Applied Mathematical Sciences
Volume 159

T0135381

Springer
New York
Berlin
Heidelberg
Hong Kong
London
Milan
Paris
Tokyo

Applied Mathematical Sciences

(continued following index)

Alexandre Ern   Jean-Luc Guermond

Theory and Practice of Finite Elements

With 89 Figures

 Springer

Alexandre Ern
CERMICS, ENPC
6 et 8, avenue Blaise Pascal
77455 Marne la Vallée cedex 2
France
ern@cermics.enpc.fr

Jean-Luc Guermond
LIMSI, CNRS
BP 133
91403 Orsay cedex
France
guermond@limsi.fr

Editors:

S.S. Antman
Department of Mathematics
and
Institute for Physical Science
  and Technology
University of Maryland
College Park, MD 20742-4015
USA
ssa@math.umd.edu

J.E. Marsden
Control and Dynamical
  Systems, 107-81
California Institute of
  Technology
Pasadena, CA 91125
USA
marsden@cds.caltech.edu

L. Sirovich
Division of Applied
  Mathematics
Brown University
Providence, RI 02912
USA
chico@camelot.mssm.edu

Mathematics Subject Classification (2000): 35A35, 65-01, 65F10, 65M60, 65N30, 65N50, 68U20, 74S05, 76M10, 80M10

Ern, Alexandre, 1967–
    Theory and practice of finite elements / Alexandre Ern, Jean-Luc Guermond.
       p. cm.
    Includes bibliographical references and index.

    1. Finite element method.   I. Guermond, Jean-Luc.   II. Title.
TA347.F5E75 2003
620′.0042—dc22   2003066022

ISBN 978-1-4419-1918-2   Printed on acid-free paper.

Printed in the United States of America.   (EB)

9 8 7 6 5 4 3 2 1

Springer-Verlag is a part of *Springer Science+Business Media*

springeronline.com

Preface

The origins of the finite element method can be traced back to the 1950s when engineers started to solve numerically structural mechanics problems in aeronautics. Since then, the field of applications has widened steadily and nowadays encompasses nonlinear solid mechanics, fluid/structure interactions, turbulent flows in industrial or geophysical settings, multicomponent reactive flows, mass transfer in porous media, viscoelastic flows in medical sciences, electromagnetism, wave scattering problems, and option pricing (to cite a few examples). Numerous commercial and academic codes based on the finite element method have been developed over the years. The method has been so successful to solve Partial Differential Equations (PDEs) that the term "Finite Element Method" nowadays refers not only to the mere interpolation technique it is, but also to a fuzzy set of PDEs and approximation techniques.

The efficiency of the finite element method relies on two distinct ingredients: the interpolation capability of finite elements (referred to as the *approximability* property in this book) and the ability of the user to approximate his model (mostly a set of PDEs) in a proper mathematical setting (thus guaranteeing *continuity*, *stability*, and *consistency* properties). Experience shows that failure to produce an approximate solution with an acceptable accuracy is almost invariably linked to departure from the mathematical foundations. Typical examples include non-physical oscillations, spurious modes, and locking effects. In most cases, a remedy can be designed if the mathematical framework is properly set up.

The starting point of the analysis is to choose a mathematical framework to set the exact problem, the goal being to state clearly in which sense the solution is stable with respect to the data, i.e., in which sense the problem in question is *well-posed*. To clarify the well-posedness issue while remaining at a graduate level, we mostly restrict ourselves to *linear* problems. Consider the following model problem: For $f \in V'$, seek a function $u \in W$ such that

$$Au = f, \tag{0.1}$$

where V and W are two Banach spaces (V being reflexive), V' is the dual space of V, and $A : W \to V'$ is a linear operator. In spite of its simplicity, (0.1) is a prototype for a broad class of engineering problems.

Many important results of functional analysis clarifying the theory of PDEs were established in the 1940s in the wake of Banach, Sobolev, Schwartz, and many others (contributing to the solution of Hilbert's 19th, 20th, and 23rd problems); however, the connection to the finite element method was not readily made, in part because of the low "diffusion constants" between mathematicians and engineers. One of the simplest well-posedness results for (0.1) that emerged in the finite element literature in the late 1950s is the now popular Lax–Milgram Lemma [LaM54]. However, the range of application of this lemma is limited since it only gives a *sufficient* condition for well-posedness known as *coercivity*. It is also limited to Hilbertian settings. *Necessary and sufficient* conditions for well-posedness are the so-called *inf-sup conditions*. These conditions were popularized by Babuška in 1972 in the context of finite element methods [BaA72, p. 112] and stated in an earlier theoretical work by Nečas in 1962 [Neč62]. From the functional analysis point of view, the inf-sup conditions are a rephrasing of two fundamental theorems by Banach: the Closed Range Theorem and the Open Mapping Theorem. For this reason, we shall refer to the well-posedness result based on the inf-sup conditions as the *Banach–Nečas–Babuška* (BNB) Theorem, although this terminology is by no means standard. One goal of the book is to go beyond the Lax–Milgram paradigm by relying systematically on the inf-sup conditions for the mathematical analysis of the finite element approximation to (0.1).

The book is organized into three parts. The first part (Chapters 1 and 2) reviews the theoretical foundations of the finite element method. The second and third parts are devoted to the practice of finite elements. The second part (Chapters 3 to 6) deals with PDE-based applications of finite elements. The third part (Chapters 7 to 10) covers implementation aspects. Two appendices summarize the main mathematical concepts used in this book: Banach and Hilbert spaces; Banach operators; distributions; and Sobolev spaces.

Part I. Chapter 1 defines finite elements and the associated *interpolation operators*. Numerous examples of scalar- and vector-valued finite elements are analyzed. The basic notions to construct meshes and approximation spaces are introduced. The last sections of this chapter are devoted to the analysis of interpolation errors and inverse inequalities.

Chapter 2 introduces the class of *well-posed problems*, states fundamental results to establish the well-posedness of (0.1), and presents the basic concepts to approximate (0.1) using *Galerkin-type methods*. In its most general form, the Galerkin method involves different solution and test spaces. These spaces may not be subsets of the original Banach spaces V and W, and both the operator A and the data f may have to be modified to account for the discrete setting. Chapter 2 investigates the well-posedness of the discrete problem and analyzes the convergence of the approximate solution to that of (0.1).

We have chosen to present the interpolation theory before the abstract results on well-posedness and approximation to stress the fact that the finite element method is foremost an interpolation technique. Nevertheless, Chapters 1 and 2 can be read independently.

Part II. Chapter 3 is centered on problems endowed with a *coercivity* property (e.g., the Laplace operator, scalar elliptic PDEs, and linear elasticity models). These problems provided the first field of applications for the finite element method and can be analyzed by relying solely on the Lax–Milgram Lemma. The last section of this chapter discusses coercivity loss (observed, in general, when the model involves parameters taking extreme values).

A first widening of the perspective occurs in Chapter 4 where non-coercive problems of *saddle-point* type are investigated. Applications include incompressible fluid flows and incompressible continuum mechanics. Because no coercivity property holds, the well-posedness of the exact problem is based on the more general inf-sup conditions. The key difference with the coercive situation is that these inf-sup conditions are not transferred automatically to the discrete setting; in other words, *discrete inf-sup conditions* must be ascertained to ensure well-posedness of the approximate problem. Violating these conditions usually leads to spurious oscillations in the discrete solution. Various finite element settings satisfying discrete inf-sup conditions are analyzed. A *Galerkin/Least-Squares* (GaLS) formulation working with any type of finite element is also studied.

Chapter 5 is dedicated to first-order PDEs. Examples include the advection equation, Darcy's and Maxwell-like equations and, more generally, Friedrichs' systems. In the same manner as it had long been thought that finite elements could not solve flow problems, the idea that finite elements can solve only PDEs dominated by a second-order coercive term is still widespread. This idea is rooted in the fact that the *standard Galerkin* technique (in which solution and test spaces are identical) cannot approximate satisfactorily first-order PDEs since the inf-sup constant involved in this type of approximation goes to zero with the mesh size. This phenomenon leads to spurious oscillations in the discrete solution. Chapter 5 analyzes various alternatives to the standard Galerkin method with satisfactory approximation properties, namely the *GaLS* formulation, *subgrid viscosity* methods, *Discontinuous Galerkin* (DG) methods, and *non-standard Galerkin* methods. All the results presented in Chapter 5 are based on the BNB Theorem.

Chapter 6 studies the time-dependent version of the problems considered in Chapters 3, 4, and 5. For parabolic and Stokes-like equations, the emphasis is set on the *method of lines* where the problem is first approximated in space, and then a time-marching algorithm is employed to construct the time-approximation. For Stokes-like problems, a class of fractional-step methods often referred to as *projection methods* is investigated in detail. Chapter 6 also analyzes various finite element methods to approximate evolution problems without coercivity, namely *DG/GaLS* techniques, the *method of char-*

acteristics, and *subgrid viscosity* techniques. Various numerical examples are presented at the end of the chapter.

Part III. Chapter 7 describes *data structures* for finite element codes and introduces general principles of *mesh generation*. Some details on how to construct *Delaunay triangulations* are also given.

Chapter 8 investigates *quadrature techniques* and data structures to implement quadratures in finite element codes. Various *assembling techniques* for matrices and right-hand sides together with some examples of *storage techniques for sparse matrices* are described. Chapter 8 also contains a brief discussion on the implementation of essential (Dirichlet-type) boundary conditions.

Chapter 9 deals with *linear algebra*. It introduces the concept of matrix conditioning and investigates the conditioning of the mass matrix and the stiffness matrix. Then, it reviews *reordering techniques for sparse matrices*, *iterative solution methods*, including the conjugate gradient algorithm and its extensions to non-symmetric systems, and *preconditioning techniques*. Issues related to parallelization are briefly discussed.

Finally, Chapter 10 analyzes *a posteriori error estimation* in the finite element method. Residual-based, hierarchical, and duality-based a posteriori error estimates are investigated. This chapter also discusses practical issues related to *adaptive mesh generation*.

—

Many theoretical and practical aspects of the finite element method are covered in detail in this book. A particular emphasis is set on the inf-sup conditions and the BNB Theorem. Various recent theoretical advances are presented, e.g., subgrid viscosity methods and DG methods. The chapters comprised in Part III should give the reader most of the practical details needed to write or understand a finite element code. Still, this book is obviously not an exhaustive monograph and always tries to remain at the *graduate level*. Some aspects of the finite element method are only briefly mentioned or simply alluded to (e.g., the *p*- and *hp*-versions of the method and the use of hierarchical settings). Many bibliographic entries to the extensive literature on finite elements are given throughout the book.

Note to Instructors. This book is an expanded version of Lecture Notes published by the authors in French [ErG02]. It has been used as a textbook for graduate finite element courses at Ecole Nationale des Ponts et Chaussées (ENPC), Université Paris VI, Université d'Evry, Université Joseph Fourier at Grenoble, and the University of Texas at Austin.

The book can be used in several courses in Mathematics, Computer Science, and Engineering programs. Each section is meant to provide a coherent teaching unit, and each chapter is accompanied with exercises. To be of real interest to graduate students and to those who are not familiar with the field, numerous hints are given, and the questions are divided into tractable sub-problems.

Here are some suggestions for course titles and syllabi:

Title   Introduction to the Finite Element Method
Syllabus §1.1 to §1.5, Chapter 2, §3.1, §3.2, §4.1, and §4.2.

Title   Finite Element Approximation of PDEs
Syllabus Chapters 1–2, §3.1, §3.2, §4.1, §4.2, §4.4, and Chapter 5.

Title   Implementation of Finite Elements
Syllabus §1.1 to §1.4, §3.1, §3.2, Chapters 7–9, and §10.4.

Title   Advanced Topics in Finite Elements
Syllabus §1.2 to §1.7, Chapter 2, §3.2, §4.2, §4.3, §5.2 to §5.7, Chapter 6, and §10.1 to §10.3.

Title   Finite Elements in Solid and Fluid Mechanics
           Finite Elements in Aerospace Engineering
           Finite Elements in Mechanical Engineering
Syllabus §1.1 to §1.5, Chapters 2–4, §5.1, §5.4, §6.2, Chapters 7 and 9, and §10.4.

Acknowledgments. We are indebted to many colleagues and former students for valuable discussions and comments on the manuscript. We express our warmest thanks to Y. Achdou (Université Paris VII), P. Azerad (Université de Perpignan), J. Bazilevs (The University of Texas at Austin), M. Braack (University of Heidelberg), E. Burman (Ecole Polytechnique Fédérale de Lausanne), E. Cancès (ENPC), D. Chapelle (INRIA), J.-P. Croisille (Université de Metz), L. Dormieux (ENPC), L. El Alaoui (ENPC), J.-F. Gerbeau (INRIA), V. Giovangigli (Ecole Polytechnique), T. Lelièvre (ENPC), L. Quartapelle (Politecnico di Milano), J. Proft (ENPC), and P. Witomski (Université Joseph Fourier, Grenoble).

Paris, France   Alexandre Ern
March 2004   Jean-Luc Guermond

Contents

Part III Implementation

Part I

Theoretical Foundations

Finite Element Interpolation

This chapter introduces the concept of finite elements along with the corresponding interpolation techniques. As an introductory example, we study how to interpolate functions in one dimension. Finite elements are then defined in arbitrary dimension, and numerous examples of scalar- and vector-valued finite elements are presented. Next, the concepts underlying the construction of meshes, approximation spaces, and interpolation operators are thoroughly investigated. The last sections of this chapter are devoted to the analysis of interpolation errors and inverse inequalities.

1.1 One-Dimensional Interpolation

The scope of this section is the interpolation theory of functions defined on an interval $]a, b[$. For an integer $k \geq 0$, $\mathbb{P}_k$ denotes the space of the polynomials in one variable, with real coefficients and of degree at most k.

1.1.1 The mesh

A mesh of $\Omega =]a, b[$ is an indexed collection of intervals with non-zero measure $\{I_i = [x_{1,i}, x_{2,i}]\}_{0 \leq i \leq N}$ forming a partition of Ω, i.e.,

$$\overline{\Omega} = \bigcup_{i=0}^{N} I_i \quad \text{and} \quad \overset{\circ}{I_i} \cap \overset{\circ}{I_j} = \emptyset \quad \text{for } i \neq j. \tag{1.1}$$

The simplest way to construct a mesh is to take $(N+2)$ points in $\overline{\Omega}$ such that

$$a = x_0 < x_1 < \ldots < x_N < x_{N+1} = b, \tag{1.2}$$

and to set $x_{1,i} = x_i$ and $x_{2,i} = x_{i+1}$ for $0 \leq i \leq N$. The points in the set $\{x_0, \ldots, x_{N+1}\}$ are called the *vertices* of the mesh. The mesh may have a variable step size

$$h_i = x_{i+1} - x_i, \qquad 0 \le i \le N,$$

and we set

$$h = \max_{0 \le i \le N} h_i.$$

In the sequel, the intervals I_i are also called *elements* (or *cells*) and the mesh is denoted by $\mathcal{T}_h = \{I_i\}_{0 \le i \le N}$. The subscript h refers to the refinement level.

1.1.2 The $\mathbb{P}_1$ Lagrange finite element

Consider the vector space of continuous, piecewise linear functions

$$P_h^1 = \{v_h \in \mathcal{C}^0(\overline{\Omega}); \, \forall i \in \{0, \ldots, N\}, \, v_{h|I_i} \in \mathbb{P}_1\}. \tag{1.3}$$

This space can be used in conjunction with Galerkin methods to approximate one-dimensional PDEs; see, e.g., Chapters 2 and 3. For this reason, P_h^1 is called an *approximation space*. Introduce the functions $\{\varphi_0, \ldots, \varphi_{N+1}\}$ defined elementwise as follows: For $i \in \{0, \ldots, N+1\}$,

$$\varphi_i(x) = \begin{cases} \frac{1}{h_{i-1}}(x - x_{i-1}) & \text{if } x \in I_{i-1}, \\ \frac{1}{h_i}(x_{i+1} - x) & \text{if } x \in I_i, \\ 0 & \text{otherwise,} \end{cases} \tag{1.4}$$

with obvious modifications if $i = 0$ or $N+1$. Clearly, $\varphi_i \in P_h^1$. These functions are often called "hat functions" in reference to the shape of their graph; see Figure 1.1.

Proposition 1.1. *The set $\{\varphi_0, \ldots, \varphi_{N+1}\}$ is a basis for P_h^1.*

Proof. The proof relies on the fact that $\varphi_i(x_j) = \delta_{ij}$, the Kronecker symbol, for $0 \le i, j \le N+1$. Let $(\alpha_0, \ldots, \alpha_{N+1})^T \in \mathbb{R}^{N+2}$ and assume that the continuous function $w = \sum_{i=0}^{N+1} \alpha_i \varphi_i$ vanishes identically in Ω. Then, for $0 \le i \le N+1$, $\alpha_i = w(x_i) = 0$; hence, the set $\{\varphi_0, \ldots, \varphi_{N+1}\}$ is linearly independent. Furthermore, for all $v_h \in P_h^1$, it is clear that $v_h = \sum_{i=0}^{N+1} v_h(x_i)\varphi_i$ since, on each element I_i, the functions v_h and $\sum_{i=0}^{N+1} v_h(x_i)\varphi_i$ are affine and coincide at two points, namely x_i and x_{i+1}. □

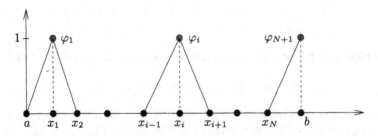

Fig. 1.1. One-dimensional hat functions.

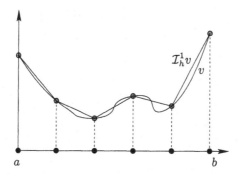

Fig. 1.2. Interpolation by continuous, piecewise linear functions.

Definition 1.2. *Choose a basis $\{\gamma_0, \ldots, \gamma_{N+1}\}$ for $\mathcal{L}(P_h^1; \mathbb{R})$; henceforth, the linear forms in this basis are called the global degrees of freedom in P_h^1. The functions in the dual basis are called the global shape functions in P_h^1.*

For $i \in \{0, \ldots, N+1\}$, choose the linear form

$$\gamma_i : \mathcal{C}^0(\overline{\Omega}) \ni v \longmapsto \gamma_i(v) = v(x_i) \in \mathbb{R}. \tag{1.5}$$

The proof of Proposition 1.1 shows that a function $v_h \in P_h^1$ is uniquely defined by the $(N+2)$-uplet $(v_h(x_i))_{0 \le i \le N+1}$. In other words, $\{\gamma_0, \ldots, \gamma_{N+1}\}$ is a basis for $\mathcal{L}(P_h^1; \mathbb{R})$. Choosing the linear forms (1.5) as the global degrees of freedom in P_h^1, the global shape functions are the functions $\{\varphi_0, \ldots, \varphi_{N+1}\}$ defined in (1.4) since $\gamma_i(\varphi_j) = \delta_{ij}$, $0 \le i, j \le N+1$.

Consider the so-called *interpolation operator*

$$\mathcal{I}_h^1 : \mathcal{C}^0(\overline{\Omega}) \ni v \longmapsto \sum_{i=0}^{N+1} \gamma_i(v)\varphi_i \in P_h^1. \tag{1.6}$$

For a function $v \in \mathcal{C}^0(\overline{\Omega})$, $\mathcal{I}_h^1 v$ is the unique continuous, piecewise linear function that takes the same value as v at all the mesh vertices; see Figure 1.2. The function $\mathcal{I}_h^1 v$ is called the *Lagrange interpolant* of v of degree 1. Note that the approximation space P_h^1 is the codomain of $\mathcal{I}_h^1$.

When approximating PDEs using finite elements, it is important to investigate the properties of $\mathcal{I}_h^1$ in Sobolev spaces; see Appendix B. In particular, recall that for an integer $m \ge 1$, $H^m(\Omega)$ denotes the space of square-integrable functions over Ω whose distributional derivatives up to order m are square-integrable. We use the following notation: $\|v\|_{0,\Omega} = \|v\|_{L^2(\Omega)}$, $|v|_{1,\Omega} = \|v'\|_{0,\Omega}$, $\|v\|_{1,\Omega} = (\|v\|_{0,\Omega}^2 + \|v'\|_{0,\Omega}^2)^{\frac{1}{2}}$, $|v|_{2,\Omega} = \|v''\|_{0,\Omega}$, etc.

Lemma 1.3. $P_h^1 \subset H^1(\Omega)$.

Proof. Let $v_h \in P_h^1$. Clearly, $v_h \in L^2(\Omega)$. Furthermore, owing to the continuity of v_h, its first-order distributional derivative is the piecewise constant function w_h such that

$$\forall I_i \in \mathcal{T}_h, \quad w_{h|I_i} = \frac{v_h(x_{i+1}) - v_h(x_i)}{h_i}. \tag{1.7}$$

Clearly, $w_h \in L^2(\Omega)$; hence, $v_h \in H^1(\Omega)$.   □

Proposition 1.4. $\mathcal{I}_h^1$ *is a linear continuous mapping from* $H^1(\Omega)$ *to* $H^1(\Omega)$, *and* $\|\mathcal{I}_h^1\|_{\mathcal{L}(H^1(\Omega);H^1(\Omega))}$ *is uniformly bounded with respect to* h.

Proof. (1) In one dimension, a function in $H^1(\Omega)$ is continuous. Indeed, for $v \in H^1(\Omega)$ and $x, y \in \overline{\Omega}$,

$$|v(y) - v(x)| \le \int_x^y |v'(s)| \, \mathrm{d}s \le |y - x|^{\frac{1}{2}} |v|_{1,\Omega}, \tag{1.8}$$

owing to the Cauchy–Schwarz inequality (this can be justified rigorously by a density argument). Furthermore, taking x to be a point where $|v|$ reaches its minimum over $\overline{\Omega}$, the above inequality implies

$$\|v\|_{L^\infty(\Omega)} \le |b - a|^{-\frac{1}{2}} \|v\|_{0,\Omega} + |b - a|^{\frac{1}{2}} |v|_{1,\Omega}, \tag{1.9}$$

since $|v(x)| \le |b - a|^{-\frac{1}{2}} \|v\|_{0,\Omega}$. Therefore, $\mathcal{I}_h^1 v$ is well-defined for $v \in H^1(\Omega)$. Moreover, Lemma 1.3 implies $\mathcal{I}_h^1 v \in H^1(\Omega)$; hence, $\mathcal{I}_h^1$ maps $H^1(\Omega)$ to $H^1(\Omega)$. (2) Let $I_i \in \mathcal{T}_h$ for $0 \le i \le N$. Owing to (1.7), $(\mathcal{I}_h^1 v)'_{|I_i} = h_i^{-1}(v(x_{i+1}) - v(x_i))$; hence, using (1.8) yields the estimate $|\mathcal{I}_h^1 v|_{1,I_i} \le |v|_{1,I_i}$. Therefore, $|\mathcal{I}_h^1 v|_{1,\Omega} \le |v|_{1,\Omega}$. Moreover, since $\|\mathcal{I}_h^1 v\|_{0,\Omega} \le |b - a|^{\frac{1}{2}} \|\mathcal{I}_h^1 v\|_{L^\infty(\Omega)}$ and $\|\mathcal{I}_h^1 v\|_{L^\infty(\Omega)} \le \|v\|_{L^\infty(\Omega)}$, we deduce from (1.9) that $\|\mathcal{I}_h^1 v\|_{0,\Omega} \le c \|v\|_{1,\Omega}$ where c is independent of h (assuming h bounded). The conclusion follows readily.   □

Proposition 1.5. *For all* h *and* $v \in H^2(\Omega)$,

$$\|v - \mathcal{I}_h^1 v\|_{0,\Omega} \le h^2 |v|_{2,\Omega} \quad and \quad |v - \mathcal{I}_h^1 v|_{1,\Omega} \le h|v|_{2,\Omega}. \tag{1.10}$$

Proof. (1) Consider an interval $I_i \in \mathcal{T}_h$. Let $w \in H^1(I_i)$ be such that w vanishes at some point ξ in I_i. Then, owing to (1.8) we infer $\|w\|_{0,I_i} \le h_i |w|_{1,I_i}$. (2) Let $v \in H^2(\Omega)$, let $i \in \{0, \ldots, N\}$, and set $w_i = (v - \mathcal{I}_h^1 v)_{|I_i}$. Note that $w_i \in H^1(I_i)$ and that w_i vanishes at some point ξ in I_i owing to the mean-value theorem. Applying the estimate derived in step 1 to w_i and using the fact that $(\mathcal{I}_h^1 v)''$ vanishes identically on I_i yields $|v - \mathcal{I}_h^1 v|_{1,I_i} \le h_i |v|_{2,I_i}$. The second estimate in (1.10) is then obtained by summing over the mesh intervals. To prove the first estimate, observe that the result of step 1 can also be applied to $(v - \mathcal{I}_h^1 v)_{|I_i}$ yielding

$$\|v - \mathcal{I}_h^1 v\|_{0,I_i} \le h_i |v - \mathcal{I}_h^1 v|_{1,I_i} \le h_i^2 |v|_{2,I_i}.$$

Conclude by summing over the mesh intervals.   □

Remark 1.6.

(i) The bound on the interpolation error involves second-order derivatives of v. This is reasonable since the larger the second derivative, the more the graph of v deviates from the piecewise linear interpolant.

(ii) If the function to be interpolated is in $H^1(\Omega)$ only, one can prove the following results:

$$\forall h, \ \|v - \mathcal{I}_h^1 v\|_{0,\Omega} \leq h|v|_{1,\Omega} \qquad \text{and} \qquad \lim_{h \to 0} |v - \mathcal{I}_h^1 v|_{1,\Omega} = 0. \qquad \square$$

The proof of Proposition 1.5 shows that the operator $\mathcal{I}_h^1$ is endowed with *local* interpolation properties, i.e., the interpolation error is controlled elementwise before being controlled globally over Ω. This motivates the introduction of local interpolation operators. Let $I_i = [x_i, x_{i+1}] \in \mathcal{T}_h$ and let $\Sigma_i = \{\sigma_{i,0}, \sigma_{i,1}\}$ where $\sigma_{i,0}, \sigma_{i,1} \in \mathcal{L}(\mathbb{P}_1; \mathbb{R})$ are such that, for all $p \in \mathbb{P}_1$,

$$\sigma_{i,0}(p) = p(x_i) \qquad \text{and} \qquad \sigma_{i,1}(p) = p(x_{i+1}). \qquad (1.11)$$

Note that Σ_i is a basis for $\mathcal{L}(\mathbb{P}_1; \mathbb{R})$. The triplet $\{I_i, \mathbb{P}_1, \Sigma_i\}$ is called a (one-dimensional) $\mathbb{P}_1$ *Lagrange finite element*, and the linear forms $\{\sigma_{i,0}, \sigma_{i,1}\}$ are the corresponding *local degrees of freedom*. The functions $\{\theta_{i,0}, \theta_{i,1}\}$ in the dual basis of Σ_i (i.e., $\sigma_{i,m}(\theta_{i,n}) = \delta_{mn}$ for $0 \leq m, n \leq 1$) are called the *local shape functions*. One readily verifies that

$$\theta_{i,0}(t) = 1 - \frac{t - x_i}{h_i} \qquad \text{and} \qquad \theta_{i,1}(t) = \frac{t - x_i}{h_i}. \qquad (1.12)$$

Finally, introduce the family $\{\mathcal{I}_{I_i}^1\}_{I_i \in \mathcal{T}_h}$ of *local interpolation operators* such that, for $i \in \{0, \ldots, N\}$,

$$\mathcal{I}_{I_i}^1 : \mathcal{C}^0(I_i) \ni v \longmapsto \sum_{m=0}^{1} \sigma_{i,m}(v)\theta_{i,m}. \qquad (1.13)$$

The proof of Propositions 1.4 and 1.5 can now be rewritten using the local interpolation operators $\mathcal{I}_{I_i}^1$. In particular, the key properties are, for $0 \leq i \leq N$ and $v \in H^2(I_i)$,

$$\|v - \mathcal{I}_{I_i}^1 v\|_{0,I_i} \leq h_i^2 |v|_{2,I_i} \qquad \text{and} \qquad |v - \mathcal{I}_{I_i}^1 v|_{1,I_i} \leq h_i |v|_{2,I_i}.$$

1.1.3 $\mathbb{P}_k$ Lagrange finite elements

The interpolation technique presented in §1.1.2 generalizes to higher-degree polynomials. Consider the mesh $\mathcal{T}_h = \{I_i\}_{0 \leq i \leq N}$ introduced in §1.1.1. Let

$$P_h^k = \{v_h \in \mathcal{C}^0(\overline{\Omega}); \ \forall i \in \{0, \ldots, N\}, \ v_{h|I_i} \in \mathbb{P}_k\}. \qquad (1.14)$$

To investigate the properties of the approximation space P_h^k and to construct an interpolation operator with codomain P_h^k, it is convenient to consider Lagrange polynomials. Recall the following:

Definition 1.7 (Lagrange polynomials). *Let $k \geq 1$ and let $\{s_0, \ldots, s_k\}$ be $(k+1)$ distinct numbers. The Lagrange polynomials $\{\mathcal{L}_0^k, \ldots, \mathcal{L}_k^k\}$ associated with the nodes $\{s_0, \ldots, s_k\}$ are defined to be*

$$\mathcal{L}_m^k(t) = \frac{\prod_{l \neq m}(t - s_l)}{\prod_{l \neq m}(s_m - s_l)}, \qquad 0 \leq m \leq k. \tag{1.15}$$

The Lagrange polynomials satisfy the important property

$$\mathcal{L}_m^k(s_l) = \delta_{ml}, \quad 0 \leq m, l \leq k.$$

Figure 1.3 presents families of Lagrange polynomials with equi-distributed nodes in the reference interval $[0, 1]$ for $k = 1, 2,$ and 3.

For $i \in \{0, \ldots, N\}$, introduce the *nodes* $\xi_{i,m} = x_i + \frac{m}{k}h_i$, $0 \leq m \leq k$, in the mesh interval I_i; see Figure 1.4. Let $\{\mathcal{L}_{i,0}^k, \ldots, \mathcal{L}_{i,k}^k\}$ be the Lagrange polynomials associated with these nodes. For $j \in \{0, \ldots, k(N+1)\}$ with $j = ki + m$ and $0 \leq m \leq k - 1$, define the function φ_j elementwise as follows: For $1 \leq m \leq k - 1$,

$$\varphi_{ki+m}(x) = \begin{cases} \mathcal{L}_{i,m}^k(x) & \text{if } x \in I_i, \\ 0 & \text{otherwise,} \end{cases}$$

and for $m = 0$,

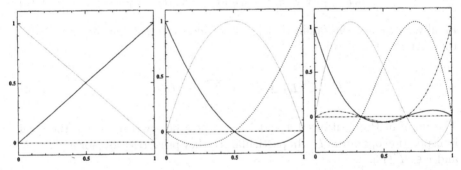

Fig. 1.3. Families of Lagrange polynomials with equi-distributed nodes in the reference interval $[0, 1]$ and of degree $k = 1$ (left), 2 (center), and 3 (right).

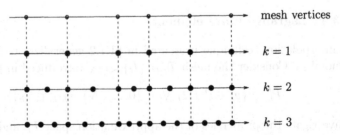

Fig. 1.4. Mesh vertices and nodes for $k = 1, 2,$ and 3.

$$\varphi_{ki}(x) = \begin{cases} \mathcal{L}_{i-1,k}^k(x) & \text{if } x \in I_{i-1}, \\ \mathcal{L}_{i,0}^k(x) & \text{if } x \in I_i, \\ 0 & \text{otherwise}, \end{cases}$$

with obvious modifications if $i = 0$ or $N + 1$. The functions φ_j are illustrated in Figure 1.5 for $k = 2$. Note the difference between the support of the functions associated with mesh vertices (two adjacent intervals) and that of the functions associated with cell midpoints (one interval).

Lemma 1.8. $\varphi_j \in P_h^k$.

Proof. Let $j \in \{0, \ldots, k(N+1)\}$ with $j = ki + m$. If $1 \le m \le k - 1$, $\varphi_j(x_i) = \varphi_j(x_{i+1}) = 0$; hence, $\varphi_j \in \mathcal{C}^0(\overline{\Omega})$. Moreover, the restrictions of φ_j to the mesh intervals are in $\mathbb{P}_k$ by construction. Therefore, $\varphi_j \in P_h^k$. Now, assume $m = 0$ (i.e., $j = ki$) and $0 < i < N + 1$. Clearly, φ_{ki} is continuous at x_i by construction and $\varphi_{ki}(x_{i-1}) = \varphi_{ki}(x_{i+1}) = 0$; hence, $\varphi_{ki} \in P_h^k$. The cases $i = 0$ and $i = N + 1$ are treated similarly. $\qquad\square$

Introduce the set of nodes $\{a_j\}_{0 \le j \le k(N+1)}$ such that $a_j = \xi_{i,m}$ where $j = ik + m$. For $j \in \{0, \ldots, k(N+1)\}$, consider the linear form

$$\gamma_j : \mathcal{C}^0(\overline{\Omega}) \ni v \longmapsto \gamma_j(v) = v(a_j). \tag{1.16}$$

Proposition 1.9. $\{\varphi_0, \ldots, \varphi_{k(N+1)}\}$ *is a basis for* P_h^k, *and* $\{\gamma_0, \ldots, \gamma_{k(N+1)}\}$ *is a basis for* $\mathcal{L}(P_h^k; \mathbb{R})$.

Proof. Similar to that of Proposition 1.1 since $\gamma_j(\varphi_{j'}) = \delta_{jj'}$ for $0 \le j, j' \le k(N+1)$. $\qquad\square$

The *global degrees of freedom* in P_h^k are chosen to be the $(k(N+1)+1)$ linear forms defined in (1.16); hence, the *global shape functions* in P_h^k are the functions $\{\varphi_0, \ldots, \varphi_{k(N+1)}\}$.

The main advantage of using high-degree polynomials is that smooth functions can be interpolated to high-order accuracy. Define the *interpolation operator* $\mathcal{I}_h^k$ to be

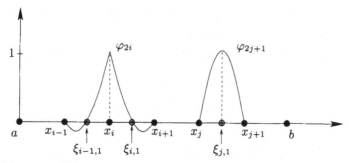

Fig. 1.5. Global shape functions in the approximation space P_h^2.

$$\mathcal{I}_h^k : \mathcal{C}^0(\overline{\Omega}) \ni v \longmapsto \sum_{j=0}^{k(N+1)} \gamma_j(v)\varphi_j \in P_h^k. \tag{1.17}$$

$\mathcal{I}_h^k v$ is called the *Lagrange interpolant* of v of degree k. Clearly, $\mathcal{I}_h^k$ is a linear operator, and $\mathcal{I}_h^k v$ is the unique function in P_h^k that takes the same value as v at all the mesh nodes. The approximation space P_h^k is the codomain of $\mathcal{I}_h^k$.

Lemma 1.10. $P_h^k \subset H^1(\Omega)$.

Proof. Similar to that of Lemma 1.3.   □

To investigate the properties of $\mathcal{I}_h^k$, it is convenient to introduce a family of local interpolation operators. On $I_i = [x_i, x_{i+1}] \in \mathcal{T}_h$, choose the *local degrees of freedom* to be the $(k+1)$ linear forms $\{\sigma_{i,0}, \ldots, \sigma_{i,k}\}$ defined as follows:

$$\sigma_{i,m} : \mathbb{P}_k \ni p \longmapsto \sigma_{i,m}(p) = p(\xi_{i,m}), \quad 0 \le m \le k. \tag{1.18}$$

The triplet $\{I_i, \mathbb{P}_k, \Sigma_i\}$ is called a (one-dimensional) $\mathbb{P}_k$ *Lagrange finite element*, and the points $\{\xi_{i,0}, \ldots, \xi_{i,k}\}$ are called the *nodes* of the finite element. Clearly, the *local shape functions* $\{\theta_{i,0}, \ldots, \theta_{i,k}\}$ are the $(k+1)$ Lagrange polynomials associated with the nodes $\{\xi_{i,0}, \ldots, \xi_{i,k}\}$, i.e., $\theta_{i,m} = \mathcal{L}_{i,m}^k$ for $0 \le m \le k$. Finally, introduce the family $\{\mathcal{I}_{I_i}^k\}_{I_i \in \mathcal{T}_h}$ of *local interpolation operators* such that, for $i \in \{0, \ldots, N\}$,

$$\mathcal{I}_{I_i}^k : \mathcal{C}^0(I_i) \ni v \longmapsto \sum_{m=0}^k \sigma_{i,m}(v)\theta_{i,m}, \tag{1.19}$$

i.e., for all $0 \le i \le N$ and $v \in \mathcal{C}^0(\overline{\Omega})$, $(\mathcal{I}_h^k v)_{|I_i} = \mathcal{I}_{I_i}^k(v_{|I_i})$.

Let us show that the family $\{\mathcal{I}_{I_i}^k\}_{I_i \in \mathcal{T}_h}$ can be generated from a single reference interpolation operator. Let $\widehat{K} = [0, 1]$ be the *unit interval*, henceforth referred to as the *reference interval*. Set $\widehat{P} = \mathbb{P}_k$, and define the $(k+1)$ linear forms $\{\widehat{\sigma}_0, \ldots, \widehat{\sigma}_k\}$ as follows:

$$\widehat{\sigma}_m : \mathbb{P}_k \ni \widehat{p} \longmapsto \widehat{\sigma}_m(\widehat{p}) = \widehat{p}(\widehat{\xi}_m), \quad 0 \le m \le k, \tag{1.20}$$

where $\widehat{\xi}_m = \frac{m}{k}$. Let $\{\widehat{\mathcal{L}}_0^k, \ldots, \widehat{\mathcal{L}}_k^k\}$ be the Lagrange polynomials associated with the nodes $\{\widehat{\xi}_0, \ldots, \widehat{\xi}_k\}$; see Figure 1.3. Set $\widehat{\theta}_m = \widehat{\mathcal{L}}_m^k$, $0 \le m \le k$, so that $\widehat{\sigma}_m(\widehat{\theta}_n) = \delta_{mn}$ for $0 \le m, n \le k$. Then, $\{\widehat{K}, \widehat{P}, \widehat{\Sigma}\}$ is a $\mathbb{P}_k$ Lagrange finite element, and the corresponding interpolation operator is

$$\mathcal{I}_{\widehat{K}}^k : \mathcal{C}^0(\widehat{K}) \ni \widehat{v} \longmapsto \sum_{m=0}^k \widehat{\sigma}_m(\widehat{v})\widehat{\theta}_m.$$

$\{\widehat{K}, \widehat{P}, \widehat{\Sigma}\}$ is called the *reference finite element* and $\mathcal{I}_{\widehat{K}}^k$ the *reference interpolation operator*. For $i \in \{0, \ldots, N\}$, consider the affine transformations

$$T_i \; : \; \widehat{K} \ni t \longmapsto x = x_i + t h_i \in I_i. \tag{1.21}$$

Since $T_i(\widehat{K}) = I_i$, the mesh $\mathcal{T}_h$ can be constructed by applying the affine transformations T_i to the reference interval $\widehat{K}$. Moreover, owing to the fact that $T_i(\widehat{\xi}_m) = \xi_{i,m}$ for $0 \leq m \leq k$, it is clear that $\theta_{i,m} \circ T_i = \widehat{\theta}_m$ and $\sigma_{i,m}(v) = \widehat{\sigma}_m(v \circ T_i)$ for all $v \in \mathcal{C}^0(I_i)$. Hence, using

$$\mathcal{I}_{I_i}^k(v)(T_i(\widehat{x})) = \sum_{m=0}^k \sigma_{i,m}(v)\theta_{i,m}(T_i(\widehat{x})) = \sum_{m=0}^k \sigma_{i,m}(v)\widehat{\theta}_m(\widehat{x}) =$$

$$= \sum_{m=0}^k \widehat{\sigma}_m(v \circ T_i)\widehat{\theta}_m(\widehat{x}) = \mathcal{I}_{\widehat{K}}^k(v \circ T_i)(\widehat{x}),$$

we infer

$$\forall v \in \mathcal{C}^0(I_i), \quad \mathcal{I}_{I_i}^k(v) \circ T_i = \mathcal{I}_{\widehat{K}}^k(v \circ T_i). \tag{1.22}$$

In other words, the family $\{\mathcal{I}_{I_i}^k\}_{I_i \in \mathcal{T}_h}$ is entirely generated by the transformations $\{T_i\}_{I_i \in \mathcal{T}_h}$ and the reference interpolation operator $\mathcal{I}_{\widehat{K}}^k$. The property (1.22) plays a key role when estimating the interpolation error; see the proof of Proposition 1.12 below.

Proposition 1.11. *$\mathcal{I}_h^k$ is a linear continuous mapping from $H^1(\Omega)$ to $H^1(\Omega)$, and $\|\mathcal{I}_h^k\|_{\mathcal{L}(H^1(\Omega);H^1(\Omega))}$ is uniformly bounded with respect to h.*

Proof. (1) To prove that $\mathcal{I}_h^k$ maps $H^1(\Omega)$ to $H^1(\Omega)$, use the argument of step 1 in the proof of Proposition 1.4.
(2) Let $v \in H^1(\Omega)$ and $I_i \in \mathcal{T}_h$. Since $\sum_{m=0}^k \theta'_{i,m} = 0$,

$$(\mathcal{I}_{I_i}^k v)' = \sum_{m=0}^k [v(\xi_{i,m}) - v(x_i)]\theta'_{i,m}.$$

Inequality (1.8) yields $|v(\xi_{i,m}) - v(x_i)| \leq h_i^{\frac{1}{2}}|v|_{1,I_i}$ for $0 \leq m \leq k$. Furthermore, changing variables in the integral, it is clear that $|\theta_{i,m}|_{1,I_i} = h_i^{-\frac{1}{2}}|\widehat{\theta}_m|_{1,\widehat{K}}$. Set $c_k = \max_{0 \leq m \leq k}|\widehat{\theta}_m|_{1,\widehat{K}}$ and observe that this quantity is mesh-independent. A straightforward calculation yields

$$|\mathcal{I}_{I_i}^k v|_{1,I_i} \leq (k+1)c_k|v|_{1,I_i},$$

showing that $|\mathcal{I}_h^k v|_{1,\Omega}$ is controlled by $|v|_{1,\Omega}$ uniformly with respect to h. In addition, since $\sum_{m=0}^k \theta_{i,m} = 1$,

$$\mathcal{I}_{I_i}^k v - v(x_i) = \sum_{m=0}^k [v(\xi_{i,m}) - v(x_i)]\theta_{i,m},$$

implying, for $x \in I_i$, $|\mathcal{I}_{I_i}^k v(x)| \leq \|v\|_{L^\infty(\Omega)} + (k+1)d_k h_i^{\frac{1}{2}} |v|_{1,I_i}$ with the mesh-independent constant $d_k = \max_{0 \leq m \leq k} \|\widehat{\theta}_m\|_{L^\infty(\widehat{K})}$. Then, using (1.9) yields $\|\mathcal{I}_h^k v\|_{L^\infty(\Omega)}$ is controlled by $\|v\|_{1,\Omega}$ uniformly with respect to h. To conclude, use the fact that $\|\mathcal{I}_h^k v\|_{0,\Omega} \leq |b-a|^{\frac{1}{2}} \|\mathcal{I}_h^k v\|_{L^\infty(\Omega)}$. □

we

Proposition 1.12. *Let* $0 \leq l \leq k$. *Then, there exists* c *such that, for all* h *and* $v \in H^{l+1}(\Omega)$,

$$\|v - \mathcal{I}_h^k v\|_{0,\Omega} + h|v - \mathcal{I}_h^k v|_{1,\Omega} \leq c\, h^{l+1} |v|_{l+1,\Omega}, \tag{1.23}$$

and for $l \geq 1$,

$$\sum_{m=2}^{l+1} h^m \left(\sum_{i=0}^{N} |v - \mathcal{I}_h^k v|_{m,I_i}^2 \right)^{\frac{1}{2}} \leq c\, h^{l+1} |v|_{l+1,\Omega}. \tag{1.24}$$

Proof. Let $0 \leq l \leq k$ and $0 \leq m \leq l+1$. Let $v \in H^{l+1}(\Omega)$.
(1) Consider a mesh interval I_i. Set $\widehat{v} = v \circ T_i$. Then, use (1.22) and change variables in the integral to obtain

$$|v - \mathcal{I}_{I_i}^k v|_{m,I_i} = h_i^{-m+\frac{1}{2}} |\widehat{v} - \mathcal{I}_{\widehat{K}}^k \widehat{v}|_{m,\widehat{K}}.$$

Similarly, $|\widehat{v}|_{l+1,\widehat{K}} = h_i^{l+\frac{1}{2}} |v|_{l+1,I_i}$.
(2) Consider the linear mapping

$$\mathcal{F} : H^{l+1}(\widehat{K}) \ni \widehat{v} \longmapsto \widehat{v} - \mathcal{I}_{\widehat{K}}^k \widehat{v} \in H^m(\widehat{K}).$$

Note that $\mathcal{I}_{\widehat{K}}^k \widehat{v}$ is meaningful since in one dimension, $\widehat{v} \in H^{l+1}(\widehat{K})$ with $l \geq 0$ implies $\widehat{v} \in C^0(\widehat{K})$. Moreover, $\mathcal{F}$ is continuous from $H^{l+1}(\widehat{K})$ to $H^m(\widehat{K})$. Indeed, one can easily adapt the proof of Proposition 1.11 to prove that $\mathcal{I}_{\widehat{K}}^k$ is continuous from $H^1(\widehat{K})$ to $H^s(\widehat{K})$ for all $s \geq 1$. Furthermore, it is clear that $\mathbb{P}_k$ is invariant under $\mathcal{F}$ since, for all $\widehat{p} \in \mathbb{P}_k$ with $\widehat{p} = \sum_{n=0}^{k} \alpha_n \widehat{\theta}_n$,

$$\mathcal{I}_{\widehat{K}}^k \widehat{p} = \sum_{m,n=0}^{k} \alpha_n \widehat{\sigma}_m(\widehat{\theta}_n) \widehat{\theta}_m = \sum_{m,n=0}^{k} \alpha_n \delta_{mn} \widehat{\theta}_m = \sum_{n=0}^{k} \alpha_n \widehat{\theta}_n = \widehat{p}.$$

(3) Since $l \leq k$, $\mathbb{P}_l$ is invariant under $\mathcal{F}$. As a result,

$$\begin{aligned}
|\widehat{v} - \mathcal{I}_{\widehat{K}}^k \widehat{v}|_{m,\widehat{K}} = |\mathcal{F}(\widehat{v})|_{m,\widehat{K}} &= \inf_{\widehat{p} \in \mathbb{P}_l} |\mathcal{F}(\widehat{v} + \widehat{p})|_{m,\widehat{K}} \\
&\leq \|\mathcal{F}\|_{\mathcal{L}(H^{l+1}(\widehat{K}); H^m(\widehat{K}))} \inf_{\widehat{p} \in \mathbb{P}_l} \|\widehat{v} + \widehat{p}\|_{l+1,\widehat{K}} \\
&\leq c \inf_{\widehat{p} \in \mathbb{P}_l} \|\widehat{v} + \widehat{p}\|_{l+1,\widehat{K}} \leq c |\widehat{v}|_{l+1,\widehat{K}},
\end{aligned}$$

the last estimate resulting from the Deny–Lions Lemma; see Lemma B.67. The identities derived in step 1 yield

$$|v - \mathcal{I}_{I_i}^k v|_{m,I_i} = h_i^{-m+\frac{1}{2}} |\widehat{v} - \mathcal{I}_{\widehat{K}}^k \widehat{v}|_{m,\widehat{K}}$$

$$\leq c\, h_i^{-m+\frac{1}{2}} |\widehat{v}|_{l+1,\widehat{K}} \leq c\, h_i^{l+1-m} |v|_{l+1,I_i}.$$

(4) To derive the estimates (1.23) and (1.24), sum over the mesh intervals. When $m = 0$ or 1, global norms over Ω can be used since $P_h^k \subset H^1(\Omega)$ owing to Lemma 1.10.   □

Remark 1.13.
(i) The proof of Proposition 1.12 shows that the interpolation properties of $\mathcal{I}_h^k$ are local.

(ii) If the function to be interpolated is smooth enough, say $v \in H^{k+1}(\Omega)$, the interpolation error is of optimal order. In particular, (1.23) yields

$$\forall h,\ \forall v \in H^{k+1}(\Omega), \quad \|v - \mathcal{I}_h^k v\|_{0,\Omega} + h|v - \mathcal{I}_h^k v|_{1,\Omega} \leq c\, h^{k+1} |v|_{k+1,\Omega}.$$

However, one should bear in mind that the order of the interpolation error may not be optimal if the function to be interpolated is not smooth. For instance, if $v \in H^s(\Omega)$ and $v \notin H^{s+1}(\Omega)$ with $s \geq 2$, considering polynomials of degree larger than $s - 1$ does not improve the interpolation error.

(iii) If the function to be interpolated is in $H^1(\Omega)$ only, one can still prove $\lim_{h\to 0} |v - \mathcal{I}_h^k v|_{1,\Omega} = 0$. To this end, use the density of $H^2(\Omega)$ in $H^1(\Omega)$ and (1.23); details are left as an exercise.   □

1.1.4 Interpolation by discontinuous functions

Let

$$P_{d,h}^k = \{v_h \in L^1(\Omega);\ \forall i \in \{0,\dots,N\},\ v_{h|I_i} \in \mathbb{P}_k\}.$$

Since the restriction of a function $v_h \in P_{d,h}^k$ to an interval I_i can be chosen independently of its restriction to the other intervals, $P_{d,h}^k$ is a vector space of dimension $(k + 1) \times (N + 1)$. However, instead of taking the Lagrange polynomials as local shape functions, it is often more convenient to consider the Legendre polynomials or modifications thereof based on the concept of hierarchical bases; see §1.1.5. Let $\widehat{K} = [0, 1]$ be the reference interval.

Definition 1.14 (Legendre polynomials). *The Legendre polynomials on the reference interval $[0, 1]$ are defined to be $\widehat{\mathcal{E}}_k(t) = \frac{1}{k!} \frac{d^k}{dt^k} (t^2 - t)^k$ for $k \geq 0$.*

The Legendre polynomial $\widehat{\mathcal{E}}_k$ is of degree k, $\widehat{\mathcal{E}}_k(0) = (-1)^k$, $\widehat{\mathcal{E}}_k(1) = 1$, and its k roots are in $\widehat{K}$. The roots of the Legendre polynomials are called *Gauß–Legendre points* and play an important role in the design of quadratures; see §8.1. The first four Legendre polynomials are (see Figure 1.6)

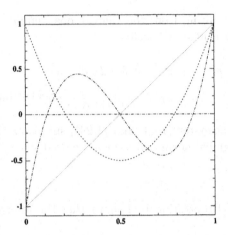

Fig. 1.6. Legendre polynomials of degree at most 3 on the reference interval $[0, 1]$.

$$\widehat{\mathcal{E}}_0(t) = 1, \qquad\qquad \widehat{\mathcal{E}}_1(t) = 2t - 1,$$
$$\widehat{\mathcal{E}}_2(t) = 6t^2 - 6t + 1, \qquad \widehat{\mathcal{E}}_3(t) = 20t^3 - 30t^2 + 12t - 1.$$

In the literature, the Legendre polynomials are sometimes defined using the reference interval $[-1, +1]$. Up to rescaling, both definitions are equivalent. In the context of finite elements, an important property of Legendre polynomials is that

$$\int_0^1 \widehat{\mathcal{E}}_m(t) \widehat{\mathcal{E}}_n(t)\, \mathrm{d}t = \tfrac{1}{2m+1}\delta_{mn}. \qquad (1.25)$$

Introduce the functions $\{\varphi_{i,m}\}_{0\leq i\leq N, 0\leq m\leq k}$ such that $\varphi_{i,m|I_j} = \delta_{ij}\, \widehat{\mathcal{E}}_m \circ T_i^{-1}$ where the geometric transformation T_i is defined in (1.21). Clearly, $\{\varphi_{i,m}\}_{0\leq i\leq N, 0\leq m\leq k}$ is a basis for $P_{\mathrm{d},h}^k$. The corresponding degrees of freedom are the linear forms $\gamma_{i,m}$, $0 \leq i \leq N$ and $0 \leq m \leq k$, such that

$$\gamma_{i,m} : L^1(\Omega) \ni v \longmapsto \gamma_{i,m}(v) = \tfrac{2m+1}{h_i} \int_{I_i} v(x)\, \widehat{\mathcal{E}}_m \circ T_i^{-1}(x)\, \mathrm{d}x,$$

since, for $0 \leq i, i' \leq N$ and $0 \leq m, m' \leq k$,

$$\gamma_{i,m}(\varphi_{i',m'}) = \tfrac{2m+1}{h_i} \int_{I_i} \varphi_{i',m'}(x)\, \widehat{\mathcal{E}}_m \circ T_i^{-1}(x)\, \mathrm{d}x$$

$$= (2m + 1)\delta_{ii'}\delta_{mm'} \int_{\widehat{K}} \widehat{\mathcal{E}}_m(t)^2\, \mathrm{d}t = \delta_{ii'}\delta_{mm'}.$$

Define the *interpolation operator* $\mathcal{I}_{\mathrm{d},h}^k$ by

$$\mathcal{I}_{\mathrm{d},h}^k : L^1(\Omega) \ni v \longmapsto \sum_{i=0}^N \sum_{m=0}^k \gamma_{i,m}(v)\varphi_{i,m} \in P_{\mathrm{d},h}^k. \qquad (1.26)$$

For instance, $\mathcal{I}_{d,h}^0 v$ is the unique piecewise constant function that takes the same mean value as v over the mesh intervals.

Let $I_i = [x_i, x_{i+1}] \in \mathcal{T}_h$ and choose for the local degrees of freedom in $\mathbb{P}_k$ the set $\Sigma_i = \{\gamma_{i,m}\}_{0 \le m \le k}$. The triplet $\{I_i, \mathbb{P}_k, \Sigma_i\}$ is often called a *modal finite element*; see §1.1.5 for further insight. The local shape functions are $\theta_{i,m} = \hat{\mathcal{E}}_m \circ T_i^{-1}$. Introduce the family $\{\mathcal{I}_{d,I_i}^k\}_{I_i \in \mathcal{T}_h}$ of *local interpolation operators* such that, for $0 \le i \le N$,

$$\mathcal{I}_{d,I_i}^k : L^1(I_i) \ni v \longmapsto \sum_{m=0}^k \sigma_{i,m}(v)\theta_{i,m}. \tag{1.27}$$

Then, it is clear that, for all $v \in L^1(\Omega)$, $(\mathcal{I}_{d,h}^k v)_{|I_i} = \mathcal{I}_{d,I_i}^k(v_{|I_i})$. Using the family $\{\mathcal{I}_{d,I_i}^k\}_{I_i \in \mathcal{T}_h}$, one easily verifies the following results:

Proposition 1.15. $\mathcal{I}_{d,h}^k$ *is a linear continuous mapping from* $L^1(\Omega)$ *to* $L^1(\Omega)$, *and* $\|\mathcal{I}_{d,h}^k\|_{\mathcal{L}(L^1(\Omega);L^1(\Omega))}$ *is uniformly bounded with respect to* h.

Proposition 1.16. *Let* $k \ge 0$ *and let* $0 \le l \le k$. *Then, there exists* c *such that, for all* h *and* $v \in H^{l+1}(\Omega)$,

$$\|v - \mathcal{I}_{d,h}^k v\|_{0,\Omega} + \sum_{m=1}^{l+1} h^m \left(\sum_{i=0}^N |v - \mathcal{I}_{d,h}^k v|_{m,I_i}^2 \right)^{\frac{1}{2}} \le c\, h^{l+1} |v|_{l+1,\Omega}.$$

Proof. Use steps 1, 2, and 3 in the proof of Proposition 1.12. □

Example 1.17. Taking $k = l = 0$ in Proposition 1.16 yields, for all h and $v \in H^1(\Omega)$, $\|v - \mathcal{I}_{d,h}^0 v\|_{0,\Omega} \le c\, h|v|_{1,\Omega}$. □

1.1.5 Hierarchical polynomial bases

Although the emphasis in this book is set on h-type finite element methods for which convergence is achieved by refining the mesh, it is also possible to consider p-type finite element methods for which convergence is achieved by increasing the polynomial degree of the interpolation in every element. The hp-type finite element method is a combination of these two strategies. The idea that the p version of the finite element method can be as efficient as the h version is rooted in a series of papers by Babuška et al. [BaS81, BaD81].

When working with high-degree polynomials, it is important to select carefully the polynomial basis. The material presented herein is set at an introductory level; see, e.g., [KaS99^b, pp. 31–59]. The following definition plays an important role in the construction of polynomial bases:

Definition 1.18 (Hierarchical modal basis). *A family* $\{\mathcal{B}_k\}_{k \ge 0}$, *where* $\mathcal{B}_k$ *is a set of polynomials, is said to be a* hierarchical modal basis *if, for all* $k \ge 0$:

(i) $\mathcal{B}_k$ *is a basis for* $\mathbb{P}_k$.

(ii) $\mathcal{B}_k \subset \mathcal{B}_{k+1}$.

Example 1.19. The simplest example of hierarchical modal basis is $\mathcal{B}_k = \{1, x, \ldots, x^k\}$. $\qquad\qquad\qquad\qquad\qquad\qquad\qquad\qquad\qquad\qquad\qquad\qquad\qquad\qquad\qquad$ □

So far, the local shape functions $\{\widehat{\theta}_0, \ldots, \widehat{\theta}_k\}$ we have used are the Lagrange polynomials $\{\widehat{\mathcal{L}}_0^k, \ldots, \widehat{\mathcal{L}}_k^k\}$ or the Legendre polynomials $\{\widehat{\mathcal{E}}_0, \ldots, \widehat{\mathcal{E}}_k\}$. Clearly, the Legendre polynomial basis is a *hierarchical modal basis*. This is not the case for the Lagrange polynomial basis, which instead has the remarkable property that $\widehat{\mathcal{L}}_l^k(\widehat{\xi}_{l'}) = \delta_{ll'}$ at the associated nodes $\{\widehat{\xi}_0, \ldots, \widehat{\xi}_k\}$. Because of this property, the Lagrange polynomial basis is said to be a *nodal basis*.

A first important criterion to select a high-degree polynomial basis is that the basis is orthogonal or nearly orthogonal with respect to an appropriate inner product. Let $\widehat{K} = [0,1]$ be the reference interval and define the matrix $\mathcal{M}_{\widehat{K}}$ of order $k+1$ with entries

$$\forall m, n \in \{0, \ldots, k\}, \quad \mathcal{M}_{\widehat{K}, mn} = \int_{\widehat{K}} \widehat{\theta}_m(t) \widehat{\theta}_n(t) \, dt. \qquad (1.28)$$

The matrix $\mathcal{M}_{\widehat{K}}$ is symmetric positive definite and is called the *elemental mass matrix*. The high-degree polynomial basis can be constructed so that $M_{\widehat{K}}$ is diagonal or "almost" diagonal. Define the condition number of $\mathcal{M}_{\widehat{K}}$ to be the ratio between its largest and smallest eigenvalue; see §9.1. Instead of diagonality, an alternative criterion to select a polynomial basis can be that the condition number of $\mathcal{M}_{\widehat{K}}$ does not increase "too much" as k grows; see Remark 1.20(i).

A second important criterion is that interface conditions between adjacent mesh elements can be imposed easily. For instance, imposing continuity at the interfaces ensures that the codomain of the global interpolation operator is in $H^1(\Omega)$; see, e.g., Lemmas 1.3 and 1.10.

Remark 1.20.

(i) The conditioning of the elemental mass matrix has important consequences on computational efficiency. For instance, in time-dependent problems discretized with explicit time-marching algorithms, this matrix has to be inverted at each time step; see, e.g., (6.27). Furthermore, for time-dependent advection problems, explicit time step restrictions are less severe when the elemental mass matrix is well-conditioned; see [KaS99^b, p. 187] and also Exercises 6.7 and 6.9.

(ii) Instead of the elemental mass matrix, one can also consider the *elemental stiffness matrix* $\mathcal{A}_{\widehat{K}}$ defined by

$$\forall m, n \in \{0, \ldots, k\}, \quad \mathcal{A}_{\widehat{K}, mn} = \int_{\widehat{K}} \frac{d}{dt} \widehat{\theta}_m(t) \frac{d}{dt} \widehat{\theta}_n(t) \, dt.$$

This matrix, which is symmetric and positive, arises when approximating the Laplace equation; see §3.1. The high-degree polynomial basis can then be constructed so that $\mathcal{A}_{\widehat{K}}$ remains relatively well-conditioned. $\qquad\qquad$ □

The Legendre polynomial basis satisfies the first criterion above. Owing to (1.25), the mass matrix is diagonal and its condition number is $(2k + 1)$. However, Legendre polynomials do not vanish at the boundary of $\widehat{K}$, making it cumbersome to enforce C^0-continuity between adjacent mesh intervals. On the other hand, the Lagrange polynomial basis satisfies the C^0-continuity criterion provided the nodal points contain the interval endpoints, but the mass matrix is dense and its condition number explodes exponentially with k; see [OlD95] for a proof and [KaS99^b, p. 44] for an illustration. We now discuss appropriate modifications of the above bases designed to better fulfill the above criteria.

Modal (C^0-continuous) basis. We first define the Jacobi polynomials.

Definition 1.21 (Jacobi polynomials). *Let $\alpha > -1$ and $\beta > -1$. The Jacobi polynomials $\{\mathcal{J}_k^{\alpha,\beta}\}_{k \geq 0}$ are defined by*

$$\mathcal{J}_k^{\alpha,\beta}(t) = \frac{(-1)^k}{k!} 2^{-\alpha-\beta} (1-t)^{-\alpha} t^{-\beta} \frac{d^k}{dt^k}\left((1-t)^{\alpha+k} t^{\beta+k} \right). \tag{1.29}$$

The Jacobi polynomials satisfy the important orthogonality property

$$\int_{\widehat{K}} (1-t)^{\alpha} t^{\beta} \mathcal{J}_m^{\alpha,\beta}(t) \mathcal{J}_n^{\alpha,\beta}(t)\, dt = c_{m,\alpha,\beta} \delta_{mn}, \tag{1.30}$$

with constant $c_{m,\alpha,\beta} = \frac{1}{2m+\alpha+\beta+1} \frac{\Gamma(m+\alpha+1)\Gamma(m+\beta+1)}{m!\Gamma(m+\alpha+\beta+1)}$. The first Jacobi polynomials for $\alpha = \beta = 1$ are $\mathcal{J}_0^{1,1}(t) = 1$, $\mathcal{J}_1^{1,1}(t) = 4t - 2$, and $\mathcal{J}_2^{1,1}(t) = 15t^2 - 15t + 3$. Note that the Legendre polynomials introduced in Definition 1.14 are Jacobi polynomials with parameters $\alpha = \beta = 0$. For more details on Jacobi polynomials, see [AbS72, Chap. 22] and [KaS99^b, p. 350].

The modal (C^0-continuous) basis is the set of functions $\{\widehat{\theta}_0, \ldots, \widehat{\theta}_k\}$ such that

$$\widehat{\theta}_l(t) = \begin{cases} 1 - t & \text{if } l = 0, \\ (1-t)t\, \mathcal{J}_{l-1}^{1,1}(t) & \text{if } 0 < l < k, \\ t & \text{if } l = k. \end{cases} \tag{1.31}$$

This basis possesses several attractive features:

(i) It is a hierarchical modal basis according to Definition 1.18.
(ii) C^0-continuity at element endpoints can be easily enforced since only the first and last basis functions do not vanish at the endpoints.
(iii) Owing to the use of Jacobi polynomials with parameters $\alpha = \beta = 1$, the elemental mass matrix $\mathcal{M}_{\widehat{K}}$ is such that $\mathcal{M}_{\widehat{K},mn} = 0$ for $|m - n| > 2$ and $0 \leq m, n \leq k$, unless $m = k$ and $n \leq 2$ or $n = k$ and $m \leq 2$. Furthermore, this matrix remains relatively well-conditioned. A precise result in arbitrary dimension d using tensor products of modal hierarchical bases is that the condition number of the elemental mass matrix (resp., stiffness matrix) is equivalent to 4^{kd} (resp., $4^{k(d-1)}$) uniformly in k; see [HuG98].

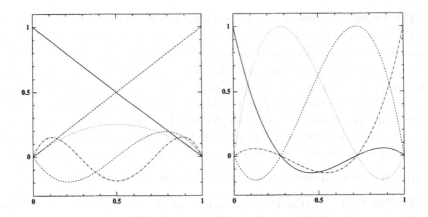

Fig. 1.7. Left: Modal ($\mathcal{C}^0$-continuous) basis functions of degree at most 4 on the reference interval $[0, 1]$. Right: Nodal ($\mathcal{C}^0$-continuous) basis functions of degree at most 3 on the same interval.

The modal ($\mathcal{C}^0$-continuous) basis functions are shown in the left panel of Figure 1.7 for $k = 5$.

Remark 1.22. Note that in the present case the degrees of freedom have no evident definition. It is more natural to define directly the local shape functions without resorting to the notion of degrees of freedom. □

Nodal ($\mathcal{C}^0$-continuous) basis. Nodal basis functions are interesting in the context of quadratures; see §8.1 for an introduction to these techniques. The principle of quadratures is to approximate the integral of a function over $\widehat{K}$ by a linear combination of the values it takes at $(k + 1)$ points in $\widehat{K}$, say $\{\widehat{\xi}_0, \ldots, \widehat{\xi}_k\}$, in the form

$$\int_{\widehat{K}} f(t)\, \mathrm{d}t \simeq \sum_{l=0}^{k} \widehat{\omega}_l f(\widehat{\xi}_l). \tag{1.32}$$

The points $\{\widehat{\xi}_0, \ldots, \widehat{\xi}_k\}$ are called the *quadrature nodes* and the numbers $\{\widehat{\omega}_0, \ldots, \widehat{\omega}_k\}$ the *quadrature weights*. For $k \geq 2$, the *Gauß–Lobatto* quadrature nodes are defined to be the two endpoints of $\widehat{K}$ and the $(k - 1)$ roots of $\widehat{\mathcal{E}}'_k$. The resulting quadrature rule is exact for polynomials up to degree $2k - 1$.

Define the local degrees of freedom $\{\widehat{\sigma}_0, \ldots, \widehat{\sigma}_k\}$ such that, for $0 \leq i \leq k$,

$$\widehat{\sigma}_i : \mathbb{P}_k \ni \widehat{p} \longmapsto \widehat{\sigma}_i(\widehat{p}) = p(\widehat{\xi}_i) \in \mathbb{R}.$$

Then, the local shape functions $\{\widehat{\theta}_0, \ldots, \widehat{\theta}_k\}$ are the Lagrange polynomials associated with the nodes $\{\widehat{\xi}_0, \ldots, \widehat{\xi}_k\}$. Using standard induction relations on the Legendre polynomials, it is possible to show that the local shape functions $\{\widehat{\theta}_0, \ldots, \widehat{\theta}_k\}$ are given by

$$\forall m \in \{0\ldots,k\}, \quad \widehat{\theta}_m(t) = \frac{(t-1)t\widehat{\mathcal{E}}_k'(t)}{k(k+1)\widehat{\mathcal{E}}_k(\widehat{\xi}_m)(t-\widehat{\xi}_m)}. \tag{1.33}$$

These functions are shown in the right panel of Figure 1.7 for $k = 4$. Although these nodal basis functions are not hierarchical, they present attractive features in the context of *spectral element methods*; see [KaS99^b, p. 51] and [Pat84] for more details. If the quadrature (1.32) is used to evaluate $\mathcal{M}_{\widehat{K}}$ in (1.28), the elemental mass matrix becomes diagonal, and each diagonal entry is equal to the row-wise sum of the entries of the exact elemental matrix. Summing row-wise the entries of the mass matrix and using the result as diagonal entries is often referred to as *lumping*.

1.2 Finite Elements: Definitions and Examples

The purpose of this section is to give a general definition of finite elements and local interpolation operators. Numerous two- and three-dimensional examples are listed.

1.2.1 Main definitions

Following Ciarlet, a finite element is defined as a triplet $\{K, P, \Sigma\}$; see, e.g., [Cia91, p. 93].

Definition 1.23. *A finite element consists of a triplet* $\{K, P, \Sigma\}$ *where:*

(i) *K is a compact, connected, Lipschitz subset of $\mathbb{R}^d$ with non-empty interior.*

(ii) *P is a vector space of functions $p : K \to \mathbb{R}^m$ for some positive integer m (typically $m = 1$ or d).*

(iii) *Σ is a set of n_{sh} linear forms $\{\sigma_1, \ldots, \sigma_{n_{\mathrm{sh}}}\}$ acting on the elements of P, and such that the linear mapping*

$$P \ni p \longmapsto \big(\sigma_1(p), \ldots, \sigma_{n_{\mathrm{sh}}}(p)\big) \in \mathbb{R}^{n_{\mathrm{sh}}}, \tag{1.34}$$

is bijective, i.e., Σ is a basis for $\mathcal{L}(P; \mathbb{R})$. The linear forms $\{\sigma_1, \ldots, \sigma_{n_{\mathrm{sh}}}\}$ are called the local degrees of freedom.

Proposition 1.24. *There exists a basis $\{\theta_1, \ldots, \theta_{n_{\mathrm{sh}}}\}$ in P such that*

$$\sigma_i(\theta_j) = \delta_{ij}, \qquad 1 \le i, j \le n_{\mathrm{sh}}.$$

Proof. Direct consequence of the bijectivity of the mapping (1.34). $\qquad \square$

Definition 1.25. $\{\theta_1, \ldots, \theta_{n_{\mathrm{sh}}}\}$ *are called the* local shape functions.

Remark 1.26. Condition (iii) in Definition 1.23 amounts to proving that

$$\forall(\alpha_1, \dots, \alpha_{n_{\rm sh}}) \in \mathbb{R}^{n_{\rm sh}}, \ \exists! p \in P, \quad \sigma_i(p) = \alpha_i \text{ for } 1 \leq i \leq n_{\rm sh},$$

which, in turn, is equivalent to

$$\begin{cases} \dim P = \operatorname{card} \Sigma = n_{\rm sh}, \\ \forall p \in P, \ (\sigma_i(p) = 0, \ 1 \leq i \leq n_{\rm sh}) \implies (p = 0). \end{cases}$$

This property is usually referred to as *unisolvence*. In the literature, the bijectivity of the mapping (1.34) is sometimes not included in the definition and, if this property holds, the finite element is said to be *unisolvent*. □

Definition 1.27 (Lagrange finite element). *Let $\{K, P, \Sigma\}$ be a finite element. If there is a set of points $\{a_1, \dots, a_{n_{\rm sh}}\}$ in K such that, for all $p \in P$, $\sigma_i(p) = p(a_i)$, $1 \leq i \leq n_{\rm sh}$, $\{K, P, \Sigma\}$ is called a* Lagrange finite element. *The points $\{a_1, \dots, a_{n_{\rm sh}}\}$ are called the* nodes *of the finite element, and the local shape functions $\{\theta_1, \dots, \theta_{n_{\rm sh}}\}$ (which are such that $\theta_i(a_j) = \delta_{ij}$ for $1 \leq i, j \leq n_{\rm sh}$) are called the* nodal basis *of P.*

Example 1.28. See §1.1.2 and §1.1.3 for one-dimensional examples of Lagrange finite elements. □

Remark 1.29. In the literature, Lagrange finite elements as defined above are also called *nodal finite elements*. □

1.2.2 Local interpolation operator

Let $\{K, P, \Sigma\}$ be a finite element. Assume that there exists a normed vector space $V(K)$ of functions $v : K \to \mathbb{R}^m$, such that:

(i) $P \subset V(K)$.
(ii) The linear forms $\{\sigma_1, \dots, \sigma_{n_{\rm sh}}\}$ can be extended to $V(K)'$.

Then, the *local interpolation operator* $\mathcal{I}_K$ can be defined as follows:

$$\mathcal{I}_K : V(K) \ni v \longmapsto \sum_{i=1}^{n_{\rm sh}} \sigma_i(v)\theta_i \in P. \tag{1.35}$$

$V(K)$ is the domain of $\mathcal{I}_K$ and P is its codomain. Note that the term "interpolation" is used in a broad sense since $\mathcal{I}_K v$ is not necessarily defined by matching point values of v.

Proposition 1.30. *P is invariant under $\mathcal{I}_K$, i.e., $\forall p \in P$, $\mathcal{I}_K p = p$.*

Proof. Letting $p = \sum_{j=1}^{n_{\rm sh}} \alpha_j \theta_j$ yields $\mathcal{I}_K p = \sum_{i,j=1}^{n_{\rm sh}} \alpha_j \sigma_i(\theta_j)\theta_i = p$. □

Example 1.31.

(i) For Lagrange finite elements, one may choose $V(K) = [\mathcal{C}^0(K)]^m$ or $V(K) = [H^s(K)]^m$ with $s > \frac{d}{2}$. The *local Lagrange interpolation operator* is defined as follows:

$$\mathcal{I}_K : V(K) \ni v \longmapsto \mathcal{I}_K v = \sum_{i=1}^{n_{\text{sh}}} v(a_i)\theta_i, \tag{1.36}$$

i.e., the Lagrange interpolant is constructed by matching the point values at the Lagrange nodes.

(ii) For the modal finite elements discussed in §1.1.4, an admissible choice is $V(K) = L^1(K)$. □

Remark 1.32. It may seem more appropriate to define a finite element as a quadruplet $\{K, P, \Sigma, V(K)\}$, where the triplet $\{K, P, \Sigma\}$ complies with Definition 1.23 and $V(K)$ satisfies properties (i)–(ii). However, for the sake of simplicity, we hereafter employ the well-established triplet-based definition, and always implicitly assume that there exists a normed vector space $V(K)$ satisfying properties (i)–(ii). In many textbooks, $V(K)$ is implicitly assumed to be of the form $\mathcal{C}^s(K)$ for some integer $s \geq 0$; see, e.g., [Cia91, p. 96] or [BrS94, p. 79]. □

1.2.3 Simplicial Lagrange finite elements

Simplices and barycentric coordinates. Let $\{a_0, \ldots, a_d\}$ be a family a points in $\mathbb{R}^d$, $d \geq 1$. Assume that the vectors $\{a_1 - a_0, \ldots, a_d - a_0\}$ are linearly independent. Then, the convex hull of $\{a_0, \ldots, a_d\}$ is called a *simplex*, and the points $\{a_0, \ldots, a_d\}$ are called the *vertices* of the simplex. The *unit simplex* of $\mathbb{R}^d$ is the set

$$\left\{ x \in \mathbb{R}^d; \ x_i \geq 0, \ 1 \leq i \leq d, \ \text{and} \ \sum_{i=1}^{d} x_i \leq 1 \right\}.$$

A simplex can be equivalently defined to be the image of the unit simplex by a bijective affine transformation. For $0 \leq i \leq d$, define F_i to be the face of K opposite to a_i, and define n_i to be the outward normal to F_i. Note that in dimension 2 a face is also called an edge, but this distinction will not be made unless necessary.

Given a simplex K in $\mathbb{R}^d$, it is often convenient to consider the associated *barycentric coordinates* $\{\lambda_0, \ldots, \lambda_d\}$ defined as follows: For $0 \leq i \leq d$,

$$\lambda_i : \mathbb{R}^d \ni x \longmapsto \lambda_i(x) = 1 - \frac{(x - a_i) \cdot n_i}{(a_j - a_i) \cdot n_i} \in \mathbb{R}, \tag{1.37}$$

where a_j is an arbitrary vertex in F_i (the definition of λ_i is clearly independent of the choice of the vertex in F_i). The barycentric coordinate λ_i is an affine function; it is equal to 1 at a_i and vanishes at F_i. Furthermore, its level-sets are hyperplanes parallel to F_i. Note that the barycenter G of K has barycentric coordinates $(\frac{1}{d+1}, \ldots, \frac{1}{d+1})$. The barycentric coordinates satisfy the following properties: For all $x \in K$, $0 \leq \lambda_i(x) \leq 1$, and for all $x \in \mathbb{R}^d$,

$$\sum_{i=1}^{d+1} \lambda_i(x) = 1 \quad \text{and} \quad \sum_{i=1}^{d+1} \lambda_i(x)(x - a_i) = 0.$$

See Exercise 1.4 for further properties in dimension 2 and 3.

Example 1.33. In the unit simplex, $\lambda_0 = 1 - x_1 - x_2$, $\lambda_1 = x_1$, and $\lambda_2 = x_2$ in dimension 2, and $\lambda_0 = 1 - x_1 - x_2 - x_3$, $\lambda_1 = x_1$, $\lambda_2 = x_2$, $\lambda_3 = x_3$ in dimension 3. □

The polynomial space $\mathbb{P}_k$. Let $x = (x_1, \ldots, x_d)$ and let $\mathbb{P}_k$ be the space of polynomials in the variables $x_1, \ldots, x_d$, with real coefficients and of global degree at most k,

$$\mathbb{P}_k = \left\{ p(x) = \sum_{\substack{0 \leq i_1, \ldots, i_d \leq k \\ i_1 + \ldots + i_d \leq k}} \alpha_{i_1 \ldots i_d} x_1^{i_1} \ldots x_d^{i_d}; \quad \alpha_{i_1 \ldots i_d} \in \mathbb{R} \right\}.$$

One readily verifies that $\mathbb{P}_k$ is a vector space of dimension

$$\dim \mathbb{P}_k = \binom{d+k}{k} = \begin{cases} k+1 & \text{if } d = 1, \\ \frac{1}{2}(k+1)(k+2) & \text{if } d = 2, \\ \frac{1}{6}(k+1)(k+2)(k+3) & \text{if } d = 3. \end{cases}$$

Proposition 1.34. *Let K be a simplex in $\mathbb{R}^d$. Let $k \geq 1$, let $P = \mathbb{P}_k$, and let $n_{\text{sh}} = \dim \mathbb{P}_k$. Consider the set of nodes $\{a_i\}_{1 \leq i \leq n_{\text{sh}}}$ with barycentric coordinates*

$$\left(\tfrac{i_0}{k}, \ldots, \tfrac{i_d}{k} \right), \quad 0 \leq i_0, \ldots, i_d \leq k, \quad i_0 + \ldots + i_d = k.$$

Let $\Sigma = \{\sigma_1, \ldots, \sigma_{n_{\text{sh}}}\}$ be the linear forms such that $\sigma_i(p) = p(a_i)$, $1 \leq i \leq n_{\text{sh}}$. Then, $\{K, P, \Sigma\}$ is a Lagrange finite element.

Proof. See Exercise 1.3. □

Table 1.1 presents examples for $k = 1, 2,$ and 3 in dimension 2 and 3. For $k = 1$, the $(d+1)$ local shape functions are the barycentric coordinates

$$\theta_i = \lambda_i, \quad 0 \leq i \leq d.$$

For $k = 2$, the local shape functions are

$$\begin{cases} \lambda_i(2\lambda_i - 1), & 0 \leq i \leq d, \\ 4\lambda_i\lambda_j, & 0 \leq i < j \leq d, \end{cases}$$

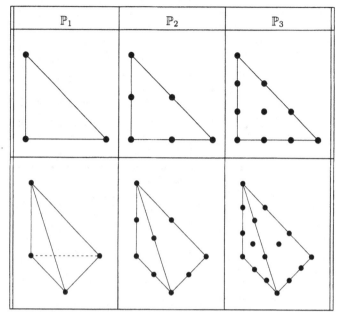

Table 1.1. Two- and three-dimensional $\mathbb{P}_1$, $\mathbb{P}_2$, and $\mathbb{P}_3$ Lagrange finite elements; in three dimensions, only visible degrees of freedom are shown.

and for $k = 3$,

$$
\begin{cases}
\frac{1}{2}\lambda_i(3\lambda_i - 1)(3\lambda_i - 2), & 0 \le i \le d, \\
\frac{9}{2}\lambda_i(3\lambda_i - 1)\lambda_j, & 0 \le i, j \le d,\ i \ne j, \\
27\lambda_i\lambda_j\lambda_k, & 0 \le i < j < k \le d.
\end{cases}
$$

1.2.4 Tensor product Lagrange finite elements

Cuboids. Given a set of d intervals $\{[c_i, d_i]\}_{1 \le i \le d}$, all with non-zero measure, the set $K = \prod_{i=1}^{d}[c_i, d_i]$ is called a *cuboid*. For $x \in K$, there exists a unique vector $(t_1, \dots, t_d) \in [0, 1]^d$ such that, for all $1 \le i \le d$, $x_i = c_i + t_i(d_i - c_i)$. The vector $(t_1, \dots, t_d)$ is called the local coordinate vector of x in K.

The polynomial space $\mathbb{Q}_k$. Let $\mathbb{Q}_k$ be the polynomial space in the variables $x_1, \dots, x_d$, with real coefficients and of degree at most k in each variable. In dimension 1, $\mathbb{Q}_k = \mathbb{P}_k$; in dimension $d \ge 2$,

$$
\mathbb{Q}_k = \left\{ q(x) = \sum_{0 \le i_1, \dots, i_d \le k} \alpha_{i_1 \dots i_d} x_1^{i_1} \dots x_d^{i_d}; \quad \alpha_{i_1 \dots i_d} \in \mathbb{R} \right\}.
$$

One readily verifies that $\mathbb{Q}_k$ is a vector space of dimension

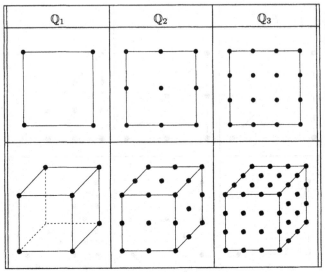

Table 1.2. Two- and three-dimensional $\mathbb{Q}_1$, $\mathbb{Q}_2$, and $\mathbb{Q}_3$ Lagrange finite elements; in three dimensions, only visible degrees of freedom are shown.

$$\dim \mathbb{Q}_k = (k+1)^d = \begin{cases} (k+1)^2 & \text{if } d = 2, \\ (k+1)^3 & \text{if } d = 3. \end{cases}$$

Note the inclusions $\mathbb{P}_k \subset \mathbb{Q}_k \subset \mathbb{P}_{kd}$.

Proposition 1.35. *Let K be a cuboid in $\mathbb{R}^d$. Let $k \geq 1$, let $P = \mathbb{Q}_k$, and let $n_{\mathrm{sh}} = \dim \mathbb{Q}_k$. Consider the set of nodes $\{a_i\}_{1 \leq i \leq n_{\mathrm{sh}}}$ with local coordinates*

$$\left(\tfrac{i_1}{k}, \ldots, \tfrac{i_d}{k} \right), \quad 0 \leq i_1, \ldots, i_d \leq k.$$

Let $\Sigma = \{\sigma_1, \ldots, \sigma_{n_{\mathrm{sh}}}\}$ be the linear forms such that $\sigma_i(p) = p(a_i)$, $1 \leq i \leq n_{\mathrm{sh}}$. Then, $\{K, P, \Sigma\}$ is a Lagrange finite element.

Table 1.2 presents examples for $k = 1$, 2, and 3 in dimension 2 and 3. For $1 \leq i \leq d$, set $\xi_{i,l} = c_i + \tfrac{l}{k}(d_i - c_i)$, $0 \leq l \leq k$, and let $\{\mathcal{L}_{i,0}^k, \ldots, \mathcal{L}_{i,k}^k\}$ be the Lagrange polynomials in the variable x_i associated with the nodes $\{\xi_{i,0}, \ldots, \xi_{i,k}\}$; see Definition 1.7. Then, the local shape functions are

$$\theta_{i_1 \ldots i_d}(x) = \mathcal{L}_{1,i_1}^k(x_1) \ldots \mathcal{L}_{d,i_d}^k(x_d), \quad 0 \leq i_1, \ldots, i_d \leq k.$$

1.2.5 Prismatic Lagrange finite elements

Prisms. For $x \in \mathbb{R}^d$, set $x' = (x_1, \ldots, x_{d-1})$. Let K' be a simplex in $\mathbb{R}^{d-1}$ and let $[a, b]$ be an interval with non-zero measure. Then, the set $K = \{x \in \mathbb{R}^d; \ x' \in K'; \ x_d \in [a, b]\}$ is called a *prism*. Let $(\lambda_0, \ldots, \lambda_{d-1})$ be the barycentric coordinates of x' in K' and let $t \in [0, 1]$ be such that $x_d = a + t(b - a)$. Then, the prismatic coordinates of $x \in K$ are defined to be $(\lambda_0, \ldots, \lambda_{d-1}; t)$.

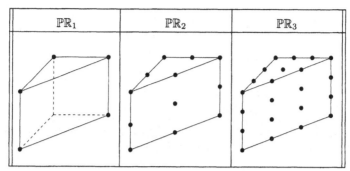

Table 1.3. Prismatic Lagrange finite elements of degree 1, 2, and 3; only visible degrees of freedom are shown.

Prismatic polynomials. Let $\mathbb{P}_k[x']$ (resp., $\mathbb{P}_k[x_d]$) be the set of polynomials with real coefficients in the variable x' (resp., x_d) of global degree at most k. Set

$$\mathbb{PR}_k = \{p(x) = p_1(x')\,p_2(x_d); \ p_1 \in \mathbb{P}_k[x'], \ p_2 \in \mathbb{P}_k[x_d]\}.$$

Clearly, $\mathbb{P}_k \subset \mathbb{PR}_k$ and $\dim \mathbb{PR}_k = \frac{1}{2}(k+1)^2(k+2)$ in dimension 3.

Proposition 1.36. *Let K be a prism in $\mathbb{R}^d$. Let $k \geq 1$, let $P = \mathbb{PR}_k$, and let $n_{\mathrm{sh}} = \dim \mathbb{PR}_k$. Consider the set of nodes $\{a_i\}_{1 \leq i \leq n_{\mathrm{sh}}}$ with prismatic coordinates*

$$\left(\tfrac{i_0}{k}, \ldots, \tfrac{i_{d-1}}{k}; \tfrac{i_d}{k}\right), \quad 0 \leq i_0, \ldots, i_{d-1}, i_d \leq k, \quad i_0 + \ldots + i_{d-1} = k.$$

Let $\Sigma = \{\sigma_1, \ldots, \sigma_{n_{\mathrm{sh}}}\}$ be the linear forms such that $\sigma_i(p) = p(a_i)$, $1 \leq i \leq n_{\mathrm{sh}}$. Then, $\{K, P, \Sigma\}$ is a Lagrange finite element.

Table 1.3 presents examples for $k = 1$, 2, and 3. The local shape functions can be expressed in tensor product form using the local shape functions on the simplex K' and the Lagrange polynomials in x_d.

1.2.6 The Crouzeix–Raviart finite element

Let K be a simplex in $\mathbb{R}^d$, set $P = \mathbb{P}_1$, and take for the local degrees of freedom the mean-value over the $(d+1)$ faces of K, i.e., for $0 \leq i \leq d$,

$$\sigma_i(p) = \frac{1}{\mathrm{meas}(F_i)} \int_{F_i} p.$$

Proposition 1.37. *Let $\Sigma = \{\sigma_i\}_{0 \leq i \leq d}$. Then, $\{K, \mathbb{P}_1, \Sigma\}$ is a finite element.*

Using the barycentric coordinates $\{\lambda_0, \ldots, \lambda_d\}$ defined in (1.37), the local shape functions are

$$\theta_i(x) = d\left(\frac{1}{d} - \lambda_i(x)\right), \quad 0 \leq i \leq d. \tag{1.38}$$

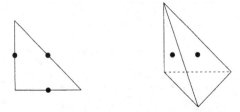

Fig. 1.8. Crouzeix–Raviart finite element in two (left) and three (right) dimensions; in three dimensions, only visible degrees of freedom are shown.

Indeed, $\theta_i \in \mathbb{P}_1$ and $\sigma_j(\theta_i) = \delta_{ij}$ for $0 \leq i, j \leq d$. Note that $\theta_{i|F_i} = 1$ and $\theta_i(a_i) = 1 - d$.

A conventional representation of the Crouzeix–Raviart finite element is shown in Figure 1.8. The dot means that the mean-value is taken over the corresponding face. This finite element has been introduced by Crouzeix and Raviart; see [CrR73] and also [BrF91^b, pp. 107–109].

An admissible choice for the domain of the local interpolation operator is $V(K) = W^{1,1}(K)$. Indeed, owing to the Trace Theorem B.52 applied with $p = 1$, the trace of a function in $W^{1,1}(K)$ is in $L^1(\partial K)$. The *local Crouzeix–Raviart interpolation operator* is then defined as follows:

$$\mathcal{I}_K^{\mathrm{CR}} : V(K) \ni v \longmapsto \mathcal{I}_K^{\mathrm{CR}} v = \sum_{i=0}^{d} \left(\tfrac{1}{\operatorname{meas} F_i} \int_{F_i} v \right) \theta_i \in \mathbb{P}_1. \tag{1.39}$$

Remark 1.38.

(i) Since a polynomial in P is linear, its mean-value over a face is equal to the value it takes at the barycenter. Therefore, another possible choice for the degrees of freedom is to take the value at the face barycenters. The resulting finite element is a Lagrange finite element according to Definition 1.27. The only difference with the Crouzeix–Raviart finite element is that it is no longer possible to take $W^{1,1}(K)$ for the domain of the local interpolation operator; an admissible choice is, for instance, $V(K) = \mathcal{C}^0(K)$.

(ii) Another choice for the local degrees of freedom is $\sigma_i(p) = \int_{F_i} p$ for $0 \leq i \leq d$; then, the local shape functions are $\theta_i = \frac{d}{\operatorname{meas}(F_i)} \left(\tfrac{1}{d} - \lambda_i \right)$.   □

1.2.7 The Raviart–Thomas finite element

Let K be a simplex in $\mathbb{R}^d$. Consider the vector space of $\mathbb{R}^d$-valued polynomials

$$\mathbb{RT}_0 = [\mathbb{P}_0]^d \oplus x\,\mathbb{P}_0. \tag{1.40}$$

Clearly, the dimension of $\mathbb{RT}_0$ is $d + 1$. For $p \in \mathbb{RT}_0$, the local degrees of freedom are chosen to be the value of the flux of the normal component of p across the faces of K, i.e., for $0 \leq i \leq d$,

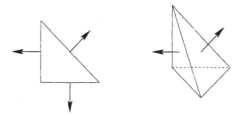

Fig. 1.9. Raviart–Thomas finite element in two (left) and three (right) dimensions; in three dimensions, only visible degrees of freedom are shown.

$$\sigma_i(p) = \int_{F_i} p \cdot n_i.$$

Proposition 1.39. *Let* $\Sigma = \{\sigma_i\}_{0 \leq i \leq d}$. *Then,* $\{K, \mathbb{RT}_0, \Sigma\}$ *is a finite element.*

The local shape functions are

$$\theta_i(x) = \frac{1}{d \operatorname{meas}(K)}(x - a_i), \quad 0 \leq i \leq d. \tag{1.41}$$

Indeed, $\theta_i \in \mathbb{RT}_0$ and $\sigma_j(\theta_i) = \delta_{ij}$ for $0 \leq i, j \leq d$. Note that the normal component of a local shape function is constant on the face with which it is associated and is zero on the other faces.

A conventional representation of the degrees of freedom of the Raviart–Thomas finite element is shown in Figure 1.9. An arrow means that the flux of the normal component is taken over the corresponding face. This finite element has been introduced by Raviart and Thomas and is often referred to as the $\mathbb{RT}_0$ finite element [RaT77]. It is used, for instance, in applications related to fluid mechanics where the functions to be interpolated are velocities.

The domain of the local interpolation operator can be taken to be $V^{\operatorname{div}}(K) = \{v \in [L^p(K)]^d; \nabla \cdot v \in L^s(K)\}$ for $p > 2$, $s \geq q$, $\frac{1}{q} = \frac{1}{p} + \frac{1}{d}$. Note that $V^{\operatorname{div}}(K) = W^{1,t}(K)$ with $t > \frac{2d}{d+2}$ is also an admissible choice. Indeed, one can show that for $v \in V^{\operatorname{div}}(K)$ and for a face F_i of K, the quantity $\int_{F_i} v \cdot n_i$ is meaningful. The *local Raviart–Thomas interpolation operator* is then defined as follows:

$$\mathcal{I}_K^{\mathrm{RT}} : V^{\operatorname{div}}(K) \ni v \longmapsto \mathcal{I}_K^{\mathrm{RT}} v = \sum_{i=0}^{d} \left(\int_{F_i} v \cdot n_i \right) \theta_i \in \mathbb{RT}_0. \tag{1.42}$$

Remark 1.40.

(i) See Exercise 1.5 for the proofs of the above results and for an alternative expression of the local shape functions in terms of barycentric coordinates. Further results can be found in [BrF91^b, p. 113] and [QuV97, p. 82].

(ii) In the spirit of Remark 1.38, the Raviart–Thomas finite element can be defined as a Lagrange finite element. Another choice for the degrees of freedom is $\sigma_i(p) = \frac{1}{\text{meas}(F_i)} \int_{F_i} p \cdot n_i$; then, the local shape functions are $\theta_i = \frac{\text{meas}(F_i)}{d \, \text{meas}(K)} (x - a_i)$.   □

Lemma 1.41. *Let* $\mathcal{I}_K^{\mathrm{RT}}$ *be defined in* (1.42). *Let* π_K^0 *be the orthogonal projection from* $L^2(K)$ *to* $\mathbb{P}_0$. *The following diagram commutes:*

$$
\begin{array}{ccc}
V^{\mathrm{div}}(K) & \xrightarrow{\ \nabla \cdot\ } & L^2(K) \\
\Big\downarrow{\scriptstyle \mathcal{I}_K^{\mathrm{RT}}} & & \Big\downarrow{\scriptstyle \pi_K^0} \\
\mathrm{RT}_0 & \xrightarrow{\ \nabla \cdot\ } & \mathbb{P}_0
\end{array}
$$

Proof. Left as an exercise.   □

1.2.8 The Nédélec (or edge) finite element

Let K be a simplex in $\mathbb{R}^d$, $d = 2$ or 3. Define the polynomial space of dimension $\frac{1}{2}d(d+1)$,

$$
\mathbb{N}_0 = [\mathbb{P}_0]^d \oplus \mathcal{R}_1, \qquad \mathcal{R}_1 = \{ p \in [\mathbb{P}_1]^d; \ x \cdot p = 0 \}. \tag{1.43}
$$

Introducing the mapping $\mathcal{R} : \mathbb{R}^2 \to \mathbb{R}^2$ such that $\mathcal{R}(x_1, x_2) = (x_2, -x_1)$, the following equivalent definition of $\mathbb{N}_0$ holds in dimension 2:

$$
\mathbb{N}_0 = [\mathbb{P}_0]^2 \oplus (\mathcal{R}(x)\mathbb{P}_0). \tag{1.44}
$$

In dimension 3, the following equivalent definition of $\mathbb{N}_0$ holds:

$$
\mathbb{N}_0 = [\mathbb{P}_0]^3 \oplus (x \times [\mathbb{P}_0]^3). \tag{1.45}
$$

For $p \in \mathbb{N}_0$, the local degrees of freedom are chosen to be the integral of the tangential component of p along the three (resp., six) edges of K in two (resp., three) dimensions. Set $n_e = 3$ if $d = 2$ and $n_e = 6$ if $d = 3$. Denote by $\{e_i\}_{1 \le i \le n_e}$ the set of edges of K and, for each edge e_i, let t_i be one of the two unit vectors parallel to e_i. For $1 \le i \le n_e$, the local degrees of freedom are

$$
\sigma_i(p) = \int_{e_i} p \cdot t_i.
$$

Proposition 1.42. *Let* $\Sigma = \{\sigma_i\}_{1 \le i \le n_e}$. *Then,* $\{K, \mathbb{N}_0, \Sigma\}$ *is a finite element.*

In two dimensions, the local shape function associated with the edge e_i, $1 \le i \le 3$, is

$$
\theta_i(x) = \frac{\mathcal{R}(x - a_i)}{t_i \cdot [\mathcal{R}(\frac{a_{i_1} + a_{i_2}}{2} - a_i)] \, \text{meas}(e_i)}, \qquad i_1, i_2 \ne i. \tag{1.46}
$$

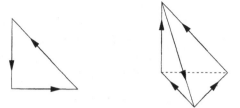

Fig. 1.10. Edge finite element in dimension 2 (left) and 3 (right); in three dimensions, only visible degrees of freedom are shown.

In three dimensions, define the mapping $j : \{1, \dots, 6\} \to \{1, \dots, 6\}$ such that $j(i)$ is the index of the edge opposite to e_i, i.e., e_i does not intersect $e_{j(i)}$. Note that $j = j^{-1}$. Let m_i be the midpoint of e_i. Then, the local shape function associated with the edge e_i, $1 \le i \le 6$, is

$$\theta_i(x) = \frac{(x - m_{j(i)}) \times t_{j(i)}}{t_i \cdot \left[(m_i - m_{j(i)}) \times t_{j(i)} \right] \operatorname{meas}(e_i)}. \tag{1.47}$$

In both cases, the tangential component of a local shape function is constant along the edge with which it is associated and vanishes along the other edges.

A conventional representation of the edge finite element is shown in Figure 1.10. An arrow means that the integral of the component parallel to this direction is taken over the corresponding edge. This finite element has been introduced by Nédélec [Néd80, Néd86]; see also [Whi57]. It is used, for instance, in electromagnetism and in magneto-hydrodynamics; see [Bos93, Chap. 3].

In two dimensions, the domain of the local interpolation operator can be taken to be $V^{\mathrm{curl}}(K) = \{v = (v_1, v_2) \in [L^p(K)]^2; \partial_2 v_1 - \partial_1 v_2 \in L^p(K)\}$ for $p > 2$. Indeed, one can show that for $v \in V^{\mathrm{curl}}(K)$ and for an edge e_i of K, the quantity $\int_{e_i} v \cdot t_i$ is meaningful. In three dimensions, a suitable choice is $V^{\mathrm{curl}}(K) = \{v \in [H^s(K)]^3; \nabla \times v \in [L^p(K)]^3\}$ for $s > \frac{1}{2}$ and $p > 2$; see, e.g., [AmB98]. The *local Nédélec interpolation operator* is then defined as follows:

$$\mathcal{I}_K^{\mathrm{N}} : V^{\mathrm{curl}}(K) \ni v \longmapsto \mathcal{I}_K^{\mathrm{N}} v = \sum_{i=1}^{n_e} \left(\int_{e_i} v \cdot t_i \right) \theta_i \in \mathbb{N}_0. \tag{1.48}$$

Remark 1.43.

(i) See Exercise 1.6 for the proofs of the above results and for an alternative expression of the local shape functions in terms of barycentric coordinates.

(ii) In the spirit of Remark 1.38, the Nédélec finite element can be defined as a Lagrange finite element. Another choice for the degrees of freedom is $\sigma_i(p) = \frac{1}{\operatorname{meas}(e_i)} \int_{e_i} p \cdot t_i$ for $1 \le i \le n_e$; the local shape functions are then readily derived from (1.46) and (1.47). □

Lemma 1.44. *Assume $d = 3$. Let $\mathcal{I}_K^{\mathrm{RT}}$ and $\mathcal{I}_K^{\mathrm{N}}$ be defined in (1.42) and (1.48), respectively. The following diagram commutes:*

$$V^{\mathrm{curl}}(K) \xrightarrow{\ \nabla\times\ } V^{\mathrm{div}}(K)$$

$$\Big\downarrow \mathcal{I}_K^N \qquad\qquad\qquad \Big\downarrow \mathcal{I}_K^{RT}$$

$$N_0 \xrightarrow{\ \nabla\times\ } RT_0$$

Proof. Let $v \in V^{\mathrm{curl}}(K)$. It is clear that $\nabla\times\mathcal{I}_K^N v \in [\mathbb{P}_0]^3 \subset RT_0$. Let F be a face of K and n_F be the corresponding outward normal. Then,

$$\int_F (\nabla\times(\mathcal{I}_K^N v))\cdot n_F = \sum_{e \subset \partial F} \int_e \mathcal{I}_K^N v\cdot t_e = \sum_{e \subset \partial F} \int_e v\cdot t_e$$

$$= \int_F (\nabla\times v)\cdot n_F = \int_F (\mathcal{I}_K^{RT}(\nabla\times v))\cdot n_F,$$

where t_e is a unit vector parallel to the edge e so that the edge integrals are taken anti-clockwise along ∂F. The above equality implies $\mathcal{I}_K^{RT}(\nabla\times v) = \nabla\times(\mathcal{I}_K^N v)$, since these two functions are in RT_0 and their fluxes across the faces of K are identical. $\qquad\square$

Lemma 1.45. *Assume $d = 2$ or 3. Let $\mathcal{I}_K^1$ be the interpolation operator associated with the $\mathbb{P}_1$ Lagrange finite element and let $V^1(K) = H^s(K)$ be its domain, $s > \frac{d}{2}$. The following diagram commutes:*

$$V^1(K) \xrightarrow{\ \nabla\ } V^{\mathrm{curl}}(K)$$

$$\Big\downarrow \mathcal{I}_K^1 \qquad\qquad\qquad \Big\downarrow \mathcal{I}_K^N$$

$$\mathbb{P}_1 \xrightarrow{\ \nabla\ } N_0$$

Proof. Le $v \in V^1(K)$. Let e be an edge of K and denote by a_1, a_2 the two vertices of e. Set $t = \frac{a_2 - a_1}{\|a_2 - a_1\|_d}$ to obtain

$$\int_e \nabla(\mathcal{I}_K^1 v)\cdot t = \mathcal{I}_K^1 v(a_2) - \mathcal{I}_K^1 v(a_1) = v(a_2) - v(a_1)$$

$$= \int_e (\nabla v)\cdot t = \int_e \mathcal{I}_K^N (\nabla v)\cdot t.$$

Conclude using the fact that both $\mathcal{I}_K^N(\nabla v)$ and $\nabla(\mathcal{I}_K^1 v)$ belong to N_0. $\qquad\square$

1.2.9 High-order finite elements

As in the one-dimensional case, basis functions must be selected carefully when working with high-degree polynomials.

Nodal finite elements. When K is a simplex in $\mathbb{R}^d$ and $P = \mathbb{P}_k$ with k large, it is possible to define sets of quadrature points with near optimal interpolation properties: the so-called *Fekete points*; see [ChB95, TaW00]. Then, these points can be used as *Lagrange nodes* to define nodal bases. Finite element methods using the Fekete points as Lagrange nodes when k is large are often referred to as *spectral element methods*; see, e.g., [KaS99^b].

When K is a cuboid in $\mathbb{R}^d$ and $P = \mathbb{Q}_k$, one can use the tensor product of Gauß–Lobatto nodes instead of equi-distributing the Lagrange nodes in each space direction. Then, the local shape functions are

$$\theta_{i_1,\ldots,i_d}(x_1,\ldots,x_d) = \theta_{i_1}(x_1)\ldots\theta_{i_d}(x_d), \quad 0 \leq i_1,\ldots,i_d \leq k, \qquad (1.49)$$

where the functions $\{\theta_i\}_{0 \leq i \leq k}$ are the images by suitable mappings of the *nodal (C^0-continuous) basis functions* defined in (1.33). An interesting property of the Gauß–Lobatto points is that they are the Fekete points for the d-dimensional cuboid, i.e., these points have near optimal interpolation properties; see [BoT01].

Modal finite elements. When K is a cuboid, hierarchical modal bases can be constructed using tensor products of one-dimensional hierarchical modal bases. For instance, we can consider the basis functions defined in (1.49), where the functions $\{\theta_i\}_{0 \leq i \leq k}$ are now the images by suitable mappings of the *modal (C^0-continuous) basis functions* defined in (1.31).

When K is a simplex or a prism, the construction of hierarchical bases is more technical. The idea is to introduce a nonlinear transformation mapping K to a square or a cube and to use tensor products of one-dimensional bases. See [KaS99^b, pp. 70–94] for a detailed presentation.

1.3 Meshes: Basic Concepts

This section presents the general principles governing the construction of a mesh. Implementation aspects are investigated in Chapter 7.

1.3.1 Domains and meshes

Throughout this book, we shall use the following:

Definition 1.46 (Domain). *In dimension 1, a domain is an open, bounded interval. In dimension $d \geq 2$, a domain is an open, bounded, connected set in $\mathbb{R}^d$ such that its boundary $\partial\Omega$ satisfies the following property: There are $\alpha > 0$, $\beta > 0$, a finite number R of local coordinate systems $x^r = (x^{r\prime}, x_d^r)$, $1 \leq r \leq R$, where $x^{r\prime} \in \mathbb{R}^{d-1}$ and $x_d^r \in \mathbb{R}$, and R local maps ϕ^r that are Lipschitz on their definition domain $\{x^{r\prime} \in \mathbb{R}^{d-1}; |x^{r\prime}| < \alpha\}$ and such that*

$$\partial\Omega = \bigcup_{r=1}^{R} \{(x^{r'}, x_d^r); \ x_d^r = \phi^r(x^{r'}); \ |x^{r'}| < \alpha\},$$

$$\{(x^{r'}, x_d^r); \ \phi^r(x^{r'}) < x_d^r < \phi^r(x^{r'}) + \beta; \ |x^{r'}| < \alpha\} \subset \Omega, \qquad \forall r,$$

$$\{(x^{r'}, x_d^r); \ \phi^r(x^{r'}) - \beta < x_d^r < \phi^r(x^{r'}); \ |x^{r'}| < \alpha\} \subset \mathbb{R}^d \backslash \overline{\Omega}, \qquad \forall r,$$

where $|x^{r'}| \leq \alpha$ means that $|x_i^{r'}| \leq \alpha$ for all $1 \leq i \leq d - 1$. For $m \geq 1$, Ω is said to be of class $\mathcal{C}^m$ (resp., piecewise of class $\mathcal{C}^m$) if all the local maps Φ^r are of class $\mathcal{C}^m$ (resp., piecewise of class $\mathcal{C}^m$).

Definition 1.47 (Polygon, polyhedron). *In dimension 2, a polygon is a domain whose boundary is a finite union of segments. In dimension 3, a polyhedron is a domain whose boundary is a finite union of polygons. When the distinction is not relevant, the term polyhedron is also employed for polygons.*

Remark 1.48.

(i) Definition 1.46 implies that a domain is necessarily located on one side of its boundary $\partial\Omega$, i.e., it excludes sets with slits. This assumption can be weakened, but this involves technical complexities that go beyond the scope of this book; see, e.g., [CoD02].

(ii) For a domain Ω in $\mathbb{R}^d$ with $d \geq 2$, the outward normal, say n, is defined for a.e. $x \in \partial\Omega$. For a domain of class $\mathcal{C}^m$, $m \geq 1$, n is defined for all $x \in \partial\Omega$ and is a function of class $\mathcal{C}^{m-1}$.

(iii) Definition 1.47 can be extended to arbitrary dimension d by induction: a polyhedron in $\mathbb{R}^d$ is a domain whose boundary is a finite union of polyhedra in $\mathbb{R}^{d-1}$. □

Definition 1.49 (Mesh). *Let Ω be a domain in $\mathbb{R}^d$. A mesh is a union of a finite number N_{el} of compact, connected, Lipschitz sets K_m with non-empty interior such that $\{K_m\}_{1 \leq m \leq N_{el}}$ forms a partition of Ω, i.e.,*

$$\overline{\Omega} = \bigcup_{m=1}^{N_{el}} K_m \qquad and \qquad \overset{\circ}{K}_m \cap \overset{\circ}{K}_n = \emptyset \qquad for \ m \neq n. \qquad (1.50)$$

The subsets K_m are called mesh cells *or* mesh elements *(or simply* elements *when there is no ambiguity).*

Figure 1.11 presents an example of a mesh of the unit square in $\mathbb{R}^2$ involving triangles and quadrangles. In the sequel, a mesh $\{K_m\}_{1 \leq m \leq N_{el}}$ is denoted by $\mathcal{T}_h$. The subscript h refers to the level of refinement of the mesh. Setting

$$\forall K \in \mathcal{T}_h, \quad h_K = \text{diam}(K) = \max_{x_1, x_2 \in K} \|x_1 - x_2\|_d,$$

where $\|\cdot\|_d$ is the Euclidean norm in $\mathbb{R}^d$, the parameter h is defined by

$$h = \max_{K \in \mathcal{T}_h} h_K.$$

A sequence of successively refined meshes is denoted by $\{\mathcal{T}_h\}_{h>0}$.

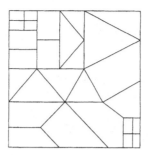

Fig. 1.11. Example of a mesh of the unit square in $\mathbb{R}^2$.

1.3.2 Mesh generation

In practice, a mesh is generated from a *reference cell*, say $\widehat{K}$, and a set of geometric transformations mapping $\widehat{K}$ to the actual mesh cells. We shall henceforth assume that the geometric transformations are $\mathcal{C}^1$-diffeomorphisms. For $K \in \mathcal{T}_h$, denote by $T_K : \widehat{K} \to K$ the corresponding transformation. Usually T_K is specified using a Lagrange finite element $\{\widehat{K}, \widehat{P}_{\text{geo}}, \widehat{\Sigma}_{\text{geo}}\}$. Let $n_{\text{geo}} = \text{card}(\widehat{\Sigma}_{\text{geo}})$, let $\{\widehat{g}_1, \ldots, \widehat{g}_{n_{\text{geo}}}\}$ be the nodes of $\widehat{K}$ associated with $\widehat{\Sigma}_{\text{geo}}$, and let $\{\widehat{\psi}_1, \ldots, \widehat{\psi}_{n_{\text{geo}}}\}$ be the local shape functions.

Definition 1.50. $\{\widehat{K}, \widehat{P}_{\text{geo}}, \widehat{\Sigma}_{\text{geo}}\}$ *is called the* geometric (reference) finite element, $\{\widehat{g}_1, \ldots, \widehat{g}_{n_{\text{geo}}}\}$ *the* geometric (reference) nodes, *and* $\{\widehat{\psi}_1, \ldots, \widehat{\psi}_{n_{\text{geo}}}\}$ *the* geometric (reference) shape functions.

For the sake of simplicity, assume that all the mesh cells are generated using the *same* geometric reference finite element. This assumption can be easily lifted. When $\widehat{K}$ is a simplex, $\mathcal{T}_h$ is called a *simplicial mesh*.

A mesh generator usually provides a list of n_{geo}-uplets

$$\{g_1^m, \ldots, g_{n_{\text{geo}}}^m\}_{1 \le m \le N_{\text{el}}},$$

where $g_i^m \in \mathbb{R}^d$ and N_{el} is the number of mesh elements. The points $\{g_1^m, \ldots, g_{n_{\text{geo}}}^m\}$ are called the *geometric nodes* of the m-th element. For $1 \le m \le N_{\text{el}}$, define the *geometric transformation*

$$T_m : \widehat{K} \ni \widehat{x} \longmapsto T_m(\widehat{x}) = \sum_{i=1}^{n_{\text{geo}}} g_i^m \widehat{\psi}_i(\widehat{x}) \in \mathbb{R}^d, \qquad (1.51)$$

so that $T_m(\widehat{g}_i) = g_i^m$ for $1 \le i \le n_{\text{geo}}$, and set $K_m = T_m(\widehat{K})$.

Remark 1.51. The hypothesis that the geometric transformation T_m is a $\mathcal{C}^1$-diffeomorphism requires that the numbering of the nodes $\{g_1^m, \ldots, g_{n_{\text{geo}}}^m\}$ and that employed in the reference element are compatible; see Figure 1.12. An usual convention is to impose the additional requirement that the numbering

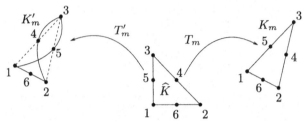

Fig. 1.12. The numbering of the nodes of K_m (right) is compatible with that of $\widehat{K}$; the numbering in K'_m (left) is not.

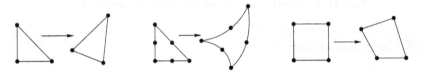

Fig. 1.13. Examples of geometric transformations: $\mathbb{P}_1$ transformation of a triangle (left); $\mathbb{P}_2$ transformation of a triangle (center); $\mathbb{Q}_1$ transformation of a quadrangle (right).

is such that the Jacobian determinant of the transformation T_m is positive. For instance, in two dimensions, there are three compatible ways of numbering the nodes of a triangle, and there are four compatible ways of numbering the nodes of a square. In three dimensions, there are 3×4 compatible ways of numbering the nodes of a tetrahedron, and there are 4×6 compatible ways of numbering the nodes of a cube.   □

Example 1.52. Figure 1.13 presents three examples in dimension 2:

(i) A transformation based on the Lagrange finite element $\mathbb{P}_1$ maps the unit simplex to a non-degenerate triangle.

(ii) A transformation based on the Lagrange finite element $\mathbb{P}_2$ maps the unit simplex to a curved triangle.

(iii) A transformation based on the Lagrange finite element $\mathbb{Q}_1$ maps the unit square to a non-degenerate quadrangle.   □

Definition 1.53 (Affine meshes). *When the transformations $\{T_m\}_{1 \le m \le N_{el}}$ are affine, the mesh is said to be* affine. *In dimension 2, when the reference cell $\widehat{K}$ is a simplex, an affine mesh is also called a* triangulation. *This terminology is used henceforth in any dimension for an affine, simplicial mesh.*

Examples of affine meshes include the following:

(i) When the geometric reference finite element is the Lagrange finite element $\mathbb{P}_1$, all the mesh elements are triangles in dimension 2 and tetrahedra in dimension 3.

(ii) When the geometric reference finite element is the Lagrange finite element $\mathbb{Q}_1$, all the mesh elements are parallelograms in dimension 2 and parallelepipeds in dimension 3.

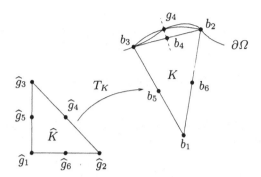

Fig. 1.14. Geometric construction of a curved triangle.

Domains with curved boundary. For domains with curved boundary, the use of affine meshes generates an interpolation error in the neighborhood of the boundary. Hence, when high-order accuracy is required, it is necessary to generate the mesh with geometric transformations of degree k_{geo} larger than one; the mesh then contains curved elements.

A relatively straightforward way to proceed is the following:

(i) Construct an affine mesh $\widetilde{\mathcal{T}}_h$ so that all the vertices of the resulting polyhedron lie on the curved boundary $\partial\Omega$.
(ii) For each element $\widetilde{K} \in \widetilde{\mathcal{T}}_h$ having a non-empty intersection with $\partial\Omega$, design a polynomial transformation (of degree larger than 1) that approximates the boundary more accurately than the first-order interpolation. The resulting element K replaces $\widetilde{K}$ in the mesh.

Example 1.54. The following example illustrates a simple technique relying on $\mathbb{P}_2$ Lagrange finite elements to approximate a curved boundary in $\mathbb{R}^2$ (see Figure 1.14):

(i) Let $\widetilde{K}$ be an element having an edge whose vertices lie on $\partial\Omega$. Let $\{b_1, \ldots, b_{n_{\text{geo}}}\}$ be the geometric nodes of $\widetilde{K}$ ($n_{\text{geo}} = 6$).
(ii) For each b_i, $1 \le i \le n_{\text{geo}}$, construct a new node g_i as follows:
  • If b_i is located on an edge whose vertices lie on $\partial\Omega$, g_i is defined as the intersection with $\partial\Omega$ of the line normal to the corresponding edge and passing through b_i.
  • Otherwise, set $g_i = b_i$.
(iii) Replace $\{b_1, \ldots, b_{n_{\text{geo}}}\}$ by $\{g_1, \ldots, g_{n_{\text{geo}}}\}$ in the list of n_{geo}-uplets provided by the mesh generator. In other words, replace $\widetilde{K}$ by the curved triangle $K = T_K(\widehat{K})$ where

$$\forall \widehat{x} \in \widehat{K}, \quad T_K(\widehat{x}) = \sum_{i=1}^{n_{\text{geo}}} g_i \widehat{\psi}_i(\widehat{x}),$$

and $\{\widehat{\psi}_1, \ldots, \widehat{\psi}_{n_{\text{geo}}}\}$ are the $\mathbb{P}_2$ Lagrange local shape functions.   □

Fig. 1.15. Examples of reference elements.

Note that a mesh consisting of curved triangles may not necessarily cover the domain Ω; see Figure 1.14. In other words, the open set Ω_h such that

$$\overline{\Omega}_h = \bigcup_{K \in \mathcal{T}_h} K, \qquad (1.52)$$

does not necessarily coincide with Ω; the domain Ω_h is called a *geometric interpolation* of Ω. For the sake of simplicity, $\mathcal{T}_h$ is said to be a mesh of Ω even though it may happen that $\Omega \neq \Omega_h$.

1.3.3 Geometrically conformal meshes

Henceforth we assume that the reference element $\widehat{K}$ used to generate the mesh is a polyhedron. Classical examples include the following (see Figure 1.15):

(i) $\widehat{K}$ is the *unit interval* $[0, 1]$ in dimension 1.
(ii) $\widehat{K}$ is either the *unit simplex* with vertices $(0,0)$, $(1,0)$, $(0,1)$ or the *unit square* $[0, 1]^2$ in dimension 2.
(iii) $\widehat{K}$ is either the *unit simplex* with vertices $(0,0,0)$, $(1,0,0)$, $(0,1,0)$, $(0,0,1)$, or the *unit cube* $[0, 1]^3$, or the *unit prism* with vertices $(0,0,0)$, $(1,0,0)$, $(0,1,0)$, $(0,0,1)$, $(1,0,1)$, $(0,1,1)$ in dimension 3.

For a given mesh cell $K = T_K(\widehat{K})$, the vertices, edges, and faces are defined to be the image by the geometric transformation T_K of the vertices, edges, and faces of the reference element $\widehat{K}$.

Definition 1.55 (Geometrically conformal meshes). *Let Ω be a domain in $\mathbb{R}^d$ and let $\mathcal{T}_h = \{K_m\}_{1 \le m \le N_{el}}$ be a mesh of Ω. The mesh $\mathcal{T}_h$ is said to be* geometrically conformal *if the following matching criterion is satisfied: For all K_m and K_n having a non-empty $(d-1)$-dimensional intersection, say $F = K_m \cap K_n$, there is a face $\widehat{F}$ of $\widehat{K}$ and renumberings of the geometric nodes of K_m and K_n such that $F = T_m(\widehat{F}) = T_n(\widehat{F})$ and*

$$T_{m|\widehat{F}} = T_{n|\widehat{F}}. \qquad (1.53)$$

If more than one geometric reference element is used to generate the mesh, say $\widehat{K}_1$ and $\widehat{K}_2$, (1.53) is replaced by the following statement: $T_{m|F}^{-1}(F)$ is a face of $\widehat{K}_1$, $T_{n|F}^{-1}(F)$ is a face of $\widehat{K}_2$, and there is a bijective affine transformation mapping $T_{m|F}^{-1}(F)$ to $T_{n|F}^{-1}(F)$.

Fig. 1.16. Example and counterexample of a geometrically conformal mesh.

Remark 1.56. If Ω_h is connected, Definition 1.55 implies that for any cell pair $\{K_m, K_n\}$ with $m \neq n$, the intersection $K_m \cap K_n$ is:

(i) either empty or a common vertex in dimension 1;

(ii) either empty, or a common vertex, or a common edge in dimension 2;

(iii) either empty, or a common vertex, or a common edge, or a common face in dimension 3.

An example and a counterexample of a geometrically conformal mesh are shown in Figure 1.16. □

Geometrically conformal meshes form a particular class of meshes that are convenient to generate H^1-conformal approximation spaces; see §1.4.5. Moreover, on such meshes, the Euler relations provide useful means to count global degrees of freedom.

Lemma 1.57 (Euler relations). *Let $\mathcal{T}_h$ be a geometrically conformal mesh and let Ω_h be defined in (1.52).*

(i) *In dimension 2, let I be the degree of multiple-connectedness[1] of Ω_h, N_{el} the number of cells (or elements), N_{ed} the number of edges, N_v the number of vertices, N_{ed}^{∂} the number of boundary edges, and N_v^{∂} the number of boundary vertices; then,*

$$\begin{cases} N_{el} - N_{ed} + N_v = 1 - I, \\ N_v^{\partial} - N_{ed}^{\partial} = 0. \end{cases}$$

Furthermore, if the mesh cells are polygons with ν vertices,

$$2N_{ed} - N_{ed}^{\partial} = \nu N_{el}.$$

In particular, $2N_{ed} - N_{ed}^{\partial} = 3N_{el}$ for triangles and $2N_{ed} - N_{ed}^{\partial} = 4N_{el}$ for quadrangles.

(ii) *In dimension 3, let I be the degree of multiple-connectedness of Ω_h, J the number of connected components of the boundary of Ω_h, N_{el} the number of elements, N_f the number of faces, N_{ed} the number of edges, N_v the*

---

[1] I is the number of holes in Ω_h.

number of vertices, N_f^∂ the number of boundary faces, N_{ed}^∂ the number of boundary edges, and N_v^∂ the number of boundary vertices; then,

$$\begin{cases} N_{el} - N_f + N_{ed} - N_v = -1 + I - J, \\ N_f^\partial - N_{ed}^\partial + N_v^\partial = 2(J - I). \end{cases}$$

Furthermore, if the mesh cells are polyhedra with ν faces,

$$2N_f - N_f^\partial = \nu N_{el}.$$

In particular, $2N_f - N_f^\partial = 4N_{el}$ for tetrahedra and $2N_f - N_f^\partial = 6N_{el}$ for hexahedra.

1.3.4 Faces, edges, and jumps

Henceforth, we denote by $\mathcal{F}_h^i$ the set of interior faces (or interfaces), i.e., $F \in \mathcal{F}_h^i$ if F is a $(d-1)$-manifold and there are K_1, $K_2 \in \mathcal{T}_h$ such that $F = K_1 \cap K_2$. We denote by $\mathcal{F}_h^\partial$ the set of the faces that separate the mesh from the exterior of Ω_h, i.e., $F \in \mathcal{F}_h^\partial$ if F is a $(d-1)$-manifold and there is $K \in \mathcal{T}_h$ such that $F = K \cap \partial\Omega_h$. Finally, we set $\mathcal{F}_h = \mathcal{F}_h^i \cup \mathcal{F}_h^\partial$. In all dimensions we refer to the elements of $\mathcal{F}_h$ as faces. In dimension 2, faces are also called edges, but this distinction will not be made unless necessary. In dimension $d \geq 3$, we define $\mathcal{E}_h^i$, $\mathcal{E}_h^\partial$, and $\mathcal{E}_h = \mathcal{E}_h^i \cup \mathcal{E}_h^\partial$ to be the sets of internal edges (i.e., one-dimensional manifolds), boundary edges, and edges, respectively.

Let $F \in \mathcal{F}_h^i$ with $F = K_1 \cap K_2$, and denote by n_1 and n_2 the outward normal to K_1 and K_2, respectively. Let v be a *scalar-valued* function defined on all cells K of the mesh. Assume that v is smooth enough to have limits on both sides of F (these limits being not necessarily the same). Set $v_1 = v_{|K_1}$ and $v_2 = v_{|K_2}$. Then, the *jump* of v across F is defined to be

$$[\![v]\!]_F = v_1 n_1 + v_2 n_2. \tag{1.54}$$

Note that $[\![v]\!]_F$ is an $\mathbb{R}^d$-valued function defined on F. When there is no ambiguity, the subscript F is dropped. When v is an $\mathbb{R}^d$-*valued* function, we use the notation

$$[\![v \cdot n]\!]_F = v_1 \cdot n_1 + v_2 \cdot n_2, \tag{1.55}$$

for the jump of the *normal component* of v. In dimension 3, we also use the notation

$$[\![v \times n]\!]_F = v_1 \times n_1 + v_2 \times n_2, \tag{1.56}$$

for the jump of the *tangential component* of v.

1.4 Approximation Spaces and Interpolation Operators

This section reviews approximation spaces and global interpolation operators that can be used in conjunction with Galerkin methods to approximate PDEs.

1.4.1 Finite element generation

Let $\{\widehat{K}, \widehat{P}, \widehat{\Sigma}\}$ be a fixed finite element. Denote by $\{\widehat{\sigma}_1, \ldots, \widehat{\sigma}_{n_{\rm sh}}\}$ the local degrees of freedom and by $\{\widehat{\theta}_1, \ldots, \widehat{\theta}_{n_{\rm sh}}\}$ the local ($\mathbb{R}^m$-valued) shape functions. Let $V(\widehat{K})$ be the domain of the *local interpolation operator* $\mathcal{I}_{\widehat{K}}$ associated with $\{\widehat{K}, \widehat{P}, \widehat{\Sigma}\}$, i.e.,

$$\mathcal{I}_{\widehat{K}} : V(\widehat{K}) \ni \widehat{v} \longmapsto \sum_{i=1}^{n_{\rm sh}} \widehat{\sigma}_i(\widehat{v}) \widehat{\theta}_i \in \widehat{P}. \tag{1.57}$$

Definition 1.58. $\{\widehat{K}, \widehat{P}, \widehat{\Sigma}\}$ *is called the* reference finite element *and* $\mathcal{I}_{\widehat{K}}$ *the* reference interpolation operator.

Let $\mathcal{T}_h$ be a mesh generated as described in §1.3.2. Recall that a cell $K \in \mathcal{T}_h$ is constructed using the $\mathcal{C}^1$-diffeomorphism $T_K : \widehat{K} \to K$ defined in (1.51).

Definition 1.59 (Iso- and subparametric interpolation). *Let* $\{\widehat{K}, \widehat{P}, \widehat{\Sigma}\}$ *be the reference finite element and let* $\{\widehat{K}, \widehat{P}_{\rm geo}, \widehat{\Sigma}_{\rm geo}\}$ *be the geometric reference finite element used to define* T_K. *When the two finite elements are identical, the interpolation is said to be* isoparametric, *whereas it is said to be* subparametric *whenever* $\widehat{P}_{\rm geo} \subsetneq \widehat{P}$.

Example 1.60. For scalar-valued finite elements, the most common example of subparametric interpolation is $\mathbb{P}_1 \subset \widehat{P}_{\rm geo} \neq \mathbb{P}_2 \subset \widehat{P}$. □

Elementary generation of finite elements. For all $K \in \mathcal{T}_h$, one must first define the counterpart of $V(\widehat{K})$, i.e., a Banach space $V(K)$ of $\mathbb{R}^m$-valued functions and a linear bijective mapping

$$\psi_K : V(K) \longrightarrow V(\widehat{K}).$$

Then, a set of $\mathcal{T}_h$-based finite elements can be defined as follows:

Proposition 1.61. *For* $K \in \mathcal{T}_h$, *the triplet* $\{K, P_K, \Sigma_K\}$ *defined by*

$$\begin{cases} K = T_K(\widehat{K}); \\ P_K = \{\psi_K^{-1}(\widehat{p}); \widehat{p} \in \widehat{P}\}; \\ \Sigma_K = \{\{\sigma_{K,i}\}_{1 \le i \le n_{\rm sh}}; \sigma_{K,i}(p) = \widehat{\sigma}_i(\psi_K(p)), \forall p \in P_K\}; \end{cases} \tag{1.58}$$

is a finite element. The local shape functions are $\theta_{K,i} = \psi_K^{-1}(\widehat{\theta}_i)$, $1 \le i \le n_{\rm sh}$, *and the associated* local interpolation operator *is*

$$\mathcal{I}_K : V(K) \ni v \longmapsto \mathcal{I}_K v = \sum_{i=1}^{n_{\rm sh}} \sigma_{K,i}(v) \theta_{K,i} \in P_K. \tag{1.59}$$

Proof. Left as an exercise. □

Proposition 1.62. *Let $\mathcal{I}_K$ be defined in* (1.59). *Then, the following diagram commutes:*

$$
\begin{array}{ccc}
V(K) & \xrightarrow{\ \psi_K\ } & V(\widehat{K}) \\
\Big\downarrow{\scriptstyle \mathcal{I}_K} & & \Big\downarrow{\scriptstyle \mathcal{I}_{\widehat{K}}} \\
P_K & \xrightarrow{\ \psi_K\ } & \widehat{P}
\end{array}
$$

Proof. Let v in $V(K)$. The definition (1.58) for $\{K, P_K, \Sigma_K\}$ implies

$$
\mathcal{I}_{\widehat{K}}(\psi_K(v)) = \sum_{i=1}^{n_{\mathrm{sh}}} \widehat{\sigma}_i(\psi_K(v))\,\widehat{\theta}_i = \sum_{i=1}^{n_{\mathrm{sh}}} \sigma_{K,i}(v)\,\psi_K(\theta_{K,i}) = \psi_K(\mathcal{I}_K(v)),
$$

owing to the linearity of ψ_K. $\square$

Proposition 1.62 plays an important role in the analysis of the interpolation error; see, e.g., the proof of Theorem 1.103. This result is the main motivation for the construction (1.58).

Example 1.63.
(i) Let $\{\widehat{K}, \widehat{P}, \widehat{\Sigma}\}$ be a Lagrange finite element. Then, one may choose $V(\widehat{K}) = [\mathcal{C}^0(\widehat{K})]^m$. Defining $V(K)$ similarly and setting

$$
\psi_K : V(K) \ni v \longmapsto \psi_K(v) = v \circ T_K \in V(\widehat{K}), \tag{1.60}
$$

yields a linear bijective mapping. Then, for all $K \in \mathcal{T}_h$, the finite element $\{K, P_K, \Sigma_K\}$ constructed in Proposition 1.61 is a Lagrange finite element. Indeed, owing to

$$
\sigma_i(v) = \widehat{\sigma}_i(\psi_K(v)) = \psi_K(v)(\widehat{a}_i) = v \circ T_K(\widehat{a}_i),
$$

and setting $a_{K,i} = T_K(\widehat{a}_i)$ for $1 \le i \le n_{\mathrm{sh}}$, we infer that $\{a_{K,i}\}_{1 \le i \le n_{\mathrm{sh}}}$ are the nodes of $\{K, P_K, \Sigma_K\}$.

(ii) For the Raviart–Thomas finite element (see §1.2.7), set $V(\widehat{K}) = \{v \in [L^p(\widehat{K})]^d;\ \nabla \cdot v \in L^s(\widehat{K})\}$ for $p > 2$, $s \ge q$, $\frac{1}{q} = \frac{1}{p} + \frac{1}{d}$, and define $V(K)$ similarly. The transformation $p \mapsto p \circ T_K$ does not map $V(K)$ to $V(\widehat{K})$. A suitable choice for ψ_K is the so-called Piola transformation; see §1.4.7. $\square$

Remark 1.64. In the literature the notation $\psi_K(v) = \widehat{v}$ is often used. Then, the relation $\mathcal{I}_{\widehat{K}}(\psi_K(v)) = \psi_K(\mathcal{I}_K(v))$ resulting from Proposition 1.62 can be rewritten in the form $\mathcal{I}_{\widehat{K}}\widehat{v} = \widehat{\mathcal{I}_K(v)}$. This notation can sometimes be misleading; in particular, it must not be confused with the notation $\widehat{x} = T_K^{-1}(x)$. $\square$

Finite element generation with rescaling. The technique described in Proposition 1.61 to generate finite elements is generally sufficient to construct approximation spaces. However, in the most general situation, a more sophisticated technique must be designed. To understand the nature of the problem, observe that the degrees of freedom in $\widehat{\Sigma}$ are constrained only locally by unisolvence. When constructing approximation spaces, one often wishes to enforce interface conditions between adjacent elements, thus introducing a new constraint on the degrees of freedom. Accordingly, we allow for a rescaling of the degrees of freedom in Σ_K.

Proposition 1.65. *For $K \in \mathcal{T}_h$, let $\alpha_K \in \mathbb{R}^{n_{\mathrm{sh}}}$ be such that $\alpha_{K,i} \neq 0$ for all $1 \leq i \leq n_{\mathrm{sh}}$. Define the triplet $\{K, P_K, \Sigma_K^\alpha\}$ by taking K and P_K as in (1.58) and by choosing the local degrees of freedom $\Sigma_K^\alpha = \{\sigma_{K,1}, \ldots, \sigma_{K,n_{\mathrm{sh}}}\}$ such that, for all $1 \leq i \leq n_{\mathrm{sh}}$,*

$$\sigma_{K,i} : P_K \ni p \longmapsto \sigma_{K,i}(p) = \alpha_{K,i}\widehat{\sigma}_i(\psi_K(p)). \tag{1.61}$$

Then, $\{K, P_K, \Sigma_K^\alpha\}$ is a finite element. Furthermore, the local shape functions on K are given by $\theta_{K,i} = \frac{1}{\alpha_{K,i}}\psi_K^{-1}(\widehat{\theta}_i)$, $1 \leq i \leq n_{\mathrm{sh}}$, and the associated local interpolation operator $\mathcal{I}_K^\alpha$ is defined as in (1.59).

Proof. Left as an exercise.   □

Proposition 1.66. *Let $\mathcal{I}_K^\alpha$ be the local interpolation operator associated with $\{K, P_K, \Sigma_K^\alpha\}$. Then, the diagram in Proposition 1.62 commutes.*

Proof. Straightforward verification.   □

Example 1.67. An example where a rescaling of the degrees of freedom is needed is the Hermite finite element discussed in §1.4.6; see also Remark 1.72(i), Remark 1.88, and Remark 1.94 for further examples.   □

1.4.2 Global interpolation operator

Using the $\mathcal{T}_h$-based family of finite elements $\{K, P_K, \Sigma_K\}_{K \in \mathcal{T}_h}$ generated in Proposition 1.61 or Proposition 1.65, a global interpolation operator $\mathcal{I}_h$ can be constructed as follows: First, choose its domain to be

$$D(\mathcal{I}_h) = \{v \in [L^1(\Omega_h)]^m; \forall K \in \mathcal{T}_h, v_{|K} \in V(K)\}, \tag{1.62}$$

where Ω_h is the geometric interpolation of Ω defined in (1.52). For a function $v \in D(\mathcal{I}_h)$, the quantities $\sigma_{K,i}(v_{|K})$ are meaningful on all the mesh elements and for all $1 \leq i \leq n_{\mathrm{sh}}$. Then, the global interpolant $\mathcal{I}_h v$ can be specified elementwise using the local interpolation operators defined in (1.59), i.e.,

$$\forall K \in \mathcal{T}_h, \quad (\mathcal{I}_h v)_{|K} = \mathcal{I}_K(v_{|K}) = \sum_{i=1}^{n_{\mathrm{sh}}} \sigma_{K,i}(v_{|K})\theta_{K,i}.$$

Note that the function $\mathcal{I}_h v$ is defined on Ω_h. It may happen that $\mathcal{I}_h v$ is multi-valued at the interfaces of the elements. This is not a major difficulty since $\mathcal{F}_h$ is of zero Lebesgue measure. The *global interpolation operator* is defined as follows:

$$\mathcal{I}_h : D(\mathcal{I}_h) \ni v \longmapsto \sum_{K \in \mathcal{T}_h} \sum_{i=1}^{n_{\mathrm{sh}}} \sigma_{K,i}(v_{|K}) \theta_{K,i} \in W_h, \qquad (1.63)$$

where W_h, the codomain of $\mathcal{I}_h$, is

$$W_h = \{ v_h \in [L^1(\Omega_h)]^m ; \forall K \in \mathcal{T}_h, \; v_{|K} \in P_K \}. \qquad (1.64)$$

The space W_h is called an *approximation space*. In (1.63), we abuse the notation by implicitly extending $\theta_{K,i}$ by zero outside K.

One often wishes to impose additional regularity properties on the functions of W_h. At this stage, we only state the following general definition:

Definition 1.68 (Conformal approximation). *Let W_h be defined in (1.64) and let V be a Banach space. W_h is said to be V-conformal if $W_h \subset V$.*

Practical examples are investigated in §1.4.5, §1.4.6, §1.4.7, and §1.4.8.

1.4.3 Totally discontinuous spaces

Totally discontinuous spaces play an important role in the so-called Discontinuous Galerkin (DG) method; see §3.2.4, §5.6, and §6.3.2. Functions in such spaces only satisfy the simplest regularity requirement, namely to be integrable over Ω_h.

For the sake of simplicity, assume that $\{\widehat{K}, \widehat{P}, \widehat{\Sigma}\}$ is such that the local degrees of freedom are of the form

$$\widehat{\sigma}_i : V(\widehat{K}) \ni \widehat{v} \longmapsto \widehat{\sigma}_i(\widehat{v}) = \frac{1}{\mathrm{meas}(\widehat{K})} \int_{\widehat{K}} \widehat{v} \, \widehat{\mathcal{K}}_i, \quad 1 \le i \le n_{\mathrm{sh}},$$

where $n_{\mathrm{sh}} = \dim(\widehat{P})$ and $\widehat{\mathcal{K}}_i$ is a smooth function on $\widehat{K}$; hence, $V(\widehat{K}) = L^1(\widehat{K})$ is an admissible choice. Define $V(K)$ similarly and choose the mapping defined in (1.60), i.e., $\psi_K(v) = v \circ T_K$. Construct the family $\{K, P_K, \Sigma_K\}_{K \in \mathcal{T}_h}$ using Proposition 1.61. Then, for each $K \in \mathcal{T}_h$, setting $\mathcal{K}_{K,i} = \widehat{\mathcal{K}}_i \circ T_K^{-1}$, we infer

$$\sigma_{K,i}(v) = \widehat{\sigma}_i(\psi_K(v)) = \frac{1}{\mathrm{meas}(K)} \int_K v \, \mathcal{K}_{K,i}.$$

The local shape functions are $\theta_{K,i} = \widehat{\theta}_i \circ T_K^{-1}$, $1 \le i \le n_{\mathrm{sh}}$, where $\{\widehat{\theta}_1, \ldots, \widehat{\theta}_{n_{\mathrm{sh}}}\}$ are the local shape functions associated with $\{\widehat{\sigma}_1, \ldots, \widehat{\sigma}_{n_{\mathrm{sh}}}\}$.

Consider the approximation space

$$Z_{\mathrm{td},h} = \{ v_h \in L^1(\Omega_h) ; \forall K \in \mathcal{T}_h, \; v_{|K} \in P_K \}. \qquad (1.65)$$

Because local degrees of freedom can be taken independently on each mesh cell, $Z_{\text{td},h}$ is of dimension $N_{\text{el}} \times n_{\text{sh}}$ where N_{el} is the number of mesh cells. For $v \in L^1(\Omega_h)$, the quantities $\sigma_{K,i}(v_{|K})$ are meaningful for $K \in \mathcal{T}_h$ and $1 \leq i \leq n_{\text{sh}}$. Then, the global interpolation operator is constructed as follows:

$$\mathcal{I}_{\text{td},h} : L^1(\Omega_h) \ni v \longmapsto \sum_{K \in \mathcal{T}_h} \sum_{i=1}^{n_{\text{sh}}} \frac{1}{\text{meas}(K)} \left(\int_K v \, \mathcal{K}_{K,i} \right) \theta_{K,i} \in Z_{\text{td},h}. \quad (1.66)$$

Example 1.69.

(i) Choosing $\widehat{P} = \mathbb{P}_k$ and assuming that the mesh is affine, we infer $P_K = \mathbb{P}_k$, so that the approximation space defined in (1.65) is

$$P_{\text{td},h}^k = \{ v_h \in L^1(\Omega_h); \, \forall K \in \mathcal{T}_h, \, v_{h|K} \in \mathbb{P}_k \}. \quad (1.67)$$

For instance, the space $P_{\text{td},h}^0 = \{ v_h \in L^1(\Omega_h); \, \forall K \in \mathcal{T}_h, \, v_{h|K} \in \mathbb{P}_0 \}$ is of dimension N_{el} and spanned by $\{ 1_K \}_{K \in \mathcal{T}_h}$ where 1_K is the characteristic function of K. The global interpolation operator associated with $P_{\text{td},h}^0$ is

$$\mathcal{I}_{\text{td},h}^0 : L^1(\Omega_h) \ni v \longmapsto \mathcal{I}_{\text{td},h}^0 v = \sum_{K \in \mathcal{T}_h} \left(\frac{1}{\text{meas}(K)} \int_K v \right) 1_K \in P_{\text{td},h}^0.$$

(ii) A similar construction is possible with $\mathbb{Q}_k$ polynomials. For instance, on quadrangular meshes, the local shape functions can be taken to be tensor products of Legendre polynomials, i.e.,

$$\widehat{\theta}_{l_1 \ldots l_d}(\widehat{x}) = \widehat{\mathcal{E}}_{l_1}(\widehat{x}_1) \ldots \widehat{\mathcal{E}}_{l_d}(\widehat{x}_d), \quad 0 \leq l_1, \ldots, l_d \leq k.$$

This choice naturally yields hierarchical bases; see, e.g., §1.1.5 and §1.2.9. □

1.4.4 Discontinuous spaces with patch-test

In this section, we assume that $\mathcal{T}_h$ is a simplicial, affine, and geometrically conformal mesh.

The Crouzeix–Raviart approximation space. Let $\{\widehat{K}, \widehat{P}, \widehat{\Sigma}\}$ be the Crouzeix–Raviart finite element introduced in §1.2.6. Set $V(\widehat{K}) = W^{1,1}(\widehat{K})$, define $V(K)$ similarly, and choose the mapping defined in (1.60), i.e., $\psi_K(v) = v \circ T_K$. Construct the family $\{K, P_K, \Sigma_K\}_{K \in \mathcal{T}_h}$ using Proposition 1.61. Then, for each $K \in \mathcal{T}_h$, letting $F_{K,i} = T_K(\widehat{F}_i)$, $0 \leq i \leq d$, where $\{\widehat{F}_0, \ldots, \widehat{F}_d\}$ are the faces of $\widehat{K}$, the local degrees of freedom are

$$\sigma_{K,i}(v) = \widehat{\sigma}_i(\psi_K(v)) = \frac{1}{\text{meas}(\widehat{F}_i)} \int_{\widehat{F}_i} \psi_K(v) = \frac{1}{\text{meas}(F_{K,i})} \int_{F_{K,i}} v. \quad (1.68)$$

In addition, since the mesh is affine, $P_K = \mathbb{P}_1$. As a result, $\{K, P_K, \Sigma_K\}$ is a Crouzeix–Raviart finite element.

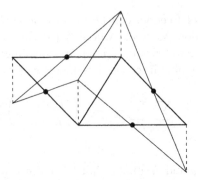

Fig. 1.17. Global shape function for the Crouzeix–Raviart approximation space. The support is materialized by thick lines and the graph by thin lines.

Consider the so-called *Crouzeix–Raviart approximation space*

$$P_{\mathrm{pt},h}^1 = \{v_h \in L^1(\Omega_h); \forall K \in \mathcal{T}_h, v_{h|K} \in \mathbb{P}_1;$$

$$\forall F \in \mathcal{F}_h^i, \int_F [\![v_h]\!] = 0\}. \tag{1.69}$$

Recall that $\mathcal{F}_h^i$ denotes the set of interior faces (interfaces) in the mesh and $[\![v_h]\!]$ the jump of v_h across interfaces. For $F \in \mathcal{F}_h$, consider the function φ_F with support consisting of the one or two simplices to which F belongs and such that on each of these simplices, say K, the function $\varphi_{F|K}$ is the local shape function of $\{K, P_K, \Sigma_K\}$ associated with the face F. The graph of a function φ_F is shown in Figure 1.17.

Lemma 1.70. $\varphi_F \in P_{\mathrm{pt},h}^1$.

Proof. Let $F \in \mathcal{F}_h^i$, say $F = K_1 \cap K_2$. Since $\varphi_{F|K_1}$ is the local shape function of $\{K_1, P_{K_1}, \Sigma_{K_1}\}$ associated with F, (1.68) implies

$$\int_F \varphi_{F|K_1} = \mathrm{meas}(F).$$

Similarly, $\int_F \varphi_{F|K_2} = \mathrm{meas}(F)$, proving that $\int_F [\![\varphi_F]\!] = 0$. Use the same argument to prove $\int_{F'} [\![\varphi_F]\!] = 0$ for all faces $F' \neq F$. Since the restriction of φ_F to any mesh element is in $\mathbb{P}_1$, $\varphi_F \in P_{\mathrm{pt},h}^1$. □

For $F \in \mathcal{F}_h$, define the linear form $\gamma_F : P_{\mathrm{pt},h}^1 \ni v_h \mapsto \frac{1}{\mathrm{meas}(F)} \int_F v_h$. Although $v_h \in P_{\mathrm{pt},h}^1$ may be multi-valued at F, the quantity $\gamma_F(v_h)$ is single-valued since $\int_F [\![v_h]\!] = 0$.

Proposition 1.71. $\{\varphi_F\}_{F \in \mathcal{F}_h}$ *is a basis for* $P_{\mathrm{pt},h}^1$, *and* $\{\gamma_F\}_{F \in \mathcal{F}_h}$ *is a basis for* $\mathcal{L}(P_{\mathrm{pt},h}^1; \mathbb{R})$.

Proof. The proof is based on the fact that (with obvious notation) $\gamma_{F'}(\varphi_F) = \delta_{FF'}$ for $F, F' \in \mathcal{F}_h$. Consider a set of real numbers $\{\alpha_F\}_{F \in \mathcal{F}_h}$, and assume that the function $w = \sum_{F \in \mathcal{F}_h} \alpha_F \varphi_F$ vanishes identically. Then, $\alpha_F = \gamma_F(w) = 0$; hence, the set $\{\varphi_F\}_{F \in \mathcal{F}_h}$ is linearly independent. Let $v_h \in P^1_{\text{pt},h}$ and set

$$w_h = \sum_{F \in \mathcal{F}_h} \left(\frac{1}{\text{meas}(F)} \int_F v_h \right) \varphi_F.$$

Then, for all $K \in \mathcal{T}_h$, $v_{h|K}$ and $w_{h|K}$ are in P_K, and for all $\sigma \in \Sigma_K$, $\sigma(v_{h|K}) = \sigma(w_{h|K})$. Unisolvence implies $v_{h|K} = w_{h|K}$. This shows that $\{\varphi_F\}_{F \in \mathcal{F}_h}$ is a basis for $P^1_{\text{pt},h}$. The proof is easily completed. □

Proposition 1.71 implies that $P^1_{\text{pt},h}$ is a space of dimension N_{ed} in two dimensions and N_f in three dimensions. The linear forms $\{\gamma_F\}_{F \in \mathcal{F}_h}$ are called the *global degrees of freedom* in $P^1_{\text{pt},h}$, and $\{\varphi_F\}_{F \in \mathcal{F}_h}$ are called the *global shape functions*.

For a function $v \in W^{1,1}(\Omega_h)$, the quantity $\gamma_F(v)$ is meaningful (and single-valued) for all $F \in \mathcal{F}_h$. The so-called *global Crouzeix–Raviart interpolation operator* is constructed as follows:

$$\mathcal{I}^{\text{CR}}_h : W^{1,1}(\Omega_h) \ni v \longmapsto \mathcal{I}^{\text{CR}}_h v = \sum_{F \in \mathcal{F}_h} \left(\frac{1}{\text{meas}(F)} \int_F v \right) \varphi_F \in P^1_{\text{pt},h}. \quad (1.70)$$

Note that $P^1_{\text{pt},h}$ is the codomain of $\mathcal{I}^{\text{CR}}_h$.

Remark 1.72.
(i) If the degrees of freedom in $\{\widehat{K}, \widehat{P}, \widehat{\Sigma}\}$ are chosen to be the integral over the faces instead of the mean-value (see Remark 1.38(ii)), Proposition 1.65 must be used to generate the family $\{K, P_K, \Sigma_K\}_{K \in \mathcal{T}_h}$. Indeed, taking $\alpha_{K,i} = \frac{\text{meas}(F_{K,i})}{\text{meas}(\widehat{F}_i)}$ for $0 \le i \le d$ in (1.61) yields $\sigma_{K,i}(v) = \int_{F_{K,i}} v$. Then, constructing φ_F as before, Lemma 1.70 holds. If Proposition 1.61 had been used instead, then $\sigma_{K,i}(v) = \frac{\text{meas}(\widehat{F}_i)}{\text{meas}(F_{K,i})} \int_{F_{K,i}} v$, yielding $\theta_{K,i} = \frac{1}{\text{meas}(\widehat{F}_i)}(1 - \frac{\lambda_{K,i}}{d})$ where $\lambda_{K,i}$ is the i-th barycentric coordinate of K. Then, $\theta_{K,i} = \frac{1}{\text{meas}(\widehat{F}_i)}$ on $F_{K,i}$; hence, $\int_F \llbracket \varphi_F \rrbracket \ne 0$, i.e., $\varphi_F \notin P^1_{\text{pt},h}$ (unless $\widehat{K}$ is equilateral).
(ii) Since $\llbracket v_h \rrbracket_F : F \ni x \mapsto \llbracket v_h \rrbracket(x) \in \mathbb{R}$ is linear, the condition $\int_F \llbracket v_h \rrbracket = 0$ in (1.69) is equivalent to the continuity of v_h at the center of gravity of F. □

Extension to high-degree polynomials. The extension of the Crouzeix–Raviart approximation space $P^1_{\text{pt},h}$ to higher-degree polynomials is somewhat technical. When approximating PDEs, one often wishes to impose the continuity of the moments, up to order $k - 1$, of the functions in $P^k_{\text{pt},h}$ on any interface of the mesh. This condition is known as the *patch-test*; see [IrR72]. The space $P^k_{\text{pt},h}$ is thus defined as follows:

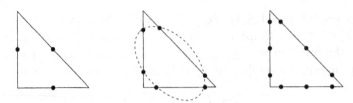

Fig. 1.18. Continuity points for functions in $P_{\mathrm{pt},h}^k$: $k = 1$ (left); $k = 2$ (center); and $k = 3$ (right). For $k = 2$, the six points lie on an ellipse.

$$P_{\mathrm{pt},h}^k = \{v_h \in L^1(\Omega_h); \forall K \in \mathcal{T}_h, \ v_{h|K} \in \mathbb{P}_k;$$

$$\forall F \in \mathcal{F}_h^i, \forall q \in \mathbb{P}_{k-1}, \int_F [\![v_h]\!]\, q = 0\}. \qquad (1.71)$$

In two dimensions, the patch-test is equivalent to the continuity of v_h at the k Gauß points located on each face of K; see Figure 1.18 and Definition 8.1. These points (completed with internal points for $k \geq 3$) can be used to define local Lagrange degrees of freedom, say Σ, on the simplex K if k is odd, but this construction is not possible if k is even. For instance, if $k = 2$, the six Gauß points lie on the ellipse of equation $2 - 3(\lambda_{0,K}^2 + \lambda_{1,K}^2 + \lambda_{2,K}^2) = 0$, where $\{\lambda_{0,K}, \lambda_{1,K}, \lambda_{2,K}\}$ are the barycentric coordinates of the simplex K. This means that the so-called *Fortin–Soulié bubble*

$$b_K = 2 - 3(\lambda_{0,K}^2 + \lambda_{1,K}^2 + \lambda_{2,K}^2), \qquad (1.72)$$

vanishes at these six Gauß points. Then, because $b_K \in \mathbb{P}_2$, the linear mapping (1.34) associated with the triplet $\{K, \mathbb{P}_2, \Sigma\}$ is not bijective; hence, $\{K, \mathbb{P}_2, \Sigma\}$ is not a finite element.

In the three-dimensional case, a similar construction is possible. However, the patch-test no longer implies point-continuity except for $k = 1$.

Remark 1.73. The space $P_{\mathrm{pt},h}^2$ can be used to approximate PDEs owing to a decomposition involving H^1-conformal quadratics and Fortin-Soulié bubbles; see [FoS83] and Exercise 1.12.   □

1.4.5 H^1-conformal spaces based on Lagrange finite elements

The goal of this section is to construct a H^1-conformal subspace of the approximation space W_h defined in (1.64). We assume that the mesh is *geometrically conformal* (but not necessarily affine) and that the reference finite element is a *Lagrange finite element*. Hence, setting $V(\widehat{K}) = [\mathcal{C}^0(\widehat{K})]^m$, defining $V(K)$ similarly, and choosing the mapping ψ_K defined in (1.60), the family $\{K, P_K, \Sigma_K\}_{K \in \mathcal{T}_h}$ constructed as in Proposition 1.61 is a family of Lagrange finite elements; see Example 1.63(i).

Consider the space

$$V_h = \{v_h \in W_h; \forall F \in \mathcal{F}_h^i, \llbracket v_h \rrbracket_F = 0\}. \tag{1.73}$$

The main motivation for introducing V_h is the following:

Proposition 1.74. $V_h \subset [H^1(\Omega_h)]^m$.

Proof. Assume $m = 1$. For vector-valued functions, the proof below is simply applied component by component. Let $v_h \in V_h$. Since its restriction to every $K \in \mathcal{T}_h$ is a polynomial, it is differentiable in the classical sense. For $1 \leq j \leq d$, consider the function $w_j \in L^2(\Omega_h)$ defined on $K \in \mathcal{T}_h$ by $w_{j|K} = \partial_j(v_{h|K})$. Let $\phi \in \mathcal{D}(\Omega_h)$. Using the Green formula yields

$$\int_{\Omega_h} w_j \phi = \sum_{K \in \mathcal{T}_h} \int_K w_j \phi = -\sum_{K \in \mathcal{T}_h} \int_K v_{h|K}\, \partial_j \phi + \sum_{K \in \mathcal{T}_h} \int_{\partial K} \phi v_{h|K} n_{K,j},$$

where ∂K is the boundary of K and $n_{K,j}$ is the j-th component of the outer normal to K. Use the fact that ϕ vanishes at the boundary of Ω_h, regroup interface terms, and employ the notation of §1.3.4 to infer

$$\int_{\Omega_h} w_j \phi = -\int_{\Omega_h} v_h \partial_j \phi + \sum_{F \in \mathcal{F}_h^i} \int_F \phi e_j \cdot \llbracket v_h \rrbracket_F,$$

where $\{e_1, \ldots, e_d\}$ is the canonical basis of $\mathbb{R}^d$. Owing to $\llbracket v_h \rrbracket_F = 0$, $\int_{\Omega_h} w_j \phi = -\int_{\Omega_h} v_h \partial_j \phi$. Therefore, for $1 \leq j \leq d$, the distributional derivative of v_h with respect to the j-th coordinate is w_j. Since $w_j \in L^2(\Omega_h)$, $v_h \in H^1(\Omega_h)$. $\quad\square$

The next question is to determine how the zero-jump condition in (1.73) can be enforced using the local degrees of freedom of adjacent cells. For $K \in \mathcal{T}_h$, denote by $\{a_{K,1}, \ldots, a_{K,n_{sh}}\}$ the Lagrange nodes (not to be confused with the geometric nodes of K). Assume that:

(sc1) All the faces of $\widehat{K}$ have the same number of nodes, say $n_{n_{sh}}^\partial$.

(sc2) Consider a face $\widehat{F}$ of $\widehat{K}$ and let $\{a_{1,\widehat{F}}, \ldots, a_{n_{sh}^\partial,\widehat{F}}\}$ be its nodes. Define $\widehat{P}_{\widehat{F}} = \{\widehat{q}; \exists \widehat{p} \in \widehat{P}, \widehat{q} = \widehat{p}_{|\widehat{F}}\}$ and $\widehat{\Sigma}_{\widehat{F}} = \{\widehat{\sigma}_1, \ldots, \widehat{\sigma}_{n_{sh}^\partial}\}$ such that $\widehat{\sigma}_i(\widehat{q}) = \widehat{q}(a_{i,\widehat{F}})$ for $\widehat{q} \in \widehat{P}_{\widehat{F}}$ and $1 \leq i \leq n_{sh}^\partial$. Then, $\{\widehat{F}, \widehat{P}_{\widehat{F}}, \widehat{\Sigma}_{\widehat{F}}\}$ is a finite element.

(sc3) For all $F \in \mathcal{F}_h^i$ with $F = K_1 \cap K_2$, assume that there are renumberings of the Lagrange nodes of K_1 and K_2 such that (see Figure 1.19):

$$\forall i \in \{1, \ldots, n_{n_{sh}}^\partial\}, \quad a_{K_1,i} = a_{K_2,i}.$$

Lemma 1.75. *Assume* (sc1)–(sc3). *Let* $v_h \in W_h$. *Then,* $\llbracket v_h \rrbracket_F = 0$ *for all* $F \in \mathcal{F}_h^i$ *if and only if, for all* $F \in \mathcal{F}_h^i$ *such that* $F = K_1 \cap K_2$,

$$\forall i \in \{1, \ldots, n_{n_{sh}}^\partial\}, \quad v_{h|K_1}(a_{K_1,i}) = v_{h|K_2}(a_{K_2,i}). \tag{1.74}$$

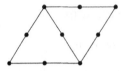 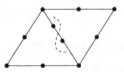

Fig. 1.19. Compatible (left) and incompatible (right) position of nodes at an interface for a geometrically conformal mesh.

Proof. The direct statement is evident. To prove the converse, let $v_h \in W_h$, let $F \in \mathcal{F}_h^i$ with $F = K_1 \cap K_2$, and assume (1.74). Let T_1 and T_2 be the geometric transformations associated with K_1 and K_2, respectively. Set $v_1 = v_{h|K_1}$ and $v_2 = v_{h|K_2}$. Since the mesh is geometrically conformal, there are renumberings of the geometric nodes of K_1 and K_2 such that (1.53) holds. Owing to (sc3), $\widehat{a}_{i,\widehat{F}} = T_{1|F}^{-1}(a_{K_1,i}) = T_{2|F}^{-1}(a_{K_2,i})$ for $1 \leq i \leq n_{n_{\mathrm{sh}}}^\partial$. Define $\widehat{v}_{1|\widehat{F}} = v_{1|F} \circ T_{1|F}$ and $\widehat{v}_{2|\widehat{F}} = v_{2|F} \circ T_{2|F}$. Then, (1.74) implies

$$\forall i \in \{1, \ldots, n_{n_{\mathrm{sh}}}^\partial\}, \quad \widehat{v}_{1|\widehat{F}}(\widehat{a}_{i,\widehat{F}}) = \widehat{v}_{2|\widehat{F}}(\widehat{a}_{i,\widehat{F}}).$$

Owing to (sc2), $\widehat{v}_{1|\widehat{F}} = \widehat{v}_{2|\widehat{F}}$, and since the geometric transformations are bijective, this readily implies $v_{1|F} = v_{2|F}$. $\qquad\square$

Remark 1.76. All the Lagrange finite elements introduced in §1.2.3–§1.2.5 satisfy assumption (sc2). This is not the case for the Crouzeix–Raviart finite element considered as a Lagrange finite element. $\qquad\square$

Let $\{a_1, \ldots, a_N\} = \bigcup_{K \in \mathcal{T}_h} \{a_{K,1}, \ldots, a_{K,n_{\mathrm{sh}}}\}$ be the set of all the *Lagrange nodes*. For $K \in \mathcal{T}_h$ and $m \in \{1, \ldots, n_{\mathrm{sh}}\}$, let $\mathrm{j}(K,m) \in \{1, \ldots, N\}$ be the corresponding index of the Lagrange node. Let $\{\varphi_1, \ldots, \varphi_N\}$ be the set of functions in W_h defined elementwise by $\varphi_{i|K}(a_{K,m}) = \delta_{mn}$ if there is $n \in \{1, \ldots, n_{\mathrm{sh}}\}$ such that $i = \mathrm{j}(K,n)$ and 0 otherwise. This implies $\varphi_i(a_j) = \delta_{ij}$ for $1 \leq i, j \leq N$.

Lemma 1.77. *Under the assumptions of Lemma 1.75, $\varphi_i \in V_h$.*

Proof. Use the converse statement in Lemma 1.75. $\qquad\square$

For $1 \leq i \leq N$, define the linear form $\gamma_i : V_h \ni v_h \mapsto v_h(a_i) \in \mathbb{R}$.

Proposition 1.78. $\{\varphi_1, \ldots, \varphi_N\}$ *is a basis for V_h, and $\{\gamma_1, \ldots, \varphi_N\}$ is a basis for $\mathcal{L}(V_h; \mathbb{R})$.*

Proof. The family $\{\varphi_1, \ldots, \varphi_N\}$ is linearly independent; indeed, if the function $\sum_{j=1}^N \alpha_j \varphi_j$ vanishes identically, evaluating it at the node a_i yields $\alpha_i = 0$. Now, let $v_h \in V_h$. Owing to the direct statement of Lemma 1.75, v_h is single-valued at all the Lagrange nodes. Set $w_h = \sum_{i=1}^N v_h(a_i)\varphi_i$. Then, for all $K \in \mathcal{T}_h$, $v_{h|K}$ and $w_{h|K}$ are in P_K and coincide at the nodes $\{a_{K,1}, \ldots, a_{K,n_{\mathrm{sh}}}\}$. Unisolvence implies $v_{h|K} = w_{h|K}$. Hence, $\{\varphi_1, \ldots, \varphi_N\}$ is a basis for V_h. Proving that $\{\gamma_1, \ldots, \varphi_N\}$ is a basis for $\mathcal{L}(V_h; \mathbb{R})$ is then straightforward. $\qquad\square$

	dim.	$k = 1$	$k = 2$	$k = 3$
$P_{c,h}^k$	2	N_v	$N_v + N_{ed}$	$N_v + 2N_{ed} + N_{el}$
$Q_{c,h}^k$	2	N_v	$N_v + N_{ed} + N_{el}$	$N_v + 2N_{ed} + 4N_{el}$
$P_{c,h}^k$	3	N_v	$N_v + N_{ed}$	$N_v + 2N_{ed} + N_f$
$Q_{c,h}^k$	3	N_v	$N_v + N_{ed} + N_f + N_{el}$	$N_v + 2N_{ed} + 4N_f + 8N_{el}$

Table 1.4. Dimension of H^1-conformal spaces constructed using a geometrically conformal mesh and various Lagrange finite elements. The second column indicates the space dimension. N_{el} denotes the number of cells in the mesh, N_f the number of faces, N_{ed} the number of edges, and N_v the number of vertices.

Proposition 1.78 implies that V_h is a space of dimension N. The linear forms $\{\gamma_1, \ldots, \gamma_N\}$ are called the *global degrees of freedom* in V_h, and $\{\varphi_1, \ldots, \varphi_N\}$ are called the *global shape functions*. The *global Lagrange interpolation operator* is defined as follows:

$$\mathcal{I}_h : \mathcal{C}^0(\overline{\Omega}_h) \ni v \longmapsto \sum_{i=1}^N v(a_i)\varphi_i \in V_h. \tag{1.75}$$

Note that the domain of $\mathcal{I}_h$ can also be taken to be $H^s(\Omega_h)$ for $s > \frac{d}{2}$.

We shall often consider the approximation spaces

$$P_{c,h}^k = \{v_h \in \mathcal{C}^0(\overline{\Omega}_h); \forall K \in \mathcal{T}_h, v_h \circ T_K \in \mathbb{P}_k\}, \tag{1.76}$$

$$Q_{c,h}^k = \{v_h \in \mathcal{C}^0(\overline{\Omega}_h); \forall K \in \mathcal{T}_h, v_h \circ T_K \in \mathbb{Q}_k\}. \tag{1.77}$$

The dimension of these spaces is given in Table 1.4 for the first values of k. The subscript 'c' refers to the continuity condition across mesh interfaces (for simplicity, it was not used in the one-dimensional cases treated in §1.1).

Example 1.79. Assume that $\mathcal{T}_h$ is composed of triangles in dimension 2.

(i) Let $\{S_1, \ldots, S_{N_v}\}$ be the mesh vertices. For $1 \leq i \leq N_v$, the global shape functions in $P_{c,h}^1$ satisfy $\varphi_i(S_j) = \delta_{ij}$ for $1 \leq i, j \leq N_v$; see the left panel of Figure 1.20. Owing to Proposition 1.78, the set $\{\varphi_1, \ldots, \varphi_{N_v}\}$ is a basis for $P_{c,h}^1$.

(ii) Let $\{T_1, \ldots, T_{N_{ed}}\}$ be the edge midpoints. For $1 \leq i \leq N_v$, let $\varphi_{i,0} \in P_{c,h}^2$ be such that $\varphi_{i,0}(S_j) = \delta_{ij}$ and $\varphi_{i,0}(T_j) = 0$. In addition, for $1 \leq i \leq N_{ed}$, let $\varphi_{i,1} \in P_{c,h}^2$ be such that $\varphi_{i,1}(S_j) = 0$ and $\varphi_{i,1}(T_j) = \delta_{ij}$. The functions $\varphi_{i,0}$ and $\varphi_{i,1}$ are illustrated in the central and right panels of Figure 1.20. Owing to Proposition 1.78, $\{\varphi_{1,0}, \ldots, \varphi_{N_v,0}, \varphi_{1,1}, \ldots, \varphi_{N_{ed},1}\}$ is a basis for $P_{c,h}^2$. $\qquad\square$

Remark 1.80. Lemma 1.77 can be easily extended to $\mathbb{R}^m$-valued functions by considering the functions $\varphi_{i,n}$ for $1 \leq i \leq N$ and $1 \leq n \leq m$, such that $\varphi_{i,n}(a_j) = \delta_{ij}e_n$, where e_n is the n-th vector of the canonical basis of $\mathbb{R}^m$. $\qquad\square$

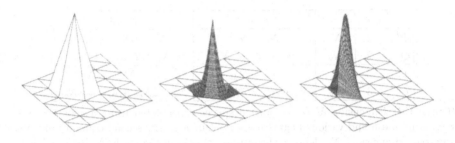

Fig. 1.20. Global shape functions for H^1-conformal spaces in two dimensions: $P^1_{c,h}$ (left) and $P^2_{c,h}$ (center and right).

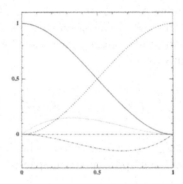

Fig. 1.21. Local shape functions for the Hermite finite element in the reference interval $[0, 1]$.

1.4.6 H^2-conformal spaces

In dimension 1, a H^2-conformal space can be constructed using Hermite finite elements. Let $\widehat{K} = [0, 1]$ be the reference interval, set $\widehat{P} = \mathbb{P}_3$, and define the local degrees of freedom $\widehat{\Sigma} = \{\widehat{\sigma}_1, \widehat{\sigma}_2, \widehat{\sigma}_3, \widehat{\sigma}_4\}$ to be

$$\widehat{\sigma}_1(\widehat{p}) = \widehat{p}(0), \qquad \widehat{\sigma}_2(\widehat{p}) = \widehat{p}'(0), \qquad \widehat{\sigma}_3(\widehat{p}) = \widehat{p}(1), \qquad \widehat{\sigma}_4(\widehat{p}) = \widehat{p}'(1).$$

One readily verifies that $\{\widehat{K}, \widehat{P}, \widehat{\Sigma}\}$ is a finite element; it is called a *Hermite finite element*. The local shape functions $\{\widehat{\theta}_1, \widehat{\theta}_2, \widehat{\theta}_3, \widehat{\theta}_4\}$ are (see Figure 1.21)

$$\widehat{\theta}_1(t) = (2t + 1)(t - 1)^2, \qquad \widehat{\theta}_2(t) = t(t - 1)^2,$$
$$\widehat{\theta}_3(t) = (3 - 2t)t^2, \qquad \widehat{\theta}_4(t) = (t - 1)t^2.$$

Owing to the choice of the local degrees of freedom, an admissible choice for $V(\widehat{K})$ is $\mathcal{C}^1(\widehat{K})$ (or $H^s(\widehat{K})$ with $s > \frac{3}{2}$).

Let $\Omega = \,]a, b[$ and let $\mathcal{T}_h = \{\widehat{I}_i\}_{0 \leq i \leq N}$ be the one-dimensional mesh of Ω introduced in §1.1.1. Consider the affine transformation T_i defined in

(1.21), i.e., $T_i : \widehat{K} \ni t \mapsto x = x_i + th \in I_i$. The goal is to generate a family of Hermite finite elements over the mesh intervals. To this end, one must use Proposition 1.65 since the degrees of freedom in $\widehat{\Sigma}$ are of different dimensionality. Specifically, set $V(I_i) = \mathcal{C}^1(I_i)$ and choose the mapping $\psi_{I_i} : V(I_i) \ni v \mapsto \psi_{I_i}(v) = v \circ T_i \in V(\widehat{K})$. Set $\alpha_{i,1} = \alpha_{i,3} = 1$, $\alpha_{i,2} = \alpha_{i,4} = \frac{1}{h_i}$, and $\alpha_i = (\alpha_{i,1}, \alpha_{i,2}, \alpha_{i,3}, \alpha_{i,4})$. Using Proposition 1.65 to generate the family $\{I_i, P_i, \Sigma_i\}_{0 \le i \le N}$, we infer $P_i = \mathbb{P}_3$ and that the local degrees of freedom are

$$\sigma_{i,1}(p) = p(x_i), \qquad \sigma_{i,2}(p) = p'(x_i),$$
$$\sigma_{i,3}(p) = p(x_{i+1}), \qquad \sigma_{i,4}(p) = p'(x_{i+1}).$$

The local shape functions are

$$\theta_{i,1} = \widehat{\theta}_1 \circ T_i^{-1}, \qquad \theta_{i,2} = h_i \widehat{\theta}_2 \circ T_i^{-1},$$
$$\theta_{i,3} = \widehat{\theta}_3 \circ T_i^{-1}, \qquad \theta_{i,4} = h_i \widehat{\theta}_4 \circ T_i^{-1},$$

and the *local Hermite interpolation operator* is defined as follows:

$$\mathcal{I}_{I_i}^{\mathrm{H}} : \mathcal{C}^1(I_i) \ni v \longmapsto \sum_{m=1}^{4} \sigma_{i,m}(v)\theta_{i,m} \in \mathbb{P}_3. \tag{1.78}$$

Consider the so-called *Hermite approximation space*

$$H_h = \{v_h \in \mathcal{C}^1(\overline{\Omega}); \forall i \in \{0, \dots, N\}, v_{h|I_i} \in \mathbb{P}_3\}. \tag{1.79}$$

The main motivation for introducing H_h is the following:

Proposition 1.81. $H_h \subset H^2(\Omega)$.

Proof. Adapt the proof of Lemma 1.3. $\qquad\square$

Introduce the functions $\{\varphi_{0,0}, \dots, \varphi_{N+1,0}, \varphi_{0,1}, \dots, \varphi_{N+1,1}\}$ such that

$$\varphi_{i,0}(x) = \begin{cases} \theta_{i-1,3}(x) & \text{if } x \in I_{i-1}, \\ \theta_{i,1}(x) & \text{if } x \in I_i, \\ 0 & \text{otherwise,} \end{cases} \qquad \varphi_{i,1}(x) = \begin{cases} \theta_{i-1,4}(x) & \text{if } x \in I_{i-1}, \\ \theta_{i,2}(x) & \text{if } x \in I_i, \\ 0 & \text{otherwise,} \end{cases}$$

with obvious modifications if $i = 0$ or $N+1$.

Lemma 1.82. $\varphi_{i,0} \in H_h$ and $\varphi_{i,1} \in H_h$.

Proof. Left as an exercise. $\qquad\square$

For $i \in \{0, \dots, N\}$, consider the linear forms

$$\gamma_{i,0} : \mathcal{C}^1(\overline{\Omega}) \ni v \longmapsto \gamma_{i,0}(v) = v(x_i),$$
$$\gamma_{i,1} : \mathcal{C}^1(\overline{\Omega}) \ni v \longmapsto \gamma_{i,1}(v) = v'(x_i).$$

Proposition 1.83. $\{\varphi_{i,l}\}_{0 \le i \le N, 0 \le l \le 1}$ *is a basis for* H_h, *and* $\{\gamma_{i,l}\}_{0 \le i \le N, 0 \le l \le 1}$ *is a basis for* $\mathcal{L}(H_h; \mathbb{R})$.

Proof. Use the fact that $\gamma_{i,l}(\varphi_{i'l'}) = \delta_{ii'}\delta_{ll'}$ and that on each interval I_i, a function in H_h is a polynomial of degree at most 3 and is, therefore, uniquely determined by its value and that of its first derivative at the endpoints x_i and x_{i+1}; details are left as an exercise.   □

Proposition 1.83 implies that H_h is a space of dimension $2(N+2)$. The linear forms $\{\gamma_{i,l}\}_{0 \le i \le N, 0 \le l \le 1}$ are called the *global degrees of freedom* in H_h, and the functions $\{\varphi_{i,l}\}_{0 \le i \le N, 0 \le l \le 1}$ are called the *global shape functions*.

Define the *global Hermite interpolation operator* $\mathcal{I}_h^{\mathrm{H}}$ with codomain H_h as follows:

$$\mathcal{I}_h^{\mathrm{H}} : C^1(\overline{\Omega}) \ni v \longmapsto \sum_{i=0}^{N+1} \gamma_{i,0}(v)\varphi_{i,0} + \sum_{i=0}^{N+1} \gamma_{i,1}(v)\varphi_{i,1} \in H_h. \qquad (1.80)$$

$\mathcal{I}_h^{\mathrm{H}}$ is a linear operator, and $\mathcal{I}_h^{\mathrm{H}} v$ is the unique function in H_h that coincides with v and its derivatives at all the mesh points.

In dimension 2, the construction of H^2-conformal spaces is more technical. A classical example uses Argyris finite elements; see, e.g., [Cia91, p. 88].

1.4.7 $H(\mathrm{div})$-conformal spaces

Let $\{\widehat{K}, \widehat{P}, \widehat{\Sigma}\}$ be the Raviart–Thomas finite element introduced in §1.2.7. Choose $V(\widehat{K}) = \{v \in [L^p(\widehat{K})]^d; \nabla \cdot v \in L^s(\widehat{K})\}$, with $p > 2$ and $s \ge q$, $\frac{1}{q} = \frac{1}{p} + \frac{1}{d}$, and define $V(K)$ similarly. Since $\psi_K(v) = v \circ T_K$ does not map $V(K)$ to $V(\widehat{K})$, one introduces the so-called *Piola transformation*

$$\psi_K : V(K) \ni v \longmapsto \psi_K(v)(\widehat{x}) = \det(J_K) J_K^{-1} \left[v \circ T_K(\widehat{x}) \right] \in \widehat{V}(\widehat{K}), \quad (1.81)$$

where J_K is the Jacobian matrix of T_K.

Lemma 1.84. *Let* $v \in V(K)$ *and set* $\widehat{v} = \psi_K(v)$. *Then, whenever the left-hand sides are meaningful, the following identities hold:*

(i) $\nabla_x v = \frac{1}{\det(J_K)} J_K \left[\nabla_{\widehat{x}} \psi_K(v) \right] J_K^{-1}$ *and* $\int_F v \cdot n = \int_{\widehat{F}} \widehat{v} \cdot \widehat{n}$.

(ii) $\int_K q \nabla_x \cdot v = \int_{\widehat{K}} \widehat{q} \nabla_{\widehat{x}} \cdot \widehat{v}$ *and* $\int_K v \cdot \nabla_x q = \int_{\widehat{K}} \widehat{v} \cdot \nabla_{\widehat{x}} \widehat{q}$ *with* $\widehat{q} = q \circ T_K$.

Proof. Observe that $\nabla_x q = (J_K^{-1})^T \nabla_{\widehat{x}} \widehat{q}$ and $\nabla_x \cdot v(x) = \frac{1}{\det(J_K)} \nabla_{\widehat{x}} \cdot \widehat{v}(\widehat{x})$.   □

Construct the family $\{K, P_K, \Sigma_K\}_{K \in \mathcal{T}_h}$ using Proposition 1.61. Then, for each $K \in \mathcal{T}_h$, letting $F_{K,i} = T_K(\widehat{F}_i)$ with $0 \le i \le d$ where $\{\widehat{F}_0, \ldots, \widehat{F}_d\}$ are the faces of $\widehat{K}$, Lemma 1.84(i) implies that the local degrees of freedom are

$$\sigma_{K,i}(v) = \int_{F_{K,i}} v \cdot n_i, \qquad (1.82)$$

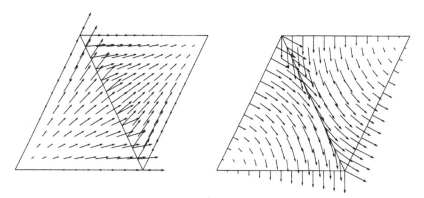

Fig. 1.22. Global shape functions associated with the Raviart–Thomas (left) and the Nédélec (right) finite elements in dimension 2. The normal (resp., tangential) component of the Raviart–Thomas (resp., Nédélec) global shape function is continuous across the interface, but since the triangles are not isosceles, the tangential (resp., normal) component is not antisymmetric.

where n_i is the outward normal to $F_{K,i}$. Furthermore, since the mesh is affine, $T_K(\widehat{x}) = J_K\widehat{x} + b_K$ where $J_K \in \mathbb{R}^{d,d}$ and $b_K \in \mathbb{R}^d$. Hence, for $p \in P_K$, $\psi_K(p) = \widehat{x}_0 + \alpha\widehat{x}$, where $\widehat{x}_0 \in [\mathbb{P}_0]^d$ and $\alpha \in \mathbb{R}$, yielding $p = \psi_K^{-1}(\widehat{x}_0 + \alpha\widehat{x}) = \frac{1}{\det(J_K)}J_K(\widehat{x}_0 + \alpha\widehat{x})$. Then, using $\widehat{x} = J_K^{-1}(x - b_K)$ yields $p \in \mathbb{RT}_0$. As a result, $P_K = \mathbb{RT}_0$ and $\{K, P_K, \Sigma_K\}$ is a Raviart–Thomas finite element.

Consider the so-called *Raviart–Thomas approximation space*

$$D_h = \{v_h \in [L^1(\Omega_h)]^d; \forall K \in \mathcal{T}_h, v_{h|K} \in \mathbb{RT}_0,$$
$$\forall F \in \mathcal{F}_h^i, [\![v_h \cdot n]\!]_F = 0\}, \tag{1.83}$$

where $[\![v_h \cdot n]\!]_F$ denotes the jump of the normal component of v_h across the interface F. The main motivation for introducing D_h is the following:

Proposition 1.85. $D_h \subset H(\text{div}; \Omega_h) = \{v \in [L^2(\Omega_h)]^d; \nabla \cdot v \in L^2(\Omega_h)\}$.

Proof. Proceed as in the proof of Proposition 1.74. □

Now, let us specify the global shape functions in D_h. For $F \in \mathcal{F}_h$, let n_F be a normal unit vector to F (its direction is irrelevant). Consider the function φ_F with support consisting of the one or two simplices to which F belongs and such that on one simplex, say K, the function $\varphi_{F|K}$ is the local shape function of $\{K, P_K, \Sigma_K\}$ associated with the face F and on the other simplex, say K', $\varphi_{F|K'}$ is the opposite of the local shape function associated with F on K'; see the left panel of Figure 1.22.

Lemma 1.86. $\varphi_F \in D_h$.

Proof. Adapt the proof of Lemma 1.70 and use the fact that $\varphi \cdot n_F$ is constant on F. □

Proposition 1.87. $\{\varphi_F\}_{F\in\mathcal{F}_h}$ *is a basis for* D_h, *and defining the linear forms* $\gamma_F : D_h \ni v_h \mapsto \int_F v_h \cdot n_F \in \mathbb{R}$, $\{\gamma_F\}_{F\in\mathcal{F}_h}$ *is a basis for* $\mathcal{L}(D_h; \mathbb{R})$.

Proof. Left as an exercise.   □

Proposition 1.87 implies that D_h is a space of dimension N_{ed} in two dimensions and N_f in three dimensions. The linear forms $\{\gamma_F\}_{F\in\mathcal{F}_h}$ are called the *global degrees of freedom* in D_h, and $\{\varphi_F\}_{F\in\mathcal{F}_h}$ the *global shape functions*.

For a function v in the space

$$V^{\text{div}} = \{v \in [L^p(\Omega_h)]^d; \nabla \cdot v \in L^s(\Omega_h)\}, \qquad (1.84)$$

with $p > 2$ and $s \geq q$, $\frac{1}{q} = \frac{1}{p} + \frac{1}{d}$, the quantity $\gamma_F(v)$ is meaningful (and single-valued) for all $F \in \mathcal{F}_h$. The so-called *global Raviart–Thomas interpolation operator* is constructed as follows:

$$\mathcal{I}_h^{\text{RT}} : V^{\text{div}} \ni v \longmapsto \mathcal{I}_h^{\text{RT}} v = \sum_{F\in\mathcal{F}_h} \left(\int_F v \cdot n_F \right) \varphi_F \in D_h. \qquad (1.85)$$

Note that D_h is the codomain of $\mathcal{I}_h^{\text{RT}}$. See [BrF91^b, RaT77] for further results on $H(\text{div})$-conformal spaces.

Remark 1.88. If the degrees of freedom in $\{\widehat{K}, \widehat{P}, \widehat{\Sigma}\}$ are chosen to be the mean-value of the flux (see Remark 1.40(ii)), Proposition 1.65 must be used to construct the family $\{K, P_K, \Sigma_K\}_{K\in\mathcal{T}_h}$; see Remark 1.72(i).   □

1.4.8 $H(\text{curl})$-conformal spaces

We consider a three-dimensional setting, but a similar construction is possible in two dimensions. Let $\{\widehat{K}, \widehat{P}, \widehat{\Sigma}\}$ be the Nédélec finite element introduced in §1.2.8. Choose $V(\widehat{K}) = \{v \in [L^p(\widehat{K})]^3; \nabla \times v \in [L^s(\widehat{K})]^3\}$ with $p > 2$ and $s > \frac{1}{2}$, and define $V(K)$ similarly. Introduce the mapping

$$\psi_K : V(K) \ni v \longmapsto \psi_K(v)(\widehat{x}) = J_K^T[v \circ T_K(\widehat{x})] \in \widehat{V}(\widehat{K}). \qquad (1.86)$$

Lemma 1.89. *Let* $\mathcal{C}(v) = \nabla v - (\nabla v)^T$. *For all* $v \in V(K)$, *the following identities hold:*

(i) *For all* $\beta \in \mathbb{R}^3$, $\mathcal{C}(v) \cdot \beta = (\nabla \times v) \times \beta$.
(ii) $\|\mathcal{C}(v)\|_{\mathbb{R}^{3,3}} = \|\nabla \times v\|_{\mathbb{R}^3}$.
(iii) $\mathcal{C}[\psi_K(v)] = (J_K)^T \mathcal{C}(v) J_K$.

Construct the family $\{K, P_K, \Sigma_K\}_{K\in\mathcal{T}_h}$ using Proposition 1.61. Denote by $\{\widehat{e}_1, \ldots, \widehat{e}_6\}$ the edges of $\widehat{K}$ and, for $1 \leq i \leq 6$, let $e_{K,i} = T_K(\widehat{e}_i)$ be the corresponding edge of K. Let $\widehat{t}_i$ (resp. $t_{K,i}$) be one of the two unit vectors parallel to $\widehat{e}_i$ (resp. $e_{K,i}$). Since $J_K \widehat{t}_i = \frac{\text{meas}(e_{K,i})}{\text{meas}(\widehat{e}_i)} t_{K,i}$,

$$\sigma_i(v) = \int_{\widehat{e}_i} \psi_K(v){\cdot}\widehat{t}_i = \int_{e_{K,i}} v{\cdot}t_{K,i}. \tag{1.87}$$

Furthermore, since the mesh is affine, $T_K(\widehat{x}) = J_K\widehat{x} + b_K$ where $J_K \in \mathbb{R}^{d,d}$ and $b_K \in \mathbb{R}^d$. Hence, for $p \in P_K$, $\psi_K(p) = J_K^T[p \circ T_K] = \alpha + \beta{\times}\widehat{x}$, yielding $p = \alpha' + (J_K^T)^{-1}[\beta{\times}J_K^{-1}x]$. Then, it is clear that $((J_K^T)^{-1}[\beta{\times}J_K^{-1}x]){\cdot}x = (\beta{\times}J_K^{-1}x){\cdot}J_K^{-1}x = 0$, i.e., $p \in \mathbb{N}_0$. As a result, $P_K = \mathbb{N}_0$ and $\{K, P_K, \Sigma_K\}$ is a Nédélec finite element.

Consider the so-called *Nédélec approximation space*

$$R_h = \{v_h \in [L^1(\Omega_h)]^3; \forall K \in \mathcal{T}_h, v_{h|K} \in \mathbb{N}_0;$$
$$\forall F \in \mathcal{F}_h^i, [\![v_h{\times}n]\!]_F = 0\}, \tag{1.88}$$

where $[\![v_h{\times}n]\!]_F$ denotes the jump of the tangential component of v_h across the interface F. The main motivation for introducing R_h is the following:

Proposition 1.90. $R_h \subset H(\mathrm{curl}; \Omega_h) = \{v \in [L^2(\Omega_h)]^3; \nabla{\times}v \in [L^2(\Omega_h)]^3\}$.

Proof. Proceed as in the proof of Proposition 1.74. ☐

To derive the global shape functions in R_h, we first state the following:

Lemma 1.91. *Let* $F = K_1 \cap K_2$ *and let* v_h *be such that* $v_{h|K_1} \in \mathbb{N}_0$ *and* $v_{h|K_2} \in \mathbb{N}_0$. *Then,* $[\![v_h{\times}n]\!]_F = 0$ *if and only if* $\int_e v_{h|K_1}{\cdot}t_e = \int_e v_{h|K_2}{\cdot}t_e$ *for the three edges of* F.

Proof. Write $v_{h|K_1} = \alpha_1 + \beta_1{\times}x$ and $v_{h|K_2} = \alpha_2 + \beta_2{\times}x$. Let n_F be one of the two unit vectors that are normal to F. Clearly, $v_{h|K_1}{\times}n_F = \alpha_1{\times}n_F + (\beta_1{\cdot}n_F)x - (x{\cdot}n_F)\beta_1$. Since $x{\cdot}n_F$ is constant on F, $v_{h|K_1}{\times}n_F = s + tx$ where $s \in \mathbb{R}^3$ and $t \in \mathbb{R}$; that is to say, $v_{h|K_1}{\times}n_F \in \mathbb{RT}_0$; see (1.40). Let e_1, e_2, and e_3 be the three edges of F. Denote by n_1, n_2, and n_3 the three unit vectors that are parallel to F, are normal to e_1, e_2, and e_3, and point outward. It is clear that $t_i = n_F \times n_i$ is a unit vector parallel to the edge e_i. Let $\{\theta_1, \theta_2, \theta_3\}$ be the two-dimensional Raviart-Thomas shape functions on F. It is readily checked that

$$v_{h|K_1}{\times}n_F = \sum_{i=1}^{3}\left(\int_{e_i}(v_{h|K_1}{\times}n_F){\cdot}n_i\right)\theta_i = \sum_{i=1}^{3}\left(\int_{e_i}v_{h|K_1}{\cdot}t_i\right)\theta_i.$$

Since the set $\{\theta_1, \theta_2, \theta_3\}$ is linearly independent, it is clear that $[\![v_h{\times}n]\!]_F = 0$ if and only if $\int_{e_i}v_{h|K_1}{\cdot}t_i = \int_{e_i}v_{h|K_2}{\cdot}t_i$ for all $i \in \{1,2,3\}$. ☐

For an edge $e \in \mathcal{E}_h$, choose one of the two unit vectors parallel to e, say t_e. Consider the function φ_e with support consisting of the simplices to which e belongs and such that on each of these simplices, say K, the function $\varphi_{e|K}$ is the local shape function of $\{K, P_K, \Sigma_K\}$ associated with the edge e oriented by t_e; see the right panel of Figure 1.22.

Lemma 1.92. $\varphi_e \in R_h$.

Proof. Let $e \in \mathcal{E}_h$.
(1) Let K_1 and K_2 be two elements sharing the edge e. Then, owing to (1.87), $\int_e \varphi_{e|K_1} \cdot t_e = 1 = \int_e \varphi_{e|K_2} \cdot t_e$ and for $e' \neq e$, $\int_{e'} \varphi_{e|K_1} \cdot t_{e'} = 0 = \int_{e'} \varphi_{e|K_2} \cdot t_{e'}$.
(2) Let $F \in \mathcal{F}_h^i$, say $F = K_1 \cap K_2$. Owing to step 1, the converse statement of Lemma 1.91 implies $[\![\varphi_e \times n]\!]_F = 0$. The conclusion follows easily. $\qquad\square$

Proposition 1.93.

(i) *For all $e \in \mathcal{E}_h$, the linear form $\gamma_e : R_h \ni v_h \mapsto \int_e v_h \cdot t_e$ is single-valued.*
(ii) *$\{\varphi_e\}_{e\in\mathcal{E}_h}$ is a basis for R_h, and $\{\gamma_e\}_{e\in\mathcal{E}_h}$ is a basis for $\mathcal{L}(R_h; \mathbb{R})$.*

Proof. (1) Let $e \in \mathcal{E}_h$ and let K_1 and K_2 be two elements sharing the edge e. Then, there exists a finite family of elements $\{K_{j_1}, \dots, K_{j_J}\}$ such that $K_{j_1} = K_1$, $K_{j_J} = K_2$, and $K_{j_l} \cap K_{j_{l+1}}$ is a face containing e. Owing to the direct statement of Lemma 1.91 for each pair $\{K_{j_l}, K_{j_{l+1}}\}$, the quantity $\int_e v_h \cdot t_e$ is single-valued for all edge $e \in \mathcal{E}_h$ and all $v_h \in R_h$.
(2) The family $\{\varphi_e\}_{e\in\mathcal{E}_h}$ is linearly independent since $\gamma_{e'}(\varphi_e) = \delta_{ee'}$ (with obvious notation). Let $v_h \in R_h$. Owing to step 1, it is legitimate to consider the function

$$w_h = \sum_{e\in\mathcal{E}_h} \left(\int_e v_h \cdot t_e \right) \varphi_e.$$

Then, it is clear that for all $K \in \mathcal{T}_h$, $v_{h|K}$ and $w_{h|K}$ are in $\mathbb{N}_0$ and that $\int_e v_{h|K} \cdot t_e = \int_e w_{h|K} \cdot t_e$ for all edge $e \in \partial K$. Unisolvence implies $v_{h|K} = w_{h|K}$. Hence, $\{\varphi_e\}_{e\in\mathcal{E}_h}$ is a basis for R_h. Proving that $\{\gamma_e\}_{e\in\mathcal{E}_h}$ is a basis for $\mathcal{L}(R_h; \mathbb{R})$ is then straightforward. $\qquad\square$

Proposition 1.93 implies that R_h is a space of dimension N_{ed}. The linear forms $\{\gamma_e\}_{e\in\mathcal{E}_h}$ are called the *global degrees of freedom* in R_h, and $\{\varphi_e\}_{e\in\mathcal{E}_h}$ are called the *global shape functions*.

For a function v in the space

$$V^{\mathrm{curl}} = \{v \in [L^p(\Omega_h)]^3; \; \nabla \times v \in [L^s(\Omega_h)]^3\}, \qquad (1.89)$$

with $p > 2$ and $s > \frac{1}{2}$, the quantity $\gamma_e(v)$ is meaningful (and single-valued) for all $e \in \mathcal{E}_h$. The so-called *global Nédélec interpolation operator* is constructed as follows:

$$\mathcal{I}_h^N : V^{\mathrm{curl}} \ni v \longmapsto \mathcal{I}_h^N v = \sum_{e\in\mathcal{E}_h} \left(\int_e v \cdot t_e \right) \varphi_e \in R_h. \qquad (1.90)$$

Note that R_h is the codomain of $\mathcal{I}_h^N$. For further results on $H(\mathrm{curl})$-conformal spaces, see, e.g., [Néd80, Néd86, Mon92, Bos93].

Remark 1.94. If the degrees of freedom in $\{\widehat{K}, \widehat{P}, \widehat{\Sigma}\}$ are chosen to be the mean-value of the integral over the edges (see Remark 1.43(ii)), Proposition 1.65 must be used to construct the family $\{K, P_K, \Sigma_K\}_{K\in\mathcal{T}_h}$; see Remark 1.72(i). $\qquad\square$

1.4.9 A link between H^1-, $H(\mathrm{curl})$-, and $H(\mathrm{div})$-conformal spaces

When the mesh $\mathcal{T}_h$ consists of affine simplices, an interesting relation exists between the spaces $P_{c,h}^1$, R_h, and D_h in three dimensions. To formalize this relation, we introduce the concept of exact sequence; see, e.g., [God71]. Let $\{E_j\}_{j\in J}$ be a sequence of vector spaces on the same field and indexed by an interval J of $\mathbb{N}$. For $j \in J$ such that $j + 1 \in J$, let $h_j : E_j \to E_{j+1}$ be a homomorphism.

Definition 1.95. *The sequence*

$$\ldots \xrightarrow{h_{j-1}} E_j \xrightarrow{h_j} E_{j+1} \xrightarrow{h_{j+1}} E_{j+2} \xrightarrow{h_{j+2}} \ldots$$

is said to be exact *if for all $j \in J$ such that $j + 2 \in J$, $\mathrm{Ker}(h_{j+1}) = \mathrm{Im}(h_j)$.*

Consider a domain Ω in $\mathbb{R}^3$. Let $H_0(\mathrm{curl}; \Omega)$ be the subspace of $H(\mathrm{curl}; \Omega)$ consisting of the vector fields whose tangential components vanish at $\partial\Omega$. Let also $H_0(\mathrm{div}; \Omega)$ be the subspace of $H(\mathrm{div}; \Omega)$ consisting of the vector fields whose normal component vanishes at $\partial\Omega$. Let i be the canonical injection and let m be the averaging operator over Ω.

Proposition 1.96. *If Ω is simply connected and $\partial\Omega$ is connected, the following sequence is exact:*

$$\{0\} \xrightarrow{i} H_0^1(\Omega) \xrightarrow{\nabla} H_0(\mathrm{curl}; \Omega) \xrightarrow{\nabla\times} H_0(\mathrm{div}; \Omega) \xrightarrow{\nabla\cdot} L^2(\Omega) \xrightarrow{m} \mathrm{span}\{1\}.$$

Let Ω_h be a geometric interpolate of the domain Ω based on a mesh $\mathcal{T}_h$. Define the approximation spaces $P_{c,h,0}^1 = P_{c,h}^1 \cap H_0^1(\Omega_h)$, $R_{h,0} = R_h \cap H_0(\mathrm{curl}; \Omega_h)$, and $D_{h,0} = D_h \cap H_0(\mathrm{div}; \Omega_h)$. Let also $P_{\mathrm{td},h}^0$ be the space of piecewise constant functions on the mesh $\mathcal{T}_h$. As a discrete counterpart of Proposition 1.96, one easily proves the following:

Proposition 1.97. *If Ω_h is simply connected and $\partial\Omega_h$ is connected, the following sequence is exact:*

$$\{0\} \xrightarrow{i} P_{c,h,0}^1 \xrightarrow{\nabla} R_{h,0} \xrightarrow{\nabla\times} D_{h,0} \xrightarrow{\nabla\cdot} P_{\mathrm{td},h}^0 \xrightarrow{m} \mathrm{span}\{1\}.$$

Assume $\Omega_h \subset \Omega$ for the sake of simplicity. Set $V^1 = H^s(\Omega)$ with $s > \frac{d}{2}$ and $V^0 = L^1(\Omega)$. Let V^{div} and V^{curl} be defined in (1.84) and (1.89), respectively. Let $\mathcal{I}_h^1$, $\mathcal{I}_h^N$, $\mathcal{I}_h^{RT}$, and $\mathcal{I}_{\mathrm{td},h}^0$ the interpolation operators associated with the finite element spaces $P_{c,h}^1$, R_h, D_h, and $P_{\mathrm{td},h}^0$, respectively. The following striking property holds:

Proposition 1.98. *The following diagram commutes:*

$$
\begin{array}{ccccccc}
V^1 & \xrightarrow{\nabla} & V^{\mathrm{curl}} & \xrightarrow{\nabla\times} & V^{\mathrm{div}} & \xrightarrow{\nabla\cdot} & V^0 \\
\downarrow{\scriptstyle \mathcal{I}_h^1} & & \downarrow{\scriptstyle \mathcal{I}_h^N} & & \downarrow{\scriptstyle \mathcal{I}_h^{RT}} & & \downarrow{\scriptstyle \mathcal{I}_{\mathrm{td},h}^0} \\
P_{c,h}^1 & \xrightarrow{\nabla} & R_h & \xrightarrow{\nabla\times} & D_h & \xrightarrow{\nabla\cdot} & P_{\mathrm{td},h}^0
\end{array}
$$

Proof. This is a simple corollary of Lemmas 1.41, 1.44, and 1.45.   □

Remark 1.99.

(i) Propositions 1.97 and 1.98 can be extended to higher-order finite element spaces; see the de Rham diagram theory developed in [DeM00, Bof01].

(ii) Proposition 1.97 provides an efficient means of constructing all the fields in $R_{h,0}$ with vanishing curl and all the solenoidal fields in $D_{h,0}$. For further results, see [Bos93].   □

1.5 Interpolation of Smooth Functions

Letting $\mathcal{I}_h$ be one of the interpolation operators constructed in §1.4, the goal of this section is to estimate the interpolation error $v - \mathcal{I}_h v$ assuming that the function v is smooth enough to be in the domain of $\mathcal{I}_h$. First, we investigate thoroughly the interpolation of scalar- and vector-valued functions on affine meshes. Then, we briefly discuss non-affine transformations.

1.5.1 Interpolation in $W^{s,p}(\Omega)$

In this section, we establish local and global interpolation error estimates on affine meshes for scalar-valued functions living in Sobolev spaces; see Appendix B for a definition of these spaces and the corresponding norms. Interpolation error estimates in vector-valued Sobolev spaces are readily derived by applying the scalar-valued interpolation error estimates componentwise.

Since the mesh is affine, the transformation T_K takes the form

$$T_K : \widehat{K} \ni \widehat{x} \longmapsto J_K \widehat{x} + b_K \in K, \tag{1.91}$$

where $J_K \in \mathbb{R}^{d,d}$ and $b_K \in \mathbb{R}^d$. The Jacobian matrix J_K is invertible since T_K is bijective. Let $\| \cdot \|_d$ be the Euclidean norm in $\mathbb{R}^d$ as well as the associated matrix norm. Throughout this section, we assume that the mapping $\psi_K : V(K) \to \widehat{V}(\widehat{K})$ in Proposition 1.62 is $\psi_K(v) = v \circ T_K$, and we set $\widehat{v} = v \circ T_K$.

Lemma 1.100. *Let ρ_K be the diameter of the largest ball that can be inscribed in K. Then,*

$$|\det(J_K)| = \frac{\operatorname{meas}(K)}{\operatorname{meas}(\widehat{K})}, \qquad \|J_K\|_d \leq \frac{h_K}{\rho_{\widehat{K}}}, \qquad and \qquad \|J_K^{-1}\|_d \leq \frac{h_{\widehat{K}}}{\rho_K}. \tag{1.92}$$

Proof. The first property in (1.92) is classical. Furthermore,

$$\|J_K\|_d = \sup_{\widehat{x} \neq 0} \frac{\|J_K \widehat{x}\|_d}{\|\widehat{x}\|_d} = \frac{1}{\rho_{\widehat{K}}} \sup_{\|\widehat{x}\|_d = \rho_{\widehat{K}}} \|J_K \widehat{x}\|_d.$$

Write $\widehat{x} = \widehat{x}_1 - \widehat{x}_2$ with $\widehat{x}_1$ and $\widehat{x}_2$ in $\widehat{K}$ and use $J_K \widehat{x} = T_K \widehat{x}_1 - T_K \widehat{x}_2 = x_1 - x_2$ to obtain $\|J_K \widehat{x}\|_d \leq h_K$. This proves the first inequality in (1.92). The second inequality is obtained by exchanging the roles of K and $\widehat{K}$.   □

Lemma 1.101. *Let $s \geq 0$ and let $1 \leq p \leq \infty$. There exists c such that, for all K and $w \in W^{s,p}(K)$,*

$$|\widehat{w}|_{s,p,\widehat{K}} \leq c \, \|J_K\|_d^s \, |\det(J_K)|^{-\frac{1}{p}} \, |w|_{s,p,K}, \qquad (1.93)$$

$$|w|_{s,p,K} \leq c \, \|J_K^{-1}\|_d^s \, |\det(J_K)|^{\frac{1}{p}} \, |\widehat{w}|_{s,p,\widehat{K}}, \qquad (1.94)$$

with $\widehat{w} = w \circ T_K$ and with the convention that, for $p = \infty$ and any positive real x, $x^{\pm \frac{1}{p}} = 1$.

Proof. Let α be a multi-index with length $|\alpha| = s$. Use the chain-rule and the fact that the transformation T_K is affine to obtain

$$\|\partial^\alpha \widehat{w}\|_{L^p(\widehat{K})} \leq c \, \|J_K\|_d^s \sum_{|\beta|=s} \|\partial^\beta w \circ T_K\|_{L^p(\widehat{K})}.$$

Changing variables in the right-hand side yields

$$\|\partial^\alpha \widehat{w}\|_{L^p(\widehat{K})} \leq c \, \|J_K\|_d^s \, |\det(J_K)|^{-\frac{1}{p}} \, |w|_{s,p,K}.$$

We deduce (1.93) upon summing over α. The proof of (1.94) is similar. □

Remark 1.102. The upper bounds in (1.93) and (1.94) involve only seminorms because affine transformations are considered. □

Theorem 1.103 (Local interpolation). *Let $\{\widehat{K}, \widehat{P}, \widehat{\Sigma}\}$ be a finite element with associated normed vector space $V(\widehat{K})$. Let $1 \leq p \leq \infty$ and assume that there exists an integer k such that*

$$\mathbb{P}_k \subset \widehat{P} \subset W^{k+1,p}(\widehat{K}) \subset V(\widehat{K}). \qquad (1.95)$$

Let $T_K : \widehat{K} \to K$ be an affine bijective mapping and let $\mathcal{I}_K^k$ be the local interpolation operator on K defined in (1.59). Let l be such that $0 \leq l \leq k$ and $W^{l+1,p}(\widehat{K}) \subset V(\widehat{K})$ with continuous embedding. Then, setting $\sigma_K = \frac{h_K}{\rho_K}$, there exists $c > 0$ such that, for all $m \in \{0, \dots, l+1\}$,

$$\forall K, \; \forall v \in W^{l+1,p}(K), \quad |v - \mathcal{I}_K^k v|_{m,p,K} \leq c \, h_K^{l+1-m} \sigma_K^m \, |v|_{l+1,p,K}. \qquad (1.96)$$

Proof. Let $\mathcal{I}_{\widehat{K}}^k$ be the local interpolation operator on $\widehat{K}$ defined in (1.57). Let $\widehat{w} \in W^{l+1,p}(\widehat{K})$. Since $W^{l+1,p}(\widehat{K}) \subset V(\widehat{K})$ with continuous embedding, the linear operator

$$\mathcal{F} : W^{l+1,p}(\widehat{K}) \ni \widehat{w} \longmapsto \widehat{w} - \mathcal{I}_{\widehat{K}}^k \widehat{w} \in W^{m,p}(\widehat{K}),$$

is continuous from $W^{l+1,p}(\widehat{K})$ to $W^{m,p}(\widehat{K})$ for all $m \in \{0, \dots, l+1\}$. Since $l \leq k$, $\mathbb{P}_l \subset \widehat{P}$ and, therefore, $\mathbb{P}_l$ is invariant under $\mathcal{I}_{\widehat{K}}^k$ owing to Proposition 1.30. Hence, $\mathcal{F}$ vanishes on $\mathbb{P}_l$. As a consequence,

$$|\widehat{w} - \mathcal{I}_{\widehat{K}}^k \widehat{w}|_{m,p,\widehat{K}} = |\mathcal{F}(\widehat{w})|_{m,p,\widehat{K}} = \inf_{\widehat{p} \in \mathbb{P}_l} |\mathcal{F}(\widehat{w} + \widehat{p})|_{m,p,\widehat{K}}$$

$$\leq \|\mathcal{F}\|_{\mathcal{L}(W^{l+1,p}(\widehat{K});W^{m,p}(\widehat{K}))} \inf_{\widehat{p} \in \mathbb{P}_l} \|\widehat{w} + \widehat{p}\|_{l+1,p,\widehat{K}}$$

$$\leq c \inf_{\widehat{p} \in \mathbb{P}_l} \|\widehat{w} + \widehat{p}\|_{l+1,p,\widehat{K}} \leq c \,|\widehat{w}|_{l+1,p,\widehat{K}},$$

the last estimate resulting from the Deny–Lions Lemma; see Lemma B.67. Now let $v \in W^{l+1,p}(K)$ and set $\widehat{v} = \psi_K(v) = v \circ T_K$. Owing to Proposition 1.62, $[\mathcal{I}_K^k v] \circ T_K = \mathcal{I}_{\widehat{K}}^k \widehat{v}$. Using Lemma 1.101 yields

$$|v - \mathcal{I}_K^k v|_{m,p,K} \leq c \|J_K^{-1}\|_d^m \,|\det(J_K)|^{\frac{1}{p}} \,|\widehat{v} - \mathcal{I}_{\widehat{K}}^k \widehat{v}|_{m,p,\widehat{K}}$$

$$\leq c \|J_K^{-1}\|_d^m \,|\det(J_K)|^{\frac{1}{p}} \,|\widehat{v}|_{l+1,p,\widehat{K}}$$

$$\leq c \|J_K^{-1}\|_d^m \,\|J_K\|_d^{l+1} \,|v|_{l+1,p,K}$$

$$\leq c (\|J_K\|_d \|J_K^{-1}\|_d)^m \,\|J_K\|_d^{l+1-m} \,|v|_{l+1,p,K}.$$

Conclude using (1.92).   □

Definition 1.104 (Degree of a finite element). *The largest integer k such that (1.95) holds is called the* degree *of the finite element $\{\widehat{K}, \widehat{P}, \widehat{\Sigma}\}$.*

Remark 1.105. If the interpolated function is in $W^{k+1,p}(K)$, one can take $l = k$ in Theorem 1.103. The resulting error estimate is *optimal*, i.e., for $m \in \{0, \dots, k+1\}$,

$$\forall K, \ \forall v \in W^{k+1,p}(K), \quad |v - \mathcal{I}_K^k v|_{m,p,K} \leq c \, h_K^{k+1-m} \sigma_K^m \,|v|_{k+1,p,K}. \qquad \square$$

Example 1.106.
 (i) For a Lagrange finite element of degree k, $V(\widehat{K}) = \mathcal{C}^0(\widehat{K})$; hence, the condition on l in Theorem 1.103 is $\frac{d}{p} - 1 < l \leq k$. Indeed, owing to Theorem B.46, $W^{l+1,p}(\widehat{K}) \subset V(\widehat{K})$ provided $l + 1 > \frac{d}{p}$. More generally, for a finite element with $V(\widehat{K}) = \mathcal{C}^t(\widehat{K})$ (for instance, $t = 1$ for the Hermite finite element), the condition on l is $\frac{d}{p} - 1 + t < l \leq k$; see also [BrS94, p. 104].
 (ii) For the Crouzeix–Raviart finite element, $k = 1$ and $V(\widehat{K}) = W^{1,1}(\widehat{K})$; as a result, the condition on l is $0 \leq l \leq k = 1$.   □

To obtain global interpolation error estimates on Ω and to prove that these estimates converge to zero as $h \to 0$, the quantity σ_K appearing in (1.96) must be controlled independently of K and h. This leads to the following:

Definition 1.107 (Shape-regularity). *A family of meshes $\{\mathcal{T}_h\}_{h>0}$ is said to be* shape-regular *if there exists σ_0 such that*

$$\forall h, \ \forall K \in \mathcal{T}_h, \quad \sigma_K = \frac{h_K}{\rho_K} \leq \sigma_0.$$

Remark 1.108.

(i) Let K be a triangle and denote by θ_K the smallest of its angles. One readily sees that

$$\frac{h_K}{\rho_K} \leq \frac{2}{\sin \theta_K}.$$

Therefore, in a shape-regular family of triangulations, the triangles cannot become too flat as $h \to 0$.

(ii) In dimension 1, $h_K = \rho_K$; hence, any mesh family is shape-regular.

(iii) Lemma 1.100 shows that for a shape-regular family of meshes, there is c such that, for all h and $K \in \mathcal{T}_h$, $\|J_K\|_d \|J_K^{-1}\|_d \leq c$. The quantity $\|J_K\|_d \|J_K^{-1}\|_d$ is called the Euclidean *condition number* of J_K. □

Corollary 1.109 (Global interpolation). *Let p, k, and l satisfy the assumptions of Theorem 1.103. Let Ω be a polyhedron and let $\{\mathcal{T}_h\}_{h>0}$ be a shape-regular family of affine meshes of Ω. Denote by V_h^k the approximation space based on $\mathcal{T}_h$ and $\{\widehat{K}, \widehat{P}, \widehat{\Sigma}\}$. Let $\mathcal{I}_h^k$ be the corresponding global interpolation operator. Then, there exists c such that, for all h and $v \in W^{l+1,p}(\Omega)$,*

$$\|v - \mathcal{I}_h^k v\|_{L^p(\Omega)} + \sum_{m=1}^{l+1} h^m \left(\sum_{K \in \mathcal{T}_h} |v - \mathcal{I}_h^k v|_{m,p,K}^p \right)^{\frac{1}{p}} \leq c \, h^{l+1} |v|_{l+1,p,\Omega}, \quad (1.97)$$

for $p < \infty$, and for $p = \infty$

$$\|v - \mathcal{I}_h^k v\|_{L^\infty(\Omega)} + \sum_{m=1}^{l+1} h^m \max_{K \in \mathcal{T}_h} |v - \mathcal{I}_h^k v|_{m,\infty,K} \leq c \, h^{l+1} |v|_{l+1,\infty,\Omega}. \quad (1.98)$$

Furthermore, for $p < \infty$ and $v \in L^p(\Omega)$, the following density result holds:

$$\lim_{h \to 0} \left(\inf_{v_h \in V_h^k} \|v - v_h\|_{L^p(\Omega)} \right) = 0. \quad (1.99)$$

Proof. Since the family $\{\mathcal{T}_h\}_{h>0}$ is shape-regular, estimates (1.97) and (1.98) result from (1.96). Let $v \in L^p(\Omega)$ and $\epsilon > 0$. Since $W^{l+1,p}(\Omega)$ is dense in $L^p(\Omega)$ for $p < \infty$, there is $v^\epsilon \in W^{l+1,p}(\Omega)$ such that $\|v - v^\epsilon\|_{L^p(\Omega)} \leq \epsilon$. Furthermore, (1.97) yields $\|v^\epsilon - \mathcal{I}_h^k v^\epsilon\|_{L^p(\Omega)} \leq ch^{l+1} |v^\epsilon|_{l+1,p,\Omega}$. Hence,

$$\inf_{v_h \in V_h^k} \|v - v_h\|_{L^p(\Omega)} \leq \|v - \mathcal{I}_h^k v^\epsilon\|_{L^p(\Omega)} \leq \|v - v^\epsilon\|_{L^p(\Omega)} + \|v^\epsilon - \mathcal{I}_h^k v^\epsilon\|_{L^p(\Omega)}.$$

That is to say, $\limsup_{h \to 0} (\inf_{v_h \in V_h^k} \|v - v_h\|_{L^p(\Omega)}) \leq \epsilon$, and (1.99) follows from the fact that ϵ is arbitrary. □

Corollary 1.110 (Interpolation in $W^{s,p}(\Omega)$). *Let the hypotheses of Corollary 1.109 hold and assume that V_h^k is $W^{1,p}$-conformal. Then, there is c such that, for all h and $v \in W^{l+1,p}(\Omega)$,*

$$|v - \mathcal{I}_h^k v|_{1,p,\Omega} \le c\, h^l |v|_{l+1,p,\Omega}. \tag{1.100}$$

For $p < \infty$, the following density result holds:

$$\forall v \in W^{1,p}(\Omega), \quad \lim_{h \to 0} \left(\inf_{v_h \in V_h^k} |v - v_h|_{1,p,\Omega} \right) = 0. \tag{1.101}$$

Example 1.111.

(i) Consider a Lagrange finite element of degree k. Take $p = 2$ and assume $d \le 3$. Then, owing to Example 1.106(i), one can take $1 \le l \le k$, and (1.97) yields, for all $v \in H^{l+1}(\Omega)$,

$$\|v - \mathcal{I}_h^k v\|_{0,\Omega} + h|v - \mathcal{I}_h^k v|_{1,\Omega} \le c\, h^{l+1} |v|_{l+1,\Omega}. \tag{1.102}$$

This estimate is optimal if v is smooth enough, i.e., $v \in H^{k+1}(\Omega)$. However, if v is in $H^s(\Omega)$ and not in $H^{s+1}(\Omega)$ for some $s \ge 2$, increasing the degree of the finite element beyond $s - 1$ does not improve the interpolation error. This phenomenon is illustrated in §3.2.5. Note also that the same asymptotic order is obtained for $\mathbb{P}_k$ and $\mathbb{Q}_k$ Lagrange finite elements. For $\mathbb{Q}_k$ Lagrange finite elements, a sharper interpolation error estimate can be derived using a different norm for v in the right-hand side of (1.97); see, e.g., [BrS94, p. 112].

(ii) Consider the Hermite finite element; see §1.4.6. Take $p = 2$; since $d = 1$ and $k = 3$, Example 1.106(i) shows that one can take $2 \le l \le 3$. Owing to (1.97), we infer, for all $v \in H^{l+1}(\Omega)$,

$$\|v - \mathcal{I}_h^k v\|_{0,\Omega} + h|v - \mathcal{I}_h^k v|_{1,\Omega} + h^2 |v - \mathcal{I}_h^k v|_{2,\Omega} \le c\, h^{l+1} |v|_{l+1,\Omega}. \tag{1.103}$$

If $l = 3$, i.e., if $v \in H^4(\Omega)$, the error estimate is optimal. □

Remark 1.112. Estimate (1.97) also applies when the parameter l is not an integer. As a simple example, consider a Lagrange finite element of degree $k \ge 1$ in dimension $d \le 3$. Since $W^{k+1-\frac{d}{2},\infty}(\widehat{K}) \subset C^0(\widehat{K}) = V(\widehat{K})$ with continuous embedding (i.e., $k+1-\frac{d}{2} > 0$), (1.98) can be applied with $l = k - \frac{d}{2}$ and $p = \infty$ to obtain $\|v - \mathcal{I}_h^k v\|_{L^\infty(\Omega)} \le c\, h^{k+1-\frac{d}{2}} |v|_{k+1-\frac{d}{2},\infty,\Omega}$ for $v \in W^{k+1-\frac{d}{2},\infty}(\Omega)$. Therefore, using the fact that $H^{k+1}(\Omega) \subset W^{k+1-\frac{d}{2},\infty}(\Omega)$ with continuous embedding yields

$$\forall h, \ \forall v \in H^{k+1}(\Omega), \quad \|v - \mathcal{I}_h^k v\|_{L^\infty(\Omega)} \le c\, h^{k+1-\frac{d}{2}} |v|_{k+1,\Omega}.$$

Obviously, if $v \in W^{k+1,\infty}(\Omega)$, (1.98) implies the sharper estimate

$$\forall h, \ \forall v \in W^{k+1,\infty}(\Omega), \quad \|v - \mathcal{I}_h^k v\|_{L^\infty(\Omega)} \le c\, h^{k+1} |v|_{k+1,\infty,\Omega}. \quad □$$

1.5.2 Interpolation in $H(\mathrm{div}; \Omega)$

We analyze in this section the interpolation properties of the Raviart–Thomas finite element introduced in §1.2.7.

We assume that the mapping $T_K : \widehat{K} \to K$ is linear, i.e., $T_K(\widehat{x}) = J_K\widehat{x} + b_K$ with $J_K \in \mathbb{R}^{d,d}$ and $b_K \in \mathbb{R}^d$. For a vector-field $v \in [W^{s,p}(K)]^d$, set $\widehat{v}(\widehat{x}) = \det(J_K)J_K^{-1}v(x)$, i.e., $\widehat{v} = \psi_K(v)$ where ψ_K is the Piola transformation defined by (1.81).

Lemma 1.113. *Let $s \geq 0$ and $1 \leq p \leq \infty$ (with $x^{\pm\frac{1}{p}} = 1$ for all $x > 0$ if $p = \infty$). Then, there is c such that, for all K and $w \in [W^{s,p}(K)]^d$ with $\nabla\cdot w \in W^{s,p}(K)$,*

$$|w|_{s,p,K} \leq c \|J_K^{-1}\|_d^s \|J_K\|_d \, |\det(J_K)|^{-\frac{1}{p'}} \, |\widehat{w}|_{s,p,\widehat{K}}, \qquad (1.104)$$

$$|\nabla\cdot w|_{s,p,K} \leq c \|J_K^{-1}\|_d^s |\det(J_K)|^{-\frac{1}{p'}} \, |\nabla\cdot\widehat{w}|_{s,p,\widehat{K}}. \qquad (1.105)$$

Proof. The proof is similar to that of Lemma 1.101; note however the different factors appearing in (1.94) and (1.104) resulting from the fact that a different mapping ψ_K has been used. $\qquad\square$

Let $\{K, \mathbb{RT}_0, \Sigma\}$ be the Raviart–Thomas finite element and let $\mathcal{I}_K^{\mathrm{RT}}$ be the associated local interpolation operator defined in (1.42).

Theorem 1.114. *Let $p > \frac{2d}{d+2}$. There is c such that, for all $v \in [W^{1,p}(K)]^d$ with $\nabla\cdot v \in W^{1,p}(K)$,*

$$\|\mathcal{I}_K^{\mathrm{RT}}v - v\|_{0,p,K} \leq c\,\sigma_K h_K |v|_{1,p,K},$$

$$\|\nabla\cdot(\mathcal{I}_K^{\mathrm{RT}}v - v)\|_{0,p,K} \leq c\,h_K |\nabla\cdot v|_{1,p,K}.$$

Proof. Set $V(\widehat{K}) = [W^{1,p}(\widehat{K})]^d$ with $p > \frac{2d}{d+2}$. The operator

$$\mathcal{F} : [W^{1,p}(\widehat{K})]^d \ni \widehat{w} \longmapsto \widehat{w} - \mathcal{I}_{\widehat{K}}^{\mathrm{RT}}\widehat{w} \in [L^p(\widehat{K})]^d,$$

is continuous. Since $[\mathbb{P}_0]^d \subset \mathbb{RT}_0$ and $\mathcal{F}$ vanishes on $[\mathbb{P}_0]^d$, it is clear that, for all $\widehat{w} \in V(\widehat{K})$,

$$\|\widehat{w} - \mathcal{I}_{\widehat{K}}^{\mathrm{RT}}\widehat{w}\|_{0,p,\widehat{K}} = \|\mathcal{F}(\widehat{w})\|_{0,p,\widehat{K}} - \inf_{\widehat{p}\in[\mathbb{P}_0]^d} \|\mathcal{F}(\widehat{w} + \widehat{p})\|_{0,p,\widehat{K}}$$

$$\leq \|\mathcal{F}\|_{[W^{1,p}(\widehat{K})]^d,[L^p(\widehat{K})]^d} \inf_{\widehat{p}\in[\mathbb{P}_0]^d} \|\widehat{w} + \widehat{p}\|_{1,p,\widehat{K}}$$

$$\leq c \inf_{\widehat{p}\in[\mathbb{P}_0]^d} \|\widehat{w} + \widehat{p}\|_{1,p,\widehat{K}} \leq c\,|\widehat{w}|_{1,p,\widehat{K}},$$

the last estimate resulting from the Deny–Lions Lemma applied componentwise. Let $v \in [W^{1,p}(K)]^d$ and set $\widehat{v} = \psi_K(v)$. Lemma 1.113 implies

$$\|v - \mathcal{I}_K^{\mathrm{RT}}v\|_{0,p,K} \leq c\|J_K\|_d |\det(J_K)|^{-\frac{1}{p'}} \|\widehat{v} - \mathcal{I}_{\widehat{K}}^{\mathrm{RT}}\widehat{v}\|_{0,p,\widehat{K}}$$

$$\leq c\|J_K\|_d |\det(J_K)|^{-\frac{1}{p'}} |\widehat{v}|_{1,p,\widehat{K}}$$

$$\leq c\|J_K\|_d^2 \|J_K^{-1}\|_d |v|_{1,p,K}$$

$$\leq c(\|J_K\|_d \|J_K^{-1}\|_d) \|J_K\|_d |v|_{1,p,K}.$$

The estimate on $\|\mathcal{I}_K^{\text{RT}} v - v\|_{0,p,K}$ then results from (1.92). To prove the estimate on the divergence of the interpolation error, use Lemma 1.41, yielding

$$\|\nabla\cdot(\mathcal{I}_K^{\text{RT}} v) - \nabla\cdot v\|_{0,p,K} = \|\pi_K^0[\nabla\cdot v] - \nabla\cdot v\|_{0,p,K} \leq c\, h_K |\nabla\cdot v|_{1,p,K}.$$

Since $\nabla\cdot v$ is scalar-valued, the technique to prove the last inequality is identical to that used in the proof of Theorem 1.103. $\qquad\square$

Corollary 1.115. *Let the assumptions of Theorem 1.114 hold. Let Ω be a polyhedron and let $\{\mathcal{T}_h\}_{h>0}$ be a shape-regular family of affine meshes of Ω. Let $\mathcal{I}_h^{\text{RT}}$ be the global Raviart–Thomas interpolation operator defined in (1.85). Let $p > \frac{2d}{d+2}$. Then, there is c such that, for all h and $v \in [W^{1,p}(\Omega)]^d$ with $\nabla\cdot v \in W^{1,p}(\Omega)$,*

$$\|v - \mathcal{I}_h^{\text{RT}} v\|_{0,p,\Omega} + \|\nabla\cdot(v - \mathcal{I}_h^{\text{RT}} v)\|_{0,p,\Omega} \leq c\, h(\|v\|_{1,p,\Omega} + \|\nabla\cdot v\|_{1,p,\Omega}). \quad (1.106)$$

1.5.3 Interpolation in $H(\text{curl}; \Omega)$

The purpose of this section is to analyze the interpolation properties of the Nédélec finite element introduced in §1.2.8.

The space dimension is $d = 2$ or 3. The results are stated for $d = 3$, those for $d = 2$ being similar. As in the previous section, we assume that the mapping $T_K : \widehat{K} \to K$ is linear, i.e., $T_K(\widehat{x}) = J_K \widehat{x} + b_K$ with $J_K \in \mathbb{R}^{d,d}$ and $b_K \in \mathbb{R}^d$. For a vector-field $v \in [W^{s,p}(K)]^3$ with $s \geq 0$ and $p \geq 1$, we set $\widehat{v}(\widehat{x}) = J_K v(T_K(x))$, i.e., $\widehat{v} = \psi_K(v)$ where ψ_K is the transformation defined in (1.86). Denote by $\|\cdot\|_{\mathbb{R}^3}$ the Euclidean vector norm in $\mathbb{R}^3$ and by $\|\cdot\|_{\mathbb{R}^{3,3}}$ the associated matrix norm.

Lemma 1.116. *Let $s \geq 0$ and $1 \leq p \leq \infty$ (with $x^{\pm\frac{1}{p}} = 1$ for all $x > 0$ if $p = \infty$). There is c such that, for all K and $w \in [W^{s,p}(K)]^3$ with $\nabla\times w \in [W^{s,p}(K)]^3$,*

$$|w|_{s,p,K} \leq c\, \|J_K^{-1}\|_{\mathbb{R}^{3,3}}^{s+1} |\det(J_K)|^{\frac{1}{p}} |\widehat{w}|_{s,p,\widehat{K}},$$

$$|\nabla\times w|_{s,p,K} \leq c\, \|J_K^{-1}\|_{\mathbb{R}^{3,3}}^{s+2} |\det(J_K)|^{\frac{1}{p}} |\nabla\times\widehat{w}|_{s,p,\widehat{K}}.$$

Proof. The proof is similar to that of Lemma 1.101 and uses Lemma 1.89. Let us prove the second inequality with $s = 1$. Observe that

$$\|\partial_{x_i}\nabla\times v\|_{[L^p(K)]^3}^p = \|\nabla\times(\partial_{x_i} v)\|_{[L^p(K)]^3}^p = \|\mathcal{C}(\partial_{x_i} v)\|_{[L^p(K)]^{3,3}}^p$$

$$= |\det(J_K)| \int_{\widehat{K}} \left\| \sum_{j=1}^3 \partial_{x_i}\widehat{x}_j\, (J_K^{-1})^T \mathcal{C}(\partial_{\widehat{x}_j}\widehat{v})\, (J_K^{-1}) \right\|_{\mathbb{R}^{3,3}}^p$$

$$\leq |\det(J_K)| \|(J_K^{-1})^T\|_{\mathbb{R}^{3,3}}^p \|J_K^{-1}\|_{\mathbb{R}^{3,3}}^p \left(\sum_{j=1}^3 |\partial_{x_i}\widehat{x}_j|^2 \right)^{\frac{p}{2}} \int_{\widehat{K}} \left(\sum_{j=1}^3 \|\partial_{\widehat{x}_j}\nabla\times\widehat{v}\|_{\mathbb{R}^3}^2 \right)^{\frac{p}{2}}.$$

Then, since $\|(J_K^{-1})^T\|_{\mathbb{R}^{3,3}} = \|J_K^{-1}\|_{\mathbb{R}^{3,3}}$ and $\sum_{j=1}^3 |\partial_{x_i}\widehat{x}_j|^2 \leq \|J_K^{-1}\|_{\mathbb{R}^{3,3}}^2$, the desired result is obtained. $\qquad\square$

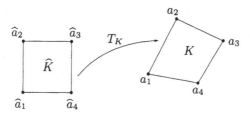

Fig. 1.23. Non-affine transformation mapping the unit square to a quadrangle.

Let $\{K, \mathbb{N}_0, \Sigma\}$ be the Nédélec finite element and let $\mathcal{I}_K^N$ be the associated local interpolation operator defined in (1.48).

Theorem 1.117. *Let $p > 2$. There is c such that, for all $v \in [W^{1,p}(K)]^3$ with $\nabla \times v \in [W^{1,p}(K)]^3$,*

$$\|\mathcal{I}_K^N v - v\|_{0,p,K} \le c\,\sigma_K h_K |v|_{1,p,K},$$
$$\|\nabla \times (\mathcal{I}_K^N v - v)\|_{0,p,K} \le c\,h_K |\nabla \times v|_{1,p,K}.$$

Proof. The proof is similar to that of Theorem 1.114. □

Corollary 1.118. *Let the assumptions of Theorem 1.117 hold. Let Ω be a polyhedron and let $\{\mathcal{T}_h\}_{h>0}$ be a shape-regular family of affine meshes of Ω. Let $\mathcal{I}_h^N$ be the global Nédélec interpolation operator defined in (1.90). Let $p > 2$. Then, there is c such that, for all h and $v \in [W^{1,p}(\Omega)]^3$ with $\nabla \times v \in [W^{1,p}(\Omega)]^3$,*

$$\|v - \mathcal{I}_h^N v\|_{0,p,\Omega} + \|\nabla \times (v - \mathcal{I}_h^N v)\|_{0,p,\Omega} \le c\,h(\|v\|_{1,p,\Omega} + \|\nabla \times v\|_{1,p,\Omega}). \quad (1.107)$$

1.5.4 Interpolation in $W^{s,p}(\Omega)$ on non-affine meshes

Interpolation on general quadrangles. This section contains a brief introduction to error estimates applicable to finite elements on quadrangles. For the sake of simplicity, we assume that Ω is a polygonal domain in $\mathbb{R}^2$ and that the reference cell $\widehat{K}$ is the unit square. For proofs and further insight, see, e.g., [GiR86, p. 104].

Let K be a non-degenerate, convex quadrangle in $\mathbb{R}^2$. We readily see that there exists a unique bijective transformation $T_K \in [\mathbb{Q}_1(\widehat{K})]^2$ such that $T_K(\widehat{K}) = K$ (see Figure 1.23); T_K maps the edges of $\widehat{K}$ to the edges of K, but unless K is a parallelogram T_K is not affine.

In this section, we assume again that $\psi_K(v) = v \circ T_K$.

Lemma 1.119. *Let K be a convex quadrangle in $\mathbb{R}^2$ and let T_K be the unique bijective transformation in $[\mathbb{Q}_1(\widehat{K})]^2$ mapping the unit square $\widehat{K}$ to K. Let J_K be the Jacobian matrix of T_K. Then, there exists c such that*

$$\|\det(J_K)\|_{L^\infty(\widehat{K})} \le c\,h_K^2, \qquad \|\det(J_K^{-1})\|_{L^\infty(K)} \le c\,\tfrac{1}{\rho_K^2},$$

$$\|J_K\|_{[L^\infty(\widehat{K})]^{2,2}} \le c\,h_K, \qquad \|J_K^{-1}\|_{[L^\infty(K)]^{2,2}} \le c\,\tfrac{h_K}{\rho_K^2},$$

where $h_K = \operatorname{diam}(K)$ and $\rho_K = \min_{1 \le i \le 4}\rho_i$, ρ_i being the diameter of the circle inscribed in the triangle formed by the three vertices $(a_j)_{j \ne i}$ of K.

Theorem 1.120 (Local interpolation). *Let $\{\widehat{K}, \widehat{P}, \widehat{\Sigma}\}$ be the reference finite element with $\widehat{K} = [0,1]^2$ and associated normed vector space $V(\widehat{K})$. Assume that there exists an integer k such that $\mathbb{Q}_k \subset \widehat{P}$ and $H^{k+1}(\widehat{K}) \subset V(\widehat{K})$. Let K be a quadrangle in $\mathbb{R}^2$ and let $\mathcal{I}_K^k$ be the local interpolation operator in K defined in (1.59). Then, setting $\sigma_K = \tfrac{h_K}{\rho_K}$, there exists c such that, for all $m \in \{0, \ldots, k+1\}$ and $v \in H^{k+1}(K)$,*

$$\begin{cases} \|v - \mathcal{I}_K^k v\|_{0,K} \le c\,\sigma_K h_K^{k+1}|v|_{k+1,K}, \\ |v - \mathcal{I}_K^k v|_{m,K} \le c\,\sigma_K^{4m-1} h_K^{k+1-m}|v|_{k+1,K}. \end{cases} \tag{1.108}$$

Definition 1.121 (Shape-regularity). *Let ρ_K be as in Lemma 1.119. A family $\{\mathcal{T}_h\}_{h>0}$ of quadrangular meshes is said to be* shape-regular *if there exists σ_0 such that*

$$\forall h, \ \forall K \in \mathcal{T}_h, \quad \sigma_K = \tfrac{h_K}{\rho_K} \le \sigma_0.$$

Corollary 1.122 (Global interpolation). *Let the assumptions of Theorem 1.120 hold. Let Ω be a polygonal domain in $\mathbb{R}^2$. Let $\{\mathcal{T}_h\}_{h>0}$ be a family of quadrangular meshes of Ω and assume that $\{\mathcal{T}_h\}_{h>0}$ is shape-regular according to Definition 1.121. Denote by V_h^k the approximation space based on $\mathcal{T}_h$ and $\{\widehat{K}, \widehat{P}, \widehat{\Sigma}\}$. Let $\mathcal{I}_h^k$ be the corresponding interpolation operator. Then, there exists c such that, for all h and $v \in H^{k+1}(\Omega)$,*

$$\|v - \mathcal{I}_h^k v\|_{0,\Omega} + \sum_{m=1}^{k+1} h^m \left(\sum_{K \in \mathcal{T}_h} \|v - \mathcal{I}_h^k v\|_{m,K}^2\right)^{\frac{1}{2}} \le c\,h^{k+1}|v|_{k+1,\Omega}.$$

In particular, if V_h^k is H^1-conformal,

$$\forall h, \ \forall v \in H^{k+1}(\Omega), \quad |v - \mathcal{I}_h^k v|_{1,\Omega} \le c\,h^k|v|_{k+1,\Omega}.$$

Remark 1.123.

(i) In Theorem 1.120, the exponent on σ_K is larger than that obtained in (1.96) for affine meshes.

(ii) We deduce from Lemma 1.119 that for a shape-regular family of quadrangular meshes, the condition number of J_K is controlled uniformly with respect to h and $K \in \mathcal{T}_h$. □

Interpolation on domains with curved boundary. The goal of this section is to highlight an important practical result, namely that using a high-order reference finite element on a domain with curved boundary only pays off if the boundary is accurately represented. In particular, if a domain with curved boundary is approximated geometrically with affine meshes, using finite elements of degree larger than one is not asymptotically more accurate than using first-order finite elements.

For the sake of simplicity, we restrict the discussion to Lagrange geometric finite elements on simplices (see §1.3.2), and we consider *isoparametric* interpolation. For proofs and further insight, see [Ber89, BrS94, Cia91, CiR72^b, Len86, Zlá73, Zlá74].

Let $\{\widetilde{T}_h\}_{h>0}$ be a family of *affine* meshes of Ω and set $\widetilde{\Omega}_h = \bigcup_{\widetilde{K}\in\widetilde{T}_h} \widetilde{K}$. Let $k_{\mathrm{geo}} \geq 2$ and let $F_h : \widetilde{\Omega}_h \to \Omega_h = F_h(\widetilde{\Omega}_h)$ be a mapping such that $\forall \widetilde{K} \in \widetilde{T}_h$, $F_{h|\widetilde{K}} \in [\mathbb{P}_{k_{\mathrm{geo}}}]^d$. Using the mapping F_h, a new triangulation is constructed from $\widetilde{T}_h$ by setting $T_h = \{F_h(\widetilde{K})\}_{\widetilde{K}\in\widetilde{T}_h}$. The concept of shape-regular family of meshes can be extended as follows:

Definition 1.124. *The family of meshes* $\{T_h\}_{h>0}$ *is said to be* shape-regular *if the affine family* $(\widetilde{T}_h)_h$ *is shape-regular according to Definition 1.107 and if the mappings* $\{F_h\}_{h>0}$ *satisfy the following properties:*

(i) F_h *is the identity away from* $\partial\Omega_h$; *that is,* $F_{h|\widetilde{K}} = \mathcal{I}$ *if* $\partial\widetilde{K} \cap \partial\widetilde{\Omega}_h = \emptyset$.

(ii) $\sup_{x\in\partial\Omega} \mathrm{dist}(x, \partial\Omega_h) \leq c\, h^{k_{\mathrm{geo}}+1}$ *with* c *independent of* h.

(iii) *The norm of the Jacobian matrix of* F_h *and the norm of its inverse are bounded uniformly in* $[W^{k_{\mathrm{geo}},\infty}(\Omega_h)]^{d,d}$ *with respect to* h.

Theorem 1.125. *Let* $\{\widehat{K}, \widehat{P}, \widehat{\Sigma}\}$ *be a Lagrange finite element of degree* k *with* $k+1 > \frac{d}{2}$. *Let* Ω *be a domain in* $\mathbb{R}^d$ *and let* $\{T_h\}_{h>0}$ *be a shape-regular family of meshes according to Definition 1.124 with* $k_{\mathrm{geo}} = k$. *Let*

$$V_h^k = \{v \in \mathcal{C}^0(\overline{\Omega}_h);\ v \circ F_h \in \widetilde{V}_h^k\},$$

where $\widetilde{V}_h^k$ *is the approximation space based on the mesh* $\widetilde{T}_h$ *and the reference finite element* $\{\widehat{K}, \widehat{P}, \widehat{\Sigma}\}$. *Let* $\mathcal{I}_h^k$ *be the interpolation operator on* V_h^k. *Then, there exists* c *such that*

$$\forall h,\ \forall v \in H^{k+1}(\Omega_h), \quad \|v - \mathcal{I}_h^k v\|_{0,\Omega_h} + h\,|v - \mathcal{I}_h^k v|_{1,\Omega_h} \leq c\, h^{k+1}|v|_{k+1,\Omega_h}.$$

Moreover,

$$\forall v \in H^1(\Omega_h), \quad \lim_{h\to 0}\left(\inf_{v_h\in V_h^k} |v - v_h|_{1,\Omega_h}\right) = 0.$$

Proof. See, e.g., [BrS94, p. 117]. □

Remark 1.126. A different approach to extend the concept of shape-regularity is presented in [Cia91, p. 227]. Assume, for instance, that the geometric finite element is the Lagrange finite element $\mathbb{P}_2$. Let $\widehat{a}_i$, $0 \le i \le d$, be the vertices of $\widehat{K}$ and let $\widehat{a}_l$, $d + 1 \le l \le \frac{1}{2}d(d+3)$, be the other nodes. Consider a similar notation for the nodes a_i and a_l of K. Let K° be the convex hull of the $(d+1)$ vertices of K and denote by a_l°, $d + 1 \le l \le \frac{1}{2}d(d+3)$, the nodes located at the midpoints of the edges of K°. The shape-regularity criterion considered in [Cia91, p. 241] involves two conditions:

(i) The family of meshes formed by the simplices K° is shape-regular according to Definition 1.107.
(ii) There exists c such that, for all $l \in \{d+1, \ldots, \frac{1}{2}d(d+3)\}$,

$$\forall h, \ \forall K, \quad \|a_l^\circ - a_l\|_d \le c \, h^2.$$

This definition can be extended to the Lagrange finite element $\mathbb{P}_3$ [Cia91, p. 247]. A general theory is presented in [CiR72^b]. $\qquad\qquad\square$

1.6 Interpolation of Non-Smooth Functions

This section is concerned with the problem of interpolating non-smooth functions, e.g., functions that are too rough to be in the domain of the Lagrange interpolation operator. This situation occurs, for instance, when interpolating discontinuous functions, e.g., in $L^2(\Omega)$ or in $H^1(\Omega)$ in dimension $d \ge 2$. Throughout this section, Ω is a polyhedron and $\{\mathcal{T}_h\}_{h>0}$ is a shape-regular family of affine, simplicial, geometrically conformal meshes.

1.6.1 Clément interpolation

An interpolation technique to handle functions in L^1 using H^1-conformal Lagrange finite elements was first analyzed by Clément [Clé75]. The main ingredient is a regularization operator based on *macroelements* consisting of element patches. Let $P_{c,h}^k$ be the H^1-conformal approximation space based on the $\mathbb{P}_k$ Lagrange finite element; see (1.76). Let $\{a_1, \ldots, a_N\}$ be the Lagrange nodes and let $\{\varphi_1, \ldots, \varphi_N\}$ be the global shape functions in $P_{c,h}^k$. Associate with each node a_i the macroelement A_i consisting of the simplices containing a_i. Examples of macroelements are shown in Figure 1.24. Clearly, the macroelements can only assume a finite number of configurations, say n_{cf}. Denote by $\{\widehat{A}_n\}_{1 \le n \le n_{\mathrm{cf}}}$ the list of reference configurations. Define the application $j : \{1, \ldots, N\} \to \{1, \ldots, n_{\mathrm{cf}}\}$ such that $j(i)$ is the index of the reference configuration associated with the macroelement A_i. Define a $\mathcal{C}^0$-diffeomorphism F_{A_i} from $\widehat{A}_{j(i)}$ to A_i such that $\forall \widehat{K} \in \widehat{A}_{j(i)}$, $F_{A_i|\widehat{K}}$ is affine. The Clément interpolation operator $\mathcal{C}_h$ is then defined by local L^2-projections onto the macroelements. More precisely, for a reference macroelement $\widehat{A}_n$ and

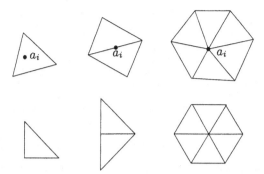

Fig. 1.24. Examples of macroelements A_i (top) and reference configuration $\widehat{A}_i$ (bottom) associated with a node a_i.

a function $\widehat{v} \in L^1(\widehat{A}_n)$, let $\widehat{C}_n\widehat{v}$ be the unique polynomial in $\mathbb{P}_k$ such that $\int_{\widehat{A}_n}(\widehat{C}_n\widehat{v} - \widehat{v})p = 0$ for all $p \in \mathbb{P}_k$. Then, the Clément interpolation operator is defined as follows:

$$\mathcal{C}_h : L^1(\Omega) \ni v \longmapsto \mathcal{C}_h v = \sum_{i=1}^N \widehat{\mathcal{C}}_{j(i)}(v \circ F_{A_i})(F_{A_i}^{-1}(a_i))\, \varphi_i \in P_{c,h}^k. \quad (1.109)$$

The stability and interpolation properties of the Clément operator are stated in the following:

Lemma 1.127 (Clément). *Under the above assumptions, the following properties hold:*

(i) *Stability: Let $1 \le p < +\infty$ and $0 \le m \le 1$. There is c such that*

$$\forall h,\ \forall v \in W^{m,p}(\Omega), \quad \|\mathcal{C}_h v\|_{W^{m,p}(\Omega)} \le c\, \|v\|_{W^{m,p}(\Omega)}. \quad (1.110)$$

(ii) *Approximation: For $K \in \mathcal{T}_h$, denote by Δ_K the set of elements in $\mathcal{T}_h$ sharing at least one vertex with K. Let F be an interface between two elements of $\mathcal{T}_h$, and denote by Δ_F the set of elements in $\mathcal{T}_h$ sharing at least one vertex with F; see Figure 1.25. Let l, m, and p satisfy $1 \le p < +\infty$ and $0 \le m \le l \le k+1$. Then, there is c such that*

$$\forall h,\ \forall K \in \mathcal{T}_h,\ \forall v \in W^{l,p}(\Delta_K), \quad \|v - \mathcal{C}_h v\|_{m,p,K} \le c\, h_K^{l-m}\|v\|_{l,p,\Delta_K}.$$

Similarly, if $m + \frac{1}{p} \le l \le k+1$,

$$\forall h,\ \forall K \in \mathcal{T}_h,\ \forall v \in W^{l,p}(\Delta_F), \quad \|v - \mathcal{C}_h v\|_{m,p,F} \le c\, h_F^{l-m-\frac{1}{p}}\|v\|_{l,p,\Delta_F}.$$

Proof. See [Clé75, Ber89, BeG98]. $\square$

An easy consequence of this result is the following:

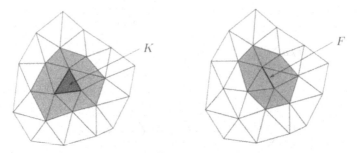

Fig. 1.25. Left: the shaded zone illustrates the set Δ_K of simplices sharing at least one vertex with the simplex K. Right: the shaded zone illustrates the set Δ_F of simplices sharing at least one vertex with the interface F.

Corollary 1.128. *Let the assumptions of Lemma* 1.127 *hold, let* $0 \leq l \leq k+1$, *and let* $0 \leq m \leq \min(1, l)$. *Then, there is* c *such that*

$$\forall h, \ \forall v \in W^{l,p}(\Omega), \quad \inf_{v_h \in P^k_{c,h}} \|v - v_h\|_{m,p,\Omega} \leq c\, h^{l-m} \|v\|_{l,p,\Omega}. \tag{1.111}$$

Remark 1.129.

(i) One difficulty with the Clément interpolation operator is that it does not preserve homogeneous boundary conditions, i.e., if v vanishes at the boundary, this is generally not the case for $\mathcal{C}_h v$. This problem is usually solved by setting boundary nodal values to zero. It can be shown that the Clément interpolant thus modified satisfies the estimates of Lemma 1.127.

(ii) The technique presented above can be generalized to other finite elements and to domains with curved boundaries; see, e.g., [Ber89, BeG98]. □

1.6.2 Scott–Zhang interpolation

Besides the fact that the Clément operator does not preserve boundary conditions, another difficulty is that it is not a projection. In [ScZ90], Scott and Zhang have addressed these two issues and defined an alternative interpolation operator.

Consider the notation and assumptions of the previous section. With each node a_i in the approximation space $P^k_{c,h}$ we associate either a d-simplex or a $(d-1)$-simplex, say Ξ_i, as follows: If a_i is in the interior of a d-simplex, say K, we simply set $\Xi_i = K$. If a_i is on a face, i.e., a $(d-1)$-simplex, say F, we set $\Xi_i = F$. Whenever a_i is at the boundary and in the intersection of many faces, it is important to pick the one face such that $F \subset \partial\Omega$. Let n_i be the number of nodes belonging to Ξ_i and denote by $\{\varphi_{i,q}\}_{1 \leq q \leq n_i}$ the restrictions to Ξ_i of the local shape functions associated with the nodes lying in Ξ_i; see Figure 1.26. Conventionally set $\varphi_{i,1} = \varphi_i$. We now construct a family $\{\gamma_{i,q}\}_{1 \leq q \leq n_i}$ as follows: For an integer q, $1 \leq q \leq n_i$, define $\gamma_{i,q} \in \text{span}\{\varphi_{i,1}, \dots, \varphi_{i,n_i}\}$ to be the unique function such that

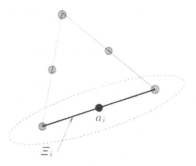

Fig. 1.26. Example of a node a_i with associated $(d-1)$-simplex Ξ_i $(d = 2)$ containing $n_i = 3$ nodes.

$$\int_{\Xi_i} \gamma_{i,q} \varphi_{i,r} = \delta_{qr}, \quad 1 \leq q, r \leq n_i. \tag{1.112}$$

Then, the Scott–Zhang interpolation operator is defined as follows:

$$\mathcal{SZ}_h : W^{l,p}(\Omega) \ni v \longmapsto \mathcal{SZ}_h v(x) = \sum_{i=1}^N \varphi_i \int_{\Xi_i} \gamma_{i,1} v \in P_{c,h}^k. \tag{1.113}$$

It is clear that $\mathcal{SZ}_h$ preserves homogeneous boundary conditions, i.e., $v_{|\partial\Omega} = 0$ implies $\mathcal{SZ}_h v_{|\partial\Omega} = 0$. Furthermore, (1.112) implies $\mathcal{SZ}_h v_h = v_h$ for all $v_h \in P_{c,h}^k$. The interpolation properties of the Scott–Zhang interpolation operator are stated in the following:

Lemma 1.130 (Scott–Zhang). *Let p and l satisfy $1 \leq p < +\infty$ and $l \geq 1$ if $p = 1$, and $l > \frac{1}{p}$ otherwise. Then, there is c such that the following properties hold:*

(i) *Stability: for all $0 \leq m \leq \min(1, l)$,*

$$\forall h, \ \forall v \in W^{l,p}(\Omega), \quad \|\mathcal{SZ}_h v\|_{m,p,\Omega} \leq c \|v\|_{l,p,\Omega}. \tag{1.114}$$

(ii) *Approximation: provided $l \leq k + 1$, for all $0 \leq m \leq l$,*

$$\forall h, \ \forall K \in \mathcal{T}_h, \ \forall v \in W^{l,p}(\Delta_K), \quad \|v - \mathcal{SZ}_h v\|_{m,p,K} \leq c\, h_K^{l-m} |v|_{l,p,\Delta_K},$$

where Δ_K is defined in Lemma 1.127.

1.6.3 Orthogonal projections

Projection onto H^1-conformal spaces. Let $P_{c,h}^k$ be the H^1-conformal approximation space based on the $\mathbb{P}_k$ Lagrange finite element; see (1.76). The results presented in this section also hold for tensor product finite element spaces, e.g., the approximation space $Q_{c,h}^k$ defined in (1.77). Consider the following orthogonal projection operators:

$$\Pi_{c,h}^{0,k} : L^2(\Omega) \longrightarrow P_{c,h}^k \qquad \text{and} \qquad \Pi_{c,h}^{1,k} : H^1(\Omega) \longrightarrow P_{c,h}^k,$$

with scalar products $(u, v)_{0,\Omega} = \int_\Omega uv$ and $(u, v)_{1,\Omega} = \int_\Omega uv + \int_\Omega \nabla u \cdot \nabla v$, respectively. Recall that

$$\forall v_h \in P_{c,h}^k, \qquad (\Pi_{c,h}^{0,k}(u), v_h)_{0,\Omega} = (u, v_h)_{0,\Omega},$$

$$\forall v_h \in P_{c,h}^k, \qquad (\Pi_{c,h}^{1,k}(u), v_h)_{1,\Omega} = (u, v_h)_{1,\Omega},$$

and that $\Pi_{c,h}^{0,k} v$ (resp., $\Pi_{c,h}^{1,k} v$) is the closest function to v in $P_{c,h}^k$ for the L^2-norm (resp., H^1-norm). The operator $\Pi_{c,h}^{1,k}$ is often called the *elliptic projector* or the *Riesz projector*.

Lemma 1.131 (Stability). *Let $k \geq 1$. The following estimates hold:*

$$\forall v \in L^2(\Omega), \quad \|\Pi_{c,h}^{0,k} v\|_{0,\Omega} \leq \|v\|_{0,\Omega}, \tag{1.115}$$

$$\forall v \in H^1(\Omega), \quad \|\Pi_{c,h}^{1,k} v\|_{1,\Omega} \leq \|v\|_{1,\Omega}. \tag{1.116}$$

Moreover, if the family $\{\mathcal{T}_h\}_{h>0}$ is quasi-uniform, there exists c such that

$$\forall h, \forall v \in H^1(\Omega), \quad \|\Pi_{c,h}^{0,k} v\|_{1,\Omega} \leq c \|v\|_{1,\Omega}. \tag{1.117}$$

Proof. The stability estimates (1.115)–(1.116) directly follow from the definition of orthogonal projections. Indeed, using the Pythagoras identity yields

$$\forall v \in L^2(\Omega), \quad \|v\|_{0,\Omega} = \|\Pi_{c,h}^{0,k} v\|_{0,\Omega} + \|v - \Pi_{c,h}^{0,k} v\|_{0,\Omega},$$

and a similar identity holds for $\Pi_{c,h}^{1,k}$. The proof of (1.117) is the subject of Exercise 1.17. □

Remark 1.132. Under reasonable assumptions, the stability estimate (1.116) can be substantially improved. In particular, the elliptic projector is stable in $W^{1,p}(\Omega)$; see Theorem 3.21 and [BrS94, p. 170]. □

For $s \geq 1$ and $v \in L^2(\Omega)$, we define the so-called *negative-norm*

$$\|v\|_{-s,\Omega} = \sup_{w \in H^s(\Omega) \cap H_0^1(\Omega)} \frac{(v, w)_{0,\Omega}}{\|w\|_{s,\Omega}}.$$

Note that this is not the norm considered to define the dual space $H^{-s}(\Omega)$, except in the particular case $s = 1$. Here, the norm $\|\cdot\|_{-s,\Omega}$ is simply used as a quantitative measure for functions in $L^2(\Omega)$.

Proposition 1.133. *Let $k \geq 1$ and $1 \leq s \leq k+1$. Then, there is c such that*

$$\forall h, \forall v \in L^2(\Omega), \quad \|v - \Pi_{c,h}^{0,k} v\|_{-s,\Omega} \leq c h^s \inf_{v_h \in P_{c,h}^k} \|v - v_h\|_{0,\Omega}. \tag{1.118}$$

Proof. Let $v \in L^2(\Omega)$ and $w \in H^s(\Omega) \cap H_0^1(\Omega)$. Since $s \leq k+1$, Lemma 1.127 implies

$$\|w - \mathcal{C}_h w\|_{0,\Omega} \leq c h^s |w|_{s,\Omega}.$$

Furthermore, since $v - \Pi_{c,h}^{0,k} v$ is L^2-orthogonal to $P_{c,h}^k$,

$$(v - \Pi_{c,h}^{0,k} v, w)_{0,\Omega} = (v - \Pi_{c,h}^{0,k} v, w - \mathcal{C}_h w)_{0,\Omega} \leq c h^s \|v - \Pi_{c,h}^{0,k} v\|_{0,\Omega} \|w\|_{s,\Omega}.$$

The result follows easily. $\qquad\square$

Finally, we state approximation properties for smooth functions.

Proposition 1.134. *Let $k \geq 1$ and $1 \leq l \leq k$.*

(i) *There exists c such that, for all h and $v \in H^{l+1}(\Omega)$,*

$$\|v - \Pi_{c,h}^{0,k}(v)\|_{0,\Omega} \leq c h^{l+1} |v|_{l+1,\Omega}, \tag{1.119}$$

$$\|v - \Pi_{c,h}^{1,k}(v)\|_{1,\Omega} \leq c h^l |v|_{l+1,\Omega}. \tag{1.120}$$

(ii) *If Ω is convex, there exists c such that, for all h and $v \in H^{l+1}(\Omega)$,*

$$\|v - \Pi_{c,h}^{1,k}(v)\|_{0,\Omega} \leq c h^{l+1} |v|_{l+1,\Omega}. \tag{1.121}$$

(iii) *If the family $\{\mathcal{T}_h\}_{h>0}$ is quasi-uniform, there exists c such that, for all h and $v \in H^{l+1}(\Omega)$,*

$$\|v - \Pi_{c,h}^{0,k}(v)\|_{1,\Omega} \leq c h^l |v|_{l+1,\Omega}. \tag{1.122}$$

Proof. See Exercise 1.18. $\qquad\square$

Projection onto totally discontinuous spaces. Let $k \geq 0$ and consider the L^2-orthogonal projection $\Pi_{td,h}^{0,k}$ from $L^2(\Omega)$ to the space $P_{td,h}^k$ defined in (1.67). Clearly, for $v \in L^2(\Omega)$ and $K \in \mathcal{T}_h$,

$$\begin{cases} (\Pi_{td,h}^{0,k} v)_{|K} \in \mathbb{P}_k, \\ \int_K (\Pi_{td,h}^{0,k} v - v) q = 0, \quad \forall q \in \mathbb{P}_k. \end{cases} \tag{1.123}$$

In particular, $\Pi_{td,h}^{0,0} v$ is the piecewise constant function equal to $\frac{1}{\text{meas}(K)} \int_K v$ on all cells $K \in \mathcal{T}_h$. The approximation properties of $\Pi_{td,h}^{0,k}$ are stated in the following:

Proposition 1.135. *There exists c, independent of h, such that, for all $0 \leq l \leq k+1$, $1 \leq p \leq \infty$, and $v \in W^{l,p}(\Omega)$,*

$$\|v - \Pi_{td,h}^{0,k} v\|_{0,p,\Omega} \leq c h^l |v|_{l,p,\Omega}. \tag{1.124}$$

Proof. Straightforward verification. $\qquad\square$

Remark 1.136.

(i) The shape-regularity assumption on the mesh is not required for (1.124).

(ii) An estimate similar to (1.118) holds for $\Pi_{td,h}^{0,k} v$. $\qquad\square$

1.6.4 The discrete commutator property

The so-called discrete commutator property is a powerful tool to analyze nonlinear problems; see Bertoluzza [Ber99] and [JoS87]. As a corollary of Lemma 1.130, we infer the following:

Lemma 1.137 (Bertoluzza). *Let the hypotheses of Lemma* 1.130 *hold. Then, there is* c *such that, for all* h, v_h *in* $P^k_{c,h}$, ϕ *in* $W^{s+1,\infty}(\Omega)$, *and* $0 \le m \le s \le 1$,

$$\|\phi v_h - \mathcal{SZ}_h(\phi v_h)\|_{m,p,\Omega} \le c\,h^{1+s-m}\|v_h\|_{s,p,\Omega}\|\phi\|_{s+1,\infty,\Omega}.$$

Proof. We prove the result locally. Let K be a cell in the mesh $\mathcal{T}_h$. Denote by x_K some point in K, say the barycenter of K. Let ϕ be a function in $W^{s+1,\infty}(\Omega)$. Define $R_K = \phi - \phi(x_K)$. It is clear that $R_K \in W^{1,\infty}(\Omega)$ and

$$\|R_K\|_{0,\infty,\Delta_K} \le c\,h_K\|\phi\|_{1,\infty,\Omega},$$
$$\|R_K\|_{1,\infty,\Delta_K} \le c\,\|\phi\|_{1,\infty,\Omega}.$$

Let $\overline{v}_h$ be the mean value of v_h on Δ_K. Then, one readily verifies that

$$\|\overline{v}_h\|_{0,p,\Delta_K} \le c\,\|v_h\|_{0,p,\Delta_K},$$
$$\|v_h - \overline{v}_h\|_{m,p,\Delta_K} \le c\,h_K^{s-m}\|v_h\|_{s,p,\Delta_K}, \qquad 0 \le m \le s \le 1.$$

Furthermore, observe that

$$\|\phi v_h - \mathcal{SZ}_h(\phi v_h)\|_{m,p,K} \le \|(\mathcal{I} - \mathcal{SZ}_h)(\phi\overline{v}_h)\|_{m,p,K}$$
$$+ \|(\mathcal{I} - \mathcal{SZ}_h)(\phi(v_h - \overline{v}_h))\|_{m,p,K},$$

and denote by R_1 and R_2 the two residuals in the right-hand side. Since $s \ge 0$, $1 + s \ge \frac{1}{p}$ if $p = 1$ and $1 + s > \frac{1}{p}$ if $p > 1$; moreover, $s \le 1 \le k$. As a result, one can use Lemma 1.130 to control R_1 as follows:

$$R_1 \le c\,h_K^{1+s-m}\|\phi\overline{v}_h\|_{s+1,p,\Delta_K} \le c\,h_K^{1+s-m}\|\overline{v}_h\|_{0,p,\Delta_K}\|\phi\|_{s+1,\infty,\Omega}$$
$$\le c\,h_K^{1+s-m}\|v_h\|_{0,p,\Delta_K}\|\phi\|_{s+1,\infty,\Omega}.$$

For the other residual, use the fact that $\mathcal{SZ}_h$ is linear, $P^k_{c,h}$ is invariant under $\mathcal{SZ}_h$, and $\mathcal{SZ}_h(\overline{v}_h) = \overline{v}_h$ on K to obtain

$$(\mathcal{I} - \mathcal{SZ}_h)(\phi(v_h - \overline{v}_h)) = (\mathcal{I} - \mathcal{SZ}_h)((\phi - \phi(x_K))(v_h - \overline{v}_h)).$$

As a result,

$$R_2 = \|(\mathcal{I} - \mathcal{SZ}_h)(R_K(v_h - \overline{v}_h))\|_{m,p,K}$$
$$\le c\,h_K^{1-m}|R_K(v_h - \overline{v}_h)|_{1,p,\Delta_K}$$
$$\le c\,h_K^{1-m}(\|R_K\|_{0,\infty,\Delta_K}|v_h - \overline{v}_h|_{1,p,\Delta_K}$$
$$+ |R_K|_{1,\infty,\Delta_K}\|v_h - \overline{v}_h\|_{0,p,\Delta_K})$$
$$\le c\,h_K^{1-m}(h_K|v_h - \overline{v}_h|_{1,p,\Delta_K} + \|v_h - \overline{v}_h\|_{0,p,\Delta_K})\|\phi\|_{1,\infty,\Omega}$$
$$\le c\,h_K^{1+s-m}\|v_h\|_{s,p,\Delta_K}\|\phi\|_{1,\infty,\Omega}.$$

Then, the desired result follows easily from the shape-regularity of the mesh, which implies that $\sup_{K' \in \mathcal{T}_h}(\mathrm{card}\{K \in \mathcal{T}_h; K' \subset \Delta_K\})$ is a fixed constant independent of h. $\square$

1.7 Inverse Inequalities

The goal of this section is to compare various functional norms on approximation spaces. Such spaces being finite-dimensional, all the norms therein are equivalent. The purpose of inverse inequalities is to specify how the equivalence constants depend on h. For the sake of simplicity, we restrict ourselves to affine meshes and to finite elements for which $\psi_K(v) = v \circ T_K$.

Lemma 1.138 (Local inverse inequalities). *Let $\{\widehat{K}, \widehat{P}, \widehat{\Sigma}\}$ be a finite element. Let $l \geq 0$ be such that $\widehat{P} \subset W^{l,\infty}(\widehat{K})$. Let $\{\mathcal{T}_h\}_{h>0}$ be a shape-regular family of affine meshes in $\mathbb{R}^d$ with $h \leq 1$. Let $0 \leq m \leq l$ and $1 \leq p, q \leq \infty$. Then, there is c, independent of h, K, p, and q, such that, for all $v \in P_K = \{\widehat{p} \circ T_K^{-1}; \widehat{p} \in \widehat{P}\}$,*

$$\|v\|_{l,p,K} \leq c\, h_K^{m-l+d(\frac{1}{p}-\frac{1}{q})}\|v\|_{m,q,K}. \tag{1.125}$$

Proof. (1) Since all the norms in $\widehat{P} \subset W^{l,\infty}(\widehat{K})$ are equivalent, there exists c, only depending on $\widehat{K}$ and l, such that, for all $\widehat{v} \in \widehat{P}$, $\|\widehat{v}\|_{l,\infty,\widehat{K}} \leq c\|\widehat{v}\|_{0,1,\widehat{K}}$; hence,

$$\forall \widehat{v} \in \widehat{P}, \qquad \|\widehat{v}\|_{l,p,\widehat{K}} \leq c\|\widehat{v}\|_{0,q,\widehat{K}}. \tag{1.126}$$

(2) Let $v \in P_K$ and $0 \leq j \leq l$. Using (1.93), (1.94), (1.126), and the shape-regularity of the family $\{\mathcal{T}_h\}_{h>0}$ yields

$$|v|_{j,p,K} \leq c\, h_K^{-j}|\det(J_K)|^{\frac{1}{p}}\|\widehat{v}\|_{j,p,\widehat{K}} \leq c\, h_K^{-j}|\det(J_K)|^{\frac{1}{p}}\|\widehat{v}\|_{0,q,\widehat{K}}$$
$$\leq c\, h_K^{-j}|\det(J_K)|^{\frac{1}{p}-\frac{1}{q}}\|v\|_{0,q,K} \leq c\, h_K^{-j+d(\frac{1}{p}-\frac{1}{q})}\|v\|_{0,q,K}.$$

Since $h_K \leq h \leq 1$ by assumption,

$$\forall v \in P_K,\ \forall j \in \{0, \dots, l\}, \quad \|v\|_{j,p,K} \leq c\, h_K^{-j+d(\frac{1}{p}-\frac{1}{p})}\|v\|_{0,q,K}.$$

Taking $j = l$ yields (1.125) for $m = 0$.

(3) Let $0 \leq m \leq l$. Let α be a multi-index such that $0 \leq |\alpha| \leq l$. If $|\alpha| < l - m$,

$$\|\partial^\alpha v\|_{0,p,K} \leq \|v\|_{l-m,p,K} \leq c h_K^{m-l+d(\frac{1}{p}-\frac{1}{p})}\|v\|_{0,q,K} \leq c h_K^{m-l+d(\frac{1}{p}-\frac{1}{p})}\|v\|_{m,q,K}.$$

If $l - m \leq |\alpha| \leq l$, one can find two multi-indices β and γ such that $\alpha = \beta + \gamma$ and $|\beta| = l - m$. Hence,

$$\|\partial^\alpha v\|_{0,p,K} = \|\partial^\beta(\partial^\gamma v)\|_{0,p,K} \leq \|\partial^\gamma v\|_{l-m,p,K}$$
$$\leq c h_K^{m-l+d(\frac{1}{p}-\frac{1}{q})}\|\partial^\gamma v\|_{0,q,K} \leq c h_K^{m-l+d(\frac{1}{p}-\frac{1}{q})}\|v\|_{m,q,K},$$

since $|\gamma| \le m$. This proves that for all multi-index α such that $0 \le |\alpha| \le l$, $\|\partial^\alpha v\|_{0,p,K} \le c\, h_K^{m-l+d\left(\frac{1}{p}-\frac{1}{q}\right)} \|v\|_{m,q,K}$. The conclusion follows readily. $\square$

Example 1.139. For $p = q$, $l = 1$ and $m = 0$, Lemma 1.138 yields $\|v\|_{1,p,K} \le c\, h_K^{-1} \|v\|_{0,p,K}$ for all h, $K \in \mathcal{T}_h$, and $v \in P_K$. $\square$

To obtain global inverse inequalities, the quantity h_K^{-1} must be controlled. This observation leads to the following:

Definition 1.140 (Quasi-uniformity). *A family of meshes $\{\mathcal{T}_h\}_{h>0}$ is said to be quasi-uniform if and only if it is shape-regular and there is c such that*

$$\forall h,\ \forall K \in \mathcal{T}_h, \quad h_K \ge c\, h. \tag{1.127}$$

Corollary 1.141 (Global inverse inequalities). *Along with the hypotheses of Lemma 1.138, assume that the family $\{\mathcal{T}_h\}_{h>0}$ is quasi-uniform. Set $W_h = \{v_h;\ \forall K \in \mathcal{T}_h,\ v_h \circ T_K \in \widehat{P}\}$. Then, using the usual convention if $p = \infty$ or $q = \infty$, there is c, independent of h, such that, for all $v_h \in W_h$ and $0 \le m \le l$,*

$$\left(\sum_{K \in \mathcal{T}_h} \|v_h\|_{l,p,K}^p\right)^{\frac{1}{p}} \le c\, h^{m-l+\min\left(0,\frac{d}{p}-\frac{d}{q}\right)} \left(\sum_{K \in \mathcal{T}_h} \|v_h\|_{m,q,K}^q\right)^{\frac{1}{q}}. \tag{1.128}$$

Proof. Let $v_h \in W_h$. Assume $p \ne \infty$ and $q \ne \infty$ (these two cases are treated similarly).
(1) Assume $p \ge q$. Then, (1.125) implies

$$\sum_{K \in \mathcal{T}_h} \|v_h\|_{l,p,K}^p \le c\, h^{p(m-l+d(\frac{1}{p}-\frac{1}{q}))} \sum_{K \in \mathcal{T}_h} \|v_h\|_{m,q,K}^p.$$

To conclude, use the inequality $\left(\sum_{i\in I} a_i^p\right)^{\frac{1}{p}} \le \left(\sum_{i\in I} a_i^q\right)^{\frac{1}{q}}$ which holds for $p \ge q \ge 0$ and for all finite sequence of non-negative numbers $\{a_i\}_{i\in I}$; see Exercise 1.20. (2) Assume $p \le q$. Then, (1.125) implies

$$\sum_{K \in \mathcal{T}_h} \|v_h\|_{l,p,K}^p \le c\, h^{p(m-l)} \sum_{K \in \mathcal{T}_h} h_K^{dp(\frac{1}{p}-\frac{1}{q})} \|v_h\|_{m,q,K}^p$$

$$\le c\, h^{p(m-l)} \left(\sum_{K \in \mathcal{T}_h} h_K^{dp(\frac{1}{p}-\frac{1}{q})\frac{q}{q-p}}\right)^{\frac{q-p}{q}} \left(\sum_{K \in \mathcal{T}_h} \|v_h\|_{m,q,K}^q\right)^{\frac{p}{q}}$$

$$\le c\, h^{p(m-l)} \operatorname{meas}(\Omega)^{\frac{q-p}{q}} \left(\sum_{K \in \mathcal{T}_h} \|v_h\|_{m,q,K}^q\right)^{\frac{p}{q}}. \qquad \square$$

The following result is often used when dealing with nonlinear problems:

Lemma 1.142. *Along with the hypotheses of Lemma 1.138, assume that the family $\{\mathcal{T}_h\}_{h>0}$ is quasi-uniform. Set $W_h = \{v_h;\ \forall K \in \mathcal{T}_h,\ v_h \circ T_K \in \widehat{P}\}$. Then, there is c, independent of h, such that, for all $v_h \in W_h \cap H^1(\Omega)$,*

$$\|v_h\|_{L^\infty(\Omega)} \leq \begin{cases} c\,(1 + |\log h|)\|v_h\|_{1,\Omega} & \text{in dimension 2,} \\ c\,h^{-\frac{1}{2}}\|v_h\|_{1,\Omega} & \text{in dimension 3.} \end{cases} \tag{1.129}$$

Proof. See Exercise 1.21 for a proof in dimension 3. □

Remark 1.143.

(i) A simple consequence of Corollary 1.141 is that for all $v_h \in W_h \cap W^{1,p}(\Omega)$, $\|v_h\|_{1,p,\Omega} \leq c\,h^{-1}\|v_h\|_{0,p,\Omega}$.

(ii) A necessary and sufficient condition for quasi-uniformity is that there exists τ such that $\rho_K \geq \tau h$ for all h and $K \in \mathcal{T}_h$. Indeed, if $\{\mathcal{T}_h\}_{h>0}$ satisfies the above property, then $\frac{h_K}{\rho_K} \leq \tau^{-1}\frac{h_K}{h} \leq \tau^{-1}$ for all h and $K \in \mathcal{T}_h$, thus showing that the family $\{\mathcal{T}_h\}_{h>0}$ is shape-regular. Furthermore, $h_K \geq \rho_K \geq \tau h$ implies (1.127). Conversely, if $\{\mathcal{T}_h\}_{h>0}$ is a quasi-uniform mesh family, $\rho_K \geq \frac{1}{\sigma}h_K \geq \frac{c}{\sigma}h$ for all $h > 0$ and $K \in \mathcal{T}_h$.

(iii) In two dimensions, one can construct a finer triangulation from an initial triangulation by connecting all the edge midpoints. Repeating this procedure yields a quasi-uniform family of meshes; see [Zha95].

(iv) See, e.g., [GiR86, p. 103] and [BrS94, p. 109] for further insight. □

1.8 Exercises

Exercise 1.1. Let $\mathcal{I}_h^1$ be the one-dimensional $\mathbb{P}_1$ Lagrange interpolation operator defined in (1.6).

(i) Prove that for all h and $v \in \mathcal{C}^0(\overline{\Omega})$, $\|\mathcal{I}_h^1 v\|_{\mathcal{C}^0(\overline{\Omega})} \leq \|v\|_{\mathcal{C}^0(\overline{\Omega})}$.

(ii) Prove that for all h and $v \in \mathcal{C}^1(\overline{\Omega})$, $\|v - \mathcal{I}_h^1 v\|_{\mathcal{C}^0(\overline{\Omega})} \leq h\|v\|_{\mathcal{C}^1(\overline{\Omega})}$. (*Hint:* Use the mean-value theorem.)

Exercise 1.2 (Hermite finite element).

(i) Prove Lemma 1.82 and Proposition 1.83.

(ii) Prove (1.103) for $v \in H^4(\Omega)$ without using the results of §1.5. (*Hint:* Adapt the proof of Proposition 1.5 by showing that on a mesh interval I_i, $(v - \mathcal{I}_h^H v_{|I_i})'''$ vanishes at least at one point of I_i.)

Exercise 1.3 ($\mathbb{P}_k$ Lagrange finite element).

(i) Let $p \in \mathbb{P}_k$ with $k \geq 1$. Assume that p vanishes on the $\mathbb{R}^d$-hyperplane of equation $\lambda = 0$. Prove that there is $q \in \mathbb{P}_{k-1}$ such that $p = \lambda q$. Then, prove Proposition 1.34. (*Hint:* By induction on k.)

(ii) Prove that if $k \leq d$ and $p \in \mathbb{P}_k$ vanishes at all the faces of K, then $p = 0$.

(iii) Prove that the number of nodes of a $\mathbb{P}_k$ Lagrange finite element located on any edge of K is $(k + 1)$ in arbitrary dimension $d \geq 2$. Prove that the number of nodes located on any face of K is the dimension of $\mathbb{P}_k$ in dimension $(d - 1)$. Justify Remark 1.76.

Exercise 1.4. Let K be a simplex in $\mathbb{R}^d$ with vertices $\{a_0, \ldots, a_d\}$ and barycentric coordinates $\{\lambda_0, \ldots, \lambda_d\}$ defined in (1.37).

(i) In dimension 2, for a point $x \in K$, let $K_i(x)$ be the triangle obtained by joining x to the two vertices a_j with $j \neq i$. Show that $\lambda_i = \frac{\text{meas}(K_i(x))}{\text{meas}(K)}$.

(ii) In dimension 3, let F be a face of K with outer normal n_F. Let a_r, a_s, and a_t be the vertices of F and assume that the ordering is such that $[(a_r - a_s) \times (a_s - a_t)] \cdot n_F > 0$. Prove that $x = \lambda_r(a_r - a_i) + \lambda_s(a_s - a_i) + \lambda_t(a_t - a_i)$, $\nabla \lambda_r = \frac{1}{6 \, \text{meas}(K)}(a_s - a_i) \times (a_t - a_i)$, and similar formulas for $\nabla \lambda_s$ and $\nabla \lambda_t$.

Exercise 1.5 (Raviart–Thomas finite element). Consider the Raviart–Thomas finite element introduced in §1.2.7.

(i) Prove Proposition 1.39 and equation (1.41).

(ii) Prove that a suitable domain for the local interpolation operator is $V(K) = \{v \in [L^p(K)]^d;\ \nabla \cdot v \in L^s(K)\}$ for $p > 2$ and $\frac{1}{s} \leq \frac{1}{p} + \frac{1}{d}$. (*Hint*: Set $\frac{1}{p} + \frac{1}{p'} = 1$ and use the fact that for a face $F \subset \partial K$, there exists $w \in W^{1,p'}(K)$ such that $w_{|F} = 1$ and $w_{|\partial \Omega \setminus F} = 0$. Conclude using (B.20) and Corollary B.43.)

(iii) Prove Lemma 1.41.

(iv) Let K be a three-dimensional simplex and let n be the outer normal. Let F_i be a face whose vertices are a_r, a_s, and a_t so that $[(a_r - a_s) \times (a_s - a_t)] \cdot n > 0$. Prove that

$$\theta_i = 2(\lambda_r \nabla \lambda_s \times \nabla \lambda_t + \lambda_s \nabla \lambda_t \times \nabla \lambda_r + \lambda_t \nabla \lambda_r \times \nabla \lambda_s).$$

(*Hint*: To prove $\theta_i \in \mathbb{RT}_0$, use Exercise 1.4(ii).)

(v) Find the counterpart of the above formula in dimension 2.

Exercise 1.6 (Nédélec finite element). Consider the Nédélec finite element introduced in §1.2.8.

(i) Verify that definitions (1.43) and (1.44) in two dimensions (resp., (1.45) in three dimensions) are equivalent.

(ii) Prove Proposition 1.42 and equations (1.46) and (1.47).

(iii) Prove that in dimension 2, a suitable domain for the local interpolation operator is $V(K) = \{v \in [L^p(K)]^2;\ \nabla \times v \in L^p(K)\}$ for $p > 2$. (*Hint*: Adapt the technique of Exercise 1.5.)

(iv) Let a_l, $0 \leq l \leq 3$, be the four vertices of $K \subset \mathbb{R}^3$. Let λ_l, $0 \leq l \leq 3$, be the corresponding barycentric coordinates in K. Denote by $e(l, l')$ the index of the edge whose endpoints are the two vertices a_l and $a_{l'}$. Choose $t_{e(l,l')}$ so that $t_{e(l,l')}$ points from a_l to $a_{l'}$. Prove that

$$\theta_{e(l,l')} = \lambda_l \nabla \lambda_{l'} - \lambda_{l'} \nabla \lambda_l.$$

(*Hint*: To prove $\theta_{e(l,l')} \in \mathbb{N}_0$, use Exercise 1.4(ii).)

(v) Find the counterpart of the above formula in dimension 2.

Exercise 1.7. Let K be a rectangle and denote by $\{a_1, a_2, a_3, a_4\}$ the midpoints of its sides. Set $P = \mathbb{Q}_1$ and $\Sigma = \{\sigma_1, \sigma_2, \sigma_3, \sigma_4\}$ such that $\sigma_i(p) = p(a_i)$, $1 \leq i \leq 4$. Show that $\{K, P, \Sigma\}$ is not a finite element, i.e., unisolvence does not hold.

Exercise 1.8 (Bicubic Hermite finite element). Let K be a rectangle and denote by $\{a_1, a_2, a_3, a_4\}$ the midpoints of its sides. Set $P = \mathbb{Q}_3$ and

$$\Sigma = \{p(a_i), \partial_{x_1}p(a_i), \partial_{x_2}p(a_i), \partial^2_{x_1 x_2}p(a_i)\}_{1 \leq i \leq 4}.$$

Show that $\{K, P, \Sigma\}$ is a finite element. Does this finite element yield H^2-conformal approximation spaces on rectangular meshes?

Exercise 1.9 (Brezzi–Douglas–Marini finite element). Let K be a triangle and set $P = [\mathbb{P}_1]^2$. On each face F of K with outer normal n_F, consider the two linear forms $\sigma_{F,1} : P \ni p \mapsto \int_F p(s) \cdot n_F \, ds$ and $\sigma_{F,2} : P \ni p \mapsto \int_F p(s) \cdot n_F \, s \, ds$. Set $\Sigma = \{\sigma_{F,1}, \sigma_{F,2}\}_{F \in \partial K}$. Prove that $\{K, P, \Sigma\}$ is a finite element; see [BrD86, BrD87].

Exercise 1.10. Prove Proposition 1.61.

Exercise 1.11. Let $\{\widehat{K}, \widehat{P}, \widehat{\Sigma}\}$ be the reference finite element, let $\mathcal{T}_h$ be a mesh, and let $K \in \mathcal{T}_h$.

(i) Assume that the geometric transformation T_K is affine and $\widehat{P} = \mathbb{P}_k$ for some integer k. Show that $v_h \circ T_K \in \mathbb{P}_k$ if and only if $v_h \in \mathbb{P}_k$.
(ii) Does the equivalence still hold for non-affine transformations?
(iii) Does the equivalence still hold for affine transformations, but with $\widehat{P} = \mathbb{Q}_k$? What happens if the transformations T_K are diagonal, i.e., if K is a rectangle in dimension 2 and a cuboid in dimension 3?

Exercise 1.12. Let $\mathcal{T}_h$ be a mesh. For $K \in \mathcal{T}_h$, let b_K be the Fortin–Soulié bubble defined in (1.72). Let $B = \text{span}\{b_K\}_{K \in \mathcal{T}_h}$.

(i) Let $P^2_{c,h}$ be the H^1-conformal approximation space based on the Lagrange finite element $\mathbb{P}_2$; see (1.76) for $k = 2$. Set $\Phi = \sum_{K \in \mathcal{T}_h} b_K$. Prove that $P^2_{c,h} \cap B = \text{span} \, \Phi$.
(ii) Set $P^2_{\text{pt},h,0} = \{v_h \in P^2_{\text{pt},h}; \forall F \in \mathcal{F}^\partial_h, \int_F v_h = 0\}$ with $P^2_{\text{pt},h}$ defined in (1.71). Let $P^2_{c,h,0} = P^2_{c,h} \cap H^1_0(\Omega)$. Prove that $P^2_{\text{pt},h,0} = P^2_{c,h,0} \oplus B$. (*Hint:* Prove $P^2_{c,h,0} \oplus B \subset P^2_{\text{pt},h,0}$ and use a dimensional argument.)

Exercise 1.13. Prove Proposition 1.87.

Exercise 1.14. Prove Proposition 1.93. (*Hint:* Use Lemma 1.91.)

Exercise 1.15. Use Proposition 1.97 to characterize all the solenoidal fields in $D_{h,0}$ and all the fields in $R_{h,0}$ whose curl is zero.

Exercise 1.16. Let $\mathcal{I}_h^1$ be the Lagrange interpolation operator associated with continuous, piecewise linears on triangles. Prove $\sup_{u \in C^0(\overline{\Omega})} \frac{\|\mathcal{I}_h^1 u\|_{L^\infty}}{\|u\|_{L^\infty}} \leq 1$. Is this true for piecewise quadratics?

Exercise 1.17. Prove (1.117). (*Hint:* For $v \in H^1(\Omega)$, consider its Scott–Zhang interpolant $\mathcal{SZ}_h v$ defined in (1.113) and write

$$\|\nabla \Pi_{c,h}^{0,k} v\|_{0,\Omega} \leq \|\nabla(\Pi_{c,h}^{0,k} v - \mathcal{SZ}_h v)\|_{0,\Omega} + \|\nabla \mathcal{SZ}_h v\|_{0,\Omega}.$$

Then, use an inverse inequality and the stability properties of the Scott–Zhang interpolant to conclude.)

Exercise 1.18. The goal of this exercise is to prove Proposition 1.134.

(i) Prove Proposition 1.134(i).
(ii) For $w \in L^2(\Omega)$, let $\varsigma(w)$ be the unique solution in $H^1(\Omega)$ to the problem: $(\varsigma(w), v)_{1,\Omega} = (w, v)_{0,\Omega}, \forall v \in H^1(\Omega)$. Show that for arbitrary $v_h \in P_{c,h}^k$,

$$\|v - \Pi_{c,h}^{1,k}(v)\|_{0,\Omega} \leq \|v - \Pi_{c,h}^{1,k}(v)\|_{1,\Omega} \sup_{f \in L^2(\Omega)} \frac{\|\varsigma(f) - v_h\|_{1,\Omega}}{\|f\|_{0,\Omega}}.$$

Then, use Theorem 3.12 to prove (1.121).
(iii) Denoting by $\mathcal{SZ}_h v$ the Scott–Zhang interpolant of v, show that

$$|v - \Pi_{c,h}^{0,k}(v)|_{1,\Omega} \leq |v - \mathcal{SZ}_h v|_{1,\Omega} + |\Pi_{c,h}^{0,k}[v - \mathcal{SZ}_h v]|_{1,\Omega}.$$

Then, use an inverse inequality to prove (1.122).

Exercise 1.19. Justify Remark 1.136(ii).

Exercise 1.20. Let $\{a_i\}_{i \in I}$ be a finite sequence of non-negative numbers. Prove that $(\sum_{i \in I} a_i^p)^{\frac{1}{p}} \leq (\sum_{i \in I} a_i^q)^{\frac{1}{q}}$ whenever $0 \leq q \leq p$. (*Hint:* Prove $\sum_{i \in I}(a_i^q / \sum_{i \in I} a_i^q)^{\frac{p}{q}} \leq 1$.)

Exercise 1.21. Prove the three-dimensional inverse estimate in Lemma 1.142. (*Hint:* Use an inverse inequality between $L^\infty(K)$ and $L^6(K)$, then use a Sobolev inequality.)

2

Approximation in Banach Spaces by Galerkin Methods

In this chapter, we consider an abstract linear problem which serves as a generic model for engineering applications. Our first goal is to specify the conditions under which this problem is well-posed. We use the definition proposed by Hadamard [Had32]: a problem is well-posed if it admits a unique solution and if it is endowed with a stability property, namely the solution is controlled by the data. Two important results asserting well-posedness are presented: the Lax–Milgram Lemma and the Banach–Nečas–Babuška Theorem. The former provides a *sufficient* condition for well-posedness, whereas the latter, relying on slightly more sophisticated assumptions, gives *necessary and sufficient* conditions. Then, we study approximation techniques based on the so-called Galerkin method. Both conformal and non-conformal settings are considered. We investigate under which conditions the stability properties of the abstract problem are transferred to the approximate problem, and we obtain a priori estimates for the approximation error. The last section of this chapter investigates a particular form of the Banach–Nečas–Babuška Theorem relevant to problems endowed with a saddle-point structure.

2.1 The Banach–Nečas–Babuška (BNB) Theorem

In this section, we introduce an abstract problem and determine the conditions under which this problem is well-posed.

2.1.1 Well-posedness

Consider the following (abstract) problem:

$$\begin{cases} \text{Seek } u \in W \text{ such that} \\ a(u,v) = f(v), \quad \forall v \in V, \end{cases} \tag{2.1}$$

where:

(i) W and V are vector spaces equipped with norms denoted by $\|\cdot\|_W$ and $\|\cdot\|_V$, respectively. In many applications, W and V are *Hilbert spaces*, but a more general case where V is a reflexive Banach space and W a Banach space can be considered. See Appendix A for an introduction to Banach and Hilbert spaces. Unless stated otherwise, we henceforth assume that W and V are Banach spaces and that V is reflexive. W is called the *solution space*, and V is called the *test space*.

(ii) a is a *continuous* bilinear form on $W \times V$, i.e., $a \in \mathcal{L}(W \times V; \mathbb{R})$; henceforth, we shall also say that a is *bounded* on $W \times V$.

(iii) f is a *continuous* linear form on V, i.e., $f \in V' = \mathcal{L}(V; \mathbb{R})$. To simplify the notation, we write $f(v)$ instead of $\langle f, v \rangle_{V',V}$.

Henceforth, well-posedness is understood in the sense introduced by Hadamard [Had32].

Definition 2.1 (Hadamard). *Problem* (2.1) *is said to be* well-posed *if it admits one and only one solution and if the following a priori estimate holds:*

$$\exists c > 0, \ \forall f \in V', \quad \|u\|_W \le c \|f\|_{V'}.$$

In many applications, the bilinear form a results from the weak formulation of PDEs posed on a domain $\Omega \subset \mathbb{R}^d$ with boundary conditions enforced on $\partial\Omega$. The linearity of a with respect to v directly results from the weak formulation whereas the linearity with respect to u is a consequence of the linearity of the model problem itself. The elements of W and V are scalar- or vector-valued functions defined on Ω.

Three important examples falling into the framework of the abstract problem (2.1) are the Laplace equation, the Stokes equations, and the advection equation. These problems (and many variants thereof) are thoroughly investigated in Chapters 3, 4, and 5, respectively. They are now briefly introduced for the sake of illustration.

The Laplace equation. Consider the PDE $-\Delta u = f$ in Ω supplemented with the homogeneous Dirichlet condition $u_{|\partial\Omega} = 0$. This problem can be reformulated in the form (2.1) by setting

$$\begin{cases} W = V = H_0^1(\Omega), \\ a(u, v) = \displaystyle\int_\Omega \nabla u \cdot \nabla v, \end{cases}$$

and $f(v) = \int_\Omega fv$ if $f \in L^2(\Omega)$ or $f(v) = \langle f, v \rangle_{H^{-1}, H_0^1}$ if $f \in H^{-1}(\Omega)$; see §3.1.1.

The Stokes equations. Consider the PDEs $-\Delta u + \nabla p = s$ and $\nabla \cdot u = g$ in Ω supplemented with the homogeneous Dirichlet condition $u_{|\partial\Omega} = 0$. This problem falls into the above framework by setting

$$\begin{cases} W = V = [H_0^1(\Omega)]^d \times L_{\int=0}^2(\Omega), \\ a((u,p),(v,q)) = \int_\Omega \nabla u{:}\nabla v - \int_\Omega p\nabla{\cdot}v + \int_\Omega q\nabla{\cdot}u, \end{cases}$$

and $f(v,q) = \int_\Omega(s{\cdot}v + gq)$ provided $s \in [L^2(\Omega)]^d$ and $g \in L^2(\Omega)$ or $f(v,q) = \langle s, v\rangle_{H^{-1},H_0^1} + \int_\Omega gq$ if $s \in [H^{-1}(\Omega)]^d$; here, $L_{\int=0}^2(\Omega)$ is the space of square-summable functions with zero mean in Ω. One important difference with the Laplace equation is that the solution and the test functions are now *vector-valued*; see §4.1.2.

The advection equation. Let $\beta \in [\mathcal{C}^1(\overline{\Omega})]^d$ be a given vector field and denote by $\partial\Omega^- = \{x \in \partial\Omega; (\beta{\cdot}n)(x) < 0\}$ the so-called inflow boundary, n being the outward normal to $\partial\Omega$. Consider the PDE $\beta{\cdot}\nabla u = f$ in Ω supplemented with the boundary condition $u_{|\partial\Omega^-} = 0$. This problem can be reformulated in the form (2.1) by setting

$$\begin{cases} W = \{u \in L^2(\Omega); \beta{\cdot}\nabla u \in L^2(\Omega); u = 0 \text{ on } \partial\Omega^-\}, \quad V = L^2(\Omega), \\ a(u,v) = \int_\Omega v(\beta{\cdot}\nabla u), \end{cases}$$

and $f(v) = \int_\Omega fv$ provided $f \in L^2(\Omega)$. The main difference with the Laplace and the Stokes equations is that *the solution space and the test space are different*; see §5.2.3.

2.1.2 The Lax–Milgram Lemma

Consider the case where the solution space and the test space are identical. Thus, the model problem is:

$$\begin{cases} \text{Seek } u \in V \text{ such that} \\ a(u,v) = f(v), \quad \forall v \in V. \end{cases} \tag{2.2}$$

Lemma 2.2 (Lax–Milgram). *Let V be a Hilbert space, let $a \in \mathcal{L}(V \times V; \mathbb{R})$, and let $f \in V'$. Assume that the bilinear form a is coercive, i.e.,*

(LM) $$\exists \alpha > 0, \forall u \in V, \quad a(u,u) \geq \alpha\|u\|_V^2.$$

Then, problem (2.2) is well-posed with a priori estimate

$$\forall f \in V', \quad \|u\|_V \leq \tfrac{1}{\alpha}\|f\|_{V'}. \tag{2.3}$$

Proof. Since this lemma is a consequence of the BNB Theorem, the proof is postponed to Lemma 2.8; see also Exercise 2.11 for a direct proof. □

Remark 2.3. The Lax–Milgram Lemma holds in Hilbert spaces only (i.e., not in Banach spaces) since coercivity is essentially an Hilbertian property; see Exercise 2.8. □

In the particular case where the bilinear form a is symmetric and positive, problem (2.2) can be interpreted as an optimization problem.

Proposition 2.4. *Along with the hypotheses of Lemma 2.2, assume that:*

(i) a *is symmetric:* $a(u,v) = a(v,u)$, $\forall u, v \in V$;
(ii) a *is positive:* $a(u,u) \geq 0$, $\forall u \in V$.

Then, setting $J(v) = \frac{1}{2}a(v,v) - f(v)$, u *solves* (2.2) *if and only if* u *minimizes* J *over* V.

Proof. The proof relies on the following identity: for all $u, v \in V$ and $t \in \mathbb{R}$,

$$J(u + tv) = J(u) + t(a(u,v) - f(v)) + \tfrac{t^2}{2}a(v,v), \tag{2.4}$$

which results from the symmetry of a. Assume that u solves (2.2). Then, owing to the positivity of a, (2.4) implies that u minimizes J over V. Conversely, assume that u minimizes J over V. Let $v \in V$ with $a(v,v) \neq 0$ and take $t = -\frac{a(u,v)-f(v)}{a(v,v)}$ in (2.4). A straightforward calculation yields

$$0 \geq J(u) - J(u + tv) = \frac{(a(u,v) - f(v))^2}{2a(v,v)}.$$

Owing to the positivity of a, this implies $a(u,v) = f(v)$. If $a(v,v) = 0$, one can conclude similarly by taking $t = -a(u,v) + f(v)$. □

Remark 2.5.

(i) When a is symmetric and coercive, the Lax–Milgram Lemma implies that the optimization problem $\inf_{v \in V} J(v)$ has a unique solution. The coercivity of a can be interpreted as a strong convexity property of the functional J. Problem (2.2) is termed a *variational formulation*.

(ii) In several applications, the functional J represents an *energy*. For instance, in continuum mechanics, consider an elastic body deformed under an externally applied load; see §3.4. Then, $\frac{1}{2}a(v,v)$ is the elastic deformation energy in the equilibrium configuration, and $-f(v)$ is the potential energy under the external load. □

2.1.3 The BNB Theorem: inf-sup conditions

The BNB Theorem plays a fundamental role in this book. Although it is by no means standard, we have adopted the terminology "BNB Theorem" since, to our knowledge, the result in the form below was first stated by Nečas in 1962 [Neč62] and popularized by Babuška in 1972 in the context of finite element methods [BaA72, p. 112]. From a functional analysis point of view, this theorem is a rephrasing of two fundamental results by Banach: the Closed Range Theorem and the Open Mapping Theorem.

Theorem 2.6 (Banach–Nečas–Babuška). *Let W be a Banach space and let V be a reflexive Banach space. Let $a \in \mathcal{L}(W \times V; \mathbb{R})$ and $f \in V'$. Then, problem (2.1) is well-posed if and only if:*

(BNB1) $$\exists \alpha > 0, \quad \inf_{w \in W} \sup_{v \in V} \frac{a(w,v)}{\|w\|_W \|v\|_V} \geq \alpha,$$

(BNB2) $$\forall v \in V, \quad (\forall w \in W, \; a(w,v) = 0) \implies (v = 0).$$

Moreover, the following a priori estimate holds:

$$\forall f \in V', \quad \|u\|_W \leq \tfrac{1}{\alpha} \|f\|_{V'}. \tag{2.5}$$

Proof. Owing to Corollary A.46, the conditions (BNB1) and (BNB2) are equivalent to the well-posedness of (2.1). Moreover, the a priori estimate (2.5) directly results from the inequalities

$$\alpha \|u\|_W \leq \sup_{v \in V} \frac{a(u,v)}{\|v\|_V} = \sup_{v \in V} \frac{f(v)}{\|v\|_V} = \|f\|_{V'}. \qquad \square$$

Remark 2.7. Let $A \in \mathcal{L}(W; V')$ be defined by

$$\forall w \in W, \; \forall v \in V, \quad \langle Aw, v \rangle_{V',V} = a(w,v). \tag{2.6}$$

Then, problem (2.1) amounts to seeking $u \in W$ such that $Au = f$ in V'. From the results of Appendix A, we infer

$$\text{(BNB1)} \iff (\operatorname{Ker}(A) = \{0\} \text{ and } \operatorname{Im}(A) \text{ closed}) \iff (A^T \text{ is surjective}),$$
$$\text{(BNB2)} \iff \qquad (\operatorname{Ker}(A^T) = \{0\}) \qquad \iff (A^T \text{ is injective}). \qquad \square$$

We are now in the position to prove the Lax–Milgram Lemma. It suffices to verify that condition (LM) implies conditions (BNB1) and (BNB2).

Lemma 2.8. *Assume $W = V$. Then, (LM) implies (BNB1) and (BNB2).*

Proof. Assume (LM) and let $w \in V$. Condition (BNB1) is readily deduced from

$$\alpha \|w\|_V \leq \frac{a(w,w)}{\|w\|_V} \leq \sup_{v \in V} \frac{a(w,v)}{\|v\|_V}.$$

Let now $v \in V$. Taking $w = v$ yields

$$\sup_{w \in W} a(w,v) \geq a(v,v) \geq \alpha \|v\|_V^2.$$

Therefore, $\sup_{w \in W} a(w,v) = 0$ implies $v = 0$, thus proving (BNB2). $\qquad \square$

Remark 2.9.

(i) The BNB Theorem is sometimes referred to in the literature as the "generalized Lax–Milgram Theorem." The reason we shall not use this terminology is that, as already mentioned in Remark 2.3, coercivity is essentially a Hilbertian property, i.e., the Lax–Milgram Lemma is meaningful only in Hilbert spaces, whereas the proper setting for the BNB Theorem is that of Banach spaces. Hence, calling this theorem the "generalized Lax–Milgram Theorem" would amount to calling Banach spaces "generalized Hilbert spaces" which would be somewhat misleading.

(ii) Condition (BNB1) is usually termed an *inf-sup condition*. It can be recast in the following form, which will often be used in the sequel:

$$\forall w \in W, \quad \alpha \|w\|_W \leq \sup_{v \in V} \frac{a(w,v)}{\|v\|_V}.$$

(iii) The reciprocal of Lemma 2.8 is wrong: conditions (BNB1) and (BNB2) do not imply property (LM). Hence, (LM) is not optimal in general (recall that (BNB1)–(BNB2) are necessary and sufficient for well-posedness). However, when the bilinear form a is symmetric and positive, coercivity is both necessary and sufficient; see Corollary A.55.

(iv) In finite dimension, (LM) is equivalent to stating that the matrix associated with the operator A is positive definite, whereas the conditions (BNB1) and (BNB2) are equivalent to its invertibility; see Remark 2.20(i) and Proposition 2.21.   □

2.1.4 Non-homogeneous Dirichlet boundary conditions

This section analyzes a particular form of problem (2.1) arising when non-homogeneous Dirichlet boundary conditions are enforced. This type of boundary condition is often encountered in engineering applications. It takes the form $u_{|\partial\Omega} = g$ where g is a given function on $\partial\Omega$.

Let B be a vector space of functions defined on $\partial\Omega$ and assume that there exists a trace operator $\gamma_0 \in \mathcal{L}(W; B)$ mapping functions of W to their restriction to $\partial\Omega$. The non-homogeneous version of problem (2.1) is:

$$\begin{cases} \text{Seek } u \in W \text{ such that} \\ a(u,v) = f(v), \quad \forall v \in V, \\ \gamma_0(u) = g, \qquad \text{in } B. \end{cases} \tag{2.7}$$

The main result of this section is the following:

Proposition 2.10. *Let W, V, and B be three Banach spaces, V being reflexive. Let $\gamma_0 \in \mathcal{L}(W; B)$ and let $a \in \mathcal{L}(W \times V; \mathbb{R})$. Assume that γ_0 is surjective and that the restriction of a to $W_0 \times V$, where $W_0 = \mathrm{Ker}(\gamma_0)$, satisfies the conditions of the BNB Theorem. Then, problem (2.7) is well-posed, and there exists $c > 0$ such that, for all $f \in V'$ and $g \in B$,*

$$\|u\|_W \leq c \left(\|f\|_{V'} + \|g\|_B \right).$$

Proof. Since γ_0 is continuous and surjective, the Open Mapping Theorem implies that there exists $c > 0$ such that, for all $g \in B$, there is $u_g \in W$ satisfying $\gamma_0 u_g = g$ and $\|u_g\|_W \le c\|g\|_B$. Clearly, (2.7) is equivalent to setting $\phi = u - u_g$ and considering the following problem:

$$\begin{cases} \text{Seek } \phi \in W_0 \text{ such that} \\ a(\phi, v) = f(v) - a(u_g, v), \quad \forall v \in V. \end{cases} \tag{2.8}$$

From the inequalities

$$|f(v) - a(u_g, v)| \le (\|f\|_{V'} + \|a\| \|u_g\|_W) \|v\|_V$$
$$\le (\|f\|_{V'} + c\|a\| \|g\|_B) \|v\|_V,$$

we deduce that the linear form $f - a(u_g, \cdot)$ is continuous on V. Since the restriction of a to $W_0 \times V$ fulfills the conditions of the BNB Theorem, problem (2.8) has a unique solution. Therefore, problem (2.7) is also well-posed. Finally, the a priori estimate directly results from the above inequalities. □

Remark 2.11. The function u_g is called a *lifting of the boundary condition*. From a theoretical viewpoint, non-homogeneous Dirichlet boundary conditions are thus handled very simply: after lifting the boundary condition and changing variable, the problem falls into the framework of the BNB Theorem. However, the argument presented above is not constructive since it does not provide an explicit expression for the lifting. In §3.2.2, we investigate a finite element approximation to (2.7) that accounts explicitly for non-homogeneous Dirichlet boundary conditions. □

Example 2.12.
   (i) Consider the Laplace equation $-\Delta u = f$ in a domain Ω with the non-homogeneous Dirichlet boundary condition $u_{|\partial\Omega} = g$. The weak formulation of this problem is:

$$\begin{cases} \text{Seek } u \in H^1(\Omega) \text{ such that} \\ \int_\Omega \nabla u \cdot \nabla v = \int_\Omega fv, \quad \forall v \in H_0^1(\Omega), \\ \gamma_0(u) = g, \quad \text{in } H^{\frac{1}{2}}(\partial\Omega). \end{cases} \tag{2.9}$$

This problem clearly falls into the framework of (2.7) by setting $W = H^1(\Omega)$, $V = H_0^1(\Omega)$, $\gamma_0(v) = v_{|\partial\Omega}$, $W_0 = H_0^1(\Omega)$, and $B = H^{\frac{1}{2}}(\partial\Omega)$. The operator γ_0 is indeed bounded and surjective from $H^1(\Omega)$ into $H^{\frac{1}{2}}(\partial\Omega)$ (see §B.3.5), and the homogeneous version of problem (2.9) is well-posed (see Remark 3.9(ii)).
   (ii) The non-homogeneous Stokes problem can be recast in the framework of (2.7) by setting $W = [H^1(\Omega)]^d \times L_{j=0}^2(\Omega)$, $V = [H_0^1(\Omega)]^d \times L_{j=0}^2(\Omega)$, $W_0 = [H_0^1(\Omega)]^d \times L_{j=0}^2(\Omega)$, $\gamma_0(u, p) = u_{|\partial\Omega}$, and $B = [H^{\frac{1}{2}}(\partial\Omega)]^d$.
   (iii) The non-homogeneous advection equation can be recast in the framework of (2.7) by setting $W = \{v \in L^2(\Omega); \beta \cdot \nabla v \in L^2(\Omega)\}$, $V = L^2(\Omega)$, $\gamma_0(v) = v_{|\partial\Omega^-}$, where $\partial\Omega^-$ is the inflow boundary, and $W_0 = \{v \in V; v_{|\partial\Omega^-} =$

0}. The characterization of the space B is somewhat technical in this case. It is possible to take $B = L^2_{loc}(\partial\Omega^-, |\beta\cdot n|)$, the space of locally square-integrable functions over $\partial\Omega^-$ for the surface measure $|\beta\cdot n|$.   □

2.2 Galerkin Methods

This section is concerned with the approximation of the abstract problem (2.1) using Galerkin methods.

2.2.1 The setting

The key idea underlying Galerkin methods is to replace the spaces W and V by *finite-dimensional spaces* W_h and V_h. The space W_h is termed the *solution space* or the *trial space*, and the space V_h is termed the *test space*. The finite element interpolation techniques presented in Chapter 1 provide practical means to construct such spaces, the index h referring to the mesh size. Henceforth, we assume that W_h and V_h are equipped with some norms, say $\|\cdot\|_{W_h}$ and $\|\cdot\|_{V_h}$, respectively. Setting

$$W(h) = W + W_h, \tag{2.10}$$

we make the important assumption that this space can be equipped with a norm, say $\|\cdot\|_{W(h)}$, such that:

(i) $\|w_h\|_{W(h)} = \|w_h\|_{W_h}$ for all $w_h \in W_h$;
(ii) $\|w\|_{W(h)} \leq c\|w\|_W$ for all $w \in W$; this property means that W is continuously embedded in $W(h)$.

In its most general form, the Galerkin method constructs an approximation of u by solving the following approximate problem:

$$\begin{cases} \text{Seek } u_h \in W_h \text{ such that} \\ a_h(u_h, v_h) = f_h(v_h), \quad \forall v_h \in V_h. \end{cases} \tag{2.11}$$

Problem (2.11) involves an approximation a_h to the bilinear form a and an approximation f_h to the linear form f.

A particular case of (2.11) is one in which the same approximation space V_h is chosen for the solution and the test functions, leading to the approximate problem:

$$\begin{cases} \text{Seek } u_h \in V_h \text{ such that} \\ a_h(u_h, v_h) = f_h(v_h), \quad \forall v_h \in V_h. \end{cases} \tag{2.12}$$

In this case, we say that a *standard Galerkin method* is used to approximate (2.1). When the solution and test spaces are different, the approximation method is sometimes called a *Petrov–Galerkin method*, but in this book we refer to it as a *non-standard Galerkin method*.

Definition 2.13 (Conformity). *The approximation setting is said to be conformal if $W_h \subset W$ and $V_h \subset V$; it is said to be* non-conformal *otherwise.*

Definition 2.14 (Approximability). *The approximation setting is said to have the* approximability property *if*

$$\forall w \in W, \quad \lim_{h \to 0} \left(\inf_{w_h \in W_h} \| w - w_h \|_{W(h)} \right) = 0. \tag{2.13}$$

Definition 2.15 (Consistency and asymptotic consistency). *Let u solve* (2.1).

(i) *The approximation setting is said to be* consistent *if a_h can be extended to $W(h) \times V_h$ and if the exact solution u satisfies the approximate problem (2.11), i.e., if*

$$\forall v_h \in V_h, \quad a_h(u, v_h) = f_h(v_h). \tag{2.14}$$

It is said to be non-consistent *otherwise.*

(ii) *When a_h is uniformly bounded on $W_h \times V_h$, the approximation method is said to be* asymptotically consistent *if there is an operator $\Pi_h : W \to W_h$ such that, for all $w \in W$, $\| \Pi_h w - w \|_{W(h)} \leq c \inf_{w_h \in W_h} \| w - w_h \|_{W(h)}$ where c is independent of w, and*

$$\lim_{h \to 0} \left(\sup_{v_h \in V_h} \frac{|f_h(v_h) - a_h(\Pi_h u, v_h)|}{\| v_h \|_{V_h}} \right) = 0. \tag{2.15}$$

The consistency error *$R_h(u)$ is defined to be*

$$R_h(u) = \sup_{v_h \in V_h} \frac{|f_h(v_h) - a_h(\Pi_h u, v_h)|}{\| v_h \|_{V_h}}. \tag{2.16}$$

Remark 2.16. The definition of asymptotic consistency is independent of the operator Π_h provided the approximation setting has the approximability property. Indeed, assume there is an operator Π_h^1 as in Definition 2.15(ii). Let $\Pi_h^2 : W \to W_h$ be an operator such that, for all $w \in W$, $\| \Pi_h^2 w - w \|_{W(h)} \leq c \inf_{w_h \in W_h} \| w - w_h \|_{W(h)}$. Then, for all $v_h \in V_h$,

$$|f_h(v_h) - a_h(\Pi_h^2 u, v_h)| \leq |f_h(v_h) - a_h(\Pi_h^1 u, v_h)| + |a_h(\Pi_h^1 u - \Pi_h^2 u, v_h)|.$$

Denote by $R_{1h}(u)$ and $R_{2h}(u)$ the consistency errors using Π_h^1 and Π_h^2, respectively. Using the uniform boundedness of a_h, the triangle inequality, and the approximation property of Π_h^1 and Π_h^2 yields

$$R_{2h}(u) \leq R_{1h}(u) + c \| a_h \|_{W_h, V_h} \inf_{w_h \in V_h} \| u - w_h \|_{W(h)},$$

implying that the method is asymptotically consistent using Π_h^2; in other words, up to a quantity controlled by the approximability property, the consistency error does not depend on the operator Π_h chosen to measure it. $\quad\square$

As an immediate consequence of Definition 2.15, we state the following:

Lemma 2.17 (Galerkin orthogonality). *If the approximation setting is consistent, the so-called* Galerkin orthogonality *holds:*

$$\forall v_h \in V_h, \quad a_h(u - u_h, v_h) = 0. \tag{2.17}$$

Remark 2.18. In practice, non-consistent methods must be considered for various reasons. For instance, quadratures are often employed to evaluate the integrals defining the exact forms a and f; see §8.1. Another important example arises in the context of non-conformal methods where the exact forms a and f are no longer defined on the approximation spaces W_h and V_h; see, e.g., §3.2.3, §3.2.4, §4.2.8, §5.6, and §5.7. Non-consistent methods are also considered in conformal stabilized finite element approximations; see, e.g., the subgrid viscosity method described in §5.5.   □

2.2.2 The linear system

The approximate problem (2.11) is simply a linear system. To see this, let

$$M = \dim W_h \quad \text{and} \quad N = \dim V_h.$$

Let $\{\psi_1, \ldots, \psi_M\}$ be a basis of W_h and let $\{\varphi_1, \ldots, \varphi_N\}$ be a basis of V_h. In the framework of finite element methods, the functions $\{\psi_1, \ldots, \psi_M\}$ (resp., $\{\phi_1, \ldots, \phi_N\}$) can be taken to be the *global shape functions* in W_h (resp., V_h); see Chapter 1. Consider the expansion of u_h in the basis of W_h,

$$u_h = \sum_{i=1}^{M} U_i \psi_i,$$

and introduce the coordinate vector of u_h, $U = (U_i)_{1 \leq i \leq M} \in \mathbb{R}^M$, relative to the basis $\{\psi_1, \ldots, \psi_M\}$. Let $\mathcal{A} \in \mathbb{R}^{N,M}$ be the *stiffness matrix* with entries

$$\mathcal{A}_{ij} = a_h(\psi_j, \varphi_i), \quad 1 \leq i \leq N, \ 1 \leq j \leq M,$$

and let $F \in \mathbb{R}^N$ be the vector with components

$$F_i = f_h(\varphi_i), \quad 1 \leq i \leq N.$$

It is readily verified that

$$(u_h \text{ solves } (2.11)) \iff (\mathcal{A}U = F).$$

2.2.3 Well-posedness

The goal of this section is to investigate the well-posedness of the approximate problem (2.11). We shall see that this property is automatically granted in the coercive case for a conformal, consistent approximation. However, in the general case, there is no guarantee that the conditions of the BNB Theorem are automatically transferred from the exact problem to the approximate problem. As a result, non-trivial inf-sup conditions must be proven at the discrete level; see the numerous examples investigated in Chapters 4 and 5.

A particular case: Conformity, consistency, and coercivity. Consider the following approximation of problem (2.2):

$$\begin{cases} \text{Seek } u_h \in V_h \text{ such that} \\ a(u_h, v_h) = f(v_h), \quad \forall v_h \in V_h, \end{cases} \tag{2.18}$$

with an approximation space $V_h \subset V$. Note that (2.18) involves the *same* bilinear form a and the *same* linear form f as (2.2).

Proposition 2.19. *Let V be a Hilbert space, let $a \in \mathcal{L}(V \times V; \mathbb{R})$, and let $f \in V'$. Let V_h be a finite-dimensional space. Assume that:*

(i) *a is coercive on V;*
(ii) *$V_h \subset V$.*

Then, the approximate problem (2.18) is well-posed. In particular, for all $f \in V'$, the a priori estimate $\|u_h\|_V \leq \frac{1}{\alpha}\|f\|_{V'}$ holds.

Proof. Since $V_h \subset V$, the bilinear form a is coercive on V_h with constant α. To conclude, use the Lax–Milgram Lemma. $\qquad\square$

Remark 2.20.

(i) For a conformal, consistent approximation of a coercive problem, the stiffness matrix is *positive definite*. Indeed, let $N = \dim V_h$ and for $X \in \mathbb{R}^N$ with $X = (X_i)_{1 \leq i \leq N}$, set $\xi = \sum_{i=1}^N X_i \varphi_i \in V_h$, where $\{\varphi_1, \dots, \varphi_N\}$ is a basis of V_h. A straightforward calculation yields

$$\forall X \in \mathbb{R}^N, \quad (\mathcal{A}X, X)_N = a(\xi, \xi) \geq \alpha \|\xi\|_V^2,$$

implying that $(\mathcal{A}X, X)_N \geq 0$. Moreover, $(\mathcal{A}X, X)_N = 0$ implies $\xi = 0$, whence we deduce $X = 0$ since $\{\varphi_1, \dots, \varphi_N\}$ is a basis of V_h.

(ii) If a is symmetric, the stiffness matrix is also symmetric. $\qquad\square$

The general case. Consider the approximate problem (2.11). Owing to the BNB Theorem, the well-posedness of (2.11) is equivalent to the two following discrete conditions:

$$(\text{BNB1}_h) \qquad \exists \alpha_h > 0, \quad \inf_{w_h \in W_h} \sup_{v_h \in V_h} \frac{a_h(w_h, v_h)}{\|w_h\|_{W_h}\|v_h\|_{V_h}} \geq \alpha_h,$$

$$(\text{BNB2}_h) \qquad \forall v_h \in V_h, \quad (\forall w_h \in W_h, \; a_h(w_h, v_h) = 0) \implies (v_h = 0).$$

Condition (BNB1_h) is often termed a *discrete inf-sup condition*. Let us interpret conditions (BNB1_h) and (BNB2_h) in terms of the stiffness matrix.

Proposition 2.21.

(i) $(\text{BNB1}_h) \iff (\text{Ker}(\mathcal{A}) = \{0\})$.
(ii) $(\text{BNB2}_h) \iff (\text{rank}\,\mathcal{A} = \dim V_h)$.
(iii) *If $\dim W_h = \dim V_h$, $(\text{BNB1}_h) \iff (\text{BNB2}_h)$.*

Proof. (i) The following equivalences hold:

$$(X \in \mathrm{Ker}(\mathcal{A})) \iff \left(\forall i, \sum_{j=1}^{M} \mathcal{A}_{ij} X_j = 0 \right) \iff (\forall i, a_h(\xi, \varphi_i) = 0)$$

$$\iff \left(\sup_{v \in V_h} a_h(\xi, v) = 0 \right),$$

with $\xi = \sum_{i=1}^{M} X_i \psi_i$; hence,

$$(\mathrm{BNB1_h}) \implies \left(\forall \xi \in W_h, \left(\sup_{v \in V_h} a_h(\xi, v) = 0 \right) \Rightarrow (\xi = 0) \right)$$

$$\implies (\mathrm{Ker}(\mathcal{A}) = \{0\}).$$

Conversely, assume that $\mathrm{Ker}(\mathcal{A}) = \{0\}$. Reasoning by contradiction, consider a sequence $w_{hn} \in W_h$ with $\|w_{hn}\|_{W_h} = 1$ and such that

$$\sup_{v \in V_h} \frac{a_h(w_{hn}, v)}{\|v\|_{V_h}} \leq \frac{1}{n}.$$

Since the unit sphere in W_h is compact, there exists a subsequence, still denoted by w_{hn}, that converges to w_h as $n \to \infty$. The limit w_h satisfies $\|w_h\|_{W_h} = 1$ and $\sup_{v \in V_h} a_h(w_h, v) = 0$. This implies that $X \in \mathrm{Ker}(\mathcal{A})$ where X is the coordinate vector of w_h, i.e., $w_h = \sum_{i=1}^{M} X_i \psi_i$. Since $\mathrm{Ker}(\mathcal{A}) = \{0\}$, $X = 0$, thus contradicting $\|w_h\|_{W_h} = 1$.
(ii) The proof of (ii) is similar to that of (i) (consider $\mathcal{A}^T$ instead of $\mathcal{A}$).
(iii) Direct consequence of (i), (ii), and the Rank Theorem.   □

Theorem 2.22. *Let V_h and W_h be two finite-dimensional spaces equipped with the norms $\| \cdot \|_{W_h}$ and $\| \cdot \|_{V_h}$, respectively. Assume that:*

(i) *a_h is bounded on $W_h \times V_h$ and f_h is continuous on V_h.*
(ii) *The discrete inf-sup condition (BNB1$_h$) is fulfilled.*
(iii) *V_h and W_h have the same dimension.*

Then, the approximate problem (2.11) is well-posed, and the a priori estimate $\|u_h\|_{W_h} \leq \frac{1}{\alpha_h} \|f_h\|_{V_h'}$ holds.

Proof. Use Proposition 2.21(iii) and the BNB Theorem.   □

Remark 2.23.
(i) Even in the case of a consistent, conformal approximation, neither condition (BNB1) nor condition (BNB2) implies its discrete counterpart.
(ii) By comparing Proposition 2.21(i) and (ii) with Remark 2.7, we realize that the interpretation of (BNB1$_h$) and (BNB2$_h$) in matrix terms is almost identical to that of (BNB1) and (BNB2) in operator terms. The only difference is that in finite dimension, the range of $\mathcal{A}$ is automatically closed.

(iii) In the linear algebra framework, the inf-sup condition has a very simple interpretation. Given a matrix $\mathcal{A} \in \mathbb{R}^{N,M}$, set

$$\alpha = \min_{W \in \mathbb{R}^M} \max_{V \in \mathbb{R}^N} \frac{(\mathcal{A}W, V)_N}{\|W\|_M \|V\|_N},$$

where $(\cdot, \cdot)_N$ denotes the Euclidean scalar product in $\mathbb{R}^N$ with associated norm $\| \cdot \|_N$. Then, a straightforward calculation shows that

$$\alpha = \min_{\|W\|_M = 1} \|\mathcal{A}W\|_N = \lambda_{\min}(\mathcal{A}^T \mathcal{A})^{\frac{1}{2}},$$

where $\lambda_{\min}(\mathcal{A}^T \mathcal{A})$ is the smallest eigenvalue of $\mathcal{A}^T \mathcal{A}$, i.e., α is the smallest *singular value* of $\mathcal{A}$.   □

2.3 Error Analysis

In this section, we derive estimates for the approximation error $u - u_h$, where u solves the exact problem (2.1) and u_h solves the approximate problem (2.11).

2.3.1 The general case

Theorem 2.24. *Assume the following:*

(i) *Condition* (BNB1$_h$) *holds uniformly in h and* $\dim(W_h) = \dim(V_h)$.
(ii) *The bilinear form a_h is uniformly bounded on $W_h \times V_h$.*
(iii) *The approximation setting is asymptotically consistent.*
(iv) *The approximation setting has the approximability property.*

Then, denoting by $R_h(u)$ the consistency error, the following estimate holds:

$$\|u - u_h\|_{W(h)} \leq \tfrac{1}{\alpha_h} R_h(u) + c \inf_{w_h \in W_h} \|u - w_h\|_{W(h)}, \qquad (2.19)$$

and $\lim_{h \to 0} \|u - u_h\|_{W(h)} = 0$.

Proof. Let $\mathit{\Pi}_h$ be an operator involved in the definition of the asymptotic consistency of a_h. (Recall that the uniform boundedness of a_h allows to measure the consistency error by any operator $\mathit{\Pi}_h$ satisfying the assumptions of Definition 2.15; see Remark 2.16.) Clearly,

$$a_h(u_h - \mathit{\Pi}_h u, v_h) = f_h(v_h) - a_h(\mathit{\Pi}_h u, v_h).$$

Condition (BNB1$_h$) yields $\alpha_h \|u_h - \mathit{\Pi}_h u\|_{W_h} \leq R_h(u)$. Since the norms $\| \cdot \|_{W_h}$ and $\| \cdot \|_{W(h)}$ coincide on W_h, the triangle inequality implies

$$\|u_h - u\|_{W(h)} \leq \tfrac{1}{\alpha_h} R_h(u) + \|u - \mathit{\Pi}_h u\|_{W(h)}.$$

Estimate (2.19) then results from $\|u - \mathit{\Pi}_h u\|_{W(h)} \leq c \inf_{w_h \in W_h} \|u - w_h\|_{W(h)}$. Finally, the asymptotic consistency and the approximability property readily yield $\lim_{h \to 0} \|u - u_h\|_{W(h)} = 0$.   □

This theorem clearly indicates the four properties which are required to prove the convergence of the approximate solution: uniform stability, uniform boundedness, asymptotic consistency, and approximability. A loose principle in numerical analysis, known as the *Lax Principle*, is that stability and consistency imply convergence. The fact that this principle does not mention continuity and approximability does not mean that these two properties should be taken for granted. For instance, the counterexample discussed in §2.3.3 shows that the approximability property may not hold in some circumstances.

2.3.2 Particular cases

The non-consistent, non-conformal case. We assume that the bilinear form $a_h(\cdot, \cdot)$ can be extended to $W(h) \times V_h$ so that $a_h(w, v_h)$ makes sense for $w \in W$ and $v_h \in V_h$. The following result is known as the Second Strang Lemma [Str72]:

Lemma 2.25 (Strang 2). *Assume the following:*

(i) *Condition* (BNB1_h) *holds and* $\dim(W_h) = \dim(V_h)$.
(ii) *The bilinear form* a_h *is bounded on* $W(h) \times V_h$.

Then, the following error estimate holds:

$$\|u - u_h\|_{W(h)} \leq \left(1 + \frac{\|a_h\|_{W(h), V_h}}{\alpha_h}\right) \inf_{w_h \in W_h} \|u - w_h\|_{W(h)}$$

$$+ \frac{1}{\alpha_h} \sup_{v_h \in V_h} \frac{|f_h(v_h) - a_h(u, v_h)|}{\|v_h\|_{V_h}}. \tag{2.20}$$

Proof. Let $w_h \in W_h$. Then,

$$a_h(u_h - w_h, v_h) = a_h(u_h - u, v_h) + a_h(u - w_h, v_h)$$
$$= f_h(v_h) - a_h(u, v_h) + a_h(u - w_h, v_h).$$

Condition (BNB1_h) implies

$$\alpha_h \|u_h - w_h\|_{W(h)} \leq \sup_{v_h \in V_h} \frac{|f_h(v_h) - a_h(u, v_h)|}{\|v_h\|_{V_h}} + \|a_h\|_{W(h), V_h} \|u - w_h\|_{W(h)}.$$

Estimate (2.20) then results from the triangle inequality. □

Remark 2.26. When the method is consistent, (2.20) simplifies into

$$\|u - u_h\|_{W(h)} \leq \left(1 + \frac{\|a_h\|_{W(h), V_h}}{\alpha_h}\right) \inf_{w_h \in W_h} \|u - w_h\|_{W(h)}. \quad □$$

The non-consistent, conformal case. We now assume that $W_h \subset W$ and $V_h \subset V$, i.e., the approximation setting is conformal. As a result, $W(h) = W$. However, we do not assume that the extended norm $\|\cdot\|_{W(h)}$ is the same as that of W. Furthermore, as opposed to the previous case, we do not assume that $a_h(\cdot, \cdot)$ can be extended to $W \times V_h$, i.e., we accept the fact that $a_h(w, v_h)$ may not make sense for $(w, v_h) \in W \times V_h$. This is the case, for instance, when $a_h(u_h, \cdot)$ involves point values of u_h, which may not necessarily exist for functions in W. It is also the case when a_h involves a direct decomposition of u_h in W_h, which may not make sense for functions in W. A case corresponding to the second situation is investigated in §5.5; see Remark 5.55(i). The following result is known as the First Strang Lemma [Str72]:

Lemma 2.27 (Strang 1). *Assume the following:*

(i) $W_h \subset W$ and $V_h \subset V$.
(ii) *Condition* ($\mathrm{BNB1_h}$) *holds and* $\dim(W_h) = \dim(V_h)$.
(iii) *The bilinear form* a_h *is bounded on* $W_h \times V_h$, *and a is bounded on* $W \times V_h$ *when W is equipped with the extended norm* $\|\cdot\|_{W(h)}$.

Then, the following error estimate holds:

$$\|u - u_h\|_{W(h)} \le \frac{1}{\alpha_h} \sup_{v_h \in V_h} \frac{|f(v_h) - f_h(v_h)|}{\|v_h\|_{V_h}} \tag{2.21}$$

$$+ \inf_{w_h \in W_h} \left[\left(1 + \frac{\|a\|_{W(h), V_h}}{\alpha_h} \right) \|u - w_h\|_{W(h)} + \frac{1}{\alpha_h} \sup_{v_h \in V_h} \frac{|a(w_h, v_h) - a_h(w_h, v_h)|}{\|v_h\|_{V_h}} \right].$$

Proof. Let $w_h \in W_h$. Condition ($\mathrm{BNB1_h}$) implies

$$\alpha_h \|u_h - w_h\|_{W(h)} \le \sup_{v_h \in V_h} \frac{a_h(u_h - w_h, v_h)}{\|v_h\|_{V_h}}.$$

A straightforward calculation yields

$$a_h(u_h - w_h, v_h) = a(u - w_h, v_h) + a(w_h, v_h) - a_h(w_h, v_h) + f_h(v_h) - f(v_h).$$

Therefore,

$$\alpha_h \|u_h - w_h\|_{W(h)} \le \|a\|_{W(h), V_h} \|u - w_h\|_{W(h)}$$

$$+ \sup_{v_h \in V_h} \frac{|a(w_h, v_h) - a_h(w_h, v_h)|}{\|v_h\|_{V_h}} + \sup_{v_h \in V_h} \frac{|f(v_h) - f_h(v_h)|}{\|v_h\|_{V_h}}.$$

Conclude using the triangle inequality. $\qquad\square$

The consistent, conformal case. For a consistent, conformal approximation with $a_h = a$ and $f_h = f$, the approximate problem is:

$$\begin{cases} \text{Seek } u_h \in W_h \text{ such that} \\ a(u_h, v_h) = f(v_h), \quad \forall v_h \in V_h. \end{cases} \tag{2.22}$$

The following result is known as Céa's Lemma [Céa64]:

Lemma 2.28 (Céa). *Let the hypotheses of Lemma 2.25 hold with $V_h \subset V$, $W_h \subset W$, $a_h = a$, and $f_h = f$. Let u_h solve the approximate problem (2.22). Then, the following error estimate holds:*

$$\|u - u_h\|_W \leq \left(1 + \tfrac{\|a\|_{W,V}}{\alpha_h}\right) \inf_{w_h \in W_h} \|u - w_h\|_W. \qquad (2.23)$$

Proof. For completeness, we present a direct proof without using (2.20). Let $w_h \in W_h$. Galerkin orthogonality implies

$$\forall v_h \in V_h, \quad a(u_h - w_h, v_h) = a(u - w_h, v_h).$$

Using property (BNB1$_h$) and the continuity of a yields

$$\alpha_h \|u_h - w_h\|_W \leq \sup_{v_h \in V_h} \frac{a(u_h - w_h, v_h)}{\|v_h\|_V} = \sup_{v_h \in V_h} \frac{a(u - w_h, v_h)}{\|v_h\|_V} \leq \|a\|_{W,V} \|u - w_h\|_W.$$

Conclude using the triangle inequality.   □

The coercive case. Assume that $W_h = V_h$, $W = V$, $a_h = a$, and $f_h = f$. Assume that the bilinear form a is coercive with coercivity constant α and continuity constant $\|a\|$. Galerkin orthogonality implies

$$\forall v_h \in V_h, \quad a(u - u_h, u - u_h) = a(u - u_h, u - v_h).$$

Using the coercivity and the continuity of a yields the estimate (2.23) with the constant $\frac{\|a\|}{\alpha}$ instead of $1 + \frac{\|a\|}{\alpha}$.

This estimate can be sharpened even further if a is also symmetric. In this case, define the scalar product $a(\cdot, \cdot)$ with associated norm $\|u\|_e^2 = a(u, u)$ for $u \in V$. The norm $\|\cdot\|_e$ is called the *energy norm* and is equivalent to the norm $\|\cdot\|_V$ since

$$\forall u \in V, \quad \alpha^{\frac{1}{2}} \|u\|_V \leq \|u\|_e \leq \|a\|^{\frac{1}{2}} \|u\|_V.$$

Let $\mathcal{P}_{e,V_h} : V \to V_h$ be the orthogonal projection for the scalar product $a(\cdot, \cdot)$. Owing to Galerkin orthogonality, it is clear that

$$u_h = \mathcal{P}_{e,V_h}(u). \qquad (2.24)$$

The Pythagoras identity yields, for all $w_h \in V_h$,

$$a(u - u_h, u - u_h)\|u - u_h\|_e^2 \leq \|u - w_h\|_e^2 = a(u - w_h, u - w_h).$$

As a result,

$$\alpha\|u - u_h\|_V^2 \leq a(u - w_h, u - w_h) \leq \|a\| \, \|u - w_h\|_V^2.$$

Hence, $\|u - u_h\|_V \leq \left(\tfrac{\|a\|}{\alpha}\right)^{\frac{1}{2}} \inf_{w_h \in V_h} \|u - w_h\|_V.$

2.3.3 A counterexample to the approximability hypothesis

The approximability property (2.13) may seem unnecessary to verify, and one may believe that it is likely to be always satisfied for polynomial-based finite elements. Generally, practitioners do not bother to check this hypothesis. However, there are situations of engineering interest where this hypothesis may fail. It may happen, for instance, in electromagnetism.

In the computational electromagnetism literature, there is a debate pitting proponents of edge finite elements against those of the all-purpose Lagrange finite elements. Both techniques work perfectly well in many cases, but there are a few situations where the two methods yield different answers, irrespective of the level of mesh refinement. It turns out that the origin of the discrepancy lies in the approximability property. This fact has been clarified by Costabel [Cos91].

Let Ω be a domain in $\mathbb{R}^3$ with boundary $\partial\Omega$ and outward normal n. In many electromagnetism problems governed by the Maxwell equations, a natural solution space is $H_0(\text{curl}; \Omega) \cap H(\text{div}; \Omega)$ where $H_0(\text{curl}; \Omega) = \{v \in H(\text{curl}; \Omega); v \times n_{|\partial\Omega} = 0\}$.

Lemma 2.29 (Costabel). *Assume that Ω is a polyhedron. If Ω is not convex, $H_0(\text{curl}; \Omega) \cap [H^1(\Omega)]^3$ is a closed proper subspace of $H_0(\text{curl}; \Omega) \cap H(\text{div}; \Omega)$.*

This result has the following striking consequence:

Corollary 2.30. *Let $\{W_h\}_{h>0}$ be a family of finite element spaces conformal in $[H^1(\Omega)]^3$ and set $W_{h0} = \{w_h \in W_h; w_h \times n_{|\partial\Omega} = 0\}$. Then, under the hypotheses of Lemma 2.29, $\{W_{h0}\}_{h>0}$ cannot have the approximability property in $H_0(\text{curl}; \Omega) \cap H(\text{div}; \Omega)$.*

Proof. It is clear that $W_{h0} \subset [H^1(\Omega)]^3$ and $W_{h0} \subset H_0(\text{curl}; \Omega)$; hence, $W_{h0} \subset H_0(\text{curl}; \Omega) \cap [H^1(\Omega)]^3$. Since $H_0(\text{curl}; \Omega) \cap [H^1(\Omega)]^3$ is closed in $H_0(\text{curl}; \Omega) \cap H(\text{div}; \Omega)$, the limit of all the Cauchy sequences in W_{h0} are in $H_0(\text{curl}; \Omega) \cap [H^1(\Omega)]^3$. Moreover, since $H_0(\text{curl}; \Omega) \cap [H^1(\Omega)]^3$ is a proper subspace of $H_0(\text{curl}; \Omega) \cap H(\text{div}; \Omega)$, there are functions of $H_0(\text{curl}; \Omega) \cap H(\text{div}; \Omega)$ that lie at a positive distance from $H_0(\text{curl}; \Omega) \cap [H^1(\Omega)]^3$. Therefore, Cauchy sequences in W_{h0} cannot reach these functions, i.e., $\{W_{h0}\}_{h>0}$ does not have the approximability property in $H_0(\text{curl}; \Omega) \cap H(\text{div}; \Omega)$. □

In the light of Corollary 2.30, we now understand why Lagrange finite elements may fail in electromagnetism. If the solution to be approximated is so rough as to be only in $H_0(\text{curl}; \Omega) \cap H(\text{div}; \Omega)$ and not in more regular spaces, then Lagrange finite elements cannot interpolate it, whereas edge finite elements can. However, if, by some argument, it is known a priori that the solution is somewhat smoother, i.e., lives in a space that is slightly more regular than $H_0(\text{curl}; \Omega) \cap H(\text{div}; \Omega)$, say $H_0(\text{curl}; \Omega) \cap [H^1(\Omega)]^3$, then Lagrange finite elements yield the approximability property. In particular, if Ω is convex, the necessary extra regularity holds.

The above counterexample shows that the *approximability property is not a hypothesis to be forgotten* or to be treated too lightly.

2.3.4 The Aubin–Nitsche Lemma

The goal of this section is to derive an error estimate in a weaker norm than that of $W(h)$. For the sake of simplicity, we restrict the analysis to the approximation of problem (2.2) in a standard, conformal, and consistent setting, i.e., $W_h = V_h$ and the discrete problem is (2.18); see, e.g., [Bra97, p. 108] for non-conformal approximation settings. Problems (2.2) and (2.18) are assumed to be well-posed. Furthermore, we make the following additional assumptions:

(AN1) There exists a Hilbert space L into which V can be continuously embedded. We assume that L is equipped with a continuous, symmetric, and positive bilinear form $l(\cdot,\cdot)$, and we denote by $|\cdot|_L = \sqrt{l(\cdot,\cdot)}$ the corresponding seminorm. We further assume that there exists a Banach space $Z \subset V$ and a stability constant $c_S > 0$ such that, for all $g \in L$, the solution $\varsigma(g)$ to the following adjoint problem:

$$\begin{cases} \text{Seek } \varsigma(g) \in V \text{ such that} \\ a\big(v,\varsigma(g)\big) = l(g,v), \quad \forall v \in V, \end{cases} \tag{2.25}$$

satisfies the a priori estimate $\|\varsigma(g)\|_Z \le c_S |g|_L$.

(AN2) There exists an interpolation constant $c_i > 0$ such that

$$\forall h, \ \forall v \in Z, \quad \inf_{v_h \in V_h} \|v - v_h\|_V \le c_i h \|v\|_Z.$$

Whenever property (AN1) holds, problem (2.25) is said to be *regularizing*. The following lemma yields an error estimate in the seminorm $|\cdot|_L$ [Aub87]:

Lemma 2.31 (Aubin–Nitsche). *Under the above assumptions,*

$$\forall h, \quad |u - u_h|_L \le c\,h \|u - u_h\|_V,$$

where $c = c_i c_S \|a\|_{W,V}$.

Proof. Setting $e_h = u - u_h$, it is clear that

$$|e_h|_L = \sup_{g \in L} \frac{l(g, e_h)}{|g|_L} = \sup_{g \in L} \frac{a\big(e_h, \varsigma(g)\big)}{|g|_L}.$$

Galerkin orthogonality implies $a\big(e_h, \varsigma(g)\big) = a(e_h, \varsigma(g) - v_h)$ for all $v_h \in V_h$. Hence,

$$\begin{aligned} a\big(e_h, \varsigma(g)\big) &\le \|a\|_{W,V} \|e_h\|_V \inf_{v_h \in V_h} \|\varsigma(g) - v_h\|_V \\ &\le \|a\|_{W,V} \|e_h\|_V\, c_i h \|\varsigma(g)\|_Z && \text{from (AN2)} \\ &\le \|a\|_{W,V} \|e_h\|_V\, c_i h\, c_S |g|_L && \text{from (AN1).} \end{aligned}$$

The conclusion is straightforward. $\square$

Example 2.32.

(i) For a model problem with the Laplace operator, set

$$Z = H^2(\Omega), \qquad V = H^1(\Omega), \qquad L = L^2(\Omega), \qquad |\cdot|_L = \|\cdot\|_{0,\Omega}.$$

Assumption (AN1) is not straightforward but can be proven when Ω satisfies some regularity properties; see §3.1.3. Assumption (AN2) is a direct consequence of Corollary 1.109.

(ii) Lemma 2.31 can also be applied to the Stokes problem. In this case, $|\cdot|_L$ is only a seminorm; see §2.4.2. □

2.4 Saddle-Point Problems

This section treats a particular form of problem (2.1) encountered, for instance, when dealing with the Stokes problem. Owing to the particular form of this problem (a saddle-point problem), we give a more precise, although equivalent, characterization of well-posedness. Then, we analyze the approximation of saddle-point problems using Galerkin methods.

2.4.1 Well-posedness

Let X and M be two reflexive Banach spaces, $f \in X'$, $g \in M'$, and consider two bilinear forms $a \in \mathcal{L}(X \times X; \mathbb{R})$ and $b \in \mathcal{L}(X \times M; \mathbb{R})$. The abstract problem we investigate is:

$$\begin{cases} \text{Seek } u \in X \text{ and } p \in M \text{ such that} \\ a(u,v) + b(v,p) = f(v), \quad \forall v \in X, \\ b(u,q) = g(q), \qquad \forall q \in M. \end{cases} \qquad (2.26)$$

Example 2.33. The prototype example for (2.26) is the Stokes problem; see Chapter 4 for a thorough presentation. In this case, $X = [H_0^1(\Omega)]^d$, $M = L^2_{\int=0}(\Omega)$, $a(u,v) = \int_\Omega \nabla u{:}\nabla v$, $b(v,p) = -\int_\Omega p\nabla{\cdot}v$, $f(v) = \int_\Omega f{\cdot}v$, and $g(q) = -\int_\Omega gq$. □

Another way of looking at problem (2.26) consists of introducing $W = X \times M$, $c((u,p),(v,q)) = a(u,v) + b(v,p) + b(u,q)$, and $k(v,q) = f(v) + g(q)$. One can then consider the following problem:

$$\begin{cases} \text{Seek } (u,p) \in W \text{ such that} \\ c((u,p),(v,q)) = k(v,q), \quad \forall (v,q) \in W. \end{cases} \qquad (2.27)$$

It is clear that (2.26) and (2.27) are equivalent. As a result, necessary and sufficient conditions for the well-posedness of (2.26) are the two conditions (BNB1) and (BNB2) for the bilinear form c. However, owing to the particular structure of (2.26), it is possible to reformulate (BNB1) and (BNB2) in terms of conditions on the bilinear forms a and b. The goal of this section is to explore this point of view.

Introduce the operators A and B such that $A : X \to X'$ with $\langle Au, v \rangle_{X',X} = a(u,v)$ and $B : X \to M'$ (and $B^T : M = M'' \to X'$ since M is reflexive) with $\langle Bv, q \rangle_{M',M} = b(v,q)$. Problem (2.26) is equivalent to:

$$\begin{cases} \text{Seek } u \in X \text{ and } p \in M \text{ such that} \\ Au + B^T p = f, \\ Bu = g. \end{cases}$$

Let $\text{Ker}(B) = \{v \in X; \forall q \in M, b(v,q) = 0\}$ be the nullspace of B and let $\pi A : \text{Ker}(B) \to \text{Ker}(B)'$ be such that $\langle \pi Au, v \rangle_{X',X} = \langle Au, v \rangle_{X',X}$ for all $u, v \in \text{Ker}(B)$.

Theorem 2.34. *Under the above framework, problem (2.26) is well-posed if and only if*

$$\begin{cases} \exists \alpha > 0, \quad \inf_{u \in \text{Ker}(B)} \sup_{v \in \text{Ker}(B)} \dfrac{a(u,v)}{\|u\|_X \|v\|_X} \geq \alpha, \\ \forall v \in \text{Ker}(B), \quad (\forall u \in \text{Ker}(B), a(u,v) = 0) \Rightarrow (v = 0), \end{cases} \tag{2.28}$$

and

$$\exists \beta > 0, \quad \inf_{q \in M} \sup_{v \in X} \dfrac{b(v,q)}{\|v\|_X \|q\|_M} \geq \beta. \tag{2.29}$$

Furthermore, the following a priori estimates hold:

$$\begin{cases} \|u\|_X \leq c_1 \|f\|_{X'} + c_2 \|g\|_{M'}, \\ \|p\|_M \leq c_3 \|f\|_{X'} + c_4 \|g\|_{M'}, \end{cases} \tag{2.30}$$

with $c_1 = \frac{1}{\alpha}$, $c_2 = \frac{1}{\beta}(1 + \frac{\|a\|}{\alpha})$, $c_3 = \frac{1}{\beta}(1 + \frac{\|a\|}{\alpha})$, and $c_4 = \frac{\|a\|}{\beta^2}(1 + \frac{\|a\|}{\alpha})$.

Proof. Problem (2.26) is well-posed if and only if the conditions (i) and (ii) of Theorem A.56 are satisfied. Owing to Corollary A.45 and the fact that $\text{Ker}(B)$ is reflexive, the two inequalities in (2.28) are equivalent to the fact that πA is an isomorphism. Furthermore, inequality (2.29) is equivalent to the fact that B is surjective owing to condition (A.9) of Lemma A.40 and the fact that M is reflexive. Therefore, the well-posedness of problem (2.26) is equivalent to conditions (2.28) and (2.29). Let us now prove the a priori estimates (2.30). From condition (2.29) and Lemma A.42 (since M is reflexive), we deduce that there exists $u_g \in X$ such that $Bu_g = g$ and $\beta \|u_g\|_X \leq \|g\|_{M'}$. Setting $\phi = u - u_g$ yields

$$\forall v \in \text{Ker}(B), \quad a(\phi, v) = f(v) - a(u_g, v).$$

Noting that

$$|f(v) - a(u_g, v)| \leq (\|f\|_{X'} + \|a\| \|u_g\|_X) \|v\|_X$$

$$\leq \left(\|f\|_{X'} + \frac{\|a\|}{\beta} \|g\|_{M'} \right) \|v\|_X,$$

where $\|a\| = \|a\|_{X,X}$, and taking the supremum for v in $\mathrm{Ker}(B)$ yields

$$\alpha \|\phi\|_X \leq \|f\|_{X'} + \frac{\|a\|}{\beta} \|g\|_{M'},$$

owing to condition (2.28). The estimate on u then results from this inequality and the triangle inequality $\|u\|_X \leq \|u - u_g\|_X + \|u_g\|_X$. To prove the estimate on p, deduce from condition (2.29) and Lemma A.40 that $\beta \|p\|_M \leq \|B^T p\|_{X'}$, yielding

$$\beta \|p\|_M \leq \|a\| \|u\|_X + \|f\|_{X'}.$$

The estimate on $\|p\|_M$ then results from that on $\|u\|_X$. $\square$

Remark 2.35.

(i) If the bilinear form a is coercive on $\mathrm{Ker}(B)$, the conditions in (2.28) are clearly fulfilled. These conditions are also fulfilled if a is coercive on the whole space X.

(ii) Saddle-point problems are historically important in the engineering literature since they contributed to the popularization of inf-sup conditions. In particular, (2.29) is known as the *Babuška–Brezzi condition* [Bab73^a, Bre74]. $\square$

To stress the fact that (2.28) and (2.29) are nothing more than a restatement of the conditions (BNB1) and (BNB2) for problem (2.27), we state the following:

Proposition 2.36. *Equip the space $W = X \times M$ with the norm $\|(u, p)\|_W = \|u\|_X + \|p\|_M$. Then, the bilinear form c satisfies* (BNB1) *and* (BNB2) *if and only if* (2.28) *and* (2.29) *hold.*

Proof. Let us prove that (2.28) and (2.29) imply (BNB1). Let $(u, p) \in W$. Let $\widehat{u} \in X$ be such that $B\widehat{u} = Bu$ and $\beta \|\widehat{u}\|_X \leq \|Bu\|_{M'}$. Clearly,

$$\sup_{(v,q) \in W} \frac{c((\widehat{u}, p), (v, q))}{\|(v, q)\|_W} \geq \sup_{q \in M} \frac{b(\widehat{u}, q)}{\|q\|_M} = \|B\widehat{u}\|_{M'} \geq \beta \|\widehat{u}\|_X.$$

Moreover, owing to the fact that $u - \widehat{u}$ is in $\mathrm{Ker}(B)$,

$$\alpha \|u - \widehat{u}\|_X \leq \sup_{v \in \mathrm{Ker}(B)} \frac{a(u - \widehat{u}, v)}{\|v\|_X} = \sup_{v \in \mathrm{Ker}(B)} \frac{a(u - \widehat{u}, v) + b(v, p) + b(u, 0)}{\|(v, 0)\|_W}$$

$$\leq \sup_{(v,q) \in W} \frac{c((u, p), (v, q))}{\|(v, q)\|_W} + \|a\| \|\widehat{u}\|_X$$

$$\leq \left(1 + \frac{\|a\|}{\beta} \right) \sup_{(v,q) \in W} \frac{c((u, p), (v, q))}{\|(v, q)\|_W}.$$

Using the triangle inequality yields the following bound on $\|u\|_X$:

$$\|u\|_X \le \|\hat{u}\|_X + \|u - \hat{u}\|_X \le \left(\tfrac{1}{\beta} + \tfrac{1}{\alpha}\left(1 + \tfrac{\|a\|}{\beta}\right)\right) \sup_{(v,q)\in W} \frac{c((u,p),(v,q))}{\|(v,q)\|_W}.$$

To bound $\|p\|_M$, proceed as follows:

$$\beta\|p\|_M \le \sup_{v\in X} \frac{b(v,p)}{\|v\|_X} \le \sup_{v\in X} \frac{a(u,v) + b(v,p) + b(u,0)}{\|(v,0)\|_W} + \sup_{v\in X} \frac{a(u,v)}{\|v\|_X}$$

$$\le \sup_{(v,q)\in W} \frac{c((u,p),(v,q))}{\|(v,q)\|_W} + \|a\|\,\|u\|_X,$$

implying

$$\|p\|_M \le \tfrac{1}{\beta}\left(1 + \|a\|\left(\tfrac{1}{\beta} + \tfrac{1}{\alpha}\left(1 + \tfrac{\|a\|}{\beta}\right)\right)\right) \sup_{(v,q)\in W} \frac{c((u,p),(v,q))}{\|(v,q)\|_W}.$$

This proves (BNB1); the rest of the proof is left as an exercise. $\square$

One can generalize Proposition 2.4 to the abstract problem (2.26) assuming that the bilinear form a is symmetric and positive. In particular, one can prove that problem (2.26) is equivalent to a saddle-point problem. Recall the following:

Definition 2.37. *Given two sets X and M, consider a mapping $\mathcal{L}: X \times M \to \mathbb{R}$. A pair (u, p) is said to be a* saddle-point *of $\mathcal{L}$ if*

$$\forall (v, q) \in X \times M, \quad \mathcal{L}(u, q) \le \mathcal{L}(u, p) \le \mathcal{L}(v, p). \tag{2.31}$$

Lemma 2.38. *(u, p) is a saddle-point of $\mathcal{L}$ if and only if*

$$\inf_{v\in X} \sup_{q\in M} \mathcal{L}(v, q) = \sup_{q\in M} \mathcal{L}(u, q) = \mathcal{L}(u, p) = \inf_{v\in X} \mathcal{L}(v, p) = \sup_{q\in M} \inf_{v\in X} \mathcal{L}(v, q). \tag{2.32}$$

Proof. Definition 2.37 implies

$$\inf_{v\in X} \sup_{q\in M} \mathcal{L}(v, q) \le \sup_{q\in M} \mathcal{L}(u, q) \le \mathcal{L}(u, p) \le \inf_{v\in X} \mathcal{L}(v, p) \le \sup_{q\in M} \inf_{v\in X} \mathcal{L}(v, q).$$

Moreover, for all pairs $(v, q) \in X \times M$,

$$\inf_{v'\in X} \mathcal{L}(v', q) \le \mathcal{L}(v, q) \le \sup_{q'\in M} \mathcal{L}(v, q'),$$

yielding

$$\sup_{q\in M} \inf_{v\in X} \mathcal{L}(v, q) \le \inf_{v\in X} \sup_{q\in M} \mathcal{L}(v, q).$$

Therefore,

$$\inf_{v\in X} \sup_{q\in M} \mathcal{L}(v, q) = \sup_{q\in M} \mathcal{L}(u, q) = \mathcal{L}(u, p) = \inf_{v\in X} \mathcal{L}(v, p) = \sup_{q\in M} \inf_{v\in X} \mathcal{L}(v, q).$$

Note that the first equality means that the infimum over v is reached at u, and the last equality means that the supremum over q is reached at p. $\square$

Proposition 2.39. *Assume that a is symmetric and positive. Then, the pair* (u,p) *solves* (2.26) *if and only if* (u,p) *is a saddle-point of the* Lagrangian *functional*

$$\mathcal{L}(v,q) = \tfrac{1}{2}a(v,v) + b(v,q) - f(v) - g(q). \tag{2.33}$$

Proof. Let (u,p) be an arbitrary pair in $X \times M$. Clearly,

$$(\forall q \in M, \ \mathcal{L}(u,q) \leq \mathcal{L}(u,p)) \iff (\forall q \in M, \ b(u,q-p) \leq g(q-p))$$
$$\iff (\forall q \in M, \ b(u,q) = g(q)).$$

(In the last equivalence, the fact that M is a vector space has been used.) Therefore, the first inequality in (2.31) is equivalent to stating that u satisfies the second equality in problem (2.26). For $p \in M$, consider now the functional $J_p(v) = \tfrac{1}{2}a(v,v) + b(v,p) - f(v)$. One readily verifies that

$$(\forall v \in X, \ \mathcal{L}(u,p) \leq \mathcal{L}(v,p)) \iff \left(J_p(u) = \min_{v \in X} J_p(v) \right)$$
$$\iff (\forall v \in X, \ a(u,v) + b(v,p) = f(v)),$$

where the last equivalence is a direct consequence of Proposition 2.4. Therefore, the second inequality in (2.31) is equivalent to stating that the pair (u,p) satisfies the first equality in problem (2.26). □

Remark 2.40. When a is symmetric and positive, (2.26) is often termed a *saddle-point problem.* □

Corollary 2.41. *Assume that a is symmetric and positive and that the two conditions* (2.28) *and* (2.29) *are fulfilled. Then:*

(i) *Problem* (2.26) *admits a unique solution.*
(ii) *This solution is the unique saddle-point of the functional* (2.33).
(iii) *This solution satisfies* (2.32).

2.4.2 Approximation

This section studies conformal approximations to problem (2.26). Let X_h be a subspace of X and let M_h be a subspace of M. Assume that X_h and M_h are finite-dimensional and consider the approximate problem:

$$\begin{cases} \text{Seek } u_h \in X_h \text{ and } p_h \in M_h \text{ such that} \\ a(u_h, v_h) + b(v_h, p_h) = f(v_h), & \forall v_h \in X_h, \\ b(u_h, q_h) = g(q_h), & \forall q_h \in M_h. \end{cases} \tag{2.34}$$

Let $B_h : X_h \to M_h'$ be the operator induced by b such that $\langle B_h v_h, q_h \rangle_{M_h', M_h} = b(v_h, q_h)$. Let $\mathrm{Ker}(B_h)$ be the nullspace of B_h, i.e.,

$$\mathrm{Ker}(B_h) = \{v_h \in X_h; \forall q_h \in M_h, \ b(v_h, q_h) = 0\}.$$

We first address the well-posedness of the approximate problem (2.34).

Proposition 2.42. *Problem* (2.34) *is well-posed if and only if*

$$\exists \alpha_h > 0, \quad \inf_{u_h \in \mathrm{Ker}(B_h)} \sup_{v_h \in \mathrm{Ker}(B_h)} \frac{a(u_h, v_h)}{\|u_h\|_X \|v_h\|_X} \geq \alpha_h, \qquad (2.35)$$

$$\exists \beta_h > 0, \quad \inf_{q_h \in M_h} \sup_{v_h \in X_h} \frac{b(v_h, q_h)}{\|v_h\|_X \|q_h\|_M} \geq \beta_h. \qquad (2.36)$$

Proof. Apply Theorem 2.34 and use the fact that in finite dimension, condition (2.35) implies both conditions in (2.28); see Proposition 2.21(iii). □

Remark 2.43. Condition (2.36) is equivalent to assuming that B_h is surjective; see Lemma A.40. □

Our next goal is to estimate the approximation errors $u - u_h$ and $p - p_h$. We first derive an a priori estimate similar to Céa's Lemma.

Lemma 2.44. *Under the assumptions* (2.35) *and* (2.36), *letting* $\|a\| = \|a\|_{X,X}$ *and* $\|b\| = \|b\|_{X,M}$, *the solution* (u_h, p_h) *to* (2.34) *satisfies the estimates*

$$\|u - u_h\|_X \leq c_{1h} \inf_{v_h \in X_h} \|u - v_h\|_X + c_{2h} \inf_{q_h \in M_h} \|p - q_h\|_M,$$

$$\|p - p_h\|_M \leq c_{3h} \inf_{v_h \in X_h} \|u - v_h\|_X + c_{4h} \inf_{q_h \in M_h} \|p - q_h\|_M,$$

with $c_{1h} = (1 + \frac{\|a\|}{\alpha_h})(1 + \frac{\|b\|}{\beta_h})$, $c_{2h} = \frac{\|b\|}{\alpha_h}$ *if* $\mathrm{Ker}(B_h) \not\subset \mathrm{Ker}(B)$ *and* $c_{2h} = 0$ *otherwise,* $c_{3h} = c_{1h} \frac{\|a\|}{\beta_h}$, *and* $c_{4h} = 1 + \frac{\|b\|}{\beta_h} + c_{2h} \frac{\|a\|}{\beta_h}$.

Proof. Introduce the notation

$$Z_h(g) = \{w_h \in X_h; \, \forall q_h \in M_h, \, b(w_h, q_h) = g(q_h)\}. \qquad (2.37)$$

Clearly, $Z_h(g)$ is non-empty because the operator B_h is surjective. Let v_h be arbitrary in X_h. Since B_h verifies (2.36), the reciprocal of Lemma A.42 implies the existence of r_h in X_h such that

$$\forall q_h \in M_h, \, b(r_h, q_h) = b(u - v_h, q_h) \quad \text{and} \quad \beta_h \|r_h\|_X \leq \|b\| \, \|u - v_h\|_X.$$

It is clear that $b(r_h + v_h, q_h) = g(q_h)$, showing that $r_h + v_h$ is in $Z_h(g)$. Let $w_h = r_h + v_h$. Since w_h is in $Z_h(g)$, $u_h - w_h$ is in $\mathrm{Ker}(B_h)$, yielding

$$\begin{aligned}
\alpha_h \|u_h - w_h\|_X &\leq \sup_{y_h \in \mathrm{Ker}(B_h)} \frac{a(u_h - w_h, y_h)}{\|y_h\|_X} \\
&\leq \sup_{y_h \in \mathrm{Ker}(B_h)} \frac{a(u_h - u, y_h) + a(u - w_h, y_h)}{\|y_h\|_X} \\
&\leq \sup_{y_h \in \mathrm{Ker}(B_h)} \frac{b(y_h, p - p_h) + a(u - w_h, y_h)}{\|y_h\|_X}.
\end{aligned}$$

If $\mathrm{Ker}(B_h) \subset \mathrm{Ker}(B)$, then $b(y_h, p - p_h) = 0$ for $y_h \in \mathrm{Ker}(B_h)$; hence,

$$\alpha_h \|u_h - w_h\|_X \leq \|a\| \|u - w_h\|_X.$$

Using the triangle inequality yields

$$\|u - u_h\|_X \leq \left(1 + \frac{\|a\|}{\alpha_h}\right) \|u - w_h\|_X.$$

In the general case, $b(y_h, p_h) = 0 = b(y_h, q_h)$ for all $q_h \in M_h$ since y_h is in $\mathrm{Ker}(B_h)$, implying

$$\alpha_h \|u_h - w_h\|_X \leq \|a\| \|u - w_h\|_X + \|b\| \|p - q_h\|_M.$$

Using the triangle inequality yields

$$\|u - u_h\|_X \leq \left(1 + \frac{\|a\|}{\alpha_h}\right) \|u - w_h\|_X + \frac{\|b\|}{\alpha_h} \|p - q_h\|_M.$$

The estimate on $\|u - u_h\|_X$ then results from the inequality

$$\|u - w_h\|_X \leq \|u - v_h\|_X + \|r_h\|_X \leq \left(1 + \frac{\|b\|}{\beta_h}\right) \|u - v_h\|_X.$$

We now estimate $\|p - p_h\|_M$. Since $b(v_h, p - p_h) = a(u_h - u, v_h)$ for all v_h in X_h, we can introduce an arbitrary $q_h \in M_h$ to obtain

$$\forall v_h \in X_h, \quad b(v_h, q_h - p_h) = a(u_h - u, v_h) + b(v_h, q_h - p).$$

Condition (2.36) then implies

$$\beta_h \|q_h - p_h\|_M \leq \|a\| \|u - u_h\|_X + \|b\| \|p - q_h\|_M.$$

The final result readily follows from the triangle inequality. $\square$

We now establish an error estimate based on the Aubin–Nitsche Lemma. To this end, we introduce the following assumptions:

(ANM1) There exists a Hilbert space H into which X can be continuously embedded. Denote by $\|\cdot\|_H$ and $(\cdot, \cdot)_H$ the norm and the scalar product in H, respectively. We further assume that there exist two Banach spaces $Y \subset X$ and $N \subset M$ and a stability constant $c_S > 0$ such that, for all $g \in H$, the solution to the adjoint problem:

$$\begin{cases} \text{Seek } \varphi(g) \in X \text{ and } \vartheta(g) \in M \text{ such that} \\ a(v, \varphi(g)) + b(v, \vartheta(g)) = (g, v)_H, & \forall v \in X, \\ b(\varphi(g), q) = 0, & \forall q \in M, \end{cases}$$

satisfies the a priori estimate $\|\varphi(g)\|_Y + \|\vartheta(g)\|_N \leq c_S \|g\|_H$.

(ANM2) There exists an interpolation constant $c_i > 0$ such that, for all h and $(v, q) \in Y \times N$,

$$\inf_{(v_h, q_h) \in X_h \times M_h} (\|v - v_h\|_X + \|q - q_h\|_M) \leq c_i h(\|v\|_Y + \|q\|_N).$$

Lemma 2.45. *Under the assumptions* (ANM1)–(ANM2), *there is c such that*

$$\forall h, \quad \|u - u_h\|_H \le c\, h(\|u - u_h\|_X + \|p - p_h\|_M).$$

Proof. Set $V = X \times M$, $Z = Y \times N$, and $L = H \times M$ equipped with the product norms. Define the symmetric positive bilinear form $l((v,q),(w,r)) = (v,w)_H$ and the seminorm $|(v,q)|_L = \|v\|_H$. To conclude, apply Lemma 2.31 using the bilinear form $c((u,p),(v,q)) = a(u,v) + b(v,p) + b(u,q)$.   □

2.5 Exercises

Exercise 2.1. Let V and W be two Banach spaces and let $a \in \mathcal{L}(W \times V; \mathbb{R})$. Let $A : W \to V'$ be the mapping defined in (2.6). Show that $\|A\|_{\mathcal{L}(W;V')} = \|a\|_{W,V}$.

Exercise 2.2. Use Proposition 2.4 to prove Proposition A.31.

Exercise 2.3. Prove Lemmas A.39 and A.40. (*Hint*: Use the Closed Range Theorem and the Open Mapping Theorem.)

Exercise 2.4. Let V be a real Hilbert space equipped with the scalar product $(\cdot,\cdot)_V$ and norm $\|\cdot\|_V$. Let U be a nonempty, closed, and convex subset of V.

(i) Let $f \in V$. Show that there is a unique u in V such that $\|f - u\|_V = \min_{v \in V} \|f - v\|_V$. (*Hint*: Consider a minimizing sequence and show that it is a Cauchy sequence.)

(ii) Show that u is the solution to the above minimization problem if and only if $(f - u, v - u)_V \le 0$ for all $v \in U$.

(iii) Let a be a continuous, symmetric, and V-coercive bilinear form. Let L be a continuous linear form on V. Set $J(v) = \frac{1}{2}a(v,v) - L(v)$. Show that there is a unique $u \in V$ such that $J(u) = \min_{v \in V} J(v)$ and that u is a minimizing solution if and only if $a(u, v - u) \ge L(v - u)$ for all $v \in U$.

Exercise 2.5. Use the notation and results of Exercise 2.4. Let u be the unique element in V such that $a(u, v - u) \ge L(v - u)$ for all $v \in U$. Let V_h be a finite-dimensional subspace of V, and let U_h be a nonempty, closed, and convex subset of V_h. Owing to Exercise 2.4, there is a unique u_h in V_h such that $a(u_h, v_h - u_h) \ge L(v_h - u_h)$ for all $v_h \in U_h$.

(i) Show that there is $c_1(u)$ such that, for all $v \in U$,

$$\|u - u_h\|_V^2 \le c_1(u)\big(\|u - v_h\|_V + \|u_h - v\|_V + \|u - u_h\|_V\|u - v_h\|_V\big).$$

(*Hint*: Prove $\alpha\|u - u_h\|_V^2 \le a(u, v - u_h) - L(v - u_h) + a(u_h, v_h - u) - L(v_h - u).$)

(ii) Show that there is $c_2(u)$ such that

$$\|u - u_h\|_V \le c_2(u)\Big(\inf_{v_h \in U_h}\big(\|u - v_h\|_V + \|u - v_h\|_V^2\big) + \inf_{v \in U}\|u_h - v\|_V\Big)^{\frac{1}{2}}.$$

Exercise 2.6. Prove Lemmas A.39 and A.40.

Exercise 2.7. Let $\mathcal{A} \in \mathbb{R}^{N,N}$ be a non-singular matrix. Show that

$$\min_{w \in \mathbb{R}^N} \max_{v \in \mathbb{R}^N} \frac{(\mathcal{A}w, v)_N}{\|v\|_N \|w\|_N} = \min_{v \in \mathbb{R}^N} \max_{w \in \mathbb{R}^N} \frac{(\mathcal{A}^T v, w)_N}{\|v\|_N \|w\|_N} > 0.$$

Does this property still hold when $A \in \mathcal{L}(W; V')$ is a bijective Banach operator?

Exercise 2.8. Let V be a Banach space. Prove that V can be equipped with a Hilbert structure with the same topology if and only if there is a coercive operator in $\mathcal{L}(V; V')$. (*Hint:* Think of $\langle Au, v\rangle_{V',V} + \langle Av, u\rangle_{V',V}$.)

Exercise 2.9. Let V be a reflexive Banach space and let $A \in \mathcal{L}(V; V')$ be a monotone self-adjoint operator; see §A.2.4. Prove that A is bijective if and only if A is coercive. (*Hint:* Prove that if A is monotone and self-adjoint, the following inequality holds:

$$\forall v, w \in V, \quad \langle Av, w\rangle_{V',V} \leq \langle Av, v\rangle_{V',V}^{\frac{1}{2}} \langle Aw, w\rangle_{V',V}^{\frac{1}{2}};$$

then, use this inequality in the inf-sup condition satisfied by A.)

Exercise 2.10. Let $a \in \mathcal{L}(V \times V; \mathbb{R})$ be a symmetric coercive bilinear form on a Hilbert space V. Explain why the Lax–Milgram Lemma is nothing more than a rephrasing of the Riesz–Fréchet Theorem.

Exercise 2.11. The goal of this exercise is to prove the Lax–Milgram Lemma without using the BNB Theorem. Assume the hypotheses of the Lax–Milgram Lemma and let

$$A : V \ni u \longmapsto a(u, \cdot) \in V'.$$

(i) Prove that coercivity implies $\|Au\|_{V'} \geq \alpha \|u\|_V$.
(ii) Prove that A is injective and $\mathrm{Im}(A)$ is closed.
(iii) Prove that $\mathrm{Im}(A)$ is dense in V'. (*Hint:* Use Corollary A.18.)
(iv) Conclude.

Exercise 2.12. Complete the proof of Proposition 2.36.

Exercise 2.13. Prove Proposition 6.55.

Exercise 2.14. Let X_1, X_2, M_1, and M_2 be four reflexive Banach spaces, $f \in X_2'$, $g \in M_2'$. Let $A \in \mathcal{L}(X_1; X_2')$, $B_1 \in \mathcal{L}(X_2; M_1')$, and $B_2 \in \mathcal{L}(X_1; M_2')$. Consider the problem:

$$\begin{cases} \text{Seek } u \in X_1 \text{ and } p \in M_1 \text{ such that} \\ Au + B_1^T p = f, \\ B_2 u = g. \end{cases}$$

Prove that this problem is well-posed if and only if

$$
\begin{cases}
\exists \alpha > 0, \quad \inf_{u \in \mathrm{Ker}(B_2)} \sup_{v \in \mathrm{Ker}(B_1)} \dfrac{\langle Au, v \rangle_{X_2', X_2}}{\|u\|_{X_1} \|v\|_{X_2}} \geq \alpha, \\[2mm]
\forall v \in \mathrm{Ker}(B_1), \ (\forall u \in \mathrm{Ker}(B_2), \ \langle Au, v \rangle_{X_2', X_2} = 0) \Rightarrow (v = 0), \\[2mm]
\exists \beta_1 > 0, \quad \inf_{q \in M_1} \sup_{v \in X_2} \dfrac{\langle B_1 v, q \rangle_{M_1', M_1}}{\|v\|_{X_2} \|q\|_{M_1}} \geq \beta_1, \\[2mm]
\exists \beta_2 > 0, \quad \inf_{q \in M_2} \sup_{v \in X_1} \dfrac{\langle B_2 v, q \rangle_{M_2', M_2}}{\|v\|_{X_1} \|q\|_{M_2}} \geq \beta_2.
\end{cases}
$$

Part II

Approximation of PDEs

3

Coercive Problems

This chapter deals with problems whose weak formulation is endowed with a coercivity property. The key examples investigated henceforth are scalar elliptic PDEs, spectral problems associated with the Laplacian, and PDE systems derived from continuum mechanics. The goal is twofold: First, to set up a mathematical framework for well-posedness; then, to investigate conformal and non-conformal finite element approximations based on Galerkin methods. Error estimates are derived from the theoretical results of Chapters 1 and 2 and are illustrated numerically. The last section of this chapter is concerned with coercivity loss and is meant to be a transition to Chapters 4 and 5.

3.1 Scalar Elliptic PDEs: Theory

Let Ω be a domain in $\mathbb{R}^d$. Consider a differential operator $\mathcal{L}$ in the form

$$\mathcal{L}u = -\nabla\cdot(\sigma\cdot\nabla u) + \beta\cdot\nabla u + \mu u, \qquad (3.1)$$

where σ, β, and μ are functions defined over Ω and taking their values in $\mathbb{R}^{d,d}$, $\mathbb{R}^d$, and $\mathbb{R}$, respectively. Given a function $f : \Omega \to \mathbb{R}$, consider the problem of finding a function $u : \Omega \to \mathbb{R}$ such that

$$\begin{cases} \mathcal{L}u = f & \text{in } \Omega, \\ \mathcal{B}u = g & \text{on } \partial\Omega, \end{cases} \qquad (3.2)$$

where the operator $\mathcal{B}$ accounts for boundary conditions. The model problem (3.2) arises in several applications:

(i) *Heat transfer:* u is the temperature, $\sigma = \kappa\mathcal{I}$ where κ is the thermal conductivity, β is the flow field, $\mu = 0$, and f is the externally supplied heat per unit volume.

(ii) *Advection–diffusion*: u is the concentration of a solute transported in a flow field β. The matrix σ models the solute diffusivity resulting from either molecular diffusion or turbulent mixing by the carrier flow. Solute production or destruction by chemical reaction is accounted for by the linear term μu, and the right-hand side f models fixed sources or sinks.

Henceforth, the following assumptions are made on the data: $f \in L^2(\Omega)$, $\sigma \in [L^\infty(\Omega)]^{d,d}$, $\beta \in [L^\infty(\Omega)]^d$, $\nabla \cdot \beta \in L^\infty(\Omega)$, and $\mu \in L^\infty(\Omega)$. Furthermore, the operator $\mathcal{L}$ is assumed to be *elliptic* in the following sense:

Definition 3.1. *The operator $\mathcal{L}$ defined in (3.1) is said to be* elliptic *if there exists $\sigma_0 > 0$ such that*

$$\forall \xi \in \mathbb{R}^d, \qquad \sum_{i,j=1}^d \sigma_{ij} \xi_i \xi_j \geq \sigma_0 \|\xi\|_d^2 \quad a.e.\ in\ \Omega. \tag{3.3}$$

Equation (3.2) is then called an elliptic PDE.

Example 3.2. A fundamental example of an elliptic operator is the *Laplacian*, $\mathcal{L} = -\Delta$, which is obtained for $\sigma = \mathcal{I}$, $\beta = 0$, and $\mu = 0$.   □

3.1.1 Review of boundary conditions and their weak formulation

We first proceed formally and then specify the mathematical framework for the weak formulation.

Homogeneous Dirichlet boundary condition. We want to enforce $u = 0$ on $\partial\Omega$. Multiplying the PDE $\mathcal{L}u = f$ by a (sufficiently smooth) test function v vanishing at the boundary, integrating over Ω, and using the Green formula

$$\int_\Omega -\nabla \cdot (\sigma \cdot \nabla u)\, v = \int_\Omega \nabla v \cdot \sigma \cdot \nabla u - \int_{\partial\Omega} v\, (n \cdot \sigma \cdot \nabla u), \tag{3.4}$$

yields

$$\int_\Omega \nabla v \cdot \sigma \cdot \nabla u + v(\beta \cdot \nabla u) + \mu u v = \int_\Omega f v.$$

A possible regularity requirement on u and v for the integrals over Ω to be meaningful is

$$u \in H^1(\Omega) \qquad \text{and} \qquad v \in H^1(\Omega).$$

Since $u \in H^1(\Omega)$, Theorem B.52 implies that u has a trace at the boundary. Because of the boundary condition $u_{|\partial\Omega} = 0$, the solution is sought in $H_0^1(\Omega)$. Test functions are also taken in $H_0^1(\Omega)$, leading to the following weak formulation:

$$\begin{cases} \text{Seek } u \in H_0^1(\Omega) \text{ such that} \\ a_{\sigma,\beta,\mu}(u, v) = \int_\Omega f v, \quad \forall v \in H_0^1(\Omega), \end{cases} \tag{3.5}$$

with the bilinear form

$$a_{\sigma,\beta,\mu}(u, v) = \int_\Omega \nabla v \cdot \sigma \cdot \nabla u + v(\beta \cdot \nabla u) + \mu u v. \tag{3.6}$$

Proposition 3.3. *If u solves (3.5), then $\mathcal{L}u = f$ a.e. in Ω and $u = 0$ a.e. on $\partial\Omega$.*

Proof. Let $\varphi \in \mathcal{D}(\Omega)$ and let u be a solution to (3.5). Hence,

$$\langle -\nabla\cdot(\sigma\cdot\nabla u), \varphi\rangle_{\mathcal{D}',\mathcal{D}} = \langle \sigma\cdot\nabla u, \nabla\varphi\rangle_{\mathcal{D}',\mathcal{D}} = \int_\Omega \nabla\varphi\cdot\sigma\cdot\nabla u$$

$$= \int_\Omega (f - \beta\cdot\nabla u - \mu u)\,\varphi,$$

yielding $\langle \mathcal{L}u, \varphi\rangle_{\mathcal{D}',\mathcal{D}} = \int_\Omega f\varphi$. Owing to the density of $\mathcal{D}(\Omega)$ in $L^2(\Omega)$, $\mathcal{L}u = f$ in $L^2(\Omega)$. Therefore, $\mathcal{L}u = f$ a.e. in Ω. Moreover, $u = 0$ a.e. on $\partial\Omega$ by definition of $H_0^1(\Omega)$; see Theorem B.52. □

Non-homogeneous Dirichlet boundary condition. We want to enforce $u = g$ on $\partial\Omega$, where $g : \partial\Omega \to \mathbb{R}$ is a given function. We assume that g is sufficiently smooth so that there exists a lifting u_g of g in $H^1(\Omega)$, i.e., a function $u_g \in H^1(\Omega)$ such that $u_g = g$ on $\partial\Omega$; see §2.1.4. We obtain the weak formulation:

$$\begin{cases} \text{Seek } u \in H^1(\Omega) \text{ such that} \\ u = u_g + \phi, \quad \phi \in H_0^1(\Omega), \\ a_{\sigma,\beta,\mu}(\phi, v) = \int_\Omega fv - a_{\sigma,\beta,\mu}(u_g, v), \quad \forall v \in H_0^1(\Omega). \end{cases} \quad (3.7)$$

Proposition 3.4. *Let $g \in H^{\frac{1}{2}}(\partial\Omega)$. If u solves (3.7), then $\mathcal{L}u = f$ a.e. in Ω and $u = g$ a.e. on $\partial\Omega$.*

Proof. Similar to that of Proposition 3.3. □

When the operator $\mathcal{L}$ is the Laplacian, (3.7) is called a *Poisson problem*.

Neumann boundary condition. Given a function $g : \partial\Omega \to \mathbb{R}$, we want to enforce $n\cdot\sigma\cdot\nabla u = g$ on $\partial\Omega$. Note that in the case $\sigma = \mathcal{I}$, the Neumann condition specifies the normal derivative of u since $n\cdot\nabla u = \partial_n u$. Proceeding as before and using the Neumann condition in the surface integral in (3.4) yields the weak formulation:

$$\begin{cases} \text{Seek } u \in H^1(\Omega) \text{ such that} \\ a_{\sigma,\beta,\mu}(u, v) = \int_\Omega fv + \int_{\partial\Omega} gv, \quad \forall v \in H^1(\Omega). \end{cases} \quad (3.8)$$

Proposition 3.5. *Let $g \in L^2(\partial\Omega)$. If u solves (3.8), then $\mathcal{L}u = f$ a.e. in Ω and $n\cdot\sigma\cdot\nabla u = g$ a.e. on $\partial\Omega$.*

Proof. Taking test functions in $\mathcal{D}(\Omega)$ readily implies $\mathcal{L}u = f$ a.e. in Ω. Therefore, $-\nabla\cdot(\sigma\cdot\nabla u) \in L^2(\Omega)$. Corollary B.59 implies $n\cdot\sigma\cdot\nabla u \in H^{\frac{1}{2}}(\partial\Omega)' = H^{-\frac{1}{2}}(\partial\Omega)$ since

$$\forall \phi \in H^{\frac{1}{2}}(\partial\Omega), \quad \langle n \cdot \sigma \cdot \nabla u, \phi \rangle_{H^{-\frac{1}{2}}, H^{\frac{1}{2}}} = \int_\Omega -\nabla \cdot (\sigma \cdot \nabla u) u_\phi + \int_\Omega \nabla u_\phi \cdot \sigma \cdot \nabla u,$$

where $u_\phi \in H^1(\Omega)$ is a lifting of ϕ in $H^1(\Omega)$. Then, (3.8) yields

$$\forall \phi \in H^{\frac{1}{2}}(\partial\Omega), \quad \langle n \cdot \sigma \cdot \nabla u, \phi \rangle_{H^{-\frac{1}{2}}, H^{\frac{1}{2}}} = \int_{\partial\Omega} g\phi,$$

showing that $n \cdot \sigma \cdot \nabla u = g$ in $H^{-\frac{1}{2}}(\partial\Omega)$ and, therefore, in $L^2(\partial\Omega)$ since g belongs to this space.   □

Mixed Dirichlet–Neumann boundary conditions. Consider a partition of the boundary in the form $\partial\Omega = \partial\Omega_D \cup \partial\Omega_N$. Impose a Dirichlet condition on $\partial\Omega_D$ and a Neumann condition on $\partial\Omega_N$. If the Dirichlet condition is non-homogeneous, assume that $\partial\Omega_D$ is smooth enough so that, for all $g \in H^{\frac{1}{2}}(\partial\Omega_D)$, there exists an extension $\widetilde{g} \in H^{\frac{1}{2}}(\partial\Omega)$ such that $\widetilde{g}_{|\partial\Omega_D} = g$ and $\|\widetilde{g}\|_{H^{\frac{1}{2}}(\partial\Omega)} \leq c\|g\|_{H^{\frac{1}{2}}(\partial\Omega_D)}$ uniformly in g. Then, using the lifting of $\widetilde{g}$ in $H^1(\Omega)$, one can assume that the Dirichlet condition is homogeneous. The boundary conditions are thus

$$\begin{cases} u = 0 & \text{on } \partial\Omega_D, \\ n \cdot \sigma \cdot \nabla u = g & \text{on } \partial\Omega_N, \end{cases}$$

with a given function $g : \partial\Omega_N \to \mathbb{R}$.

Proceeding as before, we split the boundary integral in (3.4) into its contributions over $\partial\Omega_D$ and $\partial\Omega_N$. Taking the solution and the test function in the functional space

$$H^1_{\partial\Omega_D}(\Omega) = \{ u \in H^1(\Omega); \, u = 0 \text{ on } \partial\Omega_D \},$$

the surface integral over $\partial\Omega_D$ vanishes. Furthermore, using the Neumann condition in the surface integral over $\partial\Omega_N$ yields the weak formulation:

$$\begin{cases} \text{Seek } u \in H^1_{\partial\Omega_D}(\Omega) \text{ such that} \\ a_{\sigma,\beta,\mu}(u,v) = \int_\Omega fv + \int_{\partial\Omega_N} gv, \quad \forall v \in H^1_{\partial\Omega_D}(\Omega). \end{cases} \tag{3.9}$$

Proposition 3.6. *Let $\partial\Omega_D \subset \partial\Omega$, assume $\mathrm{meas}(\partial\Omega_D) > 0$, and set $\partial\Omega_N = \partial\Omega \backslash \partial\Omega_D$. Let $g \in L^2(\partial\Omega_N)$. If u solves (3.9), then $\mathcal{L}u = f$ a.e. in Ω, $u = 0$ a.e. on $\partial\Omega_D$, and $(n \cdot \sigma \cdot \nabla u) = g$ a.e. on $\partial\Omega_N$.*

Proof. Proceed as in the previous proofs.   □

Robin boundary condition. Given two functions $g, \gamma : \partial\Omega \to \mathbb{R}$, we want to enforce $\gamma u + n \cdot \sigma \cdot \nabla u = g$ on $\partial\Omega$. Using this condition in the surface integral in (3.4) yields the weak formulation:

$$\begin{cases} \text{Seek } u \in H^1(\Omega) \text{ such that} \\ a_{\sigma,\beta,\mu}(u,v) + \int_{\partial\Omega} \gamma uv = \int_\Omega fv + \int_{\partial\Omega} gv, \quad \forall v \in H^1(\Omega). \end{cases} \tag{3.10}$$

Problem	V	$a(u,v)$	$f(v)$
Homogeneous Dirichlet	$H_0^1(\Omega)$	$a_{\sigma,\beta,\mu}(u,v)$	$\int_\Omega fv$
Neumann	$H^1(\Omega)$	$a_{\sigma,\beta,\mu}(u,v)$	$\int_\Omega fv + \int_{\partial\Omega} gv$
Dirichlet–Neumann	$H_{\partial\Omega_D}^1(\Omega)$	$a_{\sigma,\beta,\mu}(u,v)$	$\int_\Omega fv + \int_{\partial\Omega_N} gv$
Robin	$H^1(\Omega)$	$a_{\sigma,\beta,\mu}(u,v) + \int_{\partial\Omega} \gamma uv$	$\int_\Omega fv + \int_{\partial\Omega} gv$

Table 3.1. Weak formulation corresponding to the various boundary conditions for the second-order PDE (3.2). The bilinear form $a_{\sigma,\beta,\mu}(u,v)$ is defined in (3.6).

Proposition 3.7. *Let* $g \in L^2(\partial\Omega)$ *and let* $\gamma \in L^\infty(\partial\Omega)$. *If* u *solves* (3.10), *then* $\mathcal{L}u = f$ *a.e. in* Ω *and* $\gamma u + n\cdot\sigma\cdot\nabla u = g$ *a.e. on* $\partial\Omega$.

Proof. Proceed as in the previous proofs. □

Summary. Except for the non-homogeneous Dirichlet problem, all the problems considered herein take the generic form:

$$\begin{cases} \text{Seek } u \in V \text{ such that} \\ a(u,v) = f(v), \quad \forall v \in V, \end{cases} \tag{3.11}$$

where V is a Hilbert space satisfying

$$H_0^1(\Omega) \subset V \subset H^1(\Omega).$$

Moreover, a is a bilinear form defined on $V \times V$, and f is a linear form defined on V; see Table 3.1. For the non-homogeneous Dirichlet problem, $u \in H^1(\Omega)$, $u = u_g + \phi$ where u_g is a lifting of the boundary data and ϕ solves a problem of the form (3.11).

Essential and natural boundary conditions. It is important to observe the different treatment between Dirichlet conditions and Neumann or Robin conditions. The former are imposed explicitly in the functional space where the solution is sought, and the test functions vanish on the corresponding part of the boundary. For this reason, Dirichlet conditions are often termed *essential boundary conditions*. Neumann and Robin conditions are not imposed by the functional setting but by the weak formulation itself. The fact that test functions have degrees of freedom on the corresponding part of the boundary is sufficient to enforce the boundary conditions in question. For this reason, these conditions are often termed *natural boundary conditions*. Note that it is also possible to treat Dirichlet conditions as natural boundary conditions by using a penalty method; see §8.4.3.

3.1.2 Coercivity

Theorem 3.8. *Let* $f \in L^2(\Omega)$, *let* $\sigma \in [L^\infty(\Omega)]^{d,d}$ *be such that* (3.3) *holds, let* $\beta \in [L^\infty(\Omega)]^d$ *with* $\nabla\cdot\beta \in L^\infty(\Omega)$, *and let* $\mu \in L^\infty(\Omega)$. *Set*

$p = \inf \operatorname{ess}_{x \in \Omega} \left(\mu - \frac{1}{2} \nabla \cdot \beta \right)$ and let c_Ω be the constant in the Poincaré inequality (B.23).

(i) Both the homogeneous Dirichlet problem (3.5) and the non-homogeneous Dirichlet problem (3.7) are well-posed if

$$\sigma_0 + \min\left(0, \tfrac{p}{c_\Omega}\right) > 0. \qquad (3.12)$$

(ii) The Neumann problem (3.8) is well-posed if

$$p > 0 \qquad \text{and} \qquad \inf_{x \in \partial\Omega} \operatorname{ess}(\beta \cdot n) \geq 0. \qquad (3.13)$$

(iii) The mixed Dirichlet–Neumann problem (3.9) is well-posed if (3.12) holds, $\operatorname{meas}(\partial\Omega_D) > 0$, and $\partial\Omega^- = \{x \in \partial\Omega; \, (\beta \cdot n)(x) < 0\} \subset \partial\Omega_D$.

(iv) Set $q = \inf \operatorname{ess}_{x \in \partial\Omega}(\gamma + \frac{1}{2}\beta \cdot n)$. The Robin problem (3.10) is well-posed if

$$p \geq 0, \quad q \geq 0, \quad \text{and} \quad pq \neq 0. \qquad (3.14)$$

Proof. We prove (i) and (iv) only, leaving the remaining items as an exercise.
(1) Proof of (i). Using the ellipticity of $\mathcal{L}$ and the identity

$$\int_\Omega u(\beta \cdot \nabla u) = -\tfrac{1}{2} \int_\Omega (\nabla \cdot \beta) u^2 + \tfrac{1}{2} \int_{\partial\Omega} (\beta \cdot n) u^2,$$

which is a direct consequence of the divergence formula (B.19), yields

$$\forall u \in H_0^1(\Omega), \quad a_{\sigma,\beta,\mu}(u,u) \geq \sigma_0 |u|_{1,\Omega}^2 + p\|u\|_{0,\Omega}^2.$$

Setting $\delta = \min(0, \frac{p}{c_\Omega})$ and using the Poincaré inequality (B.23) yields

$$\forall u \in H_0^1(\Omega), \quad a_{\sigma,\beta,\mu}(u,u) \geq \left(\sigma_0 + \frac{\delta}{c_\Omega}\right) |u|_{1,\Omega}^2 \geq \alpha\|u\|_{1,\Omega}^2,$$

with $\alpha = \frac{c_\Omega(c_\Omega\sigma_0 + \delta)}{1 + c_\Omega^2}$, showing that the bilinear form $a_{\sigma,\beta,\mu}$ is coercive on $H_0^1(\Omega)$. The well-posedness of the homogeneous Dirichlet problem then results from the Lax–Milgram Lemma, while that of the non-homogeneous Dirichlet problem results from Proposition 2.10.
(2) Proof of (iv). Let $a(u,v) = a_{\sigma,\beta,\mu}(u,v) + \int_{\partial\Omega} \gamma uv$. A straightforward calculation shows that

$$\forall u \in H^1(\Omega), \quad a(u,u) \geq \sigma_0 |u|_{1,\Omega}^2 + p\|u\|_{0,\Omega}^2 + q\|u\|_{0,\partial\Omega}^2.$$

If $p > 0$ and $q \geq 0$, the bilinear form a is clearly coercive on $H^1(\Omega)$ with constant $\alpha = \min(\sigma_0, p)$. If $p \geq 0$ and $q > 0$, the coercivity of a is readily deduced from Lemma B.63. In both cases, well-posedness then results from the Lax–Milgram Lemma. $\qquad \square$

Remark 3.9.

(i) For the homogeneous and the non-homogeneous Dirichlet problem, f can be taken in $H^{-1}(\Omega) = (H_0^1(\Omega))'$. In this case, the right-hand side in (3.11) becomes $f(v) = \langle f, v \rangle_{H^{-1}, H_0^1}$, and the problem is still well-posed. The stability estimate takes the form $\|u\|_{1,\Omega} \leq c \|f\|_{-1,\Omega}$.

(ii) Consider the Laplacian with homogeneous Dirichlet boundary conditions, i.e., given $f \in H^{-1}(\Omega)$, solve $-\Delta u = f$ in Ω with the boundary condition $u_{|\partial \Omega} = 0$. Then, the weak formulation of this problem amounts to seeking $u \in H_0^1(\Omega)$ such that $\int_\Omega \nabla u \cdot \nabla v = \langle f, v \rangle_{H^{-1}, H_0^1}$ for all $v \in H_0^1(\Omega)$. Owing to Theorem 3.8(i) with $\beta = 0$, $\sigma = \mathcal{I}$, and $\mu = 0$, this problem is well-posed. This means that the operator $(-\Delta)^{-1} : H^{-1}(\Omega) \to H_0^1(\Omega)$ is an isomorphism.

(iii) Uniqueness is not a trivial property in spaces larger than $H^1(\Omega)$. For instance, one can construct domains in which this property does not hold in L^2 for the Dirichlet problem; see Exercise 3.4.

(iv) Consider problem (3.11). If the advection field β vanishes and if the diffusion matrix σ is symmetric a.e. in Ω, the bilinear form a is symmetric and positive. Therefore, owing to Proposition 2.4, (3.11) can be reformulated into a *variational form*. For the homogeneous Dirichlet problem, the variational form in question is

$$\min_{v \in H_0^1(\Omega)} \left(\frac{1}{2} \int_\Omega \nabla v \cdot \sigma \cdot \nabla v + \frac{1}{2} \int_\Omega \mu v^2 - \int_\Omega f v \right).$$

The case of other boundary conditions is left as an exercise.

(v) When μ and β vanish, the solution to the Neumann problem (3.8) is defined up to an additive constant. Therefore, we decide to seek a solution with zero-mean over Ω. Accordingly, we introduce the space

$$H_{\int=0}^1(\Omega) = \left\{ v \in H^1(\Omega); \int_\Omega v = 0 \right\}.$$

To ensure the existence of a solution, the data f and g must satisfy a compatibility relation. Owing to the fact that $\int_\Omega f = -\int_\Omega \nabla \cdot (\sigma \cdot \nabla u) = -\int_{\partial \Omega} n \cdot \sigma \cdot \nabla u = -\int_{\partial \Omega} g$, the compatibility condition is

$$\int_\Omega f + \int_{\partial \Omega} g = 0. \tag{3.15}$$

Thus, the weak formulation of the purely diffusive Neumann problem is:

$$\begin{cases} \text{Seek } u \in H_{\int=0}^1(\Omega) \text{ such that} \\ \int_\Omega \nabla v \cdot \sigma \cdot \nabla u = \int_\Omega f v + \int_{\partial \Omega} g v, \quad \forall v \in H_{\int=0}^1(\Omega). \end{cases} \tag{3.16}$$

Test functions have also been restricted to the functional space $H_{\int=0}^1(\Omega)$. Indeed, owing to (3.15), a constant test function leads to the trivial equation "$0 = 0$." Moreover, under the conditions (3.3) and (3.15), assuming that the

data satisfy $f \in L^2(\Omega)$ and $g \in L^2(\partial\Omega)$, and using Lemma B.66, one readily verifies that problem (3.16) is well-posed with a stability estimate of the form
$$\forall f \in L^2(\Omega), \ \forall g \in L^2(\partial\Omega), \ \|u\|_{1,\Omega} \le c \left(\|f\|_{0,\Omega} + \|g\|_{0,\partial\Omega} \right). \qquad \square$$

3.1.3 Smoothing properties

We have seen that the natural functional space V in which to seek the solution to (3.11) is such that $H_0^1(\Omega) \subset V \subset H^1(\Omega)$. For sufficiently smooth data, stronger regularity results can be derived. The interest of these results stems from the fact that in the framework of finite element methods, the regularity of the exact solution directly controls the convergence rate of the approximate solution; see §3.2.5 for numerical illustrations. In this section, it is implicitly assumed that the hypotheses of Theorem 3.8 hold so that the problems considered henceforth are well-posed. This section is set at an introductory level; see, e.g., [Gri85, Gri92, CoD02] for further insight.

Theorem 3.10 (Domain with smooth boundary). *Let $m \ge 0$, let Ω be a domain of class C^{m+2}, and let $f \in H^m(\Omega)$. Assume that the coefficients σ_{ij} are in $C^{m+1}(\overline{\Omega})$ and that the coefficients β_i and μ are in $C^m(\overline{\Omega})$. Then:*

(i) *The solution to the homogeneous Dirichlet problem (3.5) is in $H^{m+2}(\Omega)$.*

(ii) *Assuming $g \in H^{m+\frac{3}{2}}(\partial\Omega)$, the solution to the non-homogeneous Dirichlet problem (3.7) is in $H^{m+2}(\Omega)$.*

(iii) *Assuming $g \in H^{m+\frac{1}{2}}(\partial\Omega)$, the solution to the Neumann problem (3.8) is in $H^{m+2}(\Omega)$.*

(iv) *Assuming $g \in H^{m+\frac{1}{2}}(\partial\Omega)$ and $\gamma \in C^{m+1}(\partial\Omega)$, the solution to the Robin problem (3.10) is in $H^{m+2}(\Omega)$.*

Remark 3.11.

(i) The reader who is not familiar with Sobolev spaces involving fractional exponents may replace an assumption such as $g \in H^{m+\frac{3}{2}}(\partial\Omega)$ by $g \in C^{m+1}(\partial\Omega)$ and $g^{(m+1)} \in C^{0,1}(\partial\Omega)$; see Example B.32(ii).

(ii) There is no regularity result for the mixed Dirichlet–Neumann problem. Indeed, even if f, g, and the domain Ω are smooth, the solution u may not necessarily belong to $H^2(\Omega)$. For instance, in two dimensions, the solution to $-\Delta u = 0$ on the upper half-plane $\{x_2 > 0\}$ with the mixed Dirichlet–Neumann conditions

$$\partial_2 u = 0, \qquad \text{for } x_1 \le 0 \text{ and } x_2 = 0,$$

$$u = r^{\frac{1}{2}} \sin(\tfrac{1}{2}\theta), \quad \text{otherwise},$$

is $u(x_1, x_2) = r^{\frac{1}{2}} \sin(\tfrac{1}{2}\theta)$. Clearly, $u \notin H^2$ owing to the singularity at the origin.

(iii) Theorem 3.10 can be extended to more general Sobolev spaces; see, e.g., [GiR86, pp. 12–15]. For instance, let p be a real satisfying $1 < p < \infty$ and let $m \ge 0$. Let $f \in W^{m,p}(\Omega)$ and $g \in W^{m+2-\frac{1}{p},p}(\partial\Omega)$. Then, the solution to the non-homogeneous Dirichlet problem (3.7) is in $W^{m+2,p}(\Omega)$. $\qquad \square$

Theorem 3.12 (Convex polyhedron). *Let Ω be a convex polyhedron and denote by $\bigcup_{j=1}^{J} \partial\Omega_j$ the set of boundary faces (edges in two dimensions). Assume that the coefficients σ_{ij} are in $\mathcal{C}^1(\overline{\Omega})$ and that the coefficients β_i and μ are in $\mathcal{C}^0(\overline{\Omega})$. Then:*

(i) *The solution to the homogeneous Dirichlet problem (3.5) is in $H^2(\Omega)$.*

(ii) *In dimension 2, if $g \in H^{\frac{3}{2}}(\partial\Omega)$, the solution to the non-homogeneous Dirichlet problem (3.7) is in $H^2(\Omega)$.*

(iii) *In dimension 2, if $g_{|\partial\Omega_j} \in H^{\frac{1}{2}}(\partial\Omega_j)$ for $1 \le j \le J$, the solution to the Neumann problem (3.8) is in $H^2(\Omega)$. In dimension 3, the conclusion still holds if $g = 0$.*

Remark 3.13.

(i) When the polyhedron Ω is not convex, the best regularity result is $u \in H^{\frac{3}{2}}(\Omega)$. In particular, it can be shown (see [Gri85, Gri92]) that in the neighborhood of a vertex S with an interior angle $\omega > \pi$, the solution u to the homogeneous Dirichlet problem can be decomposed into the form

$$u = \Upsilon + \widetilde{u},$$

where $\widetilde{u} \in H^2(\Omega)$ and Υ is a singular function behaving like $r^{\frac{\pi}{\omega}}$ in the neighborhood of S, r being the distance to S.

(ii) Theorem 3.12 can be extended to more general Sobolev spaces. For instance, let p be a real satisfying $1 < p < \infty$, and let $f \in L^p(\Omega)$. Then, the solution to the homogeneous Dirichlet problem (3.5) posed on a convex polyhedron is in $W^{2,p}(\Omega)$.

(iii) The assumption on g in Theorem 3.12(ii) can be weakened as follows: Denote by $\{S_j\}_{1\le j\le J}$ the vertices of $\partial\Omega$ so that $\partial\Omega_j$ is the segment $S_j S_{j+1}$, and conventionally set $S_{J+1} = S_1$ and $\partial\Omega_{J+1} = \partial\Omega_1$. Then, if $g_{|\partial\Omega_j} \in H^{\frac{3}{2}}(\partial\Omega_j)$ and $g_{|\partial\Omega_j}(S_j) = g_{|\partial\Omega_{j+1}}(S_{j+1})$ for all $1 \le j \le J$, the solution to the non-homogeneous Dirichlet problem (3.7) is in $H^2(\Omega)$.

(iv) A regularity result analogous to Theorem 3.12(iii) is valid for the purely diffusive Neumann problem (3.16). □

Definition 3.14 (Smoothing property). *Problem (3.11) is said to have smoothing properties in Ω if assumption (AN1) in §2.3.4 is satisfied with $Z = H^2(\Omega) \cap H_0^1(\Omega)$, $L = L^2(\Omega)$, and $l(\cdot,\cdot) = (\cdot,\cdot)_{0,\Omega}$, i.e., if there exists c_S such that, for all $\varphi \in L^2(\Omega)$, the solution w to the adjoint problem:*

$$\begin{cases} Seek\ w \in V\ such\ that \\ a(v, w) = \int_\Omega \varphi v, \quad \forall v \in V, \end{cases} \tag{3.17}$$

satisfies $\|w\|_{2,\Omega} \le c_S \|\varphi\|_{0,\Omega}$.

Remark 3.15. Because the Laplace operator is self-adjoint, the Laplacian has *smoothing properties in Ω* if the unique solution to the homogeneous Dirichlet problem with $f \in L^2(\Omega)$ is in $H^2(\Omega) \cap H_0^1(\Omega)$, i.e., if the operator $(-\Delta)^{-1} : L^2(\Omega) \to H^2(\Omega) \cap H_0^1(\Omega)$ is an isomorphism. □

3.2 Scalar Elliptic PDEs: Approximation

This section reviews various finite element methods to approximate second-order, scalar, elliptic PDEs. Assume that the well-posedness conditions stated in Theorem 3.8 hold and denote by $u \in V$ the unique solution to (3.11).

3.2.1 H^1-conformal approximation

Let Ω be a polyhedron in $\mathbb{R}^d$, let $\{T_h\}_{h>0}$ be a family of meshes of Ω, and let $\{\widehat{K}, \widehat{P}, \widehat{\Sigma}\}$ be a reference Lagrange finite element of degree $k \geq 1$. Let $L_{c,h}^k$ be the H^1-conformal approximation space defined by

$$L_{c,h}^k = \{v_h \in C^0(\overline{\Omega}); \forall K \in T_h, v_h \circ T_K \in \widehat{P}\}. \tag{3.18}$$

For instance, $L_{c,h}^k = P_{c,h}^k$ or $Q_{c,h}^k$ defined in (1.76) and (1.77), respectively, if a $\mathbb{P}_k$ or $\mathbb{Q}_k$ Lagrange finite element is used. To obtain a V-conformal approximation space, we must account for the boundary conditions, i.e., we set

$$V_h = L_{c,h}^k \cap V. \tag{3.19}$$

This yields $V_h = \{v_h \in L_{c,h}^k; v_h = 0 \text{ on } \partial\Omega\}$ for the homogeneous Dirichlet problem and $V_h = L_{c,h}^k$ for the Neumann and the Robin problems. For the mixed Dirichlet–Neumann problem, we assume, for the sake of simplicity, that $\partial\Omega_D$ is a union of mesh faces; in this case, a suitable approximation space is $V_h = \{v_h \in L_{c,h}^k; v_h = 0 \text{ on } \partial\Omega_D\}$.

Consider the approximate problem:

$$\begin{cases} \text{Seek } u_h \in V_h \text{ such that} \\ a(u_h, v_h) = f(v_h), \quad \forall v_h \in V_h. \end{cases} \tag{3.20}$$

Our goal is to estimate the error $u - u_h$, first in the H^1-norm, then in the L^2-norm, and finally in more general norms.

Theorem 3.16 (H^1-estimate). *Let Ω be a polyhedron in $\mathbb{R}^d$ and let $\{T_h\}_{h>0}$ be a shape-regular family of geometrically conformal meshes of Ω. Let V_h be defined in (3.19). Then, $\lim_{h\to 0} \|u - u_h\|_{1,\Omega} = 0$. Furthermore, if $u \in H^s(\Omega)$ with $\frac{d}{2} < s \leq k+1$, there exists c such that*

$$\forall h, \quad \|u - u_h\|_{1,\Omega} \leq c\, h^{s-1} |u|_{s,\Omega}. \tag{3.21}$$

Proof. Since $s > \frac{d}{2}$, Corollary B.43 implies that u is in the domain of the Lagrange interpolation operator $\mathcal{I}_h^k$ associated with $L_{c,h}^k$. Moreover, $\mathcal{I}_h^k u \in V_h$ since the Lagrange interpolant preserves Dirichlet boundary conditions. As a result, Céa's Lemma yields

$$\|u - u_h\|_{1,\Omega} \leq c \left(\inf_{v_h \in V_h} \|u - v_h\|_{1,\Omega} \right) \leq c \|u - \mathcal{I}_h^k u\|_{1,\Omega}.$$

Owing to Corollary 1.110 (with $p = 2$) and since $s \leq k + 1$,

$$\|u - \mathcal{I}_h^k u\|_{1,\Omega} \leq c\, h^{s-1} |u|_{s,\Omega}.$$

Combining the above inequalities yields (3.21). If $u \in H^1(\Omega)$ only, the convergence of u_h results from the density of $H^s(\Omega) \cap V$ in V. □

Remark 3.17. The assumption $s > \frac{d}{2}$ in Theorem 3.16 can be lifted on simplicial meshes by considering the Clément or the Scott–Zhang interpolation operator instead of the Lagrange interpolation operator; details are left as an exercise. □

For the sake of simplicity, we shall henceforth restrict ourselves to homogeneous Dirichlet conditions.

Theorem 3.18 (L^2-estimate). *Along with the hypotheses of Theorem 3.16, assume $V = H_0^1(\Omega)$, $V_h = L_{c,h}^k \cap H_0^1(\Omega)$, and that problem (3.11) has smoothing properties. Then, there exists c such that*

$$\forall h, \quad \|u - u_h\|_{0,\Omega} \leq c\, h |u - u_h|_{1,\Omega}. \tag{3.22}$$

Proof. Apply the Aubin–Nitsche Lemma. □

Example 3.19. Consider the homogeneous Dirichlet problem posed on a convex polyhedron, say Ω. Owing to Theorem 3.12, the Laplacian has smoothing properties in Ω. Therefore, using $\mathbb{P}_1$ finite elements yields the estimates

$$\forall h, \quad \|u - u_h\|_{0,\Omega} + h\|u - u_h\|_{1,\Omega} \leq c\, h^2 \|f\|_{0,\Omega}. \quad □$$

Using again duality techniques, it is possible to derive negative-norm estimates for the error, provided Lagrange finite elements of degree 2 at least are employed. For $s \geq 1$, we define the norm

$$\|v\|_{-s,\Omega} = \sup_{z \in H^s(\Omega) \cap H_0^1(\Omega)} \frac{(v, z)_{0,\Omega}}{\|z\|_{s,\Omega}}.$$

Recall that this is not the norm considered to define the dual space $H^{-s}(\Omega)$, except in the particular case $s = 1$. Here, the norm $\|\cdot\|_{-s,\Omega}$ is simply used as a quantitative measure for functions in $L^2(\Omega)$.

Theorem 3.20 (Negative-norm estimates). *Along with the hypotheses of Theorem 3.16, assume $V_h \subset H_0^1(\Omega)$. Assume $k \geq 2$ and let $1 \leq s \leq k - 1$. Assume that there exists a stability constant $c_S > 0$ such that, for all $\varphi \in H^s(\Omega)$, the solution w to the adjoint problem (3.17) satisfies $\|w\|_{s+2,\Omega} \leq c_S \|\varphi\|_{s,\Omega}$. Then, there exists c such that*

$$\forall h, \quad \|u - u_h\|_{-s,\Omega} \leq c\, h^{s+1} \|u - u_h\|_{1,\Omega}. \tag{3.23}$$

Proof. Let $1 \leq s \leq k - 1$, let $z \in H^s(\Omega) \cap H_0^1(\Omega)$, and let $w \in H^{s+2}$ be the solution to the adjoint problem (3.17) with data z. Then, for any $w_h \in V_h$, Galerkin orthogonality implies

$$
\begin{aligned}
(u - u_h, z)_{0,\Omega} &= a(u - u_h, w) \\
&= a(u - u_h, w - w_h) \\
&\leq \|a\| \, \|u - u_h\|_{1,\Omega} \|w - w_h\|_{1,\Omega}.
\end{aligned}
$$

Since $w \in H^{s+2} \cap H_0^1(\Omega)$, it is legitimate to take for w_h the Lagrange interpolant of w in V_h (if $s + 2 \leq \frac{d}{2}$, the Clément or the Scott–Zhang interpolation operator must be considered). Corollary 1.109 implies

$$
\|w - w_h\|_{1,\Omega} \leq c \, h^{s+1} |w|_{s+2,\Omega},
$$

and, therefore, $\|w - w_h\|_{1,\Omega} \leq c \, h^{s+1} \|z\|_{s,\Omega}$. Hence,

$$
(u - u_h, z)_{0,\Omega} \leq c \, h^{s+1} \|u - u_h\|_{1,\Omega} \|z\|_{s,\Omega},
$$

and taking the supremum over z yields the desired estimate. □

Error estimates in the Sobolev norms $\| \cdot \|_{1,p,\Omega}$ are useful in the context of nonlinear problems; see [BrS94, p. 188] for an example. For second-order, elliptic PDEs, the main result is a stability property for the discrete problem (3.20) in the $W^{1,p}$-norm. The result requires some technical assumptions on the discretization and some regularity properties for the exact problem. For the sake of brevity, the former are not restated here. These assumptions hold for the Lagrange finite elements introduced in §1.2.3–§1.2.5 and for quasi-uniform families of geometrically conformal meshes.

Theorem 3.21 ($W^{1,p}$-**stability**). *Let Ω be a polyhedron in $\mathbb{R}^d$ with $d \leq 3$. Assume that:*

(i) *The bilinear form a is elliptic and coercive on $H_0^1(\Omega)$.*
(ii) *The assumptions of [BrS94, p. 170] on the finite element space V_h hold.*
(iii) *The diffusion coefficients are such that $\sigma \in [W^{1,p}(\Omega)]^{d,d}$ for $p > 2$ if $d = 2$ and for $p \geq \frac{12}{5}$ if $d = 3$.*
(iv) *There exists $\delta > d$ such that for all $q \in \,]1, \delta[$ and for all $f \in L^q(\Omega)$, the unique solution to the exact problem (3.11) posed on $H_0^1(\Omega)$ is in $W^{2,q}(\Omega)$. Assume also that the adjoint problem (3.17) satisfies the same regularity property.*

Then, there exist c and $h_0 > 0$ such that

$$
\forall h \leq h_0, \; \forall 1 < p \leq \infty, \quad \|u_h\|_{1,p,\Omega} \leq c \, \|u\|_{1,p,\Omega}. \tag{3.24}
$$

Proof. See [RaS82] and [BrS94, p. 169]. □

Remark 3.22. Owing to assumption (iv) and Corollary B.43, the solution to (3.11) is in $W^{1,\infty}(\Omega)$ whenever $f \in L^q(\Omega)$ with $q > d$. □

Corollary 3.23 ($W^{1,p}$-estimate). *Under the assumptions of Theorem 3.21,*

$$\lim_{h \to 0} \|u - u_h\|_{1,p,\Omega} = 0. \tag{3.25}$$

Furthermore, if $u \in W^{s,p}(\Omega)$ for some $s \geq 2$,

$$\forall h, \quad \|u - u_h\|_{1,p,\Omega} \leq c\, h^l |u|_{l+1,p,\Omega}, \tag{3.26}$$

with $l = \min(k, s-1)$ and k is the degree of the finite element.

Proof. Let $v_h \in V_h$ and $1 < p \leq \infty$. Since $a(u_h - v_h, w_h) = a(u - v_h, w_h)$ for all $w_h \in V_h$, Theorem 3.21 implies $\|u_h - v_h\|_{1,p,\Omega} \leq c \|u - v_h\|_{1,p,\Omega}$. Using the triangle inequality readily yields the estimate

$$\|u - u_h\|_{1,p,\Omega} \leq c \inf_{v_h \in V_h} \|u - v_h\|_{1,p,\Omega}.$$

Equations (3.25) and (3.26) then result from (1.100) and (1.101).   □

Using duality techniques, one can obtain an L^p-norm estimate.

Proposition 3.24 (L^p-estimate). *Under the assumptions of Theorem 3.21, there exist c and $h_0 > 0$ such that*

$$\forall h \leq h_0, \; \forall \delta' < p < \infty, \quad \|u - u_h\|_{L^p(\Omega)} \leq c\, h \|u - u_h\|_{1,p,\Omega}, \tag{3.27}$$

where $\frac{1}{\delta} + \frac{1}{\delta'} = 1$ and δ is defined in assumption (iv) of Theorem 3.21.

Proof. The proof uses duality techniques; see Exercise 3.8.   □

The derivation of L^∞-norm estimates is more technical; see [Nit76, Sco76]. In the framework of the above assumptions, one can show that for finite elements of degree 2 at least,

$$\forall h \leq h_0, \quad \|u - u_h\|_{L^\infty(\Omega)} \leq c\, h \|u - u_h\|_{1,\infty,\Omega}.$$

However, for piecewise linear approximations in two dimensions, the best error estimate in the L^∞-norm is

$$\forall h \leq h_0, \quad \|u - u_h\|_{L^\infty(\Omega)} \leq c\, h |\ln h| \, \|u - u_h\|_{1,\infty,\Omega}.$$

Remark 3.25.

(i) Let x_i be a mesh node, let $\delta_{x=x_i}$ be the Dirac mass at x_i, and assume that the following problem:

$$\begin{cases} \text{Seek } G_i \in V \text{ such that} \\ a(v, G_i) = \langle \delta_{x=x_i}, v \rangle_{\mathcal{D}',\mathcal{D}}, \quad \forall v \in V, \end{cases}$$

is well-posed. Its solution G_i is said to be the *Green function* at point x_i. If it happens that $G_i \in V_h$, Galerkin orthogonality implies

$$0 = a(u - u_h, G_i) = \langle \delta_{x=x_i}, u - u_h \rangle_{\mathcal{D}', \mathcal{D}} = u(x_i) - u_h(x_i),$$

showing that the error vanishes identically at the mesh nodes. This situation occurs when approximating the Laplacian in one dimension with Lagrange finite elements since, in this case, the Green function is continuous and piecewise linear; see also Example 3.90 for the Green function associated with a beam flexion problem.

(ii) When the solution u is not smooth enough, error estimates in weaker norms can be derived. For instance, under the assumptions of Theorem 3.18 and assuming that the family of meshes $\{\mathcal{T}_h\}_{h>0}$ is quasi-uniform, one can show (see, e.g., [QuV97, p. 174]) that there exists c such that

$$\forall h, \quad \|u - u_h\|_{L^\infty(\Omega)} \leq c\, h^{l+1-\frac{d}{2}} |u|_{l+1,\Omega},$$

with $l \leq k$. For instance, if the solution u is in $H^2(\Omega)$, the convergence in the L^∞-norm is first-order in dimension 2, and of order $\frac{1}{2}$ in dimension 3. It would scale like $h^2|\ln h|$ provided $u \in W^{2,\infty}(\Omega)$ and $\mathbb{P}_1$ finite elements are used.

(iii) Consider the purely diffusive version of problem (3.11). When the diffusion coefficients do not satisfy assumption (iii) of Theorem 3.21, but are only measurable and bounded, it is still possible to prove a stability result in $W^{1,p}(\Omega)$ if $|p - 2|$ is small enough. The proof uses the inf-sup condition to express the stability of the exact problem; see [BrS94, p. 184].   □

3.2.2 Non-homogeneous Dirichlet boundary conditions

Given $f \in L^2(\Omega)$ and $g \in H^{\frac{1}{2}}(\partial\Omega)$, the non-homogeneous version of problem (3.11) is:

$$\begin{cases} \text{Seek } u \in H^1(\Omega) \text{ such that} \\ a(u, v) = \int_\Omega fv, \quad \forall v \in H_0^1(\Omega), \\ \gamma_0(u) = g, \qquad \text{in } H^{\frac{1}{2}}(\partial\Omega), \end{cases} \qquad (3.28)$$

where γ_0 is the trace operator defined in §B.3.5. We assume that problem (3.28) is well-posed, namely that the bilinear form a satisfies the assumptions of the BNB Theorem on $H_0^1(\Omega) \times H_0^1(\Omega)$; see §2.1.4 for the theoretical background. For instance, a may be coercive on $H_0^1(\Omega)$. Henceforth, the reader unfamiliar with fractional Sobolev spaces may replace the assumption $g \in H^{\frac{1}{2}}(\partial\Omega)$ by $g \in \mathcal{C}^{0,1}(\partial\Omega)$ (since $\mathcal{C}^{0,1}(\partial\Omega) \subset H^{\frac{1}{2}}(\partial\Omega)$ with continuous embedding; see Example B.32(ii)).

We seek an approximate solution to (3.28) in the discrete space $V_h = L_{c,h}^k$ defined in (3.18). Let N be the dimension of V_h. Denote by $\{\varphi_1, \dots, \varphi_N\}$ the nodal basis of V_h and by $\{a_1, \dots, a_N\}$ the associated nodes. Recall that the Lagrange interpolant of a continuous function u on Ω is defined as

$$\mathcal{I}_h u = \sum_{i=1}^N u(a_i)\varphi_i.$$

Assuming that g is continuous on $\partial\Omega$, we introduce its Lagrange interpolant

$$\mathcal{I}_h^\partial g = \sum_{a_i \in \partial\Omega} g(a_i)\gamma_0(\varphi_i).$$

Since $\{\varphi_1, \ldots, \varphi_N\}$ is a nodal basis,

$$(a_i \notin \partial\Omega) \implies (\gamma_0(\varphi_i) = 0). \tag{3.29}$$

As a result, for $u \in \mathcal{C}^0(\overline{\Omega}) \cap H^1(\Omega)$,

$$\gamma_0(\mathcal{I}_h u) = \gamma_0 \left(\sum_{i=1}^N u(a_i)\varphi_i \right) = \sum_{i=1}^N u(a_i)\gamma_0(\varphi_i)$$
$$= \sum_{a_i \in \partial\Omega} u(a_i)\gamma_0(\varphi_i) = \mathcal{I}_h^\partial\big(\gamma_0(u)\big),$$

so that $\gamma_0 \circ \mathcal{I}_h = \mathcal{I}_h^\partial \circ \gamma_0$, i.e., the trace of the interpolant of a sufficiently smooth function coincides with the interpolant of its trace.

Consider the approximate problem :

$$\begin{cases} \text{Seek } u_h \in V_h \text{ such that} \\ a(u_h, v_h) = \int_\Omega f v_h, \quad \forall v_h \in V_{h0}, \\ \gamma_0(u_h) = \mathcal{I}_h^\partial g, \quad \text{on } \partial\Omega, \end{cases} \tag{3.30}$$

where $V_{h0} = \{v_h \in V_h; \gamma_0(v_h) = 0\} \subset H_0^1(\Omega)$. Assume that the bilinear form a satisfies the condition (BNB1$_h$) on $V_{h0} \times V_{h0}$.

Proposition 3.26. *If g is smooth enough to have a lifting in $\mathcal{C}^0(\overline{\Omega}) \cap H^1(\Omega)$, problem (3.30) is well-posed.*

Proof. Let u_g be a lifting of g in $\mathcal{C}^0(\overline{\Omega}) \cap H^1(\Omega)$. Clearly,

$$\gamma_0(\mathcal{I}_h u_g) = \mathcal{I}_h^\partial\big(\gamma_0(u_g)\big) = \mathcal{I}_h^\partial(g) = \gamma_0(u_h).$$

Therefore, setting $\phi_h = u_h - \mathcal{I}_h u_g$ yields $\phi_h \in V_{h0}$ and $a(\phi_h, v_h) = \int_\Omega f v_h - a(\mathcal{I}_h u_g, v_h)$ for all $v_h \in V_{h0}$. Since the bilinear form a satisfies the condition (BNB1$_h$) on $V_{h0} \times V_{h0}$, problem (3.30) is well-posed. $\square$

The approximate problem (3.30) being well-posed, our goal is now to estimate the approximation error $u - u_h$ in the H^1- and L^2-norms, where u and u_h solve (3.28) and (3.30), respectively. The results below generalize Céa's and Aubin–Nitsche Lemmas; see Exercises 3.9 and 3.10 for proofs.

Lemma 3.27. *Along with the hypotheses of Proposition 3.26, assume that the exact solution u is sufficiently smooth for its Lagrange interpolant $\mathcal{I}_h u$ to be well-defined. Set $\|a\| := \|a\|_{H^1(\Omega), H^1(\Omega)}$. Then,*

$$\|u - u_h\|_{1,\Omega} \leq \left(1 + \frac{\|a\|}{\alpha_h}\right) \|u - \mathcal{I}_h u\|_{1,\Omega}.$$

Lemma 3.28. *Along with the hypotheses of Lemma 3.27, assume that:*

(i) *Problem (3.11) has smoothing properties.*

(ii) *The bilinear form a satisfies the following continuity property: there exists c such that, for all $v \in H^1(\Omega)$ and $w \in H^2(\Omega)$,*

$$|a(v,w)| \leq c\,(\|v\|_{0,\Omega} + \|\gamma_0(v)\|_{0,\partial\Omega})\|w\|_{2,\Omega}.$$

(iii) *There exists an interpolation constant $c > 0$ such that*

$$\forall h,\ \forall \theta \in H^2(\Omega),\quad \|\theta - \mathcal{I}_h\theta\|_{1,\Omega} \leq c\,h\|\theta\|_{2,\Omega}.$$

Then, there exists c such that

$$\forall h,\quad \|u - u_h\|_{0,\Omega} \leq c\,(h\|\mathcal{I}_h u - u\|_{1,\Omega} + \|\mathcal{I}_h u - u\|_{0,\Omega} + \|\mathcal{I}_h g - g\|_{0,\partial\Omega}).$$

Corollary 3.29. *Let Ω be a polyhedron, let $\{\mathcal{T}_h\}_{h>0}$ be a shape-regular family of geometrically conformal meshes of Ω, and let V_h be a H^1-conformal approximation space based on $\mathcal{T}_h$ and a Lagrange finite element of degree $k \geq 1$. Along with the hypotheses of Lemma 3.28, assume that the exact solution u is in $H^{k+1}(\Omega)$. Then, there is c such that*

$$\forall h,\quad \|u - u_h\|_{0,\Omega} + h\|u - u_h\|_{1,\Omega} \leq c\,h^{k+1}\|u\|_{k+1,\Omega}. \tag{3.31}$$

Proof. Direct consequence of Lemmas 3.27 and 3.28.   □

Example 3.30. Assumptions (i)–(iii) of Lemma 3.28 are satisfied for the Poisson problem posed in dimension 2 or 3 on either a convex polyhedron or a domain of class $\mathcal{C}^2$ and for a Lagrange finite element of degree $k \geq 1$ using a shape-regular family of meshes. More precisely, assumption (i) is stated in §3.1.3. Assumption (ii) results from the identity

$$\forall v \in H^1(\Omega),\ \forall w \in H^2(\Omega),\quad a(v,w) = \int_\Omega \nabla v \cdot \nabla w = -\int_\Omega v\Delta w + \int_{\partial\Omega} v\,\partial_n w,$$

together with the continuity of the normal derivative operator $\gamma_1 : H^2(\Omega) \to L^2(\partial\Omega)$; see Theorem B.54. Assumption (iii) is a direct consequence of Corollary 1.109.   □

3.2.3 Crouzeix–Raviart non-conformal approximation

In this section, we present an example of non-conformal approximation for the Laplacian based on the Crouzeix–Raviart finite element. Let Ω be a polyhedron in $\mathbb{R}^d$ and let u be the solution to the homogeneous Dirichlet problem with data $f \in L^2(\Omega)$. Assume that $u \in H^2(\Omega)$. This property holds, for instance, if Ω is convex; see Theorem 3.12.

Let $\{\mathcal{T}_h\}_{h>0}$ be a shape-regular family of geometrically conformal, affine meshes of Ω. Let $P^1_{\mathrm{pt},h}$ be the Crouzeix–Raviart finite element space defined in (1.69). Let

$$P^1_{\text{pt},h,0} = \left\{ v_h \in P^1_{\text{pt},h}; \ \forall F \in \mathcal{F}^\partial_h, \ \int_F v_h = 0 \right\},\qquad (3.32)$$

where $\mathcal{F}^\partial_h$ denotes the set of faces of the mesh located at the boundary. Recall that $\dim P^1_{\text{pt},h,0} = N^i_f$, the number of internal faces (edges in two dimensions) in the mesh. Since functions in $P^1_{\text{pt},h,0}$ can be discontinuous, the bilinear form $\int_\Omega \nabla u \cdot \nabla v$ must be broken over the elements, yielding:

$$\begin{cases} \text{Seek } u_h \in P^1_{\text{pt},h,0} \text{ such that} \\ a_h(u_h, v_h) = f(v_h), \quad \forall v_h \in P^1_{\text{pt},h,0}, \end{cases} \qquad (3.33)$$

with

$$a_h(u_h, v_h) = \sum_{K \in \mathcal{T}_h} \int_K \nabla u_h \cdot \nabla v_h \qquad \text{and} \qquad f(v_h) = \int_\Omega f v_h. \qquad (3.34)$$

Set $V(h) = P^1_{\text{pt},h,0} + H^1_0(\Omega)$ and for $v_h \in V(h)$ define the broken H^1-seminorm

$$|v_h|_{h,1,\Omega} = \left(\sum_{K \in \mathcal{T}_h} \|\nabla v_h\|^2_{0,K} \right)^{\frac{1}{2}}.$$

Equip the space $V(h)$ with the norm $\|\cdot\|_{V(h)} = \|\cdot\|_{0,\Omega} + |\cdot|_{h,1,\Omega}$.

Our goal is to investigate the convergence of the solution to the approximate problem (3.33) in the norm $\|\cdot\|_{V(h)}$. To this end, we must exhibit stability, continuity, consistency, and approximability properties; see §2.3.1. To obtain a stability property for problem (3.33), we would like to establish the coercivity of a_h on $P^1_{\text{pt},h,0}$. Since $P^1_{\text{pt},h,0} \not\subset H^1_0(\Omega)$, this is a non-trivial result.

Lemma 3.31 (Extended Poincaré inequality). *There exists c depending only on Ω such that, for all $h \le 1$,*

$$\forall u \in V(h), \quad c \|u\|_{0,\Omega} \le |u|_{h,1,\Omega}. \qquad (3.35)$$

Proof. We restate the proof given in [Tem77, Prop. 4.13]; see also [CrG02]. Let $u \in V(h)$; then

$$\|u\|_{0,\Omega} \le \sup_{v \in L^2(\Omega)} \frac{(u,v)_{0,\Omega}}{\|v\|_{0,\Omega}}.$$

For $v \in L^2(\Omega)$, there exists $p \in [H^1(\Omega)]^d$ such that $\nabla \cdot p = v$ and $\|p\|_{1,\Omega} \le c \|v\|_{0,\Omega}$, where c depends only on Ω. Integration by parts yields

$$(u,v)_{0,\Omega} = (u, \nabla \cdot p)_{0,\Omega} = -\sum_{K \in \mathcal{T}_h} (\nabla u, p)_{0,K} + \sum_{K \in \mathcal{T}_h} \sum_{F \in \partial K} \int_F (p \cdot n_K) u,$$

where F is a face of K and n_K is the outward normal to K. Consider the second term in the right-hand side of the above equality. If F is an interface,

$F = K_m \cap K_n$, it appears twice in the sum, and since $\int_F u_{|K_m} = \int_F u_{|K_n}$ for $u \in V(h)$, we can subtract from $p \cdot n_K$ a constant function on F that we take equal to $\bar{p} \cdot n_K$ with $\bar{p} = \frac{1}{\text{meas}(F)} \int_F p$. The same conclusion is valid for faces located at the boundary since $\int_F u = 0$ on such faces. Therefore,

$$\sum_{K \in \mathcal{T}_h} \sum_{F \in \partial K} \int_F (p \cdot n_K) u = \sum_{K \in \mathcal{T}_h} \sum_{F \in \partial K} \int_F (p - \bar{p}) \cdot n_K u$$

$$= \sum_{K \in \mathcal{T}_h} \sum_{F \in \partial K} \int_F (p - \bar{p}) \cdot n_K (u - \bar{u}),$$

and using Lemma 3.32 below, this yields

$$(u, v)_{0,\Omega} \leq \|p\|_{0,\Omega} |u|_{h,1,\Omega} + \sum_{K \in \mathcal{T}_h} c \, h_K^{\frac{1}{2}} |p|_{1,K} h_K^{\frac{1}{2}} |u|_{1,K}$$

$$\leq \|p\|_{0,\Omega} |u|_{h,1,\Omega} + c \, h |p|_{1,\Omega} |u|_{h,1,\Omega}.$$

Since $h \leq 1$, $(u, v)_{0,\Omega} \leq c \, \|v\|_{0,\Omega} |u|_{h,1,\Omega}$ and, hence, (3.35) holds.   □

Lemma 3.32. *Let $\{\mathcal{T}_h\}_{h>0}$ be a shape-regular family of geometrically conformal affine meshes. Let $m \geq 1$ be a fixed integer. For $K \in \mathcal{T}_h$, $\psi \in [H^1(K)]^m$, and a face $F \in \partial K$, set $\bar{\psi} = \frac{1}{\text{meas}(F)} \int_F \psi$. Then, there exists c such that*

$$\forall h, \, \forall K \in \mathcal{T}_h, \, \forall F \in \partial K, \, \forall \psi \in [H^1(K)]^m, \quad \|\psi - \bar{\psi}\|_{0,F} \leq c \, h_K^{\frac{1}{2}} |\psi|_{1,K}. \quad (3.36)$$

Proof. Let $K \in \mathcal{T}_h$, let $\psi \in [H^1(K)]^m$, and consider a face $F \in \partial K$. Let $\hat{K}$ be the reference simplex and let $T_K : \hat{K} \to K$ be the corresponding affine transformation with Jacobian J_K. Letting $\hat{F} = T_K^{-1}(F)$, it is clear that

$$\|\psi - \bar{\psi}\|_{0,F} \leq \left(\frac{\text{meas } F}{\text{meas } \hat{F}} \right)^{\frac{1}{2}} \|\hat{\psi} - \overline{\hat{\psi}}\|_{0,\hat{F}} \leq c \left(\frac{\text{meas } F}{\text{meas } \hat{F}} \right)^{\frac{1}{2}} \|\hat{\psi} - \overline{\hat{\psi}}\|_{1,\hat{K}},$$

owing to the Trace Theorem B.52. The Deny–Lions Lemma implies

$$\|\hat{\psi} - \overline{\hat{\psi}}\|_{1,\hat{K}} \leq c \, |\hat{\psi}|_{1,\hat{K}}.$$

Returning to element K and using the shape-regularity of the mesh yields

$$\|\psi - \bar{\psi}\|_{0,F} \leq c \left(\frac{\text{meas } F}{\text{meas } \hat{F}} \right)^{\frac{1}{2}} \|J_K^{-1}\|_d \left(\frac{\text{meas } \hat{K}}{\text{meas } K} \right)^{\frac{1}{2}} |\psi|_{1,K}$$

$$\leq c \, h_K^{\frac{d-1}{2}} h_K h_K^{-\frac{d}{2}} |\psi|_{1,K} \leq c \, h_K^{\frac{1}{2}} |\psi|_{1,K},$$

thereby completing the proof.   □

Corollary 3.33 (Stability). *The bilinear form a_h defined in (3.34) is coercive on $P_{\text{pt},h,0}^1$.*

Proof. Direct consequence of the extended Poincaré inequality (3.35).   □

Lemma 3.34 (Continuity). *The bilinear form a_h defined in (3.34) is uniformly bounded on $V(h) \times V(h)$.*

Proof. Use the fact that, for all $u_h \in V(h)$, $|u_h|_{h,1,\Omega} \leq \|u_h\|_{V(h)}$.   □

Corollary 3.35 (Well-Posedness). *Problem (3.33) is well-posed.*

Proof. Direct consequence of the Lax–Milgram Lemma.   □

Lemma 3.36 (Asymptotic consistency). *Let u be the solution to the homogeneous Dirichlet problem with data $f \in L^2(\Omega)$. Assume that $u \in H^2(\Omega)$. Then, there exists c such that*

$$\forall h, \ \forall w_h \in P^1_{pt,h,0}, \quad \frac{|f(w_h) - a_h(u, w_h)|}{\|w_h\|_{V(h)}} \leq c \, h |u|_{2,\Omega}. \qquad (3.37)$$

Proof. Let $w_h \in P^1_{pt,h,0}$. Since $f = -\Delta u$,

$$a_h(u, w_h) - f(w_h) = \sum_{K \in \mathcal{T}_h} \int_K (\nabla u \cdot \nabla w_h - f w_h) = \sum_{K \in \mathcal{T}_h} \sum_{F \in \partial K} \int_F \nabla u \cdot n_K \, w_h.$$

Since each face F of an element K located inside Ω appears twice in the above sum, we can subtract from w_h its mean-value on the face, $\overline{w_h}$. If F is on $\partial \Omega$, it is clear that $\overline{w_h} = 0$. Therefore,

$$a_h(u, w_h) - f(w_h) = \sum_{K \in \mathcal{T}_h} \sum_{F \in \partial K} \int_F \nabla u \cdot n_K (w_h - \overline{w_h}).$$

We can also subtract from ∇u its mean-value on F, $\overline{\nabla u}$, yielding

$$a_h(u, w_h) - f(w_h) = \sum_{K \in \mathcal{T}_h} \sum_{F \in \partial K} \int_F (\nabla u - \overline{\nabla u}) \cdot n_K (w_h - \overline{w_h}).$$

The Cauchy–Schwarz inequality implies

$$|a_h(u, w_h) - f(w_h)| \leq \sum_{K \in \mathcal{T}_h} \sum_{F \in \partial K} \|\nabla u - \overline{\nabla u}\|_{0,F} \|w_h - \overline{w_h}\|_{0,F}.$$

Lemma 3.32 yields

$$|a_h(u, w_h) - f(w_h)| \leq \sum_{K \in \mathcal{T}_h} c \, h_K^{\frac{1}{2}} |u|_{2,K} h_K^{\frac{1}{2}} |w_h|_{1,K}$$

$$\leq c \, h \left(\sum_{K \in \mathcal{T}_h} |u|^2_{2,K} \cdot \sum_{K \in \mathcal{T}_h} |w_h|^2_{1,K} \right)^{\frac{1}{2}} \leq c \, h |u|_{2,\Omega} \|w_h\|_{V(h)},$$

leading to (3.37).   □

Lemma 3.37 (Approximability). *There exists c such that*

$$\forall h, \ \forall u \in H^2(\Omega) \cap H^1_0(\Omega), \quad \inf_{v_h \in P^1_{\mathrm{pt},h,0}} \|u - v_h\|_{V(h)} \leq c\,h|u|_{2,\Omega}. \qquad (3.38)$$

Proof. Use $P^1_{\mathrm{c},h,0} = P_{\mathrm{c},h} \cap H^1_0(\Omega) \subset P^1_{\mathrm{pt},h,0}$ and Corollary 1.109.   □

Theorem 3.38 (Convergence). *Under the assumptions of Lemma 3.36, there exists c such that*

$$\forall h, \quad \|u - u_h\|_{V(h)} \leq c\,h|u|_{2,\Omega}. \qquad (3.39)$$

Proof. Direct consequence of Lemma 2.25 and the above results.   □

Finally, an error estimate in the L^2-norm can be obtained by generalizing the Aubin–Nitsche Lemma to non-conformal approximation spaces.

Theorem 3.39 (L^2-estimate). *Along with the assumptions of Theorem 3.38, assume that the Laplacian has smoothing properties in Ω. Then, there exists c such that*

$$\forall h, \quad \|u - u_h\|_{0,\Omega} \leq c\,h|u - u_h|_{h,1,\Omega}. \qquad (3.40)$$

Proof. See [Bra97, p. 108].   □

3.2.4 Discontinuous Galerkin (DG) Approximation

In the previous section, we have investigated a first example of non-conformal method to approximate second-order elliptic PDEs. Because the degrees of freedom in the finite element space were located at the faces of the mesh, the method can be viewed as a *face-centered approximation*. In this section, we continue the investigation of non-conformal methods for elliptic problems by analyzing *cell-centered approximations* in which the degrees of freedom in the finite element space are defined independently on each cell. In the literature, such methods are often termed Discontinuous Galerkin (DG) methods, and this terminology will be employed henceforth.

For the sake of simplicity, we restrict ourselves to the approximation of the Laplacian with homogeneous Dirichlet conditions and data $f \in L^2(\Omega)$. As in the previous section, we assume that the domain Ω is a polyhedron in $\mathbb{R}^d$ in which the Laplacian has smoothing properties; hence, the exact solution u is in $H^2(\Omega)$. The material presented below is adapted from [ArB01].

Mixed formulation. We recast the problem in the form of a mixed system of first-order PDEs

$$\sigma = \nabla u, \quad -\nabla{\cdot}\sigma = f \quad \text{in } \Omega, \qquad u = 0 \quad \text{on } \partial\Omega. \qquad (3.41)$$

From a physical viewpoint, the auxiliary unknown σ plays the role of a flux, and the PDE $-\nabla{\cdot}\sigma = f$ expresses a conservation property. The unknown u is called the *primal variable*. Multiplying the first and second equations in (3.41) by test functions τ and v, respectively, and integrating formally over a subset K of Ω yields the weak formulation

$$\begin{cases} \int_K \sigma \cdot \tau = -\int_K u\, \nabla \cdot \tau + \int_{\partial K} u\, \tau \cdot n_K, \\ \int_K \sigma \cdot \nabla v = \int_K fv + \int_{\partial K} v\, \sigma \cdot n_K, \end{cases} \qquad (3.42)$$

where n_K is the outward normal to ∂K.

Let $\{\mathcal{T}_h\}_{h>0}$ be a shape-regular family of simplicial meshes of the domain Ω, and for $k \geq 1$, consider the finite element spaces

$$\begin{cases} V_h = \{v \in L^1(\Omega);\, \forall K \in \mathcal{T}_h,\, v_{|K} \in \mathbb{P}_k\}, \\ \Sigma_h = \{\tau \in [L^1(\Omega)]^d;\, \forall K \in \mathcal{T}_h,\, \tau_{|K} \in [\mathbb{P}_k]^d\}. \end{cases}$$

Note that V_h coincides with the space $P^k_{\mathrm{td},h}$ introduced in §1.4.3. For $v \in V_h$ and $\tau \in \Sigma_h$, let $\nabla_h v$ and $\nabla_h \cdot \tau$ be the functions whose restriction to each element $K \in \mathcal{T}_h$ is equal to ∇v and $\nabla \cdot \tau$, respectively. Following [CoS98], a discrete mixed formulation is derived by summing (3.42) over the mesh elements:

$$\begin{cases} \text{Seek } u_h \in V_h \text{ and } \sigma_h \in \Sigma_h \text{ such that} \\ \int_\Omega \sigma_h \cdot \tau = -\int_\Omega u_h\, \nabla_h \cdot \tau + \sum_{K \in \mathcal{T}_h} \int_{\partial K} \phi_u\, \tau \cdot n_K, \quad \forall \tau \in \Sigma_h, \\ \int_\Omega \sigma_h \cdot \nabla_h v = \int_\Omega fv + \sum_{K \in \mathcal{T}_h} \int_{\partial K} v\, \phi_\sigma \cdot n_K, \qquad \forall v \in V_h, \end{cases} \qquad (3.43)$$

where the *numerical fluxes* ϕ_u and ϕ_σ are approximations to the double-valued traces at the mesh interfaces of u_h and σ_h, respectively. The numerical fluxes need not be single-valued at the mesh interfaces.

To specify the numerical fluxes, we introduce an appropriate functional setting. For an integer $l \geq 1$, let $H^l(\mathcal{T}_h)$ be the space of functions on Ω whose restriction to each element $K \in \mathcal{T}_h$ belongs to $H^l(K)$. Recall that $\mathcal{F}_h^i$ denotes the set of interior faces, $\mathcal{F}_h^\partial$ the set of boundary faces, and $\mathcal{F}_h = \mathcal{F}_h^i \cup \mathcal{F}_h^\partial$. The traces on element boundaries of functions in $H^1(\mathcal{T}_h)$ belong to a space denoted by $T(\mathcal{F}_h)$. Functions in $T(\mathcal{F}_h)$ are double-valued on $\mathcal{F}_h^i$ and single-valued on $\mathcal{F}_h^\partial$. Denote by $L^2(\mathcal{F}_h)$ the space of single-valued functions on $\mathcal{F}_h$ whose restriction to each face $F \in \mathcal{F}_h$ is in $L^2(F)$.

Using the above notation, the numerical fluxes are chosen to be linear functions

$$\phi_u : H^1(\mathcal{T}_h) \longrightarrow T(\mathcal{F}_h), \qquad \phi_\sigma : H^2(\mathcal{T}_h) \times [H^1(\mathcal{T}_h)]^d \longrightarrow [T(\mathcal{F}_h)]^d.$$

In the present setting, ϕ_u depends only on u_h, while ϕ_σ depends on both u_h and σ_h; other settings can be considered as well.

Two properties of the numerical fluxes are important in the analysis of DG methods: consistency and conservativity.

Definition 3.40 (Consistency). *The numerical fluxes ϕ_u and ϕ_σ are said to be* consistent *if for any smooth function $v \in H^2(\Omega) \cap H_0^1(\Omega)$,*

$$\phi_u(v) = v_{|\mathcal{F}_h} \quad \text{and} \quad \phi_\sigma(v, \nabla v) = \nabla v_{|\mathcal{F}_h}.$$

Proposition 3.41. *If the numerical fluxes ϕ_u and ϕ_σ are consistent, the exact solution u and its gradient ∇u satisfy (3.43).*

Proof. Straightforward verification.   □

Definition 3.42 (Conservativity). *The numerical fluxes ϕ_u and ϕ_σ are said to be* conservative *if they are single-valued on $\mathcal{F}_h$.*

Proposition 3.43. *Assume that the numerical fluxes are conservative. Let ω be the union of any collection of elements. Then, if (u_h, σ_h) solves (3.43),*

$$\int_\omega f + \int_{\partial\omega} \phi_\sigma(u_h, \sigma_h) \cdot n_\omega = 0,$$

where n_ω is the outward normal to $\partial\omega$.

Proof. Take v to be the characteristic function of ω.   □

Primal formulation. A primal formulation is a discrete problem in which u_h is the only unknown.

To derive a primal formulation, the discrete unknown σ_h must be eliminated through a *flux reconstruction formula*, that is, a formula expressing the discrete flux σ_h in terms of the discrete primal variable u_h only. It is convenient to define averages and jumps across faces. Let F be an interior face shared by elements K_1 and K_2, and let n_1 and n_2 be the normal vectors to F pointing toward the exterior of K_1 and K_2, respectively. For $v \in V_h$, setting $v_i = v_{|F \cap K_i}$, $i = 1, 2$, define the average $\{\cdot\}$ and jump $[\![\cdot]\!]$ operators as

$$\{v\} = \tfrac{1}{2}(v_1 + v_2) \quad \text{and} \quad [\![v]\!] = v_1 n_1 + v_2 n_2 \quad \text{on each } F \in \mathcal{F}_h^i.$$

Using a similar notation for $\tau \in \Sigma_h$, set

$$\{\tau\} = \tfrac{1}{2}(\tau_1 + \tau_2) \quad \text{and} \quad [\![\tau]\!] = \tau_1 \cdot n_1 + \tau_2 \cdot n_2 \quad \text{on each } F \in \mathcal{F}_h^i.$$

Note that the jump of a scalar-valued function is vector-valued, and *vice versa* (to alleviate the notation, we write $[\![\tau]\!]$ instead of $[\![\tau \cdot n]\!]$). For $F \in \mathcal{F}_h^\partial$, set $[\![v]\!] = vn$ and $\{\tau\} = \tau$ where n is the outward normal. Owing to the identity

$$\int_\Omega \nabla_h \cdot \tau\, v + \int_\Omega \tau \cdot \nabla_h v = \sum_{K \in \mathcal{T}_h} \int_{\partial K} v\, \tau \cdot n_K = \int_{\mathcal{F}_h} [\![v]\!] \cdot \{\tau\} + \int_{\mathcal{F}_h^i} \{v\} [\![\tau]\!], \quad (3.44)$$

holding for all $v \in V_h$ and $\tau \in \Sigma_h$, (3.43) is recast into the form

$$\begin{cases} \int_\Omega \sigma_h \cdot \tau = -\int_\Omega u_h \, \nabla_h \cdot \tau + \int_{\mathcal{F}_h} [\![\phi_u(u_h)]\!] \cdot \{\tau\} + \int_{\mathcal{F}_h^i} \{\phi_u(u_h)\} [\![\tau]\!], \\[2mm] \int_\Omega \sigma_h \cdot \nabla_h v - \int_{\mathcal{F}_h} \{\phi_\sigma(u_h, \sigma_h)\} \cdot [\![v]\!] - \int_{\mathcal{F}_h^i} [\![\phi_\sigma(u_h, \sigma_h)]\!] \{v\} = \int_\Omega fv, \end{cases} \tag{3.45}$$

for all $\tau \in \Sigma_h$ and $v \in V_h$. Using (3.44) to eliminate the term $\int_\Omega u_h \, \nabla_h \cdot \tau$ in the first equation of (3.45) yields

$$\int_\Omega \sigma_h \cdot \tau = \int_\Omega \nabla_h u_h \cdot \tau + \int_{\mathcal{F}_h} [\![\phi_u(u_h) - u_h]\!] \cdot \{\tau\} + \int_{\mathcal{F}_h^i} \{\phi_u(u_h) - u_h\} [\![\tau]\!]. \tag{3.46}$$

Introduce the lifting operators $l_1 : L^2(\mathcal{F}_h^i) \to \Sigma_h$ and $l_2 : [L^2(\mathcal{F}_h)]^d \to \Sigma_h$ such that, for $q \in L^2(\mathcal{F}_h^i)$ and $\rho \in [L^2(\mathcal{F}_h)]^d$,

$$\forall \tau \in \Sigma_h, \quad \int_\Omega l_1(q) \cdot \tau = -\int_{\mathcal{F}_h^i} q [\![\tau]\!], \quad \int_\Omega l_2(\rho) \cdot \tau = -\int_{\mathcal{F}_h} \rho \cdot \{\tau\}. \tag{3.47}$$

These lifting operators involve local L^2-projections. For instance, for $F \in \mathcal{F}_h$, define the operator $l_F : [L^1(F)]^d \to \Sigma_h$ such that, for $\rho \in [L^1(F)]^d$,

$$\forall \tau \in \Sigma_h, \quad \int_\Omega l_F(\rho) \cdot \tau = -\int_F \rho \cdot \{\tau\}.$$

Clearly, the support of $l_F(\rho)$ consists of the one or two simplices sharing F as a face. For $\rho \in [L^2(\mathcal{F}_h)]^d$, it is clear that $l_2(\rho) = \sum_{F \in \mathcal{F}_h} l_F(\rho)$. A similar construction is possible for the lifting operator l_1.

Recalling that $\nabla_h V_h \subset \Sigma_h$ and using the above lifting operators, we deduce from (3.46) the flux reconstruction formula

$$\sigma_h = \nabla_h u_h - l_1(\{\phi_u(u_h) - u_h\}) - l_2([\![\phi_u(u_h) - u_h]\!]). \tag{3.48}$$

Taking now $\tau = \nabla_h v$ in (3.46), the second equation in (3.45) yields $a_h(u_h, v) = \int_\Omega fv$, where

$$\begin{aligned} a_h(u_h, v) = {} & \int_\Omega \nabla_h u_h \cdot \nabla_h v \\ & + \int_{\mathcal{F}_h} [\![\phi_u(u_h) - u_h]\!] \cdot \{\nabla_h v\} - \{\phi_\sigma(u_h, \sigma_h)\} \cdot [\![v]\!] \\ & + \int_{\mathcal{F}_h^i} \{\phi_u(u_h) - u_h\} [\![\nabla_h v]\!] - [\![\phi_\sigma(u_h, \sigma_h)]\!] \{v\}, \end{aligned} \tag{3.49}$$

with σ_h evaluated from (3.48). The bilinear form a_h is defined on $H^2(\mathcal{T}_h) \times H^2(\mathcal{T}_h)$. The primal formulation is thus:

$$\begin{cases} \text{Seek } u_h \in V_h \text{ such that} \\ a_h(u_h, v) = \int_\Omega fv, \quad \forall v \in V_h. \end{cases} \tag{3.50}$$

Clearly, if $(u_h, \sigma_h) \in V_h \times \Sigma_h$ solves (3.45), then u_h solves (3.50) provided the flux σ_h is reconstructed using (3.48).

Remark 3.44. If the fluxes are conservative, (3.49) simplifies into

$$a_h(u_h, v) = \int_\Omega \nabla_h u_h \cdot \nabla_h v - \int_{\mathcal{F}_h} [\![u_h]\!] \cdot \{\nabla_h v\} + \{\phi_\sigma(u_h, \sigma_h)\} \cdot [\![v]\!]$$

$$+ \int_{\mathcal{F}_h^i} (\phi_u(u_h) - \{u_h\}) [\![\nabla_h v]\!].$$

□

Error analysis. To estimate the error induced by the approximate problem (3.50), it is convenient to introduce the space $V(h) = V_h + H^2(\Omega) \cap H_0^1(\Omega)$. For $v \in V(h)$, set

$$|v|_{h,1,\Omega}^2 = \sum_{K \in \mathcal{T}_h} |v|_{1,K}^2, \qquad |v|_j^2 = \sum_{F \in \mathcal{F}_h} \|l_F([\![v]\!])\|_{0,\Omega}^2,$$

and let

$$\|v\|_{V(h)}^2 = |v|_{h,1,\Omega}^2 + |v|_j^2 + \sum_{K \in \mathcal{T}_h} h_K^2 |v|_{2,K}^2. \tag{3.51}$$

This choice will appear more clearly in the examples presented below.

Lemma 3.45. *If Ω has smoothing properties, there exists c, independent of h, such that*

$$\forall v \in V(h), \quad c\|v\|_{0,\Omega} \le |v|_{h,1,\Omega} + |v|_j.$$

Proof. (1) Using inverse inequalities, one can prove that there exist positive constants c_1 and c_2 such that

$$\forall \rho \in [\mathbb{P}_k(F)]^d, \quad c_1 \|l_F(\rho)\|_{0,\Omega}^2 \le h_F^{-1} \|\rho\|_{0,F}^2 \le c_2 \|l_F(\rho)\|_{0,\Omega}^2.$$

These inequalities can be applied to $\rho = [\![v]\!]$ for $v \in V(h)$, yielding

$$\forall v \in V(h), \quad c_1 |v|_j^2 \le \sum_{F \in \mathcal{F}_h} h_F^{-1} \|[\![v]\!]\|_{0,F}^2 \le c_2 |v|_j^2. \tag{3.52}$$

(2) Let $v \in V(h)$ and let $\psi \in H_0^1(\Omega)$ solve $-\Delta\psi = v$. Since Ω has smoothing properties, there is $c > 0$ such that $\|\psi\|_{2,\Omega} \le c\|v\|_{0,\Omega}$. Then,

$$\|v\|_{0,\Omega}^2 = -\int_\Omega v\Delta\psi = \int_\Omega \nabla\psi \cdot \nabla_h v - \int_{\mathcal{F}_h} \nabla\psi \cdot [\![v]\!]$$

$$\le c|v|_{1,h,\Omega}\|v\|_{0,\Omega} + \left(\sum_{F \in \mathcal{F}_h} h_F^{-1} \|[\![v]\!]\|_{0,F}^2\right)^{\frac{1}{2}} \left(\sum_{F \in \mathcal{F}_h} h_F |\psi|_{1,F}^2\right)^{\frac{1}{2}}.$$

Using a trace theorem and a scaling argument yields

$$h_F |\psi|_{1,F}^2 \le c\left(|\psi|_{1,K}^2 + h_F^2 |\psi|_{2,K}^2\right) \le c'\|\psi\|_{2,K}^2. \tag{3.53}$$

Hence,

$$\|v\|_{0,\Omega}^2 \le c_1 |v|_{1,h,\Omega}\|v\|_{0,\Omega} + c_2 |v|_j \|v\|_{0,\Omega},$$

and this completes the proof.

□

Remark 3.46. Lemma 3.45 is a discrete Poincaré-type inequality.   □

Proposition 3.47 (Well-posedness). *Assume that the bilinear form* a_h *defined in (3.49) satisfies the following properties:*

(i) *Uniform boundedness on* $V(h)$: *there exists* $c_b > 0$, *independent of* h, *such that*

$$\forall v, w \in V(h), \quad a_h(w, v) \le c_b \|w\|_{V(h)} \|v\|_{V(h)}. \tag{3.54}$$

(ii) *Coercivity on* V_h: *there exists* $c_s > 0$, *independent of* h, *such that*

$$\forall v \in V_h, \quad a_h(v, v) \ge c_s \|v\|_{V(h)}^2. \tag{3.55}$$

Then, problem (3.50) is well-posed.

Proof. Direct consequence of the Lax–Milgram Lemma.   □

Proposition 3.48 (Consistency). *Assume that the numerical fluxes* ϕ_u *and* ϕ_σ *are consistent. Then, the exact solution* u *satisfies*

$$\forall v \in V_h, \quad a_h(u, v) = \int_\Omega fv.$$

Proof. Since $u \in H^2(\Omega)$, taking $\tau = \nabla_h u$ in (3.44) yields, for all $v \in V_h$,

$$\int_\Omega \nabla_h u \cdot \nabla_h v = -\int_\Omega \Delta u\, v + \int_{\mathcal{F}_h} [\![v]\!] \cdot \{\nabla_h u\} + \int_{\mathcal{F}_h^i} \{v\} [\![\nabla_h u]\!].$$

Since $\{u\} = u$, $[\![u]\!] = 0$, $\{\nabla_h u\} = \nabla u$, $[\![\nabla_h u]\!] = 0$, and $-\Delta u = f$,

$$a_h(u, v) = \int_\Omega fv + \int_{\mathcal{F}_h} [\![\phi_u(u)]\!] \cdot \{\nabla_h v\} + (\nabla u - \{\phi_\sigma(u, \sigma_h(u))\}) \cdot [\![v]\!]$$

$$+ \int_{\mathcal{F}_h^i} \{\phi_u(u) - u\} [\![\nabla_h v]\!] - [\![\phi_\sigma(u, \sigma_h(u))]\!] \{v\}.$$

Owing to the consistency of the numerical flux ϕ_u, $\phi_u(u) = u$. Moreover, the reconstruction formula (3.48) implies $\sigma_h(u) = \nabla u$. Since the numerical flux ϕ_σ is also consistent, $\{\phi_\sigma(u, \sigma_h(u))\} = \nabla u$ and $[\![\phi_\sigma(u, \sigma_h(u))]\!] = 0$. Therefore, all the face integrals vanish.   □

Lemma 3.49 (Approximability). *There exists* c *such that, for all* $1 \le s \le k+1$,

$$\forall h, \forall u \in H^s(\Omega) \cap H_0^1(\Omega), \quad \inf_{v \in V_h} \|u - v\|_{V(h)} \le c\, h^{s-1} |u|_{s,\Omega}.$$

Proof. Let $1 \le s \le k+1$. Since V_h contains the H^1-conformal Scott–Zhang interpolant $\mathcal{SZ}_h u$ of u and since the face jumps of $u - \mathcal{SZ}_h u$ vanish,

$$\|u - \mathcal{SZ}_h u\|_{V(h)}^2 = |u - \mathcal{SZ}_h u|_{1,\Omega}^2 + \sum_{K \in \mathcal{T}_h} h_K^2 |u - \mathcal{SZ}_h u|_{2,K}^2 \le c\, h^{2(s-1)} |u|_{s,\Omega}^2,$$

the last inequality being a direct consequence of Lemma 1.130.   □

Theorem 3.50 (Convergence). *Let u be the solution to the homogeneous Dirichlet problem with data $f \in L^2(\Omega)$. Assume that the Laplacian has smoothing properties in Ω and that $u \in H^s(\Omega)$ for some $s \in \{2, \ldots, k+1\}$. Let u_h be the solution to (3.50). Along with hypotheses (i)–(ii) of Proposition 3.47, assume that the numerical fluxes ϕ_u and ϕ_σ are consistent. Then, there exists c such that*

$$\forall h, \quad \|u - u_h\|_{V(h)} \le c\, h^{s-1}|u|_{s,\Omega}. \tag{3.56}$$

Proof. Direct consequence of Lemma 2.25 and the above results. □

An L^2-norm error estimate can be obtained using duality techniques.

Definition 3.51 (Adjoint-consistency). *The bilinear form a_h is said to be adjoint-consistent if, for all $w \in H^2(\Omega) \cap H_0^1(\Omega)$,*

$$\forall v \in V(h), \quad a_h(v, w) = -\int_\Omega \Delta w\, v. \tag{3.57}$$

Lemma 3.52. *Assume that the numerical fluxes ϕ_u and ϕ_σ are conservative. Then, the bilinear form a_h is adjoint-consistent.*

Proof. Let $w \in H^2(\Omega) \cap H_0^1(\Omega)$ and let $v \in V(h)$. Note that $[\![w]\!] = 0$, $[\![\nabla_h w]\!] = 0$, and $\{\nabla_h w\} = \nabla w$. Using (3.44) yields

$$\int_\Omega \nabla_h v \cdot \nabla_h w = -\int_\Omega \Delta w\, v + \int_{\mathcal{F}_h} [\![v]\!] \cdot \nabla w.$$

Since w is smooth, Remark 3.44 implies $a_h(v, w) = \int_\Omega \nabla_h v \cdot \nabla_h w - \int_{\mathcal{F}_h} [\![v]\!] \cdot \nabla w$. The conclusion follows readily. □

Theorem 3.53 (L^2-convergence). *Under the hypotheses of Theorem 3.50, assuming that the numerical fluxes ϕ_u and ϕ_σ are conservative, there exists c such that*

$$\forall h, \quad \|u - u_h\|_{0,\Omega} \le c\, h^s|u|_{s,\Omega}. \tag{3.58}$$

Proof. Let $\psi \in H_0^1(\Omega)$ be such that $-\Delta\psi = u - u_h$. Since the Laplacian has smoothing properties in Ω, $|\psi|_{2,\Omega} \le c\|u - u_h\|_{0,\Omega}$. Furthermore, since the approximate fluxes ϕ_u and ϕ_σ are conservative, Lemma 3.52 implies

$$\forall v \in V(h), \quad a_h(v, \psi) = \int_\Omega (u - u_h)v.$$

Since $u - u_h \in V(h)$ and the numerical fluxes are consistent,

$$\|u - u_h\|_{0,\Omega}^2 = a_h(u - u_h, \psi) = a_h(u - u_h, \psi - \mathcal{SZ}_h\psi)$$
$$\le c_b\|u - u_h\|_{V(h)}\|\psi - \mathcal{SZ}_h\psi\|_{V(h)} \le c\, h|\psi|_{2,\Omega}\|u - u_h\|_{V(h)},$$

where $\mathcal{SZ}_h\psi$ is the Scott–Zhang interpolant of ψ. Conclude using (3.56). □

Example 1 (LDG). The so-called Local Discontinuous Galerkin (LDG) method has been introduced by Cockburn and Shu in 1998 [CoS98] to approximate time-dependent convection–diffusion problems. Written within the above framework, it consists of taking the numerical fluxes

$$\phi_u(u_h) = \begin{cases} \{u_h\} - \beta\cdot[\![u_h]\!] & \text{on } \mathcal{F}_h^i, \\ 0 & \text{on } \mathcal{F}_h^\partial, \end{cases} \tag{3.59}$$

and

$$\phi_\sigma(u_h, \sigma_h) = \begin{cases} \{\sigma_h\} + \beta\cdot[\![\sigma_h]\!] - \eta_F h_F^{-1}[\![u_h]\!] & \text{on } \mathcal{F}_h^i, \\ \{\sigma_h\} - \eta_F h_F^{-1}[\![u_h]\!] & \text{on } \mathcal{F}_h^\partial. \end{cases} \tag{3.60}$$

Here, $\beta \in [L^\infty(\mathcal{F}_h^i)]^d$ is a vector-valued function that is constant on each interior face, η_F is a given positive parameter on the face F, and h_F denotes the diameter of F. A straightforward calculation yields the following:

Proposition 3.54. *The numerical fluxes ϕ_u and ϕ_σ defined by (3.59)–(3.60) are consistent and conservative.*

In the LDG method, the flux reconstruction formula (3.48) takes the form

$$\sigma_h = \nabla_h u_h + l_1(\beta\cdot[\![u_h]\!]) + l_2([\![u_h]\!]),$$

and the bilinear form a_h is given by

$$
\begin{aligned}
a_h(u_h, v) = &\int_\Omega \nabla_h u_h\cdot\nabla_h v - \int_{\mathcal{F}_h} [\![u_h]\!]\cdot\{\nabla_h v\} + \{\nabla_h u_h\}[\![v]\!] \\
&+ \int_{\mathcal{F}_h} \eta_F h_F^{-1}[\![u_h]\!]\,[\![v]\!] + \int_{\mathcal{F}_h^i} \beta\cdot[\![u_h]\!]\,[\![v]\!] + [\![\nabla_h u_h]\!]\,\beta\cdot[\![v]\!] \quad (3.61) \\
&+ \int_\Omega \big(l_1(\beta\cdot[\![u_h]\!]) + l_2([\![u_h]\!])\big)\cdot\big(l_1(\beta\cdot[\![v]\!]) + l_2([\![v]\!])\big).
\end{aligned}
$$

Proposition 3.55. *The bilinear form a_h defined by (3.61) is continuous on $V(h)$ and, provided $\inf_F \eta_F$ is large enough, it is also coercive on V_h.*

Proof. The proof is only sketched; see [ArB01] and the references therein.
(i) To prove continuity, i.e., property (3.54), the various terms appearing in the right-hand side of (3.61) must be bounded. Let $w, v \in V(h)$. First, it is clear that $\int_\Omega \nabla_h w\cdot\nabla_h v \le |w|_{h,1,\Omega}|v|_{h,1,\Omega}$. Owing to (3.52), $\int_{\mathcal{F}_h} \eta_F h_F^{-1}[\![u_h]\!]\,[\![v]\!] \le c_3|w|_j|v|_j$ with $c_3 = c_2 \sup_F \eta_F$. Next, for $w \in H^2(K)$ and a face F of K, (3.53) implies

$$\|\nabla w\cdot n\|_{0,F}^2 \le c_4\big(h_F^{-1}|w|_{1,K}^2 + h_F|w|_{2,K}^2\big).$$

This in turn implies

$$
\int_{\mathcal{F}_h} \{\nabla_h w\}\cdot[\![v]\!] \le c_5 \left(\sum_{K\in T_h} |w|_{1,K}^2 + h_K^2|w|_{2,K}^2\right)^{\frac{1}{2}} \left(\sum_{F\in\mathcal{F}_h} h_F^{-1}\|[\![v]\!]\|_{0,F}^2\right)^{\frac{1}{2}}
$$

$$\le c_5\|w\|_{V(h)}|v|_j.$$

The remaining face integrals in (3.61) are bounded similarly. Finally, one can readily show that

$$\forall v \in V(h), \quad \|l_1(\beta \cdot [\![v]\!])\|_{0,\Omega} \le c_6 \|\beta\|_{L^\infty(\mathcal{F}_h^i)}^{\frac{1}{2}} |v|_j,$$

and

$$\forall v \in V(h), \quad \|l_2([\![v]\!])\|_{0,\Omega} \le c_7 |v|_j.$$

Using the above estimates, one easily bounds the second integral over Ω in the right-hand side of (3.61).

(ii) Let us prove the coercivity of a_h, i.e., property (3.55). Consider $v \in V_h$. It is clear that

$$a_h(v,v) = |v|_{h,1,\Omega}^2 + \int_{\mathcal{F}_h} \eta_F h_F^{-1} [\![v]\!]^2 + b(v,v),$$

where the bilinear form b gathers all the remaining terms. It follows from the first part of the proof that

$$\int_{\mathcal{F}_h} \eta_F h_F^{-1} [\![v]\!]^2 \ge c_8 \left(\inf_F \eta_F \right) |v|_j^2 \quad \text{and} \quad b(v,v) \le c_9 \|v\|_{V(h)} |v|_j.$$

Therefore,

$$a_h(v,v) \ge |v|_{h,1,\Omega}^2 + c_8 \left(\inf_F \eta_F \right) |v|_j^2 - c_9 \|v\|_{V(h)} |v|_j,$$

and the last term in the right-hand side can be lower bounded in the form $-c_9 \|v\|_{V(h)} |v|_j \ge -\epsilon \|v\|_{V(h)}^2 - \frac{c_9^2}{4\epsilon} |v|_j^2$ for any positive ϵ. Moreover, using an inverse inequality on V_h yields

$$\|v\|_{V(h)}^2 \le c_{10}(|v|_{h,1,\Omega}^2 + |v|_j^2).$$

Coercivity follows by taking ϵ small enough and $\inf_F \eta_F$ large enough.   □

The above results show that the LDG method approximates the exact solution to $\mathcal{O}(h^k)$ in the H^1-norm and to $\mathcal{O}(h^{k+1})$ in the L^2-norm.

Example 2 (NIPG). The so-called Non-symmetric Interior Penalty Galerkin (NIPG) method has been derived in [OdB98, BaO99] and further investigated in [RiW99]. Written within the above framework, it consists of taking the numerical fluxes

$$\phi_u(u_h) = \begin{cases} \{u_h\} + n_K \cdot [\![u_h]\!] & \text{on } \mathcal{F}_h^i, \\ 0 & \text{on } \mathcal{F}_h^\partial, \end{cases} \tag{3.62}$$

and

$$\phi_\sigma(u_h, \sigma_h) = \{\nabla_h u_h\} - \eta_F h_F^{-1} [\![u_h]\!] \quad \text{on } \mathcal{F}_h. \tag{3.63}$$

Note that ϕ_u is not single-valued on $\mathcal{F}_h^i$. A straightforward calculation yields the following:

Proposition 3.56. *The numerical fluxes ϕ_u and ϕ_σ given by (3.62)–(3.63) are consistent, but not conservative.*

In the NIPG method, the bilinear form a_h is given by

$$a_h(u_h, v) = \int_\Omega \nabla_h u_h \cdot \nabla_h v + \int_{\mathcal{F}_h} \eta_F h_F^{-1} [\![u_h]\!] [\![v]\!]$$
$$+ \int_{\mathcal{F}_h} [\![u_h]\!] \cdot \{\nabla_h v\} - \{\nabla_h u_h\} [\![v]\!]. \tag{3.64}$$

Proposition 3.57. *The bilinear form a_h given by (3.64) is continuous on $V(h)$ and coercive on V_h.*

Proof. Similar to that of Proposition 3.55.   □

The above results show that the NIPG method approximates the exact solution to $\mathcal{O}(h^k)$ in the H^1-norm. However, because of the lack of conservativity in the numerical fluxes, an improved error estimate in the L^2-norm cannot be derived in general.

Remark 3.58.

(i) Because of the skew-symmetric form of the face integrals in (3.64), $\inf_F \eta_F$ needs not be large to ensure the coercivity of a_h. However, skew-symmetry is at the origin of the lack of adjoint-consistency, thus preventing optimal convergence order in the L^2-norm.

(ii) For a face $F \in \mathcal{F}_h$, one can choose the penalty parameter η_F to be proportional to a negative power of h_F, leading to the so-called superpenalty procedure. It is then possible to recover optimal convergence order for the error in the L^2-norm. The NIPG method with superpenalty is analyzed in [RiW99].   □

3.2.5 Numerical illustrations

This section presents two examples of finite element approximations to elliptic PDEs. The purpose of the first example is to illustrate the link between the convergence order of the finite element approximation and the regularity of the exact solution. The purpose of the second example is to illustrate qualitatively the behavior of the solution of advection–diffusion equations depending on whether advection effects dominate or not.

Convergence tests. Consider the Laplace equation in the domain $\Omega = \,]0,1[\times\,]0,1[$ and a positive parameter α. Choose the right-hand side f and the non-homogeneous Dirichlet conditions so that the exact solution is $u(x_1, x_2) = (x_1^2 + x_2^2)^{\frac{\alpha}{2}}$. Note that $u \in H^1(\Omega)$ if $0 < \alpha \leq 1$, $u \in H^2(\Omega)$ if $1 < \alpha \leq 2$, and $u \in H^3(\Omega)$ if $2 < \alpha \leq 3$. In the numerical experiments, we consider the values $\alpha = 0.25$, 1.25, and 2.25. A H^1-conformal Lagrange finite element approximation of degree $k = 1$ or 2 is implemented. The triangulation

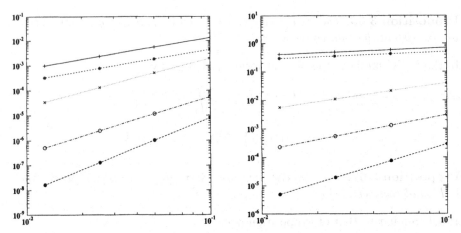

Fig. 3.1. Errors in the L^2-norm (left) and H^1-norm (right) as a function of the mesh step size h: $\mathbb{P}_1$ finite element and $\alpha = 0.25$ (+); $\mathbb{P}_2$ finite element and $\alpha = 0.25$ (*); $\mathbb{P}_1$ finite element and $\alpha = 1.25$ (×); $\mathbb{P}_2$ finite element and $\alpha = 1.25$ (○); $\mathbb{P}_2$ finite element and $\alpha = 2.25$ (•).

of Ω is uniform with vertices of the triangles given by (ih, jh), $0 \leq i, j \leq N+1$, where $h = \frac{1}{N+1}$ and N is a given integer.

Figure 3.1 presents the error in the L^2- and H^1-norms as a function of h. Results are presented in a log-log scale so that the slopes indicate orders of convergence. For $\alpha = 0.25$ and $k = 1$, the error converges "slowly" to zero as $h \to 0$, with a slope lower than 1 in the H^1-norm and lower than 2 in the L^2-norm. For $\alpha = 1.25$ and still $k = 1$, the slope is equal to 1 in the H^1-norm and to 2 in the L^2-norm. Moreover, using a higher-order method ($k = 2$) does not improve the convergence order. Finally, for $\alpha = 2.25$ and with a second-order finite element, the slopes in both the H^1-norm and the L^2-norm are one order higher than those obtained with the first-order finite element method, in agreement with theoretical predictions. As a conclusion, only when the exact solution is smooth enough does it pay off to use a high-order finite element method.

Advection–diffusion equation. Consider a two-dimensional flow through a heated pipe. The flow velocity is assumed to be known, and we want to evaluate the temperature u inside the pipe at steady-state. The temperature is governed by the advection–diffusion equation

$$\beta \cdot \nabla u - \epsilon \Delta u = 0. \tag{3.65}$$

The pipe is modeled by a rectangular domain Ω with sides numbered clockwise from 1 to 4 starting from the left-most side. The flow enters the pipe through $\partial \Omega_1$ and flows out through $\partial \Omega_3$ while the sides $\partial \Omega_2$ and $\partial \Omega_4$ are solid boundaries. Spatial coordinates are denoted by (x_1, x_2) with the x_1-axis parallel to the pipe axis. Temperature boundary conditions are $u = 0$

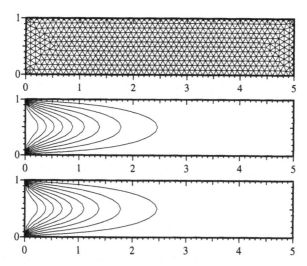

Fig. 3.2. Heat transfer problem through a two-dimensional pipe: computational mesh (top); temperature field for dominant diffusion (center); and temperature field for dominant advection (bottom).

on $\partial\Omega_1$ (cold upstream flow), $u = 1$ on $\partial\Omega_2$ and $\partial\Omega_4$ (heated boundaries), and $\partial_1 u = 0$ on $\partial\Omega_3$ (outflow condition). The flow velocity is taken to be $\beta = (4x_2(1 - x_2), 0)$. The solution to (3.65) is approximated on the mesh shown in the top panel of Figure 3.2 using continuous $\mathbb{P}_1$ finite elements. The central panel of Figure 3.2 presents isotherms for a diffusion-dominated case ($\epsilon = 10^{-1}$); the peak temperature is quickly reached on the symmetry axis. The bottom panel displays isotherms resulting from a moderate diffusion coefficient ($\epsilon = 10^{-3}$). Advection effects are dominant, i.e., the boundary layer in which the temperature undergoes significant variations remains localized near the top and bottom boundaries. If advection effects become even more dominant, the approximation method needs to be stabilized to avoid spurious oscillations in the solution profile; see Chapter 5.

3.3 Spectral Problems

This section contains a brief introduction to spectral problems and their approximation by finite element methods. Spectral problems occur when analyzing the response of buildings, vehicles, or aircrafts to vibrations. Henceforth, we restrict the presentation to a simple model problem: the Laplace operator with homogeneous Dirichlet conditions. Although this problem is somewhat simple, it is representative of a large class of engineering applications. As such, it models membrane and string vibrations.

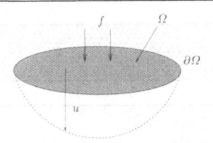

Fig. 3.3. Elastic deformation of a membrane: reference configuration Ω, externally applied load f, and equilibrium displacement u. The boundary $\partial\Omega$ of the membrane is kept fixed.

3.3.1 Modeling a vibrating membrane

Figure 3.3 presents an elastic homogeneous membrane. In the reference configuration, the membrane occupies the domain Ω in $\mathbb{R}^2$ and is tightened according to a two-dimensional stress tensor $\sigma \in \mathbb{R}^{2,2}$. For the sake of simplicity, we assume that σ is uniform and isotropic, i.e., $\sigma = \tau\mathcal{I}$ where τ is the membrane tension. Apply now a transverse load f and assume first that f is *time-independent*. If the strains in the membrane are sufficiently small, the equilibrium configuration is described by a transverse displacement which is a function $u : \Omega \to \mathbb{R}$ governed by the PDE

$$-\tau\Delta u = f \quad \text{in } \Omega. \tag{3.66}$$

We assume that the boundary of the membrane is kept fixed, yielding the homogeneous Dirichlet condition $u = 0$ on $\partial\Omega$.

Consider now the *time-dependent* load $f(x,t) = g(x)\cos(\omega t)$ for $(x,t) \in Q$, where $g : \Omega \to \mathbb{R}$ is a given function, ω a real parameter representing the angular velocity of the excitation, $Q = \Omega \times]0,T[$, and T a given time. Assuming again that the strains in the membrane remain sufficiently small, the (time-dependent) displacement $u : Q \to \mathbb{R}$ is governed by the PDE

$$\rho\partial_{tt}u - \tau\Delta u = g(x)\cos(\omega t) \quad \text{in } Q, \tag{3.67}$$

where ρ is the membrane density. Equation (3.67) is a *wave equation* with celerity $c = (\tau\rho^{-1})^{\frac{1}{2}}$. It has to be supplemented with initial and boundary conditions. The initial data comprises the initial value of the displacement $u_0(x)$ and its time-derivative $u_1(x)$, i.e., the initial membrane velocity. We assume that the membrane boundary is kept fixed at all times., i.e., we enforce a homogeneous Dirichlet boundary condition.

3.3.2 The spectral problem

Consider the *spectral problem*:

$$\begin{cases} \text{Seek } \psi \in H_0^1(\Omega), \ \psi \neq 0, \text{ and } \lambda \in \mathbb{R} \text{ such that} \\ -\Delta\psi = \lambda\psi, \end{cases}$$

for which a weak formulation is

$$\begin{cases} \text{Seek } \psi \in H_0^1(\Omega), \ \psi \neq 0, \text{ and } \lambda \in \mathbb{R} \text{ such that} \\ \int_\Omega \nabla\psi \cdot \nabla v = \lambda \int_\Omega \psi v, \quad \forall v \in H_0^1(\Omega). \end{cases} \tag{3.68}$$

Definition 3.59. *Let $\{\lambda, \psi\}$ be a solution to (3.68). The real λ is called an eigenvalue of the Laplacian (with homogeneous Dirichlet conditions) and the function ψ an eigenfunction.*

Theorem 3.60 (Spectral decomposition). *Let Ω be a domain in $\mathbb{R}^d$. Then, the spectral problem (3.68) admits infinitely many solutions. These solutions form a sequence $\{\lambda_n, \psi_n\}_{n>0}$ such that:*

(i) *$\{\lambda_n\}_{n>0}$ is an increasing sequence of positive numbers, and $\lambda_n \to \infty$.*
(ii) *$\{\psi_n\}_{n>0}$ is an orthonormal Hilbert basis of $L^2(\Omega)$.*

Proof. This is a consequence of the fact that the injection $H_0^1(\Omega) \subset L^2(\Omega)$ is compact; see Theorem B.46 and [Yos80, p. 284] or [Bre91, p. 192]. □

Example 3.61. For $\Omega =]0,1[$, the eigenvalues of the Laplacian are $\lambda_n = n^2\pi^2$ with corresponding eigenfunctions $\psi_n(x) = \sin(n\pi x)$. These functions become more and more oscillatory as n grows. □

The solution u to the wave equation (3.67) can be written as a series in terms of the Laplacian eigenfunctions. Indeed, set $\omega_n = (\lambda_n \tau \rho^{-1})^{\frac{1}{2}}$ and assume $\omega \neq \omega_n$. Denote by $g_n = \int_\Omega g\psi_n$ the coordinates of g relative to the orthonormal basis $\{\psi_n\}_{n>0}$ and by α_n and β_n the coordinates of the initial data u_0 and u_1, respectively. A straightforward calculation shows that for $\omega \neq \omega_n$,

$$u(x,t) = \sum_{n=1}^\infty \Bigg(\alpha_n \cos(\omega_n t) + \beta_n \sin(\omega_n t)$$

$$+ \frac{g_n}{\rho(\omega + \omega_n)} \frac{\sin\left(\frac{\omega - \omega_n}{2}t\right)}{\frac{\omega - \omega_n}{2}} \sin\left(\frac{\omega + \omega_n}{2}t\right) \Bigg) \psi_n(x).$$

As ω draws closer to one of the ω_n's, a resonance phenomenon occurs. In particular, when $\omega = \omega_n$, $u(x,t)$ grows linearly in time.

3.3.3 The Rayleigh quotient

Set $a(u,v) = (\nabla u, \nabla v)_{0,\Omega}$ for all u, v in $H_0^1(\Omega)$. This bilinear form is symmetric, continuous, and coercive on $H_0^1(\Omega)$. The *Rayleigh quotient* of a function $u \in H_0^1(\Omega)$, $u \neq 0$, is defined to be

$$R(u) = \frac{a(u,u)}{\|u\|_{0,\Omega}^2}.$$

Proposition 3.62. *Let λ_1 be the smallest eigenvalue of the spectral problem (3.68) and let ψ_1 be a corresponding eigenfunction. Then,*

$$\lambda_1 = R(\psi_1) = \inf_{v \in H_0^1(\Omega)} R(v).$$

Proof. Clearly, $\lambda_1 = R(\psi_1) \geq \inf_{v \in H_0^1(\Omega)} R(v)$. Furthermore, for $v \in H_0^1(\Omega)$, the identity $v = \sum_{n=1}^{\infty} v_n \psi_n$ yields

$$R(v) = \frac{\sum_{n=1}^{\infty} \lambda_n v_n^2}{\sum_{n=1}^{\infty} v_n^2} \geq \lambda_1. \qquad \square$$

Proposition 3.63. *Let λ_m be the m-th eigenvalue of problem (3.68) (eigenvalues are counted with their multiplicity and ordered increasingly). Let V_m denote the set of subspaces of $H_0^1(\Omega)$ having dimension m. Then,*

$$\lambda_m = \min_{E_m \in V_m} \max_{v \in E_m} R(v). \tag{3.69}$$

Proof. Let $E_m = \text{span}\{\psi_1, \ldots, \psi_m\}$ be the space spanned by the m first eigenfunctions. For all $v = \sum_{n=1}^{m} v_n \psi_n$ in E_m,

$$R(v) = \frac{\sum_{n=1}^{m} \lambda_n v_n^2}{\sum_{n=1}^{m} v_n^2} \leq \lambda_m,$$

which yields

$$\lambda_m \geq \min_{E_m \in V_m} \max_{v \in E_m} R(v).$$

Consider now $E_m \in V_m$. A simple dimensional argument shows that there exists $v \neq 0$ in $E_m \cap E_{m-1}^{\perp}$. Since v can be written in the form $v = \sum_{n=m}^{\infty} v_n \psi_n$, it is clear that $R(v) \geq \lambda_m$. As a result, $\max_{v \in E_m} R(v) \geq \lambda_m$; hence,

$$\lambda_m \leq \min_{E_m \in V_m} \max_{v \in E_m} R(v). \qquad \square$$

3.3.4 H^1-conformal approximation

The spectral problem (3.68) can be solved analytically only in a limited number of remarkable cases when the domain Ω has a very simple shape. In the general case, eigenvalues and eigenfunctions must be approximated using, for instance, a finite element method.

Let $\{\mathcal{T}_h\}_{h>0}$ be a family of geometrically conformal meshes of Ω and let $\{V_h\}_{h>0}$ be the corresponding family of H^1-conformal approximation spaces. Denote by N the dimension of V_h. The approximate spectral problem we consider is the following:

$$\begin{cases} \text{Seek } \psi_h \in V_h, \ \psi_h \neq 0, \text{ and } \lambda_h \in \mathbb{R} \text{ such that} \\ \int_{\Omega} \nabla \psi_h \cdot \nabla v_h = \lambda_h \int_{\Omega} \psi_h v_h, \quad \forall v_h \in V_h. \end{cases} \tag{3.70}$$

Let $\{\varphi_1, \ldots, \varphi_N\}$ be a basis of V_h and let $\Psi_h \in \mathbb{R}^N$ be the coordinate vector of Ψ_h relative to this base. The approximate problem (3.70) is recast in the form:

$$\begin{cases} \text{Seek } \Psi_h \in \mathbb{R}^N, \ \Psi_h \neq 0, \text{ and } \lambda_h \in \mathbb{R} \text{ such that} \\ \mathcal{A}\Psi_h = \lambda_h \mathcal{M}\Psi_h, \end{cases} \tag{3.71}$$

where the *stiffness matrix* $\mathcal{A}$ and the *mass matrix* $\mathcal{M}$ have entries

$$\mathcal{A}_{ij} = \int_\Omega \nabla\varphi_i \cdot \nabla\varphi_j \qquad \text{and} \qquad \mathcal{M}_{ij} = \int_\Omega \varphi_i \varphi_j. \tag{3.72}$$

Because the matrix $\mathcal{M}$ is not the identity matrix, problem (3.71) is often called a generalized eigenvalue problem.

Proposition 3.64. *The matrices $\mathcal{A}$ and $\mathcal{M}$ defined in (3.72) are symmetric positive definite. Furthermore, the spectral problem (3.71) admits N (positive) eigenvalues (counted with their multiplicity).*

Proof. The symmetry and positive definiteness of the matrices $\mathcal{A}$ and $\mathcal{M}$ directly results from the fact that they are Gram matrices; see also Remark 2.20. Orthogonalizing the quadratic form associated with $\mathcal{A}$ with respect to the scalar product induced by $\mathcal{M}$ yields N positive reals $\{\lambda_{h1}, \ldots, \lambda_{hN}\}$ and a basis $\{\Psi_{h1}, \ldots, \Psi_{hN}\}$ of $\mathbb{R}^N$ such that, for $1 \leq i, j \leq N$,

$$(\Psi_{hi}, \mathcal{A}\Psi_{hj})_N = \lambda_{hi}\delta_{ij}, \qquad (\Psi_{hi}, \mathcal{M}\Psi_{hj})_N = \delta_{ij},$$

where $(\cdot, \cdot)_N$ denotes the Euclidean product in $\mathbb{R}^N$. As a result,

$$\mathcal{A}\Psi_{hi} = \lambda_{hi}\mathcal{M}\Psi_{hi}, \qquad 1 \leq i \leq N,$$

showing that the λ_{hi}'s are the eigenfunctions of the generalized eigenvalue problem (3.71) and that the Ψ_{hi}'s are the corresponding eigenvectors. $\qquad \square$

3.3.5 Error analysis

Let $\{\psi_{h1}, \ldots, \psi_{hN}\}$ be an orthonormal basis of eigenvectors in V_h, i.e., $(\psi_{hi}, \psi_{hj})_{0,\Omega} = \delta_{ij}$ for $1 \leq i, j \leq N$, and assume that the enumeration of these vectors is such that $\lambda_{h1} \leq \ldots \leq \lambda_{hN}$.

Henceforth, $m \geq 1$ denotes a fixed number, and we assume that h is small enough so that $m \leq N$. Set $V_m = \mathrm{span}\{\psi_1, \ldots, \psi_m\}$, and define S_m to be the unit sphere of V_m in $L^2(\Omega)$. Introduce the elliptic projector $\Pi_h : H_0^1(\Omega) \to V_h$ such that $a(\Pi_h u - u, v_h) = 0$ for all v_h in V_h, and define

$$\sigma_{hm} = \inf_{v \in S_m} \|\Pi_h v\|_{0,\Omega}. \tag{3.73}$$

Lemma 3.65. *Let $1 \leq m \leq N$. Assume $\sigma_{hm} \neq 0$. Then,*

$$\lambda_m \leq \lambda_{hm} \leq \lambda_m \sigma_{hm}^{-2}. \tag{3.74}$$

Proof. The first inequality is a simple consequence of Proposition 3.63. Furthermore, since $\sigma_{hm} \neq 0$, $\text{Ker}(\Pi_h) \cap V_m = \{0\}$; hence, the Rank Theorem implies $\dim(\Pi_h V_m) = m$. Adapting the proof of Proposition 3.63, one readily infers

$$\lambda_{hm} \leq \max_{v_h \in \Pi_h V_m} \frac{a(v_h, v_h)}{\|v_h\|_{0,\Omega}^2} = \max_{v \in V_m} \frac{a(\Pi_h v, \Pi_h v)}{\|\Pi_h v\|_{0,\Omega}^2}.$$

Hence,

$$\lambda_{hm} \leq \max_{v \in V_m} \frac{a(v, v)}{\|\Pi_h v\|_{0,\Omega}^2} \leq \max_{v \in V_m} R(v) \max_{v \in V_m} \frac{\|v\|_{0,\Omega}^2}{\|\Pi_h v\|_{0,\Omega}^2} = \frac{1}{\sigma_{hm}^2} \max_{v \in S_m} R(v).$$

Then, use $\lambda_m = \max_{v \in S_m} R(v)$ to conclude. $\square$

Remark 3.66. It is remarkable that, independently of the approximation space (provided conformity holds), the N eigenvalues of the approximate problem (3.71) are larger than the corresponding eigenvalues of the exact problem (3.68). Eigenvalues are thus approximated from above. $\square$

Lemma 3.67. *Let $1 \leq m \leq N$. There is $c(m)$, independent of h, such that*

$$\sigma_{hm}^2 \geq 1 - c(m) \max_{v \in S_m} \|v - \Pi_h v\|_{1,\Omega}^2. \tag{3.75}$$

Proof. Let $v \in S_m$. Let $(V_i)_{1 \leq i \leq m}$ be the coordinate vector of v relative to the basis $\{\psi_1, \ldots, \psi_m\}$. It is clear that $\|v\|_{0,\Omega}^2 = \sum_{1 \leq i \leq m} V_i^2 = 1$. In addition, $\|\Pi_h v\|_{0,\Omega}^2$ is bounded from below as follows:

$$\|\Pi_h v\|_{0,\Omega}^2 \geq \|v\|_{0,\Omega}^2 - 2(v, v - \Pi_h v)_{0,\Omega}. \tag{3.76}$$

Using the symmetry of a and the definition of $\Pi_h v$ yields

$$(v, v - \Pi_h v)_{0,\Omega} = \sum_{1 \leq i \leq m} V_i (\psi_i, v - \Pi_h v)_{0,\Omega} = \sum_{1 \leq i \leq m} \frac{V_i}{\lambda_i} a(\psi_i, v - \Pi_h v)$$

$$= \sum_{1 \leq i \leq m} \frac{V_i}{\lambda_i} a(\psi_i - \Pi_h \psi_i, v - \Pi_h v)$$

$$\leq \frac{\|a\|}{\lambda_1} \|v - \Pi_h v\|_{1,\Omega} \left(\sum_{1 \leq i \leq m} \|\psi_i - \Pi_h \psi_i\|_{1,\Omega}^2 \right)^{\frac{1}{2}}$$

$$\leq \sqrt{m} \frac{\|a\|}{\lambda_1} \|v - \Pi_h v\|_{1,\Omega} \sup_{w \in S_m} \|w - \Pi_h w\|_{1,\Omega}$$

$$\leq \sqrt{m} \frac{\|a\|}{\lambda_1} \sup_{w \in S_m} \|w - \Pi_h w\|_{1,\Omega}^2.$$

Then, the desired estimate is obtained by inserting this bound into (3.76) and setting $c(m) = 2\sqrt{m} \frac{\|a\|}{\lambda_1}$. $\square$

Lemma 3.68. *Assume that the sequence of approximation spaces $\{V_h\}_{h>0}$ is endowed with the following approximability property:*

$$\forall v \in H_0^1(\Omega), \quad \lim_{h \to 0} \left(\inf_{v_h \in V_h} \|v - v_h\|_{1,\Omega} \right) = 0. \tag{3.77}$$

Then, for all $m \geq 1$, there is $h_0(m)$ such that, for all $h \leq h_0(m)$,

$$0 \leq \lambda_{hm} - \lambda_m \leq 2\lambda_m c(m) \max_{v \in S_m} \inf_{v_h \in V_h} \|v - v_h\|_{1,\Omega}^2. \tag{3.78}$$

Proof. Let $m \geq 1$ be a fixed number, and assume that h is small enough so that $m \leq N$. Since S_m is compact, there is v_0 in S_m such that $\sup_{v \in S_m} \|v - \Pi_h v\|_{1,\Omega}^2 = \|v_0 - \Pi_h v_0\|_{1,\Omega}^2$. Owing to (2.24),

$$\|v_0 - \Pi_h v_0\|_{1,\Omega} \leq \left(\frac{\|a\|}{\alpha} \right)^{\frac{1}{2}} \inf_{v_h \in V_h} \|v_0 - v_h\|_{1,\Omega}.$$

Since m is fixed, (3.77) implies that there is $h_0(m)$ such that, for all $h \leq h_0(m)$, $c(m)\|v_0 - \Pi_h v_0\|_{1,\Omega}^2 \leq \frac{1}{2}$. Then, observing that $1 + 2x \geq \frac{1}{1-x}$ for all $0 \leq x \leq \frac{1}{2}$ and using (3.75) yields

$$1 + 2c(m)\|v_0 - \Pi_h v_0\|_{1,\Omega}^2 = 1 + 2c(m) \sup_{v \in S_m} \|v - \Pi_h v\|_{1,\Omega}^2 \geq \sigma_{hm}^{-2}.$$

Conclude using (3.74). □

To analyze the approximation error for eigenvectors, we assume, for the sake of simplicity, that the eigenvalues are simple.

Lemma 3.69. *Let $1 \leq m \leq N$ and set $\rho_{hm} = \max_{1 \leq i \neq m \leq N} \frac{\lambda_m}{|\lambda_m - \lambda_{hi}|}$. If λ_m is simple, there is $h_0(m)$ and a choice of eigenvector such that, for all $h \leq h_0(m)$,*

$$\|\psi_m - \psi_{hm}\|_{0,\Omega} \leq 2(1 + \rho_{hm})\|\psi_m - \Pi_h \psi_m\|_{0,\Omega}. \tag{3.79}$$

Proof. (1) Note that owing to Lemma 3.68, $\lambda_{hi} \to \lambda_i$ as $h \to 0$. Hence, since λ_m is simple, ρ_{hm} is uniformly bounded when h is small enough.
(2) Define $v_{hm} = (\Pi_h \psi_m, \psi_{hm})_{0,\Omega} \psi_{hm}$ and let us evaluate $\|\Pi_h \psi_m - v_{hm}\|_{0,\Omega}$. Note first that

$$(\Pi_h \psi_m, \psi_{hi})_{0,\Omega} = \frac{1}{\lambda_{hi}} a(\psi_{hi}, \Pi_h \psi_m) = \frac{1}{\lambda_{hi}} a(\psi_m, \psi_{hi}) = \frac{\lambda_m}{\lambda_{hi}}(\psi_m, \psi_{hi})_{0,\Omega}.$$

Hence, $(\Pi_h \psi_m, \psi_{hi})_{0,\Omega} = \frac{\lambda_m}{\lambda_{hi} - \lambda_m}(\psi_m - \Pi_h \psi_m, \psi_{hi})_{0,\Omega}$. As a result,

$$\|\Pi_h \psi_m - v_{hm}\|_{0,\Omega}^2 = \sum_{1 \leq i \neq m \leq N} (\Pi_h \psi_m, \psi_{hi})_{0,\Omega}^2 \leq \rho_{hm}^2 \|\psi_m - \Pi_h \psi_m\|_{0,\Omega}^2. \tag{3.80}$$

(3) Let us now estimate $\|\psi_{hm} - v_{hm}\|_{0,\Omega}$. Since

$$\|\psi_m\|_{0,\Omega} - \|\psi_m - v_{hm}\|_{0,\Omega} \le \|v_{hm}\|_{0,\Omega} \le \|\psi_m\|_{0,\Omega} + \|\psi_m - v_{hm}\|_{0,\Omega},$$

and $\|\psi_m\|_{0,\Omega} = 1$, we infer $\left| \|v_{hm}\|_{0,\Omega} - 1 \right| \le \|\psi_m - v_{hm}\|_{0,\Omega}$. But,

$$\|\psi_{hm} - v_{hm}\|_{0,\Omega} = |(\Pi_h\psi_m - \psi_{hm}, \psi_{hm})_{0,\Omega}| = |(\Pi_h\psi_m, \psi_{hm})_{0,\Omega} - 1|.$$

Assume that ψ_{hm} is chosen so that $(\Pi_h\psi_m, \psi_{hm})_{0,\Omega} \ge 0$. Then, $\|v_{hm}\|_{0,\Omega} = (\Pi_h\psi_m, \psi_{hm})_{0,\Omega}$, yielding

$$\|\psi_{hm} - v_{hm}\|_{0,\Omega} \le \|\psi_m - v_{hm}\|_{0,\Omega}. \tag{3.81}$$

(4) To conclude, use the triangle inequality together with (3.80) and (3.81):

$$\|\psi_m - \psi_{hm}\|_{0,\Omega} \le \|\psi_m - \Pi_h\psi_m\|_{0,\Omega} + \|\Pi_h\psi_m - v_{hm}\|_{0,\Omega} + \|v_{hm} - \psi_{hm}\|_{0,\Omega}$$
$$\le 2(\|\psi_m - \Pi_h\psi_m\|_{0,\Omega} + \|\Pi_h\psi_m - v_{hm}\|_{0,\Omega}).$$

The conclusion follows from (3.80).   □

Theorem 3.70. *Let $1 \le m \le N$. If λ_m is simple, there is $h_0(m)$ and a choice of eigenvector such that, for all $h \le h_0(m)$,*

$$\|\psi_m - \psi_{hm}\|_{0,\Omega} \le c_2(m)\|\psi_m - \Pi_h\psi_m\|_{0,\Omega}, \tag{3.82}$$

$$\|\psi_m - \psi_{hm}\|_{1,\Omega} \le c_1(m) \max_{v \in S_m} \inf_{v_h \in V_h} \|v - v_h\|_{1,\Omega}. \tag{3.83}$$

Proof. Estimate (3.82) is a direct consequence of Lemma 3.69. To control $\|\psi_m - \psi_{hm}\|_{1,\Omega}$, use the coercivity of a as follows:

$$\alpha\|\psi_m - \psi_{hm}\|^2_{1,\Omega} \le a(\psi_m - \psi_{hm}, \psi_m - \psi_{hm})$$
$$= \lambda_{hm} + \lambda_m - 2\lambda_m(\psi_m, \psi_{hm})_{0,\Omega}$$
$$= \lambda_{hm} - \lambda_m + \lambda_m\|\psi_m - \psi_{hm}\|^2_{0,\Omega}.$$

Then, (3.83) is a consequence of the above equality, together with Lemmas 3.68 and 3.69.   □

Corollary 3.71. *Let $1 \le m \le N$. Assume that the approximation setting is such that there is $k \ge 1$ and $c_1(m)$ so that $\inf_{v \in S_m} \|\Pi_h v - v\|_{0,\Omega} + h\|\Pi_h v - v\|_{1,\Omega} \le c_1(m)h^{k+1}$. Then, there are $c_2(m)$, $c_3(m)$, $c_4(m)$, independent of h, such that, if h is sufficiently small, the following estimates hold:*

$$\lambda_m \le \lambda_{hm} \le \lambda_m + c_2(m) h^{2k}\lambda_m^2. \tag{3.84}$$

Moreover, if the eigenvalue λ_m is simple,

$$\begin{cases} \|\psi_m - \psi_{hm}\|_{0,\Omega} \le c_3(m) h^{k+1}\lambda_m, \\ \|\psi_m - \psi_{hm}\|_{1,\Omega} \le c_4(m) h^k\lambda_m, \end{cases} \tag{3.85}$$

and the constants $c_2(m)$, $c_3(m)$, $c_4(m)$ grow unboundedly as $m \to +\infty$. If λ_m is multiple, ψ_m can be chosen so that (3.85) still holds.

Proof. Simple consequence of Lemma 3.68 and Theorem 3.70.   □

Remark 3.72. The above corollary shows that when h is fixed, the accuracy of the approximation decreases as m increases since $c_2(m)$, $c_3(m)$, and $c_4(m)$ grow unboundedly as $m \to +\infty$; see §3.3.6 for an illustration.   □

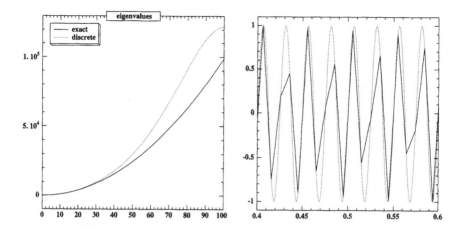

Fig. 3.4. Left: Finite element approximation to the eigenvalues of the Laplacian in one dimension. Right: Eightieth eigenfunction for the exact problem (dashed line) and for the approximate problem (solid line).

3.3.6 Numerical illustrations

In one dimension. Consider the spectral problem for the Laplacian posed in the domain $\Omega =]0, 1[$, whose solutions are the pairs

$$\{\lambda_m, \psi_m\} = \{m^2\pi^2, \sin(m\pi x)\} \qquad \text{for } m \geq 1.$$

Consider now a uniform mesh of Ω with step size $h = \frac{1}{N+1}$ and a $\mathbb{P}_1$ Lagrange finite element approximation. A straightforward calculation shows that the matrices $\mathcal{A}$ and $\mathcal{M}$ are tridiagonal and given by

$$\mathcal{A} = \frac{1}{h}\text{tridiag}(-1, 2, -1), \qquad \mathcal{M} = \frac{h}{6}\text{tridiag}(1, 4, 1).$$

The eigenvalues of the approximate problem (3.71) are easily shown to be

$$\lambda_{hm} = \frac{6}{h^2}\left(\frac{1 - \cos(m\pi h)}{2 + \cos(m\pi h)}\right), \qquad 1 \leq m \leq N.$$

The left panel in Figure 3.4 presents the first 100 eigenvalues of both the exact and the approximate problems, the latter being obtained with a mesh containing $N = 100$ points. The exact eigenvalues are approximated from above, as predicted by the theory. We also observe that only the first eigenvalues are approximated accurately. Eigenfunctions corresponding to large eigenvalues oscillate too much to be represented accurately on the mesh; see the right panel in Figure 3.4. To approximate the m-th eigenvalue with a relative accuracy of ϵ, i.e., $|\lambda_{hm} - \lambda_m| < \epsilon\lambda_m$, a mesh with step size lower than $\frac{\sqrt{\epsilon}}{m}$ must be used. In the present example, only the first 10 eigenvalues are approximated within 1% accuracy.

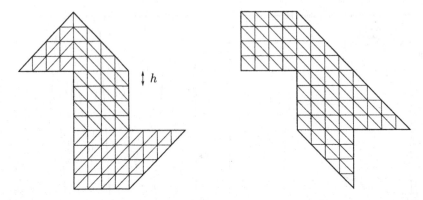

Fig. 3.5. Two domains on which the Laplacian has the same spectrum: the hen-shaped domain (left) and the arrow-shaped domain (right). The coarsest meshes used for the finite element approximation are shown. The length scale is such that the area of the two domains is equal to $\frac{7}{2}$ and that the meshes correspond to $h = \frac{1}{4}$.

Shape	Hen			Arrow		
Mesh size	$\frac{1}{4}$	$\frac{1}{8}$	$\frac{1}{16}$	$\frac{1}{4}$	$\frac{1}{8}$	$\frac{1}{16}$
Eigenvalue 1	11.16	10.44	10.24	11.03	10.42	10.24
Eigenvalue 2	16.37	15.09	14.76	16.19	15.06	14.75
Eigenvalue 3	24.45	21.67	20.98	24.07	21.64	20.98

Table 3.2. First three eigenvalues for the hen- and arrow-shaped domains obtained with a first-order finite element method on three meshes of increasing refinement.

In two dimensions. Relating the spectrum and the shape of a two-dimensional membrane is a nontrivial task. For instance, knowing the spectrum $\{\lambda_m\}_{m\geq 1}$, is it possible to reconstruct the shape of the domain Ω (or, in other words, can we hear the shape of a drum)? The answer is negative, as proven recently by Gordon and Webb [GoW96] who discovered two domains in $\mathbb{R}^2$ having exactly the same spectrum. These domains take on the shape of a "hen" and an "arrow" as depicted in Figure 3.5. We verify numerically that the first eigenvalues of these two domains indeed coincide. Eigenvalues are computed using the $\mathbb{P}_1$ Lagrange finite element on a sequence of three meshes that are successively refined. The coarsest meshes are displayed in Figure 3.5; results are presented in Table 3.2. Both sets of eigenvalues converge to a common limit as $h \to 0$. The first two eigenfunctions are shown in Figure 3.6.

3.4 Continuum Mechanics

This section is concerned with PDE systems endowed with a multicomponent coercivity property. Important examples include those arising in continuum mechanics. Hereafter we restrict ourselves to *linear isotropic elasticity*. The

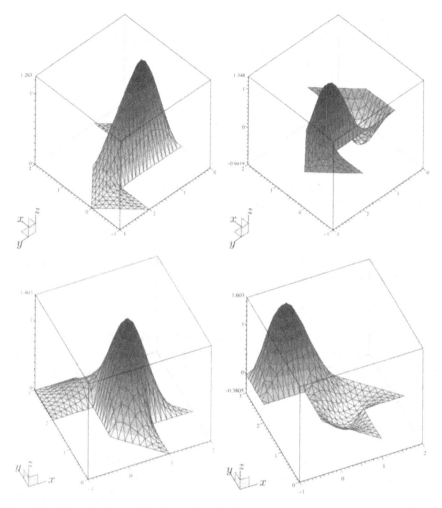

Fig. 3.6. Two first eigenfunctions for the hen-shaped domain (top) and for the arrow-shaped domain (bottom). Courtesy of E. Cancès (ENPC).

first part of this section introduces a setting for the mathematical analysis and the finite element approximation of continuum mechanics problems in this framework. The second part focuses on some problems related to beam flexion.

3.4.1 Model problems and their weak formulation

The physical model. The domain $\Omega \subset \mathbb{R}^3$ represents a deformable medium initially at equilibrium and to which an external load $f : \Omega \to \mathbb{R}^3$ is applied. Our goal is to determine the displacement field $u : \Omega \to \mathbb{R}^3$ induced by f once the system has reached equilibrium again. We assume that the deformations are small enough so that the linear elasticity theory applies.

Let $\sigma : \Omega \to \mathbb{R}^{3,3}$ be the *stress tensor* in the medium. The equilibrium conditions under the external load f can be expressed as

$$\nabla{\cdot}\sigma + f = 0 \quad \text{in } \Omega. \tag{3.86}$$

Let $\varepsilon(u) : \Omega \to \mathbb{R}^{3,3}$ be the (linearized) *strain rate tensor* defined as

$$\varepsilon(u) = \tfrac{1}{2}(\nabla u + \nabla u^T). \tag{3.87}$$

In the framework of linear isotropic elasticity, the stress tensor is related to the strain rate tensor by the relation

$$\sigma(u) = \lambda \operatorname{tr}(\varepsilon(u))\mathcal{I} + 2\mu\varepsilon(u),$$

where λ and μ are the so-called *Lamé coefficients*, and $\mathcal{I}$ is the identity matrix. Using (3.87), the above relation yields

$$\sigma(u) = \lambda(\nabla{\cdot}u)\mathcal{I} + \mu(\nabla u + \nabla u^T). \tag{3.88}$$

The Lamé coefficients λ and μ are phenomenological coefficients. Owing to thermodynamic stability, these coefficients are constrained to be such that $\mu > 0$ and $\lambda + \tfrac{2}{3}\mu \geq 0$. Moreover, for the sake of simplicity, we shall henceforth assume that λ and μ are constant and that $\lambda \geq 0$. In this case, owing to the identity $\nabla{\cdot}\big(\varepsilon(u)\big) = \tfrac{1}{2}\big(\Delta u + \nabla(\nabla{\cdot}u)\big)$, (3.86) and (3.88) yield

$$-\mu\Delta u - (\lambda + \mu)\nabla(\nabla{\cdot}u) = f \quad \text{in } \Omega.$$

The model problem (3.86)–(3.88) must be supplemented with boundary conditions. We investigate two cases: a *mixed problem* in which the displacement is imposed on part of the boundary, and a *pure-traction problem* in which the normal component of the stress tensor is imposed on the entire boundary. The *pure-displacement problem* in which the displacement is imposed on the entire boundary can be treated as a special case of the mixed problem.

Remark 3.73.

(i) The coefficient $\lambda + \tfrac{2}{3}\mu$ describes the compressibility of the medium; very large values correspond to almost incompressible materials.

(ii) Instead of using λ and μ, it is sometimes more convenient to consider the *Young modulus* E and the *Poisson coefficient* ν. These quantities are related to the Lamé coefficients by

$$E = \mu\frac{3\lambda + 2\mu}{\lambda + \mu} \quad \text{and} \quad \nu = \tfrac{1}{2}\frac{\lambda}{\lambda + \mu}.$$

The Poisson coefficient is such that $-1 \leq \nu < \tfrac{1}{2}$, and owing to the assumption $\lambda \geq 0$, we infer $\nu \geq 0$. An almost incompressible material corresponds to a

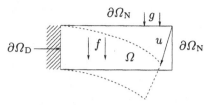

Fig. 3.7. Example of a mixed problem in continuum mechanics.

Poisson coefficient very close to $\frac{1}{2}$.

(iii) The linear isotropic elasticity model is in general valid for problems involving infinitesimal strains. In this case, the medium responds linearly to externally applied loads so that one can normalize the problem and consider arbitrary loads.

(iv) The finite element method originated in the 1950s when engineers developed it to solve continuum mechanics problems in aeronautics; see, e.g., [Lev53, ArK67] and the references cited in [Ode91]. These problems involved complex geometries that could not be easily handled by classical finite difference techniques. At the same time, theoretical researches on the approximation of linear elasticity equations were carried out [TuC56]. In 1960, Clough coined the terminology "finite elements" in a paper dealing with linear elasticity in two dimensions [Clo60].   □

Mixed problem and its weak formulation. Consider the partition $\partial\Omega = \partial\Omega_D \cup \partial\Omega_N$ illustrated in Figure 3.7. The boundary $\partial\Omega_D$ is clamped, whereas a normal load $g : \partial\Omega_N \to \mathbb{R}^3$ is imposed on $\partial\Omega_N$. The model problem we consider is the following:

$$
\begin{cases}
\nabla\cdot\sigma(u) + f = 0 & \text{in } \Omega, \\
\sigma(u) = \lambda(\nabla\cdot u)\mathcal{I} + \mu(\nabla u + \nabla u^T) & \text{in } \Omega, \\
u = 0 & \text{on } \partial\Omega_D, \\
\sigma(u)\cdot n = g & \text{on } \partial\Omega_N.
\end{cases}
\tag{3.89}
$$

To derive a weak formulation for (3.89), take the scalar product of the equilibrium equation with a test function $v : \Omega \to \mathbb{R}^3$. Since $\int_\Omega -(\nabla\cdot\sigma(u))\cdot v = \int_\Omega \sigma(u){:}\nabla v - \int_{\partial\Omega} v\cdot\sigma(u)\cdot n$ and $\sigma(u){:}\nabla v = \sigma(u){:}\varepsilon(v)$ owing to the symmetry of $\sigma(u)$,

$$
\int_\Omega \sigma(u){:}\varepsilon(v) - \int_{\partial\Omega} v\cdot\sigma(u)\cdot n = \int_\Omega f\cdot v.
$$

The displacement u and the test function v are taken in the functional space

$$
V_{DN} = \{v \in [H^1(\Omega)]^3; \ v = 0 \text{ on } \partial\Omega_D\},
\tag{3.90}
$$

equipped with the norm $\|v\|_{1,\Omega} = \sum_{i=1}^3 \|v_i\|_{1,\Omega}$ where $v = (v_1, v_2, v_3)^T$. The weak formulation of (3.89) is thus:

$$\begin{cases} \text{Seek } u \in V_{\text{DN}} \text{ such that} \\ a(u,v) = \int_\Omega f{\cdot}v + \int_{\partial\Omega_{\text{N}}} g{\cdot}v, \quad \forall v \in V_{\text{DN}}, \end{cases} \tag{3.91}$$

with the bilinear form

$$a(u,v) = \int_\Omega \sigma(u){:}\varepsilon(v) = \int_\Omega \lambda\,\nabla{\cdot}u\,\nabla{\cdot}v + \int_\Omega 2\mu\,\varepsilon(u){:}\varepsilon(v). \tag{3.92}$$

In continuum mechanics, the test function v plays the role of a virtual displacement and the weak formulation (3.91) expresses the *principle of virtual work*.

Proposition 3.74. *Let Ω be a domain in $\mathbb{R}^3$, consider the partition $\partial\Omega = \partial\Omega_{\text{D}} \cup \partial\Omega_{\text{N}}$, and assume that the measure of $\partial\Omega_{\text{D}}$ is positive. Let λ and μ be two coefficients satisfying $\mu > 0$ and $\lambda \geq 0$. Let $f \in [L^2(\Omega)]^3$ and $g \in [L^2(\partial\Omega_{\text{N}})]^3$. Then, the solution u to (3.91) satisfies*

$$-\mu\Delta u - (\lambda + \mu)\nabla(\nabla{\cdot}u) = f \quad \text{a.e. in } \Omega, \tag{3.93}$$

$u = 0$ a.e. on $\partial\Omega_{\text{D}}$, and $\sigma{\cdot}n = g$ a.e. on $\partial\Omega_{\text{N}}$.

Proof. Straightforward verification. □

Pure-traction problem and its weak formulation. The pure-traction problem consists of the following equations:

$$\begin{cases} \nabla{\cdot}\sigma(u) + f = 0 & \text{in } \Omega, \\ \sigma(u) = \lambda(\nabla{\cdot}u)\mathcal{I} + \mu(\nabla u + \nabla u^T) & \text{in } \Omega, \\ \sigma(u){\cdot}n = g & \text{on } \partial\Omega. \end{cases} \tag{3.94}$$

It is natural to seek the solution and take the test functions in $[H^1(\Omega)]^3$. Proceeding as before yields the problem:

$$\begin{cases} \text{Seek } u \in [H^1(\Omega)]^3 \text{ such that} \\ a(u,v) = \int_\Omega f{\cdot}v + \int_{\partial\Omega} g{\cdot}v, \quad \forall v \in [H^1(\Omega)]^3. \end{cases} \tag{3.95}$$

The bilinear form a is still defined by (3.92). The difficulty is that a becomes singular on $[H^1(\Omega)]^3$. To see this, introduce the set $\mathcal{R} = \{u \in [H^1(\Omega)]^3; u(x) = \alpha + \beta \times x\}$, where α and β are vectors in $\mathbb{R}^3$ and where $\times$ denotes the cross-product in $\mathbb{R}^3$. A function in $\mathcal{R}$ is called a *rigid displacement* field since it corresponds to a global motion consisting of a translation and a rotation.

Lemma 3.75. *The following equivalence holds:*

$$(u \in \mathcal{R}) \iff (\forall v \in [H^1(\Omega)]^3,\ a(u,v) = 0).$$

Proof. Let $u \in \mathcal{R}$. Clearly, $\nabla \cdot u = 0$ and $\varepsilon(u) = 0$. Therefore, $a(u, v) = 0$ for all $v \in [H^1(\Omega)]^3$. Conversely, if $a(u, v) = 0$ for all $v \in [H^1(\Omega)]^3$, take $v = u$ to obtain

$$a(u, u) = \int_\Omega \lambda (\nabla \cdot u)^2 + \int_\Omega 2\mu \varepsilon(u){:}\varepsilon(u) = 0,$$

implying that $\varepsilon(u) = 0$. Moreover, the fact that, for all j, k with $1 \leq j, k \leq 3$,

$$\partial_{jk} u_i = \partial_k (\partial_j u_i) = \partial_k (2\varepsilon_{ij}) - \partial_i \partial_k u_j = \partial_j (2\varepsilon_{ik}) - \partial_i \partial_j u_k$$
$$= \partial_k \varepsilon_{ij} + \partial_j \varepsilon_{ik} - \partial_i \varepsilon_{jk} = 0,$$

implies that all the components u_i of u are first-order polynomials. Hence,

$$u(x) = \alpha + Bx,$$

with $\alpha \in \mathbb{R}^3$ and $B \in \mathbb{R}^{3,3}$. Moreover, $\varepsilon(u) = 0$ implies $B + B^T = 0$, showing that the matrix B is skew-symmetric. Therefore, there exists a vector $\beta \in \mathbb{R}^3$ such that $Bx = \beta \times x$. This shows that $u \in \mathcal{R}$. $\qquad\square$

Taking $v \in \mathcal{R}$ in (3.95), Lemma 3.75 shows that a necessary condition for the existence of a solution to (3.94) is that the data f and g satisfy the compatibility relation

$$\forall v \in \mathcal{R}, \quad \int_\Omega f \cdot v + \int_{\partial\Omega} g \cdot v = 0. \tag{3.96}$$

Note that (3.96) expresses that the sum of the externally applied forces and their moments vanish. Furthermore, it is clear that the solution u, if it exists, is defined only up to a rigid displacement. Conventionally, we choose to seek the solution u such that $\int_\Omega u = \int_\Omega \nabla \times u = 0$ (note that both quantities are meaningful if $u \in [H^1(\Omega)]^3$). This leads to the following weak formulation:

$$\begin{cases} \text{Seek } u \in V_N \text{ such that} \\ a(u, v) = \int_\Omega f \cdot v + \int_{\partial\Omega} g \cdot v, \quad \forall v \in V_N, \end{cases} \tag{3.97}$$

with

$$V_N = \left\{ u \in [H^1(\Omega)]^3; \int_\Omega u = 0; \int_\Omega \nabla \times u = 0 \right\}, \tag{3.98}$$

equipped with the norm $\| \cdot \|_{1,\Omega}$.

Proposition 3.76. *Let Ω be a domain in $\mathbb{R}^3$. Let λ and μ be two coefficients satisfying $\mu > 0$ and $\lambda \geq 0$. Let $f \in [L^2(\Omega)]^3$ and let $g \in [L^2(\partial\Omega)]^3$. Assume that the compatibility condition (3.96) is satisfied. Then, the solution u to (3.97) satisfies (3.93) and $\sigma \cdot n = g$ a.e. on $\partial\Omega$.*

Proof. Straightforward verification. $\qquad\square$

3.4.2 Well-posedness

The *coercivity* of the bilinear form a defined in (3.92) relies on the following Korn inequalities:

Theorem 3.77 (Korn's first inequality). *Let Ω be a domain in $\mathbb{R}^3$. Set* $\|\varepsilon(v)\|_{0,\Omega} = (\int_\Omega \varepsilon(v){:}\varepsilon(v))^{\frac{1}{2}}$. *Then, there exists c such that*

$$\forall v \in [H_0^1(\Omega)]^3, \quad c\|v\|_{1,\Omega} \leq \|\varepsilon(v)\|_{0,\Omega}. \tag{3.99}$$

Proof. Let $v \in [H_0^1(\Omega)]^3$. Since v vanishes at the boundary,

$$\int_\Omega \nabla v{:}\nabla v^T = \sum_{i,j} \int_\Omega (\partial_i v_j)(\partial_j v_i) = -\sum_{i,j} \int_\Omega (\partial_{ij}^2 v_j) v_i$$

$$= \sum_{i,j} \int_\Omega (\partial_i v_i)(\partial_j v_j) = \int_\Omega (\nabla{\cdot}v)^2.$$

A straightforward calculation yields

$$\int_\Omega \varepsilon(v){:}\varepsilon(v) = \tfrac{1}{4}\int_\Omega (\nabla v + \nabla v^T){:}(\nabla v + \nabla v^T)$$

$$= \tfrac{1}{2}\int_\Omega \nabla v{:}\nabla v + \tfrac{1}{2}\int_\Omega \nabla v{:}\nabla v^T$$

$$= \tfrac{1}{2}\int_\Omega \nabla v{:}\nabla v + \tfrac{1}{2}\int_\Omega (\nabla{\cdot}v)^2 \geq \tfrac{1}{2}\int_\Omega \nabla v{:}\nabla v = \tfrac{1}{2}|v|_{1,\Omega}^2.$$

Hence, $|v|_{1,\Omega}^2 \leq 2\|\varepsilon(v)\|_{0,\Omega}^2$. Inequality (3.99) then results from the Poincaré inequality applied componentwise. $\square$

Theorem 3.78 (Korn's second inequality). *Let Ω be a domain in $\mathbb{R}^3$. Then, there exists c such that*

$$\forall v \in [H^1(\Omega)]^3, \quad c\|v\|_{1,\Omega} \leq \|\varepsilon(v)\|_{0,\Omega} + \|v\|_{0,\Omega}. \tag{3.100}$$

Proof. See [Cia97, p. 11] or [DuL72, p. 110]. $\square$

Proposition 3.79 (Mixed problem). *Let Ω be a domain in $\mathbb{R}^3$ and let $\partial\Omega_D \subset \partial\Omega$ have positive measure. Let $f \in [L^2(\Omega)]^3$ and let $g \in [L^2(\partial\Omega_N)]^3$. Then, problem (3.91) is well-posed and there exists c such that*

$$\forall f \in [L^2(\Omega)]^3, \ \forall g \in [L^2(\partial\Omega_N)]^3, \quad \|u\|_{1,\Omega} \leq c(\|f\|_{0,\Omega} + \|g\|_{0,\partial\Omega_N}).$$

Moreover, (3.91) is equivalent to the variational formulation

$$\min_{u \in V_{DN}} \left(\tfrac{1}{2}\lambda\int_\Omega (\nabla{\cdot}u)^2 + \tfrac{1}{2}\mu\int_\Omega \varepsilon(u){:}\varepsilon(u) - \int_\Omega f{\cdot}u - \int_{\partial\Omega_N} g{\cdot}u\right).$$

Proof. If $\partial\Omega_D = \partial\Omega$, $V_{DN} = [H_0^1(\Omega)]^3$. Coercivity then results from Korn's first inequality since

$$\forall u \in [H_0^1(\Omega)]^3, \quad a(u,u) \geq 2\mu \int_\Omega \varepsilon(u){:}\varepsilon(u) \geq c\,\|u\|_{1,\Omega}^2.$$

If $\partial\Omega_D \subsetneq \partial\Omega$, coercivity results from Korn's second inequality and a compacity argument; see the proof of Proposition 3.81. Conclude using the Lax–Milgram Lemma and Proposition 2.4. □

Remark 3.80. Given a displacement u, the quantity $J(u)$ represents the total energy of the deformed medium Ω. The quadratic terms correspond to the elastic deformation energy and the linear terms to the potential energy associated with external loads. □

Proposition 3.81 (Pure-traction problem). *Let Ω be a domain in $\mathbb{R}^3$. Assume that $f \in [L^2(\Omega)]^3$ and $g \in [L^2(\partial\Omega)]^3$ satisfy the compatibility condition (3.96). Then, problem (3.97) is well-posed and there exists c such that*

$$\forall f \in [L^2(\Omega)]^3, \ \forall g \in [L^2(\partial\Omega)]^3, \quad \|u\|_{1,\Omega} \leq c\,(\|f\|_{0,\Omega} + \|g\|_{0,\partial\Omega}).$$

Moreover, (3.97) is equivalent to the variational formulation

$$\min_{u \in V_N} \left(\tfrac{1}{2}\lambda \int_\Omega (\nabla{\cdot}u)^2 + \tfrac{1}{2}\mu \int_\Omega \varepsilon(u){:}\varepsilon(u) - \int_\Omega f{\cdot}u - \int_{\partial\Omega} g{\cdot}u \right).$$

Proof. Coercivity results from Korn's second inequality and from the Petree–Tartar Lemma. Indeed, set $X = V_N$, $Y = [L^2(\Omega)]^{3,3}$, and $A : X \ni u \mapsto \varepsilon(u) \in Y$. Lemma 3.75 implies that the operator A is injective. Set $Z = [L^2(\Omega)]^3$ and let T be the compact injection from X into Z. Korn's second inequality yields

$$\forall u \in X, \quad \|u\|_X \leq c\,(\|Au\|_Y + \|Tu\|_Z).$$

Applying the Petree–Tartar Lemma yields $\|u\|_X \leq c\|Au\|_Y$ for all $u \in X$, i.e.,

$$\forall u \in V_N, \quad \|u\|_{1,\Omega} \leq c\,\|\varepsilon(u)\|_{0,\Omega}.$$

This inequality shows that the bilinear form a is coercive on V_N. To complete the proof, use the Lax–Milgram Lemma and Proposition 2.4. □

3.4.3 Finite element approximation

For the sake of simplicity, we assume that Ω is a polyhedron.

H^1-**conformal approximation.** We consider a H^1-conformal finite element approximation of problems (3.91) and (3.97) based on a family of affine, geometrically conformal meshes $\{T_h\}_{h>0}$ and a Lagrange finite element of degree $k \geq 1$ denoted by $\{\widehat{K}, \widehat{P}, \widehat{\Sigma}\}$.

To approximate the mixed problem, we assume, for the sake of simplicity, that $\partial\Omega_D$ is a union of mesh faces. Hence, the approximation space

$$V_h^k = \{v_h \in [\mathcal{C}^0(\overline{\Omega})]^3; \forall K \in T_h, v_h \circ T_K \in [\widehat{P}]^3; v_h = 0 \text{ on } \partial\Omega_D\},$$

is V_{DN}-conformal. Consider the discrete problem:

$$\begin{cases} \text{Seek } u_h \in V_h^k \text{ such that} \\ a(u_h, v_h) = \int_\Omega f \cdot v_h + \int_{\partial\Omega_N} g \cdot v_h, \quad \forall v_h \in V_h^k. \end{cases} \tag{3.101}$$

Proposition 3.82 (Mixed problem). *Let u solve (3.91) and let u_h solve (3.101). In the above setting, $\lim_{h\to 0} \|u - u_h\|_{1,\Omega} = 0$. Furthermore, if $u \in [H^{l+1}(\Omega)]^3 \cap V_{DN}$ for some $l \in \{1, \dots, k\}$, there exists c such that*

$$\forall h, \quad \|u - u_h\|_{1,\Omega} \leq c\, h^l |u|_{l+1,\Omega}.$$

Proof. Direct consequence of Céa's Lemma and Corollary 1.109 applied componentwise.   □

Remark 3.83. It is not possible to apply the Aubin–Nitsche Lemma to derive an error estimate in the $[L^2(\Omega)]^3$-norm because the mixed problem is not endowed with a suitable smoothing property.   □

For the pure-traction problem, one possible way to eliminate the arbitrary rigid displacement is the following:

(i) Impose the displacement of a node, say a_0, to be zero.
(ii) Choose three additional nodes a_1, a_2, a_3, and three unit vectors τ_1, τ_2, τ_3 such that the set $\{(a_i - a_0) \times \tau_i\}_{1\leq i\leq 3}$ forms a basis of $\mathbb{R}^3$, and impose the displacement of the node a_i along the direction τ_i to be zero.

This procedure leads to the approximation space

$$W_h^k = \{v_h \in [\mathcal{C}^0(\overline{\Omega})]^3; \forall K \in T_h, v_h \circ T_K \in [\widehat{P}]^3;$$
$$v_h(a_0) = 0; v_h(a_i) \cdot \tau_i = 0, i = 1, 2, 3\},$$

and to the discrete problem:

$$\begin{cases} \text{Seek } u_h \in W_h^k \text{ such that} \\ a(u_h, v_h) = \int_\Omega f \cdot v_h + \int_{\partial\Omega} g \cdot v_h, \quad \forall v_h \in W_h^k. \end{cases} \tag{3.102}$$

Proposition 3.84 (Pure-traction problem). *Let u solve (3.91) and let u_h solve (3.102). In the above setting, $\lim_{h\to 0} \|u - u_h\|_{1,\Omega} = 0$. Furthermore, if $u \in [H^{l+1}(\Omega)]^3 \cap V_N$ for some $l \in \{1, \dots, k\}$, there exists c such that*

$$\forall h, \quad \|u - u_h\|_{1,\Omega} \leq c\, h^l |u|_{l+1,\Omega}.$$

In addition, if Ω is convex and $g = 0$, there is c such that

$$\forall h, \quad \|u - u_h\|_{0,\Omega} \leq c\, h^{l+1} |u|_{l+1,\Omega}.$$

Proof. Use Céa's Lemma, together with Corollary 1.109, to obtain the H^1-error estimate. Furthermore, the homogeneous pure-traction problem posed over a convex polyhedron is endowed with a smoothing property [Gri92, p. 135]. The L^2-error estimate then results from the Aubin–Nitsche Lemma. $\qquad \square$

Crouzeix–Raviart approximation. Non-conformal finite element approximations to the equations of elasticity can be considered using the Crouzeix–Raviart finite element introduced in §1.2.6. For pure-traction problems, the main difficulty in the analysis is to prove an appropriate version of Korn's second inequality. This result can be established for non-conformal piecewise quadratic or cubic finite elements, but is false for piecewise linear interpolation. For Crouzeix–Raviart interpolation, appropriate modifications of the method are discussed in [Fal91, Rua96].

One important advantage of non-conformal approximations is that they yield optimal-order error estimates that are uniform in the Poisson coefficient ν. Such a property is particularly useful when modeling almost incompressible materials since it is well-known that, in this case, H^1-conformal finite elements suffer from a severe deterioration in the convergence rate; see §3.5.3 for an illustration.

Numerical illustrations. As a first example, consider the horizontal deformations of a two-dimensional, rectangular plate with a circular hole. The triangulation of the plate is depicted in the left panel of Figure 3.8. The left side is clamped, the displacement $(1,0)$ is imposed on the right side, and zero normal stress is imposed on the three remaining sides. There is no external load, and the Lamé coefficients are such that $\frac{\lambda}{\mu} = 1$. The plate in its equilibrium configuration is shown in the right panel of Figure 3.8. $\mathbb{P}_1$ Lagrange finite elements have been used.

The second example deals with the three-dimensional body illustrated in Figure 3.9. A transverse load is imposed at the forefront of the body. The approximate solution has been obtained using first-order prismatic Lagrange

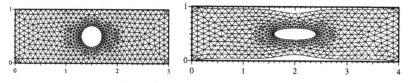

Fig. 3.8. Deformation of an elastic plate with a hole: reference configuration (left); equilibrium configuration (right).

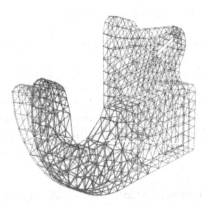

Fig. 3.9. Three-dimensional continuum mechanics problem in which a transverse load is applied to the forefront of the body; reference and equilibrium configurations are presented; approximation with prismatic Lagrange finite elements of degree 1. Courtesy of D. Chapelle (INRIA).

finite elements. Figure 3.9 presents the reference and the equilibrium configurations.

3.4.4 Beam flexion and fourth-order problems

The physical model. We investigate a model for beam flexion due to *Timoshenko*; see, e.g., [Bat96]. Consider the horizontal beam of length L shown in Figure 3.10. The x-coordinate is set so as to coincide with the beam axis. The beam is clamped into a rigid wall at $x = 0$. Impose a distributed load $f = (f_x, f_y)$ in the (x, y)-plane and a distributed momentum m parallel to the z-axis. Impose further a point force $F = (F_x, F_y)$ and a point momentum M at the beam extremity located at $x = L$. Assuming that the axis of the beam remains in the (x, y)-plane, the beam flexion can be described by the displacement $u = (u_x, u_y)$ of the points along the axis and by the rotation angle θ of the corresponding transverse sections.

In the Timoshenko model, the tangential displacement u_x uncouples from the unknowns u_y and θ. Setting $\Omega =]0, L[$, u_x solves $-u_x'' = \frac{1}{ES}f_x$ in Ω with boundary conditions $u_x(0) = 0$ and $u_x'(L) = \frac{1}{ES}F_x$, where E is the Young modulus and S is the area of the beam section. Thus, a one-dimensional second-order PDE with mixed boundary conditions is recovered.

To alleviate the notation, we now write u instead of u_y, f instead of f_y, and F instead of F_y. The displacement u and the rotation angle θ satisfy the PDEs

$$-(u'' - \theta') = \frac{\gamma}{EI}f \quad \text{and} \quad -\gamma\theta'' - (u' - \theta) = \frac{\gamma}{EI}m, \qquad (3.103)$$

where I is the inertia moment of the beam, $\gamma = \frac{2(1+\nu)I}{S\kappa}$, and κ is an empirical correction factor (usually set to $\frac{5}{6}$). Boundary conditions for u and θ are

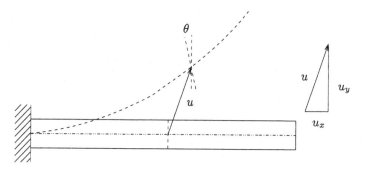

Fig. 3.10. Timoshenko model for beam flexion.

$$u(0) = 0, \quad \theta(0) = 0, \qquad (u' - \theta)(L) = \frac{\gamma}{EI}F, \quad \theta'(L) = \frac{1}{EI}M. \quad (3.104)$$

Weak formulation and coercivity. Let v be a test function for the normal displacement u and let ω be a test function for the rotation angle θ. Multiply the first equation in (3.103) by v, the second by ω, and integrate by parts over Ω to obtain the weak formulation:

$$\begin{cases} \text{Seek } (u, \theta) \in X \times X \text{ such that } \forall (v, \omega) \in X \times X, \\ a\big((u, \theta), (v, \omega)\big) = \frac{\gamma}{EI}[\int_\Omega (fv + m\omega) + Fv(L) + M\omega(L)], \end{cases} \quad (3.105)$$

where

$$a\big((u, \theta), (v, \omega)\big) = \int_\Omega \gamma \theta' \omega' + \int_\Omega (u' - \theta)(v' - \omega), \quad (3.106)$$

and $X = \{v \in H^1(\Omega); v(0) = 0\}$. Equip the product space $X \times X$ with the norm $\|(u, \theta)\|_{X \times X} = \|u\|_{1,\Omega} + \|\theta\|_{1,\Omega}$. One readily verifies the following:

Proposition 3.85. *Let f and $m \in L^2(\Omega)$. If the couple (u, θ) solves (3.105), it satisfies (3.103) a.e. in Ω and the boundary conditions (3.104).*

Theorem 3.86 (Coercivity). *Let $\gamma > 0$, let $f, m \in L^2(\Omega)$, and let $F, M \in \mathbb{R}$. Then, problem (3.105) is well-posed. Moreover, (u, θ) solves (3.105) if and only if it minimizes over $X \times X$ the energy functional*

$$J(u, \theta) = \frac{1}{2} \int_\Omega \gamma (\theta')^2 + \frac{1}{2} \int_\Omega (u' - \theta)^2 - \frac{\gamma}{EI} \left[\int_\Omega (fu + m\theta) + Fu(L) + M\theta(L) \right].$$

Proof. The key point is to verify the coercivity of the bilinear form a defined by (3.106). A straightforward calculation yields

$$a\big((u, \theta), (u, \theta)\big) = \int_\Omega \gamma (\theta')^2 + \int_\Omega (u')^2 + \int_\Omega \theta^2 - 2 \int_\Omega \theta u'.$$

Let $\mu > 0$. Use inequality (A.3) with parameter μ, together with the Poincaré inequality $c_\Omega \|v\|_{0,\Omega} \leq \|v'\|_{0,\Omega}$ valid for all $v \in X$, to obtain

$$a\big((u,\theta),(u,\theta)\big) \geq \gamma|\theta|_{1,\Omega}^2 + |u|_{1,\Omega}^2 + \|\theta\|_{0,\Omega}^2 - \mu\|\theta\|_{0,\Omega}^2 - \frac{1}{\mu}|u|_{1,\Omega}^2$$

$$\geq \left(1 - \frac{1}{\mu}\right)|u|_{1,\Omega}^2 + \frac{\gamma}{2}|\theta|_{1,\Omega}^2 + \left(\frac{\gamma}{2}c_\Omega^2 + 1 - \mu\right)\|\theta\|_{0,\Omega}^2.$$

Taking $\mu = 1 + \frac{\gamma}{2}c_\Omega^2$ yields

$$a\big((u,\theta),(u,\theta)\big) \geq \frac{\frac{\gamma}{2}c_\Omega^2}{1 + \frac{\gamma}{2}c_\Omega^2}|u|_{1,\Omega}^2 + \frac{\gamma}{2}|\theta|_{1,\Omega}^2 \geq \alpha(\gamma)\|(u,\theta)\|_{X \times X}^2,$$

with $\alpha(\gamma) = \frac{\gamma}{4}\frac{c_\Omega^2}{1+c_\Omega^2}\inf\big(1, c_\Omega^2/(1 + \frac{\gamma}{2}c_\Omega^2)\big) > 0$; since $\gamma > 0$, a is coercive. Conclude using the Lax–Milgram Lemma and Proposition 2.4. $\qquad\square$

Discrete approximation. Let $\mathcal{T}_h$ be a mesh of Ω with vertices $0 = x_0 < x_1 < \ldots < x_N < x_{N+1} = L$ where N is a given integer. Consider a conformal $\mathbb{P}_k$ Lagrange finite element approximation for both u and θ. The approximation space we consider is thus

$$X_h = \{v_h \in C^0(\overline{\Omega}); \forall i \in \{0, \ldots, N\}, v_{h|[x_i,x_{i+1}]} \in \mathbb{P}_k; v_h(0) = 0\},$$

yielding the approximate problem:

$$\begin{cases} \text{Seek } (u_h, \theta_h) \in X_h \times X_h \text{ such that, } \forall(v_h, \omega_h) \in X_h \times X_h, \\ a\big((u_h, \theta_h),(v_h, \omega_h)\big) = \frac{\gamma}{EI}[\int_\Omega (fv_h + m\omega_h) + Fv_h(L) + M\omega_h(L)]. \end{cases} \quad (3.107)$$

Theorem 3.87. *Let $\mathcal{T}_h$ be a mesh of Ω. Along with the assumptions of Theorem 3.86, assume that u and $\theta \in H^s(\Omega)$ for some $s \geq 2$. Then, setting $l = \min(k, s - 1)$, there exists c such that, for all h,*

$$|u - u_h|_{1,\Omega} + |\theta - \theta_h|_{1,\Omega} \leq c\,h^l \max(|u|_{l+1,\Omega}, |\theta|_{l+1,\Omega}),$$

$$\|u - u_h\|_{0,\Omega} + \|\theta - \theta_h\|_{0,\Omega} \leq c\,h^{l+1} \max(|u|_{l+1,\Omega}, |\theta|_{l+1,\Omega}).$$

Proof. The estimate in the H^1-norm results from Céa's Lemma and from Proposition 1.12 applied to u and θ. The estimate in the L^2-norm results from the Aubin–Nitsche Lemma. Indeed, one easily checks that the adjoint problem is endowed with the required smoothing property. $\qquad\square$

Navier–Bernoulli model and fourth-order problems. A case often encountered in applications arises when the parameter γ becomes extremely small. In the limit $\gamma \to 0$, the Navier–Bernoulli model is recovered

$$u' - \theta = 0 \quad \text{on } \Omega,$$

meaning that the sections of the bended beam remain orthogonal to the axis. Assuming that $m = 0$, $EI = 1$, and that the beam is clamped at its two extremities, the normal displacement u is governed by the fourth-order PDE

$u'''' = f$ in Ω with boundary conditions $u(0) = u(L) = u'(0) = u'(L) = 0$, leading to the weak formulation:

$$\begin{cases} \text{Seek } u \in H_0^2(\Omega) \text{ such that} \\ \int_0^L u''v'' = \int_0^L fv, \quad \forall v \in H_0^2(\Omega). \end{cases} \tag{3.108}$$

Proposition 3.88. *Let $f \in L^2(\Omega)$. Then, problem (3.108) is well-posed. Moreover, problem (3.108) is equivalent to minimizing over $H_0^2(\Omega)$ the energy functional $J(v) = \frac{1}{2}\int_\Omega (v'')^2 - \int_\Omega fv$.*

Proof. Left as an exercise.   □

We consider a H^2-*conformal* approximation to problem (3.108) using a Hermite finite element approximation. Taking the boundary conditions into account leads to the approximation space

$$X_{h0}^3 = \{v_h \in C^1(\overline{\Omega}); \forall i \in \{0, \ldots, N\}, v_{h|[x_i, x_{i+1}]} \in \mathbb{P}_3;$$
$$v_h(0) = v_h'(0) = v_h(L) = v_h'(L) = 0\},$$

and the discrete problem:

$$\begin{cases} \text{Seek } u_h \in X_{h0}^3 \text{ such that} \\ \int_0^1 u_h'' v_h'' = \int_0^1 fv_h, \quad \forall v_h \in X_{h0}^3. \end{cases} \tag{3.109}$$

Proposition 3.89. *Let $\mathcal{T}_h$ be a mesh of Ω. Let $f \in L^2(\Omega)$, let u solve (3.108), and let u_h solve (3.109). Then, there exists c such that, for all h,*

$$\|u - u_h\|_{0,\Omega} + h|u - u_h|_{1,\Omega} + h^2|u - u_h|_{2,\Omega} \le ch^4\|f\|_{0,\Omega}.$$

Proof. Left as an exercise.   □

Example 3.90. Consider a unit-length beam clamped at its two extremities. Apply a unit load $f \equiv 1$. Approximate problem (3.109) using uniform meshes with step size $h = \frac{1}{10}, \frac{1}{20}, \frac{1}{40}$, and $\frac{1}{80}$. The left panel in Figure 3.11 presents the error along the beam. We observe that the error vanishes at the mesh points. This is because, in this simple one-dimensional problem, the Green function associated with (3.108) belongs to the approximation space X_{h0}^3; see Remark 3.25 for a justification. The right panel in Figure 3.11 presents the error in the L^2-norm, H^1-seminorm, and H^2-seminorm. Convergence orders are 4, 3, and 2, respectively, as predicted by the theory.   □

Remark 3.91. The two-dimensional version of problem (3.108) is to seek $u \in H_0^2(\Omega)$ such that

$$\int_\Omega \Delta u \, \Delta v = \int_\Omega fv, \quad \forall v \in H_0^2(\Omega). \tag{3.110}$$

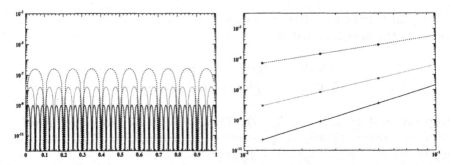

Fig. 3.11. Hermite finite element approximation for a beam flexion problem. Left: Error distribution along the beam for various mesh sizes; $h = \frac{1}{10}$ (dashed), $\frac{1}{20}$ (dotted), and $\frac{1}{40}$ (solid). Right: Error in the L^2-norm (solid), H^1-seminorm (dotted), and H^2-seminorm (dashed) as a function of mesh size.

This problem models, for instance, the bending of a clamped plate submitted to a transverse load; see [Des86, Cia97]. Regularity results for problem (3.110) are found in [GiR86, p. 17], [Cia91, p. 297], and [Gri92, p. 109]. Finite element approximations are discussed, e.g., in [Cia91, p. 273]; see also [GiR86, p. 204] for a related mixed formulation of problem (3.110) in the context of the Stokes equations in dimension 2. $\square$

3.5 Coercivity Loss

Coercivity loss occurs when some model parameters take extreme values. In this case, although the exact problem is well-posed, discrete stability is observed only if very fine meshes are employed. The examples addressed in this section are:

(i) Advection–diffusion problems of the form (3.2) with dominant advection.

(ii) Elastic deformations of a quasi-incompressible material.

(iii) Elastic bending of a very thin Timoshenko beam.

The scope of this section is not to fix the above-mentioned problems, but to highlight the mathematical background related to coercivity loss. We identify the model parameter taking extreme values, and by letting this parameter approach zero, we derive formally a problem with no coercivity, i.e., typically involving a saddle-point or a first-order PDE. Such problems are thoroughly investigated in Chapters 4 and 5.

3.5.1 The setting

Consider the problem:

$$\begin{cases} \text{Seek } u \in V \text{ such that} \\ a_\eta(u, v) = f(v), \quad \forall v \in V, \end{cases} \tag{3.111}$$

where V is a Hilbert space, $f \in V'$, and a_η is a continuous, *coercive*, bilinear form on $V \times V$. The form a_η depends on the phenomenological parameter η that will subsequently take arbitrarily small values. Set $\|a_\eta\| := \|a_\eta\|_{V,V}$ and denote by α_η the coercivity constant of a_η, i.e.,

$$\alpha_\eta = \inf_{u \in V} \frac{a_\eta(u,u)}{\|u\|_V^2}.$$

Definition 3.92. Coercivity loss *occurs in* (3.111) *if*

$$\lim_{\eta \to 0} \frac{\|a_\eta\|}{\alpha_\eta} = \infty.$$

Remark 3.93. By analogy with the terminology adopted for linear systems in §9.1, coercivity loss amounts to the ill-conditioning of the form a. □

Let V_h be a V-conformal approximation space and assume, as is often the case in practice, that V_h is endowed with the optimal interpolation property

$$\forall u \in W, \qquad \inf_{v_h \in V_h} \|u - v_h\|_V \le c_i h^k \|u\|_W,$$

where W is a dense subspace of V and c_i is an interpolation constant. Let u_h be the solution to the approximate problem:

$$\begin{cases} \text{Seek } u_h \in V_h \text{ such that} \\ a_\eta(u_h, v_h) = f(v_h), \quad \forall v_h \in V_h. \end{cases}$$

Assuming that the exact solution u is in W yields the error estimate

$$\|u - u_h\|_V \le \frac{\|a_\eta\|}{\alpha_\eta} c_i h^k \|u\|_W.$$

If problem (3.111) suffers from coercivity loss, this estimate does not yield any practical control of the error. Obviously, keeping η fixed and letting $h \to 0$, convergence is achieved. However, the mesh size is limited from below by the available computer resources. Therefore, it is not always possible in practice to compensate coercivity losses by systematic mesh refinement. Some explicit examples where this situation occurs are detailed below.

3.5.2 Advection–diffusion with dominant advection

Let Ω be a domain in $\mathbb{R}^d$. Consider the advection–diffusion equation

$$-\nu\Delta u + \beta{\cdot}\nabla u = f \quad \text{in } \Omega, \tag{3.112}$$

where $\nu > 0$ is the diffusion coefficient, $\beta : \Omega \to \mathbb{R}^d$ the advection velocity, and $f : \Omega \to \mathbb{R}$ the source term. Following §3.1, we consider the bilinear form

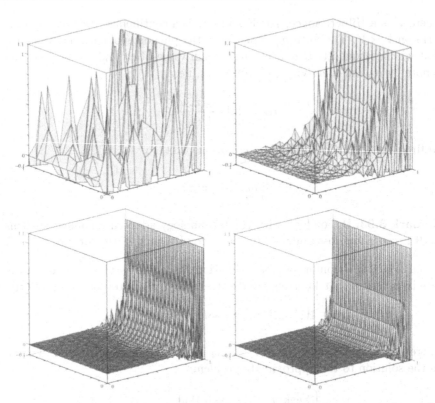

Fig. 3.12. Finite element approximation of an advection–diffusion equation with dominant advection: $h \approx \frac{1}{10}$ and $\mathbb{P}_1$ approximation (top left); $h \approx \frac{1}{20}$ and $\mathbb{P}_1$ approximation (top right); $h \approx \frac{1}{40}$ and $\mathbb{P}_1$ approximation (bottom left); $h \approx \frac{1}{20}$ and $\mathbb{P}_2$ approximation (bottom right).

$$a_\eta(u,v) = \int_\Omega \nu \nabla u \cdot \nabla v + \int_\Omega v(\beta \cdot \nabla u).$$

The parameter $\eta = \frac{\nu}{\|\beta\|_{[L^\infty(\Omega)]^d}}$ measures the relative importance of advective and diffusive effects. Assuming $\eta \ll 1$ implies

$$\frac{\|a_\eta\|}{\alpha_\eta} = O\left(\frac{\|\beta\|_{[L^\infty(\Omega)]^d}}{\nu}\right) = O\left(\frac{1}{\eta}\right) \gg 1,$$

leading to *coercivity loss*.

Figure 3.12 presents various approximate solutions to the advection–diffusion equation (3.112). The domain Ω is the unit square in $\mathbb{R}^2$. We impose $u = 1$ on the right side, $u = 0$ on the left side, and $\partial_{x_2} u = 0$ on the two other sides. The diffusion coefficient is set to $\nu = 0.002$, the advection velocity is constant and equal to $\beta = (1,0)$, and the source term f is zero. The exact solution is

$$u(x_1, x_2) = \frac{e^{\frac{x_1}{\nu}} - 1}{e^{\frac{1}{\nu}} - 1}.$$

Since the diffusion coefficient ν takes very small values, the exact solution u is almost identically zero in Ω except in a boundary layer of width ν located near the right side where u sharply goes from 0 to 1. Three unstructured triangulations of the domain Ω are considered: a coarse mesh containing 238 triangles (triangle size $h \approx \frac{1}{10}$); an intermediate mesh containing 932 triangles (triangle size $h \approx \frac{1}{20}$); and a fine mesh containing 3694 triangles (triangle size $h \approx \frac{1}{40}$). The $\mathbb{P}_1$ Galerkin solution is computed on the three meshes: $h \approx \frac{1}{10}$, top left panel; $h \approx \frac{1}{20}$, top right panel; and $h \approx \frac{1}{40}$, bottom left panel. The $\mathbb{P}_2$ Galerkin solution computed on the intermediate mesh is shown in the bottom right panel. We observe that *spurious oscillations* pollute the approximate solution in the four cases presented. Oscillations are larger on the two coarser meshes and for the $\mathbb{P}_1$ approximation.

In the limit $\eta \to 0$, the diffusion term is negligible and the solution u is governed by a first-order PDE. Hence, to understand and fix the problems associated with coercivity loss, it is important to analyze the limit first-order PDE; this is the purpose of Chapter 5.

3.5.3 Almost incompressible materials

Almost incompressible materials, such as rubber, are characterized by Lamé coefficients λ and μ with a very large ratio $\frac{\lambda}{\mu}$. Another equivalent characterization is that the Poisson coefficient ν is very close to $\frac{1}{2}$. In §3.4.1 we introduced the bilinear form

$$a_\eta(u,v) = \int_\Omega \lambda \, \nabla\!\cdot\!u \, \nabla\!\cdot\!v + \int_\Omega 2\mu \, \varepsilon(u){:}\varepsilon(v),$$

where $\varepsilon(u)$ is the strain rate tensor. When the ratio $\eta = \frac{\mu}{\lambda}$ is very small, one verifies that

$$\frac{\|a_\eta\|}{\alpha_\eta} = O\left(\frac{\lambda}{\mu}\right) = O\left(\frac{1}{\eta}\right) \gg 1,$$

leading to *coercivity loss*.

Consider a horizontal elastic flat plate with three internal holes; see Figure 3.13. The left side is kept fixed, the displacement $(1,0)$ is imposed on the right side, and zero normal stress is imposed on the remaining external sides as well as on the three internal sides. No internal load is applied, and the

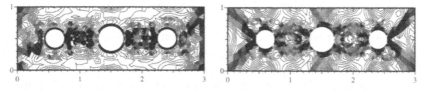

Fig. 3.13. Deformations of a horizontal, flat plate with three holes: maximal stresses (left); Tresca stresses (right).

ratio of the Lamé coefficients is $\frac{\lambda}{\mu} = 100$. Figure 3.13 presents Tresca stresses and maximal stresses obtained with a $\mathbb{P}_1$ Lagrange finite element approximation. We observe that *spurious oscillations* pollute the discrete solution; in the literature, this phenomenon is often referred to as *locking*.

When $\frac{\lambda}{\mu} \gg 1$, one can show that $\nabla \cdot u \to 0$. Introducing a new scalar unknown p in place of the product $-\lambda \nabla \cdot u$ yields

$$\begin{cases} \sigma = -p\mathcal{I} + 2\mu\varepsilon(u), \\ \nabla \cdot u = 0. \end{cases}$$

Since $\Delta u = 2\nabla \cdot \varepsilon(u)$ when $\nabla \cdot u = 0$, the governing equations of an incompressible medium in the framework of linear elasticity become

$$\begin{cases} -\mu \Delta u + \nabla p = f, \\ \nabla \cdot u = 0. \end{cases}$$

Formally, we recover the Stokes equations often considered to model steady, incompressible flows of creeping fluids. The new unknown p can be identified with a pressure. The Stokes equations are endowed with a saddle-point structure. The analysis of this class of problems is the purpose of Chapter 4.

3.5.4 Very thin beams

Referring to §3.4.4 for more details, the bilinear form arising in Timoshenko's model of beam flexion is

$$a_\eta \big((u, \theta), (v, \omega) \big) = \int_\Omega \gamma \theta' \omega' + \int_\Omega (u' - \theta)(v' - \omega),$$

where u is the normal displacement of the beam axis and θ the rotation angle of the beam section. The parameter η is simply equal to γ. When $\gamma \ll 1$, the proof of Theorem 3.86 shows

$$\frac{\|a_\eta\|}{\alpha_\eta} = O\left(\frac{1}{\gamma}\right) = O\left(\frac{1}{\eta}\right) \gg 1,$$

leading to *coercivity loss*. Note that $\gamma \ll 1$ when the ratio between the inertia moment and the section of the beam is very small, as for very thin beams. In this case, the beam bends according to the Navier–Bernoulli assumption, meaning that the sections remain almost perpendicular to the beam axis.

Figure 3.14 compares analytical and approximate solutions for a beam of length $L = 1$ and parameter $EI = 1$. The flexion is induced by a force $F = 1$ applied at the extremity $x = L$. Solutions are obtained using the $\mathbb{P}_1$ finite element approximation for both the displacement u and the rotation angle θ on a uniform mesh with step size $h = \frac{1}{20}$. The left column in Figure 3.14 corresponds to the case $\gamma = 0.01$ and the right column to the case $\gamma = 0.0001$.

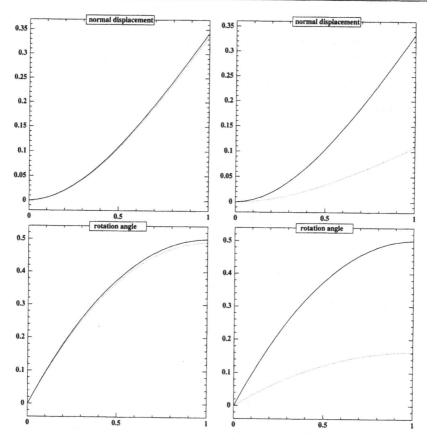

Fig. 3.14. Comparison between the analytical and finite element solutions (solid and dashed lines, respectively) for the bending of a Timoshenko beam clamped at its left extremity: $\gamma = 0.01$ (left column); $\gamma = 0.0001$ (right column).

In the second case, coercivity loss leads to very poor accuracy, indicating a *locking* phenomenon.

To pass to the limit $\gamma \twoheadrightarrow 0$ in Timoshenko's model (3.103), we introduce the auxiliary unknown

$$v = \frac{1}{\gamma}\left(u - \int_0^x \theta\right).$$

The unknowns (v, θ) satisfy the PDEs $-v'' = \frac{1}{EI}f$ and $-\theta'' - v' = \frac{1}{EI}m$ in $]0, L[$, together with the boundary conditions $v(0) = 0$, $\theta(0) = 0$, $v'(L) = \frac{1}{EI}F$, and $\theta'(L) = \frac{1}{EI}M$. One readily checks that this new problem leads to a coercive bilinear form. Furthermore, the displacement u is recovered from the first-order PDE

$$\begin{cases} u' = \gamma v' + \theta, \\ u(0) = 0. \end{cases}$$

Here, as in §3.5.2, coercivity loss is associated with the presence of a first-order PDE in the limit problem. The finite element approximation of such PDEs is investigated in Chapter 5.

3.6 Exercises

Exercise 3.1. Complete the proof of Theorem 3.8.

Exercise 3.2. Let $\Omega =]0,1[$, let $f \in L^2(\Omega)$, and let $k \in \mathbb{R}$. Consider the problem:

$$\begin{cases} \text{Seek } u \in H_0^1(\Omega) \text{ such that} \\ \int_0^1 u'v' + k \int_0^1 u'v + \int_0^1 uv = \int_0^1 fv, \quad \forall v \in H_0^1(\Omega). \end{cases}$$

(i) Write the corresponding PDE and boundary conditions.
(ii) Prove that the problem is well-posed. (*Hint*: Use the Lax–Milgram Lemma.)

Exercise 3.3. Let Ω be a domain in $\mathbb{R}^2$, let $f \in L^2(\Omega)$, and let $\sigma \in \mathbb{R}$. Show that if $|\sigma| < 1$, the following problem is well-posed:

$$\begin{cases} \text{Seek } u \in H_0^1(\Omega) \text{ such that} \\ \int_\Omega [\partial_x u \partial_x v + \sigma(\partial_x u \partial_y v + \partial_y u \partial_x v) + \partial_y u \partial_y v] = \int_\Omega fv, \quad \forall v \in H_0^1(\Omega). \end{cases}$$

Exercise 3.4. Consider the domain Ω whose definition in polar coordinates is $\Omega = \{(r,\theta); 0 < r < 1, \frac{\pi}{\alpha} < \theta < 0\}$ with $\alpha < -\frac{1}{2}$. Let $\partial\Omega_1 = \{(r,\theta); r = 1, \frac{\pi}{\alpha} < \theta < 0\}$ and $\partial\Omega_2 = \partial\Omega \setminus \partial\Omega_1$. Consider the following problem: $-\Delta u = 0$ in Ω, $u = \sin(\alpha\theta)$ on $\partial\Omega_1$, and $u = 0$ on $\partial\Omega_2$.

(i) Let $\varphi_1 = r^\alpha \sin(\alpha\theta)$ and $\varphi_2 = r^{-\alpha} \sin(\alpha\theta)$. Prove that φ_1 and φ_2 solve the above problem. (*Hint*: In polar coordinates, $\Delta\varphi = \frac{1}{r}\partial_r(r\partial_r\varphi) + \frac{1}{r^2}\partial_{\theta\theta}\varphi$.)
(ii) Prove that φ_1 and φ_2 are in $L^2(\Omega)$ if $-1 < \alpha < -\frac{1}{2}$.
(iii) Consider the following problem: Seek $u \in H^1(\Omega)$ such that $u = \sin(\alpha\theta)$ on $\partial\Omega_1$, $u = 0$ on $\partial\Omega_2$, and $\int_\Omega \nabla u \cdot \nabla v = 0$ for all $v \in H_0^1(\Omega)$. Prove that φ_2 solves this problem, but φ_1 does not. Comment.

Exercise 3.5 (Péclet number). Let $\Omega =]0,1[$, let $\nu > 0$, and let $\beta \in \mathbb{R}$. Consider the following problem:

$$\begin{cases} -\nu u'' + \beta u' = 1, \\ u(0) = u(1) = 0. \end{cases}$$

(i) Verify that the exact solution is $u(x) = \frac{1}{\beta}(x - \frac{1-e^{\lambda x}}{1-e^\lambda})$ with $\lambda = \frac{\beta}{\nu}$.
(ii) Plot the solution for $\beta = 1$ and $\nu = 1$, $\nu = 0.1$, and $\nu = 0.01$. Comment.
(iii) Write the problem in weak form and show that it is well-posed.

(iv) Consider a $\mathbb{P}_1$ H^1-conformal finite element approximation on a uniform grid $\mathcal{T}_h = \bigcup_{0 \le i < N}[ih, (i+1)h]$ where $h = \frac{1}{N+1}$. Show that the stiffness matrix is $\mathcal{A} = \frac{\nu}{h}\text{tridiag}(-1 - \frac{\gamma}{2}, 2, -1 + \frac{\gamma}{2})$, where $\gamma = \frac{\beta h}{\nu}$ is the so-called local *Péclet number*.

(v) Solve the linear system and comment. (*Hint:* If $\gamma \ne 2$, the solution is $U_i = \frac{1}{\beta}(ih - \frac{1 - \delta^i}{1 - \delta^{N+1}})$ where $\delta = \frac{2+\gamma}{2-\gamma}$.) What happens if $\gamma = 2$ or $\gamma = -2$?

(vi) Plot the approximate solution for $\gamma = 1$ and $\gamma = 10$. Comment.

Exercise 3.6. Let $\nu > 0$ and $b > 0$. Consider the equation $-\nu u'' + bu' = f$ posed on $]0, 1[$ with the boundary conditions $u(0) = 0$ and $u'(1) = 0$.

(i) Write the weak formulation of the problem.

(ii) Let $\mathcal{T}_h$ be a mesh of $]0, 1[$ and use $\mathbb{P}_1$ finite elements to approximate the problem. Let $[x_{N-1}, x_N]$ be the element such that $x_N = 1$. Let U_{N-1} and U_N be the value of the approximate solution at x_{N-1} and x_N. Write the equation satisfied by U_{N-1} and U_N when testing the weak formulation by the nodal shape function φ_N.

(iii) What is the limit of the equation derived in question (ii) when $|x_N - x_{N-1}| \to 0$? What is the limit equation when $\nu \ll |x_N - x_{N-1}|$? Comment.

Exercise 3.7. Let Ω be a domain in $\mathbb{R}^d$. Let μ be a positive constant, let β be a constant vector field, and let $f \in L^2(\Omega)$. Equip $V = H_0^1(\Omega)$ with the norm $v \mapsto \|v\|_V = \|\nabla v\|_{0,\Omega}$. Consider the problem: Seek $u \in V$ such that, for all $v \in V$, $a(u, v) = \int_\Omega fv$, where $a(u, v) = \int_\Omega \mu \nabla u \cdot \nabla v + (\beta \cdot \nabla u)v$.

(i) Explain why $v \mapsto \|\nabla v\|_{0,\Omega}$ is a norm in V.

(ii) Show that the above problem is well-posed.

(iii) Let V_h be a finite-dimensional subspace of V. Let $\lambda \ge 0$, define the bilinear form $a_h(w_h, v_h) = a(w_h, v_h) + \lambda h \int_\Omega \nabla w_h \cdot \nabla v_h$, and let $u_h \in V_h$ be such that $a_h(u_h, v_h) = \int_\Omega f v_h$ for all $v_h \in V_h$. Set $\mu_h = \mu + \lambda h$. Prove

$$\|u - u_h\|_V \le \inf_{v_h \in V_h} \left\{ \frac{\lambda h}{\mu_h} \sup_{w_h \in V_h} \frac{\int_\Omega \nabla v_h \cdot \nabla w_h}{\|w_h\|_V} + \left(1 + \frac{\|a\|}{\mu_h}\right) \|u - v_h\|_V \right\}.$$

(iv) Assume that there is an interpolation operator Π_h and an integer $k > 0$ such that $\|v - \Pi_h v\|_V \le c h^{l-1} \|v\|_{l,\Omega}$ for all $1 \le l \le k + 1$ and all $v \in H^l(\Omega) \cap V$. Prove and comment the following estimate:

$$\|u - u_h\|_V \le c \left\{ \left(1 + \frac{\|a\|}{\mu_h}\right) h^k |u|_{k+1,\Omega} + \frac{\lambda}{\mu_h} h \|\nabla u\|_{0,\Omega} \right\}.$$

Exercise 3.8. The goal of this exercise is to prove estimate (3.27) using duality techniques. Assume $p < \infty$. Let $v = |u - u_h|^{p-1}\text{sgn}(u - u_h)$ and let z be the solution to the adjoint problem (3.17) with data v.

(i) Verify that $v \in L^{p'}(\Omega)$ with $\frac{1}{p} + \frac{1}{p'} = 1$.

(ii) Using assumption (iv) of Theorem 3.21, find a constant δ' such that, for $p' > \delta'$, $z \in W^{2,p'}(\Omega)$.

(iii) Show that, for all $z_h \in V_h$,

$$\|u - u_h\|_{L^p(\Omega)}^p \le \|a\| \, \|u - u_h\|_{1,p,\Omega} \|z - z_h\|_{1,p',\Omega}.$$

(iv) Conclude.

Exercise 3.9 (Proof of Lemma 3.27).

(i) Explain why $\gamma_0(\mathcal{I}_h u) = \mathcal{I}_h^\partial(\gamma_0(u)) = \mathcal{I}_h^\partial(g) = \gamma_0(u_h)$.

(ii) Show that $a(\mathcal{I}_h u - u_h, v_h) = a(\mathcal{I}_h u - u, v_h)$, for all $v_h \in V_{h0}$.

(iii) Use (BNB1$_h$) to prove $\alpha_h \|\mathcal{I}_h u - u_h\|_{1,\Omega} \le \|a\| \, \|\mathcal{I}_h u - u\|_{1,\Omega}$.

(iv) Conclude.

Exercise 3.10 (Proof of Lemma 3.28).

(i) Prove that there is $\theta \in H^2(\Omega) \cap H_0^1(\Omega)$ such that $a(v, \theta) = \int_\Omega (\mathcal{I}_h u - u_h) v$ for all $v \in H_0^1(\Omega)$. Show that

$$\|\mathcal{I}_h u - u_h\|_{0,\Omega}^2 \le \|a\| \, \|u - u_h\|_{1,\Omega} \|\theta - w_h\|_{1,\Omega} + a(\mathcal{I}_h u - u, \theta).$$

(ii) Using Lemma 3.27 to estimate $\|u - u_h\|_{1,\Omega}$ and using assumption (ii) in Lemma 3.27, show that

$$\|\mathcal{I}_h u - u_h\|_{0,\Omega}^2 \le c \|u - \mathcal{I}_h u\|_{1,\Omega} \inf_{w_h \in V_{h0}} \|\theta - w_h\|_{1,\Omega}$$
$$+ c \left(\|u - \mathcal{I}_h u\|_{0,\Omega} + \|g - \mathcal{I}_h g\|_{0,\partial\Omega} \right) \|\theta\|_{2,\Omega}.$$

(iii) Show that $\inf_{w_h \in V_{h0}} \|\theta - w_h\|_{1,\Omega} \le c h \|\theta\|_{2,\Omega}$ and that

$$\|\mathcal{I}_h u - u_h\|_{0,\Omega} \le c \left(h \|u - \mathcal{I}_h u\|_{1,\Omega} + \|u - \mathcal{I}_h u\|_{0,\Omega} + \|g - \mathcal{I}_h g\|_{0,\partial\Omega} \right).$$

(iv) Conclude.

Exercise 3.11. Prove Propositions 3.88 and 3.89.

Exercise 3.12. Assume that Ω is a bounded domain of class C^2 in $\mathbb{R}^2$. Using the notation of Lemma B.69, prove that $\nabla \cdot : [H_0^1(\Omega)]^2 \to L_{\int=0}^2(\Omega)$ is continuous and surjective. (*Hint:* For $g \in L_{\int=0}^2(\Omega)$, construct $[H_0^1(\Omega)]^2 \ni u = \nabla q + \nabla \times \psi$ such that $\nabla \cdot u = g$ and q solves a Poisson problem, ψ solves a biharmonic problem, and $\nabla \times \psi := (\partial_2 \psi, -\partial_1 \psi)$.)

Exercise 3.13. Let Ω be a domain in $\mathbb{R}^d$. Prove that $C^{0,1}(\partial\Omega) \subset H^{\frac{1}{2}}(\partial\Omega)$ with continuous embedding.

Exercise 3.14. Let $\Omega = \,]0,1[^2$. Consider the problem $-\Delta u + u = 1$ in Ω and $u_{|\partial\Omega} = 0$. Approximate its solution with $\mathbb{P}_1$ H^1-conformal finite elements.

(i) Let $\{\lambda_0, \lambda_1, \lambda_2\}$ be the barycentric coordinates in the triangle K_h shown in the figure. Compute the entries of the elementary stiffness matrix $\mathcal{A}_{ij} = \int_{K_h} \nabla\lambda_i \cdot \nabla\lambda_j + \int_{K_h} \lambda_i\lambda_j$, and the right-hand side vector $\int_{K_h} \lambda_i$. (*Hint*: Use a quadrature from Table 8.2.)

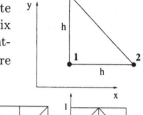

(ii) Consider the two meshes shown in the figure. Assemble the stiffness matrix and the right-hand side in both cases and compute the solution. For a fine mesh composed of 800 elements, $u_h(\frac{1}{2}, \frac{1}{2}) \approx 0,0702$. Comment.

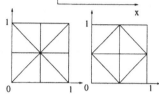

Exercise 3.15. Let $\Omega =]0, 3[\times]0, 2[$. Consider the problem $-\Delta u = 1$ in Ω and $u_{|\partial\Omega} = 0$. Approximate its solution with $\mathbb{P}_1$ H^1-conformal finite elements.

(i) Consider the reference simplex $\widehat{T}$ and the reference square $\widehat{K}$ shown in the figure. The nodes are numbered anticlockwise from $(0,0)$. Let $\{\widehat{\lambda}_1, \widehat{\lambda}_2, \widehat{\lambda}_3\}$ and $\{\widehat{\theta}_1, \widehat{\theta}_2, \widehat{\theta}_3, \widehat{\theta}_4\}$ be the local shape functions on $\widehat{T}$ and $\widehat{K}$, respectively. Compute the matrices $\left(\int_{\widehat{T}} \nabla\widehat{\lambda}_i \cdot \nabla\widehat{\lambda}_j\right)_{1 \le i,j \le 3}$ and $\left(\int_{\widehat{K}} \nabla\widehat{\theta}_i \cdot \nabla\widehat{\theta}_j\right)_{1 \le i,j \le 4}$.

(ii) Consider the meshes shown in the figure. Assemble the stiffness matrix for each of these three meshes.

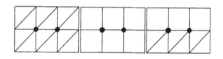

Exercise 3.16. Let Ω be a two-dimensional domain and let $\{\mathcal{T}_h\}_{h>0}$ be a shape-regular family of meshes composed of affine simplices. Let $P^2_{\mathrm{pt},h}$ be the finite element space defined in (1.71). Let

$$P^2_{\mathrm{pt},h,0} = \left\{ v_h \in P^2_{\mathrm{pt},h}; \forall F \in \mathcal{F}^\partial_h, \int_F v_h = 0 \right\}.$$

Prove that the extended Poincaré inequality (3.35) holds in $P^2_{\mathrm{pt},h,0}$. (*Hint*: Proceed as in the proof of Lemma 3.31.)

Exercise 3.17 (Discrete maximum principle). Let Ω be a polygonal domain in $\mathbb{R}^2$ and let $\mathcal{T}_h$ be an affine simplicial mesh of Ω. Assume that all the angles of the triangles in $\mathcal{T}_h$ are acute. Let $P^1_{c,h}$ be the approximation space constructed on $\mathcal{T}_h$ using continuous, piecewise linears. Let $\{\varphi_1, \ldots, \varphi_N\}$ be the global shape functions and let $\mathcal{A}$ be the stiffness matrix associated with the Laplace operator, i.e., $\mathcal{A}_{ij} = \int_\Omega \nabla\varphi_i \cdot \nabla\varphi_j$ for $1 \le i, j \le N$.

(i) Show that $\mathcal{A}$ is an *M-matrix*, i.e., all its off-diagonal entries are non-positive and its row-wise sums are non-negative.

(ii) Prove the following discrete maximum principle: If $f \in L^2(\Omega)$ is such that $f \leq 0$ in Ω, the finite element solution u_h to the homogeneous Dirichlet problem with right-hand side f is such that $u_h \leq 0$ in Ω.

4

Mixed Problems

This chapter is concerned with the study of model problems in incompressible fluid and solid mechanics. As opposed to the previous chapter, we now deal with PDEs that are not endowed with a coercivity property. The prototypical situation is that of the Stokes problem modeling incompressible fluid flows at low Reynolds number. More generally, we are concerned with mixed problems in which two dependent variables (say pressure and velocity) play different roles. As a consequence, we are led to approximate the two variables in different finite element spaces.

This chapter is organized into four sections. The mixed formulation and the constrained formulation of the Stokes problem are introduced in the first section. In addition to the well-posedness analysis of these two formulations, various non-standard boundary conditions are addressed. In the second section, the attention is focused on the approximation theory of the Stokes problem in mixed form. The standard Galerkin technique is studied in details, and it is shown that a necessary and sufficient condition for the discrete problem to be well-posed is that the velocity and the pressure spaces satisfy an inf-sup compatibility condition. Examples and counterexamples of finite elements satisfying this condition are reviewed. The third section investigates a technique tailored to solve the Stokes problem with any type of finite element, in particular those that do not satisfy the above mentioned inf-sup condition. Various solution techniques to solve the linear system arising from the discretization of the Stokes problem are analyzed in the last section.

4.1 Mathematical Study of the Stokes Problem

This section briefly describes the physical model underlying the Stokes problem and two mathematical formulations thereof, namely the mixed formulation and the constrained formulation. Various non-standard boundary conditions are also reviewed.

4.1.1 Physical model

Let Ω be a domain in $\mathbb{R}^d$ in the sense of Definition 1.46. We are interested in incompressible fluid flows in Ω. We assume that the flow is stationary and that inertial forces are negligible. Given two functions $f : \Omega \to \mathbb{R}^d$ (the body force acting on the fluid) and $g : \Omega \to \mathbb{R}$ (the mass production rate), the *Stokes problem* consists of seeking two functions $u : \Omega \to \mathbb{R}^d$ and $p : \Omega \to \mathbb{R}$ such that

$$\begin{cases} -\Delta u + \nabla p = f & \text{in } \Omega, \\ \nabla \cdot u = g & \text{in } \Omega, \\ u = 0 & \text{on } \partial\Omega. \end{cases} \tag{4.1}$$

The first equation in (4.1) is termed the *momentum equation* and the second one the *mass conservation equation*. The unknown functions u and p are the velocity and pressure, respectively. The function g is constrained by the relation $\int_\Omega g = 0$, since the divergence formula implies $\int_\Omega \nabla \cdot u = \int_{\partial\Omega} u \cdot n = 0$. Although many applications involve incompressible flows for which g is zero, we treat here a slightly more general case.

4.1.2 Mixed formulation

We first derive heuristically a *mixed weak formulation* of the Stokes problem (4.1) and then analyze its well-posedness.

Let v be a (sufficiently smooth) $\mathbb{R}^d$-valued test function. Since the value of the velocity u is enforced at the boundary, proceed as for elliptic PDEs by taking test functions that vanish on $\partial\Omega$ (see §3.1.1). Multiply the first equation in (4.1) by v and integrate over Ω. Integrating by parts yields

$$-\int_\Omega v \cdot \Delta u = \sum_{i=1}^d -\int_\Omega v_i \, \Delta u_i = \sum_{i=1}^d \int_\Omega \nabla u_i \cdot \nabla v_i = \int_\Omega \nabla u : \nabla v,$$

since v vanishes at the boundary. For the same reason, the term $\int_\Omega v \cdot \nabla p$ is equal to $-\int_\Omega p \nabla \cdot v$. Therefore,

$$\int_\Omega \nabla u : \nabla v - \int_\Omega p \nabla \cdot v = \int_\Omega f \cdot v.$$

The three integrals are well-defined if u and v are in $[H_0^1(\Omega)]^d$, p in $L^2(\Omega)$, and $f \in [L^2(\Omega)]^d$. It is also possible to consider $f \in [H^{-1}(\Omega)]^d$; the right-hand side must then be replaced by $\langle f, v \rangle_{H^{-1}, H_0^1}$.

To derive a weak formulation for the second equation in (4.1), consider a (sufficiently smooth) scalar-valued test function q. Proceeding as before yields

$$\int_\Omega q \nabla \cdot u = \int_\Omega g q.$$

Since u is in $[H_0^1(\Omega)]^d$, the left-hand side is well-defined provided $q \in L^2(\Omega)$. However, since $\int_\Omega \nabla \cdot u = \int_\Omega g = 0$, the equation needs not be tested by constant functions. The test functions q are then restricted to those in $L^2(\Omega)$ with zero-mean over Ω. The corresponding functional space is denoted by

$$L_{f=0}^2(\Omega) = \left\{ q \in L^2(\Omega); \ \int_\Omega q = 0 \right\}.$$

Observe that if a constant c is added to the pressure field p, the pair $(u, p+c)$ is also a solution to (4.1). Therefore, the pressure can be restricted to have zero-mean over Ω.

Introducing the bilinear forms

$$a(u, v) = \int_\Omega \nabla u : \nabla v, \qquad b(v, p) = -\int_\Omega p \nabla \cdot v,$$

and the linear forms $f(v) = \langle f, v \rangle_{H^{-1}, H_0^1}$ and $g(q) = -\int_\Omega gq$, we obtain the following weak formulation: Given $f \in [H^{-1}(\Omega)]^d$ and $g \in L_{f=0}^2(\Omega)$,

$$\begin{cases} \text{Seek } u \in [H_0^1(\Omega)]^d \text{ and } p \in L_{f=0}^2(\Omega) \text{ such that} \\ a(u, v) + b(v, p) = f(v), \quad \forall v \in [H_0^1(\Omega)]^d, \\ b(u, q) = g(q), \quad \forall q \in L_{f=0}^2(\Omega). \end{cases} \qquad (4.2)$$

The spaces $[H_0^1(\Omega)]^d$ and $[H^{-1}(\Omega)]^d$ are equipped with the norms $\|u\|_{1,\Omega}^2 = \sum_{i=1}^d \|u_i\|_{1,\Omega}^2$ and $\|f\|_{-1,\Omega}^2 = \sum_{i=1}^d \|f_i\|_{-1,\Omega}^2$, where $u = (u_i)_{1 \le i \le d} \in [H_0^1(\Omega)]^d$ and $f = (f_i)_{1 \le i \le d} \in [H^{-1}(\Omega)]^d$.

Proposition 4.1. *If f and g are in $[L^2(\Omega)]^d$ and $L_{f=0}^2(\Omega)$, respectively, the solution (u, p) to (4.2) satisfies*

$$\begin{cases} -\Delta u + \nabla p = f & \text{a.e. in } \Omega, \\ \nabla \cdot u = g & \text{a.e. in } \Omega, \\ u = 0 & \text{a.e. on } \partial\Omega. \end{cases} \qquad (4.3)$$

Proof. Consider functions in $[\mathcal{D}(\Omega)]^d$ to test the momentum equation. Integration by parts shows that in the distribution sense,

$$\forall v \in [\mathcal{D}(\Omega)]^d, \quad \langle -\Delta u + \nabla p, v \rangle_{\mathcal{D}', \mathcal{D}} = \int_\Omega f \cdot v,$$

since $f \in [L^2(\Omega)]^d$. By density of $[\mathcal{D}(\Omega)]^d$ in $[L^2(\Omega)]^d$, the equality $-\Delta u + \nabla p = f$ holds in $[L^2(\Omega)]^d$, which also implies equality a.e. in Ω. Proceed similarly for the mass conservation equation. The boundary condition on u is a natural consequence of u being in $[H_0^1(\Omega)]^d$ together with the Trace Theorem B.52. $\qquad \square$

Remark 4.2. If f and g are in $[\mathcal{C}^0(\overline{\Omega})]^d$ and $\mathcal{C}^0(\overline{\Omega})$, respectively, if Ω is $\mathcal{C}^2$, and if the solution (u, p) to (4.2) is such that $u \in [\mathcal{C}^2(\overline{\Omega})]^d$ and $p \in \mathcal{C}^1(\overline{\Omega})$, the pair (u, p) solves the Stokes problem in the classical sense.   □

Theorem 4.3. *Problem* (4.2) *is well-posed and there exist* c_1 *and* c_2 *such that, for all* $f \in [H^{-1}(\Omega)]^d$ *and* $g \in L^2_{\int=0}(\Omega)$,

$$\|u\|_{1,\Omega} + \|p\|_{0,\Omega} \le c_1\|f\|_{-1,\Omega} + c_2\|g\|_{0,\Omega}. \tag{4.4}$$

Proof. We apply Theorem 2.34. Set $B = \nabla\cdot\ :\ [H^1_0(\Omega)]^d \to L^2_{\int=0}(\Omega)$ and $\mathrm{Ker}(B) = V_0 = \{v \in [H^1_0(\Omega)]^d;\ \nabla\cdot v = 0\}$. Note that V_0 is a Hilbert space since it is a closed subspace of $[H^1_0(\Omega)]^d$.
(1) The bilinear form $\int_\Omega \nabla u{:}\nabla v$ is coercive on $[H^1_0(\Omega)]^d$ owing to Poincaré's Lemma; hence, it is also coercive on V_0. As a result, the two conditions in (2.28) hold.
(2) Inequality (2.29) amounts to

$$\inf_{q \in L^2_{\int=0}(\Omega)} \sup_{v \in [H^1_0(\Omega)]^d} \frac{\int_\Omega q\nabla\cdot v}{\|v\|_{1,\Omega}\|q\|_{0,\Omega}} \ge \beta. \tag{4.5}$$

The existence of $\beta > 0$ is a consequence of Corollary B.71.   □

Remark 4.4. The well-posedness of the mixed formulation results from two facts: the coercivity of a and the surjectivity of the operator $\nabla\cdot\ :\ [H^1_0(\Omega)]^d \to L^2_{\int=0}(\Omega)$. The reader may already foresee one major difficulty: although one can imagine quite easily that any reasonable discrete formulation of the mixed problem (4.2) will naturally inherit the coercivity of a provided space-conformity is satisfied, there is no a priori reason for the discrete problem to inherit the surjectivity of the divergence operator.   □

4.1.3 Constrained formulation

An alternative formulation to (4.2) consists of including the constraint on the divergence of u in the solution space. Since the operator $\nabla\cdot\ :\ [H^1_0(\Omega)]^d \to L^2_{\int=0}(\Omega)$ is surjective (see Lemma B.69 applied with $p = 2$), the Open Mapping Theorem ensures the existence of a constant c such that, for all $g \in L^2_{\int=0}(\Omega)$, there is $u_g \in [H^1_0(\Omega)]^d$ satisfying the equality $\nabla\cdot u_g = g$ and $\|u_g\|_{1,\Omega} \le c\|g\|_{0,\Omega}$; see Lemma A.36. Let us now define the space

$$V_0 = \{v \in [H^1_0(\Omega)]^d;\ \nabla\cdot v = 0\}. \tag{4.6}$$

Upon introducing the new function $u' = u - u_g$, it is clear that $\nabla\cdot u' = 0$, i.e., $u' \in V_0$. By restricting the test functions v in (4.2) to those in V_0 only, one enforces $b(v, p)$ to be zero. Furthermore, the weak form of the mass conservation equation is automatically satisfied if u' is sought in V_0. Hence, we are led to consider the weak formulation:

$$\begin{cases} \text{Seek } u' \in V_0 \text{ such that} \\ a(u', v) = f(v) - a(u_g, v), \quad \forall v \in V_0. \end{cases} \tag{4.7}$$

This problem is called the *constrained formulation* of the Stokes problem. Note that the pressure is no longer involved.

Proposition 4.5. *Problem* (4.7) *is well-posed.*

Proof. Apply the Lax–Milgram Lemma.
(i) V_0 is a Hilbert space.
(ii) The bilinear form a is both continuous and coercive on $V_0 \times V_0$.
(iii) The linear form $f(\cdot) - a(u_g, \cdot)$ is continuous in V_0 since

$$\forall v \in V_0, \quad |f(v) - a(u_g, v)| \leq (\|f\|_{-1,\Omega} + c\|g\|_{0,\Omega})\|v\|_{1,\Omega}. \qquad \square$$

The connection between the mixed formulation (4.2) and the constrained formulation (4.7) is emphasized by the following:

Proposition 4.6. *Let u be a function in $[H_0^1(\Omega)]^d$ and define $u' = u - u_g$. The following statements are equivalent:*

(i) *There is p in $L_{\int=0}^2(\Omega)$ such that the pair (u, p) solves* (4.2).
(ii) *u' solves* (4.7).

Proof. The implication (i) $\Rightarrow$ (ii) is evident. The converse (ii) $\Rightarrow$ (i) is a consequence of de Rham's Theorem. The linear form $a(u, \cdot) - f(\cdot)$ is continuous on $[H_0^1(\Omega)]^d$ and is zero on V_0. As a result, owing to de Rham's Theorem, there is p in $L_{\int=0}^2(\Omega)$ such that $a(u, \cdot) - f(\cdot) = \langle \nabla p, \cdot \rangle_{H^{-1}, H_0^1}$ on $[H_0^1(\Omega)]^d$. $\qquad \square$

The formulation (4.7) is of interest on many theoretical accounts, one of them being the coercivity of the bilinear form a on V_0. However, this formulation is not very easy to implement since it requires using divergence-free finite elements which are cumbersome to program; see [BrF91^b] for other details on this type of finite element.

4.1.4 Variations on boundary conditions

In this section, we broaden the scope by considering various types of boundary conditions. We restrict ourselves to the Stokes problem

$$\begin{cases} -\Delta u + \nabla p = f, \\ \nabla \cdot u = 0, \end{cases} \tag{4.8}$$

with $f \in [L^2(\Omega)]^d$.

Conditions based on $\Delta u = \nabla{\cdot}(\nabla u)$. Let v be a test function in $[\mathcal{C}^\infty(\Omega)]^d$. Take the Euclidean scalar product of the momentum equation with v and integrate the result over Ω. Integration by parts yields

$$\int_\Omega \nabla u{:}\nabla v - \int_\Omega p\nabla{\cdot}v - \int_{\partial\Omega} \partial_n u{\cdot}v + \int_{\partial\Omega} p(v{\cdot}n) = \int_\Omega f{\cdot}v.$$

Since at the boundary, every vector field v can be decomposed into the form $v_{|\partial\Omega} = (v{\cdot}n)n + n \times (v{\times}n)$, we obtain

$$\int_\Omega \nabla u{:}\nabla v - \int_\Omega p\nabla{\cdot}v + \int_{\partial\Omega} (p - \partial_n u{\cdot}n)\,(v{\cdot}n) - \int_{\partial\Omega} (\partial_n u \times n) \cdot (v{\times}n) = \int_\Omega f{\cdot}v.$$

To extract from this equation a bilinear form of the type $a(u,v) + b(v,p)$ where a is coercive, the boundary integrals need to be transformed. Three strategies are possible to this end: we can enforce the integrals to be zero, or we can put them on the right-hand side of the equation by eliminating the dependence on u, and finally we can enforce these integrals to be positive when $v = u$. To illustrate the various options, we introduce the partition $\partial\Omega = \partial\Omega_1 \cup \partial\Omega_2 \cup \partial\Omega_3 \cup \partial\Omega_4$.

(i) Setting $v_{|\partial\Omega_1} = 0$ enforces the boundary integrals to be zero on $\partial\Omega_1$. Since we want the solution and test spaces for the velocity to be identical, this condition amounts to enforcing

$$u_{|\partial\Omega_1} = 0. \tag{4.9a}$$

(ii) We enforce the first boundary integral to be zero on $\partial\Omega_2$ by choosing $v{\cdot}n_{|\partial\Omega_2} = 0$. The second integral can be transformed into a positive bilinear form by setting $\partial_n u \times n = g_{24} - \mu_{24}(u \times n)$, where g_{24} and μ_{24} are smooth functions on $\partial\Omega_2$, and μ_{24} is positive. That is, on $\partial\Omega_2$ we enforce

$$\begin{cases} u{\cdot}n_{|\partial\Omega_2} = 0, \\ (\partial_n u \times n + \mu_{24}(u \times n))_{|\partial\Omega_2} = g_{24}, \quad \mu_{24}(x) \geq 0. \end{cases} \tag{4.9b}$$

(iii) By setting $(v{\times}n)_{|\partial\Omega_3} = 0$, the second boundary integral is zero on $\partial\Omega_3$. We transform the first integral into a positive form by setting $p - (\partial_n u){\cdot}n = \mu_{34}(u{\cdot}n) - g_{34}$, where g_{34} and μ_{34} are smooth functions on $\partial\Omega_3$, and μ_{34} is positive. In other words, on $\partial\Omega_3$ we enforce

$$\begin{cases} u \times n_{|\partial\Omega_3} = 0, \\ (p - \partial_n u{\cdot}n - \mu_{34}(u{\cdot}n))_{|\partial\Omega_3} = -g_{34}, \quad \mu_{34}(x) \geq 0. \end{cases} \tag{4.9c}$$

(iv) A fourth possibility consists of enforcing nothing on both the solution u and the test function v on the boundary $\partial\Omega_4$. To ensure that the two boundary integrals can be transformed into a positive bilinear form, we consider the conditions

$$\begin{cases} (\partial_n u \times n + \mu_{24}(u \times n))_{|\partial\Omega_4} = g_{24}, & \mu_{24}(x) \geq 0, \\ (p - \partial_n u \cdot n - \mu_{34}(u \cdot n))_{|\partial\Omega_4} = -g_{34}, & \mu_{34}(x) \geq 0. \end{cases} \tag{4.9d}$$

To formalize the above choices for the boundary conditions, we introduce the Hilbert space

$$X = \{v \in [H^1(\Omega)]^d; \, v_{|\partial\Omega_1} = 0; \, v \cdot n_{|\partial\Omega_2} = 0; \, v \times n_{|\partial\Omega_3} = 0\},$$

and define the bilinear and linear forms

$$\begin{cases} a(u,v) = \displaystyle\int_\Omega \nabla u : \nabla v + \int_{\partial\Omega_2 \cup \partial\Omega_4} \mu_{24}(u \times n) \cdot (v \times n) + \int_{\partial\Omega_3 \cup \partial\Omega_4} \mu_{34}(u \cdot n)(v \cdot n), \\ b(v,p) = -\displaystyle\int_\Omega p \nabla \cdot v, \\ h(v) = \displaystyle\int_\Omega f \cdot v + \int_{\partial\Omega_2 \cup \partial\Omega_4} g_{24} \cdot (v \times n) + \int_{\partial\Omega_3 \cup \partial\Omega_4} g_{34}(v \cdot n). \end{cases}$$

The mixed formulation of the problem is:

$$\begin{cases} \text{Seek } u \in X \text{ and } p \in M = L^2(\Omega) \text{ such that} \\ a(u,v) + b(v,p) = h(v), & \forall v \in X, \\ b(u,q) = 0, & \forall q \in M. \end{cases} \tag{4.10}$$

Proposition 4.7. *If* meas$(\partial\Omega_3 \cup \partial\Omega_4) \neq 0$, *problem* (4.10) *is well-posed.*

Proof. The form a is clearly coercive. The operator $\nabla \cdot : X \to L^2(\Omega)$ is surjective according to Lemma 4.9. Apply Theorem 2.34 to conclude. □

Remark 4.8. Note that if meas$(\partial\Omega_3 \cup \partial\Omega_4) \neq 0$, the pressure is well-defined in $L^2(\Omega)$, i.e., the pressure mean-value is controlled by the natural boundary conditions enforced on $\partial\Omega_3 \cup \partial\Omega_4$. □

Lemma 4.9. *If* meas$(\partial\Omega_3 \cup \partial\Omega_4) \neq 0$, *the operator* $\nabla \cdot : X \to L^2(\Omega)$ *is surjective.*

Proof. Let q be in $L^2(\Omega)$. Let $\mathcal{O}$ be a smooth subset of $\partial\Omega_3 \cup \partial\Omega_4$ of non-zero measure and let ρ be a smooth positive function compactly supported in $\mathcal{O}$. Let $g = c\rho n$ be a vector-valued function in $\mathcal{O}$ where the constant c is chosen so that $\int_{\mathcal{O}} g \cdot n = \int_\Omega q$ (this is possible since meas$(\partial\Omega_3 \cup \partial\Omega_4) \neq 0$). Extend g by zero on $\partial\Omega$. Since $\mathcal{O}$ is smooth, g is in $[H^{\frac{1}{2}}(\partial\Omega)]^d$; moreover, $g = 0$ on $\partial\Omega \backslash (\partial\Omega_3 \cup \partial\Omega_4)$. As a result, it is possible to find a function w in $[H^1(\Omega)]^d$ such that $w_{|\partial\Omega} = g$. Clearly, $w_{|\partial\Omega_1 \cup \partial\Omega_2} = 0$ and $w \times n_{|\partial\Omega_3} = 0$. Set $q_0 = \nabla \cdot w - q$. Clearly, $q_0 \in L^2(\Omega)$ and $\int_\Omega q_0 = 0$. Hence, q_0 is in $L^2_{\int=0}(\Omega)$. Since the operator $\nabla \cdot : [H^1_0(\Omega)]^d \to L^2_{\int=0}(\Omega)$ is surjective according to Lemma B.69, there is $w_0 \in [H^1_0(\Omega)]^d$ such that $\nabla \cdot w_0 = q_0$. In conclusion, the function $w - w_0$ is in X and satisfies $\nabla \cdot (w - w_0) = q$. □

Conditions based on $\Delta u = \nabla(\nabla \cdot u) - \nabla \times \nabla \times u$. The technique developed above can also be applied to derive a weak formulation based on the vector identity $\Delta u = \nabla(\nabla \cdot u) - \nabla \times \nabla \times u$. Let v be a test function in $[\mathcal{C}^\infty(\Omega)]^d$. Test the momentum equation of the Stokes problem by v. Using (B.20) and (B.21), integration by parts yields

$$
\int_\Omega (\nabla \cdot u \nabla \cdot v + \nabla \times u \cdot \nabla \times v) - \int_\Omega p \nabla \cdot v - \int_{\partial\Omega} (v \cdot n)\nabla \cdot u
$$
$$
+ \int_{\partial\Omega} (v \times n) \cdot \nabla \times u + \int_{\partial\Omega} p(v \cdot n) = \int_\Omega f \cdot v.
$$

All the terms involving $\nabla \cdot u$ are zero since u is solenoidal. On the other hand, since at the boundary $\nabla \times u$ can be decomposed into

$$
\nabla \times u_{|\partial\Omega} = (\nabla \times u \cdot n)n + n \times (\nabla \times u \times n),
$$

we infer

$$
\int_\Omega \nabla \times u \cdot \nabla \times v - \int_\Omega p \nabla \cdot v + \int_{\partial\Omega} p(v \cdot n) + \int_{\partial\Omega} (n \times (\nabla \times u \times n)) \cdot (v \times n) = \int_\Omega f \cdot v.
$$

To extract from this equality a bilinear form of the type $a(u,v) + b(v,p)$ where a is coercive, three strategies are possible: enforce the boundary integrals to be zero, transform them into symmetric positive bilinear forms, or move them to the right-hand side. Upon introducing the partition $\partial\Omega = \partial\Omega_1 \cup \partial\Omega_2 \cup \partial\Omega_3$, the following boundary conditions are admissible:

$$
u_{|\partial\Omega_1} = 0, \tag{4.11a}
$$
$$
u \cdot n_{|\partial\Omega_2} = 0 \quad \text{and} \quad (n \times (\nabla \times u \times n) - \mu_2(u \times n))_{|\partial\Omega_2} = -g_2, \tag{4.11b}
$$
$$
u \times n_{|\partial\Omega_3} = 0 \quad \text{and} \quad (p - \mu_3(u \cdot n))_{|\partial\Omega_3} = -g_3, \tag{4.11c}
$$

where $\mu_2 \geq 0$ and g_2 (resp., $\mu_3 \geq 0$ and g_3) are smooth functions on $\partial\Omega_2$ (resp., $\partial\Omega_3$). Introducing the Hilbert spaces

$$
X = \{v \in [H^1(\Omega)]^d;\ v_{|\partial\Omega_1} = 0;\ v \cdot n_{|\partial\Omega_2} = 0;\ v \times n_{|\partial\Omega_3} = 0\},
$$

$M = L^2(\Omega)$ if $\mathrm{meas}(\partial\Omega_3) \neq 0$ and $M = L^2_{\int=0}(\Omega)$ otherwise, and the forms

$$
\begin{cases}
a(u,v) = \int_\Omega \nabla \times u \cdot \nabla \times v + \int_{\partial\Omega_2} \mu_2(u \times n) \cdot (v \times n) + \int_{\partial\Omega_3} \mu_3(u \cdot n)(v \cdot n), \\[2mm]
b(v,p) = -\int_\Omega p \nabla \cdot v, \\[2mm]
h(v) = \int_\Omega f \cdot v + \int_{\partial\Omega_2} g_2 \cdot (v \times n) + \int_{\partial\Omega_3} g_3(v \cdot n),
\end{cases}
$$

the weak form of the Stokes equations is:

$$\begin{cases} \text{Seek } u \in X \text{ and } p \in M \text{ such that} \\ a(u,v) + b(v,p) = h(v), \quad \forall v \in X, \\ b(u,q) = 0, \qquad\qquad\quad \forall q \in M. \end{cases} \qquad (4.12)$$

Proposition 4.10. *Problem* (4.12) *is well-posed.*

Proof. Simple consequence of Theorem 2.34 and Lemma 4.11 together with either Lemma 4.9 if $\text{meas}(\partial\Omega_3) \neq 0$ or Lemma B.69 otherwise. $\qquad\square$

Lemma 4.11. *Let* $V = \{v \in X; \nabla\cdot v = 0\}$ *be the nullspace of* $\nabla\cdot : X \to L^2(\Omega)$. *Then, the bilinear form* $\int_\Omega \nabla\times u \cdot \nabla\times v$ *is* V-*coercive.*

Proof. See, e.g., [GuQ97]. $\qquad\square$

Remark 4.12.

(i) Recall that V-coercivity is sufficient to ensure that the abstract problem (4.2) is well-posed; see Remark 2.35. The reader can verify that the bilinear form $\int_\Omega \nabla\times u \cdot \nabla\times v$ is not X-coercive!

(ii) In principle one could imagine letting u and v be free on a portion of the boundary $\partial\Omega_4$ to enforce a fully natural boundary condition on $\partial\Omega_4$. However, the resulting bilinear form $\int_\Omega \nabla\times u \cdot \nabla\times v$ would not be V-coercive; see [GuQ97] for the details.

(iii) The list of boundary conditions proposed above is not exhaustive. For instance, the reader may find new ones by using the fact that for a solenoidal vector field u the following identity holds: $\Delta u = \nabla\cdot(\nabla u + (\nabla u)^T)$. $\qquad\square$

4.2 Mixed Finite Element Approximation

This section is concerned with the mixed finite element approximation of the Stokes problem. Using the theoretical results of §2.4.2, we show that the discrete spaces for the velocity and the pressure must satisfy an inf-sup compatibility condition to ensure the well-posedness of the discrete problem.

4.2.1 The compatibility condition

Let $X_h \subset [H_0^1(\Omega)]^d$ and $M_h \subset L_{j=0}^2(\Omega)$ be finite-dimensional spaces. Using the notation of §4.1.2, consider the discrete problem:

$$\begin{cases} \text{Seek } u_h \in X_h \text{ and } p_h \in M_h \text{ such that} \\ a(u_h, v_h) + b(v_h, p_h) = f(v_h), \quad \forall v_h \in X_h, \\ b(u_h, q_h) = g(q_h), \qquad\qquad\quad \forall q_h \in M_h. \end{cases} \qquad (4.13)$$

Proposition 4.13. *The discrete problem* (4.13) *is well-posed if and only if the spaces X_h and M_h satisfy the compatibility condition*

$$\exists \beta_h > 0, \qquad \inf_{q_h \in M_h} \sup_{v_h \in X_h} \frac{\int_\Omega q_h \nabla \cdot v_h}{\|q_h\|_{0,\Omega} \|v_h\|_{1,\Omega}} \geq \beta_h. \qquad (4.14)$$

Proof. Since the bilinear form a is coercive on $[H_0^1(\Omega)]^d \times [H_0^1(\Omega)]^d$ and $X_h \subset [H_0^1(\Omega)]^d$, the statement is a direct consequence of Proposition 2.42. $\qquad \square$

Using the notation of Proposition 2.42, set $B = -\nabla \cdot : [H_0^1(\Omega)]^d \to L_{f=0}^2(\Omega)$ and $B_h = \pi_h \nabla \cdot (i_h(\cdot)) : X_h \to M_h'$, where i_h is the natural injection of X_h into $[H_0^1(\Omega)]^d$ and π_h is the L^2-projection from $L^2(\Omega)$ to M_h. Clearly,

$$\mathrm{Ker}(B_h) = \{v_h \in X_h; \forall q_h \in M_h, b(v_h, q_h) = 0\}.$$

Proposition 4.14. *Under hypothesis* (4.14), *the following estimates hold:*

$$\|u - u_h\|_{1,\Omega} \leq c_{1h} \inf_{v_h \in X_h} \|u - v_h\|_{1,\Omega} + c_{2h} \inf_{q_h \in M_h} \|p - q_h\|_{0,\Omega},$$

$$\|p - p_h\|_{0,\Omega} \leq c_{3h} \inf_{v_h \in X_h} \|u - v_h\|_{1,\Omega} + c_{4h} \inf_{q_h \in M_h} \|p - q_h\|_{0,\Omega},$$

where $c_{1h} = (1 + \frac{\|a\|}{\alpha})(1 + \frac{\|b\|}{\beta_h})$, $c_{2h} = 0$ if $\mathrm{Ker}(B_h) \subset \mathrm{Ker}(B)$ and $c_{2h} = \frac{\|b\|}{\alpha}$ otherwise, $c_{3h} = c_{1h} \frac{\|a\|}{\beta_h}$, and $c_{4h} = 1 + \frac{\|b\|}{\beta_h} + c_{2h} \frac{\|a\|}{\beta_h}$. Here, α is the coercivity constant of the bilinear form $\int_\Omega \nabla u : \nabla v$ on $[H_0^1(\Omega)]^d \times [H_0^1(\Omega)]^d$, $\|a\|$ its norm, and $\|b\|$ the norm of the bilinear form b on $[H_0^1(\Omega)]^d \times L_{f=0}^2(\Omega)$.

Proof. Direct application of Lemma 2.44. $\qquad \square$

Remark 4.15.

(i) In the literature, the condition (4.14) is often referred to as the *Babuška–Brezzi condition* [Bab73[a], Bre74]. The quantity β_h is often called the inf-sup constant; see also [Pir83, GiR86, Gun89, BrF91[b]].

(ii) The error estimates are optimal provided β_h is bounded uniformly from below as h goes to zero. Whenever possible, it is recommended to choose X_h and M_h so that the inf-sup constant does not depend on h.

(iii) In the estimate on $u - u_h$, the constants depend on $\frac{1}{\beta_h}$, whereas those in the estimate on $p - p_h$ depend on $\frac{1}{\beta_h^2}$. This means that if $\beta_h \to 0$ when $h \to 0$, the suboptimal behavior of β_h is more damaging for the convergence rate on the pressure than that on the velocity.

(iv) It is straightforward to extend Propositions (4.13) and (4.14) to non-conformal approximation settings; see §4.2.8 for an example. $\qquad \square$

Definition 4.16 (Smoothing property). *The Stokes problem* (4.3) *is said to have* smoothing properties *in Ω if assumption* (ANM1) *in §2.4.2 is satisfied with $H = [L^2(\Omega)]^d$, $Y = [H^2(\Omega)]^d \cap [H_0^1(\Omega)]^d$, and $N = H^1(\Omega) \cap L_{f=0}^2(\Omega)$.*

Lemma 4.17. *The Stokes problem has smoothing properties if one of the following statements holds:*

  (i) *Ω is a convex polygon in two dimensions.*
  (ii) *In two or three dimensions, Ω is of class $C^{1,1}$.*

Proof. The proof is quite technical; see [Cat61, AmG91]. □

Proposition 4.18. *Assume that:*

  (i) *The inf-sup condition (4.14) holds.*
  (ii) *The Stokes problem has smoothing properties.*
  (iii) *There is c_i, independent of h, such that, for all pairs $(v, q) \in ([H_0^1(\Omega)]^d \cap [H^2(\Omega)]^d) \times (L_{f=0}^2(\Omega) \cap H^1(\Omega))$,*

$$\inf_{(v_h, q_h) \in X_h \times M_h} \|v - v_h\|_{1,\Omega} + \|q - q_h\|_{0,\Omega} \leq c_i h(\|v\|_{2,\Omega} + \|q\|_{1,\Omega}).$$

Then, there is c such that

$$\forall h, \quad \|u - u_h\|_{0,\Omega} \leq c\,h \left(\inf_{v_h \in X_h} \|u - v_h\|_{1,\Omega} + \inf_{q_h \in M_h} \|p - q_h\|_{0,\Omega} \right).$$

Proof. Apply Proposition 2.45. □

4.2.2 The Fortin criterion

A powerful tool to prove the compatibility condition (4.14) is a lemma due to Fortin [For77]. This lemma is presented in an abstract form. The reader can easily rewrite it in the Stokes framework by setting $X = [H_0^1(\Omega)]^d$, $M = L_{f=0}^2(\Omega)$, and $b(v, q) = \int_\Omega q \nabla \cdot v$ for $v \in X$ and $q \in M$.

Lemma 4.19 (Fortin criterion). *Let X and M be two Banach spaces and let $b \in \mathcal{L}(X \times M; \mathbb{R})$. Assume that there is $\beta > 0$ such that the inf-sup condition $\inf_{q \in M} \sup_{v \in X} \frac{b(v,q)}{\|v\|_X \|q\|_M} \geq \beta$ holds. Let $X_h \subset X$ and $M_h \subset M$, M_h being reflexive. Then, there is $\beta_h > 0$ such that*

$$\inf_{q_h \in M_h} \sup_{v_h \in X_h} \frac{b(v_h, q_h)}{\|v_h\|_X \|q_h\|_M} \geq \beta_h, \tag{4.15}$$

iff there is $\gamma_h > 0$ such that, for all $v \in X$, there is $\Pi_h(v) \in X_h$ such that

$$\forall q_h \in M_h, \quad b(v, q_h) = b(\Pi_h(v), q_h) \quad \text{and} \quad \|\Pi_h(v)\|_X \leq \gamma_h \|v\|_X. \tag{4.16}$$

Proof. (1) Assume that (4.16) holds. Let $q_h \in M_h$. Clearly,

$$\sup_{v_h \in X_h} \frac{b(v_h, q_h)}{\|v_h\|_X} \geq \sup_{v \in X} \frac{b(\Pi_h(v), q_h)}{\|\Pi_h(v)\|_X} = \sup_{v \in X} \frac{b(v, q_h)}{\|\Pi_h(v)\|_X} \geq \sup_{v \in X} \frac{b(v, q_h)}{\gamma_h \|v\|_X}.$$

The last term is bounded from below by $\frac{\beta}{\gamma_h}\|q_h\|_M$, proving (4.15).

(2) Let us now prove the converse. Let v be in X. It is clear that $b(v,\cdot) \in M_h'$. Define $B_h : X_h \to M_h'$ to be the operator such that $\langle B_h v_h, q_h \rangle_{M_h', M_h} = b(v_h, q_h)$ for all $v_h \in X_h$ and $q_h \in M_h$. Since M_h is reflexive, owing to the converse statement in Lemma A.42, the inf-sup inequality implies that B_h is a surjective mapping and there is $\Pi_h(v) \in X_h$ such that $B_h \Pi_h(v) = b(v,\cdot)$ and $\beta_h \|\Pi_h(v)\|_X \le \|b(v,\cdot)\|_{M_h'}$. That is to say, for all $v \in X$, there is $\Pi_h(v)$ such that $b(\Pi_h(v), q_h) = b(v, q_h)$ for all $q_h \in M_h$ and $\beta_h \|\Pi_h(v)\|_X \le \|b\| \|v\|_X$. $\square$

4.2.3   Counter-examples

In this section, we study three pairs of finite element spaces that do not satisfy the inf-sup condition (4.14). This condition is not satisfied if and only if the operator $B_h^T : M_h \longrightarrow X_h'$ is not injective (or, once global shape functions have been chosen, the associated matrix has not full column rank). Equivalently, the inf-sup condition is not satisfied if and only if the operator $B_h : X_h \longrightarrow M_h'$ is not surjective. If B_h^T is not injective, a nonzero pressure field in $\text{Ker}(B_h^T)$ is called a *spurious mode*.

The $\mathbb{Q}_1/\mathbb{P}_0$ finite element: The checkerboard instability. The most well-known pair of incompatible finite element spaces is that where the velocity is approximated by means of continuous $\mathbb{Q}_1$ polynomials and the pressure by means of $\mathbb{P}_0$ polynomials, i.e., piecewise constants. This pair of spaces produces the so-called checkerboard instability.

Let us restrict ourselves to a two-dimensional setting and assume that the domain is the unit square $\Omega =]0,1[^2$. Define a uniform Cartesian mesh on Ω as follows: Let N be an integer larger than 2. Set $h = \frac{1}{N}$, and for $0 \le i, j \le N$, denote by a_{ij} the point whose coordinates are (ih, jh). Let K_{ij} be the square cell whose bottom left node is a_{ij}; see Figure 4.1. The resulting mesh is denoted by $\mathcal{T}_h = \bigcup_{i,j} K_{ij}$. Define the approximation spaces

$$X_h = \{u_h \in [\mathcal{C}^0(\overline{\Omega})]^2; \forall K_{ij} \in \mathcal{T}_h, u_h \circ T_{K_{ij}} \in [\mathbb{Q}_1]^2; u_{h|\partial\Omega} = 0\}, \quad (4.17)$$

$$M_h = \{p_h \in L^2_{f=0}(\Omega); \forall K_{ij} \in \mathcal{T}_h, p_h \circ T_{K_{ij}} \in \mathbb{P}_0\}. \quad (4.18)$$

Recall that for a mesh cell K, $T_K : \widehat{K} \to K$ denotes the $\mathcal{C}^1$-diffeomorphism that maps the reference cell $\widehat{K}$ to K; see §1.3.2. For all p_h in M_h, set $p_{i+\frac{1}{2}, j+\frac{1}{2}} = p_{h|K_{ij}}$, and for all u_h in X_h, denote by (u_{ij}, v_{ij}) the values of the two Cartesian components of u_h at the node a_{ij}.

To prove that the inf-sup constant is zero, it is sufficient to prove the existence of a nonzero pressure field $p_h \in \text{Ker}(B_h^T) = \text{Im}(B_h)^\perp$, i.e., such that $\int_\Omega p_h \nabla \cdot u_h = 0$ for all $u_h \in X_h$. By definition of M_h, p_h is constant on each cell; as a result,

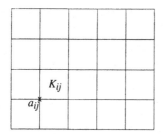

+1	−1	+1	−1	+1
−1	+1	−1	+1	−1
+1	−1	+1	−1	+1
−1	+1	−1	+1	−1

Fig. 4.1. The $\mathbb{Q}_1/\mathbb{P}_0$ finite element and the checkerboard instability: mesh (left) and spurious mode (right).

$$\int_{K_{ij}} p_h \nabla \cdot u_h = p_{i+\frac{1}{2},j+\frac{1}{2}} \int_{\partial K_{ij}} u_h \cdot n$$

$$= \tfrac{1}{2} h p_{i+\frac{1}{2},j+\frac{1}{2}} \left(u_{i+1,j} + u_{i+1,j+1} + v_{i+1,j+1} + v_{i,j+1} \right.$$

$$\left. - u_{i,j} + u_{i,j+1} - v_{i,j} + v_{i+1,j} \right).$$

Summing over all the cells and rearranging the sum yields

$$\int_{\Omega} p_h \nabla \cdot u_h = -h^2 \sum_{0<i,j<N} \left(u_{i,j} (\partial_1 p)_{ij} + v_{i,j} (\partial_2 p)_{ij} \right),$$

where

$$(\partial_1 p)_{ij} = \tfrac{1}{2h} \left(p_{i+\frac{1}{2},j+\frac{1}{2}} + p_{i+\frac{1}{2},j-\frac{1}{2}} - p_{i-\frac{1}{2},j+\frac{1}{2}} - p_{i-\frac{1}{2},j-\frac{1}{2}} \right),$$

$$(\partial_2 p)_{ij} = \tfrac{1}{2h} \left(p_{i+\frac{1}{2},j+\frac{1}{2}} + p_{i-\frac{1}{2},j+\frac{1}{2}} - p_{i+\frac{1}{2},j-\frac{1}{2}} - p_{i-\frac{1}{2},j-\frac{1}{2}} \right).$$

Hence, $\int_{\Omega} p_h \nabla \cdot u_h = 0$ for all $u_h \in X_h$ if and only if for all $1 \le i,j \le N-1$,

$$p_{i+\frac{1}{2},j+\frac{1}{2}} = p_{i-\frac{1}{2},j-\frac{1}{2}} \qquad \text{and} \qquad p_{i-\frac{1}{2},j+\frac{1}{2}} = p_{i+\frac{1}{2},j-\frac{1}{2}}.$$

The solution set of this linear system is a two-dimensional vector space. One dimension is spanned by the constant field $p_h = 1$. But, since the elements in M_h must be of zero mean, the line spanned by constant pressure fields must be excluded from the solution set. The other dimension is spanned by the field whose value is alternatively $+1$ and -1 on adjacent cells in a way similar to that of a checkerboard; see Figure 4.1. This oscillating function is usually referred to as a *spurious mode*. It is now clear that the inf-sup condition is not satisfied. Hence, the spaces X_h and M_h are incompatible to solve the Stokes problem.

Since the $\mathbb{Q}_1/\mathbb{P}_0$ finite element is very simple to program, one may be tempted to cure its deficiencies by restricting the size of M_h. For instance, one may enforce the pressure to be orthogonal (in the L^2-sense) to the space spanned by the spurious mode. Unfortunately, the cure is not strong enough to produce a healthy finite element. More precisely, it can be shown that by

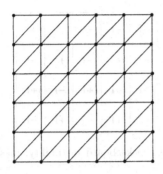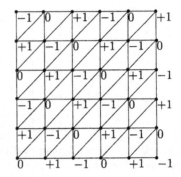

Fig. 4.2. The $\mathbb{P}_1/\mathbb{P}_1$ finite element: the mesh (left); one spurious mode (right).

doing so, one obtains a pair of finite element spaces for which the inf-sup constant β_h is such that

$$ch \le \beta_h \le c'h,$$

where the constants c and c' are positive and independent of h; see [BoN85] or [GiR86, p. 164] for further insight. This estimate shows that the method may not converge at all since $\frac{1}{\beta_h}$ comes into play in the error bound on the velocity and $\frac{1}{\beta_h^2}$ appears in the error bound on the pressure; see Proposition 4.14.

The $\mathbb{P}_1/\mathbb{P}_1$ finite element: Checkerboard-like instability. Because it is one of the simplest to program, the continuous $\mathbb{P}_1$ finite element for both velocity and pressure is a natural choice for approximating the Stokes problem. Unfortunately, the $\mathbb{P}_1/\mathbb{P}_1$ finite element does not satisfy the inf-sup condition (4.14).

To understand the origin of the problem, let us construct a two-dimensional counterexample in the square $\Omega =]0,1[^2$. Let us consider on Ω a uniform Cartesian mesh composed of squares of side h. The squares are split into triangles by cutting them along one diagonal as shown in the left panel of Figure 4.2. Denote by $\mathcal{T}_h$ the resulting triangulation and define the velocity and pressure approximation spaces to be

$$X_h = \{u_h \in [\mathcal{C}^0(\overline{\Omega})]^2; \forall K \in \mathcal{T}_h, u_h \circ T_K \in [\mathbb{P}_1]^2; u_{h|\partial\Omega} = 0\}, \qquad (4.19)$$

$$M_h = \{p_h \in L^2_{f=0}(\Omega) \cap \mathcal{C}^0(\overline{\Omega}); \forall K \in \mathcal{T}_h, p_h \circ T_K \in \mathbb{P}_1\}. \qquad (4.20)$$

Given a triangle K, denote by $\{a_{0,K}, a_{1,K}, a_{2,K}\}$ its three vertices. Now, consider a pressure field p_h such that the sum $\sum_{n=0}^2 p_h(a_{n,K})$ is zero on each triangle K. An example of such a *spurious mode* is shown in the right panel of Figure 4.2. Then,

$$\forall v_h \in X_h, \quad \int_\Omega p_h \nabla \cdot v_h = \sum_{K \in \mathcal{T}_h} (\nabla \cdot v_h)_{|K} \int_K p_h,$$

$$= \sum_{K \in \mathcal{T}_h} (\nabla \cdot v_h)_{|K} \frac{\text{meas}(K)}{3} \sum_{n=0}^2 p_h(a_{n,K}) = 0.$$

The pressure field p_h is such that $\int_\Omega p_h \nabla \cdot v_h$ is zero for all $v_h \in X_h$. In other words, this field is a spurious mode and the inf-sup constant is zero.

The $\mathbb{P}_1/\mathbb{P}_0$ finite element: The locking effect. A simple alternative to the $\mathbb{Q}_1/\mathbb{P}_0$ element consists of using the $\mathbb{P}_1/\mathbb{P}_0$ element. This element is appealing since it is very simple to program. Assuming that Ω is meshed with simplices, the velocity is approximated using continuous piecewise linear polynomials, and the pressure is approximated by means of (discontinuous) piecewise constants. Since the velocity is piecewise linear, its divergence is constant on each simplex. As a result, testing the divergence of the velocity by piecewise constants enforces the divergence to be zero everywhere. That is to say, the $\mathbb{P}_1/\mathbb{P}_0$ finite element yields a velocity approximation which is exactly divergence-free. Unfortunately, this finite element does not satisfy the inf-sup condition (4.14).

Let us produce a counterexample in two dimensions. Assume that Ω is a simply connected polygon in $\mathbb{R}^2$ and that Ω is meshed with triangles. Let N_{el}, $N_{\mathrm{v}}^{\mathrm{i}}$, and $N_{\mathrm{ed}}^{\partial}$ be the number of elements, internal vertices, and boundary edges in the triangulation, respectively. The Euler relations yield (see Lemma 1.57)

$$N_{\mathrm{el}} = 2N_{\mathrm{v}}^{\mathrm{i}} + N_{\mathrm{ed}}^{\partial} - 2.$$

It is clear that $\dim(M_h) = N_{\mathrm{el}} - 1$ and $\dim(X_h) = 2N_{\mathrm{v}}^{\mathrm{i}}$. Let $B_h : X_h \to M_h$ be the operator such that $(B_h v_h, q_h)_{0,\Omega} = (\nabla \cdot v_h, q_h)_{0,\Omega}$ for all $v_h \in X_h$ and $q_h \in M_h$. The Rank Theorem implies

$$\dim(\mathrm{Ker}(B_h^T)) = \dim(M_h) - \dim(\mathrm{Im}(B_h^T)) \geq \dim(M_h) - \dim(X_h)$$
$$= N_{\mathrm{el}} - 1 - 2N_{\mathrm{v}}^{\mathrm{i}} = N_{\mathrm{ed}}^{\partial} - 3.$$

As a result, there are at least $N_{\mathrm{ed}}^{\partial} - 3$ spurious modes. This means that the space M_h is far too rich for B_h to be surjective. Actually, in some cases it can be shown that B_h is injective, meaning that the only solution to $B_h u_h = 0$ is $u_h = 0$. In the literature, this situation is referred to as the *locking* phenomenon.

4.2.4 The $\mathbb{P}_1$-bubble/$\mathbb{P}_1$ finite element

The reason for which the $\mathbb{P}_1/\mathbb{P}_1$ element does not satisfy the inf-sup condition (4.14) is that the velocity space is not rich enough (or, conversely, the pressure space is too rich). To circumvent this difficulty, we enlarge the velocity space. The simplest idea consists of adding one degree of freedom per element associated with the barycenter of each simplex.

Assume that Ω is a polyhedron in $\mathbb{R}^d$ and consider a sequence of affine simplicial meshes $\{\mathcal{T}_h\}_{h>0}$. On the reference simplex $\widehat{K}$, define a function $\widehat{b}$ such that

$$\widehat{b} \in H_0^1(\widehat{K}), \qquad 0 \leq \widehat{b} \leq 1, \qquad \widehat{b}(\widehat{C}) = 1, \tag{4.21}$$

where $\widehat{C}$ is the barycenter of $\widehat{K}$. Then, let

$$\widehat{P} = [\mathbb{P}_1(\widehat{K}) \oplus \text{span}(\widehat{b})]^d,$$

and define $\widehat{\Sigma}$ to be the set of the linear forms on $\widehat{P}$ that map a vector-valued function $\widehat{v} \in \widehat{P}$ to the value of one of its Cartesian components at one vertex of $\widehat{K}$ or at the barycenter. The approximation spaces are defined as

$$X_h = \{u_h \in [\mathcal{C}^0(\overline{\Omega})]^d; \forall K \in \mathcal{T}_h, u_h \circ T_K \in \widehat{P}; u_{h|\partial\Omega} = 0\}, \tag{4.22}$$

$$M_h = \{p_h \in L^2_{\int=0}(\Omega) \cap \mathcal{C}^0(\overline{\Omega}); \forall K \in \mathcal{T}_h, p_h \circ T_K \in \mathbb{P}_1\}. \tag{4.23}$$

A first possible definition of the function $\widehat{b}$ consists of setting

$$\widehat{b} = (d+1)^{d+1} \prod_{i=1}^{d+1} \widehat{\lambda}_i,$$

where $\{\widehat{\lambda}_1, \ldots, \widehat{\lambda}_{d+1}\}$ are the barycentric coordinates on $\widehat{K}$. This function is usually referred to as a *bubble function* in reference to the shape of its graph; see Figure 4.3. A second possibility consists of dividing the simplex $\widehat{K}$ into $d+1$ subsimplices by connecting the $d+1$ vertices of $\widehat{K}$ to its barycenter. Then, define $\widehat{b}$ as the continuous, piecewise linear function equal to one at $\widehat{C}$ and zero at the vertices of $\widehat{K}$.

Lemma 4.20. *Let $1 < p < \infty$ and let p' the conjugate of p, i.e., $\frac{1}{p} + \frac{1}{p'} = 1$. Let X_h and M_h be defined in (4.22) and (4.23), respectively. If the mesh family $\{\mathcal{T}_h\}_{h>0}$ is shape-regular, there is β, independent of h, such that*

$$\inf_{q_h \in M_h} \sup_{v_h \in X_h} \frac{\int_\Omega q_h \nabla \cdot v_h}{\|v_h\|_{[W^{1,p}(\Omega)]^d} \|q_h\|_{L^{p'}(\Omega)}} \geq \beta > 0. \tag{4.24}$$

Proof. We apply Lemma 4.19 using Lemma B.69. Let v be a function in $[W_0^{1,p}(\Omega)]^d$. The idea is to construct $\Pi_h(v) \in X_h$ such that

$$\forall q_h \in M_h, \quad \int_\Omega q_h \nabla \cdot (\Pi_h(v)) = \int_\Omega q_h \nabla \cdot v.$$

M_h being clearly a subspace of $W^{1,p'}(\Omega)$, this amounts to proving

$$\forall q_h \in M_h, \quad \sum_{K \in \mathcal{T}_h} \int_K \Pi_h(v) \cdot \nabla q_h = \sum_{K \in \mathcal{T}_h} \int_K v \cdot \nabla q_h.$$

Since $\nabla q_h \in [\mathbb{P}_0(K)]^d$, $\Pi_h(v)$ must be such that $\int_K \Pi_h(v) = \int_K v$ for all K in $\mathcal{T}_h$.

Let us first define an interpolant of v. Since $v \in [W_0^{1,p}(\Omega)]^d$ may not be continuous, its Lagrange interpolant may not exist. However, the Clément interpolant modified to preserve homogeneous boundary conditions, $\mathcal{C}_h(v)$, is well-defined; see Lemma 1.127 and Remark 1.129(i). Hence, it is legitimate to set

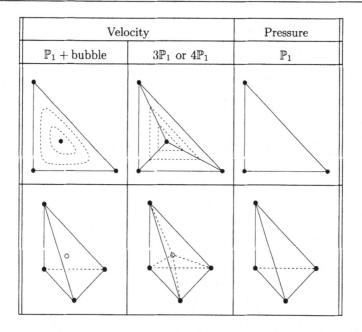

Velocity		Pressure
$\mathbb{P}_1$ + bubble	$3\mathbb{P}_1$ or $4\mathbb{P}_1$	$\mathbb{P}_1$

Fig. 4.3. Conventional representation of the $\mathbb{P}_1$-bubble/$\mathbb{P}_1$ finite element in two dimensions (top) and in three dimensions (bottom). The degrees of freedom for the velocity are shown in the first column or in the second column ($3\mathbb{P}_1$ in two dimensions and $4\mathbb{P}_1$ in three dimensions). Some isolines of the two-dimensional bubble function are drawn. The pressure degrees of freedom are shown in the third column.

$$\Pi_h(v) = \mathcal{C}_h(v) + \sum_{K \in \mathcal{T}_h} \sum_{i=1}^{d} \gamma_K^i e_i b_K,$$

where $\{e_1, \ldots, e_d\}$ is the canonical basis of $\mathbb{R}^d$ and $b_K = \widehat{b} \circ T_K^{-1}$. To enforce $\int_K \Pi_h(v) = \int_K v$, set

$$\gamma_K^i = \frac{\int_K (v^i - \mathcal{C}_h(v)^i)}{\int_K b_K},$$

where v^i and $\mathcal{C}_h(v)^i$ denote the Cartesian components of v and $\mathcal{C}_h(v)$, respectively. Since the mesh is affine, $|\det(J_K)| = \frac{\mathrm{meas}(K)}{\mathrm{meas}(\widehat{K})}$ and

$$\int_K b_K = \frac{\mathrm{meas}(K)}{\mathrm{meas}(\widehat{K})} \int_{\widehat{K}} \widehat{b} = \widehat{c} \, \mathrm{meas}(K).$$

Furthermore, since the family $\{\mathcal{T}_h\}_{h>0}$ is shape-regular,

$$\|b_K\|_{1,p,K} \leq c \, \|J_K^{-1}\|_d |\det(J_K)|^{\frac{1}{p}} \|\widehat{b}\|_{1,p,\widehat{K}}$$

$$\leq c \, \frac{\widehat{h}}{\rho_K} \frac{\mathrm{meas}(K)^{\frac{1}{p}}}{\mathrm{meas}(\widehat{K})^{\frac{1}{p}}} \|\widehat{b}\|_{1,p,\widehat{K}} \leq c' \, \mathrm{meas}(K)^{\frac{1}{p}} h_K^{-1}.$$

Moreover, letting Δ_K be defined as in Figure 1.25,

$$|\gamma_K^i| \leq \frac{\operatorname{meas}(K)^{\frac{1}{p'}}\|v - \mathcal{C}_h(v)\|_{L^p(K)}}{\widehat{c}\operatorname{meas}(K)} \leq c\operatorname{meas}(K)^{\frac{1}{p'}-1}h_K\|v\|_{W^{1,p}(\Delta_K)}.$$

In conclusion,

$$\left\|\sum_{K \in \mathcal{T}_h}\sum_{i=1}^d \gamma_K^i e^i b_K\right\|_{1,p,\Omega}^p \leq d^p \sum_{K \in \mathcal{T}_h}\|b_K\|_{1,p,K}^p \max\{|\gamma_K^1|^p, \ldots, |\gamma_K^d|^p\}$$

$$\leq c \sum_{K \in \mathcal{T}_h} h_K^{-p+p}\operatorname{meas}(K)^{1+\frac{p}{p'}-p}\|v\|_{1,p,\Delta_K}^p$$

$$\leq c \sum_{K \in \mathcal{T}_h}\operatorname{card}\{K';\, K \subset \Delta_{K'}\}\|v\|_{1,p,K}^p.$$

Since the family $\{\mathcal{T}_h\}_{h>0}$ is shape-regular, $\operatorname{card}\{K';\, K \subset \Delta_{K'}\}$ is bounded uniformly with respect to h. Hence, $\|\Pi_h(v)\|_{1,p,\Omega} \leq c\|v\|_{1,p,\Omega}$. Conclude using Lemma 4.19 together with Lemma B.69. □

Theorem 4.21. *Assume that the solution to the Stokes problem* (4.2) *is smooth enough, that is,* $u \in [H^2(\Omega) \cap H_0^1(\Omega)]^d$ *and* $p \in H^1(\Omega) \cap L_{j=0}^2(\Omega)$. *Then, the solution* (u_h, p_h) *to* (4.13) *with* X_h *and* M_h *defined in* (4.22) *and* (4.23) *satisfies*

$$\forall h, \quad \|u - u_h\|_{1,\Omega} + \|p - p_h\|_{0,\Omega} \leq c\, h(\|u\|_{2,\Omega} + \|p\|_{1,\Omega}).$$

Moreover, if the Stokes problem has smoothing properties, then

$$\forall h, \quad \|u - u_h\|_{0,\Omega} \leq c\, h^2(\|u\|_{2,\Omega} + \|p\|_{1,\Omega}).$$

Remark 4.22. The idea of using bubble functions has been introduced by Crouzeix and Raviart [CrR73]. The analysis of the $\mathbb{P}_1$-bubble/$\mathbb{P}_1$ finite element is due to Arnold, Brezzi, and Fortin [ArB84]. In the literature, this element is sometimes called the *mini-element*. □

4.2.5 The Taylor–Hood finite element and its generalizations

The Taylor–Hood element: $\mathbb{P}_2/\mathbb{P}_1$. Let us now consider a more accurate velocity approximation. We still assume that the domain Ω is a polyhedron and that $\{\mathcal{T}_h\}_{h>0}$ is a shape-regular family of affine meshes composed of simplices. We keep the continuous $\mathbb{P}_1$ approximation for the pressure, but we approximate the velocity by means of continuous $\mathbb{P}_2$ polynomials. Accordingly, we define

$$X_h = \{u_h \in [\mathcal{C}^0(\overline{\Omega})]^d;\, \forall K \in \mathcal{T}_h,\, u_h \circ T_K \in [\mathbb{P}_2]^d;\, u_{h|\partial\Omega} = 0\}, \qquad (4.25)$$

$$M_h = \{p_h \in L_{j=0}^2(\Omega) \cap \mathcal{C}^0(\overline{\Omega});\, \forall K \in \mathcal{T}_h,\, p_h \circ T_K \in \mathbb{P}_1\}. \qquad (4.26)$$

The conventional representation of this element is shown in Figure 4.4.

Velocity	Pressure	Velocity	Pressure
$\mathbb{P}_2$	$\mathbb{P}_1$	$\mathbb{Q}_2$	$\mathbb{Q}_1$

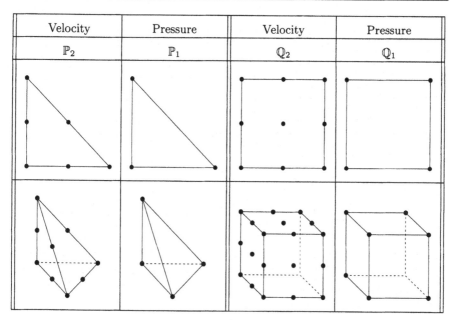

Fig. 4.4. Conventional representation of the $\mathbb{P}_2/\mathbb{P}_1$ element (left) and the $\mathbb{Q}_2/\mathbb{Q}_1$ element (right) in two dimensions (top) and in three dimensions (bottom). In three dimensions, only visible degrees of freedom are shown.

Lemma 4.23. *Assume that the space dimension d is either two or three. Assume that every mesh element has at least d edges in Ω. Then, there is c such that, for X_h and M_h defined in (4.25) and (4.26), the following inequality holds for all $1 < p < +\infty$:*

$$\sup_{v_h \in X_h} \frac{\int_\Omega q_h \nabla \cdot v_h}{\|v_h\|_{1,p,\Omega}} \geq c \left(\sum_{K \in \mathcal{T}_h} h_K^{p'} |q_h|_{1,p',K}^{p'} \right)^{\frac{1}{p'}}. \tag{4.27}$$

Proof. We give the proof in three dimensions, the proof in two dimensions being similar. Number all the internal edges of the mesh from 1 to N_{ed}^{i}. For edge i, with $1 \leq i \leq N_{\text{ed}}^{\text{i}}$, denote by d_i and f_i its two extremities and by m_i its midpoint. Set $l_i = \|f_i - d_i\|_3$ and $\tau_i = \frac{f_i - d_i}{\|f_i - d_i\|_3}$; l_i is the length of the edge and τ_i is a unit vector spanning the line passing through d_i and f_i. Let q_h be a function in M_h and let sgn be the sign function. Define a function $v_h \in X_h$ such that, for all $K \in \mathcal{T}_h$,

$$\begin{cases} v_h = 0 & \text{at the vertices of } K, \\ v_h(m_i) = -l_i^{p'} \tau_i \operatorname{sgn}(\partial_{\tau_i} q_h) |\partial_{\tau_i} q_h|^{p'-1} & \text{for the edges } i \text{ of } K. \end{cases}$$

Note that the definition of $v_h(m_i)$ is consistent, i.e., this value does not depend on K but only on the edge i. Using the quadrature formula

$$\forall \phi \in \mathbb{P}_2, \qquad \int_K \phi(x)\,\mathrm{d}x = \left(\sum_m \frac{\phi(m)}{5} - \sum_n \frac{\phi(n)}{20} \right) \mathrm{meas}(K),$$

where m spans the set of edge midpoints of K and n spans the set of nodes of K, we infer

$$
\begin{aligned}
\int_\Omega q_h \nabla \cdot v_h &= -\int_\Omega v_h \cdot \nabla q_h = -\sum_{K \in \mathcal{T}_h} \int_K v_h \cdot \nabla q_h \\
&= -\sum_{K \in \mathcal{T}_h} \sum_{m_i \in K} v_h(m_i) \cdot \nabla q_h(m_i) \frac{\mathrm{meas}(K)}{5} \\
&= \sum_{K \in \mathcal{T}_h} \sum_{m_i \in K} |\partial_{\tau_i} q_h(m_i)|^{p'} l_i^{p'} \frac{\mathrm{meas}(K)}{5} \geq c \sum_{K \in \mathcal{T}_h} h_K^{p'} |q_h|_{1,p',K}^{p'}.
\end{aligned}
$$

The last inequality results from the fact that every tetrahedron in $\mathcal{T}_h$ has at least three edges in Ω, so that the quantities $\partial_{\tau_i} q_h(m_i)$ where m_i spans the edge midpoints indeed control ∇q_h on K. Finally, inequality (4.27) results from

$$\|v_h\|_{1,p,K}^p \leq c\, h_K^{p'} |q_h|_{1,p',K}^{p'}. \qquad \qquad \square$$

Lemma 4.24. *Under the hypotheses of Lemma 4.23, the spaces X_h and M_h satisfy the inf-sup condition (4.24) uniformly with respect to h.*

Proof. Let q_h be a non-zero function in M_h. According to inequality (4.5) and the converse statement in Lemma A.42, there is $v \in [W_0^{1,p}(\Omega)]^d$ such that $\nabla \cdot v = q_h$ and $\beta \|v\|_{1,p,\Omega} \leq \|q_h\|_{0,p',\Omega}$. As a result,

$$
\begin{aligned}
\sup_{v_h \in X_h} \frac{\int_\Omega q_h \nabla \cdot v_h}{\|v_h\|_{1,p,\Omega}} &\geq \frac{\int_\Omega q_h \nabla \cdot \mathcal{C}_h(v)}{\|\mathcal{C}_h(v)\|_{1,p,\Omega}} \geq c\, \frac{\int_\Omega q_h \nabla \cdot \mathcal{C}_h(v)}{\|v\|_{1,p,\Omega}} \\
&= c\, \frac{\int_\Omega q_h \nabla \cdot v}{\|v\|_{1,p,\Omega}} + c\, \frac{\int_\Omega q_h \nabla \cdot (\mathcal{C}_h(v) - v)}{\|v\|_{1,p,\Omega}},
\end{aligned}
$$

where $\mathcal{C}_h$ is the Clément interpolation operator modified to preserve homogeneous boundary conditions; see Lemma 1.127 and Remark 1.129(i). This implies

$$
\begin{aligned}
\sup_{v_h \in X_h} \frac{\int_\Omega q_h \nabla \cdot v_h}{\|v_h\|_{1,p,\Omega}} &\geq c\beta \|q_h\|_{0,p',\Omega} + c\, \frac{\int_\Omega (\mathcal{C}_h(v) - v) \cdot \nabla q_h}{\|v\|_{1,p,\Omega}} \\
&\geq c\beta \|q_h\|_{0,p',\Omega} - c'\, \frac{\sum_{K \in \mathcal{T}_h} |q_h|_{1,p',K} h_K \|v\|_{1,p,\Delta_K}}{\|v\|_{1,p,\Omega}} \\
&\geq c\beta \|q_h\|_{0,p',\Omega} - c' \left(\sum_{K \in \mathcal{T}_h} h_K^{p'} |q_h|_{1,p',K}^{p'} \right)^{\frac{1}{p'}}.
\end{aligned}
$$

Owing to Lemma 4.23, the negative term is bounded from below as follows:

$$\sup_{v_h \in X_h} \frac{\int_\Omega q_h \nabla \cdot v_h}{\|v_h\|_{1,p,\Omega}} \geq c_1 \|q_h\|_{0,p',\Omega} - c_2 \sup_{v_h \in X_h} \frac{\int_\Omega q_h \nabla \cdot v_h}{\|v_h\|_{1,p,\Omega}}.$$

This yields the expected inequality with the constant $\frac{c_1}{1+c_2}$. □

Remark 4.25. For other details and alternative proofs, the reader is referred to [BeP79, pp. 255–257] and [GiR86, p. 176]. The main ideas of the above proof are adapted from [Ver84], but the extension to L^p is original. In the literature, the $\mathbb{P}_2/\mathbb{P}_1$ finite element is also known as the Taylor–Hood element. □

Theorem 4.26. *Assume that the solution* (u, p) *to the Stokes problem (4.2) is smooth enough, that is,* $u \in [H^3(\Omega) \cap H_0^1(\Omega)]^d$ *and* $p \in H^2(\Omega) \cap L^2_{\int=0}(\Omega)$. *Then, the solution* (u_h, p_h) *to (4.13) with* X_h *and* M_h *defined in (4.25) and (4.26) satisfies*

$$\forall h, \quad \|u - u_h\|_{1,\Omega} + \|p - p_h\|_{0,\Omega} \leq c h^2(\|u\|_{3,\Omega} + \|p\|_{2,\Omega}).$$

Moreover, if the Stokes problem has smoothing properties, then

$$\forall h, \quad \|u - u_h\|_{0,\Omega} \leq c h^3(\|u\|_{3,\Omega} + \|p\|_{2,\Omega}).$$

The $\mathbb{P}_k/\mathbb{P}_{k-1}$ **and** $\mathbb{Q}_k/\mathbb{Q}_{k-1}$ **finite elements.** Still keeping a continuous approximation of the pressure, it is possible to generalize the Taylor–Hood element to quadrangles and hexahedra. For instance, the $\mathbb{Q}_2/\mathbb{Q}_1$ finite element has the same properties as those of the Taylor–Hood element; see Figure 4.4.

It is also possible to use higher-degree polynomials. For $k \geq 2$, the $\mathbb{P}_k/\mathbb{P}_{k-1}$ finite elements ($\mathbb{P}_k$ for velocity and $\mathbb{P}_{k-1}$ for pressure) as well as the $\mathbb{Q}_k/\mathbb{Q}_{k-1}$ finite elements ($\mathbb{Q}_k$ for velocity and $\mathbb{Q}_{k-1}$ for pressure) are compatible in two and three dimensions. These elements yield the errors estimates

$$\|u - u_h\|_{0,\Omega} + h(\|u - u_h\|_{1,\Omega} + \|p - p_h\|_{0,\Omega}) \leq c h^{k+1}(\|u\|_{k+1,\Omega} + \|p\|_{k,\Omega}),$$

provided the exact solution is smooth enough. Proofs and further insight can be found in [BrF91^a, BrF91^b].

4.2.6 The $\mathbb{P}_1$-iso-$\mathbb{P}_2/\mathbb{P}_1$ finite element and its generalizations

The $\mathbb{P}_1$-**iso-**$\mathbb{P}_2/\mathbb{P}_1$ **finite element.** An alternative to the $\mathbb{P}_2/\mathbb{P}_1$ finite element is to replace the $\mathbb{P}_2$ approximation of the velocity by a $\mathbb{P}_1$ approximation on a finer mesh. Again, assume that Ω is a polyhedron and that the family $\{\mathcal{T}_h\}_{h>0}$ is shape-regular and composed of affine simplices. We construct a new mesh $\mathcal{T}_{\frac{h}{2}}$ as follows. In dimension 2, we divide each triangle of $\mathcal{T}_h$ into four new triangles by connecting the midpoint of the edges. In dimension 3, we divide each tetrahedron of $\mathcal{T}_h$ into eight new tetrahedra by dividing each face into four new triangles like in two dimensions, and by connecting the midpoints of one pair of non-intersecting edges; see bottom left panel in Figure 4.5. The approximation spaces are defined as follows:

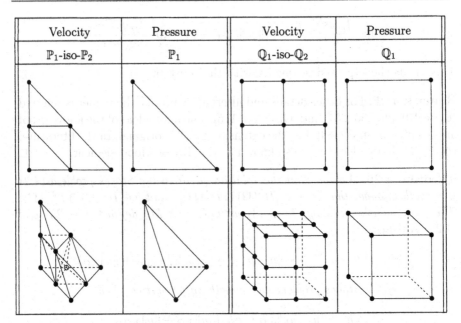

Velocity	Pressure	Velocity	Pressure
$\mathbb{P}_1$-iso-$\mathbb{P}_2$	$\mathbb{P}_1$	$\mathbb{Q}_1$-iso-$\mathbb{Q}_2$	$\mathbb{Q}_1$

Fig. 4.5. $\mathbb{P}_1$-iso-$\mathbb{P}_2/\mathbb{P}_1$ (left) and $\mathbb{Q}_1$-iso-$\mathbb{Q}_2/\mathbb{Q}_1$ (right) finite elements in two dimensions (top) and in three dimensions (bottom). In three dimensions, only visible degrees of freedom are shown.

$$X_h = \{u_h \in [\mathcal{C}^0(\overline{\Omega})]^d; \forall K \in \mathcal{T}_{\frac{h}{2}}, u_h \circ T_K \in [\mathbb{P}_1]^d; u_{h|\partial\Omega} = 0\}, \qquad (4.28)$$

$$M_h = \{p_h \in L^2_{f=0}(\Omega) \cap \mathcal{C}^0(\overline{\Omega}); \forall K \in \mathcal{T}_h, p_h \circ T_K \in \mathbb{P}_1\}. \qquad (4.29)$$

This finite element is often called $\mathbb{P}_1$-iso-$\mathbb{P}_2/\mathbb{P}_1$ or also $4\mathbb{P}_1/\mathbb{P}_1$ in two dimensions and $8\mathbb{P}_1/\mathbb{P}_1$ in three dimensions.

Lemma 4.27. *The spaces X_h and M_h defined in (4.28) and (4.29) satisfy the inf-sup condition (4.24) uniformly with respect to h.*

Proof. Easy adaptation of Lemma 4.24; see [BeP79] for the analysis in dimension 2. □

Theorem 4.28. *Assume that the solution to the Stokes problem (4.2) is smooth enough, that is, $u \in [H^2(\Omega) \cap H^1_0(\Omega)]^d$ and $p \in H^1(\Omega) \cap L^2_{f=0}(\Omega)$. Then, the solution to (4.13) with X_h and M_h defined in (4.28) and (4.29) satisfies*

$$\forall h, \quad \|u - u_h\|_{1,\Omega} + \|p - p_h\|_{0,\Omega} \leq c\,h(\|u\|_{2,\Omega} + \|p\|_{1,\Omega}).$$

Moreover, if the Stokes problem has smoothing properties, then

$$\forall h, \quad \|u - u_h\|_{0,\Omega} \leq c\,h^2(\|u\|_{2,\Omega} + \|p\|_{1,\Omega}).$$

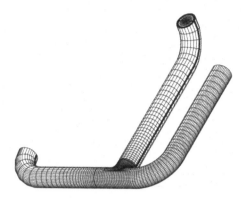

Fig. 4.6. $8\mathbb{Q}_1/\mathbb{Q}_1$ mesh of pipes. Courtesy of S. Chapuliot and J.-P. Magnaud (CEA).

The $\mathbb{Q}_1$-iso-$\mathbb{Q}_2/\mathbb{Q}_1$ finite element. It is possible to generalize the concept of the $\mathbb{P}_1$-iso-$\mathbb{P}_2/\mathbb{P}_1$ finite element to quadrangles in dimension 2 and hexahedra in dimension 3.

Assume that Ω is a polygon in $\mathbb{R}^2$ (resp., polyhedron in $\mathbb{R}^3$). Let $\{\mathcal{T}_h\}_{h>0}$ be a shape-regular family of meshes composed of quadrangles or hexahedra. Let us construct a new mesh $\mathcal{T}_{\frac{h}{2}}$ as follows. In dimension 2, we divide each quadrangle of $\mathcal{T}_h$ into four new quadrangles by connecting the midpoints of non-intersecting edges. In dimension 3, we divide each hexahedron of $\mathcal{T}_h$ into eight new hexahedra by dividing each face into four quadrangles and by connecting the midpoint nodes of non-intersecting faces; see right panels in Figure 4.5. The velocity and pressure approximation spaces are defined to be

$$X_h = \{u_h \in [\mathcal{C}^0(\overline{\Omega})]^d; \forall K \in \mathcal{T}_{\frac{h}{2}}, \, u_h \circ T_K \in [\mathbb{Q}_1]^d; \, u_{h|\partial\Omega} = 0\}, \qquad (4.30)$$

$$M_h = \{p_h \in L^2_{\int=0}(\Omega) \cap \mathcal{C}^0(\overline{\Omega}); \forall K \in \mathcal{T}_h, \, p_h \circ T_K \in \mathbb{Q}_1\}. \qquad (4.31)$$

This finite element is often called $\mathbb{Q}_1$-iso-$\mathbb{Q}_2/\mathbb{P}_1$ or $4\mathbb{Q}_1/\mathbb{Q}_1$ in dimension 2 (resp., $8\mathbb{Q}_1/\mathbb{Q}_1$ in dimension 3).

The $\mathbb{Q}_1$-iso-$\mathbb{Q}_2/\mathbb{Q}_1$ finite element satisfies the inf-sup condition uniformly with respect to h and yields the same error estimates as those of the $\mathbb{P}_1$-iso-$\mathbb{P}_2/\mathbb{P}_1$ element; see Proposition 4.28. This type of element is often used in the industry since it is simple to implement. Figure 4.6 shows an $8\mathbb{Q}_1/\mathbb{Q}_1$ mesh of pipes. Only the $\mathbb{Q}_1$ hexahedra approximating the pressure are shown.

4.2.7 Discontinuous approximation of the pressure

The $\mathbb{P}_2$-bubble/$\mathbb{P}_1$-discontinuous finite element. Consider a shape-regular family of affine simplicial meshes of Ω, say $\{\mathcal{T}_h\}_{h>0}$. Consider the bubble function $\widehat{b}$ defined in (4.21) and set

$$\widehat{P} = [\mathbb{P}_2(\widehat{K}) \oplus \mathrm{span}(\widehat{b})]^d.$$

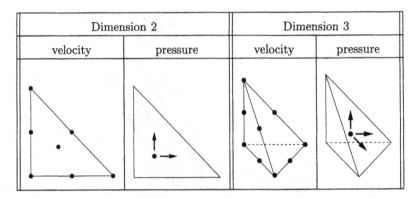

Fig. 4.7. Mixed finite element $\mathbb{P}_2$-bubble/$\mathbb{P}_1$- in two (left) and three (right) dimensions. In three dimensions, only visible degrees of freedom are shown for the velocity.

To approximate the velocity and the pressure, introduce the spaces

$$X_h = \{u_h \in [\mathcal{C}^0(\overline{\Omega})]^d; \forall K \in \mathcal{T}_h, u_h \circ T_K \in \widehat{P}; u_{h|\partial\Omega} = 0\}, \qquad (4.32)$$

$$M_h = \{p_h \in L^2_{j=0}(\Omega); \forall K \in \mathcal{T}_h, p_h \circ T_K \in \mathbb{P}_1\}. \qquad (4.33)$$

Note that the pressure is locally $\mathbb{P}_1$ on each simplex but is not necessarily continuous across the interfaces of the simplices. The local degrees of freedom for the pressure can be taken to be its mean-value and its gradient. Note also that the approximation space $X_h \times M_h$ is conformal in $[H^1_0(\Omega)]^d \times L^2_{j=0}(\Omega)$. This finite element is often called the *conformal Crouzeix–Raviart mixed finite element* [CrR73]. A conventional representation is shown in Figure 4.7.

The $\mathbb{P}_2$-bubble/$\mathbb{P}_1$-discontinuous element satisfies the inf-sup condition uniformly with respect to h and yields error estimates that are identical to those of the Taylor–Hood element; see [BrF91^b, p. 214].

The $\mathbb{Q}_2$/$\mathbb{P}_1$-discontinuous finite element. It is possible to generalize the conformal Crouzeix–Raviart mixed finite element to cuboids. Let Ω be a polyhedron in $\mathbb{R}^d$ and let $\{\mathcal{T}_h\}_{h>0}$ be a shape-regular family of meshes composed of cuboids. Introducing the spaces

$$X_h = \{u_h \in [\mathcal{C}^0(\overline{\Omega})]^d; \forall K \in \mathcal{T}_h, u_h \circ T_K \in [\mathbb{Q}_2]^d; u_{h|\partial\Omega} = 0\}, \qquad (4.34)$$

$$M_h = \{p_h \in L^2_{j=0}(\Omega); \forall K \in \mathcal{T}_h, p_h \circ T_K \in \mathbb{P}_1\}, \qquad (4.35)$$

one obtains a mixed finite element that satisfies the inf-sup condition uniformly in h, and yields the same error estimates as those of the Taylor–Hood element. As before, the local degrees of freedom for the pressure can be taken to be its mean-value over an element and that of its gradient; see [BrF91^b, pp. 216–219].

Remark 4.29. Note that the $\mathbb{Q}_2$/$\mathbb{Q}_1$-discontinuous finite element does not satisfy the inf-sup condition.   □

4.2.8 The $\mathbb{P}_1$-non-conformal/$\mathbb{P}_0$ finite element

We now present a non-conformal approximation technique for the velocity based on the non-conformal Crouzeix–Raviart finite element studied in §3.2.3.

Let $\{\mathcal{T}_h\}_{h>0}$ be a shape-regular family of affine triangulations of a domain $\Omega \subset \mathbb{R}^d$ with $d = 2$ or 3. Let $\mathcal{F}_h^i$ be the set of internal faces. Let $\mathcal{F}_h^\partial$ be the set of faces at the boundary and set $\mathcal{F}_h = \mathcal{F}_h^i \cup \mathcal{F}_h^\partial$. For $F \in \mathcal{F}_h^i$ with $F = K_1 \cap K_2$, denote by n_1 and n_2 the outward normal to K_1 and K_2, respectively. Let v be an $\mathbb{R}^d$-valued function that is smooth enough to have limits on both sides of F (these limits being not necessarily the same). Set $v_1 = v_{|K_1}$ and $v_2 = v_{|K_2}$, and define the jump of v across F to be $[\![v]\!] = v_1 \otimes n_1 + v_2 \otimes n_2$. Define the velocity and pressure spaces to be

$$X_h = \{v_h;\ \forall K \in \mathcal{T}_h,\ v_{h|K} \in [\mathbb{P}_1]^d;\ \forall F \in \mathcal{F}_h^i,\ \textstyle\int_F [\![v_h]\!] = 0;\ \forall F \in \mathcal{F}_h^\partial,\ \textstyle\int_F v_h = 0\},$$
$$M_h = \{q_h;\ \forall K \in \mathcal{T}_h,\ q_{h|K} \in \mathbb{P}_0;\ \textstyle\int_\Omega q_h = 0\}.$$

It is clear that functions in X_h are continuous at the center of the interfaces and are zero at the center of those faces that are at the boundary. The conventional representation of this element is shown in Figure 4.8. Let

$$a_h(v_h, w_h) = \sum_{K \in \mathcal{T}_h} \int_K \nabla v_h : \nabla w_h, \qquad b_h(v_h, q_h) = -\sum_{K \in \mathcal{T}_h} \int_K q_h \nabla \cdot v_h,$$

and, for $1 < p < +\infty$, equip X_h with the mesh-dependent norm $\|v_h\|_{1,p,h,\Omega}^p = \sum_{K \in \mathcal{T}_h} \|v_h\|_{1,p,K}^p$. Assuming $f \in [L^2(\Omega)]^d$, the approximate Stokes problem is:

$$\begin{cases} \text{Seek } u_h \in X_h \text{ and } p_h \text{ in } M_h \text{ such that} \\ a_h(u_h, v_h) + b_h(v_h, p_h) = \int_\Omega f v_h, & \forall v_h \in X_h, \\ b_h(u_h, q_h) = -\int_\Omega g q_h, & \forall q_h \in M_h. \end{cases} \qquad (4.36)$$

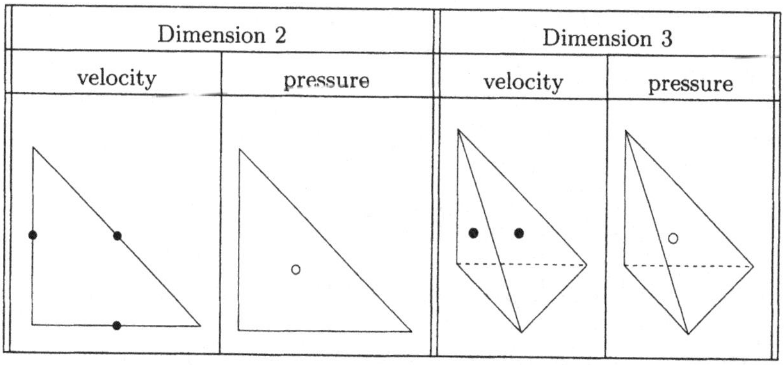

Dimension 2		Dimension 3	
velocity	pressure	velocity	pressure

Fig. 4.8. Conventional representation of the $\mathbb{P}_1$-non-conformal/$\mathbb{P}_0$ finite element in two (left) and three (right) dimensions. In three dimensions, only visible degrees of freedom for the velocity are shown. The pressure degree of freedom is its average over the mesh cell.

To prove that the spaces pair $\{X_h, M_h\}$ satisfies the inf-sup condition, we need to introduce some technicalities. Let F be a face and denote by m_F the center of F. Then, let $P_{\mathrm{pt},h}^1$ be the Crouzeix–Raviart finite element space introduced in §1.4.3, and define $\Pi_h : W^{1,p}(\Omega) \to P_{\mathrm{pt},h}^1$ to be the $\mathbb{P}_1$ interpolation operator such that

$$\forall \phi \in W^{1,p}(\Omega),\ \forall F \in \mathcal{F}_h, \quad \Pi_h(\phi)(m_F) = \frac{1}{\mathrm{meas}(F)} \int_F \phi.$$

Lemma 4.30. *Let $1 < p < \infty$. There is c, independent of h, such that*

$$\forall \phi \in W^{1,p}(\Omega), \quad \|\Pi_h(\phi)\|_{1,p,h,\Omega} \le c \|\phi\|_{1,p,\Omega}. \tag{4.37}$$

Proof. Let $\{F_0, \ldots, F_d\}$ be the faces of K and denote by $\{\phi_0, \ldots, \phi_d\}$ the mean-values of ϕ on $\{F_0, \ldots, F_d\}$, respectively. Set $\tilde{\phi} = \phi - \phi_0$. Since constants are invariant under Π_h,

$$|\Pi_h \phi|_{1,p,K} = |\Pi_h \tilde{\phi}|_{1,p,K} \le c \|J_K^{-1}\|_d |\det(J_K)|^{\frac{1}{p}} |\Pi_h \widehat{\tilde{\phi}}|_{1,p,\widehat{K}}$$

$$\le c \|J_K^{-1}\|_d |\det(J_K)|^{\frac{1}{p}} \max_{0 \le j \le d} |\widehat{\tilde{\phi}}_j|.$$

Furthermore, $|\widehat{\tilde{\phi}}_j| \le c \|\widehat{\tilde{\phi}}\|_{0,p,\partial\widehat{K}} \le c' \|\widehat{\tilde{\phi}}\|_{1,p,\widehat{K}}$. Now, recall the Poincaré–Friedrichs inequality (see Lemma B.63)

$$\exists c > 0,\ \forall \psi \in H^1(\widehat{K}), \quad c \|\psi\|_{1,p,\widehat{K}} \le |\psi|_{1,p,\widehat{K}} + \left| \int_{\widehat{F}_0} \psi \right|. \tag{4.38}$$

Using (4.38) together with the fact that $\int_{\widehat{F}_0} \widehat{\tilde{\phi}} = 0$ yields $|\widehat{\tilde{\phi}}_j| \le c |\widehat{\tilde{\phi}}|_{1,p,\widehat{K}}$. As a result,

$$|\Pi_h \phi|_{1,p,K} \le c \|J_K^{-1}\|_d |\det(J_K)|^{\frac{1}{p}} |\widehat{\tilde{\phi}}|_{1,p,\widehat{K}} \le c' |\tilde{\phi}|_{1,p,K} = c' |\phi|_{1,p,K}.$$

The rest of the proof follows easily.   □

Lemma 4.31. *There is β, independent of h, such that*

$$\inf_{q_h \in M_h} \sup_{v_h \in X_h} \frac{b_h(v_h, q_h)}{\|v_h\|_{1,p,h,\Omega} \|q_h\|_{0,p'\Omega}} \ge \beta. \tag{4.39}$$

Proof. Let $v \in [W_0^{1,p}(\Omega)]^d$, $q_h \in M_h$, and $K \in \mathcal{T}_h$. It is clear that

$$\int_K q_h \nabla \cdot v = q_h \sum_{i=0}^d n_i \cdot \int_{F_i} v = q_h \sum_{i=0}^d n_i \cdot \int_{F_i} \Pi_h(v) = \int_K q_h \nabla \cdot \Pi_h(v).$$

Hence, $b_h(v, q_h) = b_h(\Pi_h v, q_h)$. Then, using Lemma 4.30 and adapting slightly the proof of Fortin's Lemma, the conclusion follows.   □

Theorem 4.32. *Assume that the solution to the Stokes problem* (4.2) *is smooth enough, that is,* $u \in [H^2(\Omega) \cap H_0^1(\Omega)]^d$ *and* $p \in H^1(\Omega) \cap L_{j=0}^2(\Omega)$. *Then, the solution to* (4.36) *satisfies*

$$\forall h, \quad \|u - u_h\|_{1,h,\Omega} + \|p - p_h\|_{0,\Omega} \le c\, h(\|u\|_{2,\Omega} + \|p\|_{1,\Omega}).$$

Moreover, if the Stokes problem has smoothing properties, then

$$\forall h, \quad \|u - u_h\|_{0,\Omega} \le c\, h^2(\|u\|_{2,\Omega} + \|p\|_{1,\Omega}). \tag{4.40}$$

Proof. See Exercise 4.6.   □

Remark 4.33.

(i) The $\mathbb{P}_1$-non-conformal/$\mathbb{P}_0$ finite element has been introduced by Crouzeix and Raviart [CrR73] and is often called the *non-conformal Crouzeix-Raviart mixed finite element*. A quadrilateral non-conformal mixed finite element has been introduced by Rannacher and Turek [RaT92, Tur99].

(ii) Non-conformal mixed finite elements can be used to construct piecewise divergence-free approximation spaces, i.e., non-conformal approximations to the constrained problem (4.7); see Hecht [Hec81, Hec84], Braess [Bra97, p. 154], or Brezzi and Fortin [BrF91^b, p. 268] for further insight.   □

4.2.9 Numerical illustration

We conclude this section with a brief numerical illustration. Consider the two-dimensional O-shaped domain shown in the left panel of Figure 4.9. We perform time-dependent simulations of the Navier–Stokes equations with the $\mathbb{P}_1/\mathbb{P}_1$ and the Taylor–Hood finite elements. The Reynolds number is $Re = 100$ and the time step is $\delta t = 0.01$. The initial condition is the flow at rest. The flow is then driven anticlockwise by imposing a vertical velocity on the two outer faces. Pressure isolines after 100 time steps are shown in the central panel (resp., right panel) of Figure 4.9 for the $\mathbb{P}_1/\mathbb{P}_1$ (resp., Taylor–Hood) finite element. The pressure field obtained with the $\mathbb{P}_1/\mathbb{P}_1$ element is polluted by *spurious oscillations*. This example clearly shows the adverse effects of violating the inf-sup condition.

4.3 Galerkin/Least-Squares (GaLS) Approximation

Employing mixed finite elements to solve Stokes-like problems may seem a cumbersome constraint. The goal of this section is to show that it is possible to work without mixed finite elements provided the Galerkin formulation is slightly modified. We present an approximation technique known in the literature as the Galerkin/Least-Squares (GaLS) method. This method is suitable for solving Stokes-like problems with any kind of finite element. The material presented hereafter is adapted from [FrF92, ToV96].

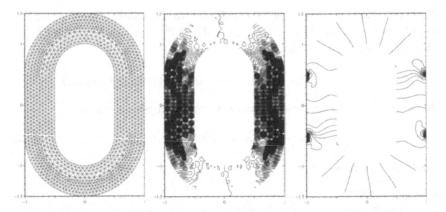

Fig. 4.9. Simulation of the Navier–Stokes equations in an O-shaped, closed channel. From left to right: the triangulation; pressure isolines obtained with the $\mathbb{P}_1/\mathbb{P}_1$ element; and pressure isolines obtained with the Taylor–Hood element.

4.3.1 The GaLS formulation

Let $V = [H_0^1(\Omega)]^d \times L_{\int=0}^2(\Omega)$ with dual space $V' = [H^{-1}(\Omega)]^d \times L_{\int=0}^2(\Omega)$ and consider the operator $A : V \to V'$ such that $A(u,p) = (\nabla p - \Delta u, \nabla{\cdot}u)$. Owing to Theorem 4.3, A is an isomorphism, and (4.2) can be recast in the form:

$$\begin{cases} \text{Seek } (u,p) \in V \text{ such that, } \forall (v,q) \in V, \\ \langle A(u,p), (v,q) \rangle_{V',V} = \langle f, v \rangle_{H^{-1}, H_0^1} + (g,q)_{0,\Omega}. \end{cases} \tag{4.41}$$

Let $(\cdot, \cdot)_{-1,\Omega}$ be a scalar product in $[H^{-1}(\Omega)]^d$ and equip V' with the scalar product $(\!(\!(v,q), (w,r))\!)\! = (v,w)_{-1,\Omega} + (q,r)_{0,\Omega}$. The Least-Squares formulation of (4.2) is as follows:

$$\begin{cases} \text{Seek } (u,p) \in V \text{ such that, } \forall (v,q) \in V, \\ (\!(\!(A(u,p), A(v,q))\!)\! = (\!(\!((f,g), A(v,q))\!)\!. \end{cases} \tag{4.42}$$

Proposition 4.34. *Problem* (4.42) *is equivalent to problem* (4.41).

Proof. Use the fact that $A : V \to V'$ is an isomorphism. $\qquad\qquad\square$

The formulation (4.42) is scarcely used in practice since it requires evaluating the $[H^{-1}(\Omega)]^d$-scalar product. An alternative idea consists of replacing this scalar product by that of $[L^2(\Omega)]^d$. The difficulty is that the quantity $A(u,p)$ cannot be controlled in $[L^2(\Omega)]^d \times L_{\int=0}^2(\Omega)$ unless u is in $[H^2(\Omega)]^d$ and p is in $H^1(\Omega)$, but this type of regularity for u cannot be guaranteed by H^1-conformal finite elements. Nevertheless, H^1-conformal functions are $\mathcal{C}^\infty$ on each element K of the mesh, i.e., the restriction to each element of the quantity $A(u_h, p_h)$ is a smooth function. This leads to the idea sustaining the

GaLS method, namely to reformulate (4.42) by replacing the scalar product $((\cdot, \cdot))$ by that of $[L^2(\Omega)]^d \times L^2_{j=0}(\Omega)$ on each element.

Let $\mathcal{T}_h$ be a mesh on Ω. Let $X_h \subset [H^1_0(\Omega)]^d$ and $M_h \subset L^2_{j=0}(\Omega)$ be two finite element spaces constructed on this mesh. Set $V_h = X_h \times M_h$. Assume $M_h \subset H^1(\Omega)$; this assumption is adopted to simplify the presentation. See [ToV96] for the analysis of the case when M_h is not H^1-conformal. Assume that the spaces X_h and M_h satisfy the following interpolation and inverse inequalities: There are two interpolation operators $\mathcal{I}_h : [H^1_0(\Omega)]^d \to X_h$, $\mathcal{J}_h : L^2_{j=0}(\Omega) \to M_h$, and an integer $k \geq 1$ such that, for all $K \in \mathcal{T}_h$,

$$\|u - \mathcal{I}_h u\|_{m,K} \leq c h_K^{k+1-m} \|u\|_{k+1,K}, \quad \forall u \in H^{k+1}(K), \quad 0 \leq m \leq 2, \quad (4.43)$$

$$\|p - \mathcal{J}_h p\|_{m,K} \leq c h_K^{k-m} \|p\|_{k,K}, \quad \forall p \in H^k(K), \quad 0 \leq m \leq 1, \quad (4.44)$$

$$\|\Delta v_h\|_{0,K} \leq c h_K^{-1} \|\nabla v_h\|_{0,K}, \quad \forall v_h \in X_h, \quad (4.45)$$

$$\|\nabla q_h\|_{0,K} \leq c h_K^{-1} \|q_h\|_{0,K}, \quad \forall q_h \in X_h, \quad (4.46)$$

with c independent of h. These properties are, for instance, satisfied if the spaces X_h and M_h have been constructed using shape-regular mesh families and Lagrange finite elements; see §1.5 and §1.7.

Remark 4.35. Note that X_h and M_h are not assumed to satisfy the inf-sup condition (4.13). □

For all $(v_h, q_h) \in V_h$, it is clear that v_h and q_h are of class $\mathcal{C}^\infty$ on each element $K \in \mathcal{T}_h$. Define the local scalar product

$$(((\phi, \psi), (\alpha, \beta)))_{0,K} = \int_K \phi \cdot \alpha + \psi \beta,$$

for $\phi, \alpha \in [L^2(K)]^d$ and $\psi, \beta \in L^2(K)$. For all $K \in \mathcal{T}_h$, the quantities $((A(u_h, p_h), A(v_h, q_h)))_{0,K}$ and $(((f,g), A(v_h, q_h)))_{0,K}$ are meaningful. To simplify the notation, we introduce the bilinear form $a_h \in \mathcal{L}(V_h \times V_h; \mathbb{R})$ such that

$$a_h((u_h, p_h), (v_h, q_h)) = \langle A(u_h, p_h), (v_h, q_h) \rangle_{V',V}$$
$$+ \sum_{K \in \mathcal{T}_h} \delta(h_K)((A(u_h, p_h), A(v_h, q_h)))_{0,K},$$

where $\delta(h_K) = c h_K^2$. This choice will be justified by the error analysis. The so-called GaLS formulation is as follows:

$$\begin{cases} \text{Seek } (u_h, p_h) \in V_h \text{ such that } \forall (v_h, q_h) \in V_h, \\ a_h((u_h, p_h), (v_h, q_h)) = \langle (f,g), (v_h, q_h) \rangle_{V',V} \\ \qquad\qquad + \sum_{K \in \mathcal{T}_h} \delta(h_K)(((f,g), A(v_h, q_h)))_{0,K}. \end{cases} \quad (4.47)$$

Remark 4.36. The full expansion of a_h is

$$a_h((u_h, p_h), (v_h, q_h)) = (\nabla u_h, \nabla v_h)_{0,\Omega} - (\nabla \cdot v_h, p_h)_{0,\Omega} + (\nabla \cdot u_h, q_h)_{0,\Omega}$$
$$+ \sum_{K \in T_h} \delta(h_K)((\nabla p_h - \Delta u_h, \nabla q_h - \Delta v_h)_{0,K} + (\nabla \cdot u_h, \nabla \cdot v_h)_{0,K}). \qquad \square$$

4.3.2 Error analysis

The new formulation is more stable than the standard Galerkin formulation. To formalize this idea, we introduce the following norm on V_h:

$$|||(u_h, p_h)|||_{h,\Omega}^2 = |u_h|_{1,\Omega}^2 + \|p_h\|_{0,\Omega}^2$$
$$+ \sum_{K \in T_h} \delta(h_K) \left(\|\nabla p_h\|_{0,K}^2 + \|\nabla p_h - \Delta u_h\|_{0,K}^2 + \|\nabla \cdot u_h\|_{0,K}^2 \right).$$

Lemma 4.37 (Continuity). *There is c, independent of h, such that*

$$\forall (v, q), (w, r) \in V_h, \quad a_h((v, q), (w, r)) \leq c |||(v, q)|||_{h,\Omega} |||(w, r)|||_{h,\Omega}. \quad (4.48)$$

Proof. Directly results from the choice of the norm $||| \cdot |||_{h,\Omega}$. $\square$

Lemma 4.38 (Stability). *Under assumptions (4.43) to (4.46), there is $c > 0$, independent of h, such that*

$$\inf_{(u_h, p_h) \in V_h} \sup_{(v_h, q_h) \in V_h} \frac{a_h((u_h, p_h), (v_h, q_h))}{|||(u_h, p_h)|||_{h,\Omega} |||(v_h, q_h)|||_{h,\Omega}} \geq c.$$

Proof. Let $(u_h, p_h) \in V_h$. The proof proceeds in three steps.
(1) A straightforward calculation yields

$$a_h((u_h, p_h), (u_h, p_h)) = |u_h|_{1,\Omega}^2 + \tfrac{1}{2} \sum_{K \in T_h} \delta(h_K)(2\|\nabla \cdot u_h\|_{0,K}^2 + \|\nabla p_h - \Delta u_h\|_{0,K}^2$$
$$+ \|\Delta u_h\|_{0,K}^2 + \|\nabla p_h\|_{0,K}^2 - 2(\Delta u_h, \nabla p_h)_{0,K}).$$

Owing to (4.45) and the fact that $\delta(h_K) = ch_K^2$, we infer $\delta(h_K)\|\Delta v_h\|_{0,K}^2 \leq c|v_h|_{1,K}^2$ for all $v_h \in X_h$ and $K \in T_h$. Let $\gamma > 1$ and use (A.3) to obtain

$$2 \sum_{K \in T_h} \delta(h_K)(\Delta u_h, \nabla p_h)_{0,K} \leq \sum_{K \in T_h} \delta(h_K) \left(\gamma \|\Delta u_h\|_{0,K}^2 + \tfrac{1}{\gamma}\|\nabla p_h\|_{0,K}^2 \right)$$
$$\leq \sum_{K \in T_h} \delta(h_K) \left((\gamma - 1)\|\Delta u_h\|_{0,K}^2 + \|\Delta u_h\|_{0,K}^2 + \tfrac{1}{\gamma}\|\nabla p_h\|_{0,K}^2 \right)$$
$$\leq (\gamma - 1)c |u_h|_{1,\Omega}^2 + \sum_{K \in T_h} \delta(h_K) \left(\|\Delta u_h\|_{0,K}^2 + \tfrac{1}{\gamma}\|\nabla p_h\|_{0,K}^2 \right).$$

Combining this estimate with the preceding yields

$$a_h((u_h, p_h), (u_h, p_h)) \geq \left(1 - \tfrac{c}{2}(\gamma - 1)\right) |u_h|_{1,\Omega}^2$$
$$+ \tfrac{1}{2} \sum_{K \in \mathcal{T}_h} \delta(h_K) \left(2\|\nabla \cdot u_h\|_{0,\Omega}^2 + \|\nabla p_h - \Delta u_h\|_{0,\Omega}^2 + (1 - \tfrac{1}{\gamma})\|\nabla p_h\|_{0,\Omega}^2\right).$$

Set $\gamma = 1 + \tfrac{1}{c} > 1$ to infer

$$a_h((u_h, p_h), (u_h, p_h)) \geq \tfrac{1}{2}|u_h|_{1,\Omega}^2$$
$$+ c \sum_{K \in \mathcal{T}_h} \delta(h_K)(\|\nabla \cdot u_h\|_{0,\Omega}^2 + \|\nabla p_h - \Delta u_h\|_{0,\Omega}^2 + \|\nabla p_h\|_{0,\Omega}^2),$$

yielding the bound from below

$$a_h((u_h, p_h), (u_h, p_h)) \geq c_1 \left(|||(u_h, p_h)|||_{h,\Omega}^2 - \|p_h\|_{0,\Omega}^2\right). \tag{4.49}$$

(2) Inequality (4.5) and the converse statement in Lemma A.42 imply that there exists a function $v \in [H_0^1(\Omega)]^d$ such that

$$\nabla \cdot v = -p_h \qquad \text{and} \qquad \|v\|_{1,\Omega} \leq c\|p_h\|_{0,\Omega}.$$

Let $v_h = \mathcal{I}_h v$. The stability of $\mathcal{I}_h$ implies $\|\mathcal{I}_h v\|_{1,\Omega} \leq c'\|p_h\|_{0,\Omega}$. Hence,

$$a_h((u_h, p_h), (v_h, 0)) = (\nabla u_h, \nabla v_h)_{0,\Omega} - (\nabla \cdot v_h, p_h)_{0,\Omega}$$
$$+ \sum_{K \in \mathcal{T}_h} \delta(h_K)\left((\nabla \cdot u_h, \nabla \cdot v_h)_{0,K} + (\Delta u_h - \nabla p_h, \Delta v_h)_{0,K}\right)$$
$$\geq -|u_h|_{1,\Omega}|v_h|_{1,\Omega} - (\nabla \cdot (v_h - v), p_h)_{0,\Omega} + \|p_h\|_{0,\Omega}^2$$
$$- c \sum_{K \in \mathcal{T}_h} \delta(h_K)\left(\|\nabla \cdot u_h\|_{0,K}|v_h|_{1,K} + \|\nabla p_h - \Delta u_h\|_{0,K}\|\Delta v_h\|_{0,K}\right).$$

Using the fact that $M_h \subset H^1(\Omega)$ and $\delta(h_K) = ch_K^2$ yields

$$a_h((u_h, p_h), (v_h, 0)) \geq \|p_h\|_{0,\Omega}^2 - c|u_h|_{1,\Omega}\|p_h\|_{0,\Omega} + \sum_{K \in \mathcal{T}_h} ch_K|v_h|_{1,K}\|\nabla p_h\|_{0,K}$$
$$- c' \sum_{K \in \mathcal{T}_h} h_K \left(\|\nabla \cdot u_h\|_{0,K} + \|\nabla p_h - \Delta u_h\|_{0,K}\right)|v_h|_{1,K}$$
$$\geq \|p_h\|_{0,\Omega}^2 - c|u_h|_{1,\Omega}\|p_h\|_{0,\Omega} - c'\|p_h\|_{0,\Omega}\left(\sum_{K \in \mathcal{T}_h} \delta(h_K)(\|\nabla \cdot u_h\|_{0,K}^2 \right.$$
$$\left. + \|\nabla p_h\|_{0,K}^2 + \|\nabla p_h - \Delta u_h\|_{0,K}^2)\right)^{\frac{1}{2}}$$
$$\geq \tfrac{1}{2}\|p_h\|_{0,\Omega}^2 - c\left(|u_h|_{1,\Omega}^2 + \sum_{K \in \mathcal{T}_h} \delta(h_K)(\|\nabla \cdot u_h\|_{0,K}^2 \right.$$
$$\left. + \|\nabla p_h\|_{0,K}^2 + \|\nabla p_h - \Delta u_h\|_{0,K}^2)\right).$$

This yields

$$a_h((u_h, p_h), (v_h, 0)) \geq \tfrac{1}{2}\|p_h\|_{0,\Omega}^2 - c_2(\||(u_h, p_h)\||_{h,\Omega}^2 - \|p_h\|_{0,\Omega}^2). \qquad (4.50)$$

(3) We now combine (4.49) and (4.50). Let $0 < \rho < 1$. Clearly,

$$a_h((u_h, p_h), ((1-\rho)u_h + \rho v_h, (1-\rho)p_h))$$
$$\geq ((1-\rho)c_1 - \rho c_2)\left(\||(u_h, p_h)\||_{h,\Omega}^2 - \|p_h\|_{0,\Omega}^2\right) + \frac{\rho}{2}\|p_h\|_{0,\Omega}^2.$$

Choosing $\rho = \frac{c_1}{c_1 + c_2 + \frac{1}{4}}$ and setting $w_h = (1-\rho)u_h + \rho v_h$ and $r_h = (1-\rho)p_h$ finally yields

$$a_h((u_h, p_h), (w_h, r_h)) \geq \tfrac{\rho}{4}\||(u_h, p_h)\||_{h,\Omega}^2 \geq c\||(u_h, p_h)\||_{h,\Omega}\||(w_h, r_h)\||_{h,\Omega},$$

completing the proof.

Lemma 4.39 (Consistency). *Let* (u, p) *be the solution to* (4.41) *and let* (u_h, p_h) *be the solution to* (4.47). *The following consistency property holds:*

$$\forall (v_h, q_h) \in V_h, \quad a_h((u - u_h, p - p_h), (v_h, q_h)) = 0. \qquad (4.51)$$

Proof. Straightforward verification.   □

Theorem 4.40 (Convergence). *Under assumptions* (4.43) *to* (4.46), *if the solution to* (4.41) *is smooth enough, the solution to* (4.47) *satisfies*

$$\forall h, \quad \|u - u_h\|_{1,\Omega} + \|p - p_h\|_{0,\Omega} \leq c\, h^k (\|u\|_{k+1,\Omega} + \|p\|_{k,\Omega}).$$

Proof. Set $v_h = \mathcal{I}_h u$ and $q_h = \mathcal{J}_h p$. Owing to Lemmas 4.37, 4.38 and 4.39,

$$\||(u_h - v_h, p_h - q_h)\||_{h,\Omega} \leq c \sup_{(w_h, r_h) \in V_h} \frac{a_h((u_h - v_h, p_h - q_h), (w_h, r_h))}{\||(w_h, r_h)\||_{h,\Omega}}$$
$$= c \sup_{(w_h, r_h) \in V_h} \frac{a_h((u - v_h, p - q_h), (w_h, r_h))}{\||(w_h, r_h)\||_{h,\Omega}}$$
$$\leq c\||(u - v_h, p - q_h)\||_{h,\Omega}.$$

Furthermore, it is clear that

$$\||(u - v_h, p - q_h)\||_{h,\Omega} \leq c\left(\sum_{K \in \mathcal{T}_h}(h_K^{2k} + \delta(h_K)h_K^{2(k-1)})(\|u\|_{k+1,K}^2 + \|p\|_{k,K}^2)\right)^{\frac{1}{2}}.$$

To conclude, use the fact that $\delta(h_K) = ch_K^2$ together with the triangle inequality.   □

Corollary 4.41. *Under the hypotheses of Theorem* 4.40, *if the Stokes problem has smoothing properties, the solution to* (4.47) *satisfies*

$$\forall h, \quad \|u - u_h\|_{0,\Omega} \leq c\, h^{k+1}(\|u\|_{k+1,\Omega} + \|p\|_{k,\Omega}).$$

Proof. Let $(e, \delta) \in V$ be the solution to the dual problem

$$\begin{cases} -\Delta e - \nabla \delta = u - u_h, \\ \nabla \cdot e = 0. \end{cases}$$

It is clear that this problem amounts to $\langle A(v, q), (e, \delta) \rangle_{V',V} = (u - u_h, v)_{0,\Omega}$ for all $(v, q) \in V$. As a result,

$$\|u - u_h\|_{0,\Omega}^2 = \langle A(u - u_h, p - p_h), (e, \delta) \rangle_{V',V}.$$

Set $v_h = \mathcal{I}_h e$ and $q_h = \mathcal{J}_h \delta$. Owing to (4.51),

$$\begin{aligned} \|u - u_h\|_{0,\Omega}^2 &= \langle A(u - u_h, p - p_h), (e - v_h, \delta - q_h) \rangle_{V',V} \\ &\quad - \sum_{K \in \mathcal{T}_h} \delta(h_K)((A(u - u_h, p - p_h), A(v_h, q_h)))_{0,K} \\ &\leq c_1 \big(|u - u_h|_{1,\Omega} + \|p - p_h\|_{0,\Omega}\big)\big(|e - v_h|_{1,\Omega} + \|\delta - q_h\|_{0,\Omega}\big) \\ &\quad + \sum_{K \in \mathcal{T}_h} c_2 h_K^2 \big(\|\Delta(u - u_h)\|_{0,K} + |p - p_h|_{1,K}\big)\big(\|v_h\|_{2,K} + \|q_h\|_{1,K}\big). \end{aligned}$$

Furthermore, $\|v_h\|_{2,K} + \|q_h\|_{1,K} \leq c(\|e\|_{2,K} + \|\delta\|_{1,K})$ and $|e - v_h|_{1,K} + \|\delta - q_h\|_{0,K} \leq c h_K(\|e\|_{2,K} + \|\delta\|_{1,K})$. Moreover, the smoothing property of the Stokes problem yields

$$\|e\|_{2,\Omega} + \|\delta\|_{1,\Omega} \leq c\|u - u_h\|_{0,\Omega}.$$

Properties (4.43) to (4.46) imply

$$\begin{aligned} h_K^2 \|\Delta(u - u_h)\|_{0,K} &\leq h_K^2 \big(\|u - \mathcal{I}_h u\|_{2,K} + \|\Delta(\mathcal{I}_h u - u_h)\|_{0,K}\big) \\ &\leq c h_K^2 \big(h_K^{k-1}\|u\|_{k+1,K} + h_K^{-1}|\mathcal{I}_h u - u_h|_{1,K}\big) \\ &\leq c h_K^2 \big(h_K^{k-1}\|u\|_{k+1,K} + h_K^{-1}|\mathcal{I}_h u - u|_{1,K} + h_K^{-1}|u - u_h|_{1,K}\big) \\ &\leq c \big(h_K^{k+1}\|u\|_{k+1,K} + h_K|u - u_h|_{1,K}\big). \end{aligned}$$

Likewise, $h_K^2 \|p - p_h\|_{1,K} \leq c\big(h_K^{k+1}\|p\|_{k,K} + h_K\|p - p_h\|_{0,K}\big)$. This yields the following bound from above:

$$\|u - u_h\|_{0,\Omega} \leq c\, h(|u - u_h|_{1,\Omega} + \|p - p_h\|_{0,\Omega}) + c\, h^{k+1}(\|u\|_{k+1,\Omega} + \|p\|_{k,\Omega}).$$

The final estimate is now a consequence of Theorem 4.40. $\qquad\square$

Remark 4.42.

(i) GaLS techniques are often used in the industry since they enable the use of finite elements that do not satisfy the inf-sup condition.

(ii) Without affecting the stability and convergence properties of the method, it is possible to remove the Least-Squares control on the divergence in (4.47).

(iii) The reader is referred to §5.4 for extensions of the GaLS method to advection equations and, more generally, to first-order PDEs. When solving the Navier–Stokes equations in the advection-dominated regime, it is necessary to combine techniques that stabilize the Stokes problem and advection equations. The reader is referred to [FrF92, ToV96] for further aspects of this question and details on the implementation.

(iv) As an alternative to the GaLS formulation, one can also consider a pressure gradient stabilization based on local projections; see [BeB01]. The technique is based on an extension of the subgrid viscosity concept introduced in [Gue99^b]; see §5.5 for a presentation of the subgrid viscosity technique for first-order PDEs.   □

4.4 Linear Algebra

In this section, we study the linear system associated with the approximate Stokes problem and exhibit some of its remarkable properties. We describe two methods for solving the system, one based on a penalty technique and the other based on a Schur complement.

4.4.1 Matrix version of the discrete problem

Consider the discrete problem:

$$\begin{cases} \text{Seek } u_h \in X_h \text{ and } p_h \in M_h \text{ such that} \\ a(u_h, v_h) + b(v_h, p_h) = f(v_h), \quad \forall v_h \in X_h, \\ b(u_h, q_h) = g(q_h), \qquad\qquad \forall q_h \in M_h, \end{cases} \qquad (4.52)$$

whose mathematical analysis has been carried out in §4.2.1. Let N_u and N_p denote the respective dimensions of the subspaces $X_h \subset X$ and $M_h \subset M$. Let $\{v_h^i\}_{1 \leq i \leq N_u}$ be a basis for X_h and let $\{q_h^k\}_{1 \leq k \leq N_p}$ be a basis for M_h. Recall that in the finite element framework, these bases consist of global shape functions.

For all $u_h = \sum_{i=1}^{N_u} u_i v_h^i$ in X_h and $p_h = \sum_{k=1}^{N_p} p_k q_h^k$ in M_h, we define the column vectors $U = (u_1, \ldots, u_{N_u})^T$ in $\mathbb{R}^{N_u}$ and $P = (p_1, \ldots, p_{N_p})^T$ in $\mathbb{R}^{N_p}$. The correspondences between u_h and U and between p_h and P are one-to-one since $\{v_h^i\}_{1 \leq i \leq N_u}$ and $\{q_h^k\}_{1 \leq k \leq N_p}$ are bases. Inserting the expansions of u_h and p_h into (4.52) and choosing as test functions the basis functions of X_h and M_h, we obtain the linear system

$$\begin{bmatrix} \mathcal{A} & \mathcal{B}^T \\ \mathcal{B} & 0 \end{bmatrix} \begin{bmatrix} U \\ P \end{bmatrix} = \begin{bmatrix} F \\ G \end{bmatrix}, \qquad (4.53)$$

where the matrices $\mathcal{A} \in \mathbb{R}^{N_u, N_u}$ and $\mathcal{B} \in \mathbb{R}^{N_p, N_u}$ are such that $\mathcal{A}_{ij} = a(v_h^j, v_h^i)$ and $\mathcal{B}_{ki} = b(v_h^i, q_h^k)$, and the vectors $F \in \mathbb{R}^{N_u}$ and $G \in \mathbb{R}^{N_p}$ are such that $F_i = f(v_h^i)$ and $G_k = g(q_h^k)$.

Example 4.43. For conformal approximations of the Stokes problem, we set $a(u_h, v_h) = \int_\Omega \nabla u_h : \nabla v_h$, $b(v_h, p_h) = -\int_\Omega p_h \nabla \cdot v_h$, $\mathcal{A}_{ij} = \int_\Omega \nabla v_h^j : \nabla v_h^i$, $\mathcal{B}_{kj} = -\int_\Omega q_h^k \nabla \cdot v_h^j$, $F_i = \langle f, v_h^i \rangle_{H^{-1}, H_0^1}$, and $G_k = -\int_\Omega g q_h^k$.   □

4.4.2 Properties of the linear system

For simplicity, assume that the bilinear form a is coercive on X, and let α be the coercivity constant. Assume also that the bilinear form b satisfies an inf-sup condition uniformly on $X_h \times M_h$, and let β be the corresponding inf-sup constant. To highlight the structure of the linear system (4.53), let us interpret in matrix terms the coercivity of a and the inf-sup condition satisfied by b. Let $\| \cdot \|_{N_u}$ and $\| \cdot \|_{N_p}$ (resp., $(\cdot, \cdot)_{N_u}$ and $(\cdot, \cdot)_{N_p}$) denote the Euclidean norms (resp., Euclidean scalar products) on $\mathbb{R}^{N_u}$ and $\mathbb{R}^{N_p}$, and consider the norms

$$\forall U \in \mathbb{R}^{N_u}, \quad \|U\|_X = \|u_h\|_X \quad \text{and} \quad \forall P \in \mathbb{R}^{N_p}, \|P\|_M = \|p_h\|_M. \quad (4.54)$$

Define the matrix $\mathcal{M}_{N_p} \in \mathbb{R}^{N_p, N_p}$ such that

$$(\mathcal{M}_{N_p} P, Q)_{N_p} = (p_h, q_h)_M,$$

where $(\cdot, \cdot)_M$ is the scalar product on M. The matrix $\mathcal{M}_{N_p}$ is called the mass matrix. This matrix is clearly symmetric positive definite. Denoting by $\mu_{p,\min}$ and $\mu_{p,\max}$ the smallest and largest eigenvalue of $\mathcal{M}_{N_p}$, respectively, we infer

$$\mu_{p,\min} \|P\|_{N_p}^2 \leq \|P\|_M^2 \leq \mu_{p,\max} \|P\|_{N_p}^2. \quad (4.55)$$

When M_h is a finite element space constructed on a quasi-uniform family of meshes, it is possible to show that the eigenvalue $\mu_{p,\min}$ is bounded from below by $c_1 h^d$ and the eigenvalue $\mu_{p,\max}$ is bounded from above by $c_2 h^d$, where both c_1 and c_2 are independent of h; see Lemma 9.7. In other words,

$$\forall P \in \mathbb{R}^{N_p}, \quad c_1 h^d \|P\|_{N_p}^2 \leq \|P\|_M^2 \leq c_2 h^d \|P\|_{N_p}^2.$$

Letting $\kappa(\mathcal{M}_{N_p}) = \frac{\mu_{p,\max}}{\mu_{p,\min}}$ be the condition number of the matrix $\mathcal{M}_{N_p}$, we infer $\kappa(\mathcal{M}_{N_p}) \leq c$ uniformly with respect to h; see Theorem 9.8. Now, let us introduce the norm $\| \cdot \|_*$ defined by

$$\forall U \in \mathbb{R}^{N_u}, \quad \|U\|_* = \sup_{V \in \mathbb{R}^{N_u}} \frac{(U, V)_{N_u}}{\|V\|_X}.$$

This norm naturally translates in matrix terms the coercivity of a and the inf-sup condition on b. Indeed, owing the coercivity of a together with the identity $(\mathcal{A}U, V)_{N_u} = a(u_h, v_h)$, holding for all U and $V \in \mathbb{R}^{N_u}$, $\mathcal{A}$ satisfies

$$\forall U \in \mathbb{R}^{N_u}, \quad (U, \mathcal{A}U)_{N_u} \geq \alpha \|U\|_X^2, \quad (4.56)$$

or, in other words,

$$\forall U \in \mathbb{R}^{N_u}, \quad \alpha \|U\|_X \le \|\mathcal{A}U\|_*. \tag{4.57}$$

Note that (4.56) shows that the matrix $\mathcal{A}$ is positive definite. Moreover, the continuity of a implies

$$\forall U \in \mathbb{R}^{N_u}, \quad \|\mathcal{A}U\|_* \le \|a\| \, \|U\|_X. \tag{4.58}$$

Since the matrix $\mathcal{B} \in \mathbb{R}^{N_p,N_u}$ is such that $(\mathcal{B}U,P)_{N_p} = b(u_h,p_h)$ for $U \in \mathbb{R}^{N_u}$ and $P \in \mathbb{R}^{N_p}$, the inf-sup inequality yields

$$\min_{\|P\|_M \ne 0} \max_{\|U\|_X \ne 0} \frac{(\mathcal{B}^T P, U)_{N_u}}{\|P\|_M \|U\|_X} \ge \beta,$$

which is equivalent to

$$\forall P \in \mathbb{R}^{N_p}, \quad \beta \|P\|_M \le \|\mathcal{B}^T P\|_*. \tag{4.59}$$

Note that this shows that $\mathcal{B}^T$ is injective, i.e., $\mathcal{B}$ is surjective. Moreover, the continuity of b implies

$$\forall P \in \mathbb{R}^{N_p}, \quad \|\mathcal{B}^T P\|_* \le \|b\| \, \|P\|_M. \tag{4.60}$$

Solving the linear system (4.53) by direct methods is out of the question when its size is large. Iterative methods are usually preferred. However, one must be careful when selecting an iterative technique, since many of them require the matrix to be positive definite. One distinctive feature of (4.53) is that the matrix is neither positive nor definite, although it is symmetric. Note that the following equivalent system yields a positive matrix:

$$\begin{bmatrix} \mathcal{A} & \mathcal{B}^T \\ -\mathcal{B} & 0 \end{bmatrix} \begin{bmatrix} U \\ P \end{bmatrix} = \begin{bmatrix} F \\ -G \end{bmatrix}. \tag{4.61}$$

However, this new matrix is neither symmetric nor definite. Hence, it is not possible to use either Gauß–Seidel-like techniques or block-Jacobi techniques. Gradient-based methods are also inefficient on this type of matrix; see Chapter 9. Actually, the saddle-point structure of the linear system (4.53), together with the inf-sup compatibility condition to be satisfied by the approximation spaces, is specific to mixed formulations.

We now describe two techniques which are frequently employed to solve (4.53). We keep track of the coefficients α and β in the error estimates since in practice these two quantities may be small, or even go to zero when the mesh is refined.

4.4.3 Penalty techniques and artificial compressibility

The principle of the penalty method is to replace (4.61) by the perturbed system

$$
\begin{bmatrix} \mathcal{A} & \mathcal{B}^T \\ -\mathcal{B} & \epsilon \mathcal{M}_{N_p} \end{bmatrix} \begin{bmatrix} U_\epsilon \\ P_\epsilon \end{bmatrix} = \begin{bmatrix} F \\ -G \end{bmatrix}, \tag{4.62}
$$

where $\epsilon > 0$ is a small penalty coefficient. Eliminating P_ϵ from the first equation yields

$$
\left(\mathcal{A} + \frac{1}{\epsilon} \mathcal{B}^T \mathcal{M}_{N_p}^{-1} \mathcal{B} \right) U_\epsilon = F + \frac{1}{\epsilon} \mathcal{B}^T \mathcal{M}_{N_p}^{-1} G. \tag{4.63}
$$

This system can be solved by means of standard techniques (e.g., conjugate gradient) since the matrix in (4.63) is symmetric positive definite. This *penalty method* is often referred to as the *artificial compressibility* technique when the underlying engineering situation is that of incompressible materials in solid mechanics or incompressible fluid mechanics. The proposition below shows that by introducing a penalty term, the perturbed solution is not too far from the original one, and that the distance between the two solutions goes to zero as the parameter ϵ goes to zero.

Proposition 4.44. *Let $\epsilon > 0$. Let (U, P) be the solution to (4.53) and (U_ϵ, P_ϵ) be the solution to the perturbed system (4.62). Then, the following error estimate holds:*

$$
\frac{\alpha\beta}{\|a\|} \|U - U_\epsilon\|_X + \frac{\alpha\beta^2}{\|a\|^2} \|P - P_\epsilon\|_M \le \epsilon \|P\|_M. \tag{4.64}
$$

Proof. Subtracting the perturbed system from (4.61) yields

$$
\mathcal{A}(U - U_\epsilon) + \mathcal{B}^T(P - P_\epsilon) = 0,
$$
$$
-\mathcal{B}(U - U_\epsilon) - \epsilon \mathcal{M}_{N_p} P_\epsilon = 0.
$$

Using inequalities (4.58) and (4.59) in the first equation yields

$$
\|P - P_\epsilon\|_M \le \tfrac{1}{\beta} \|\mathcal{B}^T(P - P_\epsilon)\|_* = \tfrac{1}{\beta} \|\mathcal{A}(U - U_\epsilon)\|_* \le \tfrac{\|a\|}{\beta} \|U - U_\epsilon\|_X.
$$

Multiply the first equation by $U - U_\epsilon$ and use the coercivity of $\mathcal{A}$ together with the second equation to infer

$$
\begin{aligned}
\alpha \|U - U_\epsilon\|_X^2 &\le (\mathcal{A}(U - U_\epsilon), U - U_\epsilon)_{N_u} = (\mathcal{B}^T(P_\epsilon - P), U - U_\epsilon)_{N_u} \\
&= (P_\epsilon - P, \mathcal{B}(U - U_\epsilon))_{N_p} = -\epsilon(P_\epsilon - P, \mathcal{M}_{N_p} P_\epsilon)_{N_p} \\
&= -\epsilon(P_\epsilon - P, \mathcal{M}_{N_p}(P_\epsilon - P))_{N_p} - \epsilon(P_\epsilon - P, \mathcal{M}_{N_p} P)_{N_p} \\
&\le -\epsilon(P_\epsilon - P, \mathcal{M}_{N_p} P)_{N_p} \le \epsilon \|P_\epsilon - P\|_M \|P\|_M.
\end{aligned}
$$

Combining these two inequalities yields (4.64). $\qquad\square$

Remark 4.45.

(i) In (4.64), the coefficient $\frac{1}{\beta}$ arises in the estimate for $\|U - U_\epsilon\|_X$ whereas $\frac{1}{\beta^2}$ arises in the estimate for $\|P - P_\epsilon\|_M$. If the spaces pair $\{X_h, M_h\}$ satisfies the inf-sup condition with $\beta \to 0$ when $h \to 0$, the singular behavior of the stability constant β has a greater effect on the convergence rate of P_ϵ than on that of U_ϵ.

(ii) Owing to (4.54), (4.64) also translates into estimates on $\|u_h - u_{\epsilon h}\|_X$ and $\|p_h - p_{\epsilon h}\|_M$, where $u_{\epsilon h}$ and $p_{\epsilon h}$ are the velocity and pressure fields reconstructed from U_ϵ and P_ϵ, respectively.

4.4.4 Uzawa matrix

A second method to solve (4.53) consists of eliminating the quantity U from (4.53), yielding

$$\mathcal{B}\mathcal{A}^{-1}\mathcal{B}^T P = \mathcal{B}\mathcal{A}^{-1}F - G. \tag{4.65}$$

The matrix $\mathcal{B}\mathcal{A}^{-1}\mathcal{B}^T$ is often called the *Uzawa matrix* in reference to the so-called Uzawa iterative algorithm. This matrix is also called the *Schur complement* of $\mathcal{A}$. Henceforth, we set

$$\mathcal{U} = \mathcal{B}\mathcal{A}^{-1}\mathcal{B}^T.$$

Some interesting properties of the matrix $\mathcal{U}$ are summarized in Proposition 4.46, the proof of which is left as an exercise.

Proposition 4.46.

(i) *If the matrix $\mathcal{A}$ is positive definite, $\mathcal{U}$ is also positive definite.*
(ii) *If $\mathcal{A}$ is symmetric, $\mathcal{U}$ is also symmetric.*

Implementation. Although the matrix $\mathcal{A}$ is generally sparse, its inverse is almost always dense. Since inverting $\mathcal{A}$ is an extremely inefficient strategy (see §9.3), (4.65) is usually solved using iterative methods.

The unit computational cost for iterative methods is that of a matrix–vector multiplication. Hence, it is desirable to compute the product $\mathcal{U}R$ as efficiently as possible for any vector $R = (r_1, \ldots, r_{N_p})^T \in R^{N_p}$. This operation is performed in three steps as follows:

1. The first step consists of evaluating the vector $\mathcal{B}^T R$. This operation can be accomplished in two different ways. If the user has already assembled the matrix $\mathcal{B}$, then he uses the computational resource as best as he can to compute $\mathcal{B}^T R$. If the matrix is not yet assembled, the following procedure can be used: Upon denoting by r_h the field of M_h such that $r_h = \sum_{k=1}^{N_p} R_k q_h^k$, the definition of $\mathcal{B}$ implies

$$(\mathcal{B}^T R)_i = b(v_h^i, r_h), \qquad 1 \le i \le N_u.$$

   In particular, for the Stokes problem, $(\mathcal{B}^T R)_i = -\int_\Omega r_h \nabla \cdot v_h^i$. The main interest of this approach is that it does not require storing the matrix $\mathcal{B}$. This memory saving can be important when dealing with large problems.

2. Set $T = \mathcal{B}^T R$. The second step consists of evaluating $S = \mathcal{A}^{-1}T$. This operation is performed by solving iteratively the system $\mathcal{A}S = T$. Here again, the unit cost is that of the multiplication of $\mathcal{A}$ by a vector. If the matrix $\mathcal{A}$ is assembled, this operation is straightforward. Otherwise, the i-th component of the vector $\mathcal{A}W$ is evaluated as

$$(\mathcal{A}W)_i = a(v_h^i, w_h).$$

For the Stokes problem, $(\mathcal{A}W)_i = \int_\Omega \nabla v_h^i : \nabla w_h$.

3. The last operation consists of evaluating $\mathcal{B}S$. If $\mathcal{B}$ is assembled, this operation is straightforward. Otherwise, the k-th component of $\mathcal{B}S$ is

$$(\mathcal{B}S)_k = b(s_h, q_h^k), \qquad 1 \le k \le N_p.$$

For the Stokes problem, $(\mathcal{B}S)_k = -\int_\Omega q_h^k \nabla \cdot s_h$.

There does not seem to be a unique strategy for choosing between assembling the matrices $\mathcal{B}$ and $\mathcal{A}$ and computing the matrix–vector multiplications on the flight. One must find a compromise between many (often conflicting) parameters: the memory space available; the number of times problem (4.65) has to be solved; the ratio between the speed to access memory and that to perform arithmetic operations; vectorization; parallelization; etc.

Condition number. Since the convergence rate of most iterative methods depends on the condition number of the matrix involved (see, e.g., Proposition 9.30), it is important to derive a bound on the condition number of the Uzawa matrix. Since $\mathcal{U}$ is symmetric, its condition number is $\kappa(\mathcal{U}) = \frac{\lambda_{\mathcal{U},\max}}{\lambda_{\mathcal{U},\min}}$, i.e., it is the ratio of the largest to the smallest eigenvalue of $\mathcal{U}$; see Proposition 9.2.

Proposition 4.47. *If $\mathcal{U}$ is symmetric, the following estimate holds:*

$$\kappa(\mathcal{U}) \le \left(\frac{\|a\|}{\alpha} \frac{\|b\|}{\beta} \right)^2 \kappa(\mathcal{M}_{N_p}). \tag{4.66}$$

Proof. (1) Let Q be an eigenvector of $\mathcal{U}$ associated with the smallest eigenvalue $\lambda_{\mathcal{U},\min}$. By definition, $\lambda_{\mathcal{U},\min}\|Q\|_{N_p}^2 = (\mathcal{U}Q, Q)_{N_p} = (\mathcal{A}^{-1}\mathcal{B}^T Q, \mathcal{B}^T Q)_{N_u}$. Let F in $\mathbb{R}^{N_u}$. Setting $U = \mathcal{A}^{-1}F$ and using (4.57) and (4.58) yields

$$(F, \mathcal{A}^{-1}F)_{N_u} = (\mathcal{A}U, U)_{N_u} \ge \alpha\|U\|_X^2 \ge \tfrac{\alpha}{\|a\|^2}\|F\|_\star^2.$$

The inf-sup inequality (4.59) implies

$$\lambda_{\mathcal{U},\min}\|Q\|_{N_p}^2 \ge \tfrac{\alpha}{\|a\|^2}\|\mathcal{B}^T Q\|_\star^2 \ge \tfrac{\alpha\beta^2}{\|a\|^2}\|Q\|_M^2.$$

Finally, the bound from below in (4.55) leads to

$$\lambda_{\mathcal{U},\min} \ge \tfrac{\alpha\beta^2}{\|a\|^2}\lambda_{\mathcal{M}_{N_p},\min}. \tag{4.67}$$

(2) Let Q be an eigenvector associated with the largest eigenvalue of $\mathcal{U}$. Then,

$$\lambda_{\mathcal{U},\max}\|Q\|_{N_p}^2 = (\mathcal{U}Q, Q)_{N_p} = (\mathcal{A}^{-1}\mathcal{B}^T Q, \mathcal{B}^T Q)_{N_u}.$$

Let F in $\mathbb{R}^{N_u}$. Setting $U = \mathcal{A}^{-1}F$ yields

$$(F, \mathcal{A}^{-1}F)_{N_u} = (F, U)_{N_u} \le \|F\|_\star\|U\|_X \le \tfrac{1}{\alpha}\|F\|_\star^2.$$

The bounds (4.55) and (4.60) imply

$$\lambda_{\mathcal{U},\max} \|Q\|_{N_p}^2 \leq \tfrac{1}{\alpha} \|\mathcal{B}^T Q\|_\star^2 \leq \tfrac{\|b\|^2}{\alpha} \|Q\|_M^2 \leq \tfrac{\|b\|^2}{\alpha} \lambda_{\mathcal{M}_{N_p},\max} \|Q\|_{N_p}^2,$$

showing that $\lambda_{\mathcal{U},\max} \leq \frac{\|b\|^2}{\alpha} \lambda_{\mathcal{M}_{N_p},\max}$. The estimate (4.66) follows easily. □

Remark 4.48.

(i) If the family $\{\mathcal{T}_h\}_{h>0}$ is quasi-uniform, the Euclidean condition number $\kappa(\mathcal{M}_{N_p})$ is bounded by a constant independent of h. As a result, estimate (4.66) shows that if $\frac{\|a\|}{\alpha}$ and $\frac{\|b\|}{\beta}$ are of order 1, the condition number of the Uzawa matrix is also of order 1. In this case, classical iterative solution techniques converge very fast without any preconditioning; see Proposition 9.30. Roughly speaking, $\mathcal{U}$ behaves like the identity.

(ii) Unfortunately, the ideal framework described above does not hold when solving the time-dependent Stokes problem. In this case, after time-discretization, one is led at each time step to solve a problem like $u - \nu \delta t \Delta u + \nabla p = f$, where δt is the time step and ν is a viscosity constant (possibly small). Then, the coercivity constant α is $\nu \delta t$ and assuming that $\frac{\|b\|}{\beta}$ is of order 1, we infer that the $\kappa(\mathcal{U})$ behaves like $\frac{c}{\nu^2 \delta t^2}$. In these circumstances, standard iterative solution methods cannot converge well without adequate preconditioning. The reader is referred to §6.2 for further insight. □

4.5 Exercises

Exercise 4.1. Find the natural boundary conditions enforced by using the formula in Remark 4.12(iii).

Exercise 4.2. For the artificial compressibility technique, use the identity matrix I_{dK} instead of the mass matrix $\mathcal{M}_K$. Does the method still converge? What is the interest of doing so?

Exercise 4.3. Prove inequality (4.38). (*Hint*: Use the Petree–Tartar Lemma.)

Exercise 4.4. Construct a counterexample showing that the $\mathbb{Q}_1/\mathbb{Q}_1$ mixed finite element (continuous on velocity and pressure, respectively) does not satisfy the inf-sup condition. (*Hint*: Consider $\Omega =]0, 1[^2$ and a mesh composed of squares; then, adapt the $\mathbb{P}_1/\mathbb{P}_1$ counterexample from §4.2.3.)

Exercise 4.5. Construct a counterexample to justify Remark 4.29. (*Hint*: Consider a uniform mesh; given an interior vertex, consider the patch composed of the four square cells sharing this vertex; and find suitable values for an oscillating pressure field.)

Exercise 4.6. Prove Theorem 4.32. (*Hint*: See the proof of Theorem 3.38.)

Exercise 4.7. Prove the statements in Proposition 4.46.

Exercise 4.8. Let $\{\Omega_1, \Omega_2\}$ be a partition of Ω. Assume $\mathrm{meas}(\Omega_1) \neq 0$ and $\mathrm{meas}(\Omega_2) \neq 0$. Let $\sigma \geq 0$, $f \in [H^{-1}(\Omega)]^d$, and $g \in L^2(\Omega_2)$.

(i) Write a weak formulation in $[H_0^1(\Omega)]^d \times L^2(\Omega_1)$ of the problem

$$\begin{cases} -\Delta u + \nabla p = f & \text{in } \Omega, \\ \nabla\cdot u = 0 & \text{in } \Omega_1, \\ \sigma\nabla\cdot u + p = g & \text{in } \Omega_2, \\ u = 0 & \text{on } \partial\Omega. \end{cases}$$

(*Hint*: Introduce the forms $a(u, v) = (\nabla u, \nabla v)_{0,\Omega} + \sigma(\nabla\cdot u, \nabla\cdot v)_{0,\Omega_2}$ and $b(v, p) = -(\nabla\cdot v, p)_{0,\Omega_1}$.)

(ii) For $q \in L^2(\Omega_1)$, prove that there exists $\tilde{q} \in L_{f=0}^2(\Omega)$ such that $\tilde{q}_{|\Omega_1} = q$ and $\|\tilde{q}\|_{0,\Omega} \leq c\|q\|_{0,\Omega_1}$.

(iii) Prove that $\exists c > 0$ such that

$$\forall q \in L^2(\Omega_1), \qquad \sup_{v \in [H_0^1(\Omega)]^d} \frac{b(v, q)}{\|v\|_{1,\Omega}} \geq c\|q\|_{0,\Omega_1}.$$

(iv) Prove that the weak problem derived in question (i) is well-posed.

(v) Propose a pair of finite elements to solve this problem. Substantiate your claim by a proof.

Exercise 4.9 (Simplified magneto-hydrodynamics). Partition $\Omega \subset \mathbb{R}^2$ into subdomains Ω_1 and Ω_2, both with non-zero measure. Let $\mu, \sigma > 0$. Let H be a vector field in $\mathbb{R}^2$ and E be a scalar field. Define $\nabla\times H = \partial_1 H_2 - \partial_2 H_1$ and $\nabla\times E = (\partial_2 E, -\partial_1 E)$. For $j \in L^2(\Omega)$, consider the problem

$$\begin{cases} \mu H = -\nabla\times E & \text{in } \Omega, \\ \nabla\times H = 0 & \text{in } \Omega_1, \\ \nabla\times H = \sigma E + j & \text{in } \Omega_2, \\ H\times n = 0 & \text{on } \partial\Omega. \end{cases}$$

(i) Give a weak formulation of this problem in the form of a saddle-point problem in $H_0(\mathrm{curl}; \Omega) \times L^2(\Omega_1)$.

(ii) For $e \in L^2(\Omega_1)$, prove that there exists $\tilde{e} \in L_{f=0}^2(\Omega)$ such that $\tilde{e}_{|\Omega_1} = e$ and $\|\tilde{e}\|_{0,\Omega} \leq c\|e\|_{0,\Omega_1}$.

(iii) Prove that $\exists c > 0$ such that

$$\forall e \in L^2(\Omega_1), \qquad \sup_{b \in H_0(\mathrm{curl};\Omega)} \frac{(\nabla\times b, e)}{\|b\|_{H(\mathrm{curl};\Omega)}} \geq c\|e\|_{0,\Omega_1}.$$

(*Hint*: Solve $-\Delta\phi = \tilde{e}$, $\partial_n\phi_{|\partial\Omega} = 0$ and set $b = \nabla\times\phi$.)

(iv) Prove that the weak problem derived in question (i) is well-posed.

(v) Propose a pair of finite elements to solve this problem.

Exercise 4.10 (Darcy equations 1). Let $f \in [L^2(\Omega)]^d$, $g \in H^{\frac{1}{2}}(\partial\Omega)$, and $k \in L^2(\Omega)$. Consider the problem:

$$
\begin{cases}
\text{Seek } u \in H(\text{div}; \Omega) \text{ and } p \in L^2(\Omega) \text{ such that} \\
(u,v)_{0,\Omega} - (\nabla \cdot v, p)_{0,\Omega} = (f,v)_{0,\Omega} + \int_{\partial\Omega} g v \cdot n, \quad \forall v \in H(\text{div}; \Omega), \\
(\nabla \cdot u, q)_{0,\Omega} = (k,q)_{0,\Omega}, \qquad\qquad\qquad\qquad \forall q \in L^2(\Omega).
\end{cases}
$$

(i) What are the corresponding PDEs and boundary conditions? Are the boundary conditions enforced naturally or essentially?
(ii) Let $q \in L^2(\Omega)$. Solve for $\phi \in H_0^1(\Omega)$ such that $(\nabla\phi, \nabla\psi)_{0,\Omega} = (q,\psi)_{0,\Omega}$ for all $\psi \in H_0^1(\Omega)$. Set $v = \nabla\phi$. Estimate $\nabla \cdot v$ and $\|v\|_{H(\text{div};\Omega)}$.
(iii) Let $q \in L^2(\Omega)$. Set $\alpha = \frac{1}{\text{meas}(\Omega)} \int_\Omega q$. Construct $v_\alpha \in [H^1(\Omega)]^d$ such that $\nabla \cdot v = \alpha$. Prove that $\nabla \cdot : [H^1(\Omega)]^d \to L^2(\Omega)$ is surjective. (*Hint:* Use the fact that $\nabla \cdot : [H_0^1(\Omega)]^d \to L_{f=0}^2(\Omega)$ is surjective.)
(iv) Prove that $\sup_{w \in H(\text{div};\Omega)} \frac{(\nabla \cdot w, q)_{0,\Omega}}{\|w\|_{H(\text{div};\Omega)}} \geq c\|q\|_{0,\Omega}$. (*Hint:* Use either question (ii) or (iii).)
(v) Prove that the weak problem is well-posed. (*Hint:* Note that $(u,v)_{0,\Omega}$ is coercive on $V = \{v \in H(\text{div}; \Omega); \nabla \cdot v = 0\}$.)

Exercise 4.11 (Darcy equations 2). Let $f \in [L^2(\Omega)]^d$, $l \in L^2(\Omega)$, and $g \in H^{-\frac{1}{2}}(\partial\Omega)$. Set $H_{f=0}^1(\Omega) = \{p \in H^1(\Omega); \int_\Omega p = 0\}$. Consider the problem:

$$
\begin{cases}
\text{Seek } u \in [L^2(\Omega)]^d \text{ and } p \in H_{f=0}^1(\Omega) \text{ such that} \\
(u,v)_{0,\Omega} + (v, \nabla p)_{0,\Omega} = (f,v)_{0,\Omega}, \qquad \forall v \in [L^2(\Omega)]^d, \qquad (4.68) \\
-(u, \nabla q)_{0,\Omega} = (l,q)_{0,\Omega} + \langle g, q \rangle_{H^{-\frac{1}{2}}, H^{\frac{1}{2}}}, \quad \forall q \in H_{f=0}^1(\Omega).
\end{cases}
$$

(i) What are the corresponding PDEs and boundary conditions?
(ii) Prove that problem (4.68) is well-posed.
(iii) Let $k \geq 1$ and let $\mathcal{T}_h$ be a mesh of affine simplices. Set $X_h = \{v_h \in [L^2(\Omega)]^d; \forall K \in \mathcal{T}_h, v_h \circ T_K \in [\mathbb{P}_{k-1}]^d\}$ and $M_h = \{q_h \in H_{f=0}^1(\Omega); \forall K \in \mathcal{T}_h, q_h \circ T_K \in \mathbb{P}_k\}$. Show that the discrete problem is well-posed.
(iv) Show that u_h can be algebraically eliminated from (4.68). Write the corresponding problem. (*Hint:* Observe that $\nabla M_h \subset X_h$.)
(v) Assume $l = 0$, meaning that there are no mass sources or sink inside Ω. Denote by $\mathcal{V}_h^i$ the set of mesh vertices inside Ω. For $\nu \in \mathcal{V}_h^i$, let $\mathcal{F}_\nu$ be the set of mesh faces (edges in two dimensions) to which ν belongs. For $F \in \mathcal{F}_\nu$, let $[\![u_h \cdot n]\!]_F$ be the jump of the normal component of u_h across F and meas(F) the measure of F. Prove that

$$
\forall \nu \in \mathcal{V}_h^i, \quad \sum_{F \in \mathcal{F}_\nu} [\![u_h \cdot n]\!]_F \, \text{meas}(F) = 0. \qquad (4.69)
$$

Exercise 4.12 (Darcy equations 3). Using the notation of Exercise 4.11, consider problem (4.68). Use the space X_h with $k = 1$ to approximate the velocity, and use the Crouzeix–Raviart finite element to approximate the pressure.

(i) Prove that the resulting discrete problem is well-posed.
(ii) Show that the discrete velocity u_h can be algebraically eliminated from the approximate problem. Write the corresponding problem.
(iii) Assume $l = 0$. Prove that the discrete velocity satisfies $[\![u_h \cdot n]\!]_F = 0$, $\forall F \in \mathcal{F}_h^i$. Compare with (4.69).

Exercise 4.13 ($\mathbb{RT}_0$). Let $\mathcal{T}_h$ be a shape-regular family of affine simplicial meshes of Ω. Let $D_h = \{v_h \in H(\mathrm{div}; \Omega); \forall K \in \mathcal{T}_h, v_{h|K} \in \mathbb{RT}_0\}$ and $M_h = \{q_h \in L^2(\Omega); \forall K \in \mathcal{T}_h, q_{h|K} \in \mathbb{P}_0\}$. For $v \in [H^1(\Omega)]^d$, let $\mathcal{I}_h^{\mathrm{RT}} v$ be the function in D_h such $\int_F (v - \mathcal{I}_h^{\mathrm{RT}} v) \cdot n = 0$ on every face F of the mesh.

(i) Let π_K be the L^2-projection from $L^2(K)$ to $\mathbb{P}_0$. Prove that $\forall v \in [H^1(\Omega)]^d$, $\nabla \cdot (\mathcal{I}_h^{\mathrm{RT}} v_{|K}) = \pi_K \nabla \cdot v$ and $\|\nabla \cdot \mathcal{I}_h^{\mathrm{RT}} v\|_{0,\Omega} \leq \|\nabla \cdot v\|_{0,\Omega}$.
(ii) Show that there exists c, independent of h_K, such that $\|\mathcal{I}_h^{\mathrm{RT}} v\|_{0,K} \leq c (\|v\|_{0,K} + h_K \|v\|_{1,K})$. (*Hint*: Work on $\widehat{K}$, use norm equivalence and a trace property, then go back to K.)
(iii) Prove that there exists $c > 0$, independent of h, such that

$$\forall q_h \in M_h, \quad \sup_{v_h \in D_h} \frac{(\nabla \cdot v_h, q_h)_{0,\Omega}}{\|v_h\|_{H(\mathrm{div};\Omega)}} \geq c \|q_h\|_{0,\Omega}.$$

(*Hint*: Use the Fortin criterion.) Comment in regard to Exercise 4.10.

Exercise 4.14. Use the notation of §4.4.1 and let $\mathcal{U} = \mathcal{B}\mathcal{A}^{-1}\mathcal{B}^T$ be the Uzawa matrix as defined in §4.4.4 assuming $X_h \subset [H_0^1(\Omega)]^d$. Assume moreover that $M_h \subset L^2(\Omega)$ and, letting 1_Ω be the characteristic function of Ω, assume that $1_\Omega = \sum_{k=1}^{N_p} q_h^k$. Let $I = (1, \ldots, 1)^T \in \mathbb{R}^{N_p}$. For $P_0 \in \mathbb{R}^{N_p}$ and $k \in \mathbb{N}$, set $\mathcal{K}(\mathcal{U}, P_0, k) = \mathrm{span}\{P_0, \mathcal{U} P_0, \ldots, \mathcal{U}^{k-1} P_0\}$.

(i) Show that $\mathcal{B}^T I = 0$.
(ii) Without assuming that $\mathcal{A}$ is symmetric, show that $I^T \mathcal{U} = 0$ and $\mathcal{U} I = 0$.
(iii) Show that if $(I, P_0)_{N_p} = 0$, then for all $P \in \mathcal{K}(\mathcal{U}, P_0, k)$, $(I, P)_{N_p} = 0$.
(iv) Let $P_0 \in \mathbb{R}^{N_p}$ and set $p_{0h} = \sum_{k=1}^{N_p} P_{0k} q_h^k$. Show that if p_{0h} is of zero mean, then $p_h \in M_h$ is also of zero mean if the coordinate vector of p_h relative to the basis $\{q_h^1, \ldots, q_h^{N_p}\}$ is in $\mathcal{K}(\mathcal{U}, P_0, k)$.

5

First-Order PDEs

This chapter deals with first-order PDEs and, more generally, with problems where solution and test spaces are different and where no coercivity property holds. The prototypical example is the advection equation.

This chapter is organized into seven sections. In the first section, we study the equation $u' = f$ in one dimension. We show that approximating this equation by means of the standard Galerkin method is not optimal. The second section sets a general framework for studying first-order PDEs in L^2 and introduces Friedrichs' systems as a general example for systems of first-order PDEs. In the third section, we introduce the *Least-Squares formulation* and show that this technique is well-suited to approximate first-order PDEs. In the fourth section, we study elliptic equations with a small coercivity constant. This situation corresponds to advection–diffusion equations with dominant advection. We show that the *Galerkin/Least-Squares* (GaLS) formulation is appropriate for approximating this type equation. In the fifth section, we introduce a *subgrid viscosity* technique and show that the domain of application of this technique is slightly larger than that of the GaLS method. The sixth section is devoted to the *Discontinuous Galerkin* (DG) method with the emphasis set on advection–reaction equations. A discontinuous GaLS method is also investigated. Finally, in the last section, we study a non-standard Galerkin technique to approximate Laplacian-type problems in mixed form.

5.1 Standard Galerkin Approximation in One Dimension

We begin with a one-dimensional model equation to demonstrate that, in general, the standard Galerkin technique is not optimal for approximating first-order PDEs.

5.1.1 The model problem

Let f be a smooth function, and consider the one-dimensional problem:

$$\begin{cases} \text{Seek } u \text{ such that} \\ u'(x) = f(x) \quad \text{in } \Omega = \,]0,1[, \\ u(0) = 0. \end{cases} \tag{5.1}$$

Before approximating the solution to problem (5.1), it is necessary to clarify the mathematical setting in which the solution is sought.

5.1.2 Formulation in $L^1(\Omega)$

Formally, the solution to (5.1) is $u(x) = \int_0^x f(t)\,dt$. To give a sense to this formula, we introduce the Banach space

$$W^{1,1}(\Omega) = \{v \in L^1(\Omega); v' \in L^1(\Omega)\},$$

where, as usual, the derivative is understood in the distribution sense, and the integral is defined in the Lebesgue sense. The $L^1(\Omega)$ setting gives a reasonable meaning to the following statement: The derivative of the antiderivative of a function is the function itself; see Lemma B.24.

Theorem 5.1. *If $f \in L^1(\Omega)$, (5.1) has a unique solution in $W^{1,1}(\Omega)$.*

Proof. (1) Let $u(x) = \int_0^x f(t)\,dt$. This is meaningful since $f \in L^1(\Omega)$.
(2) Let us show that $u \in C^0([0,1])$. Let $x \in [0,1]$ and let $\{x_n\}$ be a sequence converging to x in $[0,1]$. Clearly, $\int_0^x f - \int_0^{x_n} f = \int_{x_n}^x f = \int_0^1 1_{[x_n,x]}f$, where $1_{[x_n,x]}$ denotes the characteristic function of the interval $[x_n,x]$. Since $1_{[x_n,x]}f \to 0$ and $|1_{[x_n,x]}f| \le f$ a.e. in Ω, Lebesgue's Dominated Convergence Theorem implies $u(x_n) \to u(x)$.
(3) Step (2) implies that the boundary condition $u(0) = 0$ is meaningful.
(4) Lemma B.24 shows that $u' = f$ holds in $\mathcal{D}'(\Omega)$. Moreover, by identifying functions of $L^1(\Omega)$ with distributions, the equality holds in $L^1(\Omega)$.
(5) The uniqueness of the solution is a consequence of Lemma B.29.   □

Theorem 5.1 shows that provided f is integrable, it is meaningful to look for a solution to (5.1) in $W^{1,1}(\Omega)$. However, although this setting is coherent from a mathematical viewpoint, it is not easily amenable to approximation by duality; see Exercise 5.2. Henceforth, we shall restrict ourselves to the "friendlier" Hilbertian setting.

5.1.3 Formulation in $L^2(\Omega)$

Let us proceed by duality, i.e., we multiply $u' = f$ by a smooth test function and integrate over Ω to obtain $\int_0^1 u'v = \int_0^1 fv$. If $f \in L^2(\Omega)$, the right-hand side is meaningful if v is selected in $L^2(\Omega)$. Likewise, the left-hand side is meaningful if u is sought in $H^1(\Omega)$. Since functions in $H^1(\Omega)$ have a trace at 0, it is legitimate to define the Hilbert space

$$X = \{v \in H^1(\Omega); v(0) = 0\}. \tag{5.2}$$

Introduce the bilinear form

$$a(u, v) = \int_0^1 u'v,$$

and consider the problem:

$$\begin{cases} \text{Seek } u \text{ in } X \text{ such that} \\ a(u, v) = (f, v)_{0,\Omega}, \quad \forall v \in L^2(\Omega). \end{cases} \tag{5.3}$$

Theorem 5.2. *Problem* (5.3) *is well-posed.*

Proof. (1) It is clear that the bilinear form a is in $\mathcal{L}(X \times L^2(\Omega); \mathbb{R})$ and that the linear form $v \mapsto \int_0^1 fv$ is continuous on $L^2(\Omega)$. Let us now prove that the conditions (BNB1) and (BNB2) of the BNB Theorem hold.
(2) Let u be a function in X. Owing to the Poincaré inequality,

$$\inf_{u \in X} \sup_{v \in L^2(\Omega)} \frac{a(u, v)}{\|u\|_{1,\Omega}\|v\|_{0,\Omega}} = \inf_{u \in X} \frac{|u|_{1,\Omega}}{\|u\|_{1,\Omega}} \geq \sqrt{\tfrac{2}{3}}.$$

(3) Let v in $L^2(\Omega)$ be such that, for all u in X, $a(u, v) = 0$. Choosing u to be a function in $\mathcal{D}(\Omega)$ yields

$$\forall u \in \mathcal{D}(\Omega), \quad \int_0^1 u'v = 0,$$

which means that $v' = 0$ in the distribution sense. Lemma B.29 implies that v is a constant. Choosing $u = x$ as a test function leads to $\int_0^1 v = 0$ and, as a consequence, $v = 0$. □

5.1.4 The discrete viewpoint

Let N be a positive integer. Set $h = \frac{1}{N}$ and $x_i = ih$ for $0 \leq i \leq N$, and introduce the space

$$X_h = \{v_h \in C^0(\overline{\Omega}); \forall i \in \{0, \ldots, N-1\}, v_{h|[x_i, x_{i+1}]} \in \mathbb{P}_1; v_h(0) = 0\}. \tag{5.4}$$

It is clear that X_h is a subspace of X. Denote by $\{\varphi_1, \ldots, \varphi_N\}$ the basis of X_h such that $\varphi_i(x_j) = \delta_{ij}$, $1 \leq i, j \leq N$; see §1.1.

The simplest approach to approximate problem (5.3) in the framework of Galerkin methods is to use the discrete space X_h both as solution space and as test space. The discrete problem is then:

$$\begin{cases} \text{Seek } u_h \text{ in } X_h \text{ such that} \\ a(u_h, v_h) = (f, v_h)_{0,\Omega}, \quad \forall v_h \in X_h. \end{cases} \tag{5.5}$$

This problem falls into the framework of Theorem 2.22 where $V_h = X_h$ is equipped with the $H^1(\Omega)$-inner product, and $W_h = X_h$ is equipped with the $L^2(\Omega)$-inner product. Hence, (5.5) is well-posed iff there is $\alpha_h > 0$ such that

$$\inf_{u_h \in X_h} \sup_{v_h \in X_h} \frac{a(u_h, v_h)}{\|u_h\|_{1,\Omega} \|v_h\|_{0,\Omega}} \geq \alpha_h.$$

Furthermore, to derive optimal error estimates, it is necessary that α_h be bounded from below by a positive constant independent of h. Unfortunately, this is not the case as shown by the following:

Theorem 5.3. *There are $c_1 > 0$ and $c_2 > 0$, independent of h, such that*

$$c_1 h \leq \inf_{u_h \in X_h} \sup_{v_h \in X_h} \frac{a(u_h, v_h)}{\|u_h\|_{1,\Omega} \|v_h\|_{0,\Omega}} \leq c_2 h.$$

Proof. (1) Assume that N is even, the other case being treated similarly. We first prove the bound from above and then the bound from below.
(2) The bound from above.
(2.i) The idea consists of constructing an oscillating function u_h such that $|u_h|_{1,\Omega}$ diverges when $h \to 0$, while the L^2-projection of u_h' onto X_h is bounded. Set

$$u_h = \sum_{i=1}^{N} U_i \varphi_i \quad \text{with} \quad \begin{cases} U_{2i} = 2ih & \text{if } 1 \leq i \leq \frac{N}{2}, \\ U_{2i+1} = 1 & \text{if } 0 \leq i \leq \frac{N}{2} - 1. \end{cases}$$

Figure 5.1 shows the graph of u_h for $N = 10$. Set $U_0 = 0$ and denote by $[\cdot]$ the integer part operator. Then,

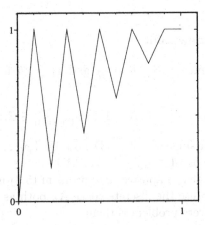

Fig. 5.1. Graph of the oscillating function u_h used in the proof of Theorem 5.3.

$$h|u_h|^2_{1,\Omega} = \sum_{i=0}^{N-1} h \int_{x_i}^{x_{i+1}} (u'_h)^2 = \sum_{i=0}^{N-1} \int_{x_i}^{x_{i+1}} h^{-1}(U_{i+1} - U_i)^2$$

$$= \sum_{i=0}^{\frac{N}{2}-1} (U_{2i+1} - U_{2i})^2 + \sum_{i=1}^{\frac{N}{2}}(U_{2i} - U_{2i-1})^2 \geq 2 \sum_{i=0}^{\frac{N}{2}-1} (1 - 2ih)^2$$

$$\geq 2 \sum_{i=0}^{[\frac{N}{4}]} (1 - 2ih)^2 \geq \tfrac{1}{2}([\tfrac{N}{4}] + 1),$$

since $1 - 2ih \geq \tfrac{1}{2}$ if $i \leq [\tfrac{N}{4}]$. Using the inequality $[\tfrac{N}{4}] + 1 > \tfrac{N}{4} = \tfrac{1}{4h}$, yields the bound from below $|u_h|_{1,\Omega} \geq \tfrac{1}{2\sqrt{2h}}$.

(2.ii) Now, we evaluate the L^2-projection of u'_h onto X_h. Let $f_h \in X_h$ be the projection in question, i.e., $(f_h, \varphi_i)_{0,\Omega} = a(u_h, \varphi_i)$ for all $1 \leq i \leq N$. Our objective is to bound $\|f_h\|_{0,\Omega}$ from above. By conventionally setting $U_{N+1} = U_N$, a simple computation shows that

$$1 \leq i \leq N, \quad \int_0^1 f_h\varphi_i = \int_0^1 u'_h\varphi_i = \tfrac{1}{2}(U_{i+1} - U_{i-1}) = \begin{cases} 0 & \text{if } i \text{ even,} \\ h & \text{if } i \text{ odd.} \end{cases}$$

(2.iii) Set $f_h = \sum_{i=1}^N F_i\varphi_i$. A bound on $\|f_h\|_{0,\Omega}$ is derived as follows:

$$\|f_h\|^2_{0,\Omega} = \sum_{i=1}^N \left(\int_0^1 f_h\varphi_i\right) F_i \leq h \sum_{i=1}^N |F_i| \leq h^{\frac{1}{2}} \left(\sum_{i=1}^N F_i^2\right)^{\frac{1}{2}}.$$

To conclude, the smallest eigenvalue of the matrix $\mathcal{M} = \left(\int_\Omega \varphi_i\varphi_j\right)_{1 \leq i,j \leq N}$ must be bounded from below. Owing to Theorem 9.8, there is a constant $c > 0$, independent of h, such that

$$\|w_h\|^2_{0,\Omega} = \sum_{i=1}^N \sum_{j=1}^N W_i W_j \int_0^1 \varphi_i\varphi_j = (W, \mathcal{M}W)_N \geq ch \sum_{i=1}^N W_i^2.$$

Actually, a direct computation shows that $c \geq \tfrac{1}{6}$; hence,

$$\|f_h\|^2_{0,\Omega} \leq h^{\frac{1}{2}} \left(\sum_{i=1}^N F_i^2\right)^{\frac{1}{2}} \leq h^{\frac{1}{2}} \left(\frac{6}{h}\right)^{\frac{1}{2}} \|f_h\|_{0,\Omega},$$

yielding the bound from above $\|f_h\|_{0,\Omega} \leq \sqrt{6}$.
(2.iv) Finally,

$$\sup_{v_h \in X_h} \frac{a(u_h, v_h)}{\|u_h\|_{1,\Omega}\|v_h\|_{0,\Omega}} \leq \frac{1}{|u_h|_{1,\Omega}} \sup_{v_h \in X_h} \frac{\int_0^1 f_h v_h}{\|v_h\|_{0,\Omega}} \leq \frac{\|f_h\|_{0,\Omega}}{|u_h|_{1,\Omega}} \leq 4\sqrt{3}h.$$

(3) The bound from below.

(3.i) Let $u_h = \sum_{i=1}^{N} U_i \varphi_i$ be a function in X_h. Set $U_0 = 0$, $U_{N+1} = U_N$, and for $1 \leq i \leq N$, set $G_i = a(u_h, \varphi_i)$. For all $i \in \{0, \ldots, \frac{N}{2} - 1\}$, $|U_{2i+2}| \leq |U_{2i}| + 2|G_{2i+1}|$ since $G_{2i+1} = \frac{1}{2}(U_{2i+2} - U_{2i})$, and, hence,

$$|U_{2i+2}| \leq |U_0| + 2\sum_{k=0}^{i} |G_{2k+1}| \leq 2\sum_{k=0}^{\frac{N}{2}-1} |G_{2k+1}|,$$

since $U_0 = 0$. Furthermore, for $1 \leq i \leq \frac{N}{2}$, it is clear that $|U_{2i-1}| \leq |U_{2i+1}| + 2|G_{2i}|$ since $G_{2i} = \frac{1}{2}(U_{2i+1} - U_{2i-1})$, and, hence,

$$|U_{2i-1}| \leq |U_{N+1}| + 2\sum_{k=i}^{\frac{N}{2}} |G_{2k}| \leq 2\sum_{k=0}^{\frac{N}{2}-1} |G_{2k+1}| + 2\sum_{k=i}^{\frac{N}{2}} |G_{2k}| \leq 2\sum_{k=1}^{N} |G_k|.$$

The above inequalities imply $\max_{1 \leq i \leq N} |U_i| \leq 2\sum_{k=1}^{N} |G_k|$.

(3.ii) Let $f_h \in X_h$ be the projection of u'_h onto X_h. Set $\Omega_i = [x_{i-1}, x_{i+1}]$ with $x_i = ih$ for $1 \leq i \leq N$, and extend f_h by zero on $[x_N, x_{N+1}]$. Then,

$$|G_i| = |(f_h, \varphi_i)_{0,\Omega}| \leq \|f_h\|_{0,\Omega_i} \|\varphi_i\|_{0,\Omega_i} \leq \left(\tfrac{2h}{3}\right)^{\frac{1}{2}} \|f_h\|_{0,\Omega_i}.$$

(3.iii) The seminorm $|u_h|_{1,\Omega}$ can now be estimated as follows:

$$|u_h|_{1,\Omega}^2 = \sum_{i=0}^{N-1} \int_{x_i}^{x_{i+1}} \frac{1}{h^2}(U_{i+1} - U_i)^2 \leq \frac{N}{h}\left(2 \max_{1 \leq i \leq N} |U_i|\right)^2 \leq \frac{N}{h}\left(4\sum_{k=1}^{N} |G_k|\right)^2$$

$$\leq 16Nh^{-1}N\sum_{k=1}^{N} |G_k|^2 \leq 16N^2h^{-1}\frac{2h}{3}\sum_{k=1}^{N} \|f_h\|_{0,\Omega_k}^2 \leq \frac{64}{3h^2}\|f_h\|_{0,\Omega}^2.$$

(3.iv) Finally, the following bound from below holds:

$$\sup_{v_h \in X_h} \frac{a(u_h, v_h)}{\|u_h\|_{1,\Omega}\|v_h\|_{0,\Omega}} = \frac{1}{\|u_h\|_{1,\Omega}} \sup_{v_h \in X_h} \frac{\int_0^1 f_h v_h}{\|v_h\|_{0,\Omega}} = \frac{\|f_h\|_{0,\Omega}}{\|u_h\|_{1,\Omega}}$$

$$\geq \sqrt{\tfrac{2}{3}} \frac{\|f_h\|_{0,\Omega}}{|u_h|_{1,\Omega}} \geq \sqrt{\tfrac{2}{3}} \tfrac{\sqrt{3}}{8} h \geq \tfrac{\sqrt{2}}{8} h. \qquad \square$$

Theorem 5.3 has very important consequences. It shows that the standard Galerkin approximation (5.5) to the advection problem (5.1) cannot produce optimal error estimates, even though it yields an invertible linear system ($c_1 \neq 0$). Indeed, Céa's Lemma yields the estimate

$$\|u - u_h\|_{1,\Omega} \leq \left(1 + \frac{\|a\|}{\alpha_h}\right) \inf_{w_h \in X_h} \|u - w_h\|_{1,\Omega}.$$

Theorem 5.3 implies $\alpha_h \sim h$. Hence, if u is in $H^2(\Omega)$, it is not reasonable to expect more than

$$\|u - u_h\|_{1,\Omega} \leq c\|u\|_{2,\Omega}.$$

If u is only in $H^1(\Omega)$, the error is not a priori bounded, i.e., the method may diverge in the H^1-norm! In practice, this problem manifests itself through the presence of spurious wiggles in the approximate solution; see Figure 5.2.

Remark 5.4.

(i) In the continuous problem (5.3), solution and test spaces are different, whereas in (5.5), we chose the same space to approximate the solution and to test the equation. Although this choice automatically ensures that the corresponding linear system has as many equations as unknowns, it is clearly not optimal. One of the reasons formulation (5.5) is not adequate is that distinct solution and test spaces have not been accounted for at the discrete level. This idea is further developed in §5.7.

(ii) The weak formulation of problem (5.5) yields a linear system that is identical to the one that would be produced by using centered finite differences. Hence, the negative conclusions that hold for finite elements extend to centered finite differences as well, i.e., centered finite differences are not suitable to approximate first-order PDEs. □

5.1.5 Numerical examples

To illustrate Theorem 5.3 and to show its consequences in higher space dimensions, consider the problem

$$\begin{cases} \partial_y u = -8\pi \sin(8\pi y) & \text{in } \Omega =]0,1[^2, \\ u_{|y=0} = 1. \end{cases} \tag{5.6}$$

The solution to (5.6) is simply $u(x,y) = \cos(8\pi y)$.

Let us approximate the solution to this problem by the standard Galerkin technique with $\mathbb{P}_1$ finite elements (resp., $\mathbb{P}_2$) on a mesh such that $h \approx \frac{1}{20}$ (resp., $h \approx \frac{1}{10}$). Figure 5.2 shows the isolines of the Lagrange interpolant of the exact solution and the isolines of the Galerkin approximation for the $\mathbb{P}_1$ and $\mathbb{P}_2$ approximations. It is clear that the solution is polluted by spurious oscillations. These oscillations are the consequence of the fact that the stability constant of the bilinear form goes to zero with h, as stated in Theorem 5.3.

5.2 First-Order PDEs in L^2

This section investigates first-order PDEs in a Hilbertian setting. The goal is to introduce an abstract counterpart to the one-dimensional problem (5.3).

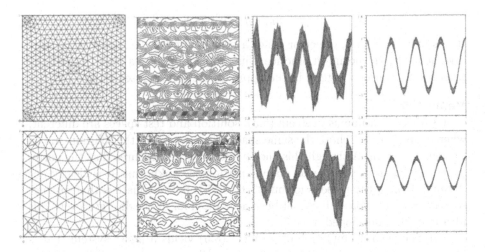

Fig. 5.2. Top: $\mathbb{P}_1$ Galerkin approximation. Bottom: $\mathbb{P}_2$ Galerkin approximation. From left to right: meshes; isolines of the approximate solution; graph of the approximate solution; and graph of the $\mathbb{P}_1$ Lagrange interpolant of the exact solution.

5.2.1 An abstract model problem

Let L be a real Hilbert space equipped with an inner product $(\cdot, \cdot)_L$. Owing to the Riesz–Fréchet Theorem, we identify L and its dual: $L \equiv L'$. Let $D(A)$ be a subspace of L and let

$$A : D(A) \subset L \longrightarrow L$$

be a linear operator of domain $D(A)$ whose graph $\bigcup_{v \in D(A)} (v, Av)$ is assumed to be closed in $L \times L$. This hypothesis means that, equipped with the *graph norm* $\|v\|_L + \|Av\|_L$, $D(A)$ is complete. Actually, equipped with the inner product $(u, v)_L + (Au, Av)_L$, $D(A)$ is a Hilbert space, and we hereafter denote it by V. Hence, $A \in \mathcal{L}(V; L)$. Henceforth, the reader unfamiliar with these notions may think of A as a first-order differential operator.

Example 5.5. For $1 \le i \le d$, the operator

$$\partial_i : V = \{v \in L^2(\Omega);\ \partial_i v \in L^2(\Omega)\} \subset L^2(\Omega) = L \longrightarrow L$$

satisfies the above hypotheses, i.e., V equipped with the norm $\|v\|_{0,\Omega} + \|\partial_i v\|_{0,\Omega}$ is a Hilbert space and $\partial_i \in \mathcal{L}(V; L^2(\Omega))$. $\qquad\square$

We assume that $A : V \to L$ is an isomorphism. Owing to the BNB Theorem, this hypothesis is equivalent to assuming that there is $\alpha > 0$ such that

$$\forall u \in V, \quad \|Au\|_L \ge \alpha \|u\|_V, \tag{5.6a}$$

$$\forall w \in L, \quad (\forall u \in V,\ (Au, w)_L = 0) \implies (w = 0). \tag{5.6b}$$

For $f \in L$, consider the abstract problem:

$$\begin{cases} \text{Seek } u \in V \text{ such that} \\ Au = f. \end{cases} \tag{5.7}$$

Problem (5.7) is well-posed owing to the above hypotheses. An equivalent viewpoint consists of introducing the bilinear form $a \in \mathcal{L}(V \times L; \mathbb{R})$ such that $a(u, w) = (A(u), w)_L$ for all $(u, w) \in V \times L$. This form is clearly continuous and satisfies the conditions (BNB1)–(BNB2) of the BNB Theorem. Problem (5.7) can be reformulated as follows: For $f \in L$,

$$\begin{cases} \text{Seek } u \in V \text{ such that} \\ a(u, w) = (f, w)_L, \quad \forall w \in L. \end{cases} \tag{5.8}$$

5.2.2 Friedrichs' systems

A wide range of first-order PDEs encountered in applications belong to the so-called class of Friedrichs' symmetric systems [Fre58].

Let m be a positive integer. Let $\mathcal{K}$, $\{\mathcal{A}^k\}_{1 \le k \le d}$ be a family of $(d+1)$ functions on Ω with values in $\mathbb{R}^{m,m}$. Assume that the Euclidean matrix norms of these matrix-valued fields together with that of $\sum_{k=1}^{d} \partial_k \mathcal{A}^k$ are in $L^\infty(\Omega)$. Define the $\mathbb{R}^m$-valued operators $A_1 = (A_1^1, \ldots, A_1^m)$ and $A_0 = (A_0^1, \ldots, A_0^m)$ such that

$$\forall i \in \{1, \ldots, m\}, \quad A_1^i u = \sum_{j=1}^{m} \sum_{k=1}^{d} \mathcal{A}_{ij}^k \frac{\partial u_j}{\partial x_k} \quad \text{and} \quad A_0^i u = \sum_{j=1}^{m} \mathcal{K}_{ij} u_j. \tag{5.9}$$

Define $A = A_1 + A_0$. Setting $L = [L^2(\Omega)]^m$ and defining the Hilbert space (equipped with the graph norm)

$$W = \{ u \in [L^2(\Omega)]^m ; A_1 u \in [L^2(\Omega)]^m \},$$

it is clear that $A \in \mathcal{L}(W; L)$. Denote the graph norm by $\|u\|_W = \|A_1 u\|_L + \|u\|_L$. Note that the zeroth-order part of A is already controlled by $\|u\|_L$.

To guarantee that A is an isomorphism, additional (sufficient) hypotheses must be made. Let $n = (n_1, \ldots, n_d)$ be the outward normal to $\partial\Omega$ and set

$$\mathcal{D} = \sum_{k=1}^{d} n_k \mathcal{A}^k.$$

Owing to the regularity assumptions on $\{\mathcal{A}^k\}_{1 \le k \le d}$, $\mathcal{D}$ is in $[L^\infty(\partial\Omega)]^{m \times m}$. Assume that there is a matrix-valued field $\mathcal{M} \in [L^\infty(\partial\Omega)]^{m \times m}$ and a positive constant μ_0 such that the following hypotheses hold:

(F1) $\mathcal{A}^k$ is symmetric for $k = 1, \ldots, d$.
(F2) $\mathcal{K} + \mathcal{K}^T - \sum_{k=1}^{d} \partial_{x_k} \mathcal{A}^k \ge 2\mu_0 \mathcal{I}$ a.e. on Ω.
(F3) $\mathcal{M} + \mathcal{M}^T \ge 0$ a.e. on $\partial\Omega$.
(F4) $\text{Ker}(\mathcal{D} - \mathcal{M}) + \text{Ker}(\mathcal{D} + \mathcal{M}) = \mathbb{R}^m$ a.e. on $\partial\Omega$.

Set $\mathcal{V} = \{v \in [\mathcal{C}^\infty(\overline{\Omega})]^m; \ (\mathcal{M} - \mathcal{D})v_{|\partial\Omega} = 0\}$ and let V be the closure of $\mathcal{V}$ in W. For any $u \in W$, the quantity $\mathcal{D}u_{|\partial\Omega}$ is meaningful in the sense that the bilinear form

$$[\mathcal{C}^\infty(\overline{\Omega})]^m \times [\mathcal{C}^\infty(\overline{\Omega})]^m \ni (u,v) \longmapsto \int_{\partial\Omega} v^T \mathcal{D}u \in \mathbb{R},$$

extends continuously to $W \times W$ (for simplicity, the extended bilinear form is still denoted by an integral sign). As a result, the space V can be interpreted as follows:

$$V = \{u \in [L^2(\Omega)]^m; \ Au \in [L^2(\Omega)]^m; \ (\mathcal{M} - \mathcal{D})u_{|\partial\Omega} = 0\}. \tag{5.10}$$

A first important consequence of the above setting is the following:

Lemma 5.6. *Assume* (F1)–(F3). *Then, for all* $u \in L$,

$$\int_\Omega u^T Au \geq \mu_0 \|u\|_L^2 + \int_{\partial\Omega} \tfrac{1}{2} u^T \mathcal{D}u. \tag{5.11}$$

As a result, A is L-coercive on V.

Proof. Let $u \in L$. Owing to the symmetry property (F1), $u^T \mathcal{A}^k \partial_{x_k} u = \tfrac{1}{2}\mathcal{A}^k{:}\partial_{x_k}(u \otimes u)$. Therefore,

$$\int_\Omega u^T Au = \int_\Omega \tfrac{1}{2}\sum_{k=1}^d \partial_{x_k}(\mathcal{A}^k{:}(u \otimes u)) + \int_\Omega u^T \left(\mathcal{K} - \tfrac{1}{2}\sum_{k=1}^d \partial_{x_k}\mathcal{A}^k\right) u$$

$$= \int_{\partial\Omega} \tfrac{1}{2}\sum_{k=1}^d n_k \mathcal{A}^k{:}(u \otimes u) + \int_\Omega u^T \left(\tfrac{1}{2}(\mathcal{K} + \mathcal{K}^T) - \tfrac{1}{2}\sum_{k=1}^d \partial_{x_k}\mathcal{A}^k\right) u$$

$$\geq \int_{\partial\Omega} \tfrac{1}{2}\mathcal{D}{:}(u \otimes u) + \mu_0\|u\|_{0,\Omega}^2 = \int_{\partial\Omega} \tfrac{1}{2}u^T \mathcal{D}u + \mu_0\|u\|_L^2,$$

where the inequality results from (F2). This proves (5.11). Furthermore, $u \in V$ implies $(\mathcal{M} - \mathcal{D})u_{|\partial\Omega} = 0$, which, owing to (F3), yields

$$\int_\Omega u^T Au \geq \int_{\partial\Omega} \tfrac{1}{4}u^T(\mathcal{M} + \mathcal{M}^T)u + \mu_0\|u\|_L^2 \geq \mu_0\|u\|_L^2. \qquad \square$$

Theorem 5.7 (Friedrichs). *Assume* (F1)–(F4). *Then, $A : V \to L$ is an isomorphism.*

Proof. (1) Let us prove (5.6a). Let $u \in V$. Then, Lemma 5.6 implies $\sup_{v \in L} \frac{(Au,v)_L}{\|v\|_L} \geq \mu_0 \|u\|_L$. Furthermore,

$$\sup_{v \in L} \frac{(Au,v)_L}{\|v\|_L} \geq \sup_{v \in L} \frac{(A_1 u,v)_L}{\|v\|_L} - \|\mathcal{K}\|_{[L^\infty(\Omega)]^{m \times m}} \|u\|_L$$

$$\geq \|A_1 u\|_L - \frac{\|\mathcal{K}\|_{[L^\infty(\Omega)]^{m \times m}}}{\mu_0} \sup_{v \in L} \frac{(Au,v)_L}{\|v\|_L}.$$

Hence,

$$\left(1 + \frac{1 + \|\mathcal{K}\|_{[L^\infty(\Omega)]^{m \times m}}}{\mu_0}\right) \sup_{v \in L} \frac{(Au, v)_L}{\|v\|_L} \geq \|A_1 u\|_L + \|u\|_L \geq \|u\|_W.$$

(2) Let us prove (5.6b). Assume that $v \in L$ is such that $(Au, v)_L = 0$ for all $u \in V$. A standard distribution argument shows that

$$-\sum_{k=1}^{d} \partial_{x_k}((A^k)^T v) + \mathcal{K}^T v = 0, \tag{5.12}$$

i.e., $v \in W$. Furthermore, $\int_{\partial\Omega} u^T \mathcal{D}^T v = 0$ for all $u \in V$, which means $\mathcal{D}^T v \perp \text{Ker}(\mathcal{D} - \mathcal{M})$ on $\partial\Omega$. Owing to hypothesis (F4), the following decomposition holds: $v_{|\partial\Omega} = v_{\text{Ker}(\mathcal{D}-\mathcal{M})} + v_{\text{Ker}(\mathcal{D}+\mathcal{M})}$. Hence, on $\partial\Omega$,

$$\begin{aligned}
v^T(\mathcal{M} + \mathcal{D})v &= v^T(\mathcal{D} + \mathcal{M})(v_{\text{Ker}(\mathcal{D}-\mathcal{M})} + v_{\text{Ker}(\mathcal{D}+\mathcal{M})}) \\
&= v^T(\mathcal{D} + \mathcal{M})v_{\text{Ker}(\mathcal{D}-\mathcal{M})} = v^T(\mathcal{D} - \mathcal{M})v_{\text{Ker}(\mathcal{D}-\mathcal{M})} + 2v^T \mathcal{M} v_{\text{Ker}(\mathcal{D}-\mathcal{M})} \\
&= 2v^T \mathcal{D} v_{\text{Ker}(\mathcal{D}-\mathcal{M})} = 2v_{\text{Ker}(\mathcal{D}-\mathcal{M})}^T \mathcal{D}^T v = 0,
\end{aligned}$$

the last equality resulting from the fact that $\mathcal{D}^T v \perp \text{Ker}(\mathcal{D} - \mathcal{M})$. Taking the L-scalar product of (5.12) with v yields

$$(Av, v)_L - \int_{\partial\Omega} v^T \mathcal{D} v = 0.$$

Lemma 5.6 implies

$$0 \geq \mu_0 \|v\|_L^2 - \tfrac{1}{2} \int_{\partial\Omega} v^T \mathcal{D} v,$$

and owing to (F3) and the boundary condition satisfied by v,

$$\begin{aligned}
0 &\geq \mu_0 \|v\|_L^2 - \tfrac{1}{2} \int_{\partial\Omega} v^T(\mathcal{D} + \mathcal{M})v + \tfrac{1}{2} \int_{\partial\Omega} v^T \mathcal{M} v \\
&= \mu_0 \|v\|_L^2 + \tfrac{1}{4} \int_{\partial\Omega} v^T(\mathcal{M} + \mathcal{M}^T)v \geq \mu_0 \|v\|_L^2,
\end{aligned}$$

i.e., v is zero. $\qquad\square$

It is possible to weaken assumption (F2) as follows: Set

$$V^* = \{v \in W; \ \mathcal{D}v_{|\partial\Omega} \perp \text{Ker}(\mathcal{D} - \mathcal{M})\},$$

and

$$A^* : V^* \ni v \longmapsto -\sum_{j=1}^{d} \partial_{x_k}(A^k v) + \mathcal{K}^T v \in L.$$

Note that (5.12) can be rewritten $v \in V^*$ and $A^* v = 0$. Assume that there is an orthogonal projection operator $\mathcal{P}$ (i.e., $\mathcal{P} = \mathcal{P}^T$, $\mathcal{P}^2 = \mathcal{P}$) and constants $\alpha > 0$, $\gamma > 0$, and λ such that:

(F2a) $\mathcal{K} + \mathcal{K}^T - \sum_{k=1}^{d} \partial_{x_k} A^k \geq 2\mu_0 \mathcal{P}$ a.e. on Ω.

(F2b) $\sup_{v \in L} \frac{(Au,v)_L}{\|v\|_L} \geq \alpha \|(\mathcal{I} - \mathcal{P})u\|_L - \lambda \|\mathcal{P}u\|_L$ for all u in V.

(F2c) $\|\mathcal{P}v\|_L \geq \gamma \|(\mathcal{I} - \mathcal{P})v\|_L$ for all v in Ker A^*.

Corollary 5.8. *If* (F2) *is replaced by* (F2a)–(F2c), A *is still an isomorphism.*

Proof. (1) Let $u \in V$. Following the proof of Lemma 5.6, using (F1), (F3), (F2a), and the fact that $\mathcal{P}$ is an orthogonal projection, we infer $(Au, u)_L \geq \mu_0 \|\mathcal{P}u\|_L^2$.

(2) Let us prove (5.6a). Step 1, the triangle inequality, and (F2b) imply

$$\mu_0 \|\mathcal{P}u\|_L^2 \leq \frac{(Au,u)_L}{\|u\|_L} (\|\mathcal{P}u\|_L + \|(\mathcal{I} - \mathcal{P})u\|_L)$$

$$\leq \sup_{v \in L} \frac{(Au,v)_L}{\|v\|_L} \left((1 + \tfrac{\lambda}{\alpha}) \|\mathcal{P}u\|_L + \tfrac{1}{\alpha} \sup_{v \in L} \frac{(Au,v)_L}{\|v\|_L} \right).$$

The arithmetic-geometric inequality $ab \leq \frac{\mu_0}{2} a^2 + \frac{1}{2\mu_0} b^2$, valid for all $\mu_0 > 0$, implies

$$\left(\frac{\mu_0}{2} \left(\frac{1}{2\mu_0} (1 + \tfrac{\lambda}{\alpha})^2 + \tfrac{1}{\alpha} \right)^{-1} \right)^{\frac{1}{2}} \|\mathcal{P}u\|_L \leq \sup_{v \in L} \frac{(Au,v)_L}{\|v\|_L}.$$

Owing to (F2b), $\sup_{v \in L} \frac{(Au,v)_L}{\|v\|_L} \geq c \|u\|_L$. The rest of the proof is then identical to that of step 1 in the proof of Theorem 5.7.

(3) Let us prove (5.6b). Assume that $v \in L$ is such that $(Au, v)_L = 0$ for all $u \in V$. Following step 2 in the proof of Theorem 5.7, we infer $v \in V^*$ and $A^*v = 0$, and replacing $\mu_0 \|v\|_L^2$ by $\mu_0 \|\mathcal{P}v\|_L^2$ leads to $\|\mathcal{P}v\|_L = 0$. Then, (F2c) yields $v = 0$, completing the proof. $\qquad\square$

5.2.3 Example 1: Advection–reaction

Let β be a vector field in $\mathbb{R}^d$, assume $\beta \in [L^\infty(\Omega)]^d$, $\nabla \cdot \beta \in L^\infty(\Omega)$, and define

$$\partial\Omega^- = \{x \in \partial\Omega;\ \beta(x) \cdot n(x) < 0\}, \qquad \partial\Omega^+ = \{x \in \partial\Omega;\ \beta(x) \cdot n(x) > 0\}.$$

$\partial\Omega^-$ is called the *inflow boundary* and $\partial\Omega^+$ the *outflow boundary*; see Figure 5.3. Note that it is possible that these two subsets of $\partial\Omega$ are empty (think of a vector field β such that $\beta \cdot n(x) = 0$ for all $x \in \partial\Omega$).

Let μ be a function in $L^\infty(\Omega)$, and consider the problem

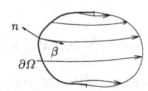

Fig. 5.3. Advection field β in a domain Ω with inflow boundary $\partial\Omega^-$.

$$\begin{cases} \mu u + \beta \cdot \nabla u = f, \\ u_{|\partial\Omega^-} = 0. \end{cases} \tag{5.13}$$

This is the so-called *advection–reaction equation*. It models the transport of the quantity u by the flow β. To give a mathematical meaning to (5.13), consider the space

$$W = \{w \in L^2(\Omega); \beta \cdot \nabla w \in L^2(\Omega)\} \subset L^2(\Omega).$$

Equipped with the norm $\|w\|_W = \|w\|_{0,\Omega} + \|\beta \cdot \nabla w\|_{0,\Omega}$, W is a Hilbert space. Now, define the differential operator

$$A : W \ni u \longmapsto \mu u + \beta \cdot \nabla u \in L^2(\Omega).$$

It is clear that A is continuous. Without additional hypotheses on μ and β, A is unlikely to be an isomorphism (think of $\partial\Omega^- = \partial\Omega^+ = \emptyset$ and $\mu = 0$).

Proposition 5.9. *Assume there is $\mu_0 > 0$ such that*

$$\mu(x) - \tfrac{1}{2}\nabla \cdot \beta(x) \geq \mu_0 > 0 \quad a.e. \ in \ \Omega, \tag{5.14}$$

and let $V = \{w \in W; w_{|\partial\Omega^-} = 0\}$. Then, $A : V \to L^2(\Omega)$ is an isomorphism.

Proof. Let us prove that A is a Friedrichs' operator, i.e., that assumptions (F1)–(F4) hold. Indeed, set $m = 1$, $\mathcal{A}^k = \beta_k$ for $1 \leq k \leq d$, $\mathcal{K} = \mu$, $\mathcal{M} = |\beta \cdot n|$, and $\mathcal{D} = \beta \cdot n$. Since $m = 1$, the symmetry of $\mathcal{A}^k$ is evident, proving (F1). Hypothesis (5.14) amounts to (F2). $\mathcal{M}$ is clearly positive on $\partial\Omega$, yielding (F3). Finally, $\mathcal{D} - \mathcal{M} = 0$ on $\partial\Omega\backslash\partial\Omega^-$ and $\mathcal{D} + \mathcal{M} = 0$ on $\partial\Omega\backslash\partial\Omega^+$. Hence, (F4) holds. Since $\mathcal{D} - \mathcal{M}$ is non-zero only on $\partial\Omega^-$, the space V defined in (5.10) is $V = \{w \in W; w_{|\partial\Omega^-} = 0\}$. Conclude using Theorem 5.7. $\square$

Remark 5.10.
    (i) See also Exercise 5.3 for a direct proof of Proposition 5.9.
    (ii) If $\mu = 0$ and $\nabla \cdot \beta = 0$, hypothesis (5.14) is not satisfied. However, the conclusions of Proposition 5.9 still hold if β is a filling field, i.e., if for almost every x in Ω, there is a characteristic line of the vector field β that starts from $\partial\Omega^-$ and reaches x in finite time; see, e.g., [Aze95]. $\square$

5.2.4 Example 2: Darcy's equations

Let Ω be a porous medium characterized by a symmetric positive definite permeability tensor S. Assume that the smallest (resp., largest) eigenvalue of S^{-1} is uniformly bounded from below (resp., from above) in Ω by $\sigma_* > 0$ (resp., by $\sigma^* < \infty$). Let $\partial\Omega = \partial\Omega_1 \cup \partial\Omega_2$ be a partition of $\partial\Omega$, and consider the problem

$$\begin{cases} S^{-1}{\cdot}u + \nabla p = f & \text{a.e. in } \Omega, \\ \nabla{\cdot}u = g & \text{a.e. in } \Omega, \\ u{\cdot}n = 0 & \text{a.e. on } \partial\Omega_1, \\ p = 0 & \text{a.e. on } \partial\Omega_2. \end{cases} \qquad (5.15)$$

The first PDE in (5.15) is Darcy's phenomenological law relating the pressure gradient ∇p to the velocity u. The second PDE is the mass conservation equation. Nonlinear variants of (5.15) are used in hydrogeology and in the oil industry to model underground storage problems, flows in porous media, or oil recovery.

One way to reformulate problem (5.15) consists of multiplying the first equation by S and applying the divergence operator to obtain

$$\begin{cases} \nabla{\cdot}(S{\cdot}\nabla p) = \nabla{\cdot}(S{\cdot}f) - g, & \text{a.e. in } \Omega, \\ n{\cdot}S{\cdot}\nabla p_{|\partial\Omega_1} = n{\cdot}S{\cdot}f, \quad p_{|\partial\Omega_2} = 0. \end{cases}$$

In this form, the problem falls into the framework of the elliptic problems developed in Chapter 3. Although this approach is quite simple, it is not systematically used in practice, since quite often the quantity $u = -S{\cdot}\nabla p$ is physically important. It is then useful to approximate it directly without having to differentiate the quantity p. For instance, the field u may transport chemical species, a polluting compound, or any other passive scalar, so that (5.15) is coupled with an advection equation of the form $\partial_t c + \nabla{\cdot}(uc) = s(c)$.

To derive a weak formulation of (5.15), introduce the spaces

$$\begin{aligned} X &= \{v \in [L^2(\Omega)]^d;\ \nabla{\cdot}v \in L^2(\Omega);\ v{\cdot}n_{|\partial\Omega_1} = 0\}, \\ Y &= \{q \in L^2(\Omega);\ \nabla q \in [L^2(\Omega)]^d;\ q_{|\partial\Omega_2} = 0\}, \end{aligned} \qquad (5.16)$$

with corresponding norms $\|v\|_X = \|v\|_{0,\Omega} + \|\nabla{\cdot}v\|_{0,\Omega}$ and $\|q\|_Y = \|q\|_{1,\Omega}$. X and Y are clearly Hilbert spaces. Set $V = X \times Y$ and $L = [L^2(\Omega)]^d \times L^2(\Omega)$, and equip these spaces with the norms $\|(v,q)\|_V = \|v\|_X + \|q\|_Y$ and $\|(v,q)\|_L = \|v\|_{0,\Omega} + \|q\|_{0,\Omega}$, respectively. Now, define

$$A : V \ni (v,q) \longmapsto (S^{-1}v + \nabla q, \nabla{\cdot}v) \in L.$$

Proposition 5.11. *Assume* $\mathrm{meas}(\partial\Omega_2) > 0$. *Then,* A *is an isomorphism.*

Proof. We apply Corollary 5.8. Let χ_1 and χ_2 be the characteristic functions of $\partial\Omega_1$ and $\partial\Omega_2$. Set $\chi = \chi_1 - \chi_2$, $m = d + 1$, and

$$\mathcal{A}^1 = \begin{bmatrix} 0 & \cdots\cdots & 0 & 1 \\ 0 & \cdots\cdots & 0 & 0 \\ \vdots & & \vdots & \vdots \\ 0 & \cdots\cdots & 0 & 0 \\ 1 & 0 & \cdots & 0 & 0 \end{bmatrix}, \quad \cdots, \quad \mathcal{A}^d = \begin{bmatrix} 0 & \cdots\cdots & 0 & 0 \\ \vdots & & \vdots & \vdots \\ 0 & \cdots\cdots & 0 & 0 \\ 0 & \cdots\cdots & 0 & 1 \\ 0 & \cdots & 0 & 1 & 0 \end{bmatrix}, \quad \mathcal{K} = \begin{bmatrix} S^{-1} & 0 \\ 0 & 0 \end{bmatrix},$$

$$
\mathcal{D} = \begin{bmatrix} 0 & \cdots & 0 & n_1 \\ \vdots & & \vdots & \vdots \\ 0 & \cdots & 0 & n_d \\ \hline n_1 & \cdots & n_d & 0 \end{bmatrix}, \quad \mathcal{M} = \begin{bmatrix} 0 & \cdots & 0 & \chi n_1 \\ \vdots & & \vdots & \vdots \\ 0 & \cdots & 0 & \chi n_d \\ \hline -\chi n_1 & \cdots & -\chi n_d & 0 \end{bmatrix}, \quad \mathcal{P} = \begin{bmatrix} \mathcal{I} & 0 \\ 0 & 0 \end{bmatrix}.
$$

Clearly, $\mathcal{A}^1, \ldots, \mathcal{A}^d$ are symmetric, $\mathcal{K} + \mathcal{K}^T + \sum_{k=1}^{d} \partial_{x_k} \mathcal{A}^k \geq 2\sigma_* \mathcal{P}$, and $\mathcal{M} + \mathcal{M}^T = 0$ is non-negative. Hence, (F1), (F2a), and (F3) are verified. Furthermore, since $(\mathcal{D}+\mathcal{M})(u,p)^T = 2(\chi_1 pn, \chi_2 u \cdot n)^T$ and $(\mathcal{D}-\mathcal{M})(u,p)^T = 2(\chi_2 pn, \chi_1 u \cdot n)^T$, it is clear that $(\chi_1 u, \chi_2 p)^T \in \mathrm{Ker}(\mathcal{D}+\mathcal{M})$ and $(\chi_2 u, \chi_1 p)^T \in \mathrm{Ker}(\mathcal{D} - \mathcal{M})$; hence, (F4) results from the unique decomposition $(u,p)^T = (\chi_1 u, \chi_2 p)^T + (\chi_2 u, \chi_1 p)^T$. The last technical point consists of verifying (F2b) and (F2c). Since $\mathrm{meas}(\partial\Omega_2) > 0$, Lemma B.66 implies that there exists $c > 0$ such that $\|\nabla p\|_{0,\Omega} \geq c \|p\|_{0,\Omega}$ for all $p \in Y$. This yields

$$
\sup_{(v,q)\in L} \frac{(A(u,p),(v,q))_L}{\|(v,q)\|_L} \geq \sup_{v\in[L^2(\Omega)]^d} \frac{(A(u,p),(v,0))_L}{\|v\|_{0,\Omega}}
$$

$$
\geq \sup_{v\in[L^2(\Omega)]^d} \frac{(\nabla p, v)_{0,\Omega}}{\|v\|_{0,\Omega}} - \sigma^* \|u\|_{0,\Omega}
$$

$$
\geq c \|p\|_{0,\Omega} - \sigma^* \|\mathcal{P}(u,p)\|_L
$$

$$
\geq c \|(\mathcal{I} - \mathcal{P})(u,p)\|_L - \sigma^* \|\mathcal{P}(u,p)\|_L,
$$

since $\|(\mathcal{I} - \mathcal{P})(u,p)\|_L = \|p\|_{0,\Omega}$ and $\|\mathcal{P}(u,p)\|_L = \|u\|_{0,\Omega}$. This proves (F2b) with $\alpha = c$ and $\lambda = \sigma^*$. Finally, noting that $A^*(v,q) = (\mathcal{S}^{-1}v - \nabla q, -\nabla \cdot v)$, $(v,q) \in \mathrm{Ker}\, A^*$ implies $\nabla q = \mathcal{S}^{-1}v$ and $q_{|\partial\Omega_2} = 0$. Using again Lemma B.66 yields $\|\mathcal{P}(v,q)\|_L \geq \frac{c}{\sigma^*} \|(\mathcal{I}-\mathcal{P})(v,q)\|_L$ for all $(v,q) \in \mathrm{Ker}\, A^*$, i.e., (F2c) holds with $\gamma = \frac{c}{\sigma^*}$. □

Remark 5.12. See Exercise 5.4 for a direct proof of Proposition 5.11. □

5.2.5 Example 3: Maxwell's equations

We consider a simplified version of Maxwell's equations. Assume that the electromagnetic field is time-harmonic with angular velocity ω and frequency ν. In the low-frequency regime (typically with ν lower than 1 MHz), the displacement currents are negligible, and one can consider the quasi-static approximation to Maxwell's equations. Let $\partial\Omega = \partial\Omega_1 \cup \partial\Omega_2$ be a partition of the boundary of a domain Ω in $\mathbb{R}^3$ and consider the problem

$$
\begin{cases}
E - \nabla \times B = f & \text{a.e. in } \Omega, \\
-i\omega B + \nabla \times E = g & \text{a.e. in } \Omega, \\
E \times n = 0 & \text{a.e. on } \partial\Omega_1, \\
B \times n = 0 & \text{a.e. on } \partial\Omega_2,
\end{cases}
\tag{5.17}
$$

where $i^2 = -1$. To formulate (5.17) in a weak sense, introduce the spaces

$$X = \{v \in [L^2(\Omega)]^3; \nabla \times v \in [L^2(\Omega)]^3; v \times n_{|\partial\Omega_1} = 0\},$$
$$Y = \{v \in [L^2(\Omega)]^3; \nabla \times v \in [L^2(\Omega)]^3; v \times n_{|\partial\Omega_2} = 0\}, \tag{5.18}$$

where, this time, $L^2(\Omega)$ is a $\mathbb{C}$-vector space of complex-valued functions equipped with the inner product

$$(f,g)_{0,\Omega} = \Re\left(\int_\Omega f\bar{g}\right),$$

where $\bar{g}$ denotes the conjugate of g and $\Re$ the real part. Equipped with the norms $\|v\|_X = \|v\|_Y = \|v\|_{0,\Omega} + \|\nabla \times v\|_{0,\Omega}$, X and Y are Hilbert spaces. Let $V = X \times Y$ and $L = [L^2(\Omega)]^3 \times [L^2(\Omega)]^3$ be equipped with the norms $\|(e,b)\|_V = \|e\|_X + \|b\|_Y$ and $\|(e,b)\|_L = \|e\|_{0,\Omega} + \|b\|_{0,\Omega}$. Define

$$A : V \ni (e,b) \longmapsto (e - \nabla \times b, -i\omega b + \nabla \times e) \in L.$$

This operator is continuous by construction. Finally, define the bilinear form $a \in \mathcal{L}(V \times L; \mathbb{R})$ by $a((E,B),(e,b)) = (A(E,B),(e,b))_L$. Assuming $(f,g) \in L$, the weak form of (5.17) is:

$$\begin{cases} \text{Seek } (E,B) \in V \text{ such that} \\ a((E,B),(e,b)) = ((f,g),(e,b))_L, \quad \forall(e,b) \in L. \end{cases} \tag{5.19}$$

Proposition 5.13. *Problem* (5.19) *is well-posed.*

Proof. Let us verify the hypotheses of the BNB Theorem. Using (B.21), the boundary conditions on E and B, and choosing the test functions $e = (1 + i)E - (1 - i)\nabla \times B$, $b = (1 - i)B + \frac{1-i}{\omega}\nabla \times E$, yields

$$a((E,B),(e,b)) = \|E\|_{0,\Omega}^2 + \omega\|B\|_{0,\Omega}^2 + \|\nabla \times B\|_{0,\Omega}^2 + \frac{1}{\omega}\|\nabla \times E\|_{0,\Omega}^2$$
$$\geq \min\left(1, \omega, \frac{1}{\omega}\right)\|(E,B)\|_V^2 \geq c\|(E,B)\|_V\|(e,b)\|_L.$$

This shows that (BNB1) holds. The proof of (BNB2) is left as an exercise. □

Remark 5.14.

(i) One can prove that A is a Friedrichs operator, i.e., that it satisfies assumptions (F1) to (F4).

(ii) It is also possible to consider a two-dimensional setting for Maxwell's equations. In this case, B is a $\mathbb{C}^2$-valued field and E is a $\mathbb{C}$-valued field; see, e.g., Exercise 4.9. □

5.3 Least-Squares Formulation and Variants

Section 5.1 has shown that the standard Galerkin method is not suited to approximate first-order PDEs. We now reformulate the problem in order to obtain a bilinear form with satisfactory stability properties.

5.3.1 Principle of the Least-Squares method

Let us take inspiration from linear algebra. Consider the linear system $\mathcal{A}U = F$ in $\mathbb{R}^N$, where $\mathcal{A}$ is an arbitrary square invertible matrix. Multiplying this system by $\mathcal{A}^T$ yields $\mathcal{A}^T\mathcal{A}U = \mathcal{A}^T F$. It is clear that the matrix $\mathcal{A}^T\mathcal{A}$ is symmetric positive definite. Hence, after multiplication by $\mathcal{A}^T$, a linear system with no particular property is transformed into a symmetric positive definite system. The new linear system can be recast in the following weak form:

$$\begin{cases} \text{Seek } U \text{ in } \mathbb{R}^N \text{ such that} \\ (\mathcal{A}U, \mathcal{A}V)_N = (\mathcal{A}V, F)_N, \quad \forall V \in \mathbb{R}^N. \end{cases} \tag{5.20}$$

This technique is usually called the Least-Squares formulation owing to the following:

Proposition 5.15. *Consider the functional* $J(V) = \frac{1}{2}(\mathcal{A}V, \mathcal{A}V)_N - (\mathcal{A}V, F)_N$. *Then, problem (5.20) is equivalent to the following optimization problem:*

$$\begin{cases} \text{Seek } U \text{ in } \mathbb{R}^N \text{ such that} \\ J(U) = \inf_{V \in \mathbb{R}^N} J(V). \end{cases} \tag{5.21}$$

Proof. Adapt the proof of Theorem 2.4. $\qquad\qquad\qquad\qquad\qquad\qquad\square$

Remark 5.16. Since $\frac{1}{2}\|\mathcal{A}V - F\|_N^2 = J(V) + \frac{1}{2}\|F\|_N^2$, problem (5.21) is equivalent to minimizing $\|\mathcal{A}V - F\|_N$ over $\mathbb{R}^N$. $\qquad\qquad\qquad\qquad\square$

5.3.2 Application to the one-dimensional model problem

Within the framework of problem (5.1), the role of the matrix $\mathcal{A}$ is played by the operator $A : X \ni u \to u' \in L^2(\Omega)$ with X defined in (5.2). By analogy with (5.20), define

$$\forall (u, v) \in X \times X, \quad \tilde{a}(u, v) = (u', v')_{0,\Omega},$$

and consider the following problem: For f in $L^2(\Omega)$,

$$\begin{cases} \text{Seek } u \text{ in } X \text{ such that} \\ \tilde{a}(u, v) = (f, v')_{0,\Omega}, \quad \forall v \in X. \end{cases} \tag{5.22}$$

The equivalence between the original problem (5.3) and the Least-Squares problem (5.22) is guaranteed by the following:

Proposition 5.17. u *solves* (5.3) *if and only if* u *solves* (5.22).

Proof. From Theorem 5.2, it follows that the operator $A : X \ni u \to u' \in L^2(\Omega)$ is an isomorphism. As a consequence, test functions spanning $L^2(\Omega)$ in (5.3) are in one-to-one correspondence with the derivatives of the functions in X that are used as test functions in (5.22). $\qquad\qquad\qquad\square$

Remark 5.18. Formally, problem (5.22) is equivalent to

$$\begin{cases} u''(x) = f'(x) & \text{in }]0,1[, \\ u(0) = 0, \quad u'(1) = f(1). \end{cases} \qquad \square$$

Theorem 5.19. *Problem* (5.22) *has a unique solution.*

Proof. It is clear that $\tilde{a}$ is continuous on $X \times X$, and Poincaré's Lemma implies coercivity: $\tilde{a}(u,u) \geq \frac{2}{3}\|u\|_{1,\Omega}^2$. Furthermore, the linear form $v \mapsto \int_0^1 fv'$ is continuous on X. The Lax–Milgram Lemma yields the conclusion. $\qquad \square$

In the light of the above proof, the interest of the new formulation becomes evident: it brings into play a symmetric coercive bilinear form. As a consequence, the finite element approximation theory which has been developed for this setting applies without restriction; see Chapter 3. In particular, setting $X_h = \{v_h \in \mathcal{C}^0(\overline{\Omega}); \forall i \in \{0, \dots, N-1\}, v_{h|[x_i, x_{i+1}]} \in \mathbb{P}_k; v_h(0) = 0\}$, we can consider the approximate problem:

$$\begin{cases} \text{Seek } u_h \in X_h \text{ such that} \\ \tilde{a}(u_h, v_h) = (f, v_h')_{0,\Omega}, \quad \forall v_h \in X_h. \end{cases} \qquad (5.23)$$

As a direct consequence of Céa's Lemma, we state the following:

Proposition 5.20. *Let* u *solve* (5.22) *and let* u_h *solve* (5.23). *If* $f \in H^k(\Omega)$ *for* $k \geq 1$, *then*

$$\forall h, \quad \|u - u_h\|_{1,\Omega} \leq c\, h^k \|f\|_{k,\Omega}.$$

The error analysis can be extended slightly further by using the Aubin–Nitsche duality argument.

Proposition 5.21. *Under the hypotheses of Proposition* 5.20,

$$\forall h, \quad \|u - u_h\|_{0,\Omega} \leq c\, h^{k+1} \|f\|_{k,\Omega}.$$

Proof. Apply the Aubin–Nitsche Lemma with $L = L^2(\Omega)$, $l(\cdot, \cdot) = (\cdot, \cdot)_{0,\Omega}$, and $Z = H^2(\Omega) \cap X$. Since, for all $g \in L^2(\Omega)$, the solution to the dual problem $\int_0^1 \varsigma'(g)v' = \int_0^1 gv$ satisfies $\varsigma''(g) = -g$, we infer $\|\varsigma(g)\|_{2,\Omega} \leq c\|g\|_{0,\Omega}$, and, hence, assumption (AN1) holds. Furthermore, assumption (AN2) is a direct consequence of Corollary 1.109. $\qquad \square$

Remark 5.22. Although Proposition 5.20 generalizes to all dimensions and to abstract settings, Proposition 5.21 holds in one dimension only since assumption (AN1) is generally false in higher dimensions. $\qquad \square$

5.3.3 Generalization of the Least-Squares method

When A is a first-order differential operator, the standard Galerkin approximation to problem (5.8) is not optimal in general, since the discrete inf-sup inequality (BNB1$_h$) has a stability constant α_h that goes to 0 with h; see Theorem 5.3. To reformulate the problem in the Least-Squares sense, define

$$\tilde{a}(u, v) = (Au, Av)_L$$

and consider the problem:

$$\begin{cases} \text{Seek } u \in V \text{ such that} \\ \tilde{a}(u, w) = (f, Aw)_L, \quad \forall w \in V. \end{cases} \tag{5.24}$$

We assume that problem (5.8) is well-posed. Recall that this implies the existence of $\alpha > 0$ such that, for all $u \in V$, $\|Au\|_L \geq \alpha\|u\|_V$.

Proposition 5.23.

(i) *Problem (5.24) is well-posed.*

(ii) *Problems (5.8) and (5.24) are equivalent.*

Proof. (1) Observe that the bilinear form $\tilde{a}$ is continuous with $\|\tilde{a}\| \leq \|A\|^2$. Furthermore, from the inequality $\tilde{a}(u, u) = \|Au\|_L^2 \geq \alpha^2\|u\|_V^2$, it is clear that $\tilde{a}$ is coercive. Since $f \in L$, the linear form $(f, A\cdot)_L$ is continuous on V. The Lax–Milgram Lemma yields the conclusion.

(2) Let $u \in V$. If u solves (5.8), u clearly solves (5.24). Conversely, assume that u solves (5.24). Since A is bijective,

$$\forall v \in L, \quad (f, v)_L = (f, A(A^{-1}v))_L = (Au, A(A^{-1}v))_L = (Au, v)_L,$$

which means that u solves (5.8). □

Now, consider a V-conformal approximation to (5.24). Let $V_h \subset V$ be a finite-dimensional space, and consider the discrete problem:

$$\begin{cases} \text{Seek } u_h \in V_h \text{ such that} \\ \tilde{a}(u_h, w_h) = (f, Aw_h)_L, \quad \forall w_h \in V_h. \end{cases} \tag{5.25}$$

Theorem 5.24. *Problem (5.25) has a unique solution u_h, and*

$$\|u - u_h\|_V \leq \frac{\|A\|}{\alpha} \inf_{w_h \in V_h} \|u - w_h\|_V. \tag{5.26}$$

Proof. The existence and uniqueness of u_h is a direct consequence of the Lax–Milgram Lemma and the conformal setting. Furthermore, the symmetry and the coercivity of $\tilde{a}$ together with the Galerkin orthogonality $\tilde{a}(u - u_h, z_h) = 0$ holding for all z_h in V_h, imply

$$\alpha^2\|u - u_h\|_V^2 \leq \tilde{a}(u - u_h, u - u_h) \leq \tilde{a}(u - w_h, u - w_h) \leq \|A\|^2\|u - w_h\|_V^2,$$

yielding the desired result. □

Remark 5.25.

(i) It is sometimes possible to improve the error estimate in the norm of L by the means of the Aubin–Nitsche duality argument, but this is not systematic since, very often, first-order PDEs are not endowed with a smoothing property.

(ii) The origins of the Least-Squares technique can be traced back to Gauß (*Theoria Motus Corporum Coelestium*, 1809). The method has gained some popularity in the numerical analysis community at the beginning of the 1970s owing to a series of papers by Bramble and Schatz [BrS70, BrS71], although it was already popular in the Russian literature [Dži68, Luč69]. For a quite theoretical introduction to this technique to solve elliptic problems, see [AzK85]. For a review of applications and implementation, see [Jia89].

(iii) Note that the estimate (5.26) is optimal, i.e., the Least Square technique is optimal to solve (5.7).   □

5.3.4 Example 1: Advection–reaction

In general, approximating (5.13) by means of the standard Galerkin method is not optimal owing to the reasons stated at the beginning of this chapter. A possible remedy consists of using the Least-Squares technique. Using the notation of §5.2.3 and setting

$$\tilde{a}(u,v) = (\mu u + \beta\cdot\nabla u, \mu v + \beta\cdot\nabla v)_{0,\Omega},$$

the Least-Squares formulation of (5.13) is:

$$\begin{cases} \text{Seek } u \in V \text{ such that} \\ \tilde{a}(u,v) = (f, \mu v + \beta\cdot\nabla v)_{0,\Omega}, \quad \forall v \in V, \end{cases} \tag{5.27}$$

with $V = \{v \in L^2(\Omega);\ \beta\cdot\nabla v \in L^2(\Omega)\}$. Let $\{\mathcal{T}_h\}_{h>0}$ be a shape-regular family of meshes of Ω and let V_h be a V-conformal space of continuous finite elements, say of type $\mathbb{P}_k$ or $\mathbb{Q}_k$. The discrete formulation is:

$$\begin{cases} \text{Seek } u_h \in V_h \text{ such that} \\ \tilde{a}(u_h, v_h) = (f, \mu v_h + \beta\cdot\nabla v_h)_{0,\Omega}, \quad \forall v_h \in V_h. \end{cases} \tag{5.28}$$

From what precedes, we infer the following:

Proposition 5.26. *Problem* (5.28) *has a unique solution* u_h, *and if* $u \in H^{k+1}(\Omega)$, *the following estimate holds:*

$$\forall h, \quad \|u - u_h\|_{0,\Omega} + \|\beta\cdot\nabla(u - u_h)\|_{0,\Omega} \leq c\,h^k \|u\|_{k+1,\Omega}.$$

Note that this estimate is optimal in the norm induced by the streamwise derivative but is not optimal in the L^2-norm. It is not clear whether it is possible to improve the L^2-estimate. The Aubin–Nitsche duality argument does not hold here since the dual problem has no smoothing property.

Remark 5.27. Formally, problem (5.27) is equivalent to solving

$$\mu(\mu u + \beta\!\cdot\!\nabla u) - \nabla\!\cdot\!(\beta(\mu u + \beta\!\cdot\!\nabla u)) = \mu f - \nabla\!\cdot\!(\beta f) \quad \text{a.e. in } \Omega,$$

with the boundary conditions $u = 0$ a.e. on $\partial\Omega^-$ and $\mu u + \beta\!\cdot\!\nabla u = f$ a.e. on $\partial\Omega^+$. $\qquad\qquad\square$

5.3.5 Example 2: Darcy's equations

The approximation of (5.15) by the Galerkin method in its standard form is not optimal in general. Let us reformulate this problem in the Least-Squares sense. Using the notation of §5.2.4, set, for all $((u,p),(v,q)) \in V \times V$,

$$\tilde{a}((u,p),(v,q)) = (\mathcal{S}^{-1}\!\cdot\!u + \nabla p, \mathcal{S}^{-1}\!\cdot\!v + \nabla q)_{0,\Omega} + (\nabla\!\cdot\!u, \nabla\!\cdot\!v)_{0,\Omega},$$

where $V = X \times Y$ with X and Y defined in (5.16). The Least-Squares formulation of (5.15) is:

$$\begin{cases} \text{Seek } (u,p) \in V \text{ such that, } \forall(v,q) \in V, \\ \tilde{a}((u,p),(v,q)) = (f, \mathcal{S}^{-1}\!\cdot\!v + \nabla q)_{0,\Omega} + (g, \nabla\!\cdot\!v)_{0,\Omega}. \end{cases} \tag{5.29}$$

Let $X_h \times Y_h = V_h$ be a V-conformal approximation space; for instance, a space of H^1-conformal Lagrange finite elements based on a shape-regular family of meshes $\{\mathcal{T}_h\}_{h>0}$. Assume that there are two numbers $k \geq 1$ and $r \geq 1$ and two interpolation operators $\mathcal{I}_h$ and $\mathcal{J}_h$ such that

$$\forall v \in H^{r+1}(\Omega), \quad \|v - \mathcal{I}_h(v)\|_X \leq ch^r\|v\|_{r+1,\Omega},$$
$$\forall q \in H^{k+1}(\Omega), \quad \|q - \mathcal{J}_h(q)\|_{0,\Omega} + h\|q - \mathcal{J}_h(q)\|_{1,\Omega} \leq ch^{k+1}\|q\|_{k+1,\Omega}.$$

The discrete problem in the Least-Squares sense is:

$$\begin{cases} \text{Seek } (u_h,p_h) \in V_h \text{ such that, } \forall(v_h,q_h) \in V_h, \\ \tilde{a}((u_h,p_h),(v_h,q_h)) = (f, \mathcal{S}^{-1}\!\cdot\!v_h + \nabla q_h)_{0,\Omega} + (y, \nabla\!\cdot\!v_h)_{0,\Omega}. \end{cases} \tag{5.30}$$

Theorem 5.24 readily yields the following:

Proposition 5.28. *Problem* (5.30) *has a unique solution, and if u and p are sufficiently smooth, letting $s = \min(k,r)$, there is c such that, for all h,*

$$\|u - u_h\|_{0,\Omega} + \|\nabla\!\cdot\!(u - u_h)\|_{0,\Omega} + \|p - p_h\|_{1,\Omega} \leq ch^s(\|u\|_{r+1,\Omega} + \|p\|_{k+1,\Omega}).$$

A slightly stronger convergence result can be inferred using a duality argument in a simplified setting; see, e.g., [PeC94].

Proposition 5.29. *Assume:*

(i) $\partial\Omega_1 = \emptyset$.
(ii) $\mathcal{S}^{-1} = \mu(x)I$ *where I is the identity matrix, $\mu \in W^{1,\infty}(\Omega)$, and μ is uniformly bounded from below by a positive number.*

(iii) $(-\Delta)^{-1}$ *is continuous from* $L^2(\Omega)$ *to* $H^2(\Omega) \cap H_0^1(\Omega)$.

Then, if $r = k+1$ *and if* u *and* p *are sufficiently smooth, there is* c *such that, for all* h,

$$\|u - u_h\|_{0,\Omega} + \|\nabla\cdot(u - u_h)\|_{0,\Omega} + \|p - p_h\|_{0,\Omega} \leq c\,h^{k+1}(\|u\|_{k+2,\Omega} + \|p\|_{k+1,\Omega}).$$

Proof. We slightly adapt the proof of the Aubin–Nitsche Lemma. For all $q \in H_0^1(\Omega)$, introduce the quantity $P_h(q) \in Y_h$ such that

$$(\nabla(q - P_h q), \nabla q_h)_{0,\Omega} = 0, \quad \forall q_h \in Y_h.$$

Note that P_h is the elliptic projector defined in §1.6.3. Owing to the smoothing property of the Laplacian, the Aubin–Nitsche Lemma implies

$$\forall q \in H^{k+1}(\Omega), \quad \|q - P_h q\|_{0,\Omega} \leq c\,h^{k+1}\|q\|_{k+1,\Omega}.$$

Owing to Galerkin orthogonality and the definition of P_h,

$$
\begin{aligned}
c\|(\mathcal{I}_h u - u_h, P_h p - p_h)\|_V^2 &\leq \tilde{a}((\mathcal{I}_h u - u_h, P_h p - p_h), (\mathcal{I}_h u - u_h, P_h p - p_h)) \\
&= \tilde{a}((\mathcal{I}_h u - u, P_h p - p), (\mathcal{I}_h u - u_h, P_h p - p_h)) \\
&= (\mu(\mathcal{I}_h u - u), \mu(\mathcal{I}_h u - u_h))_{0,\Omega} + (\mu(\mathcal{I}_h u - u), \nabla(P_h p - p_h))_{0,\Omega} \\
&\quad - (\nabla\cdot(\mu(\mathcal{I}_h u - u_h)), P_h p - p)_{0,\Omega} + (\nabla\cdot(\mathcal{I}_h u - u), \nabla\cdot(\mathcal{I}_h u - u_h))_{0,\Omega} \\
&\leq c\,(\|\mathcal{I}_h u - u\|_X + \|P_h p - p\|_{0,\Omega})\|(\mathcal{I}_h u - u_h, P_h p - p_h)\|_V \\
&\leq c(u,p)h^{k+1}\|(\mathcal{I}_h u - u_h, P_h p - p_h)\|_V,
\end{aligned}
$$

with $c(u,p) = c(\|u\|_{k+2,\Omega} + \|p\|_{k+1,\Omega})$. Hence, $\|(\mathcal{I}_h u - u_h, P_h p - p_h)\|_V \leq c(u,p)h^{k+1}$. The triangle inequality yields the estimate on $\|u - u_h\|_{0,\Omega} + \|\nabla\cdot(u - u_h)\|_{0,\Omega}$. To prove the estimate on $\|p - p_h\|_{0,\Omega}$, observe that

$$
\begin{aligned}
\|p - p_h\|_{0,\Omega} &\leq \|p - P_h p\|_{0,\Omega} + \|P_h p - p_h\|_{0,\Omega} \\
&\leq c\,h^{k+1}\|p\|_{k+1,\Omega} + c'|P_h p - p_h|_{1,\Omega} \\
&\leq c\,h^{k+1}\|p\|_{k+1,\Omega} + c'\|(\mathcal{I}_h u - u_h, P_h p - p_h)\|_V \leq c(u,p)h^{k+1}. \quad \square
\end{aligned}
$$

Remark 5.30. Darcy's problem (5.15) can be formulated in the framework of saddle-point problems; see Chapter 4. In this spirit, it is possible to define mixed finite elements giving the same estimates as those in Proposition 5.29. For instance, the Raviart–Thomas finite element introduced in §1.2.7 belongs to this class of elements [RaT77]; see Exercises 4.10, 4.11, and 4.13. See also [BrF91^b, RoT91] for further insight. $\square$

5.3.6 Example 3: Maxwell's equations

The reader can easily reformulate problem (5.17) in the Least-Squares sense. However, the resulting problem couples the two fields E and B. To avoid this coupling, while keeping the spirit of the Least-Squares method, one interesting approach consists of introducing the sesquilinear form

$$\widetilde{a}((E,B),(e,b)) = \Re\left(-i\omega \int_\Omega (E - \nabla\times B)\cdot(\bar{e} - \nabla\times\bar{b}) \right.$$

$$\left. + \int_\Omega (-i\omega B + \nabla\times E)\cdot(i\omega\bar{b} + \nabla\times\bar{e}) \right).$$

Set $V = X \times Y$ with X and Y defined in (5.18). The modified Least-Squares formulation is:

$$\begin{cases} \text{Seek } (E,B) \in V \text{ such that, } \forall (e,b) \in V, \\ \widetilde{a}((E,B),(e,b)) = (-i\omega f, e - \nabla\times b)_{0,\Omega} + (g, -i\omega b + \nabla\times e)_{0,\Omega}. \end{cases}$$

This problem is well-posed by construction. After two integrations by parts, it takes the following equivalent form in which E and B are uncoupled:

$$\begin{cases} \text{Seek } (E,B) \in V \text{ such that, } \forall (e,b) \in V, \\ (-i\omega E, e)_{0,\Omega} + (\nabla\times E, \nabla\times e)_{0,\Omega} = (-i\omega f, e)_{0,\Omega} + (g, \nabla\times e)_{0,\Omega}, \quad (5.31) \\ (-i\omega B, b)_{0,\Omega} + (\nabla\times B, \nabla\times b)_{0,\Omega} = -(f, \nabla\times b)_{0,\Omega} + (g, b)_{0,\Omega}. \end{cases}$$

Let $X_h \times Y_h = V_h$ be a V-conformal space, e.g., based on $\mathbb{P}_k$ or $\mathbb{Q}_k$ H^1-conformal finite elements. The approximate Least-Squares problem is:

$$\begin{cases} \text{Seek } (E_h, B_h) \in V_h \text{ such that, } \forall (e_h, b_h) \in V_h, \\ (-i\omega E_h, e_h)_{0,\Omega} + (\nabla\times E_h, \nabla\times e_h)_{0,\Omega} = (-i\omega f, e_h)_{0,\Omega} + (g, \nabla\times e_h)_{0,\Omega}, \quad (5.32) \\ (-i\omega B_h, b_h)_{0,\Omega} + (\nabla\times B_h, \nabla\times b_h)_{0,\Omega} = -(f, \nabla\times b_h)_{0,\Omega} + (g, b_h)_{0,\Omega}. \end{cases}$$

Proposition 5.31. *Problem* (5.32) *is well-posed, and, for all h,*

$$\|E - E_h\|_X + \|B - B_h\|_Y \le c \inf_{(e_h, b_h) \in V_h} \|(E - e_h, B - b_h)\|_V.$$

Remark 5.32.

(i) See Exercise 5.11 for a formal interpretation of (5.31).

(ii) When the domain is not smooth, or when Maxwell's equations involve discontinuous coefficients, it is in general preferable to use edge finite elements instead of Lagrange finite elements; see §1.2.8 and §2.3.3. The reader interested by this type of technique is referred to [Néd86, Néd91, Mon92, Bos93]. □

5.3.7 Generalizations in $H^{-1}(\Omega)$

The main drawback of the Least-Squares technique is that it cannot be extended to H^1-conformal approximations of second-order differential operators. Indeed, if the operator A contains a term such as $-\Delta$, its range is no longer in $L^2(\Omega)$ but in $H^{-1}(\Omega)$ instead. As a result, expressions such as $\int_\Omega AuAv$ are no longer meaningful. Of course, these expressions become meaningful if one uses H^2-conformal approximation spaces, but this option is rarely used in practice except in one dimension.

Actually, if the range of A is in $H^{-1}(\Omega)$, the abstract theory developed above still holds by setting $L = H^{-1}(\Omega)$, that is, by using $H^{-1}(\Omega)$ as the pivot space. The main practical difficulty at this point is to compute the $H^{-1}(\Omega)$-scalar product. One possibility is to set $(f, q)_{-1, \Omega} = \langle g, (-\Delta)^{-1} f \rangle_{H^{-1}, H_0^1}$: For two distributions f and g in $H^{-1}(\Omega)$, one solves the Poisson problem $-\Delta u = f$ supplemented with homogeneous Dirichlet boundary conditions; then, one computes $\langle g, u \rangle_{H^{-1}, H_0^1}$. This strategy is interesting only if a very fast solver (or preconditioner) for the Laplace operator is available. Such solvers (or preconditioners) usually involve a hierarchical decomposition of the approximation space; see [AzK85, BrP96, BrL97, Boc97, BrS98, Boc99, BrL01].

5.4 Galerkin/Least-Squares (GaLS) Approximation

To avoid $H^{-1}(\Omega)$-scalar products when dealing with second-order (or higher-order) differential operators, we consider in this section a hybrid technique combining the Galerkin and the Least-Squares methods.

5.4.1 The model problem

Let us return to the abstract framework introduced in §5.2.1. Let m be a positive integer. Set $L = [L^2(\Omega)]^m$ and equip this space with the inner product $(u, v)_L = \int_\Omega \sum_{i=1}^m u_i v_i$. As usual, L is identified with its dual. Let $V \subset L$ be a Hilbert space such that the embedding $V \subset L$ is continuous and dense. Let $A \in \mathcal{L}(V; L)$ be an isomorphism. A is assumed to be *monotone*:

$$\forall u \in V, \quad (Au, u)_L \geq 0.$$

Henceforth, the reader may think of A as a first-order differential operator.

Let $X \subset V$ be a Hilbert space continuously and densely embedded in V and equipped with the norm $\| \cdot \|_X$. Let $d \in \mathcal{L}(X \times X; \mathbb{R})$ be a continuous bilinear form. We assume that d is X-coercive and that d has been rescaled so that

$$\forall u \in X, \quad d(u, u) \geq \|u\|_X^2. \tag{5.33}$$

We also assume that d is associated with a second-order differential operator D such that, in the distribution sense,

$$\forall u \in X, \forall v \in [\mathcal{D}(\Omega)]^m, \quad d(u, v) = \langle Du, v \rangle_{\mathcal{D}', \mathcal{D}}.$$

Let ϵ be a positive real number. Define the operator T_ϵ and the bilinear form $t_\epsilon \in \mathcal{L}(X \times X; \mathbb{R})$ by $T_\epsilon = A + \epsilon D$ and, for all $(u, v) \in X \times X$,

$$t_\epsilon(u, v) = (Au, v)_L + \epsilon d(u, v).$$

Consider the following problem: For $f \in L$,

$$\begin{cases} \text{Seek } u \in X \text{ such that} \\ t_\epsilon(u, w) = (f, w)_L, \quad \forall w \in X. \end{cases} \tag{5.34}$$

Proposition 5.33. *For all $\epsilon > 0$, problem (5.34) has a unique solution.*

Proof. Apply the Lax–Milgram Lemma. □

We are interested in the singular perturbation limit where

$$\epsilon \ll \|A\|_{\mathcal{L}(X;L)}. \tag{5.35}$$

To alleviate the notation, we henceforth assume that $\|A\|_{\mathcal{L}(X;L)}$ is of order one and loose track of this quantity in the constants. Clearly, if $\|A\|_{\mathcal{L}(X;L)} \gg 1$, problem (5.34) can be rescaled appropriately. In the context of advection–diffusion problems, ϵ is the reciprocal of the Péclet number; see §5.4.4.

The coercivity of problem (5.34) yields the stability estimate

$$\|u\|_X \le c\,\epsilon^{-1}\|f\|_L.$$

When ϵ is of order 1, the stability in the norm $\|\cdot\|_X$ is optimal and the Galerkin approximation theory of coercive problems presented in Chapters 2 and 3 applies without restriction. However, difficulties arise when ϵ is small. In this case, coercivity is no longer dominant (see §3.5) and the standard Galerkin approximation of (5.34) is no longer optimal. A possible remedy to this situation consists of combining the Galerkin and the Least-Squares formulations.

Remark 5.34. Problem (5.34) is formally equivalent to

$$\begin{cases} T_\epsilon u = f & \text{a.e. in } \Omega, \\ \Lambda u = 0 & \text{a.e. on } \partial\Omega, \end{cases} \tag{5.36}$$

where Λ is a boundary condition operator (i.e., a trace operator). The essential boundary conditions are enforced in the definition of the space X, whereas the natural conditions are enforced by the differential operator D and its associated bilinear form d. For instance, consider an advection–diffusion problem with Dirichlet conditions enforced on $\partial\Omega_D \subset \partial\Omega$ with $\text{meas}(\partial\Omega_D) > 0$. Then, $X = \{v \in H^1(\Omega); v_{|\partial\Omega_D} = 0\}$, $D = -\Delta$, and the trace operator Λ is such that $\Lambda u_{|\partial\Omega_D} = u$ and $\Lambda u_{|\partial\Omega \setminus \partial\Omega_D} = \partial_n u$. □

5.4.2 Principle of the GaLS method

Formally, the Least-Squares formulation of problem (5.34) is:

$$\begin{cases} \text{Seek } u \in H = \{u \in L; T_\epsilon u \in L; \Lambda u_{|\partial\Omega} = 0\} \text{ such that} \\ (T_\epsilon u, T_\epsilon v)_L = (f, T_\epsilon v)_L, \quad \forall v \in H. \end{cases}$$

The main difficulty at this point is that the operators D and Λ are not bounded on X. For instance, think of the Laplace operator and the normal derivative: these operators are not bounded on $X = H^1(\Omega)$. If D is a second-order differential operator, the Least-Squares formulation requires a priori controls on second derivatives and on the trace of first-order derivatives at the boundary. This type of regularity is not compatible with H^1-conformal finite elements. However, since H^1-conformal finite elements are $\mathcal{C}^\infty$ on each element K of the mesh, the quantity $T_\epsilon u = Au + \epsilon Du$ can be controlled elementwise.

Let $\{\mathcal{T}_h\}_{h>0}$ be a family of meshes of Ω, and let $X_h \subset X$ be a space of H^1-conformal $\mathbb{R}^m$-valued finite elements. Any v_h in X_h is of class $\mathcal{C}^\infty$ on each element $K \in \mathcal{T}_h$. As a consequence, letting $(u,v)_{L,K} = \int_K \sum_{i=1}^m u_i v_i$, the quantities $(T_\epsilon u_h, T_\epsilon v_h)_{L,K}$ and $(f, T_\epsilon v_h)_{L,K}$ are meaningful. Then set

$$t_{\epsilon h}(u_h, v_h) = t_\epsilon(u_h, v_h) + \sum_{K \in \mathcal{T}_h} \delta(h_K)(T_\epsilon u_h, T_\epsilon v_h)_{L,K}. \qquad (5.37)$$

A hybrid approximation of (5.34) is constructed as follows:

$$\begin{cases} \text{Seek } u_h \in X_h \text{ such that} \\ t_{\epsilon h}(u_h, v_h) = (f, v_h)_L + \sum_{K \in \mathcal{T}_h} \delta(h_K)(f, T_\epsilon v_h)_{L,K}, \quad \forall v_h \in X_h, \end{cases} \qquad (5.38)$$

where $\delta(h_K)$ is a function (depending on ϵ) of the diameter of the cell K. The Galerkin formulation corresponds to $\delta = 0$, whereas the Least-Squares formulation (if it were meaningful) would correspond to $\delta \to +\infty$. Actually, the error analysis will reveal that the function $\delta : \mathbb{R}_+ \to \mathbb{R}_+$ must be chosen such that:

(i)   there is c_1 such that $\forall(h_K, \epsilon)$, $\delta(h_K) \le c_1 h_K$;

(ii) there is c_2 such that $\forall(h_K, \epsilon)$, $\delta(h_K) \le c_2 \frac{h_K^2}{\epsilon}$;

(iii) for all $c_3 > 0$, there is $c_4 > 0$ such that $\epsilon \le c_3 h_K$ implies $\delta(h_K) \ge c_4 h_K$.

Henceforth, $\delta(h_K)$ is chosen to be such that

$$\delta(h_K) = \left(\frac{1}{h_K} + \frac{\epsilon}{h_K^2} \right)^{-1}. \qquad (5.39)$$

One readily verifies that properties (i) and (ii) hold with $c_1 = c_2 = 1$ and that property (iii) holds with $c_4 = \frac{1}{1+c_3}$.

Remark 5.35. In the literature, the formulation (5.38) is usually referred to as the GaLS method, but it is also sometimes called the *streamline-diffusion method* [JoN84]. This technique has been popularized by Hughes et al. [BrH82, HuF89]. Reviews on GaLS-like techniques for approximating advection–diffusion equations can be found in [Joh87, Cod98, QuV97].   □

5.4.3 Convergence analysis

Henceforth, c and c' denote generic constants *independent of h_K and ϵ.* The convergence analysis of (5.38) relies on the following assumptions:

(i) Localize the norm of X and the bilinear form d,

$$
\begin{cases}
\|v\|_X^2 = \displaystyle\sum_{K \in T_h} \|v\|_{X,K}^2, & \forall v \in X, \\[2mm]
d(u,v) = \displaystyle\sum_{K \in T_h} d_K(u,v), & \forall (u,v) \in X \times X.
\end{cases}
\tag{5.40}
$$

(ii) Localize the continuity hypotheses

$$
\begin{cases}
\|Av\|_{L,K} \leq c\|v\|_{X,K}, & \forall v \in X,\ \forall K \in T_h, \\[1mm]
d_K(u,v) \leq c\|u\|_{X,K}\|v\|_{X,K}, & \forall (u,v) \in X \times X,\ \forall K \in T_h.
\end{cases}
\tag{5.41}
$$

(iii) Assume the local inverse inequality

$$
\forall v_h \in X_h,\ \forall K \in T_h,\quad \|Dv_h\|_{L,K} \leq c\, h_K^{-1}\|v_h\|_{X,K}.
\tag{5.42}
$$

Define the symmetric bilinear form $a_s \in \mathcal{L}(V \times V; \mathbb{R})$ such that

$$
a_s(v,w) = \tfrac{1}{2}\big((Av,w)_L + (v,Aw)_L\big).
$$

Owing to the monotonicity of A, a_s is positive. Furthermore, define the norms

$$
\begin{cases}
\|v\|_{h,T_\epsilon}^2 = a_s(v,v) + \epsilon\|v\|_X^2 + \displaystyle\sum_{K \in T_h} \delta(h_K)\|T_\epsilon v\|_{L,K}^2, \\[2mm]
\|v\|_{h,\frac{1}{2}}^2 = \|v\|_{h,T_\epsilon}^2 + \displaystyle\sum_{K \in T_h} h_K^{-1}\|v\|_{L,K}^2.
\end{cases}
$$

Lemma 5.36 (Stability). *The following coercivity property holds:*

$$
\forall v \in X,\quad t_{\epsilon h}(v,v) \geq \|v\|_{h,T_\epsilon}^2.
\tag{5.43}
$$

Proof. Direct consequence of the choice for the norm $\|\cdot\|_{h,T_\epsilon}$. $\qquad\square$

Lemma 5.37 (Continuity). *There is $c > 0$, independent of h and ϵ, such that*

$$
\forall v \in X,\ \forall w_h \in X_h,\quad t_{\epsilon h}(v,w_h) \leq c\|v\|_{h,\frac{1}{2}}\|w_h\|_{h,T_\epsilon}.
\tag{5.44}
$$

Proof. Let (v,w_h) be in $X \times X_h$ and observe that

$$
t_{\epsilon h}(v,w_h) = (Av,w_h)_L + \epsilon d(v,w_h) + \sum_{K \in T_h} \delta(h_K)(T_\epsilon v, T_\epsilon w_h)_{L,K}.
$$

Each term in the right-hand side is bounded from above as follows: Consider the first term. Since a_s is positive and symmetric,

$$(Av, w_h)_L = 2a_s(v, w_h) - (v, Aw_h)_L$$
$$\leq 2a_s(v,v)^{\frac{1}{2}}a_s(w_h, w_h)^{\frac{1}{2}} - \sum_{K \in \mathcal{T}_h}(v, Aw_h)_{L,K}.$$

Let K be an arbitrary element. If $\epsilon \leq h_K$, the inverse inequality (5.42) together with property (iii) for the function δ implies

$$|(v, Aw_h)_{L,K}| = |(v, Aw_h + \epsilon Dw_h)_{L,K} - \epsilon(v, Dw_h)_{L,K}|$$
$$\leq \|v\|_{L,K}\left(\|T_\epsilon w_h\|_{L,K} + c\epsilon h_K^{-1}\|w_h\|_{X,K}\right)$$
$$\leq h_K^{-\frac{1}{2}}\|v\|_{L,K}\left(h_K^{\frac{1}{2}}\|T_\epsilon w_h\|_{L,K} + c\epsilon h_K^{-\frac{1}{2}}\|w_h\|_{X,K}\right)$$
$$\leq c\,h_K^{-\frac{1}{2}}\|v\|_{L,K}\left(\delta(h_K)^{\frac{1}{2}}\|T_\epsilon w_h\|_{L,K} + \epsilon^{\frac{1}{2}}\|w_h\|_{X,K}\right).$$

If $h_K \leq \epsilon$, the continuity hypothesis (5.41) yields

$$|(v, Aw_h)_{L,K}| \leq \|v\|_{L,K}\|Aw_h\|_{L,K} \leq c\epsilon^{-\frac{1}{2}}\|v\|_{L,K}\epsilon^{\frac{1}{2}}\|w_h\|_{X,K}$$
$$\leq ch_K^{-\frac{1}{2}}\|v\|_{L,K}\epsilon^{\frac{1}{2}}\|w_h\|_{X,K}.$$

In both cases, applying the Cauchy–Schwarz inequality leads to

$$(Av, w_h)_L \leq c\|v\|_{h,\frac{1}{2}}\|w_h\|_{h,T_\epsilon}.$$

For the second term, it is clear that, for all ϵ,

$$\epsilon d(v, w_h) \leq c\epsilon\|v\|_X\|w_h\|_X \leq c\|v\|_{h,\frac{1}{2}}\|w_h\|_{h,T_\epsilon}.$$

For the last term, use again the Cauchy–Schwarz inequality to obtain

$$\sum_{K\in\mathcal{T}_h}\delta(h_K)(T_\epsilon v, T_\epsilon w_h)_{L,K} \leq \|v\|_{h,\frac{1}{2}}\|w_h\|_{h,T_\epsilon}. \qquad \square$$

Lemma 5.38 (Consistency). *Let u solve (5.34) and let u_h solve (5.38). Then,*

$$\forall v_h \in X_h, \quad t_{\epsilon h}(u - u_h, v_h) = 0. \tag{5.45}$$

Proof. Since f is in $L = [L^2(\Omega)]^m$, it is clear that the equality $T_\epsilon u = f$ holds in L. That is to say, $T_\epsilon u$ is square-integrable and

$$\forall v_h \in X_h, \quad \sum_{K\in\mathcal{T}_h}\delta(h_K)(T_\epsilon u, T_\epsilon v_h)_{L,K} = \sum_{K\in\mathcal{T}_h}\delta(h_K)(f, T_\epsilon v_h)_{L,K}.$$

This yields

$$\forall v_h \in X_h, \quad t_{\epsilon h}(u, v_h) = (f, v_h) + \sum_{K\in\mathcal{T}_h}\delta(h_K)(f, T_\epsilon v_h)_{L,K}.$$

Subtracting (5.38) from this equality yields (5.45). $\square$

Theorem 5.39 (Convergence). *Under the above assumptions,*

$$\|u - u_h\|_{h,T_\epsilon} \le c \inf_{w_h \in X_h} \|u - w_h\|_{h,\frac{1}{2}}. \tag{5.46}$$

Proof. Let w_h be an arbitrary function in X_h. Apply successively Lemmas 5.43, 5.45, and 5.44 to obtain

$$\begin{aligned}
\|u_h - w_h\|_{h,T_\epsilon}^2 &\le t_{\epsilon h}(u_h - w_h, u_h - w_h) \\
&\le t_{\epsilon h}(u - w_h, u_h - w_h) \\
&\le c \|u - w_h\|_{h,\frac{1}{2}} \|u_h - w_h\|_{h,T_\epsilon}.
\end{aligned}$$

The result follows by applying the triangle inequality. Note that this proof is in the spirit of the Second Strang Lemma. $\qquad\qquad\qquad\qquad\qquad\square$

To have a further insight into the above result, assume that the space X_h has a local approximation property, i.e., there exists a dense subspace of X, say W, with localized norm $\|\cdot\|_{W,K}$ and such that, for all u in W,

$$\begin{aligned}
\inf_{w_h \in X_h} \big(\|u - w_h\|_{L,K} + h_K \|u - w_h\|_{X,K} \\
+ h_K^2 \|D(u - w_h)\|_{L,K} \big) \le c h_K^{k+1} \|u\|_{W,K}.
\end{aligned} \tag{5.47}$$

Corollary 5.40. *Along with hypothesis (5.47) and those of Theorem 5.39, assume $u \in W$. Then, for all $\gamma > 0$, there is $c_\gamma > 0$ such that, for all $\epsilon > 0$ and $\{T_h\}_{h>0}$,*

(i) *If $\epsilon \ge \gamma \max_{K \in T_h}\{h_K\}$,*

$$\|u - u_h\|_X \le c_\gamma \left(\sum_{K \in T_h} h_K^{2k} \|u\|_{W,K}^2 \right)^{\frac{1}{2}}.$$

(ii) *If $\epsilon \le \gamma \min_{K \in T_h}\{h_K\}$,*

$$\left(\sum_{K \in T_h} h_K \|A(u - u_h)\|_{L,K}^2 \right)^{\frac{1}{2}} \le c_\gamma \left(\sum_{K \in T_h} h_K^{2k+1} \|u\|_{W,K}^2 \right)^{\frac{1}{2}}.$$

Proof. (1) First, we estimate the right-hand side of (5.46) as follows: Setting $\eta = u - w_h$, it is clear that

$$\begin{aligned}
\|\eta\|_{h,\frac{1}{2}}^2 &\le \|A\eta\|_L \|\eta\|_L + \epsilon \|\eta\|_X^2 \\
&\quad + \sum_{K \in T_h} h_K^{-1} \|\eta\|_{L,K}^2 + \delta(h_K)\big(\|A\eta\|_{L,K}^2 + \epsilon^2 \|D\eta\|_{L,K}^2 \big) \\
&\le c \sum_{K \in T_h} h_K^{-1} \|\eta\|_{L,K}^2 + (\delta(h_K) + h_K + \epsilon)\|\eta\|_{X,K}^2 + \epsilon^2 \delta(h_K)\|D\eta\|_{L,K}^2.
\end{aligned}$$

Properties (i) and (ii) of the function δ imply

$$\|\eta\|_{h,\frac{1}{2}}^2 \le c \sum_{K \in \mathcal{T}_h} h_K^{-1} \|\eta\|_{L,K}^2 + (h_K + \epsilon)\|\eta\|_{X,K}^2 + \epsilon h_K^2 \|D\eta\|_{L,K}^2.$$

Hence,

$$\inf_{w_h \in X_h} \|u - w_h\|_{h,\frac{1}{2}}^2 \le c \sum_{K \in \mathcal{T}_h} (h_K + \epsilon) h_K^{2k} \|u\|_{W,K}^2,$$

and using (5.46), this yields

$$\|u - u_h\|_{h,\mathcal{T}_\epsilon} \le c \left(\sum_{K \in \mathcal{T}_h} (h_K + \epsilon) h_K^{2k} \|u\|_{W,K}^2 \right)^{\frac{1}{2}}. \qquad (5.48)$$

(2) Assume $\epsilon \ge \gamma \max_{K \in \mathcal{T}_h}\{h_K\}$. Use (5.46), (5.48), and the fact that $\|u - u_h\|_X^2 \le \frac{1}{\epsilon}\|u - u_h\|_{h,\mathcal{T}_\epsilon}^2$ to infer

$$\|u - u_h\|_X^2 \le c \sum_{K \in \mathcal{T}_h} \frac{1}{\epsilon}(h_K + \epsilon) h_K^{2k} \|u\|_{W,K}^2 \le c' \sum_{K \in \mathcal{T}_h} h_K^{2k} \|u\|_{W,K}^2.$$

(3) Assume now $\epsilon \le \gamma \min_{K \in \mathcal{T}_h}\{h_K\}$. Set $e = u - u_h$ and $\eta = u - w_h$. Then,

$$\begin{aligned}
\|Ae\|_{L,K}^2 = \|T_\epsilon e - \epsilon De\|_{L,K}^2 &\le 2\|T_\epsilon e\|_{L,K}^2 + 2\epsilon^2 \|De\|_{L,K}^2 \\
&\le 2\|T_\epsilon e\|_{L,K}^2 + c\epsilon^2 (\|D\eta\|_{L,K}^2 + \|D(w_h - u_h)\|_{L,K}^2) \\
&\le 2\|T_\epsilon e\|_{L,K}^2 + c\epsilon^2 (\|D\eta\|_{L,K}^2 + h_K^{-2}\|w_h - u_h\|_{X,K}^2) \\
&\le 2\|T_\epsilon e\|_{L,K}^2 + c\epsilon^2 (\|D\eta\|_{L,K}^2 + h_K^{-2}\|e\|_{X,K}^2 + h_K^{-2}\|\eta\|_{X,K}^2).
\end{aligned}$$

Owing to property (iii) of the function δ, this implies

$$h_K \|Ae\|_{L,K}^2 \le c(\delta(h_K)\|T_\epsilon e\|_{L,K}^2 + \epsilon\|e\|_{X,K}^2) + c'(h_K\|\eta\|_{X,K}^2 + h_K^3\|D\eta\|_{L,K}^2),$$

from which the second estimate follows easily.   □

Corollary 5.41. *Assume that the hypotheses of Corollary 5.40 hold and that the family $\{\mathcal{T}_h\}_{h>0}$ is quasi-uniform. Then, for all $\gamma > 0$, there is $c_\gamma > 0$ such that, for all $\epsilon > 0$ and $h > 0$,*

(i) *If $h \le \gamma\epsilon$, $\|u - u_h\|_X \le c_\gamma h^k \|u\|_W$.*
(ii) *If $\epsilon \le \gamma h$, $\|u - u_h\|_V \le c_\gamma h^k \|u\|_W$.*

Proof. The first estimate is a direct consequence of Corollary 5.40(i). Furthermore, since $u - u_h \in X_h \subset X \subset V$, since the family $\{\mathcal{T}_h\}_{h>0}$ is quasi-uniform (i.e., $h \le \tau h_K$), and since A is an isomorphism, the following inequalities hold:

$$\alpha \|u - u_h\|_V \le \|A(u - u_h)\|_L \le \left(\frac{\tau}{h} \sum_{K \in \mathcal{T}_h} h_K \|A(u - u_h)\|_{L,K}^2 \right)^{\frac{1}{2}}.$$

Then, the second estimate follows from Corollary 5.40(ii).   □

Remark 5.42. Corollary 5.41 shows that if the mesh is sufficiently fine, the bilinear form ϵd (i.e., the diffusion) has sufficient coercivity to ensure optimal convergence in the X-norm. In this case, the Galerkin formulation can guarantee this property alone, i.e., one could set $\delta(h) = 0$. On the other hand, if the mesh is such that $\epsilon \leq ch$, the Least-Squares part of the bilinear form $t_{\epsilon h}$ compensates for the lack of coercivity in the Galerkin part, and convergence is ensured only in the V-norm. This is not surprising since the V-norm is naturally associated with the limit equation as $\epsilon \to 0$. □

If A is L-coercive, one can obtain a convergence estimate in the L-norm that improves the result of Corollary 5.41.

Corollary 5.43. *Under the hypotheses of Theorem 5.39, if there is $c_0 > 0$ such that $(Av, v)_L \geq c_0 \|v\|_L^2$ for all $v \in V$, then*

$$\|u - u_h\|_L \leq c \left(\sum_{K \in \mathcal{T}_h} (h_K^{2k+1} + \epsilon h_K^{2k}) \|u\|_{W,K}^2 \right)^{\frac{1}{2}}. \tag{5.49}$$

Furthermore, if $\epsilon \leq ch$, then $\|u - u_h\|_L \leq c' h^{k+\frac{1}{2}} \|u\|_W$.

Proof. The L-coercivity of A together with (5.48) implies

$$c_0 \|u - u_h\|_L^2 \leq (A(u - u_h), u - u_h)_L = a_s(u - u_h, u - u_h)$$
$$\leq \|u - u_h\|_{h,T_\epsilon}^2 \leq c \sum_{K \in \mathcal{T}_h} h_K^{2k}(h_K + \epsilon) \|u\|_{W,K}^2,$$

yielding (5.49). Furthermore, if $\epsilon \leq ch$, the right-hand side is clearly controlled by $c' h^{k+\frac{1}{2}} \|u\|_W$. □

Remark 5.44. The GaLS method is slightly more accurate than the Least-Squares formulation when A is L-coercive. This property is in part responsible for its popularity. □

5.4.4 Example: Advection–diffusion–reaction

To illustrate the GaLS method, we apply it to an advection–diffusion–reaction problem. Let $\beta \in [L^\infty(\Omega)]^d$, $\nabla \cdot \beta \in L^\infty(\Omega)$, $\mu \in L^\infty(\Omega)$, and assume that (5.14) holds. Introduce the operator

$$A_{\mu,\beta} : V \ni v \longmapsto A_{\mu,\beta} v = \mu v + \beta \cdot \nabla v \in L = L^2(\Omega),$$

with $V = \{w \in L^2(\Omega); \beta \cdot \nabla w \in L^2(\Omega); \beta \cdot n_{|\partial\Omega^-} = 0\}$ and inflow boundary $\partial\Omega^-$ defined in §5.2.3. Owing to Proposition 5.9, $A_{\mu,\beta}$ is bijective and monotone.

Let $\partial\Omega_D$ be a subset of $\partial\Omega$ with positive measure such that $\partial\Omega^- \subset \partial\Omega_D$. Define the Hilbert space

$$X = \{u \in H^1(\Omega); \, u_{|\partial\Omega_D} = 0\}.$$

The condition $\partial\Omega^- \subset \partial\Omega_D$ implies $X \subset V$. Let $\mathcal{S}$ be a field in $[W^{1,\infty}(\Omega)]^{d\times d}$ such that, a.e. in Ω, the matrix $\mathcal{S}(x)$ is symmetric. Introduce the diffusion operator $D_{\mathcal{S}} = -\nabla\cdot(\mathcal{S}\cdot\nabla(\cdot))$ and assume that $D_{\mathcal{S}}$ is elliptic on X (see Definition 3.1): there exists $\nu_0 > 0$ such that, a.e. in Ω and for all $Z \in \mathbb{R}^d$,

$$Z^T\cdot\mathcal{S}\cdot Z \geq \nu_0\|Z\|_d^2.$$

As a result, setting $s(u,v) = \int_\Omega (\nabla u)^T\cdot\mathcal{S}\cdot\nabla v$, it is clear that $s \in \mathcal{L}(X \times X; \mathbb{R})$ and s is X-coercive. Given $f \in L^2(\Omega)$, the advection–diffusion–reaction equation we are interested in is the following:

$$\begin{cases} \mu u + \beta\cdot\nabla u - \nabla\cdot(\mathcal{S}\cdot\nabla u) = f \text{ in } \Omega, \\ u = 0 \text{ on } \partial\Omega_D \quad \text{and} \quad n^T\cdot\mathcal{S}\cdot\nabla u = 0 \text{ on } \partial\Omega\setminus\partial\Omega_D. \end{cases} \tag{5.50}$$

To set (5.50) in the framework of problem (5.34), we proceed as follows: First, the domain Ω is rescaled by $\ell_\Omega = c_\Omega^{-1}$, where c_Ω is the Poincaré constant; see Lemma B.61. Note that ℓ_Ω is a length. Denote by Ω' the rescaled domain. Let $\nabla' = \ell_\Omega\nabla$ be the gradient operator with respect to the rescaled coordinates $x' = \frac{1}{\ell_\Omega}x$. Define $X' = \{v' \in H^1(\Omega'); \, v'_{|\partial\Omega'_D} = 0\}$ with norm $\|\cdot\|_{X'} = \|\cdot\|_{H^1(\Omega')}$ and define $d \in \mathcal{L}(X' \times X'; \mathbb{R})$ to be $d(u',v') = \frac{2}{\nu_0}\int_{\Omega'}(\nabla'u')^T\cdot\mathcal{S}(x')\cdot\nabla'v' \, dx'$.

Dropping primes to alleviate the notation, it is clear that

$$\forall u \in X, \quad d(u,u) \geq \|u\|_{1,\Omega}^2,$$

i.e., (5.33) holds. Introduce the first-order differential operator A such that $Au = \frac{1}{\|\beta\|_{L^\infty(\Omega)}}(\ell_\Omega\mu u + \beta\cdot\nabla u)$, and observe that $\|A\|_{\mathcal{L}(X;L)}$ is of order one. Introduce the dimensionless parameter

$$\epsilon = \frac{\nu_0}{2\ell_\Omega\|\beta\|_{L^\infty(\Omega)}}. \tag{5.51}$$

This quantity is the reciprocal of the so-called *Péclet number*. Finally, set $T_\epsilon = A + \epsilon D$ and $t_\epsilon(u,v) = (Au,v)_{0,\Omega} + \epsilon d(u,v)$. For $f \in L^2(\Omega)$, consider the (rescaled) advection–diffusion–reaction problem:

$$\begin{cases} \text{Seek } u \in X \text{ such that} \\ t_\epsilon(u,v) = (f,v)_{0,\Omega}, \quad \forall v \in X. \end{cases}$$

To construct the GaLS approximation, introduce a shape-regular mesh family, say $\{\mathcal{T}_h\}_{h>0}$, and a H^1-conformal finite element space $X_h \subset X$ based on $\mathcal{T}_h$. The GaLS formulation is obtained by substituting the above definition of T_ϵ into (5.38). Clearly, the localization hypotheses (5.40) and (5.41) hold. Furthermore, the inverse inequality (5.42) results from the following:

Lemma 5.45. *If the family* $\{\mathcal{T}_h\}_{h>0}$ *is shape-regular, there is* c *such that*

$$\forall h, \ \forall v_h \in X_h, \ \forall K \in \mathcal{T}_h, \quad \|\nabla \cdot (\mathcal{S} \cdot \nabla v_h)\|_{0,K} \le c\, h_K^{-1} \|\nabla v_h\|_{0,K}.$$

Proof. Use the local inverse inequality in Lemma 1.138 to infer

$$\|\nabla \cdot (\mathcal{S} \cdot \nabla v_h)\|_{0,K} \le \|\nabla \cdot \mathcal{S}\|_{0,\infty,K} \|\nabla v_h\|_{0,K} + \|\mathcal{S}\|_{0,\infty,K} |v_h|_{2,K}$$
$$\le c\, |v_h|_{1,K} + c'\, h_K^{-1} |v_h|_{1,K} \le c''\, h_K^{-1} |v_h|_{1,K}. \qquad \square$$

In conclusion, all the hypotheses of Theorem 5.39 are satisfied. Furthermore, if $\mathbb{P}_k$ or $\mathbb{Q}_k$ finite elements are used, Theorem 1.103 implies

$$\forall K \in \mathcal{T}_h, \ \forall v \in H^{k+1}(K), \quad \inf_{v_h \in X_h} \|v - v_h\|_{m,K} \le c\, h_K^{k+1-m} |v|_{k+1,K},$$

for all $0 \le m \le k+1$. This yields the following:

Corollary 5.46. *Assume* $u \in H^{k+1}(\Omega)$. *Then:*

(i) $\forall \epsilon, \ \|u - u_h\|_{0,\Omega} \le c\, h^{k+\frac{1}{2}} \|u\|_{k+1,\Omega}.$
(ii) *If* $\epsilon \ge ch, \ \|u - u_h\|_{1,\Omega} \le c\, h^k \|u\|_{k+1,\Omega}.$
(iii) *If the family* $\{\mathcal{T}_h\}_{h>0}$ *is quasi-uniform and* $\epsilon \le ch, \ \|\beta \cdot \nabla(u - u_h)\|_{0,\Omega} \le c\, h^k \|u\|_{k+1,\Omega}.$

Remark 5.47.
(i) When $\epsilon \le ch$ and the mesh family is quasi-uniform, the above estimate shows that the GaLS formulation yields a better control on the L^2-norm of the streamwise derivative than on the seminorm associated with diffusion.
(ii) Denoting by h_K the local mesh size of elements in the mesh of the unrescaled domain, a condition such that $\epsilon \le ch_K$ is equivalent to saying that the *cell Péclet number* $\frac{2h_K \|\beta\|_{L^\infty(\Omega)}}{\nu_0}$ is large. $\qquad \square$

5.5 Subgrid Viscosity Approximation

Although the GaLS technique presented in the preceding section is widely used, it is not fully satisfactory for two reasons: it involves a function $\delta(h_K)$ that may require debatable tuning when it comes to solving complicated problems, and the extension of GaLS to time-dependent problems is not straightforward since it requires using discontinuous finite elements in time; see §6.3.2 or, e.g., [Joh87].

The goal of the present section is to present an alternative technique which allows for solving non-coercive PDEs almost optimally without resorting to a tunable function (although the technique still involves one tunable parameter), and which extends easily to time-dependent problems; see §6.3.4.

5.5.1 The model problem

Once again, consider the theoretical setting introduced in §5.2.1. Let $V \subset L$ be two real Hilbert spaces with dense and continuous embedding, and let $a \in \mathcal{L}(V \times L; \mathbb{R})$ be a bilinear form satisfying the two hypotheses of the BNB Theorem. As a result, the operator $A : D(A) = V \rightarrow L$, defined by $(Au, v)_L = a(u, v)$ for $(u, v) \in V \times L$, is an isomorphism. Henceforth, we consider the problem: For $f \in L$,

$$\begin{cases} \text{Seek } u \in V \text{ such that} \\ a(u, v) = (f, v)_L, \quad \forall v \in L. \end{cases} \qquad (5.52)$$

The bilinear form a is also assumed to be positive, i.e., A is assumed to be *monotone*. The examples presented in §5.2.3, §5.2.4, and §5.2.5 fall into this framework.

5.5.2 The discrete setting

To construct an approximate solution to (5.52), let us introduce a sequence of finite-dimensional subspaces of V, say $\{X_H\}_{H>0}$, and assume that there exist a dense subspace of V, say W, as well as numbers $k > 0$ and $c > 0$ such that

$$\forall v \in W, \quad \inf_{w_H \in X_H} \|v - w_H\|_L + H\|v - w_H\|_V \le cH^{k+1}\|v\|_W. \qquad (5.53)$$

Since, for first-order PDEs, $\inf_{u_H \in X_H} \sup_{v_h \in X_H} a(u_H, v_H)/(\|u_H\|_V\|v_H\|_L)$ behaves like H when $H \rightarrow 0$ (see Theorem 5.3), it is clear that the standard Galerkin technique is not optimal in general. A possible solution to this difficulty consists of enlarging the test space. Indeed, since a satisfies the condition (BNB1), it is clear that $\inf_{u_H \in X_H} \sup_{v \in L} a(u_H, v)/(\|u_H\|_V\|v\|_L) \ge \alpha$. Hence, while keeping the solution space to be X_H, there certainly exists between X_H and L a large collection of test spaces such that the inf-sup inequality is satisfied uniformly. Assume for the time being that we are able to exhibit such a finite-dimensional space, say X_h, such that $X_H \subsetneq X_h \subset L$ and

$$\exists c_a > 0, \quad \inf_{v_H \in X_H} \sup_{\phi_h \in X_h} \frac{a(v_H, \phi_h)}{\|v_H\|_V\|\phi_h\|_L} \ge c_a, \qquad (5.54)$$

uniformly in h. Now, consider the non-standard Galerkin approximation:

$$\begin{cases} \text{Seek } u_H \in X_H \text{ such that} \\ a(u_H, v_h) = (f, v_h)_L, \quad \forall v_h \in X_h. \end{cases}$$

Clearly, if this problem has a solution, (5.54) ascertains that the solution is stable uniformly with respect to H and h. Unfortunately, the dimension of X_h is larger than that of X_H since $X_H \subsetneq X_h$; as a result, Theorem 2.22 cannot be applied. If, to avoid the dimension problem, one takes $X_h = X_H$, then (5.54) cannot hold. In summary: Hypothesis (5.54) yields a control on $\|u_H\|_V$, but the standard Galerkin formulation in X_h does not control $\|u_h\|_V$. That is to say, an a priori control on $\|u_h - u_H\|_V$ is missing. A very simple technique to control this quantity consists of adding some artificial viscosity that only acts on $u_h - u_H$ so that (5.52) is not perturbed too much.

Let us now detail the hypotheses underlying the *hierarchical setting* needed to carry out our program:

(i) There is a pair of spaces $\{X_H, X_h\}$ with $X_H \subsetneq X_h \subset V$.

(ii) There is an L-stable direct sum $X_h = X_H \oplus X_h^H$, i.e., letting $P_H : X_h \to X_H$ be the projection from X_h to X_H parallel to X_h^H,

$$\exists c_s > 0, \ \forall v_h \in X_h, \quad \|P_H v_h\|_L \leq c_s \|v_h\|_L. \tag{5.55}$$

Henceforth, we shall use the notation $v_H = P_H v_h$ and $v_h^H = (1 - P_H)v_h$.

(iii) The following inverse inequality holds in X_h:

$$\exists c_i > 0, \ \forall v_h \in X_h, \quad \|v_h\|_V \leq c_i H^{-1} \|v_h\|_L. \tag{5.56}$$

(iv) The interpolation property (5.53) holds.

(v) There exists a norm $\| \cdot \|_b$ such that

$$\exists c_{e1}, c_{e2} > 0, \ \forall v_h^H \in X_h^H, \quad c_{e1} \|v_h^H\|_V \leq \|v_h^H\|_b \leq c_{e2} H^{-1} \|v_h^H\|_L, \tag{5.57}$$

and there exists $b_h \in \mathcal{L}(X_h^H \times X_h^H; \mathbb{R})$ such that, $\forall (v_h^H, w_h^H) \in X_h^H \times X_h^H$,

$$c_{b1} H \|v_h^H\|_b^2 \leq b_h(v_h^H, v_h^H) \quad \text{and} \quad b_h(v_h^H, w_h^H) \leq c_{b2} H \|v_h^H\|_b \|w_h^H\|_b. \tag{5.58}$$

(vi) The inf-sup condition (5.54) holds.

All the constants involved in (5.53)–(5.58) do not depend on the pair $\{H, h\}$. The discrete problem we consider henceforth is:

$$\begin{cases} \text{Seek } u_h \in X_h \text{ such that} \\ a(u_h, v_h) + b_h(u_h^H, v_h^H) = (f, v_h)_L, \quad \forall v_h \in X_h. \end{cases} \tag{5.59}$$

Remark 5.48.

(i) In (5.59), the only difference with the standard Galerkin formulation is the presence of the bilinear form b_h, i.e., the Galerkin formulation consists of taking $b_h = 0$.

(ii) The reader may think of X_H as being the space associated with the *resolved scales* of the solution, whereas X_h^H is associated with the *fluctuating scales*. The operator P_H is a *filter* that removes the space fluctuations of the solution supported in the subgrid scale space X_h^H.

(iii) Property (5.56) is an inverse inequality. If the operator A is a first-order differential operator and if X_h and X_H are finite element spaces constructed on quasi-uniform meshes of characteristic size h and H, respectively, then (5.56) holds if H and h are of similar order, i.e., $c_1 h \leq H \leq c_2 h$. In the applications described in §5.5.6, either $H = 2h$ or $H = h$.

(iv) The theory generalizes to shape-regular families of non-uniform meshes by localizing the definition of b_h as described in [Gue01^b]; see also the localization hypotheses of GaLS (5.40), (5.41), and (5.42).

(v) The notion of scale separation and subgrid scale dissipation is similar in spirit to the spectral viscosity technique introduced by [Tad89] to approximate nonlinear conservation equations by means of spectral methods.   □

Example 5.49. Denote by $(\cdot, \cdot)_V$ the inner product of V. The simplest choice to define b_h consists of setting $b_h(v_h^H, w_h^H) = H(v_h^H, w_h^H)_V$, which immediately implies $\| \cdot \|_b = \| \cdot \|_V$. For instance, in the framework of the scalar advection problem of §5.2.3, this choice yields

$$b_h(v_h^H, w_h^H) = H(\mu v_h^H + \beta \cdot \nabla v_h^H, \mu w_h^H + \beta \cdot \nabla w_h^H)_{0,\Omega}. \qquad (5.60)$$

A second possibility consists of assuming that there exists a Hilbert subspace of V, say X, dense and continuously embedded, such that the inverse inequality $\|v_h^H\|_X \le c_{e2} H^{-1} \|v_h^H\|_L$ holds for all v_h^H in X_h^H. This hypothesis means that V and X are domains of differential operators of the same order. Denote by $(\cdot, \cdot)_X$ the inner product of X and assume $X_h \subset X$. Then, it is admissible to set $b_h(v_h^H, w_h^H) = H(v_h^H, w_h^H)_X$. This setting implies $\| \cdot \|_b = \| \cdot \|_X$. In the framework of a scalar advection equation, one can set $X = H^1(\Omega) \subset V$ and

$$b_h(v_h^H, w_h^H) = H(v_h^H, w_h^H)_{0,\Omega} + H(\nabla v_h^H, \nabla w_h^H)_{0,\Omega}. \qquad (5.61)$$

The advantage of (5.61) over (5.60) is that if β is a time-dependent field, the matrix associated with (5.60) must be reassembled at each time step, whereas that associated with (5.61) needs to be assembled only once.   □

5.5.3 Convergence analysis

To simplify the notation, introduce the bilinear form

$$a_h(v_h, w_h) = a(v_h, w_h) + b_h(v_h^H, w_h^H),$$

and let a_s denote the symmetric part of a, i.e., $a_s(v, w) = \frac{1}{2}((Av, w)_L + (v, Aw)_L)$. Owing to the monotonicity of A, a_s is positive. Furthermore, define the norms

$$\begin{cases} \|v\|_{h,A}^2 = a_s(v, v) + H\|v\|_V^2, \\ \|v\|_{h,\frac{1}{2}}^2 = \|v\|_{h,A}^2 + H^{-1}\|v\|_L^2. \end{cases}$$

Owing to the inverse inequality (5.56), we infer

$$\forall w_h \in X_h, \quad \|w_h\|_{h,\frac{1}{2}} \le c(H^{-\frac{1}{2}}\|w_h\|_L + H^{\frac{1}{2}}\|w_h\|_V) \le c' H^{-\frac{1}{2}}\|w_h\|_L. \qquad (5.62)$$

Lemma 5.50 (Stability). *There is $c > 0$, independent of $\{H, h\}$, such that*

$$\inf_{v_h \in X_h} \sup_{w_h \in X_h} \frac{a_h(v_h, w_h)}{\|v_h\|_{h,A}\|w_h\|_{h,A}} \ge c. \qquad (5.63)$$

Proof. (1) Let v_h be an arbitrary element in X_h. Then, (5.58) implies

$$a_s(v_h, v_h) + c_{b1} H \|v_h^H\|_b^2 \leq a_h(v_h, v_h). \tag{5.64}$$

(2) A control on $\|v_H\|_V$ is obtained from the inf-sup condition (5.54),

$$c_a \|v_H\|_V \leq \sup_{w_h \in X_h} \frac{a(v_H, w_h)}{\|w_h\|_L} = \sup_{w_h \in X_h} \frac{a_h(v_h, w_h) - a_h(v_h^H, w_h)}{\|w_h\|_L}.$$

Use the continuity of b_h in (5.58), the inverse inequality in (5.57), and the stability of P_H in (5.55) to infer

$$\begin{aligned}
a_h(v_h^H, w_h) &= a(v_h^H, w_h) + b_h(v_h^H, w_h^H) \\
&\leq \|a\| \, \|v_h^H\|_V \|w_h\|_L + c_{b2} H \|v_h^H\|_b \|w_h^H\|_b \\
&\leq c_{e1}^{-1} \|a\| \, \|v_h^H\|_b \|w_h\|_L + c_{b2} c_{e2} \|v_h^H\|_b \|w_h^H\|_L \leq c \|v_h^H\|_b \|w_h\|_L.
\end{aligned}$$

As a result,

$$c_a \|v_H\|_V \leq \sup_{w_h \in X_h} \frac{a_h(v_h, w_h)}{\|w_h\|_L} + c_1 \|v_h^H\|_b. \tag{5.65}$$

(3) Take the square of (5.65), multiply the result by $\frac{c_{b1} H}{4 c_1^2}$, and add to (5.64) to obtain

$$a_s(v_h, v_h) + c H \|v_H\|_V^2 + \frac{c_{b1}}{2} H \|v_h^H\|_b^2 \leq a_h(v_h, v_h) + c' \sup_{w_h \in X_h} \frac{a_h(v_h, w_h)^2}{H^{-1} \|w_h\|_L^2}.$$

Since $\|v_h\|_V \leq \|v_H\|_V + c_{e1}^{-1} \|v_h^H\|_b$, the above inequality leads to

$$\|v_h\|_{h,A}^2 \leq c_1 \frac{a_h(v_h, v_h)}{\|v_h\|_{h,A}} \|v_h\|_{h,A} + c_2 \sup_{w_h \in X_h} \frac{a_h(v_h, w_h)^2}{H^{-1} \|w_h\|_L^2}.$$

The arithmetic–geometric inequality yields

$$\|v_h\|_{h,A}^2 \leq c_1 \sup_{w_h \in X_h} \frac{a_h(v_h, w_h)^2}{\|w_h\|_{h,A}^2} + c_2 \sup_{w_h \in X_h} \frac{a_h(v_h, w_h)^2}{H^{-1} \|w_h\|_L^2}.$$

Now, use (5.62) and the definitions of the norms $\| \cdot \|_{h,A}$ and $\| \cdot \|_{h,\frac{1}{2}}$ to infer $H^{-1} \|w_h\|_L^2 \geq c \|w_h\|_{h,\frac{1}{2}}^2 \geq c' \|w_h\|_{h,A}^2$. Inserting this bound into the above inequality yields the desired result.

Lemma 5.51 (Continuity). *There is c, independent of $\{H, h\}$, such that*

$$\forall (v, w) \in V \times V, \quad a(v, w) \leq c \|v\|_{h,\frac{1}{2}} \|w\|_{h,A}. \tag{5.66}$$

Proof. Let v and w be two elements in V. Use the fact that a_s is symmetric and positive together with the Cauchy–Schwarz inequality to infer

$$\begin{aligned}
a(v, w) &= 2 a_s(v, w) - a(w, v) \leq 2 a_s(v, v)^{\frac{1}{2}} a_s(w, w)^{\frac{1}{2}} + \|a\| \, \|w\|_V \|v\|_L \\
&\leq c \left(a_s(v, v) + H^{-1} \|v\|_L^2\right)^{\frac{1}{2}} \left(a_s(w, w) + H \|w\|_V^2\right)^{\frac{1}{2}} \leq c \|v\|_{h,\frac{1}{2}} \|w\|_{h,A},
\end{aligned}$$

yielding the desired result. $\qquad \square$

Lemma 5.52 (Consistency).

$$\forall w_h \in X_h, \quad a_h(u_h, w_h) = a(u, w_h), \tag{5.67}$$

$$\forall w_h \in X_h, \ \forall v_H \in X_H, \quad a_h(v_H, w_h) = a(v_H, w_h). \tag{5.68}$$

Proof. Straightforward verification: the first equality is obtained by subtracting (5.59) from (5.52) and using the V-conformity of X_h; the second equality simply results from the fact that v_H has no subgrid scales.   □

Theorem 5.53 (Convergence). *Under the assumptions* (5.54)–(5.58), *problem* (5.59) *has a unique solution, and this solution satisfies*

$$\forall \{H, h\}, \quad \|u - u_h\|_{h,A} \le c \inf_{v_H \in X_H} \|u - v_H\|_{h,\frac{1}{2}}. \tag{5.69}$$

Proof. Let v_H be an arbitrary element in X_H. Lemma 5.52 shows that $\forall w_h \in X_h$ and $\forall v_H \in X_H$, $a_h(u_h - v_H, w_h) = a(u - v_H, w_h)$. The stability and continuity properties then imply

$$\|u_h - v_H\|_{h,A} \le c \sup_{w_h \in X_h} \frac{a_h(u_h - v_H, w_h)}{\|w_h\|_{h,A}} \le c \sup_{w_h \in X_h} \frac{a(u - v_H, w_h)}{\|w_h\|_{h,A}}$$

$$\le c \|u - v_H\|_{h,\frac{1}{2}}.$$

To conclude, apply the triangle inequality, use the fact that $\|\cdot\|_{h,A} \le \|\cdot\|_{h,\frac{1}{2}}$, and take the infimum on v_H.   □

Corollary 5.54. *Under hypotheses* (5.53)–(5.58), *if u, the solution to* (5.52), *is in W, the solution u_h to* (5.59) *satisfies for all $\{H, h\}$,*

$$H^{\frac{1}{2}} \|u - u_h\|_V + a_s(u - u_h, u - u_h)^{\frac{1}{2}} \le c H^{k+\frac{1}{2}} \|u\|_W. \tag{5.70}$$

Proof. Estimate (5.69) implies

$$H\|u - u_h\|_V^2 + a_s(u - u_h, u - u_h) \le \|u - u_h\|_{h,A}^2 \le c \inf_{v_H \in X_H} \|u - v_H\|_{h,\frac{1}{2}}^2.$$

If $u \in W$, (5.70) results from (5.53) since

$$\inf_{v_H \in X_H} \|u - v_H\|_{h,\frac{1}{2}} \le c H^{k+\frac{1}{2}} \|u\|_W.$$ □

Remark 5.55.

(i) The above convergence proof uses the arguments of the First Strang Lemma since the form a_h is defined only on the discrete space $V_h \times V_h$. Indeed, the expression $a_h(u, v_h)$ does not make sense for $u \in V$ and $u \notin V_h$, since u has a priori no decomposition in $V_H \oplus V_h^H$.

(ii) Estimate (5.70) is optimal in V. If a_s is L-coercive, (5.70) yields $\|u - u_h\|_L \le c H^{k+\frac{1}{2}} \|u\|_W$. This estimate is not optimal in L since a $H^{\frac{1}{2}}$ factor is missing. Optimality can be recovered with finite elements if the mesh satisfies some geometric properties; [Zho97] for the details.

(iii) Estimate (5.70) is identical to the estimate given by the GaLS formulation with $\epsilon = 0$; see Corollaries 5.41 and 5.43.

(iv) The subgrid viscosity technique discussed in this section has been introduced in [Gue99^a, Gue01^b].   □

5.5.4 Refinement of the hypotheses

It happens frequently that the operator A can be decomposed into $A = A_0 + A_1$ where A_0 is a zeroth-order operator and A_1 is a first-order differential operator. For instance, for the advection operator considered in §5.2.3, we set $A_0 u = \mu u$ and $A_1 u = \beta \cdot \nabla u$. A more general example is given by the Friedrichs operators presented in §5.2.2.

Using the decomposition $a = a_0 + a_1$ where $a_0(u,v) = (A_0 u, v)_L$ and $a_1(u,v) = (A_1 u, v)_L$, we now assume:

1. There is a seminorm in V, which we denote by $|\cdot|_V$, such that

$$\forall (u,v) \in V \times L, \quad \begin{cases} \|u\|_V \leq c \left(a_s(u,u)^{\frac{1}{2}} + |u|_V \right), \\ a_0(u,v) \leq c_0 a_s(u,u)^{\frac{1}{2}} \|v\|_L, \\ a_1(u,v) \leq c_1 |u|_V \|v\|_L. \end{cases} \quad (5.71)$$

2. We weaken hypothesis (5.54) by replacing it as follows: there are two constants $c_{a1} > 0$, $c_\delta \geq 0$, independent of $\{H, h\}$, such that

$$\forall u_h \in X_h, \quad \sup_{v_h \in X_h} \frac{a_1(u_H, v_h)}{\|v_h\|_L} \geq c_{a1} |u_H|_V - c_\delta a_s(u_h, u_h)^{\frac{1}{2}}. \quad (5.72)$$

3. We weaken the definition of b_h: there is a seminorm $|\cdot|_b$ such that

$$\forall (v_h^H, w_h^H) \in X_h^H \times X_h^H, \quad \begin{cases} c_{e1} |v_h^H|_V \leq |v_h^H|_b \leq c_{e2} H^{-1} \|v_h^H\|_L, \\ b_h(v_h^H, v_h^H) \geq c_{b1} H |v_h^H|_b^2, \\ b_h(v_h^H, w_h^H) \leq c_{b2} H |v_h^H|_b |w_h^H|_b. \end{cases} \quad (5.73)$$

Proposition 5.56. *Under hypotheses* (5.53), (5.55), (5.56), (5.71), (5.72), *and* (5.73), *if the solution to* (5.52) *is in* W, *the solution to* (5.59) *satisfies the estimate* (5.70).

Proof. Adapt the proof of Theorem 5.53; see [Gue01^b]. □

Remark 5.57.

(i) The reason for weakening (5.54) is that (5.72) is usually simpler to prove; see Corollary 5.62.

(ii) For the advection–reaction equation $\mu u + \beta \cdot \nabla u = f$, assuming $\mu - \frac{1}{2} \nabla \cdot \beta \geq \mu_0 > 0$, the bilinear form a is $L^2(\Omega)$-coercive. Hence, instead of using (5.60), one can set $|v_h^H|_b = \|\nabla v_h^H\|_{0,\Omega}$ and use the bilinear form

$$\forall (v_h^H, w_h^H) \in X_h^H \times X_h^H, \quad b_h(v_h^H, w_h^H) = H(\nabla v_h^H, \nabla w_h^H)_{0,\Omega}. \quad \square$$

5.5.5 A singular perturbation problem

The technique developed above is tailored for problems where A is a first-order differential operator. In practice, one often deals with operators of the form $T_\epsilon = A + \epsilon D$, where A is a monotone first-order differential operator and D is second-order and coercive (for instance, for an advection–diffusion problem, ϵ is the reciprocal of the Péclet number; see §5.4.4). Owing to the monotonicity of A, the operator T_ϵ is coercive with coercivity constant ϵ. If ϵ is of order 1, the problem $T_\epsilon u = f$ is elliptic and can easily be approximated by means of the standard Galerkin technique. On the other hand, if ϵ is small, the coercivity is not strong enough to guarantee that the standard Galerkin technique works properly. Indeed, in first approximation $T_\epsilon \approx A$ when $\epsilon \to 0$.

The subgrid viscosity technique generalizes to this situation and yields optimal convergence estimates. Let the hypotheses on a, V, and L stated before hold. Moreover, introduce a Hilbert space X and assume that $X \subset V$ with dense and continuous embedding. Define $d \in \mathcal{L}(X \times X; \mathbb{R})$ and assume that the bilinear form $a + d$ is X-coercive, i.e., $\|v\|_X^2 \le a(v,v) + d(v,v)$. For $0 \le \epsilon \le 1$ and $f \in L$, consider the problem:

$$\begin{cases} \text{Seek } u \in X \text{ such that} \\ a(u,v) + \epsilon d(u,v) = (f,v)_L, \quad \forall v \in X. \end{cases} \tag{5.74}$$

Example 5.58. Advection–diffusion–reaction problems correspond to $a(u,v) = (\mu u + \beta \cdot \nabla u, v)_{0,\Omega}$ and $d(u,v) = \epsilon(\nabla u, \nabla v)_\Omega$ with $X = H_0^1(\Omega)$, $V = \{v \in L^2(\Omega); \beta \cdot \nabla v \in L^2(\Omega); w_{|\partial\Omega^-} = 0\}$, and $L = L^2(\Omega)$. ☐

Let us now approximate the solution to problem (5.74). Let $X_h \subset X_H \subset X$ satisfy hypotheses (5.53), (5.55), (5.56), (5.71), (5.72), and (5.73). Furthermore, assume that there is c, independent of $\{H, h\}$, such that

$$\forall v_h \in X_h, \quad \|v_h\|_X \le cH^{-1}\|v_h\|_L. \tag{5.75}$$

This hypothesis means that X and V are domains of differential operators of the same order. Consider the discrete problem:

$$\begin{cases} \text{Seek } u_h \in X_h \text{ such that} \\ a(u_h, v_h) + \epsilon d(u_h, v_h) + b_h(u_h^H, v_h^H) = (f, v_h)_L, \quad \forall v_h \in X_h. \end{cases} \tag{5.76}$$

Theorem 5.59. *Under hypotheses* (5.53), (5.55), (5.56), (5.71), (5.72), (5.73), *and* (5.75), *and provided* $u \in W$, *there is* c, *independent of* $\{H, h\}$, *such that the solution* u_h *to* (5.76) *satisfies:*

(i) $a_s(u - u_h, u - u_h)^{\frac{1}{2}} + \epsilon^{\frac{1}{2}}\|u - u_h\|_X \le c(H^{k+\frac{1}{2}} + H^k \epsilon^{\frac{1}{2}})\|u\|_W$.

(ii) $\|u - u_h\|_V \le cH^k\|u\|_W$.

Proof. See [Gue01^b]. ☐

Remark 5.60. The error estimates are similar to those from the GaLS technique. Note however, that contrary to GaLS where the stabilizing parameter $\delta(h_K)$ is a nontrivial function of ϵ and h_K (see (5.39)), the parameter c_{b1} (see (5.58) and (5.73)) does not depend on ϵ. This is an interesting feature especially when the subgrid viscosity method is used to solve nonlinear problems and/or vector-valued problems with anisotropic diffusion matrices. Designing the right stabilizing function $\delta(h_K)$ for GaLS may no be straightforward in this situation.   □

5.5.6 Two-level $\mathbb{P}_1$ and $\mathbb{P}_2$ interpolation

This section describes four finite element settings which satisfy the hypotheses of the subgrid viscosity technique. For the sake of simplicity, we assume that Ω is a polyhedron in $\mathbb{R}^d$ and that $\{T_H\}_{H>0}$ is a shape-regular family of affine triangulations of Ω. The reference simplex is denoted by $\widehat{K}$ and $T_{K_H} : \widehat{K} \to K_H$ is the affine transformation that maps $\widehat{K}$ to K_H.

$\mathbb{P}_1$/**bubble setting.** A $\mathbb{P}_1$ space for the resolved scales is defined as

$$X_H = \{v_H \in [H^1(\Omega)]^m; \forall K_H \in T_H, v_{H|K_H} \in [\mathbb{P}_1(K_H)]^m\}. \qquad (5.77)$$

Let $\widehat{\psi}$ be a bubble function in $H_0^1(\widehat{K})$ such that $0 \leq \widehat{\psi} \leq 1$. Letting $\psi_h = \widehat{\psi} \circ T_{K_H}^{-1}$, define the vector space $X_h^H(K_H) = [\mathrm{span}\{\psi_h\}]^m$ for all K_H in T_H, and construct the space of the subgrid scales X_h^H as follows:

$$X_h^H = \bigoplus_{K_H \in T_H} X_h^H(K_H), \qquad (5.78)$$

where $\oplus_{K_H \in T_H}$ is the direct sum on all the simplices of T_H. Letting $X_h = X_H \oplus X_h^H$, the pair $\{X_H, X_h\}$ is hereafter referred to as the $\mathbb{P}_1$/bubble setting.

$\mathbb{P}_2$/**bubble setting.** A $\mathbb{P}_2$ space for the resolved scales is defined as

$$X_H = \{v_H \in [H^1(\Omega)]^m; \forall K_H \in T_H, v_{H|K_H} \in [\mathbb{P}_2(K_H)]^m\}. \qquad (5.79)$$

To construct the subgrid scale space, introduce $d + 1$ linearly independent functions of $H_0^1(\widehat{K})$, say $\{\widehat{\psi}_0, \ldots, \widehat{\psi}_d\}$. Let $\{\widehat{a}_0, \ldots, \widehat{a}_d\}$ be the $(d + 1)$ vertices of the reference simplex $\widehat{K}$. Let R_{ij} be the symmetries of $\widehat{K}$ such that $R_{ij}(\widehat{a}_i) = \widehat{a}_j$ and $R_{ij}(\widehat{a}_l) = \widehat{a}_l$ if $l \notin \{i, j\}$. We assume that the $(d + 1)$ functions $\{\widehat{\psi}_0, \ldots, \widehat{\psi}_d\}$ satisfy the symmetry properties

$$\begin{cases} \widehat{\psi}_i \circ R_{ij} = \widehat{\psi}_j, \\ \widehat{\psi}_i \circ R_{jl} = \widehat{\psi}_i & \text{if } i \notin \{j, l\}. \end{cases} \qquad (5.80)$$

Set $\psi_{i,h} = \widehat{\psi}_i \circ T_{K_H}^{-1}$ for $0 \leq i \leq d$, $X_h^H(K_H) = [\mathrm{span}(\psi_{0,h}, \ldots, \psi_{d,h})]^m$, and, finally, define

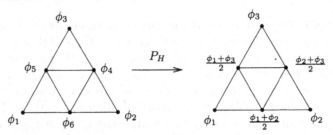

Fig. 5.4. Definition of P_H for the two-level $\mathbb{P}_1$ setting.

$$X_h^H = \bigoplus_{K_H \in \mathcal{T}_H} X_h^H(K_H) \quad \text{and} \quad X_h = X_H \oplus X_h^H. \qquad (5.81)$$

The pair $\{X_H, X_h\}$ is referred to as the $\mathbb{P}_2$/bubble setting.

Two-level $\mathbb{P}_1$ setting. The two settings described above are not really *hierarchical* since X_H and X_h^H are defined on the same mesh, i.e., in some sense $h = H$. We now present an alternative approach. We restrict ourselves to two dimensions, but three-dimensional generalizations are straightforward. First, define X_H by

$$X_H = \{v_H \in [H^1(\Omega)]^m; \forall K_H \in \mathcal{T}_H, v_{H|K_H} \in [\mathbb{P}_1(K_H)]^m\}. \qquad (5.82)$$

From each triangle $K_H \in \mathcal{T}_H$, create four new triangles by connecting the midpoints of the three edges of K_H. Set $h = \frac{H}{2}$ and denote by $\mathcal{T}_h$ the resulting triangulation. For each macrotriangle K_H, define $\mathbb{P}_{K_H}$ to be the space of functions that are continuous on K_H, vanish at the three vertices of K_H, and are piecewise $\mathbb{P}_1$ on each subtriangle of K_H. Set

$$X_h^H = \{v_h^H \in [H^1(\Omega)]^m; \forall K_H \in \mathcal{T}_H, v_{h|K_H}^H \in [\mathbb{P}_{K_H}]^m\}. \qquad (5.83)$$

Letting $X_h = X_H \oplus X_h^H$, it is clear that X_h is characterized by

$$X_h = \{v_h \in [H^1(\Omega)]^m; \forall K_h \in \mathcal{T}_h, v_{h|K_h} \in [\mathbb{P}_1(K_h)]^m\}. \qquad (5.84)$$

The pair $\{X_H, X_h\}$ is referred to as the two-level $\mathbb{P}_1$ setting. Figure 5.4 shows a schematic representation of the action of the filter $P_H : X_h \to X_H$ on a macroelement K_H of $\mathcal{T}_H$.

Two-level $\mathbb{P}_2$ setting. We now construct a two-level $\mathbb{P}_2$ setting. Set $h = \frac{H}{2}$ and let $\mathcal{T}_h$ be the triangulation obtained by dividing each macrotriangle of $\mathcal{T}_H$ into four subtriangles. For each triangle K_h, let ψ_1, ψ_2, and ψ_3 be the three nodal $\mathbb{P}_2$ functions associated with the midpoints of the edges of K_h. Set

$$X_H = \{v_H \in [H^1(\Omega)]^m; \forall K_H \in \mathcal{T}_H, v_{H|K_H} \in [\mathbb{P}_2(K_H)]^m\}, \qquad (5.85)$$

and define the space of the subgrid scales to be

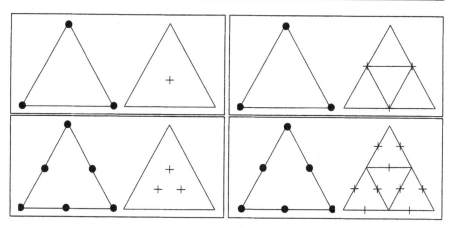

Fig. 5.5. Examples of hierarchical finite elements. In each panel, the resolved scale element is on the left and the subgrid scale element on the right: $\mathbb{P}_1$/bubble (top left); $\mathbb{P}_2$/bubble (bottom left); two-level $\mathbb{P}_1$ (top right); two-level $\mathbb{P}_2$ (bottom right).

$$X_h^H = \{v_h^H \in [H^1(\Omega)]^m; \forall K_h \in \mathcal{T}_h, v_{h|K_h}^H \in [\text{span}\{\psi_1, \psi_2, \psi_3\}]^m\}. \quad (5.86)$$

The space $X_h = X_H \oplus X_h^H$ is characterized by

$$X_h = \{v_h \in [H^1(\Omega)]^m; \forall K_h \in \mathcal{T}_h, v_{h|K_h} \in [\mathbb{P}_2(K_h)]^m\}. \quad (5.87)$$

The pair $\{X_H, X_h\}$ is called the two-level $\mathbb{P}_2$ setting.

The four interpolation settings described above are shown in Figure 5.5.

The inf-sup condition. For all the settings considered above, the decomposition $X_h = X_H \oplus X_h^H$ is L^2-stable. Furthermore, with the notation of §5.2.2, consider the Friedrichs operator A_1 defined in (5.9). Referring to [Gue99^a] for the technical details, we state the following:

Lemma 5.61. *If the fields $\mathcal{A}^k$, $1 \le k \le d$, are piecewise constant on each $K_H \in \mathcal{T}_H$, there is $c_\beta > 0$, independent of $\{H, h\}$, such that*

$$\forall u_H \in X_H, \quad \sup_{v_h \in X_h} \frac{\int_\Omega v_h^T A_1 u_H}{\|v_h\|_{[L^2(\Omega)]^m}} \ge c_\beta \|A_1 u_H\|_{[L^2(\Omega)]^m}. \quad (5.88)$$

Corollary 5.62. *If the fields $\mathcal{A}^k$, $1 \le k \le d$, are in $C^1(\overline{\Omega}; \mathcal{M}_m(\mathbb{R}))$, there are two constants $c_\beta > 0$ and $c_\delta \ge 0$, both independent of $\{H, h\}$, such that*

$$\forall u_H \in X_H, \quad \sup_{v_h \in X_h} \frac{\int_\Omega v_h^T A_1 u_H}{\|v_h\|_{[L^2(\Omega)]^m}} \ge c_\beta \|A_1 u_H\|_{[L^2(\Omega)]^m} - c_\delta \|u_H\|_{[L^2(\Omega)]^m}. \quad (5.89)$$

Remark 5.63. The stabilizing properties of bubble functions for advection-diffusion problems have been emphasized in [BrB92]. Theoretical justifications can be found in [BaB93] and [BrF97]. □

5.5.7 Some applications

We now show that for the two problems considered in §5.2.3 and §5.2.4, hypotheses (5.71), (5.72), and (5.73) are satisfied. The reader may verify that the same conclusion holds for the problem considered in §5.2.5.

Advection–reaction equation. Define $a_0(u, v) = (\mu u, v)_{0,\Omega}$, $a_1(u, v) = (\beta \cdot \nabla u, v)_{0,\Omega}$, and $|u|_V = \|\beta \cdot \nabla u\|_{0,\Omega}$. Assume that $\mu - \frac{1}{2} \nabla \cdot \beta \geq \mu_0 > 0$. Then, (5.71) is a consequence of the relation $a_s(u, u) \geq \mu_0 \|u\|_{0,\Omega}^2$ together with the definition of the seminorm $| \cdot |_V$. Moreover, (5.72) is a consequence of Corollary 5.62 together with the $L^2(\Omega)$-coercivity of a_s. Setting

$$b(v_h^H, w_h^H) = c_b H(\nabla v_h^H, \nabla w_h^H)_{0,\Omega} \qquad \text{and} \qquad |v_h^H|_b = |v_h^H|_{1,\Omega},$$

hypothesis (5.73) is obviously satisfied.

Darcy's equations. Introduce $a_0((u, p), (v, q)) = (K^{-1} \cdot u, v)_{0,\Omega}$ as well as $a_1((u, p), (v, q)) = (A_1(u, p), (v, q))_{0,\Omega} = (q, \nabla \cdot u)_{0,\Omega} + (\nabla p, v)_{0,\Omega}$. Define $|(u, p)|_V = \|A_1(u, p)\|_{0,\Omega} = \|\nabla \cdot u\|_{0,\Omega} + \|\nabla p\|_{0,\Omega}$. Hypothesis (5.71) is a consequence of the relation $a_s((u, p), (u, p)) = a_0((u, p), (u, p)) \geq \alpha' \|u\|_{0,\Omega}^2$ together with the definition of the seminorm $| \cdot |_V$ and the Poincaré inequality. Since the matrix-valued fields $\mathcal{A}^1, \ldots, \mathcal{A}^d$ are constant on Ω, hypothesis (5.72) is a consequence of Lemma 5.61. Finally, setting

$$b((v_h^H, q_h^H), (w_h^H, r_h^H)) = c_b H((\nabla v_h^H, \nabla w_h^H)_{0,\Omega} + (\nabla q_h^H, \nabla r_h^H)_{0,\Omega}),$$

and $|(v_h^H, q_h^H)|_b = |v_h^H|_{1,\Omega} + |q_h^H|_{1,\Omega}$, hypothesis (5.73) is obviously satisfied.

5.5.8 Numerical illustrations

Example 5.64 (An advection problem). Consider the problem

$$\begin{cases} \partial_y u = \frac{1}{2\epsilon} \left(1 - \tanh^2\left(\frac{1}{\epsilon}(y - 0.5)\right)\right) & \text{in } \Omega =]0, 1[^2, \\ u_{|y=0} = 0, \end{cases} \tag{5.90}$$

where $u(y) = \frac{1}{2}(\tanh(\frac{1}{\epsilon}(y - 0.5)) + 1)$ is the exact solution. In the numerical tests, we set $\epsilon = 0.04$. We use two-level $\mathbb{P}_1$ and $\mathbb{P}_2$ finite elements on a mesh $\mathcal{T}_h$ composed of 952 triangles and 517 vertices, i.e., $h \approx \frac{1}{20}$, and we set

$$b_h(v_h^H, w_h^H) = c_b \sum_{K_h \in \mathcal{T}_h} \text{meas}(K_h)^{\frac{1}{2}} \int_{K_h} \nabla v_h^H \cdot \nabla w_h^H, \tag{5.91}$$

with $c_b = 1$. Results are shown in Figure 5.6. The projections in the plane $x = 0$ of the graphs of the $\mathbb{P}_1$ and $\mathbb{P}_2$ interpolants of the exact solution are shown in the left panel of the figure. The two-level $\mathbb{P}_1$ and two-level $\mathbb{P}_2$ solutions are in the center panels. The standard Galerkin $\mathbb{P}_1$ and $\mathbb{P}_2$ solutions are in the right panels. The stabilizing effects of the subgrid viscosity method are clearly visible.   □

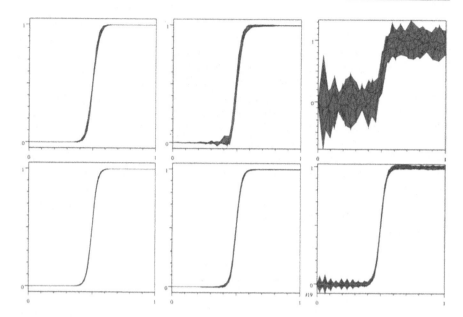

Fig. 5.6. Problem (5.90): $\mathbb{P}_1$ solution (top); $\mathbb{P}_2$ solution (bottom). From left to right: Lagrange interpolant of the exact solution; stabilized solution; and standard Galerkin solution.

Example 5.65 (A boundary layer problem). Consider the problem

$$\begin{cases} \partial_y u - \nu \nabla^2 u = 0 & \text{in } \Omega =]0,1[^2, \\ u_{|y=0} = u_{|y=1} = 0, \quad \partial_x u_{|x=0} = \partial_x u_{|x=1} = 0, \end{cases} \tag{5.92}$$

with exact solution $u(x,y) = \frac{\exp(\lambda y) - 1}{\exp(\lambda) - 1}$ with $\lambda = \frac{1}{\nu}$. In the numerical tests, we set $\nu = 0.002$. We use the same mesh as in the previous example and we approximate the solution by means of two-level $\mathbb{P}_1$ finite elements. The bilinear form b_h is the same as in (5.91).

Results are presented in Figure 5.7. The standard Galerkin solution is polluted by spurious oscillations spreading throughout the computational domain. The oscillations disappear in the stabilized solution except in the vicinity of the boundary layer where the slope of the solution is large. These residual oscillations are due to the *Gibbs phenomenon*. The Gibbs phenomenon is the manifestation of a far-reaching theorem in analysis that states that truncated Fourier series of a given function do not converge uniformly to the function in question unless the function is very smooth (continuity is not enough); see, e.g., [Rud66, Theorem 5.12].

A simple trick to eliminate these oscillations consists of adding strong dissipation in the region where the solution is rough. Of course, one does not know a priori where the solution is rough, but one may expect that in this region the quantity $\nabla u_h^H = \nabla(u_h - P_H u_h)$ is of the same order as ∇u_h. Indeed,

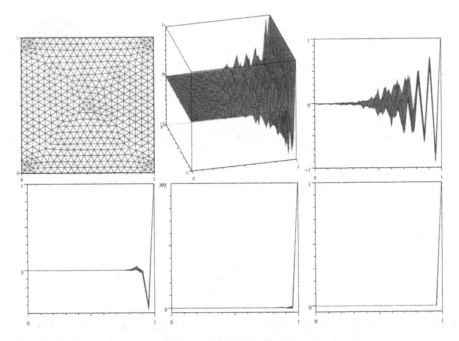

Fig. 5.7. Problem (5.92): mesh (top left); graph of standard Galerkin solution using $\mathbb{P}_1$ finite elements (top center); its projection onto the plane $x = 0$ (top right); stabilized solution (bottom left); stabilized solution with shock-capturing (bottom center); and $\mathbb{P}_1$ interpolant of the exact solution (bottom right).

if u is a smooth function, denoting by $\mathcal{I}_h u$ the Lagrange interpolant of u, the quantity $\|\mathcal{I}_h u - P_H \mathcal{I}_h u\|_{0,\Omega}$ is of order h^{k+1} and $|\mathcal{I}_h u - P_H \mathcal{I}_h u|_{1,\Omega}$ is of order h^k. Hence, we introduce the nonlinear form

$$c_h(u_h; v_h, w_h) = c_{sc} \sum_{K_H \in T_H} \operatorname{meas}(K_H)^{\frac{1}{2}} \frac{\|\nabla u_h^H\|_{0,K_H}}{\|\nabla u_h\|_{0,K_H}} \int_{K_H} \nabla v_h \cdot \nabla w_h, \qquad (5.93)$$

and consider the modified problem:

$$\begin{cases} \text{Seek } u_h \in X_h \text{ such that} \\ a(u_h, v_h) + b_h(u_h^H, v_h^H) + c_h(u_h; u_h, v_h) = (f, v_h)_L, \quad \forall v_h \in X_h. \end{cases}$$

The additional nonlinear form is usually referred to as a *shock-capturing* term. Since the nonlinearity induced by c_h is generally small, the discrete problem can be solved using a fixed-point algorithm. The solution with $c_{sc} = 0.1$ is shown in Figure 5.7 in the bottom center panel. The efficiency of the shock-capturing term is clearly visible since the boundary layer is captured within one element.

Similar shock-capturing terms must be added to the GaLS formulation when solving problems with shocks or sharp boundary layers; see, e.g.,

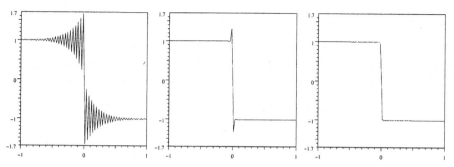

Fig. 5.8. Burgers equation (5.94) approximated on a mesh with $h \approx \frac{1}{50}$: standard Galerkin solution (left); two-level $\mathbb{P}_1$ solution with linear subgrid viscosity (center); and linear subgrid viscosity + shock-capturing (right).

[HuM86, JoS87, KnL02]. When solving nonlinear scalar conservation equations, it is shown in [JoS90] that adding such a nonlinear term guarantees that the approximate solution converges to an entropy solution. See also [BuE02] for a shock-capturing term guaranteeing rigorously a discrete maximum principle for advection–diffusion–reaction equations.   □

Example 5.66 (Burgers equation). To further compare the effects of the subgrid viscosity method and those of the subgrid shock-capturing technique, consider the *Burgers equation*

$$\begin{cases} u\partial_y u - \nu\nabla^2 u = 0 & \text{in } \Omega = \,]0,1[\,\times\,]-1,1[, \\ u_{|x=-1} = -1, \quad u_{|x=1} = 1, \quad u_{|y=0} = u_{|y=1}. \end{cases} \tag{5.94}$$

We set $\nu = 10^{-4}$ and use the two-level $\mathbb{P}_1$ approximation technique on a uniform grid with $h = \frac{1}{50}$, $H = 2h$, so that in the x-direction there are 101 nodes in the fine mesh.

The projection onto the plane $x = 0$ of the graph of the solution is plotted in Figure 5.8. The standard Galerkin solution is displayed in the left panel. The solution oscillates throughout the domain. The stabilized solution is shown in the central panel. Some overshoots and undershoots are still present in the vicinity of the shock. These remaining oscillations are symptoms of the Gibbs phenomenon. Note that except near the shock, all the spurious oscillations have disappeared. Combining subgrid viscosity with shock-capturing techniques yields the solution shown in the right panel ($c_b = 0.05$ and $c_{sc} = 1$). The result is very satisfactory considering that a quite coarse mesh has been used.   □

5.6 Discontinuous Galerkin (DG) Approximation

This section introduces the principles of DG methods to solve first-order PDEs.

5.6.1 A one-dimensional model problem

To make things simple, consider the one-dimensional problem (5.3). Let N be a positive integer. Set $h = \frac{1}{N}$ and $x_i = ih$ for $0 \le i \le N$, define the mesh $\mathcal{T}_h = \bigcup_{i=0}^{N}[x_i, x_{i+1}]$ of the interval $\Omega =]0, 1[$, and introduce the space

$$V_h = \{v_h \in L^1(\Omega);\ \forall i \in \{0, \dots, N-1\},\ v_{h|]x_i, x_{i+1}[} \in \mathbb{P}_k\}. \tag{5.95}$$

Functions in V_h can be discontinuous across element interfaces. Define

$$\forall x \in \Omega, \qquad v_h^+(x) = \lim_{\epsilon \to 0^+} v_h(x + \epsilon), \quad v_h^-(x) = \lim_{\epsilon \to 0^-} v_h(x + \epsilon).$$

Consider the discrete problem: For $f \in L^2(\Omega)$,

$$\begin{cases} \text{Seek } u_h \in V_h \text{ such that} \\ a_h(u_h, v_h) = \int_\Omega f v_h, \quad \forall v_h \in V_h, \end{cases} \tag{5.96}$$

where

$$a_h(u_h, v_h) = \sum_{K \in \mathcal{T}_h} \int_K u_h' v_h + \sum_{i=1}^{N-1} \left[u_h^+(x_i) - u_h^-(x_i) \right] v_h^+(x_i) + u_h(0) v_h(0),$$

and K is a mesh cell of the form $[x_i, x_{i+1}]$.

Example 5.67. To illustrate why the jump of u_h at x_i comes into the picture, assume $k = 0$. Set $U_i = u_{h|]x_i, x_{i+1}[} = u^+(x_i) = u^-(x_{i+1})$ for $i \in \{0, \dots, N-1\}$ and $U_{-1} = 0$. Then, (5.96) amounts to

$$\forall i \in \{0, \dots, N-1\}, \quad U_i - U_{i-1} = \int_{x_i}^{x_{i+1}} f,$$

which is a reasonable first-order approximation to $u' = f$ and $u(0) = 0$. $\quad\square$

Proposition 5.68 (Stability). *For all h and $v_h \in V_h$,*

$$a_h(v_h, v_h) \ge \frac{1}{2}\left(|v_h(0)|^2 + |v_h(1)|^2 + \sum_{i=1}^{N-1} |v_h^+(x_i) - v_h^-(x_i)|^2 \right). \tag{5.97}$$

Proof. Let $v_h \in X_h$. A straightforward calculation yields

$$a_h(v_h, v_h) = \sum_{i=0}^{N-1} \int_{x_i}^{x_{i+1}} v_h v_h' + \sum_{i=1}^{N-1} \left[v_h^+(x_i)^2 - v_h^-(x_i) v_h^+(x_i) \right] + v_h(0)^2$$

$$= v_h(0)^2 + \sum_{i=0}^{N-1} \tfrac{1}{2}\left[v_h^-(x_{i+1})^2 - v_h^+(x_i)^2 \right] + \sum_{i=1}^{N-1} \left[v_h^+(x_i)^2 - v_h^-(x_i) v_h^+(x_i) \right]$$

$$= \tfrac{1}{2} v_h(0)^2 + \tfrac{1}{2} v_h(1)^2 + \sum_{i=1}^{N-1} \tfrac{1}{2} \left[v_h^+(x_i) - v_h^-(x_i) \right]^2. \qquad\square$$

Obviously, (5.97) is not sufficient to ascertain that (5.96) is well-posed. Actually, a full inf-sup condition can be proved as shown in the next section.

5.6.2 Generalization

Let us generalize the above setting to the advection–reaction equation (5.13). Let Ω be a domain in $\mathbb{R}^d$ with outward normal n. Let μ be a function in $L^\infty(\Omega)$ and let β be a vector field in $[C^{0,\frac{1}{2}}(\overline{\Omega})]^d$ with $\nabla\cdot\beta \in L^\infty(\Omega)$. Assume that (5.14) holds. For any $S \subset \Omega$, denote $\partial S^- = \{x \in \partial S; \beta\cdot n(x) < 0\}$. For any subset $E \subset \partial\Omega$, set $L^2_\beta(E) = \{v; \int_E |\beta\cdot n|v^2 < \infty\}$. Let $f \in L^2(\Omega)$, let $g \in L^2_\beta(\partial\Omega^-)$, and set $W = \{v \in L^2(\Omega); \beta\cdot\nabla v \in L^2(\Omega)\}$. Let $u \in W$ solve

$$\begin{cases} \mu u + \beta\cdot\nabla u = f, \\ u_{|\partial\Omega^-} = g. \end{cases} \tag{5.98}$$

Let $\{\mathcal{T}_h\}_{h>0}$ be a shape-regular family of affine simplicial meshes of Ω. Assume for simplicity that Ω is a polyhedron so that $\overline{\Omega} = \bigcup_{K\in\mathcal{T}_h} K$. Set

$$\mathcal{F}_h = \bigcup_{K\in\mathcal{T}_h} \partial K \quad \text{and} \quad v^\pm(x) = \lim_{\epsilon\to\pm 0} v(x + \epsilon\beta), \ \forall x \in \mathcal{F}_h,$$

with the convention that $v^\pm(x)$ is zero whenever x is at the boundary and the limit is taken outside the domain. Consider the bilinear form

$$a_h(w,v) = \sum_{K\in\mathcal{T}_h} \int_K (\mu w + \beta\cdot\nabla w)v + \int_{\mathcal{F}^i_h} |\beta\cdot n|(w^+ - w^-)v^+ + \int_{\partial\Omega^-} |\beta\cdot n|wv,$$

where $\mathcal{F}^i_h = \mathcal{F}_h\backslash\partial\Omega$. Let k be a non-negative integer, define

$$W_h = \{v_h \in L^1(\Omega); \forall K \in \mathcal{T}_h, v_h \circ T_K \in \mathbb{P}_k\}, \tag{5.99}$$

and set $W(h) = W + W_h$. Finally, consider the discrete problem:

$$\begin{cases} \text{Seek } u_h \in W_h \text{ such that} \\ a_h(u_h, v_h) = \int_\Omega fv_h + \int_{\partial\Omega^-} |\beta\cdot n|gv_h, \quad \forall v_h \in W_h. \end{cases} \tag{5.100}$$

5.6.3 Error analysis

For all $\mathcal{E} \subset \mathcal{F}_h$, define the norm $\|v\|^2_{L^2_\beta(\mathcal{E})} = \int_\mathcal{E} |\beta\cdot n|v^2$. As a counterpart of Proposition 5.68, we state the following:

Proposition 5.69 (Coercivity). *For all h and v in $W(h)$,*

$$a_h(v,v) \geq \mu_0\|v\|^2_{L^2(\Omega)} + \tfrac{1}{2}\|v\|^2_{L^2_\beta(\partial\Omega)} + \tfrac{1}{2}\|v^+ - v^-\|^2_{L^2_\beta(\mathcal{F}^i_h)}. \tag{5.101}$$

Proof. Let v be an arbitrary element in $W(h)$. Integration by parts over each element yields

$$a_h(v,v) = \int_\Omega (\mu - \tfrac{1}{2}\nabla\cdot\beta)v^2 + \sum_{K\in T_h}\int_{\partial K}\tfrac{1}{2}\beta\cdot nv^2 + \int_{\mathcal{F}_h^i}|\beta\cdot n|(v^+ - v^-)v^+ + \|v\|_{L_\beta^2(\partial\Omega^-)}^2$$

$$\geq \mu_0\|v\|_{L^2(\Omega)}^2 + \int_{\mathcal{F}_h^i}|\beta\cdot n|(v^{+2} - v^-v^+ - \tfrac{1}{2}v^{+2} + \tfrac{1}{2}v^{-2}) + \tfrac{1}{2}\|v\|_{L_\beta^2(\partial\Omega)}^2.$$

The result follows easily. $\square$

On $W(h)$, define the norms

$$\|v\|_{h,A}^2 = \|v\|_{L^2(\Omega)}^2 + \sum_{K\in T_h}h_K\|\beta\cdot\nabla v\|_{L^2(K)}^2 + \|v^+ - v^-\|_{L_\beta^2(\mathcal{F}_h)}^2 + \|v\|_{L_\beta^2(\partial\Omega)}^2,$$

$$\|v\|_{h,\frac{1}{2}}^2 = \|v\|_{h,A}^2 + \sum_{K\in T_h}h_K^{-1}\|v\|_{L^2(K)}^2 + \|v^-\|_{L_\beta^2(\mathcal{F}_h^i)}^2.$$

Lemma 5.70 (Stability). *Let Ω be a domain in $\mathbb{R}^d$. Let μ be a function in $L^\infty(\Omega)$ and let β be a vector field in $[C^{0,\frac{1}{2}}(\overline\Omega)]^d$ with $\nabla\cdot\beta \in L^\infty(\Omega)$. Assume that (5.14) holds. Then, there is $c > 0$, independent of h, such that*

$$\inf_{v_h\in W_h}\sup_{w_h\in W_h}\frac{a_h(v_h, w_h)}{\|v_h\|_{h,A}\|w_h\|_{h,A}} \geq c. \tag{5.102}$$

Proof. (1) Let v_h be an arbitrary element in W_h. Denote by $\overline\beta_K$ the mean-value of β on K; then, $\|\beta - \overline\beta_K\|_{L^\infty(K)} \leq \|\beta\|_{[C^{0,\frac{1}{2}}(\overline\Omega)]^d}h_K^{\frac{1}{2}}$. Let π_K be the function whose restriction to K is $\overline\beta_K\cdot\nabla v_h$ and zero outside K. It is clear that π_K is in W_h. Moreover, using two inverse inequalities, namely $\|w_h\|_{L_\beta^2(\partial K^-)} \leq ch_K^{-\frac{1}{2}}\|w_h\|_{L^2(K)}$ and $\|\nabla w_h\|_{L^2(K)} \leq ch_K^{-1}\|w_h\|_{L^2(K)}$, leads to

$$\|\pi_K\|_{L_\beta^2(\partial K^-)} \leq ch_K^{-\frac{1}{2}}\|\pi_K\|_{L^2(K)} \leq c'h_K^{-\frac{1}{2}}(\|\beta\cdot\nabla v_h\|_{L^2(K)} + h_K^{-\frac{1}{2}}\|v_h\|_{L^2(K)}).$$

Hence,

$$\|\beta\cdot\nabla v_h\|_{L^2(K)}^2 = a_h(v_h, \pi_K) - (\mu v_h, \pi_K)_{0,K} - \int_{\partial K^-}|\beta\cdot n|(v_h^+ - v_h^-)\pi_K$$

$$+ (\beta\cdot\nabla v_h, (\beta - \overline\beta_K)\cdot\nabla v_h)_{0,K}$$

$$\leq a_h(v_h, \pi_K) + \|\mu v_h\|_{L^2(K)}\|\pi_K\|_{L^2(K)}$$

$$+ \|v_h^+ - v_h^-\|_{L_\beta^2(\partial K^-)}\|\pi_K\|_{L_\beta^2(\partial K^-)}$$

$$+ \|\beta\cdot\nabla v_h\|_{L^2(K)}\|\beta - \overline\beta_K\|_{L^\infty(K)}\|\nabla v_h\|_{L^2(K)}$$

$$\leq a_h(v_h, \pi_K)$$

$$+ ch_K^{-\frac{1}{2}}\|\beta\cdot\nabla v_h\|_{L^2(K)}(\|v_h\|_{L^2(K)} + \|v_h^+ - v_h^-\|_{L_\beta^2(\partial K^-)})$$

$$+ c'h_K^{-\frac{1}{2}}\|v_h\|_{L^2(K)}(\|v_h\|_{L^2(K)} + h_K^{-\frac{1}{2}}\|v_h^+ - v_h^-\|_{L_\beta^2(\partial K^-)}).$$

Setting $\pi_h = \sum_{K \in T_h} h_K \pi_K$ yields

$$\frac{1}{2} \sum_{K \in T_h} h_K \|\beta \cdot \nabla v_h\|^2_{L^2(K)} \leq a_h(v_h, \pi_h) + c\, a_h(v_h, v_h). \tag{5.103}$$

(2) Let us now prove that $\|\pi_h\|_{h,A} \leq c\|v_h\|_{h,A}$. The inverse inequality $\|\nabla v_h\|_{L^2(K)} \leq c h_K^{-1} \|v_h\|_{L^2(K)}$ implies $\|\pi_h\|_{0,\Omega} \leq c\|v_h\|_{0,\Omega}$. Using again the above inverse inequality together with $\|\beta - \bar{\beta}_K\|_{L^\infty(K)} \leq \|\beta\|_{[C^{0,\frac{1}{2}}(\overline{\Omega})]^d} h_K^{\frac{1}{2}}$ leads to

$$\sum_{K \in T_h} h_K \|\beta \cdot \nabla \pi_h\|^2_{0,K} \leq c \sum_{K \in T_h} h_K \|\bar{\beta} \cdot \nabla v_h\|^2_{0,K} \leq c \sum_{K \in T_h} (h_K \|\bar{\beta} \cdot \nabla v_h\|^2_{0,K} + \|v_h\|^2_{0,K}).$$

Moreover, owing to the inverse inequality $\|\bar{\beta} \cdot \nabla v_h\|_{L^2(\partial K)} \leq c h_K^{-\frac{1}{2}} \|\bar{\beta} \cdot \nabla v_h\|_{0,K}$, $\|\pi_h^+ - \pi_h^-\|_{L_\beta^2(\mathcal{F}_h^i)}$ is controlled as follows:

$$\|\pi_h^+ - \pi_h^-\|^2_{L_\beta^2(\mathcal{F}_h^i)} \leq c \sum_{K \in T_h} h_K^2 \|\pi_K\|^2_{L^2(\partial K)} \leq c \sum_{K \in T_h} h_K \|\bar{\beta} \cdot \nabla v_h\|^2_{0,K}$$

$$\leq c \sum_{K \in T_h} (h_K \|\bar{\beta} \cdot \nabla v_h\|^2_{0,K} + \|v_h\|^2_{0,K}).$$

Proceed similarly to control $\|\pi_h\|_{L_\beta^2(\partial \Omega)}$. In conclusion,

$$\|\pi_h\|_{h,A} \leq c\|v_h\|_{h,A}. \tag{5.104}$$

(3) Owing to (5.101) and (5.103), there is $c_1 > 0$ such that

$$\|v_h\|^2_{h,A} \leq c_1 a_h(v_h, v_h) + a_h(v_h, \pi_h) = a_h(v_h, \pi_h + c_1 v_h).$$

Then, setting $w_h = \pi_h + c_1 v_h$ and using (5.104) yields

$$\|v_h\|_{h,A} \|w_h\|_{h,A} \leq c\|v_h\|^2_{h,A} \leq c\, a_h(v_h, w_h).$$

The rest of the proof follows easily. □

Lemma 5.71 (Continuity). *Under the hypotheses of Lemma 5.70, there is c, independent of h, such that*

$$\forall v \in W(h), \ \forall w_h \in W_h, \quad a_h(v, w_h) \leq c\|v\|_{h,\frac{1}{2}} \|w_h\|_{h,A}. \tag{5.105}$$

Proof. Integration by parts over each element implies

$$a_h(v, w_h) = \sum_{K \in T_h} (v, (\mu - \nabla \cdot \beta) w_h - \beta \cdot \nabla w_h)_{0,K}$$

$$+ \int_{\mathcal{F}_h^i} |\beta \cdot n|(w_h^- - w_h^+) v^- + \int_{\partial \Omega^+} |\beta \cdot n| w_h v$$

$$\leq \sum_{K \in T_h} \|v\|_{0,K} (\|w_h\|_{0,K} + \|\beta \cdot \nabla w_h\|_{0,K})$$

$$+ \|v^-\|_{L_\beta^2(\mathcal{F}_h^i)} \|w_h^- - w_h^+\|_{L_\beta^2(\mathcal{F}_h^i)} + \|v\|_{L_\beta^2(\partial \Omega^+)} \|w_h\|_{L_\beta^2(\partial \Omega^+)}.$$

The Cauchy–Schwarz inequality yields the desired result. □

Lemma 5.72 (Consistency). *Let u solve (5.98) and let u_h solve (5.100). Then,*

$$\forall v_h \in W_h, \quad a_h(u - u_h, v_h) = 0. \tag{5.106}$$

Proof. Since u belongs to W, the integral $\int_{\mathcal{F}_h^i} |\beta \cdot n|(u^+ - u^-)v_h^+$ is zero for all v_h in W_h. Moreover, since u solves (5.98), the following equality holds:

$$\forall v_h \in W_h, \quad a_h(u, v_h) = \int_\Omega f v_h + \int_{\partial \Omega^-} |\beta \cdot n| g v_h = a_h(u_h, v_h). \qquad \square$$

Theorem 5.73 (Convergence). *Under the hypotheses of Lemma 5.70, there is c, independent of h, such that*

$$\|u - u_h\|_{h,A} \le c \inf_{v_h \in W_h} \|u - v_h\|_{h,\frac{1}{2}}. \tag{5.107}$$

Moreover, if $u \in H^{k+1}(\Omega)$,

$$\|u - u_h\|_{h,A} \le c\, h^{k+\frac{1}{2}} \|u\|_{H^{k+1}(\Omega)}. \tag{5.108}$$

Proof. Simple application of the Second Strang Lemma. $\qquad \square$

Remark 5.74.

(i) The DG method in the form presented here has been introduced and analyzed by Lesaint and Raviart [LeR74] where an $\mathcal{O}(h^k)$ estimate was proved. The $h^{\frac{1}{2}}$ improvement is due to Johnson and Pitkäranta [JoP86]. Interesting extensions of the analysis to the hp framework can be found in [BeO96, HoS00].

(ii) Note again that like the GaLS formulation and the subgrid viscosity method, DG is an L^2-based method. The basic stability estimate (5.101) is the coercivity argument in $L^2(\Omega)$. As a consequence, if f is rough, say $f \in L^1(\Omega)$ only, then the method may not work optimally. This problem is solved in practice by adding a nonlinear shock-capturing term to a_h; see Example 5.65. For additional details on this question, the reader is referred to [CoJ98]. $\quad \square$

5.6.4 Discontinuous GaLS method and other variants

It is possible to replace the inf-sup condition (5.102) by a coercivity property if the bilinear form a_h is slightly modified. Denoting by A the differential operator defined by $Av = \mu v + \beta \cdot \nabla v$, the idea is to introduce the form

$$b_h(v, w) = a_h(v, w) + \sum_{K \in \mathcal{T}_h} h_K (Av, Aw)_{0,K}.$$

Then, it is clear that b_h satisfies

$$b_h(v, v) \ge \mu_0 \|v\|^2_{L^2(\Omega)} + \tfrac{1}{2}\|v\|^2_{L^2_\beta(\partial\Omega)} + \tfrac{1}{2}\|v^+ - v^-\|^2_{L^2_\beta(\mathcal{F}_h^i)} + \sum_{K \in \mathcal{T}_h} h_K \|Av\|^2_{0,K}.$$

The discrete problem is recast in the form:

$$\begin{cases} \text{Seek } u_h \in W_h \text{ such that, } \forall v_h \in W_h, \\ b_h(u_h, v_h) = \sum_{K \in T_h} (f, v_h + h_K A v_h)_{0,K} + \int_{\partial \Omega^-} |\beta \cdot n| g v_h, \end{cases} \quad (5.109)$$

and the conclusions of Theorem 5.73 remain valid.

The method generalizes to Friedrichs' symmetric systems; see [Les75, JoN84]. For all $x \in \partial K$, set

$$[\![u_h]\!]_K(x) = u_h^i(x) - u_h^e(x), \quad u_h^i(x) = \lim_{\substack{y \to x \\ y \in K}} u_h(y), \quad u_h^e(x) = \lim_{\substack{y \to x \\ y \notin K}} u_h(y),$$

with $u_h^e(x) = 0$ if $x \in \partial \Omega$. Using the notation of §5.2.2, a_h takes the form

$$a_h(u_h, v_h) = \sum_{K \in T_h} (A u_h, v_h)_{L,K} + \tfrac{1}{2}((\mathcal{M}_K - \mathcal{D}_K)[\![u_h]\!]_K, v_h)_{L,\partial K}, \quad (5.110)$$

where $(\cdot, \cdot)_{L,K}$ and $(\cdot, \cdot)_{L,\partial K}$ denote the L^2-scalar product on K and ∂K, respectively, and where $\mathcal{D}_K$ and $\mathcal{M}_K$ are local versions of the operators $\mathcal{D}$ and $\mathcal{M}$, respectively. The field $\mathcal{D}_K$ is such that $\mathcal{D}_K = \sum_{k=1}^d n_k A_{|K}^k$. We assume that $\mathcal{M}_K$ satisfies:

(F1_DG) $\mathcal{M}_K = \mathcal{M}$ on $\partial \Omega$.
(F2_DG) $\mathcal{M}_K = \mathcal{M}_{K'}$ on $\partial K \cap \partial K'$.
(F3_DG) $\mathcal{M}_K + \mathcal{M}_K^T \geq 0$.
(F4_DG) $v^T(\mathcal{M}_K - \mathcal{D}_K)w \leq v^T \mathcal{M}_K v + c\|w\|_m^2, \forall v, w \in \mathbb{R}^m$.

Under the above assumptions together with assumptions (F1) to (F4) introduced in §5.2.2, Theorem 5.73 holds. In particular, the DG method can be used to solve reaction–advection–diffusion equations.

5.7 Non-Standard Galerkin Approximation

In this section, we present a non-standard Galerkin technique for solving Darcy's equations (5.15) and more generally *elliptic PDEs in mixed form*. For the sake of simplicity, we restrict ourselves to the situation $S = \mathcal{I}$, i.e., the permeability is isotropic. The method can be extended to anisotropic settings.

5.7.1 Principle of the method

Let the spaces X and Y be as in (5.16). Set $V = X \times Y$ and $L = [L^2(\Omega)]^d \times L^2(\Omega)$. Define the bilinear form $a((u, p), v, q)) = (u, v)_{0,\Omega} + (\nabla p, v)_{0,\Omega} + (\nabla \cdot u, q)_{0,\Omega}$ for all $(u, p) \in V$ and $(v, q) \in L$. Problem (5.15) is restated as follows: For $f \in [L^2(\Omega)]^d$ and $g \in L^2(\Omega)$,

$$\begin{cases} \text{Seek } (u, p) \in V \text{ such that} \\ a((u, p), (v, q)) = (f, v)_{0,\Omega} + (g, q)_{0,\Omega}, \quad \forall (v, q) \in L. \end{cases} \quad (5.111)$$

Proposition 5.11 guarantees that this problem is well-posed. The key difference with the mixed form of Darcy's equations investigated in Chapter 4 (see Exercises 4.10 and 4.11) is that no integration by parts has been performed in (5.111). In particular, the solution space in (5.111) assumes more regularity than the test space. Following Croisille et al. [Cro00, CrG02], we construct a non-standard Galerkin approximation to the above problem.

Assume, for the sake of simplicity, that Ω is a polyhedron in $\mathbb{R}^3$ (resp., polygon in $\mathbb{R}^2$), and consider a shape-regular family of affine simplicial meshes, say $\{\mathcal{T}_h\}_{h>0}$. Denote by $\mathcal{F}_h$ the set of faces (resp., edges) of the mesh. The faces (resp., edges) that are at the boundary are denoted by $\mathcal{F}_h^\partial$ and those that are internal are denoted by $\mathcal{F}_h^i$. Set $\mathcal{F}_{h1}^\partial = \mathcal{F}_h \cap \partial\Omega_1$, $\mathcal{F}_{h2}^\partial = \mathcal{F}_h \cap \partial\Omega_2$. Recall the definition of the Raviart–Thomas and Crouzeix–Raviart finite element spaces:

$$D_h = \{v_h; \forall K \in \mathcal{T}_h, v_{h|K} \in \mathbb{RT}_0, \forall F \in \mathcal{F}_h^i, \int_F [\![v_h \cdot n]\!] = 0\},$$

$$P_{\mathrm{pt},h}^1 = \{q_h; \forall K \in \mathcal{T}_h, q_{h|K} \in \mathbb{P}_1; \forall F \in \mathcal{F}_h^i, \int_F [\![q_h]\!] = 0\},$$

where $[\![v_h \cdot n]\!]$ denotes the jump of the normal component of v_h across interfaces and $[\![q_h]\!]$ the jump of q_h across interfaces. Define the spaces

$$X_h = \{v_h \in D_h; \forall F \in \mathcal{F}_{h1}^\partial, \int_F v_h \cdot n = 0\},$$

$$Y_h = \{q_h \in P_{\mathrm{pt},h}^1; \forall F \in \mathcal{F}_{h2}^\partial, \int_F q_h = 0\}.$$

Letting $F \in \mathcal{F}_h^i$ such that $F = K_1 \cap K_2$, observe that

$$\forall v_h \in X_h, \quad v_{h|K_1} \cdot n_1 = -v_{h|K_2} \cdot n_2 \in \mathbb{P}_0(F), \tag{5.112}$$

$$\forall v_h \in X_h, \quad v_h(x) = \bar{v}_h + \tfrac{x - g_K}{d} \nabla \cdot v_h, \quad \forall x \in K, \tag{5.113}$$

where $\bar{v}_h$ is the mean value of v_h on K and g_K is the barycenter of K. Owing to (5.113), there is $c > 0$ independent of h such that

$$\forall v_h \in X_h, \quad \|v_h\|_{0,K} \le \|\bar{v}_h\|_{0,K} + c\, h_K \|\nabla \cdot v_h\|_{0,K}. \tag{5.114}$$

Set $V_h = X_h \times Y_h$, $V(h) = V + V_h$, and equip $V(h)$ with the norm

$$\|(v,q)\|_{V(h)}^2 = \|v\|_{0,\Omega}^2 + \|\nabla \cdot v\|_{0,\Omega}^2 + \|q\|_{0,\Omega}^2 + \sum_{K \in \mathcal{T}_h} \|\nabla q\|_{0,K}^2.$$

It is clear that $\|\cdot\|_{V(h)}$ is an extension of the norm of V to $V(h)$. Furthermore, introduce the test space

$$L_h = \{(v_h, q_h); \forall K \in \mathcal{T}_h, v_{h|K} \in [\mathbb{P}_0]^d \text{ and } q_{h|K} \in \mathbb{P}_0\}. \tag{5.115}$$

Finally, define the bilinear form

$$a_h((u_h, p_h), (v_h, q_h)) = (u_h, v_h)_{0,\Omega} + (\nabla \cdot u_h, q_h)_{0,\Omega} + \sum_{K \in \mathcal{T}_h} (\nabla p_h, v_h)_{0,K}.$$

The discrete problem is formulated as follows: For $(f, g) \in L$,

$$\begin{cases} \text{Seek } (u_h, p_h) \in V_h \text{ such that} \\ a_h((u_h, p_h), (v_h, q_h)) = ((f, g), (v_h, q_h))_L, \quad \forall (v_h, q_h) \in L_h. \end{cases} \tag{5.116}$$

Note that the approximation setting is conformal on the velocity since $X_h \subset X$, but not on the pressure.

5.7.2 Convergence analysis

Let us start the convergence analysis by showing that (5.116) is well-posed. This fact is a direct consequence of the following:

Lemma 5.75 (Stability). *If h is small enough, the bilinear form a_h satisfies the conditions* (BNB1$_h$) *and* (BNB2$_h$) *on $V_h \times L_h$ uniformly in h.*

Proof. Proof of (BNB1$_h$). Let $(u_h, p_h) \in V_h$. Denote by $\bar{u}_h$ (resp., $\bar{p}_h$) the function whose restriction to each element $K \in \mathcal{T}_h$ is the mean value of u_h (resp., p_h). Denote by $\nabla_h p_h$ the function whose restriction to each element $K \in \mathcal{T}_h$ is $\nabla p_{h|K}$. Set $v_h = \bar{u}_h + \nabla_h p_h$ and $q_h = 2p_h + \nabla \cdot u_h$. Note that (v_h, q_h) is in L_h. Hence,

$$a_h((u_h, p_h), (v_h, q_h)) = (u_h, \bar{u}_h)_{0,\Omega} + \|\nabla \cdot u_h\|_{0,\Omega}^2 + \sum_{K \in \mathcal{T}_h} \|\nabla p_h\|_{0,K}^2$$

$$+ 2(\nabla \cdot u_h, p_h)_{0,\Omega} + \sum_{K \in \mathcal{T}_h} (u_h, \nabla p_h)_{0,K} + (\nabla p_h, \bar{u}_h)_{0,K}$$

$$= \|\bar{u}_h\|_{0,\Omega}^2 + \|\nabla \cdot u_h\|_{0,\Omega}^2 + \sum_{K \in \mathcal{T}_h} \|\nabla p_h\|_{0,K}^2$$

$$+ 2(\nabla \cdot u_h, p_h)_{0,\Omega} + 2 \sum_{K \in \mathcal{T}_h} (u_h, \nabla p_h)_{0,K}.$$

Denote by R the last two terms in the right-hand side. Owing to (5.112), we infer $\frac{1}{2}R = \sum_{F \in \mathcal{F}_h^i} u_h \cdot n \int_F [\![p_h]\!]$ and hence, $R = 0$ since $p_h \subset P_{\mathrm{pt},h}^1$. This result together with (5.114) implies

$$a_h((u_h, p_h), (v_h, q_h)) \geq c\|u_h\|_{0,\Omega}^2 + (1 - c'h^2)\|\nabla \cdot u_h\|_{0,\Omega}^2 + \sum_{K \in \mathcal{T}_h} \|\nabla p_h\|_{0,K}^2.$$

If h is small enough, $(1 - c'h^2)$ is bounded from below by $\frac{1}{2}$. The extended Poincaré inequality (3.35) yields

$$a_h((u_h, p_h), (v_h, q_h)) \geq c\|(u_h, p_h)\|_{V(h)}^2 \geq c'\|(u_h, p_h)\|_{V(h)}\|(v_h, q_h)\|_{L_h},$$

leading to the discrete inf-sup condition

$$\inf_{(u_h, p_h) \in V_h} \sup_{(v_h, q_h) \in L_h} \frac{a_h((u_h, p_h), (v_h, q_h))}{\|(u_h, p_h)\|_{V(h)}\|(v_h, q_h)\|_L} \geq c > 0.$$

Proof of (BNB2$_h$). Owing to Proposition 2.21(iii), we only have to check that $\dim(V_h) = \dim(L_h)$. Clearly, $\dim(L_h) = (d+1)N_{\mathrm{el}}$, where N_{el} is the number of elements. On the other hand, $\dim(V_h) = 2N_{\mathrm{ed}} - N_{\mathrm{ed}}^{\partial}$, where N_{ed} is the number edges and $N_{\mathrm{ed}}^{\partial}$ is the number boundary edges in two dimensions (resp., $2N_{\mathrm{f}} - N_{\mathrm{f}}^{\partial}$, where N_{f} is the number faces and $N_{\mathrm{f}}^{\partial}$ is the number faces at the boundary in three dimensions). The conclusion is a consequence of the Euler relations; see Lemma 1.57.   □

Lemma 5.76 (Continuity). *There is c, independent of h, such that, for all* $(w, r) \in V(h)$ *and* $(v_h, q_h) \in L_h$,

$$|a_h((w,r),(v_h,q_h))| \leq c\,\|(w,r)\|_{V(h)}\|(v_h,q_h)\|_L.$$

Lemma 5.77 (Consistency). *Let* (u, p) *solve* (5.111) *and* (u_h, p_h) *solve* (5.116). *Then,*

$$\forall (v_h, q_h) \in L_h, \quad a_h((u - u_h, p - p_h), (v_h, q_h)) = 0.$$

Theorem 5.78 (Convergence). *Let* (u, p) *and* (u_h, p_h) *solve* (5.111) *and* (5.116), *respectively. Then, for all* h,

$$\|(u - u_h, p - p_h)\|_{V(h)} \leq c \inf_{(v_h, q_h) \in V_h} \|(u - v_h, p - q_h)\|_{V(h)}.$$

Proof. Apply the Second Strang Lemma.   □

Corollary 5.79. *Assume that the solution* (u, p) *to* (5.111) *is in* $[H^2(\Omega)]^d \times H^1(\Omega)$. *Then, the solution* (u_h, p_h) *to* (5.116) *satisfies*

$$\|(u - u_h, p - p_h)\|_{V(h)} \leq c\,h(\|u\|_{2,\Omega} + \|p\|_{1,\Omega}).$$

Proof. Direct consequence of the interpolation properties of the Crouzeix–Raviart element and the Raviart–Thomas elements; see Theorem 1.103 and Theorem 1.114.   □

Theorem 5.80 (L^2-estimate). *Assume* $\partial\Omega_2 = \partial\Omega$ *and* $\partial\Omega_1 = \emptyset$. *Under the hypotheses of Corollary 5.79, if the Laplace operator with homogeneous Dirichlet conditions has smoothing properties in* Ω, *then*

$$\forall h, \quad \|p - p_h\|_{0,\Omega} \leq c\,h^2(\|u\|_{2,\Omega} + \|p\|_{1,\Omega}).$$

Proof. Apply the Aubin–Nitsche Lemma in the context of non-conformal methods; see [Cro00].   □

Remark 5.81. For other details on the above non-standard Galerkin approximation method and non-standard schemes of higher order, the reader is referred to Croisille et al. [Cro00, CrG02].   □

5.8 Exercises

Exercise 5.1. Using the setting of §5.1, denote by $\{\varphi_1, \ldots, \varphi_N\}$ the nodal basis for X_h defined in (5.4). Let u_h be the solution to problem (5.5). Expand u_h in the nodal basis and write the linear system resulting from (5.5). Compare this system with the one obtained by approximating (5.1) with second-order centered finite differences.

Exercise 5.2 (Formulation in $L^1(\Omega)$). Use the notation of §5.1.4 with X_h defined in (5.4). Set $Y_h = \{v_h \in L^1(\Omega); \forall i \in \{0, \ldots, N-1\}, v_{h|]x_i, x_{i+1}[} \in \mathbb{P}_0\}$.

(i) Prove that

$$\inf_{u_h \in X_h} \sup_{v_h \in Y_h} \frac{a(u_h, v_h)}{\|u_h\|_{W^{1,1}(\Omega)} \|v_h\|_{L^\infty(\Omega)}} \geq c > 0.$$

(ii) Assuming that $f \in L^1(\Omega)$, let $u_h \in X_h$ be the solution to $a(u_h, v_h) = \int_0^1 f v_h, \forall v_h \in Y_h$. Show that this problem is well-posed and that $\|u - u_h\|_{W^{1,1}(\Omega)} \leq c \inf_{v_h \in X_h} \|u - v_h\|_{W^{1,1}(\Omega)}$, where $u \in W^{1,1}(\Omega)$ solves (5.1).

Exercise 5.3. Prove Proposition 5.9 directly, i.e., instead of using the theory of Friedrichs' systems, prove that the two conditions of the BNB Theorem hold on $V \times L^2(\Omega)$ for the bilinear form $a(u, v) = (\mu u + \beta \cdot \nabla u, v)_{0,\Omega}$ with $V = \{w \in W; w_{|\partial\Omega^-} = 0\}$. (*Hint*: To prove (BNB1), first show that a is coercive on $L^2(\Omega)$.)

Exercise 5.4. Prove Proposition 5.11 directly, i.e., instead of using the theory of Friedrichs' systems, prove that the two conditions of the BNB Theorem hold for the bilinear form $a((u, p), (v, q)) = (K^{-1}u + \nabla p, v)_{0,\Omega} + (\nabla \cdot u, q)_{0,\Omega}$. (*Hint*: To prove (BNB1), consider $(v, q) = (u + K \cdot \nabla p, 2p + \nabla \cdot u)$.)

Exercise 5.5. Set $\Omega =]0, 1[\times]-1, +1[$ and consider the *wave equation*

$$\begin{cases} \partial_{x_1}^2 u - \partial_{x_2}^2 u = f & \text{in } \Omega, \\ u(x_1, -1) = u(x_1, 1) = 0 & 0 < x_1 < 1, \\ u(0, x_2) = \partial_{x_1} u(0, x_2) = 0 & |x_2| < 1. \end{cases}$$

Using the new dependent variable $v = (u, \partial_{x_1} u, \partial_{x_2} u)$, show that this problem can be rewritten in the form of a symmetric Friedrichs' system.

Exercise 5.6. Complete the proof of Proposition 5.13, i.e., prove that condition (BNB2) holds.

Exercise 5.7. Use the notation of §5.2.2.

(i) Prove that $\mathcal{D} \in [L^\infty(\partial\Omega)]^{m \times m}$. (*Hint*: Use the divergence formula and the fact that $L^\infty(\partial\Omega) = (L^1(\partial\Omega))'$.)

(ii) Prove that the bilinear form

$$[\mathcal{C}^\infty(\overline{\Omega})]^m \times [\mathcal{C}^\infty(\overline{\Omega})]^m \ni (u,v) \longmapsto \int_{\partial\Omega} v^T \mathcal{D}u \in \mathbb{R},$$

extends continuously to $W \times W$.

Exercise 5.8. Prove Remark 5.14(i).

Exercise 5.9. Justify Remark 5.18.

Exercise 5.10. Derive formally the PDEs and boundary conditions satisfied by the solution to the Darcy equations (5.29) written in Least-Squares form.

Exercise 5.11 (Maxwell's Equations). Assume that $(E,B) \in V$ satisfies the Maxwell equations in Least-Squares form (5.31). Verify formally that E satisfies the PDE $-i\omega E + \nabla \times \nabla \times E = -i\omega f + \nabla \times g$ a.e. in Ω with boundary conditions $E \times n = 0$ a.e. on $\partial\Omega_1$ and $(\nabla \times E) \times n = g \times n$ a.e. on $\partial\Omega_2$, while B satisfies the PDE $-i\omega B + \nabla \times \nabla \times B = g - \nabla \times f$ a.e. in Ω with boundary conditions $B \times n = 0$ a.e. on $\partial\Omega_2$ and $(\nabla \times B) \times n = -f \times n$ a.e. on $\partial\Omega_1$.

Exercise 5.12. Why is it important to use the equality

$$(Av, w_h)_L = 2a_s(v, w_h) - (v, Aw_h)_L$$

in the first step of the proof of Lemma 5.37? If A is a differential operator, what is the meaning of this equality? Check that the proof works without using this trick if $\epsilon = 0$.

Exercise 5.13. Prove Lemma 5.61 for the $\mathbb{P}_1$/bubble setting when A_1 is the advection operator $\beta \cdot \nabla(\cdot)$ and β is a constant vector field.

Exercise 5.14 (One-dimensional subgrid viscosity). Let $\Omega =]0,1[$ and let N be a positive integer. Set $h = \frac{1}{2N}$, $H = 2h$, $x_i = ih$, $\mathcal{F}_h = \bigcup_{0 \le i \le 2N-1}[x_i, x_{i+1}]$, $\mathcal{F}_H = \bigcup_{0 \le i \le N-1}[x_{2i}, x_{2(i+1)}]$, and

$$X_h = \{v_h \in \mathcal{C}^0(\overline{\Omega}); \forall i \in \{0, \dots, 2N-1\}, v_{h|[x_i, x_{i+1}]} \in \mathbb{P}_1; v_h(0) = 0\}.$$

Let $\{\varphi_H^{2i}\}_{1 \le i \le N}$ be the nodal shape functions associated with the nodes $\{x_{2i}\}_{1 \le i \le N}$ on the mesh $\mathcal{F}_H$. Let $\{\varphi_h^{2i+1}\}_{0 \le i \le N-1}$ be the nodal shape functions associated with the nodes $\{x_{2i+1}\}_{0 \le i \le N-1}$ on the mesh $\mathcal{F}_h$.

(i) Let $v_h = \sum_{1 \le i \le N} a_{2i}\varphi_H^{2i} + \sum_{0 \le i \le N-1} a_{2i+1}\varphi_h^{2i+1}$. Calculate the coefficients a_i in terms of the nodal values of v_h.

(ii) Prove that $\{\varphi_H^{2i}\}_{1 \le i \le N} \bigcup \{\varphi_h^{2i+1}\}_{0 \le i \le N-1}$ is a basis for X_h.

(iii) Set $X_H = \mathrm{span}\{\varphi_H^{2i}\}_{1 \le i \le N}$ and $X_h^H = \mathrm{span}\{\varphi_h^{2i+1}\}_{0 \le i \le N-1}$. Show that $X_h = X_H \oplus X_h^H$.

(iv) Prove that there is $\widehat{c} < 1$ such that, for all $v_H \in X_H$, $v_h^H \in X_h^H$, and $I_{2i} \in \mathcal{F}_H$, $|\int_{I_{2i}} v_H v_h^H| \le \widehat{c}\|v_H\|_{0,I_{2i}}\|v_h^H\|_{0,I_{2i}}$.

(v) Let $P_H : X_h \to X_H$ be the projector associated with the decomposition $X_h = X_H \oplus X_h^H$. Prove that $\|P_H\|_{\mathcal{L}(X_h;X_H)}$ is uniformly bounded.

(vi) Let $P_h^H = \mathcal{I} - P_H$. Write the matrix associated with the bilinear form $\int_\Omega \nabla(P_h^H u_h)\cdot\nabla(P_h^H v_h)$ using the basis $\{\varphi_H^{2i}\}_{1\leq i\leq N}\bigcup\{\varphi_h^{2i+1}\}_{0\leq i\leq N-1}$.

(vii) Let $\{\varphi_h^i\}_{1\leq i\leq 2N}$ be the nodal shape functions associated with the nodes $\{x_i\}_{1\leq i\leq 2N}$ on the mesh $\mathcal{F}_h$. Write the matrix associated with the bilinear form $\int_\Omega \nabla(P_h^H u_h)\cdot\nabla(P_h^H v_h)$ using the basis $\{\varphi_h^i\}_{1\leq i\leq 2N}$.

Exercise 5.15. Using Definition (5.110) for the bilinear form a_h, assuming (F1_DG) to (F4_DG) together with (F1) to (F4), prove Proposition 5.69, Lemma 5.70, Lemma 5.71, and Theorem 5.73.

Exercise 5.16 (Non-standard Galerkin with bubbles). Let $\Omega =]0,1[$, $f \in L^2(\Omega)$, and let ν and β be two positive numbers. Consider the one-dimensional advection–diffusion equation $-\nu u'' + \beta u' = f$ posed on Ω with boundary conditions $u(0) = u(1) = 0$. Let N be a positive integer, set $h = \frac{1}{N+1}$, and consider the mesh with vertices $x_i = ih$, $0 \leq i \leq N + 1$. Let φ_i be the hat function associated with x_i and defined in (1.4). Set $W_h = \mathrm{span}\{\varphi_1,\ldots,\varphi_N\}$. Let $b : [0,1] \mapsto \mathbb{R}$ be a smooth function such that $b(0) = b(1) = 0$ and $b \geq 0$; b is called a *bubble function*. For $1 \leq i \leq N$, set $b_i(x) = b(\frac{x-x_{i-1}}{h})$ if $x \in [x_{i-1},x_i]$, $b_i(x) = -b(\frac{x-x_i}{h})$ if $x \in [x_i,x_{i+1}]$, and $b_i(x) = 0$ otherwise, and set $\psi_i = \varphi_i + b_i$. Let $V_h = \mathrm{span}\{\psi_1,\ldots,\psi_N\}$.

(i) Write the non-standard Galerkin approximation of the advection–diffusion problem using W_h as trial space and V_h as test space.

(ii) Prove that the non-standard Galerkin formulation is equivalent to the following Galerkin formulation using the space W_h: Seek $u_h \in W_h$ such that
$$\forall\varphi_i, \quad \int_0^1 ((\nu + \nu_h)u_h'\varphi_i' + \beta u_h'\varphi_i) = \int_0^1 f\varphi_i + \int_0^1 fb_i,$$
where $\nu_h = h\beta \int_0^1 b$ is an *artificial viscosity*.

(iii) How does the above formulation simplify if $f = 1$?

(iv) How to choose $\int_0^1 b$ so that the stiffness matrix is always an M-matrix?

(v) Let u_h be the approximate solution. Prove that $|u - u_h|_{1,\Omega}$ is first-order in h. (*Hint*: Use the First Strang Lemma.)

(vi) Find a bubble function such that the error vanishes identically at all the mesh vertices. (*Hint*: Introduce a dual problem with data $\delta_{x=x_i}$.)

Exercise 5.17. Let (u_h, p_h) be the unique solution to the non-standard Galerkin approximation of Darcy's equations (5.116). Assume that $\partial\Omega_1 = \emptyset$ and $\partial\Omega_2 = \partial\Omega$, i.e., homogeneous Dirichlet conditions are enforced on the entire boundary. The goal of this exercise is to link u_h to the Crouzeix–Raviart non-conforming solution of a second-order elliptic problem and to link p_h to the mixed approximation of Darcy's equations using the Raviart–Thomas finite element; see Croisille [Cro00].

(i) Prove that u_h solves the following problem:

$$\begin{cases} \text{Seek } u_h \in P^1_{\text{pt},h,0} \text{ such that} \\ a_h(u_h, v_h) = \int_\Omega (\Pi^0_{\text{td},h} f) v_h, \quad \forall v_h \in P^1_{\text{pt},h,0}, \end{cases}$$

where $P^1_{\text{pt},h,0}$ is defined in (3.32), and $\Pi^0_{\text{td},h}$ is the L^2-orthogonal projector onto the space of piecewise constant functions. (*Hint:* Take $q_h = \nabla_h v_h$ in (5.116).)

(ii) Let $Z_h = \{v_h \in H(\text{div}; \Omega); \forall K \in \mathcal{T}_h, v_{h|K} \in \mathbb{RT}_0\}$ and $M_h = \{q_h \in L^2(\Omega); \forall K \in \mathcal{T}_h, q_{h|K} \in \mathbb{P}_0\}$. Write the standard Galerkin approximation of Darcy's equations (with homogeneous Dirichlet conditions enforced on the entire boundary) using the space $Z_h \times M_h$. Prove that the discrete problem is well-posed. (*Hint:* Use Exercise 4.13.)

(iii) Let $(\overline{u}_h, \overline{p}_h)$ be the unique solution to the discrete problem derived in the previous question. Prove that $\overline{p}_h = p_h$. (*Hint:* Prove that $(\nabla \cdot (\overline{p}_h - p_h), m_h)_{0,\Omega} = 0$ for all $m_h \in M_h$, and use (5.113).)

(iv) Prove that

$$\overline{u}_h = \Pi^0_{\text{td},h} u_h + \frac{1}{d^2} \sum_{K \in \mathcal{T}_h} (\Pi^0_{\text{td},h} f)_{|K} \rho_K^2 1_K,$$

where 1_K is the characteristic function of K, and $\rho_K = \|x - g_K\|_{[L^2(K)]^d}$. (*Hint:* for $z_h \in Z_h$, integrate by parts $(\Pi^0_{\text{td},h} u_h, \nabla \cdot z_h)_{0,\Omega}$, and use (5.113) for z_h and $\nabla_h u_h$.)

6
Time-Dependent Problems

This chapter is devoted to the approximation of time-dependent PDEs by finite elements. An exhaustive review of this vast subject goes beyond the scope of this book; therefore, we restrict ourselves to illustrating some important concepts. In particular, we emphasize the method of lines where the problem is first approximated in space, yielding a system of coupled ordinary differential equations, and then a time-marching algorithm is employed to construct the time-approximation.

This chapter is organized into three sections. The first section is concerned with parabolic equations, the prototypical example being the heat equation. The differential operator in space is the Laplace operator or, more generally, a coercive operator. This section is the time-dependent extension of Chapter 3. In the second section, we investigate the time-dependent version of the Stokes problem studied in Chapter 4. A large part of this section deals with a class of fractional-step techniques often referred to in the literature as projection methods. The last section investigates evolution problems without coercivity. A prototypical example is the time-dependent advection equation. We analyze space-time discontinuous finite element approximations, the method of characteristics, and finally an extension of the subgrid viscosity technique presented in Chapter 5.

6.1 Parabolic Problems

In this section, we introduce some basic properties of parabolic equations and we investigate various approximation techniques. For a thorough analysis of parabolic equations, the reader is strongly encouraged to consult Thomée's monograph [Tho97].

6.1.1 Mathematical preliminaries

Let us first review some basic concepts of functional analysis which are useful in dealing with time-dependent functions. Let $T > 0$ be a fixed time, let Ω be a domain in $\mathbb{R}^d$, and define the space/time domain $Q = \Omega \times]0, T[$. Let u be a function defined on Q. An alternative way of looking at u is to consider it as a function of t with values in a Banach space, say V, whose elements are functions that only depend on the space variable:

$$u :]0, T[\ni t \longmapsto u(t) \equiv u(\cdot, t) \in V.$$

In this spirit, we are led to consider the following spaces:

(i) For $j \geq 0$, $\mathcal{C}^j([0, T]; V)$ is the space of V-valued functions of class $\mathcal{C}^j$ with respect to t. We denote by $d_t u$ the time-derivative of u and by $d_t^l u$ the derivative of order l. A classical result of functional analysis states that $\mathcal{C}^j([0, T]; V)$ is a Banach space for the norm

$$\|u\|_{\mathcal{C}^j([0,T];V)} = \sup_{t \in [0,T]} \sum_{l=0}^{j} \|d_t^l u(t)\|_V.$$

(ii) For $1 \leq p \leq +\infty$, $L^p(]0, T[; V)$ is the space of V-valued functions whose norm in V is in $L^p(]0, T[)$; it is a Banach space for the norm

$$\|u\|_{L^p(]0,T[;V)} = \begin{cases} \left(\int_0^T \|u(t)\|_V^p \right)^{\frac{1}{p}} & \text{if } 1 \leq p < +\infty, \\ \text{ess sup}_{t \in]0,T[} \|u(t)\|_V & \text{if } p = +\infty. \end{cases}$$

We denote by $d_t u$ the distributional time-derivative of u.

Since time-evolution problems are initial-value problems, it is important to determine for which type of function on $Q = \Omega \times]0, T[$ it is legitimate to define the value at $\Omega \times \{0\}$. To this end, we introduce the following:

Definition 6.1. *Let $1 < p_1, p_2 < +\infty$, let $B_0 \subset B_1$ be two reflexive Banach spaces with continuous embeddings, and set*

$$\mathcal{W}(B_0, B_1) = \{v :]0, T[\to B_0; \, v \in L^{p_1}(]0, T[; B_0); \, d_t v \in L^{p_2}(]0, T[; B_1)\}.$$

Equipped with the norm $\|u\|_{\mathcal{W}(B_0, B_1)} = \|u\|_{L^{p_1}(]0,T[;B_0)} + \|d_t u\|_{L^{p_2}(]0,T[;B_1)}$, $\mathcal{W}(B_0, B_1)$ *is a Banach space.*

Lemma 6.2 (Aubin). *Let $1 < p_1, p_2 < +\infty$ and let $B_0 \subset B \subset B_1$ be three reflexive Banach spaces with continuous embeddings. Every function in $\mathcal{W}(B_0, B_1)$ is continuous on $[0, T]$ with values in B. Moreover, the embedding $\mathcal{W}(B_0, B_1) \subset \mathcal{C}^0([0, T]; B)$ is compact whenever the embedding $B_0 \subset B$ is compact.*

Proof. See [Aub63]; see also, e.g., [Ama00, Lio69, LiM68, Sim87]. □

Lemma 6.2 gives an answer to the trace question raised above. It guarantees that for every function u in $\mathcal{W}(B_0, B_1)$ the quantities $u(0)$ and $u(T)$ are meaningful in B.

Henceforth, we specialize the above setting by restricting ourselves to Hilbert spaces and taking $p_1 = p_2 = 2$ in Definition 6.1. Let $V \subset L$ be two Hilbert spaces with continuous embedding. The norm of the embedding operator is denoted by $c_P^{-\frac{1}{2}}$, i.e.,

$$\forall v \in V, \quad c_P \|v\|_L^2 \leq \|v\|_V^2. \tag{6.1}$$

We assume that V is dense in L, and we identify L with L' so that we are in the situation where $V \subset L \equiv L' \subset V'$, i.e., the duality paring $\langle \cdot, \cdot \rangle_{V',V}$ can be viewed as an extension of the scalar product in L. Note that $c_P^{\frac{1}{2}} \|f\|_{V'} \leq \|f\|_L$ for all $f \in L$. In this setting, the following result, whose proof is left as an exercise, justifies *integration by parts* with respect to time:

Lemma 6.3 (Integration by parts). *Under the above assumptions, for all $u, v \in \mathcal{W}(V, V')$, the following identity holds:*

$$\int_0^T \langle d_t u(t), v(t) \rangle_{V',V} \, dt = (u(T), v(T))_L - (u(0), v(0))_L - \int_0^T \langle d_t v(t), u(t) \rangle_{V',V} \, dt.$$

6.1.2 An abstract problem

We now state a general result for time-dependent problems which plays a similar role to that played by the Lax–Milgram Lemma for elliptic equations.

Let $V \subset L \equiv L' \subset V'$ be a Hilbertian setting as defined above. Consider a mapping $a : \,]0, T[\times V \times V \to \mathbb{R}$ such that $a(t, \cdot, \cdot)$ is bilinear for a.e. t in $]0, T[$. Moreover, assume that a satisfies the following properties:

(P1) The function $t \mapsto a(t, u, v)$ is measurable $\forall u, v \in V$.
(P2) $\exists M$ such that $|a(t, u, v)| \leq M \|u\|_V \|v\|_V$ for a.e. $t \in [0, T]$, $\forall u, v \in V$.
(P3) $\exists \alpha > 0$ and $\gamma > 0$ such that $a(t, u, u) \geq \alpha \|u\|_V^2 - \gamma \|u\|_L^2$ for a.e. $t \in [0, T]$ and for all $u \in V$.

For $f \in L^2(]0, T[; V')$ and $u_0 \in L$, consider the following problem:

$$\begin{cases} \text{Seek } u \in \mathcal{W}(V, V') \text{ such that} \\ \langle d_t u, v \rangle_{V',V} + a(t, u, v) = \langle f(t), v \rangle_{V',V}, \quad \text{a.e. } t \in \,]0, T[, \, \forall v \in V, \quad (6.2) \\ u(0) = u_0. \end{cases}$$

The initial data $u(0) = u_0$ is meaningful according to Lemma 6.2.

Remark 6.4. Since the duality paring is an extension of the inner product in L, $\langle f(t), v \rangle_{V',V} = (f(t), v)_L$ whenever $f \in L^2(]0, T[; L)$. □

Definition 6.5 (Parabolic equation). *Equation* (6.2) *is said to be* parabolic *whenever the bilinear form a satisfies the conditions* (P1), (P2), *and* (P3).

Up to a change of variable, it is always possible to modify condition (P3) so that $\gamma = 0$. Indeed, set

$$\tilde{a}(t, \phi, v) = a(t, \phi, v) + \gamma(\phi, v)_L.$$

It is clear that $\tilde{a}$ satisfies conditions (P1) and (P2), and that $\tilde{a}$ is V-coercive since $\tilde{a}(t, \phi, \phi) \geq \alpha\|\phi\|_V^2$. Furthermore, setting $\phi = e^{-\gamma t}u$ and $g = e^{-\gamma t}f$, problem (6.2) is recast in the following equivalent form:

$$\begin{cases} \text{Seek } \phi \in \mathcal{W}(V, V') \text{ such that} \\ \langle d_t\phi, v\rangle_{V',V} + \tilde{a}(t, \phi, v) = \langle g(t), v\rangle_{V',V}, \quad \text{a.e. } t \in \,]0, T[, \ \forall v \in V, \quad (6.3) \\ \phi(0) = u_0. \end{cases}$$

We shall henceforth assume that a is coercive on V, i.e., $\gamma = 0$.

Before stating the main existence and uniqueness result of this section, we reformulate (6.2) in a setting which, at this point in the book, should be more familiar to the reader, namely that of the BNB Theorem. For the sake of simplicity, assume $u_0 = 0$. Consider the Hilbert spaces

$$Y = L^2(\,]0, T[; V) \quad \text{and} \quad X = \{v \in \mathcal{W}(V, V'); \ v(0) = 0\}.$$

For $g \in Y'$ and $y \in Y$, set $\langle g, y\rangle_{Y',Y} = \int_0^T \langle g(t), y(t)\rangle_{V',V}\,dt$ and define the bilinear form $b : X \times Y \to \mathbb{R}$ such that

$$\forall(x, y) \in X \times Y, \quad b(x, y) = \int_0^T [\langle d_t x, y\rangle_{V',V} + a(t, x, y)]\,dt.$$

It is clear that b is continuous. Then, consider the following problem:

$$\begin{cases} \text{Seek } u \in X \text{ such that} \\ b(u, y) = \langle f, y\rangle_{Y',Y}, \quad \forall y \in Y. \end{cases} \quad (6.4)$$

Using the distribution theory, it can be shown that (6.4) and (6.2) are equivalent. As a result, proving that (6.2) is well-posed amounts to proving that b satisfies the two conditions of the BNB Theorem. This is the purpose of the following:

Theorem 6.6 (J.L. Lions). *Under the hypotheses* (P1), (P2), *and* (P3), *problem* (6.2) *has a unique solution.*

Proof. See [LiM68, pp. 253–258] or [DaL93]. Let us prove the theorem using the formulation (6.4) and assuming $u_0 = 0$.

(1) For a.e. $t \in \,]0, T[$, define $A(t) \in \mathcal{L}(V, V')$ such that $\langle A(t)u, v\rangle_{V',V} = a(t, u, v)$ for all $u, v \in V$. Clearly, $\|A(t)\|_{\mathcal{L}(V,V')} \leq M$. Owing to the coercivity

hypothesis (P3), $A(t)$ is an isomorphism for a.e. $t \in \,]0, T[$. Hence, $A(t)^{-1}$ is continuous and $\|A(t)^{-1}\|_{\mathcal{L}(V',V)} \le \alpha^{-1}$. Moreover, $A(t)^{-1}$ is coercive since, for all $x \in V'$,

$$\langle x, A(t)^{-1}x \rangle_{V',V} = \langle A(t)A(t)^{-1}x, A(t)^{-1}x \rangle_{V',V} \ge \alpha \|A(t)^{-1}x\|_V^2 \ge \tfrac{\alpha}{M^2}\|x\|_{V'}^2.$$

Hence, the coercivity constant of $A(t)^{-1}$ is bounded from below by αM^{-2}.
(2) Let us prove that condition (BNB1) holds. Let $u \in X$, $\mu > 0$, and set $v = A(t)^{-1}d_t u + \mu u$. Step 1 shows that $\|v\|_Y \le c\|u\|_X$. Moreover,

$$b(u, v) = \int_0^T \langle d_t u + A(t)u, A(t)^{-1}d_t u + \mu u \rangle_{V',V}$$

$$\ge \tfrac{\mu}{2}\|u(T)\|_L^2 + \tfrac{\alpha}{M^2} \int_0^T \|d_t u\|_{V'}^2 + \mu\alpha \int_0^T \|u\|_V^2 - \int_0^T \|A\|\|A^{-1}\|\|u\|_V \|d_t u\|_{V'}$$

$$\ge \tfrac{\alpha}{M^2} \int_0^T \|d_t u\|_{V'}^2 + \mu\alpha \int_0^T \|u\|_V^2 - \tfrac{M}{\alpha} \int_0^T \|u\|_V \|d_t u\|_{V'}$$

$$\ge \tfrac{\alpha}{2M^2} \int_0^T \|d_t u\|_{V'}^2 + (\mu\alpha - \tfrac{M^4}{2\alpha^3}) \int_0^T \|u\|_V^2.$$

Taking $\mu = M^4 \alpha^{-4}$ yields

$$b(u, v) \ge c\|u\|_{W(V,V')}^2 \ge c'\|u\|_X \|v\|_Y.$$

The inf-sup condition (BNB1) follows easily.
(3) Let us prove that condition (BNB2) holds. Let $v \in Y$ be such that $b(u, v) = 0$ for all $u \in X$. Testing with $u \in \mathcal{D}(]0, T[; V)$ leads to

$$\left| \int_0^T (d_t u, v)_L \right| = \left| \int_0^T \langle Au, v \rangle_{V',V} \right| \le c\|u\|_Y \|v\|_Y.$$

Use integration by parts (see Lemma 6.3) to infer $d_t v \in L^2(]0, T[; V')$. As a result,

$$\forall u \in \mathcal{D}(]0, T[; V), \quad \int_0^T -\langle d_t v, u \rangle_{V',V} + \langle A(t)^T v, u \rangle_{V',V} = 0. \qquad (6.5)$$

By density of $\mathcal{D}(]0, T[; V)$ in Y, (6.5) holds for all u in Y. Moreover, using the test function $u = t\phi$ with ϕ arbitrary in V and integrating by parts in (6.5) yields $v(T) = 0$. Then, using v as a test function in (6.5) yields $v = 0$. Hence, (BNB2) holds. $\qquad \square$

Theorem 6.7 (A priori estimates). *For $f \in L^2(]0, T[; V')$, the solution to* (6.2) *satisfies the energy estimate*

$$\begin{cases} \|u\|_{C^0([0,T];L)} \le \|u_0\|_L e^{-\frac{1}{2}\alpha c_P t} + \dfrac{1}{\sqrt{\alpha}}\|f\|_{L^2(]0,T[;V')}, \\[2mm] \|u\|_{L^2(]0,T[;V)} \le \dfrac{1}{\sqrt{\alpha}}\|u_0\|_L + \dfrac{1}{\alpha}\|f\|_{L^2(]0,T[;V')}. \end{cases} \qquad (6.6)$$

Furthermore, if $f \in L^\infty(]0, +\infty[; V')$,

$$\limsup_{t \to +\infty} \|u(t)\|_L \le \tfrac{1}{\alpha\sqrt{c_P}}\|f\|_{L^\infty(]0,+\infty[;V')}. \qquad (6.7)$$

Proof. (1) Let $t \in \,]0, T[$. Choose u as a test function. The coercivity of a together with Lemma 6.3 implies

$$\tfrac{1}{2}\tfrac{d}{dt}\int_0^t \|u\|_L^2 + \alpha \int_0^t \|u\|_V^2 \le \int_0^t \|f\|_{V'} \|u\|_V \le \tfrac{\alpha}{2}\int_0^t \|u\|_V^2 + \tfrac{1}{2\alpha}\int_0^t \|f\|_{V'}^2.$$

Hence, owing to (6.1),

$$\tfrac{d}{dt}\int_0^t \|u\|_L^2 + \alpha c_P \int_0^t \|u\|_L^2 \le \tfrac{d}{dt}\int_0^t \|u\|_L^2 + \alpha \int_0^t \|u\|_V^2 \le \tfrac{1}{\alpha}\int_0^t \|f\|_{V'}^2,$$

yielding (6.6).

(2) Under the second hypothesis, use Gronwall's Lemma 6.9 to infer

$$\|u(t)\|_L^2 \le \|u(0)\|_L^2 e^{-\alpha c_P t} + \tfrac{1}{\alpha}\int_0^t e^{-\alpha c_P (t-\tau)}\|f(\tau)\|_{V'}^2 \, d\tau.$$

Then, (6.7) follows easily.   □

Remark 6.8. Theorem 6.7 establishes the continuous dependence of u with respect to the data. It is wise to retain, or at least mimic, this stability result when seeking an approximation with respect to time and/or space.   □

Lemma 6.9 (Gronwall). *Let $\beta \in \mathbb{R}$, $\varphi \in \mathcal{C}^1([0,T];\mathbb{R})$, and $f \in \mathcal{C}^0([0,T];\mathbb{R})$ such that $d_t\varphi \le \beta\varphi + f$. Then,*

$$\forall t \in [0,T], \quad \varphi(t) \le e^{\beta t}\varphi(0) + \int_0^t e^{\beta(t-\tau)}f(t)\,d\tau.$$

Proof. Multiply the inequality by $e^{-\beta t}$ and integrate with respect to t.   □

6.1.3 The heat equation

We illustrate the notions introduced above on the *heat equation*:

$$\begin{cases} d_t u - \nabla\cdot(\kappa(x)\nabla u) = f & x \in \Omega, \ t \ge 0, \\ u(x,t) = 0 & x \in \partial\Omega, t > 0, \\ u(x,0) = u_0(x) & x \in \Omega. \end{cases} \qquad (6.8)$$

Problem (6.8) models heat transfers in Ω: the unknown $u(x,t)$ is the temperature at the point $x \in \Omega$ and time t; f is a source term; u_0 is the initial temperature; and κ is the thermal conductivity. For the sake of simplicity, we assume that κ is scalar and that homogeneous Dirichlet conditions are enforced on $\partial\Omega$, i.e., that the temperature is prescribed on $\partial\Omega$.

To formulate problem (6.8) in a weak sense, assume that $u_0 \in L^2(\Omega)$ and $f \in L^2(\,]0,T[; H^{-1}(\Omega))$. Take a test function $v \in H_0^1(\Omega)$, multiply (6.8) by v, and integrate over Ω. This yields

$$\int_\Omega d_t u(t)v + \int_\Omega \kappa(x)\nabla u(t)\cdot\nabla v = \langle f(t), v\rangle_{H^{-1}, H_0^1}. \qquad (6.9)$$

Hence, a possible weak formulation of problem (6.8) is:

$$\begin{cases} \text{Seek } u \in \mathcal{W}(H_0^1(\Omega), H^{-1}(\Omega)) \text{ such that, a.e. } t, \ \forall v \in H_0^1(\Omega), \\ \langle d_t u, v \rangle_{H^{-1}, H_0^1} + \int_\Omega \kappa(x) \nabla u(t) \cdot \nabla v = \langle f(t), v \rangle_{H^{-1}, H_0^1}, \\ u(0) = u_0. \end{cases} \qquad (6.10)$$

Setting $a(t, u(t), v) = \int_\Omega \kappa(x) \nabla u(t) \cdot \nabla v$ and $V = H_0^1(\Omega)$, Theorem 6.6 implies that problem (6.10) is well-posed if $\kappa \in L^\infty(\Omega)$ and if there is $\kappa_0 > 0$ such that $\kappa(x) \geq \kappa_0$ a.e. in Ω. Indeed, letting c_Ω be the Poincaré constant (see Lemma B.61), it is clear that $a(t, u, u) \geq \alpha \|u\|_{1,\Omega}^2$ where $\alpha = \kappa_0 \frac{c_\Omega^2}{1+c_\Omega^2}$. Theorem 6.7 yields

$$\begin{cases} \|u\|_{C^0([0,T];L^2(\Omega))} \leq \|u_0\|_{L^2(\Omega)} + \frac{1}{\sqrt{\alpha}} \|f\|_{L^2(]0,T[;H^{-1}(\Omega))}, \\ \|u\|_{L^2(]0,T[;H^1(\Omega))} \leq \frac{1}{\sqrt{\alpha}} \|u_0\|_{L^2(\Omega)} + \frac{1}{\alpha} \|f\|_{L^2(]0,T[;H^{-1}(\Omega))}. \end{cases}$$

Conversely, it can be shown using standard density arguments together with the distribution theory, that if u solves (6.10) and u is smooth enough, then u is a classical solution to (6.8).

Remark 6.10. Theorem 6.6 applies to problems that are more complicated than the heat equation. Actually, it applies to the time-dependent version of all the problems studied in Chapter 3. For instance, setting

$$a(t, u, v) = \int_\Omega (\sigma(x,t) \cdot \nabla u) \cdot \nabla v + (\beta(x,t) \cdot \nabla u)v + \mu(x,t)uv, \qquad (6.11)$$

one readily verifies that the hypotheses of Theorem 6.6 are satisfied provided the fields $\sigma(x,t)$, $\beta(x,t)$, and $\mu(x,t)$ are sufficiently smooth and are such that the bilinear form $a(t, \cdot, \cdot)$ is coercive for a.e. $t \in]0, T[$; see Theorem 3.8 for sufficient conditions yielding coercivity. Moreover, the theory can be extended to Neumann and Robin boundary conditions. $\square$

We conclude this section by stating two remarkable properties of the solutions to (6.2) when using the bilinear form (6.11).

Proposition 6.11 (Positivity). *Let $u_0 \in L^2(\Omega)$ and $f \in L^2(]0,T[; L^2(\Omega))$. Let $u \in \mathcal{W}(H_0^1(\Omega), H^{-1}(\Omega))$ solve (6.2) with $a(t, \cdot, \cdot)$ defined in (6.11). Assume $u_0(x) \geq 0$ a.e. in Ω and $f(x,t) \geq 0$ a.e. in Q. Then, $u(x,t) \geq 0$ a.e. in Q.*

Proof. One can verify that $u^- = \frac{1}{2}(|u| - u) \in \mathcal{W}(H_0^1(\Omega), H^{-1}(\Omega))$ is an admissible test function. Observe that $a(t, u, u^-) = -a(t, u^-, u^-)$ to obtain

$$\tfrac{1}{2} \tfrac{d}{dt} \|u^-\|_{0,\Omega}^2 + a(t, u^-, u^-) = -(f, u^-)_{0,\Omega} \leq 0,$$

implying $\|u^-(t)\|_{0,\Omega} \leq \|u_0^-\|_{0,\Omega} = 0$. $\square$

Proposition 6.12 (Maximum principle). *Let $u_0 \in L^\infty(\Omega)$ and assume $f = 0$. Let $u(x,t) \in \mathcal{W}(H_0^1(\Omega), H^{-1}(\Omega))$ be the solution to (6.2) with $a(t, \cdot, \cdot)$ defined in (6.11). Assume $\mu \geq 0$. Then, $\|u\|_{L^\infty(Q)} \leq \|u_0\|_{L^\infty(\Omega)}$.*

Proof. Set $M = \|u_0\|_{L^\infty(\Omega)}$ and note that $(u - M)^+ = \frac{1}{2}(|u - M| + u - M) \in \mathcal{W}(H_0^1(\Omega), H^{-1}(\Omega))$ is an admissible test function. The property

$$a(t, u, (u - M)^+) = a(t, u - M, (u - M)^+) + a(t, M, (u - M)^+)$$

$$= a(t, (u - M)^+, (u - M)^+) + \int_\Omega \mu M (u - M)^+ \geq 0,$$

implies $\frac{d}{dt}\|(u - M)^+\|_{0,\Omega}^2 \leq 0$. The desired result follows easily. $\qquad\square$

Remark 6.13. It is generally difficult to retain Maximum Principle properties in discrete settings. If $V_h \subset H_0^1(\Omega)$ is a finite element space and v_h is an arbitrary function in V_h, it is unlikely that v_h^- and $(v_h - M)^+$ are in V_h. $\quad\square$

6.1.4 Space approximation

The method of lines. In problem (6.2) the space and time variables play different roles. This observation advocates for the following approach: First, approximate the solution to (6.2) in space only so as to obtain a system of coupled ordinary differential equations (ODEs), where the time is the only independent variable. Second, construct an approximation in time by making use of the vast theory of solution techniques for ODEs. This approach is often called the *method of lines*.

The theoretical setting. To simplify the analysis, we henceforth assume that $a(t, u, v)$ is continuous with respect to t on $[0, T]$ for all u, v in V, that $a(t, \cdot, \cdot)$ is coercive on V, i.e., $a(t, u, u) \geq \alpha\|u\|_V^2$, and that $f \in C^0([0, T]; L)$.

Let $\{V_h\}_{0 \leq h \leq 1}$ be a family of finite-dimensional subspaces of V. We assume that there exists a dense subspace of V, say W, a linear interpolation operator, say $\mathcal{I}_h \in \mathcal{L}(W; V_h)$, an integer k, and a constant c, independent of h, such that

$$\forall v \in W, \quad \|v - \mathcal{I}_h v\|_L + h\|v - \mathcal{I}_h v\|_V \leq c\, h^{k+1}\|v\|_W. \tag{6.12}$$

For instance, one may think of $L = L^2(\Omega)$, $V = H_0^1(\Omega)$ where V_h is a H^1-conformal finite element space using a reference finite element $\{\widehat{K}, \widehat{P}, \widehat{\Sigma}\}$ such that $\mathbb{P}_k \subset \widehat{P}$, and $W = H^{k+1}(\Omega) \cap H_0^1(\Omega)$; see §1.4.2.

Consider the approximate problem:

$$\begin{cases} \text{Seek } u_h \in C^1([0, T]; V_h) \text{ such that} \\ (d_t u_h, v_h)_L + a(t, u_h, v_h) = (f, v_h)_L, \quad \forall t \in [0, T], \, \forall v_h \in V_h, \\ u_h(0) = u_{0h}, \end{cases} \tag{6.13}$$

where $u_{0h} \in V_h$ is an approximation of u_0, the precise nature of which will be clarified in Remark 6.15. Since (6.13) is a finite linear system of coupled ODEs, the Cauchy–Lipschitz Theorem guarantees the existence and uniqueness of a solution $u_h(t)$ in $C^1([0,T];V_h)$; see, e.g., [StB80, p. 406], [Bre91, p. 104], or [CrM84, p. 65].

Our goal is to determine whether $u_h(t)$ yields an accurate approximation of $u(t)$. It is convenient to introduce the operator $P_{ht} \in \mathcal{L}(V;V_h)$ defined for $t \in [0,T]$ and such that, for all $w \in V$, $P_{ht}(w)$ is the solution to

$$\forall v_h \in V_h, \quad a(t, P_{ht}(w), v_h) = a(t, w, v_h). \tag{6.14}$$

Clearly, P_{ht} is a projection. When the bilinear form a is associated with the Laplace operator, P_{ht} is the so-called *elliptic projector*; see §1.6.3. Furthermore, we make the following technical hypothesis:

(EP) $\begin{cases} \text{There is } c > 0 \text{ such that, for all } w \text{ in } C^1([0,T];W) \text{ and } \forall j \in \{0,1\}, \\ \|w - P_{ht}(w)\|_{C^j([0,T];L)} + h\|w - P_{ht}(w)\|_{C^j([0,T];V)} \leq c_i h^{k+1} c(w), \end{cases}$

where $c(w) = \|w\|_{C^1([0,T];W)}$. This hypothesis holds whenever the differential operator associated with a has smoothing properties so that the Aubin–Nitsche Lemma can be applied.

Theorem 6.14. *Assume* (EP) *and* $u \in C^1([0,T];W)$. *Then, for all* h,

$$\|u - u_h\|_{C^0([0,T];L)} \leq \|u_0 - u_{0h}\|_L e^{-c_P \alpha \frac{T}{2}} + c_i \left(1 + \frac{1}{\alpha c_P}\right) h^{k+1} c(u),$$

$$\frac{1}{\sqrt{T}} \|u - u_h\|_{L^2([0,T];V)} \leq \frac{1}{\sqrt{\alpha T}} \|u_0 - u_{0h}\|_L + c_i \left(1 + \frac{1}{\sqrt{T}} + \frac{1}{\alpha\sqrt{c_P}}\right) h^k c(u),$$

with $c(u) = \|u\|_{C^1([0,T];W)}$.

Proof. (1) Set $e_h(t) = P_{ht}(u) - u_h(t)$ and $\eta(t) = u(t) - P_{ht}(u)$ so that $u(t) - u_h(t) = e_h(t) + \eta(t)$. Hypothesis (EP) implies

$$\|\eta\|_{C^1([0,T];L)} + h\|\eta\|_{C^1([0,T];V)} \leq c_i h^{k+1} \|u\|_{C^1([0,T];W)}.$$

Subtracting (6.13) from (6.2) leads to

$$\forall v_h \in V_h, \quad (d_t e_h, v_h)_L + a(t, e_h, v_h) = -(d_t \eta, v_h)_L.$$

Note that the term $a(t, \eta(t), v_h)$ is zero by definition of $P_{ht}(u)$. Choosing $e_h(t)$ as test function yields

$$\tfrac{1}{2}\tfrac{d}{dt}\|e_h\|_L^2 + \alpha\|e_h\|_V^2 \leq \|d_t\eta\|_{V'}\|e_h\|_V \leq \tfrac{\alpha}{2}\|e_h\|_V^2 + \tfrac{1}{2\alpha}\|d_t\eta\|_{V'}^2.$$

The relation $\|v\|_V^2 \geq c_P\|v\|_L^2$ implies

$$\tfrac{d}{dt}\|e_h\|_L^2 + \alpha c_P\|e_h\|_L^2 \leq \tfrac{d}{dt}\|e_h\|_L^2 + \alpha\|e_h\|_V^2 \leq \tfrac{1}{\alpha c_P}\|d_t\eta\|_L^2. \tag{6.15}$$

(2) Owing to Gronwall's Lemma 6.9,

$$\|e_h(t)\|_L^2 \le \|e_h(0)\|_L^2 e^{-\alpha c_P t} + \frac{1}{\alpha c_P} \int_0^t e^{-\alpha c_P (t-\tau)} \|d_\tau \eta\|_L^2 \, d\tau$$

$$\le \|e_h(0)\|_L^2 e^{-\alpha c_P t} + \frac{1}{(\alpha c_P)^2} \left(1 - e^{-\alpha c_P t}\right) \|d_t \eta\|_{C^0([0,T];L)}^2.$$

Then use the triangle inequality $\|u - u_h\|_L \le \|e_h\|_L + \|\eta\|_L$ to infer

$$\|u(t) - u_h(t)\|_L \le \|e_h(0)\|_L e^{-\alpha c_P \frac{t}{2}} + \left(1 + \frac{1}{\alpha c_P}\right) \|\eta\|_{C^1([0,T];L)}.$$

The first desired bound from above follows easily.

(3) Integrating the second inequality in (6.15) with respect to time yields

$$\alpha \int_0^T \|e_h\|_V^2 \le \frac{1}{\alpha c_P} \int_0^T \|d_t \eta\|_L^2 + \|e_h(0)\|_L^2 \le \frac{T}{\alpha c_P} \|d_t \eta\|_{C^0([0,T];L)}^2 + \|e_h(0)\|_L^2.$$

Hence,

$$\frac{1}{T} \int_0^T \|e_h\|_V^2 \le \frac{1}{\alpha^2 c_P} \|d_t \eta\|_{C^0([0,T];L)}^2 + \frac{1}{\alpha T} \|e_h(0)\|_L^2.$$

The second bound from above follows from the triangle inequality.   □

Remark 6.15.

(i) If u_0 is in W, we can choose $u_{0h} = \mathcal{I}_h u_0$ to approximate the initial data. Property (6.12) implies $\|u_0 - u_{0h}\|_L \le ch^{k+1}\|u_0\|_W$, i.e., the two bounds from above in the theorem are optimal.

(ii) The error induced by the approximation of the initial data exponentially decreases with T in the L-norm and decreases in the mean like $T^{-\frac{1}{2}}$ in the V-norm. Insensitivity to initial data as time grows is a characteristic property of dissipative equations (i.e., parabolic). In particular, $\limsup_{t \to \infty} \left[\|u(t) - u_h(t)\|_L + \frac{h}{\sqrt{t}}\|u - u_h\|_{L^2(]0,t[;V)}\right] \le ch^{k+1}\|u\|_{C^1([0,\infty[;W)}$.

(iii) The use of the elliptic projector is essential to obtain optimal error estimates in the L-norm. The trick avoids having to use a duality argument involving the adjoint problem, i.e., a retrograde equation from T to 0. This technique has been introduced by Wheeler [Whe73].

(iv) The regularity required for u in the error estimates is not optimal. Actually, the term $\|u\|_{C^1([0,T];W)}$ in the L-estimate can be replaced by $\|u\|_{C^0([0,T];W)} + \|d_t u\|_{L^2(]0,T[;W)}$, and in the V-estimate this term can be replaced by $\|u\|_{L^2([0,T];W)} + \|d_t u\|_{L^2(]0,T[;Z)}$, where Z is any subset of V where the interpolation estimate $\|v - \mathcal{I}_h v\|_L \le ch^k\|v\|_Z$ holds (think of $W = H^{k+1}(\Omega) \cap H_0^1(\Omega)$ and $Z = H^k(\Omega) \cap H_0^1(\Omega)$ for the heat equation).

Implementation. The approximate problem (6.13) is simply a system of coupled ODEs. Indeed, let $\{\varphi_1, \ldots, \varphi_N\}$ be a basis for V_h (for instance, the global shape functions if V_h is a finite element space). For all $t \in [0,T]$, the approximate solution $u_h(t) \in V_h$ can be expanded in the form

$$u_h(x,t) = \sum_{i=1}^N U_i(t)\varphi_i.$$

For $t \in [0, T]$, set $U(t) = (U_i(t))_{1 \le i \le N} \in \mathbb{R}^N$ and $F(t) = ((f(t), \varphi_i)_L)_{1 \le i \le N} \in \mathbb{R}^N$. Moreover, introduce the stiffness matrix $\mathcal{A}(t) \in \mathbb{R}^{N,N}$ and the mass matrix $\mathcal{M} \in \mathbb{R}^{N,N}$ such that

$$\mathcal{A}_{ij}(t) = a(t, \varphi_j, \varphi_i), \qquad \mathcal{M}_{ij} = (\varphi_i, \varphi_j)_L, \qquad 1 \le i, j \le N.$$

The mass matrix is symmetric definite positive, and, for all $t \in [0, T]$, the stiffness matrix $\mathcal{A}(t)$ is positive definite; see §9.1.4. Using the above notation, (6.13) is recast in the form

$$\begin{cases} \mathcal{M} d_t U(t) = -\mathcal{A}(t) U(t) + F(t), & t \in [0, T], \\ U(0) = U_0, \end{cases} \tag{6.16}$$

where $U_0 \in \mathbb{R}^N$ is the coordinate vector of u_{0h} relative to the basis $\{\varphi_i\}_{1 \le i \le N}$.

6.1.5 Time approximation: Convergence theory

Possible strategies to approximate the solution to (6.16) are numerous. For instance, one can choose a one-step method (e.g., Runge–Kutta), a multistep method (e.g., Adams methods), a backward finite difference method, splitting techniques, or techniques based on rational approximations of the exponential function. A large number of monographs and textbooks are dedicated to these techniques; see [BuF93] for a review of implementations of time-marching techniques, [CaP91] for a review on dynamical systems, [CrM84] for an introductory textbook addressing mathematical aspects, [Gea71] for an introductory textbook with many examples, [HaW91] for issues related to stiff systems, and [Lam91] for a thorough review on solution methods for ODEs.

Henceforth, we (arbitrarily) restrict ourselves to backward and forward finite difference methods. We briefly develop an abstract framework for the convergence analysis of time-marching techniques, and we introduce the fundamental concepts of consistency, stability, and convergence.

The setting. Given a positive integer N, set $\Delta t = \frac{T}{N}$, $t^n = n \Delta t$ for $0 \le n \le N$ and consider a partitioning of the time interval in the form $[0, T] = \bigcup_{n=0}^{N-1} [t^n, t^{n+1}]$. The basic approximation technique consists of constructing a sequence

$$u_{h\Delta t} = (u_h^0, u_h^1, \ldots, u_h^N),$$

in the Cartesian product space V_h^{N+1} where it is expected that u_h^n approximates $u(t^n)$. The quantity Δt is the so-called *time step* of the approximation. To alleviate the notation, we restrict ourselves to constant time steps, but the theory developed below is readily extendable to variable time steps.

Let E_1, E_2, and E_3 be subspaces of L, $L^2([0, T]; V')$, and $\mathcal{W}(V, V')$, respectively. Equip these spaces with norms $\| \cdot \|_{E_1}$, $\| \cdot \|_{E_2}$, and $\| \cdot \|_{E_3}$, respectively; the norms will be specified later. We assume henceforth that the initial data u_0 and the source term f are chosen in E_1 and E_2, respectively. Denote by

$S : E_1 \times E_2 \to E_3$ the linear operator that maps the pair $(u_0, f) \in E_1 \times E_2$ to the solution $u \in E_3$ of problem (6.2); that is, $u = S(u_0, f)$.

To approximate u_0, f, and u, we introduce the finite-dimensional spaces $E_{1h} = V_h$, $E_{2h\Delta t} = V_h^N$, and $E_{3h\Delta t} = V_h^{N+1}$, that we equip with the norms or seminorms (to be specified later) $\|\cdot\|_{E_{1h}}$, $\|\cdot\|_{E_{2h\Delta t}}$, and $\|\cdot\|_{E_{3h\Delta t}}$, respectively. Approximations of u_0 and f are provided by the linear operators (yet to be specified) $\rho_{1h} : E_1 \to E_{1h}$ and $\rho_{2h\Delta t} : E_2 \to E_{2h\Delta t}$. To construct the approximate solution $u_{h\Delta t}$, we introduce the linear operator $S_{h\Delta t} : E_{1h} \times E_{2h\Delta t} \to E_{3h\Delta t}$ (yet to be specified; see Definition 6.17 below) and we set

$$u_{h\Delta t} = S_{h\Delta t}(\rho_{1h}(u_0), \rho_{2h\Delta t}(f)).$$

Example 6.16.

(i) If the initial data u_0 is in W, the domain of the interpolation operator $\mathcal{I}_h$ satisfying (6.12), one can choose $E_1 = W$ and $\rho_{1h} = \mathcal{I}_h$. If $u_0 \in V$, one can take $E_1 = V$ and $\rho_{1h} = P_{h0}$, where P_{ht} is the elliptic projector defined in (6.14). Finally, if u_0 is only in L, one must take $E_1 = L$ and $\rho_{1h} = P_{hL}$, where P_{hL} is the orthogonal projection of L onto V_h.

(ii) If $f \in C^0([0, T]; L)$, one can choose $E_2 = C^0([0, T]; L)$ and $\rho_{2h\Delta t}(f) = (P_{hL}(f(t^n)))_{1 \le n \le N}$. If $f \in L^2([0, T]; L)$, one can take $E_2 = L^2([0, T]; L)$ and $\rho_{2h\Delta t}(f) = (P_{hL}(\frac{1}{\Delta t} \int_{t^{n-1}}^{t^n} f(\tau)\,d\tau))_{1 \le n \le N}$. Finally, if $f \in L^2([0, T]; V')$ only, one must take $E_2 = L^2([0, T]; V')$ and, letting $\tilde{P}_{hL}$ be the continuous extension of P_{hL} to V', $\rho_{2h\Delta t}(f) = (\tilde{P}_{hL}(\frac{1}{\Delta t} \int_{t^{n-1}}^{t^n} f(\tau)\,d\tau))_{1 \le n \le N}$.   □

Henceforth, we restrict ourselves to the so-called p-step schemes.

Definition 6.17 (p-step scheme). *Let $p \ge 1$ be an integer. A p-step scheme is an indexed family of bijective operators $S_{h\Delta t} \in \mathcal{L}(E_{1h} \times E_{2h\Delta t}; E_{3h\Delta t})$ such that, for all $(u_{0h}, f_{h\Delta t}) \in E_{1h} \times E_{2h\Delta t}$, the sequence $(u_h^0, \ldots, u_h^N) = S_{h\Delta t}(u_{0h}, f_{h\Delta t})$ is inductively defined by*

$$\begin{cases} u_h^0 = u_{0h}, \\ u_h^{n+1} = H_h^{n+1}(u_h^{\max(n-p+1,0)}, \ldots, u_h^n, f_h^{n+1}), & \text{for } 0 \le n \le N-1, \end{cases}$$

where $H_h^{n+1} \in \mathcal{L}(V_h^{\min(n+1,p)} \times V_h; V_h)$ for $0 \le n \le N-1$.

The convergence theory. Since the exact solution $u = S(u_0, f)$ and the discrete solution $u_{h\Delta t} = S_{h\Delta t}(\rho_{1h}(u_0), \rho_{2h\Delta t}(f))$ do not live in the same space, we introduce a new space, say $E_{3\Delta t}$, to measure the error. Accordingly, we introduce the linear operators $i_{h\Delta t} : E_{3h\Delta t} \to E_{3\Delta t}$ and $\rho_{3\Delta t} : E_3 \to E_{3\Delta t}$ so that $\rho_{3\Delta t}(u) - i_{h\Delta t}(u_{h\Delta t})$ is a measure of the approximation error. Following the terminology of Temam [Tem77, I§3], $\rho_{3\Delta t}$ is called a *restriction operator* and $i_{h\Delta t}$ is called a *prolongation operator*.

The above definitions are summarized in the following diagram:

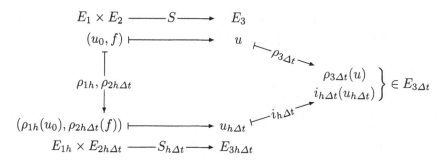

The convergence theory relies upon the fundamental concepts of consistency and stability.

Definition 6.18 (Consistency). *A p-step scheme is said to be* consistent *with problem (6.2) up to $\mathcal{O}(\epsilon(h, \Delta t))$, if there exist:*

(i) *a dense subspace of $\mathcal{W}(V, V')$, say $Z(Q)$, such that $u \in Z(Q)$ implies $u(0) \in E_1$ and $\langle d_t u, \cdot \rangle_{V', V} + a(t, u, \cdot) \in E_2$;*
(ii) *an operator $\Pi_{h\Delta t} \in \mathcal{L}(Z(Q); E_{3h\Delta t})$;*

such that if $u(t)$ solving (6.2) is in $Z(Q)$, defining $(R_{1h}, R_{2h\Delta t}) \in E_{1h} \times E_{2h\Delta t}$ such that $(R_{1h}, R_{2h\Delta t}) = S_{h\Delta t}^{-1}(\Pi_{h\Delta t}(u) - u_{h\Delta t})$, the quantity

$$\chi_{h\Delta t}(u) = \|R_{1h}\|_{E_{1h}} + \|R_{2h\Delta t}\|_{E_{2h\Delta t}} + \|\rho_{3\Delta t}(u) - i_{h\Delta t}\Pi_{h\Delta t}(u)\|_{E_{3\Delta t}},$$

is such that $\chi_{h\Delta t}(u) \leq c\,\epsilon(h, \Delta t)\|u\|_{Z(Q)}$ where c is independent of h and Δt, and $\epsilon(h, \Delta t)$ is continuous at $(0,0)$ with $\epsilon(0,0) = 0$. The quantity $\chi_{h\Delta t}(u)$ is called the consistency error.

Remark 6.19.
(i) The notion of consistency is norm-dependent, i.e., the function $\epsilon(h, \Delta t)$ depends on the norms that are chosen to measure the errors.
(ii) Note that $\Pi_{h\Delta t}(u) = S_{h\Delta t}(\rho_{1h}(u_0) + R_{1h}, \rho_{2h\Delta t}(f) + R_{2h\Delta t})$.   □

Example 6.20. If $u \in \mathcal{C}^0([0, T]; L)$, the sequence $(u(t^n))_{0 \leq n \leq N}$ is meaningful in L^{N+1}; as a result, it is reasonable to set $E_3 = \mathcal{C}^0([0, T]; L)$, $E_{3\Delta t} = L^{N+1}$, and to define $\rho_{3\Delta t}(u) = (u(t^n))_{0 \leq n \leq N}$. If $u \in \mathcal{C}^0([0, T]; V)$, we can set $E_3 = \mathcal{C}^0([0, T]; V)$, $E_{3\Delta t} = V^{N+1}$, and $\rho_{3\Delta t}(u) = (u(t^n))_{0 \leq n \leq N}$. In both cases, assuming $V_h \subset V$, we infer $E_{3h\Delta t} = V_h^{N+1} \subset V^{N+1} \subset E_{3\Delta t}$. A natural choice for $i_{h\Delta t}$ is then to take the canonical injection and to equip $E_{3h\Delta t}$ with the norm induced by that of $E_{3\Delta t}$.   □

Definition 6.21 (Stability). *Consider a consistent p-step scheme.*

(i) *The scheme is said to be* stable *if the operators $i_{h\Delta t}$ and $S_{h\Delta t}$ are uniformly bounded.*
(ii) *The scheme is said to be* conditionally stable *if there is a function $r(h) > 0$ (resp., $s(\Delta t) > 0$) such that the continuity of $i_{h\Delta t}$ and $S_{h\Delta t}$ is uniform for all $h > 0$ when $\Delta t \leq r(h)$ (resp., for all Δt when $h \leq s(\Delta t)$).*

(iii) *The scheme is said to be* completely stable *if it is stable and if the operators* ρ_{1h}, $\rho_{2h\Delta t}$, *and* $\rho_{3\Delta t}$ *are uniformly bounded.*

Remark 6.22.

(i) In general, the uniform boundedness of $S_{h\Delta t}$ is the most difficult property to establish. It amounts to proving the existence of a constant $c > 0$ such that, for all N, $h > 0$, and $(u_{0h}, f_{h\Delta t}) \in E_{1h} \times E_{2h\Delta t}$,

$$\|S_{h\Delta t}(u_{0h}, f_{h\Delta t})\|_{E_{3h\Delta t}} \le c(\|u_{0h}\|_{E_{1h}} + \|f_{h\Delta t}\|_{E_{2h\Delta t}}).$$

(ii) It is often useful to choose $\Pi_{h\Delta t}(u) = (P_{ht^n}(u))_{0 \le n \le N}$ where P_{ht} is the elliptic projector defined in (6.14); see Remark 6.15(iii). □

We are now in a position to state the major result of this section.

Theorem 6.23 (Convergence). *Consider a stable p-step scheme consistent up to* $\mathcal{O}(\epsilon(h, \Delta t))$. *Set* $u_{h\Delta t} = S_{h\Delta t}(\rho_{1h}(u_0), \rho_{2h\Delta t}(f))$. *Let* $u(t)$ *solve* (6.2).

(i) *If* $u \in Z(Q)$, *there is* c, *independent of* h *and* Δt, *such that*

$$\|\rho_{3\Delta t}(u) - i_{h\Delta t}(u_{h\Delta t})\|_{E_{3\Delta t}} \le c\epsilon(h, \Delta t)\|u\|_{Z(Q)}. \tag{6.17}$$

(ii) *If the consistency and stability hypotheses hold for* $E_1 = L$, $E_2 = L^2([0,T]; V')$, *and* $E_3 = \mathcal{W}(V, V')$, *and if the scheme is completely stable,*

$$\lim_{\substack{h \to 0 \\ \Delta t \to 0}} \|\rho_{3\Delta t}(u) - i_{h\Delta t}(u_{h\Delta t})\|_{E_{3\Delta t}} = 0. \tag{6.18}$$

Proof. (1) Set $e_{h\Delta t} = \Pi_{h\Delta t}(u) - u_{h\Delta t}$. Consistency implies

$$e_{h\Delta t} = S_{h\Delta t}(R_{1h}, R_{2h\Delta t}).$$

The stability of $S_{h\Delta t}$ yields

$$\|e_{h\Delta t}\|_{E_{3h\Delta t}} \le c(\|R_{1h}\|_{E_{1h}} + \|R_{2h\Delta t}\|_{E_{2h\Delta t}}).$$

The triangle inequality and the stability of $i_{h\Delta t}$ imply

$$\|\rho_{3\Delta t}(u) - i_{h\Delta t}u_{h\Delta t}\|_{E_{3\Delta t}} \le \|\rho_{3\Delta t}(u) - i_{h\Delta t}\Pi_{h\Delta t}(u)\|_{E_{3\Delta t}} + \|i_{h\Delta t}e_{h\Delta t}\|_{E_{3\Delta t}}$$
$$\le c\left(\|R_{1h}\|_{E_{1h}} + \|R_{2h\Delta t}\|_{E_{2h\Delta t}} + \|\rho_{3\Delta t}(u) - i_{h\Delta t}\Pi_{h\Delta t}(u)\|_{E_{3\Delta t}}\right).$$

Estimate (6.17) is a consequence of the $\mathcal{O}(\epsilon(h, \Delta t))$-consistency.
(2) Let us now prove (6.18). Assume that the scheme is completely stable. Since $\rho_{3\Delta t}$ is uniformly bounded and $Z(Q)$ is dense in $\mathcal{W}(V, V')$, we infer that for all $\xi > 0$, there is $u_\xi \in Z(Q)$ such that

$$\|\rho_{3\Delta t}(u) - \rho_{3\Delta t}(u_\xi)\|_{E_{3\Delta t}} \le c\|u - u_\xi\|_{\mathcal{W}(V,V')} \le \xi.$$

Denote by $A \in C^0([0,T]; \mathcal{L}(V, V'))$ the operator such that

$$\forall t \in [0, T], \ \forall v, w \in V, \quad \langle A(t)v, w \rangle_{V', V} = a(t, v, w).$$

This definition yields $f = d_t u + A(t)u$; hence,

$$u_{h\Delta t} = (u_h^0, \dots, u_h^N) = S_{h\Delta t}(\rho_{1h}(u(0)), \rho_{2h\Delta t}(d_t u + A(t)u)).$$

Introduce the sequence

$$u_{\xi h \Delta t} = (u_{\xi h}^0, \dots, u_{\xi h}^N) = S_{h\Delta t}(\rho_{1h}(u_\xi(0)), \rho_{2h\Delta t}(d_t u_\xi + A(t)u_\xi)),$$

and use the stability of $S_{h\Delta t}$, ρ_{1h}, and $\rho_{2h\Delta t}$ to infer

$$\begin{aligned}
\|u_{h\Delta t} - u_{\xi h \Delta t}\|_{E_{3h\Delta t}} &\leq c(\|\rho_{1h}(u(0) - u_\xi(0))\|_{E_{1h}} \\
&\quad + \|\rho_{2h\Delta t}(d_t(u - u_\xi) + A(t)(u - u_\xi))\|_{E_{2h\Delta t}}) \\
&\leq c(\|u(0) - u_\xi(0)\|_L + \|d_t(u - u_\xi) + A(t)(u - u_\xi)\|_{L^2([0,T];V')}) \\
&\leq c\|u - u_\xi\|_{W(V, V')} \leq c\xi.
\end{aligned}$$

Finally, owing to the stability of $\rho_{3\Delta t}$ and $i_{h\Delta t}$,

$$\begin{aligned}
\|\rho_{3\Delta t}(u) - i_{h\Delta t}(u_{h\Delta t})\|_{E_{3\Delta t}} &\leq \|\rho_{3\Delta t}(u - u_\xi)\|_{E_{3\Delta t}} \\
&\quad + \|\rho_{3\Delta t}(u_\xi) - i_{h\Delta t}(u_{\xi h \Delta t})\|_{E_{3\Delta t}} + \|i_{h\Delta t}(u_{\xi h \Delta t}) - i_{h\Delta t}(u_{h\Delta t})\|_{E_{3\Delta t}} \\
&\leq c_1 \|u - u_\xi\|_{W(V, V')} + c_2 \epsilon(h, \Delta t)\|u_\xi\|_{Z(Q)} + c_3 \xi.
\end{aligned}$$

Passing to the limit $h \to 0$ and $\Delta t \to 0$, yields

$$\lim_{\substack{h \to 0 \\ \Delta t \to 0}} \|\rho_{3\Delta t}(u) - i_{h\Delta t} u_{h\Delta t}\|_{E_{3\Delta t}} \leq c\xi.$$

The conclusion follows from the fact that ξ is arbitrary. □

Remark 6.24. Theorem 6.23 illustrates a general principle known in numerical analysis as the *Lax Principle*:

$$\text{Consistency} + \text{Stability} \implies \text{Convergence}.$$

While one must bear in mind that the Lax principle is valid for linear schemes only, it is not restricted to time-dependent settings; see Chapters 2, 3, 4, and 5 for numerous applications where consistency and stability properties are established to prove the convergence of a finite element approximation. □

6.1.6 Time approximation: Some examples

This section contains a small selection of time-marching algorithms: implicit Euler, explicit Euler, and the Backward Difference Formula of second order (BDF2).

Many different norms can be defined to measure the error in the space $E_{3\Delta t}$. Recall that $E_{3\Delta t} = V^{N+1}$ or L^{N+1} depending on whether the exact solution is in $C^0([0,T];V)$ or only in $C^0([0,T];L)$; see Example 6.20. In both cases, $E_{3\Delta t} = E^{N+1}$ where E is a Banach space. For convenience, we introduce the compact notation $u_{\Delta t} = (u(t^0), u(t^1), \ldots, u(t^N))$. The restriction operator $\rho_{3\Delta t} : E_3 \to E_{3\Delta t}$ is simply defined as $\rho_{3\Delta t}(u) = u_{\Delta t}$, and the prolongation operator $i_{h\Delta t}$ is simply the canonical injection of $E_{3h\Delta t}$ into $E_{3\Delta t}$. Hence, the error can be measured by some norm of $u_{\Delta t} - u_{h\Delta t}$. Let $\phi_{\Delta t} = (\phi^0, \phi^1, \ldots, \phi^N)$ be a sequence of functions in E^{N+1} and define the discrete norms

$$
\begin{cases}
\|\phi_{\Delta t}\|^2_{\ell^2([t^m,t^n];E)} = \Delta t \sum_{k=m}^{n} \|\phi^k\|^2_E, \\[4mm]
\|\phi_{\Delta t}\|_{\ell^\infty([t^m,t^n];E)} = \max_{m \le k \le n} \|\phi^k\|_E.
\end{cases}
\tag{6.19}
$$

Norms commonly used to measure the error are $\|u_{\Delta t} - u_{h\Delta t}\|_{\ell^\infty([t^0,t^N];L)}$ and $\|u_{\Delta t} - u_{h\Delta t}\|_{\ell^2([t^1,t^N];V)}$.

Example 1 (Implicit Euler). The *implicit Euler scheme* consists of constructing an approximation sequence $u_{h\Delta t} = (u_h^0, \ldots, u_h^N) \in V_h^{N+1}$ using the initial step $(u_h^0, v_h)_L = (u_0, v_h)_L$ for all $\forall v_h \in V_h$, and the time-marching algorithm

$$
\forall v_h \in V_h, \quad \tfrac{1}{\Delta t}(u_h^{n+1} - u_h^n, v_h)_L + a(t^{n+1}, u_h^{n+1}, v_h) = (f(t^{n+1}), v_h)_L. \tag{6.20}
$$

To analyze the properties of this algorithm, we introduce the operator $S_{h\Delta t} : V_h \times V_h^N \to V_h^{N+1}$ such that, for $x_h \in V_h$ and $y_{h\Delta t} \in V_h^N$, the sequence $z_{h\Delta t} = (z_h^0, \ldots, z_h^N) = S_{h\Delta t}(x_h, y_{h\Delta t})$ is inductively defined by $z_h^0 = x_h$ and

$$
\forall v_h \in V_h, \quad \tfrac{1}{\Delta t}(z_h^{n+1} - z_h^n, v_h)_L + a(t^{n+1}, z_h^{n+1}, v_h) = (y_h^{n+1}, v_h)_L. \tag{6.21}
$$

Let P_{hL} be the L-orthogonal projection from L to V_h. Set $\rho_{1h}(u_0) = P_{hL}u_0$ and $\rho_{2h\Delta t}(f) = (P_{hL}f(t^1), \ldots, P_{hL}f(t^N))$. Then,

$$
u_{h\Delta t} = S_{h\Delta t}(\rho_{1h}(u_0), \rho_{2h\Delta t}(f)),
$$

i.e., (6.20) is a one-step method where $f_h^{n+1} = P_{hL}f(t^{n+1})$; see Definition 6.17.

The operator $S_{h\Delta t}$ has stability properties (i.e., continuous dependence with respect to the data) which are similar to those of the continuous problem stated in Theorem 6.7.

Lemma 6.25 (Stability). *Assume* $\Delta t \le \frac{1}{\alpha c_P}$ *and let* $z_{h\Delta t} = S_{h\Delta t}(x_h, y_{h\Delta t})$ *be defined in* (6.21). *Then,*

$$
\|z_{h\Delta t}\|_{\ell^\infty([0,T];L)} \le
\begin{cases}
e^{-\frac{1}{4}\alpha c_P T}\|x_h\|_L + \frac{1}{\sqrt{\alpha}}\|y_{h\Delta t}\|_{\ell^2([t^1,t^n];V')}, \\[3mm]
e^{-\frac{1}{4}\alpha c_P T}\|x_h\|_L + \frac{1}{\alpha\sqrt{c_P}}\|y_{h\Delta t}\|_{\ell^\infty([0,T];V')},
\end{cases}
\tag{6.22}
$$

$$
\|z_{h\Delta t}\|_{\ell^2([t^1,t^N];V)} \le
\begin{cases}
\frac{1}{\sqrt{\alpha}}\|x_h\|_L + \frac{1}{\alpha}\|y_{h\Delta t}\|_{\ell^2([t^1,t^N];V')}, \\[3mm]
\frac{1}{\sqrt{\alpha}}\|x_h\|_L + \frac{\sqrt{T}}{\alpha}\|y_{h\Delta t}\|_{\ell^\infty([0,T];V')}.
\end{cases}
\tag{6.23}
$$

Proof. (1) Test (6.21) with $2\Delta t z_h^{n+1}$. The relation $2p(p-q) = p^2 + (p-q)^2 - q^2$ and the coercivity of a imply

$$\|z_h^{n+1}\|_L^2 + \|z_h^{n+1} - z_h^n\|_L^2 + 2\Delta t\alpha\|z_h^{n+1}\|_V^2 \leq \|z_h^n\|_L^2 + 2\Delta t\|y_h^{n+1}\|_{V'}\|z_h^{n+1}\|_V$$

$$\leq \|z_h^n\|_L^2 + \tfrac{\Delta t}{\alpha}\|y_h^{n+1}\|_{V'}^2 + \Delta t\alpha\|z_h^{n+1}\|_V^2.$$

The last inequality is obtained from the arithmetic–geometric inequality (A.3). The inequality $c_P\|z_h^{n+1}\|_L^2 \leq \|z_h^{n+1}\|_V^2$ leads to

$$(1+\Delta t\alpha c_P)\|z_h^{n+1}\|_L^2 \leq \|z_h^{n+1}\|_L^2 + \Delta t\alpha\|z_h^{n+1}\|_V^2 \leq \|z_h^n\|_L^2 + \tfrac{\Delta t}{\alpha}\|y_h^{n+1}\|_{V'}^2. \quad (6.24)$$

Owing to the Discrete Gronwall Lemma (see Lemma 6.26 below),

$$\|z_h^{n+1}\|_L^2 \leq \frac{\|x_h\|_L^2}{(1 + \Delta t\alpha c_P)^{n+1}} + \frac{\Delta t}{\alpha}\sum_{k=0}^n \frac{1}{(1 + \Delta t\alpha c_P)^{n-k+1}}\|y_h^{k+1}\|_{V'}^2. \quad (6.25)$$

Note that $(1 + \Delta t\alpha c_P)^{-1} \leq e^{-\frac{1}{2}\Delta t\alpha c_P}$ if $\Delta t\alpha c_P \leq 1$. Moreover, $(1 + \Delta t\alpha c_P)^{-1} < 1$. The first estimate in (6.22) follows easily, while the second estimate follows from (6.25) together with the inequality

$$\sum_{k=0}^n \frac{1}{(1 + \Delta t\alpha c_P)^{n-k+1}} \leq \frac{1}{\Delta t\alpha c_P}.$$

(2) Summing (6.24) from $n = 0$ to $N - 1$ leads to

$$\Delta t\alpha\sum_{n=1}^N \|z_h^n\|_V^2 \leq \|x_h\|_L^2 + \frac{1}{\alpha}\Delta t\sum_{n=1}^N \|y_h^n\|_{V'}^2,$$

whence the estimates in (6.23) are readily deduced. $\qquad\square$

Lemma 6.26 (Discrete Gronwall). *Let $\gamma \neq -1$ and β be two real numbers. Let $\{a^n\}_{n\geq 0}$ and $\{f^n\}_{n\geq 0}$ be two sequences such that $(1+\gamma)a^{n+1} \leq a^n + \beta f^n$. Then,*

$$a^{n+1} \leq \frac{a^0}{(1 + \gamma)^{n+1}} + \beta\sum_{k=0}^n \frac{f^k}{(1 + \gamma)^{n-k+1}}.$$

Remark 6.27. Since $u_{h\Delta t} = S_{h\Delta t}(P_{hL}u_0, (P_{hL}f)_{\Delta t})$, Lemma 6.25 together with the bound $\sqrt{c_P}\|f\|_V' \leq \|f\|_L$ implies that the solution to (6.20) satisfies the following stability estimates:

$$\|u_{h\Delta t}\|_{\ell^\infty([0,T];L)} \leq e^{-\frac{1}{4}\alpha c_P T}\|u_0\|_L + \tfrac{1}{\alpha c_P}\|f\|_{C^0([0,T];L)},$$

$$\tfrac{1}{\sqrt{T}}\|u_{h\Delta t}\|_{\ell^2([t^1,T];V)} \leq \tfrac{1}{\sqrt{\alpha T}}\|u_0\|_L + \tfrac{1}{\alpha\sqrt{c_P}}\|f\|_{C^0([0,T];L)}.$$

Lemma 6.28 (Consistency). *Let $E_1 = L$, $E_2 = L^2(]0,T[;V')$, $E_3 = \mathcal{W}(V,V')$, and $Z(Q) = C^1([0,T];W) \cap C^2([0,T];V')$. Introduce the norms $\|\cdot\|_{E_{1h}} = \|\cdot\|_L$ and $\|\cdot\|_{E_{2h\Delta t}} = \|\cdot\|_{\ell^\infty([0,T];V')}$. Let P_{ht} be the elliptic projector defined in (6.14); set $\Pi_{h\Delta t}(u) = (P_{ht^0}u, \ldots, P_{ht^N}u)$. Define the operator $\rho_{3\Delta t}$ by $\rho_{3\Delta t}(u) = u_{\Delta t} = (u(t^0), \ldots, u(t^N))$, and let $i_{h\Delta t} : E_{3h\Delta t} \to E_{3\Delta t}$ be the canonical injection.*

(i) Let $E_{3\Delta t} = L^{N+1}$ with $\|\cdot\|_{E_{3\Delta t}} = \|\cdot\|_{\ell^\infty([0,T];L)}$. Then, the scheme (6.20)
   is $\mathcal{O}(h^{k+1} + \Delta t)$-consistent.
(ii) Let $E_{3\Delta t} = V^{N+1}$ with $\|\cdot\|_{E_{3\Delta t}} = \|\cdot\|_{\ell^2([t^1,T];V)}$. Then, the scheme (6.20)
   is $\mathcal{O}(h^k + \Delta t)$-consistent.

Proof. (1) Set $w_h(t) = P_{ht}(u)$. By definition, $w_h^0 = P_{hL}u_0 + R_{0h}$ where $R_{0h} = P_{ht^0}(u) - P_{hL}u_0$. Furthermore, for all $v_h \in V_h$,

$$\tfrac{1}{\Delta t}(w_h^{n+1} - w_h^n, v_h)_L + a(t^{n+1}, w_h^{n+1}, v_h) = (f^{n+1}, v_h)_L + (R^{n+1}, v_h)_L, \quad (6.26)$$

where

$$R^{n+1} = \frac{1}{\Delta t}\int_{t^n}^{t^{n+1}} (d_\tau w_h(\tau) - d_\tau u(\tau))\, d\tau - \frac{1}{\Delta t}\int_{t^n}^{t^{n+1}} (\tau - t^n) d_{\tau\tau} u(\tau)\, d\tau.$$

Setting $R_{2h\Delta t} = (P_{hL}R^1, \ldots, P_{hL}R^N)$ yields

$$\varPi_{h\Delta t}(u) = S_{h\Delta t}(\rho_{1h}(u_0) + R_{0h}, \rho_{2h\Delta t}(f) + R_{2h\Delta t}).$$

It is clear that

$$\|R_{1h}\|_{E_{1h}} + \|R_{2h\Delta t}\|_{E_{2h\Delta t}} \leq c\,h^{k+1}\|u\|_{C^1([0,T];W)} + \Delta t\|u\|_{C^2([0,T];V')}.$$

Moreover,

$$\|\rho_{3\Delta t}(u) - i_{h\Delta t}\varPi_{h\Delta t}(u)\|_{E_{3\Delta t}} = \|u_{\Delta t} - \varPi_{h\Delta t}(u)\|_{\ell^\infty([0,T];L)}$$
$$\leq c\,h^{k+1}\|u\|_{C^0([0,T];W)}.$$

Hence,

$$\|R_{1h}\|_{E_{1h}} + \|R_{2h\Delta t}\|_{E_{2h\Delta t}} + \|\rho_{3\Delta t}(u) - i_{h\Delta t}\varPi_{h\Delta t}(u)\|_{E_{3\Delta t}} \leq ch^{k+1}\|u\|_{Z(Q)}.$$

Note that the constant c does not depend on T.
(2) In the second case, we obtain the estimate $\|\rho_{3\Delta t}(u) - i_{h\Delta t}\varPi_{h\Delta t}(u)\|_{E_{3\Delta t}} = \|u_{\Delta t} - \varPi_{h\Delta t}(u)\|_{\ell^2([t^1,T];V)} \leq c\sqrt{T}h^k\|u\|_{C^0([0,T];W)}$, while the bounds on the two other residuals are unchanged. Hence, for the second choice of norm in $E_{3\Delta t}$, the scheme is $\mathcal{O}(h^k + \Delta t)$-consistent. $\qquad\square$

We are now in a position to prove the main convergence result.

Theorem 6.29 (Convergence). *There is c, independent of h, Δt, and T, such that, if the solution to (6.2) is in $Z(Q) = C^1([0,T];W) \cap C^2([0,T];V')$,*

$$\|u_{\Delta t} - u_{h\Delta t}\|_{\ell^\infty([0,T];L)} \leq c(h^{k+1} + \Delta t)\|u\|_{Z(Q)},$$

$$\tfrac{1}{\sqrt{T}}\|u_{\Delta t} - u_{h\Delta t}\|_{\ell^2([t^1,T];V)} \leq c\left(1 + \tfrac{1}{\sqrt{T}}\right)(h^k + \Delta t)\|u\|_{Z(Q)}.$$

Proof. To apply Theorem 6.23, first notice that consistency has been proved in Lemma 6.28. The stability of $S_{h\Delta t}$ has been proved in Lemma 6.25. Moreover, the stability of $i_{h\Delta t}$ is obvious since $\|i_{h\Delta t}v_h\|_{E_{3h\Delta t}} = \|v_h\|_{E_{3\Delta t}}$. $\qquad\square$

Remark 6.30.

(i) If $\|u\|_{Z(Q)}$ is bounded uniformly with respect to T,

$$\limsup_{T\to+\infty} \|u(T) - u_h^N\|_L \leq c(h^{k+1} + \Delta t)\|u\|_{Z(Q)},$$

$$\limsup_{T\to+\infty} \frac{1}{\sqrt{T}}\|u_{\Delta t} - u_{h\Delta t}\|_{\ell^2([t^1,T];V)} \leq c(h^k + \Delta t)\|u\|_{Z(Q)},$$

yielding a uniform control of the error for arbitrary large times. This remarkable property is characteristic of parabolic equations: these equations loose memory of the initial data, and they also progressively loose memory of approximation errors made in the past, i.e., the approximation error does not accumulate as time grows.

(ii) With the choices we made for the approximation setting, the scheme (6.20) is not completely stable since $\rho_{2h\Delta t}$ is not uniformly bounded, i.e., $\|(P_{hL}f)_{\Delta t}\|_{\ell^\infty([0,T];V')}$ cannot be uniformly bounded by $\|f\|_{L^2(]0,T[;V')}$. A better choice could be to equip $E_{3h\Delta t}$ with the norm of $\ell^2([t^1,T];V')$ and to define $\rho_{2h\Delta t}(f) = (\tilde{P}_{hL}(\frac{1}{\Delta t}\int_{t^{n-1}}^{t^n} f(\tau)\,d\tau))_{1\leq n\leq N}$ where $\tilde{P}_{hL}$ is the extension by density of P_{hL} to V'. Note that ρ_{1h} is uniformly bounded since $\|\rho_{1h}u_0\|_{E_{1h}} = \|P_{hL}u_0\|_L \leq \|u_0\|_L = \|u_0\|_{E_1}$. If we choose $\|\cdot\|_{E_{3\Delta t}} = \|\cdot\|_{\ell^\infty([0,T];L)}$, $\rho_{3\Delta t}$ is also uniformly bounded since $\|\rho_{3\Delta t}u\|_{E_{3\Delta t}} = \|u_{\Delta t}\|_{\ell^\infty([0,T];L)} \leq \|u\|_{C^0([0,T];L)} \leq c\|u\|_{W(V,V')}$. However, if we choose $\|\cdot\|_{E_{3\Delta t}} = \|\cdot\|_{\ell^2([t^1,T];V)}$, $\rho_{3\Delta t}$ is not uniformly bounded. We could recover uniform boundedness by setting $E_{3\Delta t} = L^2(]0,T[;V)$ and $\rho_{3\Delta t}$ to be the identity operator. Then, to embed $u_{h\Delta t}$ in $E_{3\Delta t}$, we could set $i_{h\Delta t}(u_{h\Delta t})$ to be the function that is piecewise linear in time and that takes the value u_h^n at t^n for $0 \leq n \leq N$ (we could also take a piecewise constant function). In conclusion, the reader must bear in mind that there are many ways of implementing a given algorithm and measuring its accuracy.   □

Example 2 (Explicit Euler). We now define the operator $S_{h\Delta t}$ such that the sequence $z_{h\Delta t} = (z_h^0, \ldots, z_h^N) = S_{h\Delta t}(x_h, y_{h\Delta t})$ is given by $z_h^0 = x_h$ and

$$\forall v_h \in V_h, \quad \frac{1}{\Delta t}(z_h^{n+1} - z_h^n, v_h)_L + a(t^{n+1}, z_h^n, v_h) = (y_h^{n+1}, v_h)_L. \quad (6.27)$$

Set $\rho_{1h}(u_0) = P_{hL}u_0$, $\rho_{2h\Delta t}(f) = (P_{hL}f(t^1), \ldots, P_{hL}f(t^N))$, and define $u_{h\Delta t} = S_{h\Delta t}(\rho_{1h}(u_0), \rho_{2h\Delta t}(f))$ to be the approximate solution to (6.2). In a less compact form, the algorithm can be recast in the following form: initialize by solving $(u_h^0, v_h)_L = (u_0, v_h)_L$ for all $v_h \in V_h$, and then use the time-marching algorithm

$$\forall v_h \in V_h, \quad \frac{1}{\Delta t}(u_h^{n+1} - u_h^n, v_h)_L + a(t^{n+1}, u_h^n, v_h) = (f(t^{n+1}), v_h)_L. \quad (6.28)$$

This is a one-step scheme, usually referred to as the *explicit Euler scheme*.

Define

$$c_i(h) = \max_{v_h \in V_h} \frac{\|v_h\|_V}{\|v_h\|_L}. \quad (6.29)$$

This quantity is finite since V_h is finite-dimensional; see §1.7 on inverse inequalities. For a finite element setting based on a quasi-uniform mesh family, $c_i(h) \leq ch^{-1}$; see §1.7.

Lemma 6.31 (Stability). *Let* $0 < \kappa < 1$ *be a real number. Let* $z_{h\Delta t} = S_{h\Delta t}(x_h, y_{h\Delta t})$ *be defined in* (6.27). *If* $\Delta t \leq \frac{\kappa\alpha}{c_i(h)^2\|a\|^2}$, *then*

$$\|z_h^n\|_{\ell^\infty([0,T];L)} \leq \|x_h\|_L e^{-\alpha c_P(1-\kappa)T} + \frac{1}{\alpha\sqrt{c_P}\sqrt{1-\kappa}}\|y_h^n\|_{\ell^\infty([0,T];V')},$$

$$\frac{1}{\sqrt{T}}\|z_{h\Delta t}\|_{\ell^2([t^1,T];V)} \leq \frac{1}{\sqrt{\alpha T(1-\kappa)}}\|x_h\|_L + \frac{1}{\alpha\sqrt{(1-\kappa)}}\|y_h^n\|_{\ell^\infty([0,T];V')}.$$

Proof. We proceed as in the proof of Lemma 6.25. Test (6.27) with $2\Delta t z_h^{n+1}$ and use the relation $2p(p-q) = p^2 + (p-q)^2 - q^2$ together with (A.3):

$$\|z_h^{n+1}\|_L^2 + \|z_h^{n+1} - z_h^n\|_L^2 + 2\Delta t a(t^{n+1}, z_h^n, z_h^{n+1})$$
$$\leq \|z_h^n\|_L^2 + 2\Delta t\|y_h^{n+1}\|_{V'}\|z_h^{n+1}\|_V \leq \|z_h^n\|_L^2 + \frac{\Delta t}{\alpha}\|y_h^{n+1}\|_{V'}^2 + \Delta t\alpha\|z_h^{n+1}\|_V^2.$$

Furthermore, the inverse inequality (6.29) implies

$$a(t^{n+1}, z_h^n, z_h^{n+1}) = a(t^{n+1}, z_h^{n+1}, z_h^{n+1}) + a(t^{n+1}, z_h^n - z_h^{n+1}, z_h^{n+1})$$
$$\geq \alpha\|z_h^{n+1}\|_V^2 - \|a\|\,\|z_h^n - z_h^{n+1}\|_V\|z_h^{n+1}\|_V$$
$$\geq \alpha\|z_h^{n+1}\|_V^2 - c_i(h)\|a\|\,\|z_h^{n+1} - z_h^n\|_L\|z_h^{n+1}\|_V$$
$$\geq \alpha\|z_h^{n+1}\|_V^2 - \frac{\|a\|^2 c_i(h)^2}{2\kappa\alpha}\|z_h^{n+1} - z_h^n\|_L^2 - \frac{\kappa\alpha}{2}\|z_h^{n+1}\|_V^2.$$

As a result,

$$\|z_h^{n+1}\|_L^2 + \left(1 - \frac{\|a\|^2 c_i(h)^2\Delta t}{\kappa\alpha}\right)\|z_h^{n+1} - z_h^n\|_L^2 + \Delta t\alpha(1-\kappa)\|z_h^{n+1}\|_V^2$$
$$\leq \|z_h^n\|_L^2 + \frac{\Delta t}{\alpha}\|y_h^{n+1}\|_{V'}^2.$$

If $\Delta t \leq \frac{\kappa\alpha}{\|a\|^2 c_i(h)^2}$, we infer

$$\|z_h^{n+1}\|_L^2 + \Delta t\alpha(1-\kappa)\|z_h^{n+1}\|_V^2 \leq \|z_h^n\|_L^2 + \frac{\Delta t}{\alpha}\|y_h^{n+1}\|_{V'}^2. \qquad (6.30)$$

The rest of the proof is the same as that of Lemma 6.25. □

Theorem 6.32. *Assume* $\Delta t < \frac{\alpha}{c_i(h)^2\|a\|^2}$. *Then, there is* c, *independent of* h, Δt, *and* T, *such that, if the solution to* (6.2) *is in* $Z(Q) = \mathcal{C}^1([0,T];W) \cap \mathcal{C}^2([0,T];V')$,

$$\|u_{\Delta t} - u_{h\Delta t}\|_{\ell^\infty([0,T];L)} \leq c(h^{k+1} + \Delta t)\|u\|_{Z(Q)},$$

$$\frac{1}{\sqrt{T}}\|u_{\Delta t} - u_{h\Delta t}\|_{\ell^2([t^1,T];V)} \leq c\left(1 + \frac{1}{\sqrt{T}}\right)(h^k + \Delta t)\|u\|_{Z(Q)}.$$

Proof. We proceed as in the proof of Theorem 6.29. First, we prove consistency. Set $w_h(t) = P_{ht}(u)$ where P_{ht} is the elliptic projector defined in (6.14). Denote $w_h^n = w_h(t^n)$. The definition of w_h^{n+1} implies

$$\tfrac{1}{\Delta t}(w_h^{n+1} - w_h^n, v_h)_L + a(t^{n+1}, w_h^n, v_h) = (f^{n+1}, v_h)_L + R^{n+1}(v_h), \quad \forall v_h \in V_h,$$

where, for all $v_h \in V_h$,

$$R^{n+1}(v_h) = \left(\tfrac{1}{\Delta t}(w_h^{n+1} - w_h^n) - d_t u(t^{n+1}), v_h\right)_L - a(t^{n+1}, u(t^{n+1}) - u(t^n), v_h).$$

It is clear that

$$\|R^{n+1}\|_{V'} \leq c(\|w_h - u\|_{C^1([0,T];L)} + \Delta t\|u\|_{C^2([0,T];V')} + \Delta t\|u\|_{C^1([0,T];V)})$$
$$\leq c(h^{k+1} + \Delta t)\|u\|_{Z(Q)}.$$

The rest of the proof is the same as that of Theorem 6.29. □

Example 3 (BDF2). We conclude this series of examples by presenting a scheme with second-order accuracy. We define $S_{h\Delta t}$ so that the sequence $(z_h^0, \ldots, z_h^N) = S_{h\Delta t}(x_h, y_{h\Delta t})$ is generated according to

$$\begin{cases} z_h^0 = x_h, \\ \tfrac{1}{\Delta t}(z_h^1 - z_h^0, v_h)_L + \tfrac{1}{2}a(t^{\frac{1}{2}}, z_h^1 + z_h^0, v_h) = (y_h^1, v_h)_L, \\ \tfrac{1}{2\Delta t}(3z_h^{n+1} - 4z_h^n + z_h^{n-1}, v_h)_L + a(t^{n+1}, z_h^{n+1}, v_h) = (y_h^{n+1}, v_h)_L, \end{cases} \qquad (6.31)$$

where $t^{\frac{1}{2}} = \tfrac{\Delta t}{2}$ and where the second and third equations hold for all $v_h \in V_h$. Define $\rho_{1h}(u_0) = P_{hL}u_0$, $\rho_{2h\Delta t}(f) = (P_{hL}f(t^{\frac{1}{2}}), P_{hL}f(t^2), \ldots, P_{hL}f(t^N))$, and $u_{h\Delta t} = S_{h\Delta t}(\rho_{1h}(u_0), \rho_{2h\Delta t}(f))$. By definition, the sequence $u_{h\Delta t}$ is such that, for all $v_h \in V_h$,

$$\begin{cases} (u_h^0, v_h)_L = (u_0, v_h)_L, \\ \tfrac{1}{\Delta t}(u_h^1 - u_h^0, v_h)_L + \tfrac{1}{2}a(t^{\frac{1}{2}}, u_h^1 + u_h^0, v_h) = (f(t^{\frac{1}{2}}), v_h)_L, \\ \tfrac{1}{2\Delta t}(3u_h^{n+1} - 4u_h^n + u_h^{n-1}, v_h)_L + a(t^{n+1}, u_h^{n+1}, v_h) = (f(t^{n+1}), v_h)_L. \end{cases} \qquad (6.32)$$

This is a two-step method, usually referred to as *BDF2*.

Lemma 6.33 (Stability). *Let* $z_{h\Delta t} = S_{h\Delta t}(x_h, y_{h\Delta t})$ *be defined in* (6.31). *Then,*

$$\|z_{h\Delta t}\|_{\ell^\infty([0,T];L)} \leq c(\|x_h\|_V + \Delta t\|y_h^1\|_L)e^{-\frac{1}{4}\alpha c_P T} + \tfrac{1}{\alpha\sqrt{c_P}}\|y_{h\Delta t}\|_{\ell^\infty([0,T];V')},$$

$$\tfrac{1}{\sqrt{T}}\|z_{h\Delta t}\|_{\ell^2([t^1,T];V)} \leq \tfrac{c}{\sqrt{\alpha T}}(\|x_h\|_V + \Delta t\|y_h^1\|_L) + \tfrac{1}{\alpha}\|y_{h\Delta t}\|_{\ell^\infty([0,T];V')}.$$

Proof. Test the third equation in (6.31) with $4\Delta t z_h^{n+1}$. Using the relation

$$2(a^{n+1}, 3a^{n+1} - 4a^n + a^{n-1}) = |a^{n+1}|^2 + |2a^{n+1} - a^n|^2 + |\delta_{tt}a^{n+1}|^2$$
$$- |a^n|^2 - |2a^n - a^{n-1}|^2, \qquad (6.33)$$

with $\delta_{tt} a^{n+1} = a^{n+1} - 2a^n + a^{n-1}$, yields

$$\|z_h^{n+1}\|_L^2 + \|2z_h^{n+1} - z_h^n\|_L^2 + \|\delta_{tt}z_h^{n+1}\|_L^2 + 4\Delta t\alpha\|z_h^{n+1}\|_V^2 \le \|z_h^n\|_L^2$$
$$+ \|2z_h^n - z_h^{n-1}\|_L^2 + 2\Delta t\alpha\|z_h^{n+1}\|_V^2 + \tfrac{2\Delta t}{\alpha}\|y_h^{n+1}\|_{V'}^2.$$

Thus, for $n \ge 1$,

$$\|z_h^{n+1}\|_L^2 + \|2z_h^{n+1} - z_h^n\|_L^2 + 2\Delta t\alpha\|z_h^{n+1}\|_V^2 \le \|z_h^n\|_L^2$$
$$+ \|2z_h^n - z_h^{n-1}\|_L^2 + \tfrac{2\Delta t}{\alpha}\|y_h^{n+1}\|_{V'}^2.$$

Furthermore, testing the second equation in (6.31) with $\Delta t(z_h^1 + z_h^0)$ leads to

$$\|z_h^1\|_L^2 + \tfrac{\alpha\Delta t}{4}\|z_h^1 + z_h^0\|_V^2 \le \|z_h^0\|_L^2 + \tfrac{2\Delta t}{\alpha c_P}\|y_h^1\|_L^2.$$

As a result,

$$\|z_h^1\|_L^2 + \|2z_h^1 - z_h^0\|_L^2 + 2\Delta t\alpha\|z_h^1\|_V^2 \le c\left(\|z_h^0\|_V^2 + \tfrac{\Delta t}{\alpha c_P}\|y_h^1\|_L^2\right).$$

The rest of the proof is similar to that of Lemma 6.25. $\qquad\square$

Theorem 6.34. *There is c, independent of h, Δt, and T, such that, if the solution to (6.2) is in $Z(Q) = C^2([0,T]; W) \cap C^3([0,T]; V')$,*

$$\|u_{\Delta t} - u_{h\Delta t}\|_{\ell^\infty([0,T];L)} \le c(h^{k+1} + \Delta t^2)\|u\|_{Z(Q)},$$
$$\tfrac{1}{\sqrt{T}}\|u_{\Delta t} - u_{h\Delta t}\|_{\ell^2([t^1,T];V)} \le c\left(1 + \tfrac{1}{\sqrt{T}}\right)(h^k + \Delta t^2)\|u\|_{Z(Q)}.$$

Proof. Proceed as in the proof of Theorem 6.29. $\qquad\square$

6.2 Time-Dependent Mixed Problems

This section is devoted to the approximation of the time-dependent version of the steady Stokes problem investigated in Chapter 4. Most of the techniques presented hereafter are commonly applied to tackle the pressure-velocity coupling in the unsteady Navier–Stokes equations. As for parabolic problems, we consider the *method of lines*: First, space derivatives are approximated by means of finite elements while keeping time continuous. Then, time derivatives are approximated by means of a fractional-step technique.

6.2.1 The continuous problem

Consider the *time-dependent Stokes equations*

$$\begin{cases} d_t u - \nu \Delta u + \nabla p = f & \text{in } \Omega \times \,]0,T[, \\ \nabla \cdot u = 0 & \text{in } \Omega \times \,]0,T[, \\ u = 0 & \text{on } \partial\Omega \times \,]0,T[, \\ u(0) = u_0 & \text{in } \Omega. \end{cases} \qquad (6.34)$$

where $f(t)$ is a body force, u_0 is a solenoidal velocity field, and Ω is a domain in $\mathbb{R}^d$. We consider homogeneous Dirichlet boundary conditions for the sake of simplicity, but most of what follows extends easily to the boundary conditions discussed in §4.1.4.

To avoid the mean-value of the pressure to be arbitrary, we introduce the space

$$H^1_{\int=0}(\Omega) = \left\{ q \in H^1(\Omega); \int_\Omega q = 0 \right\}. \tag{6.35}$$

Moreover, to account for the incompressibility, we are led to consider the space $\mathcal{N}(\Omega) = \{v \in [\mathcal{D}(\Omega)]^d; \nabla{\cdot}v = 0\}$. We denote by H and V the closures of $\mathcal{N}(\Omega)$ in $[L^2(\Omega)]^d$ and $[H^1_0(\Omega)]^d$, respectively. These spaces are characterized by (see [Tem77, pp. 15–18])

$$H = \{v \in [L^2(\Omega)]^d; \nabla{\cdot}v = 0; v{\cdot}n_{|\partial\Omega} = 0\}, \tag{6.36}$$

$$V = \{v \in [H^1_0(\Omega)]^d; \nabla{\cdot}v = 0\}. \tag{6.37}$$

Clearly, the injection of V into H is continuous and V is dense in H. Thus, we are in the situation where $V \subset H \equiv H' \subset V'$. Henceforth, the space H is bound to play an important role through the following:

Theorem 6.35. *The following orthogonal decomposition of $[L^2(\Omega)]^d$ holds:*

$$[L^2(\Omega)]^d = H \oplus \nabla(H^1_{\int=0}(\Omega)). \tag{6.38}$$

Proof. For all v in $[L^2(\Omega)]^d$, let $p \in H^1_{\int=0}(\Omega)$ be the solution to $(\nabla p, \nabla q)_{0,\Omega} = (v, \nabla q)_{0,\Omega}$ for all $q \in H^1_{\int=0}(\Omega)$. Then, set $u = v - \nabla p$. It is clear that $u \in H$. The orthogonality of the decomposition is evident. ☐

Remark 6.36. This theorem states that for all $\tilde{u} \in [L^2(\Omega)]^d$, there is a unique $u \in H$ and a unique $\phi \in H^1_{\int=0}(\Omega)$ such that $u + \nabla\phi = \tilde{u}$, $\nabla{\cdot}u = 0$, $\int_\Omega \phi = 0$, and $u{\cdot}n_{|\partial\Omega} = 0$. Letting P_H be the L^2-orthogonal projection from $[L^2(\Omega)]^d$ to H, we obtain $u = P_H \tilde{u}$. ☐

By analogy with the stationary Stokes equations, we consider two weak forms of problem (6.34): a constrained formulation and a mixed formulation.

Constrained formulation. For all u, v in $[H^1_0(\Omega)]^d$, consider the bilinear form $a(u, v) = \nu(\nabla u, \nabla v)_{0,\Omega}$. The so-called constrained formulation is as follows: For $f \in L^2(]0, T[; [H^{-1}(\Omega)]^d)$ and $u_0 \in H$,

$$\begin{cases} \text{Seek } u \in \mathcal{W}(V, V') \text{ such that} \\ \langle d_t u, v \rangle_{V', V} + a(u, v) = \langle f, v \rangle_{H^{-1}, H^1_0}, \quad \text{a.e. } t \in]0, T[, \forall v \in V, \quad (6.39) \\ u(0) = u_0. \end{cases}$$

Proposition 6.37. *Problem (6.39) has a unique solution for all $T > 0$.*

Proof. The bilinear form a satisfies the three hypotheses (P1), (P2), and (P3) of Theorem 6.6.   □

As we have already seen for the stationary problem, it is difficult to approximate (6.39) with V-conformal finite elements, since constructing solenoidal global shape functions is a non-trivial task. Though, it is possible to construct non-conformal approximations of V; see, e.g., [Tem77, I§4], [Hec81, Hec84], [Bra97, p. 154], or [BrF91^b, p. 268]. Henceforth, we restrict ourselves to the unconstrained, or so-called mixed formulation.

Mixed formulation. In addition to the bilinear form a defined above, consider the bilinear form $b \in \mathcal{L}([H_0^1(\Omega)]^d \times L_{j=0}^2(\Omega); \mathbb{R})$ such that $b(v, p) = -(\nabla \cdot v, p)_{0,\Omega}$. The mixed formulation is as follows: For $f \in L^2(]0, T[; [L^2(\Omega)]^d)$ and $u_0 \in V$,

$$
\begin{cases}
\text{Seek } u \in \mathcal{W}([H_0^1(\Omega)]^d, [H^{-1}(\Omega)]^d) \text{ and } p \in L^2(]0, T[; L_{j=0}^2(\Omega)) \text{ such that} \\
(d_t u, v)_{0,\Omega} + a(u, v) + b(v, p) = (f, v)_{0,\Omega}, \quad \text{a.e. } t, \forall v \in [H_0^1(\Omega)]^d, \\
b(u, q) = 0, \quad\quad\quad\quad\quad\quad\quad\quad\quad\quad \text{a.e. } t, \forall q \in L_{j=0}^2(\Omega), \\
u(0) = u_0.
\end{cases}
\tag{6.40}
$$

Proposition 6.38. *Problem* (6.40) *has a unique solution for all $T > 0$.*

Proof. Owing to Proposition 6.37, there is $u \in \mathcal{W}(V, V')$ solving (6.39) with right-hand side replaced by $(f, v)_{0,\Omega}$. Use the fact that $u_0 \in V$ and $f \in L^2(]0, T[; [L^2(\Omega)]^d)$ and use formally $d_t u$ as a test function (this can be rigorously justified by a density argument) to infer $d_t u \in L^2(]0, T[; H) \subset L^2(]0, T[; [L^2(\Omega)]^d)$. Hence, the linear form $(d_t u - f, \cdot)_{0,\Omega} + a(u, \cdot)$ is in $L^2(]0, T[; [H^{-1}(\Omega)]^d)$. Furthermore,

$$
(d_t u - f, v)_{0,\Omega} + a(u, v) = 0, \quad \text{a.e. } t \in]0, T[, \forall v \in V.
$$

Owing to de Rham's Theorem, there is p in $\mathcal{D}'(]0, T[; L_{j=0}^2(\Omega))$ such that

$$
(\nabla \cdot v, p)_{0,\Omega} = (d_t u - f, v)_{0,\Omega} + a(u, v), \quad \text{a.e. } t \in]0, T[, \forall v \in [H_0^1(\Omega)]^d.
$$

Then, the inf-sup inequality (4.5) implies

$$
c \|p\|_{0,\Omega} \leq \|d_t u\|_{0,\Omega} + \|f\|_{0,\Omega} + \|a\| \|u\|_{1,\Omega}, \quad \text{a.e. } t \in]0, T[.
$$

Since $u \in \mathcal{W}([H_0^1(\Omega)]^d, [H^{-1}(\Omega)]^d)$, $p \in L^2(]0, T[; L_{j=0}^2(\Omega))$. Clearly, the pair (u, p) solves (6.40). Uniqueness is obvious.   □

Remark 6.39. When $f \in L^2(]0, T[; [H^{-1}(\Omega)]^d)$ and $u_0 \in H$, one only has $u \in \mathcal{W}(V, V')$, and it is not possible to prove $d_t u \in L^2(]0, T[; [H^{-1}(\Omega)]^d)$. Hence, one obtains only the existence of the pressure in $\mathcal{D}'(Q)$. Actually, using a Fourier technique introduced by Lions, it is possible to prove $u \in H^{\frac{1}{2}-\epsilon}(]0, T[; [L^2(\Omega)]^d)$ and $p \in H^{-\frac{1}{2}-\epsilon}(]0, T[; L_{j=0}^2(\Omega))$ for all $\epsilon > 0$; see [Lio59] and [Lio69, I§6.5].   □

6.2.2 Space approximation

We show in this section that the time-derivative in the Stokes problem can be dealt with by using the same techniques as those for the heat equation. Let $\{X_h\}_{h>0}$ and $\{M_h\}_{h>0}$ be two families of finite-dimensional vector spaces. We make the following assumptions (see §4.2):

(i) X_h is $[H_0^1(\Omega)]^d$-conformal and M_h is $L_{j=0}^2(\Omega)$-conformal.

(ii) The pairs $\{X_h, M_h\}_{h>0}$ are uniformly compatible, i.e., there exists a constant $\beta > 0$, independent of h, such that

$$\inf_{q_h \in M_h} \sup_{v_h \in X_h} \frac{\int_\Omega q_h \nabla \cdot v_h}{\|v_h\|_{1,\Omega} \|q_h\|_{0,\Omega}} \geq \beta. \tag{6.41}$$

(iii) X_h and M_h are endowed with interpolation properties, namely there is $k \geq 1$ and c, independent of h, such that, for all $v \in [H^{k+1}(\Omega)]^d \cap [H_0^1(\Omega)]^d$ and $q \in H^k(\Omega) \cap L_{j=0}^2(\Omega)$,

$$\inf_{v_h \in X_h} (\|v - v_h\|_{0,\Omega} + h\|v - v_h\|_{1,\Omega}) \leq c h^{k+1} \|v\|_{[H^{k+1}(\Omega)]^d},$$

$$\inf_{q_h \in M_h} \|q - q_h\|_{0,\Omega} \leq c h^k \|q\|_{H^k(\Omega)}.$$

(iv) The steady Stokes problem has smoothing properties; see Definition 4.16 and Lemma 4.17.

Let $u_{0h} \in X_h$ be an approximation of u_0. We approximate (6.40) as follows:

$$\begin{cases} \text{Seek } u_h \in C^1([0,T]; X_h) \text{ and } p_h \in C^0(]0,T[; M_h) \text{ such that} \\ (d_t u_h, v_h)_{0,\Omega} + a(u_h, v_h) + b(v_h, p_h) = (f, v_h)_{0,\Omega}, \quad \forall t, \forall v_h \in X_h, \\ b(u_h, q_h) = 0, \quad\quad\quad\quad\quad\quad\quad\quad\quad\quad\quad\quad \forall q_h \in M_h, \\ u_h(0) = u_{0h}. \end{cases} \tag{6.42}$$

Theorem 6.40. *Under the hypotheses* (i)–(iv), *assume that the solution* (u, p) *to* (6.40) *satisfies* $u \in C^1([0,T]; [H^{k+1}(\Omega)]^d)$ *and* $p \in C^1([0,T]; H^k(\Omega))$. *Let* (u_h, p_h) *solve* (6.42). *Then,*

$$\|u - u_h\|_{C^0([0,T];[L^2(\Omega)]^d)} \leq \|u_0 - u_{0h}\|_{0,\Omega} e^{-\frac{1}{2}c_1 T} + c_2 h^{k+1} c(u,p),$$

$$\left.\begin{array}{l} \frac{1}{\sqrt{T}} \|u - u_h\|_{L^2(]0,T[;[H^1(\Omega)]^d)} \\ \frac{1}{\sqrt{T}} \|p - p_h\|_{L^2(]0,T[;L^2(\Omega))} \end{array}\right\} \leq \frac{c_3}{\sqrt{T}} \|u_0 - u_{0h}\|_{0,\Omega} + c_4 h^k \left(1 + \frac{1}{\sqrt{T}}\right) c(u,p),$$

with constants c_1, c_2, c_3, *and* c_4 *independent of* h *and* T *and with* $c(u,p) = \|u\|_{C^1([0,T];[H^{k+1}(\Omega)]^d)} + \|p\|_{C^1([0,T];H^k(\Omega))}$.

Proof. We adapt the proof of Theorem 6.14. To this end, it is useful to define the counterpart of the elliptic projection introduced for parabolic problems in

(6.14). For $t \in [0,T]$, $u(t) \in [H_0^1(\Omega)]^d$, and $p(t) \in L_{j=0}^2(\Omega)$, let $P_{ht}(u) \in X_h$ and $Q_{ht}(p) \in M_h$ be such that

$$\forall v_h \in X_h, \quad a(P_{ht}(u), v_h) + b(v_h, Q_{ht}(p)) = a(u(t), v_h) + b(v_h, p(t)), \quad (6.43)$$

$$\forall q_h \in M_h, \quad b(P_{ht}(u), q_h) = b(u(t), q_h). \quad (6.44)$$

(1) Introduce the notation $e_h(t) = P_{ht}(u) - u_h(t)$, $\eta(t) = u(t) - P_{ht}(u)$, and $\delta_h(t) = Q_{ht}(p) - p_h(t)$. Subtracting (6.42) from (6.40) and using the definitions of $P_{ht}(u)$ and $Q_{ht}(p)$, yields, for all $t \in [0,T]$ and $v_h \in V_h$,

$$(d_t e_h, v_h)_{0,\Omega} + a(e_h, v_h) + b(v_h, \delta_h) = -(d_t \eta, v_h)_{0,\Omega}, \quad (6.45)$$

as well as $b(e_h, q_h) = 0$ for all $q_h \in M_h$. To obtain the velocity estimates, proceed as in the proof of Theorem 6.14; the relevant approximation properties of the elliptic projectors P_{ht} and Q_{ht} are stated in Lemma 6.41 below.
(2) To derive the estimate on the pressure, we use the inf-sup inequality. Condition (6.41) together with the equality

$$b(v_h, \delta_h) = -(d_t \eta, v_h)_{0,\Omega} - a(e_h, v_h) - (d_t e_h, v_h)_{0,\Omega}, \quad \forall v_h \in V_h,$$

implies

$$\beta \|\delta_h\|_{0,\Omega} \leq \|d_t \eta\|_{0,\Omega} + \|e_h\|_{1,\Omega} + \|d_t e_h\|_{0,\Omega}.$$

To conclude, $\|d_t e_h\|_{0,\Omega}$ must be controlled. Testing (6.45) with $d_t e_h$ and integrating over $]0,T[$ yields

$$\int_0^T \|d_t e\|_{0,\Omega}^2 + \nu |e_h(T)|_{1,\Omega}^2 \leq \nu |e_h(0)|_{1,\Omega}^2 + \int_0^T \|d_t \eta\|_{0,\Omega}^2.$$

The pressure estimate follows easily.   □

Lemma 6.41. *Under the hypotheses of Theorem 6.40,*

$$\|u - P_{ht}(u)\|_{C^1([0,T];[L^2(\Omega)]^d)} \leq c\, h^{k+1} c(u,p),$$

$$\|u - P_{ht}(u)\|_{C^1([0,T];[H^1(\Omega)]^d)} + \|p - Q_{ht}(p)\|_{C^1([0,T];L^2(\Omega))} \leq c\, h^k c(u,p),$$

with c independent of h and T, and $c(u,p)$ is defined in Theorem 6.40.

Proof. Apply Propositions 4.14 and 4.18.   □

Remark 6.42.
   (i) If u_{0h} is chosen so that $\|u_0 - u_{0h}\|_{0,\Omega} + h\|u_0 - u_{0h}\|_{1,\Omega} \leq c h^{k+1} \|u_0\|_{H^{k+1}}$, the error estimates in Theorem 6.40 are optimal.
   (ii) Similarly to parabolic equations, the error induced by approximating the initial data converges to zero as T grows. It exponentially decreases with T in the L^2-norm of the velocity, and it behaves like $T^{-\frac{1}{2}}$ in the mean in the H^1-norm of the velocity and in the L^2-norm of the pressure.   □

6.2.3 Time approximation

We have seen in the preceding section that the time behavior of (6.42) is very similar to that of parabolic equations. Hence, it is reasonable to expect that any time-marching algorithm working for parabolic equations should also be acceptable for the time-dependent Stokes problem.

Let us illustrate this point with the implicit Euler algorithm. Let $N \geq 1$, set $\Delta t = \frac{T}{N}$ and $t^n = n\Delta t$ for $0 \leq n \leq N$. Assume $f \in C^0([0,T];[L^2(\Omega)]^d)$ and set $f^n = f(t^n)$. We construct an approximating sequence $\{(w_h^n, q_h^n)\}_{n \geq 0}$ as follows: First, compute w_h^0 such that

$$(w_h^0, v_h)_{0,\Omega} = (u_0, v_h)_{0,\Omega}, \quad \forall v_h \in X_h.$$

Then, $\forall n \geq 0$, seek $(w_h^{n+1}, q_h^{n+1}) \in X_h \times M_h$ such that, $\forall (v_h, q_h) \in X_h \times M_h$,

$$\begin{cases} \frac{1}{\Delta t}(w_h^{n+1} - w_h^n, v_h)_{0,\Omega} + \nu a(w_h^{n+1}, v_h) + b(v_h, q_h^{n+1}) = (f^{n+1}, v_h)_{0,\Omega}, \\ b(w_h^{n+1}, q_h) = 0. \end{cases} \quad (6.46)$$

At each time step, we must solve a problem of the form

$$\begin{cases} \widetilde{a}(w_h^{n+1}, v_h) + b(v_h, \Delta t\, q_h^{n+1}) = g(v_h), & \forall v_h \in X_h, \\ b(w_h^{n+1}, q_h) = 0, & \forall q_h \in M_h, \end{cases} \quad (6.47)$$

where $\widetilde{a}(w_h^{n+1}, v_h) = (w_h^{n+1}, v_h)_{0,\Omega} + \nu\Delta t a(w_h^{n+1}, v_h)$ and $g(v_h) = (w_h^n + \Delta t f^{n+1}, v_h)_{0,\Omega}$, i.e., a steady Stokes-like problem similar to those studied in Chapter 4.

Upon choosing bases for X_h and M_h (e.g., global shape functions), we denote by $\widetilde{\mathcal{A}}$ and $\mathcal{B}$ the matrices associated with the bilinear forms $\widetilde{a}$ and b, respectively. We have seen in §4.4.4 that problem (6.47) amounts to solving the linear system

$$\mathcal{B}\widetilde{\mathcal{A}}^{-1}\mathcal{B}^T Q = G. \quad (6.48)$$

In Proposition 4.47, it is shown that the *condition number of the Uzawa matrix* $\mathcal{U} = \mathcal{B}\widetilde{\mathcal{A}}^{-1}\mathcal{B}^T$ is bounded from above by $c\frac{\kappa(\mathcal{M})}{\alpha^2\beta^2}$, where $\mathcal{M}$ is the mass matrix associated with the pressure unknown, α is the coercivity constant of $\tilde{a}$, and β is the constant in the inf-sup inequality (6.41). At this point, we are facing a major difficulty. The only bound from below we can deduce for α is $c\Delta t\nu \leq \alpha$. Hence, the bound from above on the condition number of $\mathcal{U}$ behaves like $\frac{1}{(\nu\Delta t)^2}$, i.e., we must expect that the condition number of $\mathcal{U}$ grows unboundedly like $\frac{1}{(\nu\Delta t)^2}$ when $\Delta t \to 0$. Actually, when $\nu\Delta t$ is small, $\widetilde{\mathcal{A}} \approx \mathcal{M}$, i.e., $\mathcal{U} \approx \mathcal{B}\mathcal{M}^{-1}\mathcal{B}^T$, and one can prove that $\kappa(\mathcal{U}) \sim h^{-2}$. That is to say, the condition number of $\mathcal{U}$ behaves like that of the Laplace operator. As a result, solving (6.48) by means of a standard gradient-based iterative technique entails extremely poor convergence rates as $\Delta t \to 0$; see Proposition 9.30.

There are two strategies to tackle this difficulty: either to precondition (6.48) or to reformulate the time-marching algorithm so as to uncouple the velocity and the pressure. Within the limited scope of this book, we restrict ourselves to the second solution, i.e., we shall now review some standard *fractional-step* methods and show how these techniques circumvent the difficulty raised by the condition number of the Uzawa matrix.

6.2.4 Projection methods

Standard form of projection methods. There are numerous ways to discretize the unsteady Stokes equations in time, but, undoubtedly, one of the most popular strategies is to use projection methods. This class of technique is sometimes referred to as the *Chorin–Temam method* [Cho68, Cho69, Tem69].

A projection method is a time-marching algorithm based on a fractional-step technique. It can be viewed as a predictor–corrector strategy aimed at uncoupling viscous diffusion and incompressibility effects. A time step is composed of three substeps: in the first substep, the pressure is made explicit and a provisional velocity field is computed using the momentum equation; in the second substep, the provisional velocity field is projected onto the space of incompressible (solenoidal) vector fields, i.e., H; and in the third substep, the pressure is updated.

Let $q > 0$ be an integer. To describe the implementation of projection methods with a Backward Difference Formula of order q (BDFq), conventionally set $u^n = u(t^n)$ and $p^n = p(t^n)$ for $n \geq 0$. Denote by $\frac{1}{\Delta t}(\beta_q u^{n+1} - \sum_{j=0}^{q-1} \beta_j u^{n-j})$ the qth-order backward difference formula for $d_t u(t^{n+1})$, and by $p^{\star,n+1} = \sum_{j=0}^{q-1} \gamma_j p^{n-j}$ a q-th order extrapolation of p^{n+1}. To simplify the notation, for any sequence $\phi_{\Delta t} = (\phi^0, \phi^1, \ldots, \phi^N)$, set

$$D^{(q)} \phi^{n+1} = \beta_q \phi^{n+1} - \sum_{j=0}^{q-1} \beta_j \phi^{n-j}.$$

For instance, $D^{(1)} v^{n+1} = v^{n+1} - v^n$ and $p^{\star,n+1} = p^n$ for $q = 1$, and $D^{(2)} v^{n+1} = \frac{3}{2} v^{n+1} - 2v^n + \frac{1}{2} v^{n-1}$ and $p^{\star,n+1} = 2p^n - p^{n-1}$ for $q = 2$.

Set $u^0 = u_0 \in H$ and if $q > 1$, assume that for $1 \leq r < q$, the r-th step is implemented using BDFr. For $n \geq q - 1$, seek $\tilde{u}^{n+1} \in [H_0^1(\Omega)]^d$ such that

$$\begin{cases} \frac{1}{\Delta t}\left(\beta_q \tilde{u}^{n+1} - \sum_{j=0}^{q-1} \beta_j u^{n-j} \right) - \nu \Delta \tilde{u}^{n+1} + \nabla p^{\star,n+1} = f^{n+1}, \\ \tilde{u}^{n+1}_{|\partial\Omega} = 0. \end{cases} \tag{6.49}$$

Since $\tilde{u}^{n+1}$ is unlikely to be solenoidal, this vector field is projected onto H. Owing to Theorem 6.35, we seek u^{n+1} in H and ϕ^{n+1} in $H^1_{\int=0}(\Omega)$ such that

$$\begin{cases} u^{n+1} + \nabla \phi^{n+1} = \tilde{u}^{n+1}, \\ \nabla \cdot u^{n+1} = 0, \\ u^{n+1} \cdot n_{|\partial\Omega} = 0. \end{cases} \tag{6.50}$$

Let us now interpret the auxiliary quantity ϕ^{n+1}. Multiplying the first equation in (6.50) by $\beta_q \frac{1}{\Delta t}$ and adding the result to (6.49) yields

$$\frac{D^{(q)}}{\Delta t} u^{n+1} - \nu \Delta \tilde{u}^{n+1} + \nabla \left(\frac{\beta_q}{\Delta t} \phi^{n+1} + p^{\star,n+1} \right) = f^{n+1}.$$

This equation can be interpreted as an approximation of the first equation in (6.34) if the quantity $\frac{\beta_q}{\Delta t} \phi^{n+1} + p^{\star,n+1}$ is interpreted as an approximation of the pressure. Hence, the third substep of the algorithm consists of updating the pressure by

$$p^{n+1} = \frac{\beta_q}{\Delta t} \phi^{n+1} + p^{\star,n+1}. \tag{6.51}$$

The algorithm (6.49)–(6.50)–(6.51) generates sequences denoted by $\tilde{u}_{\Delta t}$, $u_{\Delta t}$, and $p_{\Delta t}$. The exact solution to (6.40) is denoted by (u, p) with corresponding sequences $u_{\Delta t} = (u(t^0), \ldots, u(t^N))$ and $p_{\Delta t} = (p(t^0), \ldots, p(t^N))$.

The algorithm originally proposed by Chorin and Temam consists of using $p^{\star,n+1} = 0$ along with BDF1. The resulting scheme has the following convergence properties:

Theorem 6.43. *Use* BDF1 *to march in time and set* $p^{\star,n+1} = 0$. *If the solution* (u, p) *to* (6.40) *is sufficiently smooth, the sequences* $\tilde{u}_{\Delta t}$, $u_{\Delta t}$, *and* $p_{\Delta t}$ *generated by* (6.49)–(6.50)–(6.51) *satisfy*

$$\|u_{\Delta t} - u_{\Delta t}\|_{\ell^\infty([0,T];[L^2(\Omega)]^d)} + \|u_{\Delta t} - \tilde{u}_{\Delta t}\|_{\ell^\infty([0,T];[L^2(\Omega)]^d)} \le c(u, p, T)\Delta t,$$

$$\|u_{\Delta t} - \tilde{u}_{\Delta t}\|_{\ell^\infty([0,T];[H^1(\Omega)]^d)} + \|p_{\Delta t} - p_{\Delta t}\|_{\ell^\infty([0,T];L^2(\Omega))} \le c(u, p, T)\Delta t^{\frac{1}{2}}.$$

Proof. See [Ran92]. □

Despite its simplicity, the above method is not satisfactory since its convergence rate is irreducibly limited to $O(\Delta t)$. The limitation stems from the fact that the method is basically an *artificial compressibility* technique as shown in [Ran92, She92^b]. To cure this problem, numerous variants have been proposed; one of them, the so-called *pressure–correction method*, consists of setting $p^{\star,n+1} = p^n$. The idea behind pressure–correction methods can be traced back to [God79] in the form of a first-order scheme using BDF1. The following convergence result holds:

Theorem 6.44. *Use* BDF1 *to march in time and set* $p^{\star,n+1} = p^n$. *Set* $u^0 = u_0$ *and* $p^0 = p_{|t=0}$. *Then, if the solution* (u, p) *to* (6.40) *is sufficiently smooth, the sequences* $\tilde{u}_{\Delta t}$, $u_{\Delta t}$, *and* $p_{\Delta t}$ *generated by* (6.49)–(6.50)–(6.51) *satisfy*

$$\|u_{\Delta t} - u_{\Delta t}\|_{\ell^\infty([0,T];[L^2(\Omega)]^d)} + \|u_{\Delta t} - \tilde{u}_{\Delta t}\|_{\ell^\infty([0,T];[L^2(\Omega)]^d)} \le c(u, p, T)\Delta t,$$

$$\|u_{\Delta t} - \tilde{u}_{\Delta t}\|_{\ell^\infty([0,T];[H^1(\Omega)]^d)} + \|p_{\Delta t} - p_{\Delta t}\|_{\ell^\infty([0,T];L^2(\Omega))} \le c(u, p, T)\Delta t.$$

Proof. See [She92^a, GuQ98]. □

First-order pressure extrapolation became a popular strategy after it was formally shown in [vaK86] that the scheme significantly increases the accuracy when used together with a second-order time stepping method. More precisely, we state the following:

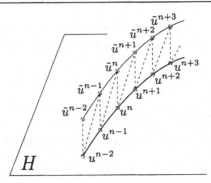

Fig. 6.1. Projection algorithm. At time t^{n+1}, compute $\tilde{u}^{n+1}$ without enforcing incompressibility; as a result, $\tilde{u}^{n+1}$ pops out of H. Then, project $\tilde{u}^{n+1}$ onto H.

Theorem 6.45. *Use* BDF2 *to march in time and set* $p^{\star,n+1} = p^n$. *Set* $u^0 = u_0$ *and* $p^0 = p_{|t=0}$. *Use* BDF1 *for the first time step. Then, if the solution* (u, p) *to* (6.40) *is sufficiently smooth, the sequences* $\tilde{u}_{\Delta t}$, $u_{\Delta t}$, *and* $p_{\Delta t}$ *generated by* (6.49)–(6.50)–(6.51) *satisfy*

$$\|\mathsf{u}_{\Delta t} - u_{\Delta t}\|_{\ell^2([0,T];[L^2(\Omega)]^d)} + \|\mathsf{u}_{\Delta t} - \tilde{u}_{\Delta t}\|_{\ell^2([0,T];[L^2(\Omega)]^d)} \leq c(\mathsf{u}, \mathsf{p}, T)\Delta t^2,$$

$$\|\mathsf{u}_{\Delta t} - \tilde{u}_{\Delta t}\|_{\ell^\infty([0,T];[H^1(\Omega)]^d)} + \|\mathsf{p}_{\Delta t} - p_{\Delta t}\|_{\ell^\infty([0,T];L^2(\Omega))} \leq c(\mathsf{u}, \mathsf{p}, T)\Delta t.$$

Proof. See [She96, Gue99^b]; see also [EwL95, StL99, BrC01] for different proofs based on normal mode analysis in the half-plane or in a periodic channel.   □

Remark 6.46.

(i) Algorithms based on the first-order extrapolation of the pressure assume more regularity than the Chorin–Temam algorithm. Indeed, they require the existence of $p_{|t=0}$ in a reasonably smooth space. Note that $p_{|t=0}$ is not an initial data for the time-dependent Stokes problem (6.40). If $u_0 \in [H^2(\Omega)]^d \cap H$, one can compute $p_{|t=0}$ by solving $(\nabla p_{|t=0}, \nabla q)_{0,\Omega} = (f(0) + \nu \Delta u_0, \nabla q)_{0,\Omega}$ for all $q \in H^1_{|=0}(\Omega)$.

(ii) Problem (6.50) is usually called a *projection step* since, owing to Theorem 6.35, it is a realization of the identity $u^{n+1} = P_H \tilde{u}^{n+1}$. The velocity $\tilde{u}^{n+1}$ is an approximation of $\mathsf{u}(t^{n+1})$ which satisfies the boundary conditions but is not solenoidal. This defect is corrected by projecting $\tilde{u}^{n+1}$ onto H (hence, the name of the method). Although u^{n+1} is divergence-free, it is not a better approximation of $\mathsf{u}(t^{n+1})$ since it does not satisfy the no-slip boundary condition, i.e., its tangential component is not zero. A schematic representation of a typical projection algorithm is shown in Figure 6.1.

(iii) The main interest of projection methods is to uncouple viscous and incompressibility effects. It is more effective computationally to solve the two independent problems (6.49) and (6.50) than to solve the coupled problem (6.46). At the present time, projection methods count among the few methods that are capable of solving the time-dependent incompressible

Navier–Stokes equations in three dimensions on fine meshes within reasonable computation times.

(iv) Some authors have tested higher-order pressure extrapolations and claimed stability and convergence. However, whether using a second-order or higher-order extrapolation on the pressure yields a stable scheme is not yet clear. To the present time no proof of stability has been found. A singular perturbation argument advanced in [She93] indicates that some of these higher-order extrapolation algorithms should not be stable for very small time steps. The issue is open. □

Rotational form of projection methods. Theorem 6.45 shows that although the scheme is second-order accurate on the velocity in the L^2-norm, it is plagued by a boundary layer that prevents it being fully second-order in the H^1-norm of the velocity and in the L^2-norm of the pressure. Actually, from (6.50) we observe that $\partial_n(p^{n+1} - p^n)_{|\partial\Omega} = 0$, which implies

$$\partial_n p^{n+1}_{|\partial\Omega} = \partial_n p^n_{|\partial\Omega} = \cdots = \partial_n p^0_{|\partial\Omega}. \tag{6.52}$$

It is this non-realistic Neumann boundary condition on the pressure that introduces the boundary layer and consequently limits the accuracy of the scheme.

To derive a more accurate algorithm, one of the main ingredients is to use the rotational form of the vector Laplacian in (6.49), namely $-\Delta\tilde{u}^{n+1} = -\nabla(\nabla\cdot\tilde{u}^{n+1}) + \nabla\times\nabla\times\tilde{u}^{n+1}$. Multiply the first equation in (6.50) by $\beta_q\frac{1}{\Delta t}$ and add the result to (6.49) to obtain

$$\frac{D^{(q)}}{\Delta t}u^{n+1} + \nu\nabla\times\nabla\times\tilde{u}^{n+1} + \nabla\left(\frac{\beta_q}{\Delta t}\phi^{n+1} + p^{\star,n+1} - \nu\nabla\cdot\tilde{u}^{n+1}\right) = f^{n+1}. \tag{6.53}$$

It is again possible to read this equation as an approximation of the momentum equation if the quantity $\frac{\beta_q}{\Delta t}\phi^{n+1} + p^{\star,n+1} - \nu\nabla\cdot\tilde{u}^{n+1}$ is interpreted as an approximation of the pressure. Hence, an alternative way of writing the third substep of the projection algorithm consists of updating the pressure as

$$p^{n+1} = \frac{\beta_q}{\Delta t}\phi^{n+1} + p^{\star,n+1} - \nu\nabla\cdot\tilde{u}^{n+1}. \tag{6.54}$$

The scheme (6.49)–(6.50)–(6.54) has been proposed in [TiM96] where it is numerically shown that the divergence correction significantly improves the pressure approximation. More precisely, the following theorem holds:

Theorem 6.47. *Use* BDF2 *to march in time and set* $p^{\star,n+1} = p^n$. *Set* $u^0 = u_0$ *and* $p^0 = p_{|t=0}$. *Use* BDF1 *for the first time step. Then, if the solution* (u, p) *to* (6.40) *is sufficiently smooth, the sequences* $\tilde{u}_{\Delta t}$, $u_{\Delta t}$, *and* $p_{\Delta t}$ *generated by* (6.49)–(6.50)–(6.54) *satisfy*

$$\|u_{\Delta t} - u_{\Delta t}\|_{\ell^2([0,T];[L^2(\Omega)]^d)} + \|u_{\Delta t} - \tilde{u}_{\Delta t}\|_{\ell^2([0,T];[L^2(\Omega)]^d)} \leq c(\mathsf{u}, \mathsf{p}, T)\Delta t^2,$$

$$\|u_{\Delta t} - \tilde{u}_{\Delta t}\|_{\ell^2([0,T];[H^1(\Omega)]^d)} + \|\mathsf{p}_{\Delta t} - p_{\Delta t}\|_{\ell^2([0,T];L^2(\Omega))} \leq c(\mathsf{u}, \mathsf{p}, T)\Delta t^{\frac{3}{2}}.$$

Proof. See [GuS01, GuS04].   □

To understand why the modified scheme performs better, observe from (6.50) that $\nabla\times\nabla\times\tilde{u}^{n+1} = \nabla\times\nabla\times u^{n+1}$. Therefore, (6.53) can be recast in the form

$$\begin{cases} \frac{D^{(q)}}{\Delta t}u^{n+1} + \nu\nabla\times\nabla\times u^{n+1} + \nabla p^{n+1} = f^{n+1}, \\ \nabla\cdot u^{n+1} = 0, \\ u^{n+1}\cdot n_{|\partial\Omega} = 0. \end{cases} \tag{6.55}$$

We deduce from (6.55) that

$$\partial_n p^{n+1}_{|\partial\Omega} = (f^{n+1} - \nu\nabla\times\nabla\times u^{n+1})\cdot n_{|\partial\Omega},$$

which, unlike (6.52), is a consistent pressure boundary condition. The splitting error now manifests itself only in the form of an inexact tangential boundary condition on the velocity.

In view of (6.55) where the operator $\nabla\times\nabla\times$ plays a key role, we refer to (6.49)–(6.50)–(6.54) as the *pressure–correction scheme in rotational form*, and to (6.49)–(6.50)–(6.51) as the *pressure–correction scheme in standard form*.

Elimination of the projected velocity. When implementing the above projection algorithms, it is not necessary to compute the sequence of projected velocities $\{u^n\}_{n\geq 0}$ since these quantities can be eliminated. To see this, set $\tilde{u}^0 = u_0$ and $\phi^l = 0$ for $0 \leq l \leq q-1$. If $q > 1$, assume that $\tilde{u}^1,\dots,\tilde{u}^{q-1}$ and $p^{\star,q}$ have been initialized properly. Now, consider (6.50) at time t^{n-j} for $0 \leq j \leq q-1$, multiply the equation by $\frac{\beta_j}{\Delta t}$, and add the result to (6.49). We obtain the following problem: For $n \geq q-1$, seek $\tilde{u}^{n+1}$ such that

$$\begin{cases} \frac{D^{(q)}}{\Delta t}\tilde{u}^{n+1} - \nu\Delta\tilde{u}^{n+1} + \nabla\left(p^{\star,n+1} + \sum_{j=0}^{q-1}\frac{\beta_j}{\Delta t}\phi^{n-j}\right) = f^{n+1}, \\ \tilde{u}^{n+1}_{|\partial\Omega} = 0. \end{cases} \tag{6.56}$$

Furthermore, taking the divergence of (6.50) yields

$$\begin{cases} \Delta\phi^{n+1} = \nabla\cdot\tilde{u}^{n+1}, \\ \partial_n\phi^{n+1}_{|\partial\Omega} = 0. \end{cases} \tag{6.57}$$

The last step consists of updating the pressure. For the rotational form of the algorithm, this step is

$$p^{n+1} = \frac{\beta_q}{\Delta t}\phi^{n+1} + p^{\star,n+1} - \nu\nabla\cdot\tilde{u}^{n+1}. \tag{6.58}$$

The scheme (6.56)–(6.57)–(6.58) is strictly equivalent to (6.49)–(6.50)–(6.54) while being somewhat easier to implement.

Finite element approximation. We now describe how the discrete setting of §6.2.2 can be used in conjunction with the projection algorithms introduced above. For the sake of brevity, we restrict ourselves to algorithm (6.56)–(6.57)–(6.58). The extension to other algorithms is straightforward.

We assume that M_h is H^1-conformal. Although this hypothesis is not required by the approximation theory of the Stokes problem, it somewhat simplifies the implementation of the method. To avoid minor technical details, we assume that the algorithm is initialized properly; see [Gue99^b]. For instance, take $\tilde{u}_h^0 = P_{h0}(u_0)$ and if the algorithm involves $p_{|t=0}$, set $p_h^0 = Q_{h0}(p_{|t=0})$, where the operators P_{ht} and Q_{ht} are defined in (6.43) and (6.44). For $n \geq q-1$, consider the following sequence of problems:

$$\begin{cases} \text{Seek } \tilde{u}_h^{n+1} \in X_h \text{ such that} \\ (\frac{D^{(q)}}{\Delta t}\tilde{u}_h^{n+1}, v_h)_{0,\Omega} + a(\tilde{u}_h^{n+1}, v_h) + b(v_h, p^{\star,n+1} + \sum_{j=0}^{q-1} \frac{\beta_j}{\Delta t}\phi^{n-j}) \qquad (6.59) \\ = (f^{n+1}, v_h)_{0,\Omega}, \quad \forall v_h \in X_h, \end{cases}$$

then:

$$\begin{cases} \text{Seek } \phi_h^{n+1} \in M_h \text{ such that} \\ (\nabla\phi_h^{n+1}, \nabla q_h)_{0,\Omega} = -(\nabla\cdot\tilde{u}_h^{n+1}, q_h)_{0,\Omega}, \quad \forall q_h \in M_h, \end{cases} \qquad (6.60)$$

and, finally, depending on whether it is the standard form or the rotational form of the algorithm which is used:

$$\begin{cases} \text{Seek } p_h^{n+1} \in M_h \text{ such that} \\ (p_h^{n+1}, q_h)_{0,\Omega} = (\frac{\beta_q}{\Delta t}\phi_h^{n+1} + p_h^{\star,n+1} - \nu\nabla\cdot\tilde{u}_h^{n+1}, q_h)_{0,\Omega}, \quad \forall q_h \in M_h, \end{cases} \qquad (6.61)$$

or

$$\text{Set: } \quad p_h^{n+1} = \frac{\beta_q}{\Delta t}\phi_h^{n+1} + p_h^{\star,n+1}. \qquad (6.62)$$

Theorem 6.48. *Use BDF2 to march in time and set $p_h^{\star,n+1} = p_h^n$. Initialize the algorithm properly. Use BDF1 for the first time step. Then, if the solution (u, p) to (6.40) is sufficiently smooth, there is $c = c(u, p, T)$ such that the sequences $\tilde{u}_{\Delta t}$, $u_{\Delta t}$, and $p_{\Delta t}$ generated by (6.59)–(6.60)–(6.62) satisfy*

$$\|u_{\Delta t} - u_{h\Delta t}\|_{\ell^2([0,T];[L^2(\Omega)]^d)} + \|u_{\Delta t} - \tilde{u}_{h\Delta t}\|_{\ell^2([0,T];[L^2(\Omega)]^d)} \leq c(h^{k+1} + \Delta t^2),$$

$$\|u_{\Delta t} - \tilde{u}_{h\Delta t}\|_{\ell^\infty([0,T];[H^1(\Omega)]^d)} + \|p_{\Delta t} - p_{h\Delta t}\|_{\ell^\infty([0,T];L^2(\Omega))} \leq c(h^k + \Delta t).$$

Proof. See [Gue99^b]; see also [Gue96, GuQ98]. $\qquad \square$

Although, at the present time, no error analysis for the fully discrete scheme (6.59)–(6.60)–(6.61) has yet been published, it is generally believed, and confirmed by numerical tests, that with this scheme the second error estimate in Theorem 6.48 should be replaced by $c(h^k + \Delta t^{\frac{3}{2}})$.

Remark 6.49. Note that the two discrete problems (6.59) and (6.60) can be solved in sequence and that none of them requires the inf-sup condition (6.41) to be well-posed (since they both involve a coercive bilinear form). This observation may lead to the conclusion that the scheme (6.59)–(6.60)–(6.62) (or (6.59)–(6.60)–(6.61)) is a clever way of solving the Navier–Stokes equations with finite elements without bothering about the inf-sup condition. This intuitive argument is false. The inf-sup condition must be satisfied (preferably uniformly) for the above algorithms to yield the expected accuracy. The reader can verify in [Gue96, GuQ98] that the convergence proof of Theorem 6.48 uses the inf-sup condition.   □

6.3 Evolution Equations Without Coercivity

This section deals with the finite element approximation of evolution equations without coercivity. The prototypical example is the time-dependent advection equation.

6.3.1 The model problem

Let us introduce a general setting for non-coercive time-dependent problems. Let L be a separable Hilbert space and let $A : D(A) \subset L \to L$ be a linear operator. Consider the following problem: For $f \in C^1([0,T]; L)$ and $u_0 \in D(A)$,

$$\begin{cases} \text{Seek } u \in C^1([0,T]; L) \cap C^0([0,T]; D(A)) \text{ such that} \\ d_t u + Au = f, \\ u(0) = u_0. \end{cases} \tag{6.63}$$

To investigate the well-posedness of problem (6.63), we restrict A to be a monotone, maximal operator. Recall the following:

Definition 6.50. *The operator A is said to be* monotone *iff*

$$\forall v \in D(A), \quad (Av, v)_L \geq 0, \tag{6.64}$$

and A is said to be maximal *iff*

$$\forall f \in L, \exists v \in D(A), \quad v + Av = f. \tag{6.65}$$

Lemma 6.51. *If $A : D(A) \subset L \to L$ is maximal and monotone, the following properties hold:*

  (i) *$D(A)$ is dense in L.*
  (ii) *The graph of A is closed.*
  (iii) *$\forall \lambda > 0$, $I + \lambda A \in \mathcal{L}(D(A); L)$ is bijective and $\|(I + \lambda A)^{-1}\|_{\mathcal{L}(L; L)} \leq 1$.*

Proof. See [Sho96, p. 22], [Yos80, p. 246], or [Bre91, p. 101].   □

Set $V = D(A)$ and equip V with the graph norm $\|v\|_V = \|v\|_L + \|Av\|_L$. Owing to Lemma 6.51, the graph of A is closed, implying that V is a Banach space. Consider the bilinear form a such that $a(u,v) = (Au,v)_L$ for all $u \in V$ and $v \in L$. Since the graph of A is closed, $A \in \mathcal{L}(V;L)$, i.e., the bilinear form $a : V \times L \to \mathbb{R}$ is continuous. Furthermore, when equipped with the scalar product $(u,v)_L + (Au, Av)_L$, V is a Hilbert space. Since $D(A) = V$ is dense in L, we are in the classical situation where $V \subset L \equiv L' \subset V'$; see also §5.2.1. Using the above notation, we can reformulate (6.63) as follows: For $f \in \mathcal{C}^1([0,T];L)$ and $u_0 \in V$,

$$\begin{cases} \text{Seek } u \text{ in } \mathcal{C}^1([0;T];L) \cap \mathcal{C}^0([0,T];V) \text{ such that} \\ (d_t u, v)_L + a(u,v) = (f,v)_L, \quad \forall v \in L, \forall t \geq 0, \qquad (6.66) \\ (u(0), v) = (u_0, v), \qquad\qquad \forall v \in L. \end{cases}$$

Clearly, problems (6.63) and (6.66) are equivalent. The main existence and uniqueness result of this section is the following:

Theorem 6.52 (Hille–Yosida). *For all $f \in \mathcal{C}^1([0,T];L)$ and $u_0 \in D(A)$, the problem (6.66) has a unique solution. Moreover, there is c, independent of T, such that*

$$\|u\|_{\mathcal{C}^0([0,T];L)} \leq c(\|u_0\|_L + T\|f\|_{\mathcal{C}^0([0,T];L)}), \qquad (6.67)$$
$$\|u\|_{\mathcal{C}^1([0,T];L)} + \|u\|_{\mathcal{C}^0([0,T];V)} \leq c(\|u_0\|_V + T\|f\|_{\mathcal{C}^1([0,T];L)}). \qquad (6.68)$$

Proof. See [Yos80, p. 248] or [Bre91, p. 110] for the existence and uniqueness result. To prove (6.67), observe that for all t, u is an admissible test function, and test (6.66) with u. Since A is monotone,

$$d_t(\|u\|_L^2) \leq 2\|f\|_L \|u\|_L.$$

The first estimate then results from Lemma 6.54 below. To prove (6.68), differentiate the evolution equation with respect to time and use the first estimate (this can be rigorously justified) to obtain

$$\|d_t u\|_{\mathcal{C}^0([0,T];L)} \leq c(\|d_t u(0)\|_L + T\|f\|_{\mathcal{C}^1([0,T];L)}).$$

Using the evolution equation at time $t = 0$ yields $\|d_t u(0)\|_L \leq \|f(0)\|_L + \|Au_0\|_L$. As a result,

$$\|d_t u\|_{\mathcal{C}^0([0,T];L)} \leq c(\|u_0\|_V + T\|f\|_{\mathcal{C}^1([0,T];L)}).$$

The desired result is a consequence of this bound along with the inequality $\|Au\|_L \leq \|d_t u\|_L + \|f\|_L$. $\qquad\square$

Remark 6.53.

(i) The Hille–Yosida Theorem is slightly more general than Theorem 6.6

for it does not require the bilinear form a to be coercive; only positivity is required.

(ii) The differential operators associated with the advection–reaction equation, Darcy's equations, and Maxwell's equations studied in §5.2.3, §5.2.4, and §5.2.5 of Chapter 5 are maximal and monotone. Hence, the time-dependent extension of these three problems falls into the above framework.   □

We now state a modified version of Gronwall's Lemma.

Lemma 6.54. *Let $\phi \in W^{1,1}(]0,T[; \mathbb{R})$ and $\psi \in L^1(]0,T[; \mathbb{R})$ be two functions such that $\phi \geq 0$ and $\psi \geq 0$. Assume that there are two numbers $a \geq 0$ and $b \geq 0$ such that $d_t \phi + \psi \leq a\phi^{\frac{1}{2}} + b$. Then,*

$$\|\phi\|_{L^\infty(]0,T[;\mathbb{R})} + \|\psi\|_{L^1(]0,T[;\mathbb{R})} \leq e \times \left(\frac{a^2}{4}T^2 + bT + \phi(0)\right).$$

Proof. First, it is clear that $d_t \phi + \psi \leq \frac{a^2 T}{4} + \frac{\phi}{T} + b$. Once multiplied by $e^{-\frac{t}{T}}$, this inequality yields

$$d_t(\phi(t)e^{-\frac{t}{T}}) + \psi e^{-\frac{t}{T}} \leq e^{-\frac{t}{T}}\left(\frac{a^2}{4}T + b\right).$$

The result is obtained by integrating this inequality over $[0,T]$ and using the fact that $1 \leq e^{\frac{\tau}{T}} \leq e$ for all $\tau \in [0,T]$.   □

Introducing the seminorm $|v|_V = \|Av\|_L$ and using the definition of the graph norm, we infer

$$\forall u \in V, \quad \sup_{v \in L} \frac{a(u,v)}{\|v\|_L} = |u|_V \geq \|u\|_V - \|u\|_L. \tag{6.69}$$

Owing to Proposition 6.55 below, this inequality is actually a consequence of the fact that A is maximal.

Proposition 6.55. *Let $E \subset F$ be two Hilbert spaces with dense and continuous embedding and let $R \in \mathcal{L}(E;F)$ be a monotone operator. The two following statements are equivalent:*

(i) *R is maximal.*
(ii) *There exist $c_1 > 0$ and $c_2 \geq 0$ such that*

$$\forall u \in V, \quad \sup_{v \in L} \frac{(Ru,v)_F}{\|v\|_F} \geq c_1 \|u\|_E - c_2 \|u\|_F. \tag{6.70}$$

The main difference between problem (6.66) and parabolic equations is that a is no longer assumed to be coercive. This point renders the task of constructing an approximate solution to (6.66) non-trivial. The following sections review some classical methods to approximate evolution equations without coercivity, namely discontinuous finite elements (GaLS/DG in time and space-time DG), the method of characteristics, and the subgrid viscosity technique.

6.3.2 Discontinuous finite elements (DG)

GaLS/DG in time. A first class of techniques is based on the GaLS method developed in §5.4. The key idea is to use GaLS in V for the space approximation and DG for the time approximation.

Let $V_h \subset V$ be a finite-dimensional subspace of V. Take $(N+1)$ points in $[0, T]$ such that $0 = t^0 < t^1 < \ldots < t^N = T$, set $I_n = [t^n, t^{n+1}]$ and $\Delta t = \max_{0 \le n \le N-1}(t^{n+1} - t^n)$, and denote by $\mathcal{I}_{\Delta t} = \{I_n\}_{0 \le n \le N-1}$ the resulting mesh of $]0, T[$. Let $k \ge 0$ be an integer. Define

$$V_{h\Delta t}^k = \{v \in L^2(]0, T[; V_h); \forall I_n \in \mathcal{I}_{\Delta t}, \, v_{|I_n} \in V_{hI_n}^k\},$$

where the space $V_{hI_n}^k = \mathbb{P}^k(I_n; V_h)$ is composed of those functions that are polynomials of degree at most k in time with values in V_h. We shall seek an approximation U of u in $V_{h\Delta t}^k$. Introduce the notation

$$\forall v \in \mathcal{W}(V; L) + V_{h\Delta t}^k, \quad v_{\pm}^n = \lim_{s \to 0^{\pm}} v(t^n + s), \qquad (6.71)$$

and set

$$U_{-}^0 = u_0. \qquad (6.72)$$

Consider the discrete problem:

$$\begin{cases} \text{Seek } U \in V_{h\Delta t}^k \text{ such that, for all } n \ge 0 \text{ and } v \in V_{hI_n}^k, \\ \int_{I_n} (d_t U + AU)(v + \delta(h)(d_t v + Av)) \, dt + (U_+^n - U_-^n, v_+^n)_L \\ \qquad\qquad = \int_{I_n} f(v + \delta(h)(d_t v + Av)) \, dt. \end{cases} \qquad (6.73)$$

The parameter $\delta(h)$ is chosen according to the following:

Lemma 6.56 (Stability). *Let $\gamma > 0$ and set $\delta(h) = \gamma \min_{v \in V_h} \|v\|_L / \|Av\|_L$. Assume that $\mathcal{I}_{\Delta t}$ is quasi-uniform. Let $c_{CFL} = \Delta t / \delta(h)$. There is c, depending only on γ and c_{CFL}, such that the following stability estimate holds:*

$$\|U\|_{L^\infty(]0,T[;L)} + \delta(h)^{\frac{1}{2}} \|d_t U + AU\|_{L^2(]0,T[;L)} < c(\|u_0\|_L + T^{\frac{1}{2}} \|f\|_{L^2(]0,T[;L)}).$$

Proof. See [JoN84] and also Exercise 6.11. □

Remark 6.57.

(i) Let $\Delta t_{\min} = \min_{0 \le n < N} |t^{n+1} - t^n|$. Quasi-uniformity of $\mathcal{I}_{\Delta t}$ implies that there is $\sigma > 0$ such that $\Delta t \le \Delta t_{\min}/\sigma$; as a result, $c_{CFL} \Delta t_{\max} \le \delta(h) \le \Delta t_{\min}/(\sigma c_{CFL})$. Assume that V_h is constructed by using a quasi-uniform mesh family $\{\mathcal{T}_h\}_h$ and $\delta(h) \sim h$. Assuming that c_{CFL} is uniformly bounded from below and above as $\Delta t \to 0$ and $h \to 0$ is equivalent to assuming that the mesh family $\{\mathcal{F}_h \times \mathcal{I}_{\Delta t}\}_{h, \Delta t}$ is quasi-uniform.

(ii) The origin of this type of technique using $\delta(h) = 0$ can be traced back to [Jam78]. The GaLS/DG method can also account for a diffusion operator as in §5.4. See, e.g., to [JoN84] for a thorough analysis of this method. □

Space-time DG. Another strategy to approximate (6.66) is to reformulate the problem in a setting that does not distinguish time and space. For instance, one could set $X = \mathcal{W}(V, L)$, $Y = L^2(]0, T[; L)$, and $b(u, v) = (d_t u + Au, v)_L$. Assume $u_0 = 0$ for simplicity. Then, problem (6.66) can be recast in the form:

$$\begin{cases} \text{Seek } u \in X \text{ such that} \\ b(u, v) = (f, v)_Y, \quad \forall v \in Y. \end{cases}$$

When A is a first-order differential operator, the operator associated with b is also a first-order differential operator. For instance, if $Au = \partial_x u$, the operator associated with b is $\partial_t(\cdot) + \partial_x(\cdot)$. Then, an appropriate approximation strategy consists of using the DG method described in §5.6. Using the notation of §5.6.2, let A be defined in (5.98), i.e., $Au = \mu u + \beta \cdot \nabla u$, and consider the *unsteady advection–reaction equation*:

$$\begin{cases} d_t u + \mu u + \beta \cdot \nabla u = f & \text{in } \Omega \times]0, T[, \\ u = g & \text{in } \partial \Omega^- \times]0, T[, \\ u(0) = u_0 & \text{in } \Omega, \end{cases} \qquad (6.74)$$

with $u_0 \in L^2(\Omega)$, $f \in L^2(]0, T[; L^2(\Omega))$, and $g \in L^2(]0, T[; L^2_\beta(\partial\Omega^-))$. Let W_h be the finite element space defined in (5.99). Let $\mathcal{I}_{\Delta t} = \{I_n\}_{0 \le n \le N-1}$ be a mesh of $]0, T[$ with $I_n = [t^n, t^{n+1}]$. Assume $\text{diam}(I_n) \sim h$ uniformly with respect to n. Let $k \ge 0$ be an integer. Define $W^k_{hI_n} = \mathbb{P}^k(I_n; W_h)$ and

$$W^k_{h\Delta t} = \{v \in L^2(]0, T[; V_h); \forall I_n \in \mathcal{I}_{\Delta t}, \, v_{|I_n} \in W^k_{hI_n}\}.$$

Consider the bilinear form a_h defined in §5.6.2. Adopting the notation (6.71) and the initialization convention (6.72), the discrete problem is as follows:

$$\begin{cases} \text{Seek } U \in W^k_{h\Delta t} \text{ such that, for all } v \in W^k_{h\Delta t}, \\ \int_{I_n} [(d_t U, v)_{0,\Omega} + a_h(U, v)] \, dt + (U^n_+ - U^n_-, v^n_+)_{0,\Omega} \\ \qquad\qquad = \int_{I_n} (f, v)_{0,\Omega} + \int_{\partial\Omega^- \times I_n} |\beta \cdot n| g v. \end{cases} \qquad (6.75)$$

The error analysis of §5.6.2 fully applies to the present setting. Set $\tilde{\Omega} = \Omega \times]0, T[$, $\tilde{\beta} = (\beta, 1) \in \mathbb{R}^{d+1}$, $\tilde{\nabla} = (\nabla, \partial_t)$, $\partial\tilde{\Omega}^- = \partial\Omega^- \times]0, T[\,\cup\, \Omega \times \{0\}$, and $\tilde{g}_{|\partial\Omega^- \times]0,T[} = g$, $\tilde{g}_{|\Omega \times \{0\}} = u_0$. Then, (6.74) is clearly equivalent to

$$\begin{cases} \mu u + \tilde{\beta} \cdot \tilde{\nabla} u = f & \text{in } \tilde{\Omega}, \\ u = \tilde{g} & \text{on } \partial\tilde{\Omega}^-. \end{cases} \qquad (6.76)$$

The scheme (6.75) is nothing but the DG approximation of (6.76); see [CoK00, LeR74, Joh87, JoN84, ErE96] for further insight.

6.3.3 The method of characteristics

The method of characteristics is tailored to solve *unsteady advection equations*

$$
\begin{cases}
d_t u + \beta \cdot \nabla u = f & \text{in } \Omega \times \,]0, T[, \\
u = g & \text{in } \partial\Omega^- \times \,]0, T[, \\
u(0) = u_0 & \text{in } \Omega.
\end{cases}
\tag{6.77}
$$

To formulate the approximation scheme, introduce the characteristic curves $X(x, s; t)$ of the vector field (i.e., the flow) $\beta(x, t)$ defined as follows:

$$
\begin{cases}
d_t X(x, s; t) = \beta(X(x, s; t), t), \\
X(x, s; s) = x.
\end{cases}
\tag{6.78}
$$

If $\beta \in C^0([0, T]; [C^{0,1}(\overline{\Omega})]^d)$, this ODE has a unique solution owing to the Cauchy–Lipschitz Theorem. When no confusion may arise, we set $X^n(x) = X(x, t^{n+1}; t^n)$. The quantity $X^n(x)$ is usually referred to as the *foot of the characteristic* passing at (x, t^{n+1}). Set $D_t u(x, t) = \partial_t u(x, t) + \beta(x, t) \cdot \nabla u(x, t)$ so that the PDE in (6.77) becomes $D_t u = f$. This quantity is called the *material derivative* of u. With this tool in hand, we write

$$
d_t \big[u(X(x, t^{n+1}; t), t) \big] = [D_t u](X(x, t^{n+1}; t), t).
\tag{6.79}
$$

Henceforth, we assume $\nabla \cdot \beta = 0$ and $\beta \cdot n_{|\partial\Omega} = 0$; note that $\beta \cdot n_{|\partial\Omega} = 0$ implies that $\partial\Omega^- = \emptyset$. The following result guarantees that the mapping $\Omega \ni x \mapsto X(x, s; t) \in \Omega$ is a homeomorphism:

Lemma 6.58. *Assume that* $\beta \in C^0([0, T]; [C^1(\overline{\Omega})]^d)$.

(i) *If* $|s - t|$ *is sufficiently small, then* $\Omega \ni x \mapsto X(x, s; t) \in \Omega$ *is a homeomorphism and its Jacobian determinant equals 1 a.e. on* Ω.

(ii) *Assume that* Δt *is small enough, say* $\Delta t \|\beta\|_{C^0([0,T]; [C^1(\overline{\Omega})]^d)} \leq \frac{1}{6}$, *then the mapping* $\Omega \ni x \mapsto (1 - \theta)x + \theta X(x, t^{n+1}; t^n) \in \Omega$ *is a homeomorphism with a Jacobian determinant* $\geq \frac{1}{2}$ *for all* θ *in* $[0, 1]$.

Proof. See, e.g., [Rus85, DoR82, Sül88]. □

The introduction of characteristics is motivated by the following approximation property:

Proposition 6.59. *Let* $1 \leq p \leq +\infty$. *Under the hypotheses of Lemma 6.58, if* u *and* β *are such that* $D_{tt} u \in L^\infty(]0, T[; L^p(\Omega))$, *then*

$$
\Big\| \tfrac{1}{\Delta t} \big[u(\cdot, t^{n+1}) - u(X^n(\cdot), t^n) \big] - D_t u(\cdot, t^{n+1}) \Big\|_{L^p(\Omega)} \leq c(u) \Delta t.
$$

Proof. See, e.g., [Pir82, DoR82, Sül88] or Exercise 6.10. □

In nonlinear problems, the field β may depend on u and must therefore be approximated on a mesh at the discrete time t^n. Denote by V_h the finite element space associated with the mesh at time t^n and denote by $\beta_h^n \in V_h$ the corresponding approximation. Define the approximate characteristic line $X_h(x, t^{n+1}; t)$ as follows:

$$\begin{cases} d_t X_h(x, t^{n+1}; t) = \beta_h^n(X_h(x, t^{n+1}; t)), \\ X_h(x, t^{n+1}; t^{n+1}) = x. \end{cases} \tag{6.80}$$

This problem has a unique solution owing to the Cauchy–Lipschitz Theorem if $V_h \subset [C^{0,1}(\overline{\Omega})]^d$. This hypothesis is satisfied with H^1-conformal Lagrange finite elements. For the sake of simplicity, set $X_h^n(x) = X_h(x, t^{n+1}; t^n)$.

The discrete counterpart of Lemma 6.58 can be stated as follows:

Lemma 6.60. *Assume that for all n, $\Delta t \|\beta_h^n\|_{1,\infty,\Omega} e^{\Delta t \|\beta_h^n\|_{1,\infty,\Omega}} \le \frac{1}{8}$.*

(i) *Then, for all t in $[t^n, t^{n+1}]$, $\Omega \ni x \mapsto X_h(x, t^{n+1}; t) \in \Omega$ is a homeomorphism and its Jacobian determinant satisfies $\frac{1}{2} \le J_h(x, t^{n+1}; t) \le \frac{3}{2}$.*

(ii) *Under the hypotheses of Lemma 6.58, there is ϵ such that, if $\Delta t \|\beta_h^n\|_{1,\infty,\Omega} \le \epsilon$, then the mapping $\Omega \ni x \mapsto (1-\theta)X_h(x, t^{n+1}; t^n) + \theta X(x, t^{n+1}; t^n) \in \Omega$ is a homeomorphism with Jacobian determinant $\ge \frac{1}{2}$ for all θ in $[0, 1]$.*

Owing to (6.79), the finite difference $\frac{1}{\Delta t}(u(x, t^{n+1}) - u(X_h^n, t^n))$ is an approximation to $[D_t u](x, t^{n+1})$, yielding the approximate equation

$$\frac{1}{\Delta t}(u(x, t^{n+1}) - u(X_h^n, t^n)) \approx f^{n+1},$$

where $f^{n+1} = f(t^{n+1})$. Then, introducing a finite element space $W_h \subset L^2(\Omega)$, the discrete sequence $(u_h^0, \dots, u_h^N) \in W_h^{N+1}$ is generated as follows: initialize u_h^0 by solving $(u_h^0, v_h)_{0,\Omega} = (u_0, v_h)_{0,\Omega}$ for all $v_h \in W_h$, and then march in time by solving the problems:

$$\begin{cases} \text{Seek } u_h^{n+1} \in W_h \text{ such that} \\ (u_h^{n+1}, v_h)_{0,\Omega} = (u_h^n \circ X_h^n + \Delta t f^{n+1}, v_h)_{0,\Omega}, \quad \forall v_h \in W_h, \end{cases} \tag{6.81}$$

where $u_h^n \circ X_h^n$ means $(u_h^n \circ X_h^n)(x) = u_h^n(X_h^n(x))$ for a.e. $x \in \Omega$.

Actually, the method of characteristics applies equally well to *unsteady advection-diffusion equations*. For instance, if the evolution equation is

$$\begin{cases} d_t u + \beta \cdot \nabla u - \nu \Delta u = f & \text{in } \Omega \times]0, T[, \\ u = 0 & \text{on } \partial\Omega \times]0, T[, \\ u(0) = u_0 & \text{in } \Omega, \end{cases} \tag{6.82}$$

then, assuming $W_h \subset H_0^1(\Omega)$, $u_0 \in H_0^1(\Omega)$, and denoting by P_h the elliptic projector, the method consists of initializing $u_h^0 = P_h u_0$ and marching in time by solving the problems:

$$\begin{cases} \text{Seek } u_h^{n+1} \in W_h \text{ such that, for all } v_h \in W_h, \\ (u_h^{n+1}, v_h)_{0,\Omega} + \Delta t \nu (\nabla u_h^{n+1}, \nabla v_h)_{0,\Omega} = (u_h^n \circ X_h^n + \Delta t f^{n+1}, v_h)_{0,\Omega}. \end{cases} \quad (6.83)$$

The method of characteristics offers two advantages. On the one hand, it yields remarkable stability properties (roughly speaking, unconditional stability on the material derivative; see [Pir82]). On the other hand, since the treatment of the advective term is explicit, the linear system is symmetric and time-invariant. If there is no diffusion, this amounts to inverting the mass matrix, which is usually an easy task. If there is a second-order term, the matrix of the linear system is $\mathcal{M} + \Delta t \nu \mathcal{A}$, where $\mathcal{M}$ is the mass matrix and $\mathcal{A}$ the stiffness matrix. For further insight, see [Pir83, ErE96, LuP98]; see also [BoM97] for a second-order implementation of the method in the context of the Navier–Stokes equations, and [MoP88] for an implementation using approximate Gaussian quadratures.

We now state a series of important approximation results which are useful in the error analysis of (6.81). These results are adapted from Russell [Rus85], Douglas–Russell [DoR82], and Süli [Sül88]. Denoting by ϵ the small parameter introduced in Lemma 6.60, we henceforth assume that

(HS1) $$\Delta t \|\beta\|_{\mathcal{C}^0([0,T];[\mathcal{C}^1(\overline{\Omega})]^d)} \le \tfrac{1}{6}.$$

(HS2) $$\Delta t \|\beta_h^n\|_{1,\infty,\Omega} \le \epsilon.$$

Assumption (HS1) is a restriction on the time step, whereas (HS2) is a stability hypothesis on the approximate velocity.

Lemma 6.61. *Let* $x \mapsto \psi_0(x)$ *and* $x \mapsto \psi_1(x)$ *be two homeomorphisms of* Ω *onto itself such that, for all* $\theta \in [0,1]$, $x \mapsto (1-\theta)\psi_0(x) + \theta\psi_1(x)$ *is a homeomorphism of* Ω *onto itself with Jacobian* $\ge \tfrac{1}{2}$. *Then, for all* p *and* q *with* $1 \le q < \infty$ *and* $1 \le p \le \infty$, *and for* $\eta \in W^{1,qp'}(\Omega)$ *with* $\tfrac{1}{p} + \tfrac{1}{p'} = 1$,

$$\|\eta \circ \psi_0 - \eta \circ \psi_1\|_{L^q(\Omega)} \le 2\|\psi_0 - \psi_1\|_{[L^{pq}(\Omega)]^d} \|\nabla \eta\|_{[L^{qp'}(\Omega)]^d}. \quad (6.84)$$

Corollary 6.62. *Assume* $\beta \in \mathcal{C}^0([0,T];[\mathcal{C}^1(\overline{\Omega})]^d)$, $d_t\beta \in L^\infty(]0,T[;[L^2(\Omega)]^d)$, *and that* (HS1) *and* (HS2) *hold. Set* $c_1 = 2\exp(\Delta t \|\beta\|_{\mathcal{C}^0([0,T];[\mathcal{C}^1(\overline{\Omega})]^d)})$ *and* $c_2 = \|d_t\beta\|_{L^\infty(]0,T[;[L^2(\Omega)]^d)}$. *Then,*

$$\|X^n - X_h^n\|_{0,\Omega} \le c_1 \Delta t (\|\beta(\cdot,t^n) - \beta_h^n\|_{0,\Omega} + c_2 \Delta t). \quad (6.85)$$

A direct consequence of Lemma 6.61 and Corollary 6.62 is the following:

Corollary 6.63. *Under the assumptions of Corollary* 6.62, *for all* $\eta \in H^1(\Omega)$,

$$\|\eta \circ X^n - \eta \circ X_h^n\|_{L^1(\Omega)} \le 2c_1 \Delta t (\|\beta(t^n) - \beta_h^n\|_{L^2(\Omega)} + c_2 \Delta t)\|\nabla \eta\|_{[L^2(\Omega)]^d}. \quad (6.86)$$

The following corollary is crucial for deriving optimal L^2-estimates; it is mainly due to Douglas–Russell [DoR82]:

Corollary 6.64. *Assume $\beta \in C^0([0,T];[C^1(\overline{\Omega})]^d)$ and that (HS1) holds. Then, for all $\eta \in L^2(\Omega)$,*

$$\|\eta - \eta \circ X^n\|_{H^{-1}(\Omega)} \le 2\Delta t \|\beta\|_{C^0([0,T];[C^1(\overline{\Omega})]^d)} \|\eta\|_{L^2(\Omega)}. \tag{6.87}$$

We are now in a position to state the main convergence result. Henceforth, c denotes a generic constant independent of h and Δt. Assume that there is a subspace $Z \subset H_0^1(\Omega)$ such that, for all $u \in Z$, the elliptic projector P_h is such that $\|u - P_h u\|_{0,\Omega} + h\|u - P_h u\|_{1,\Omega} \le c(u)h^{k+1}$. Let $D(h)$ be the constant, depending on h only, such that $\|v_h\|_{0,\infty,\Omega} \le D(h)\|v_h\|_{1,2,\Omega}$ for all $v_h \in W_h$. For finite elements, $D(h) = c(1 + |\log h|)$ in dimension 2, and $D(h) = ch^{-\frac{1}{2}}$ in dimension 3; see Lemma 1.142. Moreover, assume that the approximate vector field β_h is such that

$$\|\beta - \beta_h\|_{C^0([0,T];[L^2(\Omega)]^d)} \le c(\Delta t + h^{k+1}), \tag{6.88}$$

and $\|\beta_h(t)\|_{1,\infty,\Omega} \le c\|\beta(t)\|_{1,\infty,\Omega}$ for all $t \in [0,T]$.

Theorem 6.65. *Assume that the solution to (6.82) is smooth enough. Under the above assumptions, there are h_0 and c_s such that, for all $h \le h_0$ and $\Delta t \le c_s D(h)^{-1}$, the following estimates hold:*

$$\|u - u_h\|_{\ell^\infty([t^0,t^N];L^2(\Omega))} \le c(u)(\Delta t + h^{k+1}), \tag{6.89}$$

$$\|u - u_h\|_{\ell^2([t^1,t^N];H^1(\Omega))} \le c(u)(\Delta t + h^k), \tag{6.90}$$

with norms defined in (6.19).

Proof. In this proof, c denotes a constant that may depend on u. Set $u^n = u(\cdot,t^n)$. Let $e_h^n = P_h u^n - u_h^n$ and $\eta^n = u^n - P_h u^n$. Then $e_h^0 = 0$ and for $n \ge 0$,

$$(e_h^{n+1} - e_h^n, v_h)_{0,\Omega} + \Delta t\nu(\nabla e_h^{n+1}, \nabla v_h)_{0,\Omega} = R(v_h), \quad \forall v_h \in W_h, \tag{6.91}$$

where the residual $R(v_h)$ is decomposed as follows:

$$\begin{aligned}
R(v_h) = &-(\eta^{n+1} - \eta^n, v_h)_{0,\Omega} + (u^{n+1} - u^n \circ X^n - \Delta t D_t u^{n+1}, v_h)_{0,\Omega} \\
&- (u^n \circ X_h^n - u^n \circ X^n, v_h)_{0,\Omega} - (u^n - u_h^n - (u^n - u_h^n) \circ X^n, v_h)_{0,\Omega} \\
&- ((u^n - u_h^n) \circ X^n - (u^n - u_h^n) \circ X_h^n, v_h)_{0,\Omega}.
\end{aligned}$$

Test (6.91) with $2e_h^{n+1}$. Denote by $R_1, \ldots, R_5$ the five terms in the above equation. Let $\gamma > 0$. Then,

$$2R_1(e_h^{n+1}) \le \gamma\Delta t\|e_h^{n+1}\|_{1,\Omega}^2 + c\Delta t\|\eta_t\|_{L^2(]t^n,t^{n+1}[;L^2(\Omega))}^2.$$

Proposition 6.59 implies

$$2R_2(e_h^{n+1}) \le \gamma\Delta t\|e_h^{n+1}\|_{1,\Omega}^2 + c\Delta t^3\|D_{tt}u\|_{L^2(]t^n,t^{n+1}[;L^2(\Omega))}^2.$$

Lemma 6.61 with $q = 2$ and $p = 1$ together with Corollary 6.62 implies

$$2R_3(e_h^{n+1}) \leq c \, \|e_h^{n+1}\|_{L^2(\Omega)} \|X^n - X_h^n\|_{[L^2(\Omega)]^d} \|\nabla u^n\|_{[L^\infty(\Omega)]^d}$$
$$\leq c \, \Delta t (\Delta t + h^{k+1}) \|e_h^{n+1}\|_{1,\Omega} \leq \gamma \Delta t \|e_h^{n+1}\|_{1,\Omega}^2 + c \, \Delta t (\Delta t + h^{k+1})^2.$$

Corollary 6.64 yields

$$2R_4(e_h^{n+1}) \leq 2 \, \|e_h^{n+1}\|_{1,\Omega} \|u^n - u_h^n - (u^n - u_h^n) \circ X^n\|_{-1,\Omega}$$
$$\leq c \, \Delta t \|e_h^{n+1}\|_{1,\Omega} \|u^n - u_h^n\|_{0,\Omega}$$
$$\leq \gamma \Delta t \|e_h^{n+1}\|_{1,\Omega}^2 + c_1 \Delta t \|e_h^n\|_{0,\Omega}^2 + c_2 \Delta t \|\eta\|_{C^0([t^n,t^{n+1}];L^2(\Omega))}^2.$$

Now, use Lemma 6.61 with $q = 1$ and $p = 2$ together with Corollary 6.62:

$$2R_5(e_h^{n+1}) \leq 2 \, \|e_h^{n+1}\|_{L^\infty(\Omega)} \|(u^n - u_h^n) \circ X^n - (u^n - u_h^n) \circ X_h^n\|_{L^1(\Omega)}$$
$$\leq c \, \|e_h^{n+1}\|_{L^\infty(\Omega)} \|X^n - X_h^n\|_{[L^2(\Omega)]^d} \|\nabla(u^n - u_h^n)\|_{[L^2(\Omega)]^d}$$
$$\leq c \, D(h) \Delta t (\Delta t + h^{k+1}) \|e_h^{n+1}\|_{1,\Omega} (\|e_h^n\|_{1,\Omega} + \|\eta^n\|_{1,\Omega}).$$

If h_0 is small enough, $D(h)\|\eta^n\|_{1,\Omega} \leq 1$ for all $h \leq h_0$. Moreover, if h_0 and c_s are chosen small enough, then $c \, D(h)(\Delta t + h^{k+1}) \leq 2\gamma$ for all $\Delta t \leq c_s D(h)^{-1}$ and for all $h \leq h_0$. Hence,

$$2R_5(e_h^{n+1}) \leq 2\gamma \Delta t \|e_h^{n+1}\|_{1,\Omega}^2 + \gamma \Delta t \|e_h^n\|_{1,\Omega}^2 + c \, \Delta t (\Delta t + h^{k+1})^2.$$

Collecting the five bounds derived above leads to

$$\|e_h^{n+1}\|_{0,\Omega}^2 + \Delta t (2\alpha\nu - 6\gamma) \|e_h^{n+1}\|_{1,\Omega}^2 \leq (1 + c_1 \Delta t) \|e_h^n\|_{0,\Omega}^2$$
$$+ \gamma \Delta t \|e_h^n\|_{1,\Omega}^2 + c_2 \Delta t (\Delta t + h^{k+1})^2,$$

where $\alpha = \frac{c_\Omega^2}{1 + c_\Omega^2}$. Then, choose $\gamma = \frac{\alpha\nu}{4}$ and conclude using a discrete Gronwall Lemma. □

6.3.4 Subgrid viscosity approximation

We show in this section how the subgrid viscosity technique developed in §5.5 can be adapted to approximate evolution problems without coercivity.

Let us briefly restate the setting introduced in §5.5.2. We assume that there exists a family of pairs of finite-dimensional subspaces $\{X_H, X_h\}_{\{H,h\}}$ such that:

(i) $X_H \subsetneq X_h \subset V$.
(ii) There is $X_h^H \subset X_h$ such that $X_h = X_H \oplus X_h^H$ and the projection $P_H : X_h \to X_H$ induced by this decomposition is stable in L:

$$\exists c > 0, \ \forall v_h \in X_h, \quad \|P_H v_h\|_L \leq c \|v_h\|_L. \tag{6.92}$$

For all v_h in X_h, set $v_H = P_H v_h$ and $v_h^H = (1 - P_H)v_h$.

(iii) There is $c_i > 0$ such that

$$\forall v_h \in X_h, \quad |v_h|_V \leq c_i H^{-1} \|v_h\|_L. \tag{6.93}$$

Recall that this hypothesis holds for finite elements based on quasi-uniform mesh families provided $c_1 H \leq h \leq c_2 H$.

(iv) There exist a dense subspace of V, say W, a linear interpolation operator $I_H \in \mathcal{L}(W; X_H)$, an integer k, and a constant $c > 0$ such that

$$\forall v \in W, \quad \|v - I_H v\|_L + H\|v - I_H v\|_V \leq cH^{k+1}\|v\|_W. \tag{6.94}$$

(v) There is a subgrid viscosity bilinear form $b_h \in \mathcal{L}(X_h^H \times X_h^H; \mathbb{R})$ satisfying the following continuity and coercivity hypotheses:

$$\forall v_h^H, w_h^H \in X_h^H, \quad \begin{cases} b_h(v_h^H, w_h^H) \leq c_{b2} H |v_h^H|_b |w_h^H|_b, \\ b_h(v_h^H, v_h^H) \geq c_{b1} H |v_h^H|_b^2, \end{cases} \tag{6.95}$$

where the seminorm $|\cdot|_b$ is such that there are $c_{e1} > 0$ and $c_{e2} > 0$ for which

$$\forall v_h^H \in X_h^H, \quad c_{e1} |v_h^H|_V \leq |v_h^H|_b \leq c_{e2} H^{-1} \|v_h^H\|_L. \tag{6.96}$$

(vi) There are $c_a > 0$ and $c_\delta \geq 0$ such that

$$\forall v_h \in X_h, \quad \sup_{\phi_h \in X_h} \frac{a(v_H, \phi_h)}{\|\phi_h\|_L} \geq c_a |v_H|_V - c_\delta \|v_h\|_L. \tag{6.97}$$

All the constants involved in (6.92)–(6.97) do not depend on $\{H, h\}$. See §5.5.2 and §5.5.6 for a discussion of the above hypotheses and for examples of admissible finite elements pairs $\{X_H, X_h\}$.

Remark 6.66. Hypothesis (6.97) is the keystone of the theory that we are going to develop. It is simply the discrete counterpart of the tautology (6.69). In general, (6.69) has no uniform discrete counterpart in the standard Galerkin framework.   □

Set $a_h(u_h, v_h) = a(u_h, v_h) + b(u_h^H, v_h^H)$ and also introduce $a_s(u, v) = \frac{1}{2}(a(u, v) + a(v, u))$. Since a is positive, a_s induces a seminorm on V; let $|\cdot|_{V_s}$ be this seminorm. Assume that $u_0 \in W$ so that u_0 can be approximated by $I_H u_0$. The discrete problem we consider hereafter is the following:

$$\begin{cases} \text{Seek } u_h \in \mathcal{C}^1([0, T]; X_h) \text{ such that} \\ (d_t u_h, v_h)_L + a_h(u_h, v_h) = (f, v_h)_L, \quad \forall v_h \in X_h, \\ u_h(0) = I_H u_0. \end{cases} \tag{6.98}$$

This problem has clearly a unique solution since it is a linear system of ODEs. The major result of this section is the following:

Theorem 6.67. *Under hypotheses* (i)–(vi), *if the solution u to* (6.66) *is in* $C^2([0,T];W)$, *the solution u_h to* (6.98) *satisfies*

$$\|u - u_h\|_{C^0([0,T];L)} + |u - u_h|_{L^2(]0,T[;V_s)} \leq c_1 H^{k+\frac{1}{2}}, \qquad (6.99)$$

$$\frac{1}{\sqrt{T}}\|u - u_h\|_{L^2(]0,T[;V)} \leq c_2 H^k, \qquad (6.100)$$

where the constants c_1 and c_2 are bounded as follows

$$c_1 \leq c\left[H + T\left(1 + T\right)\right]^{\frac{1}{2}} \|u\|_{C^2([0,T];W)}, \qquad c_2 \leq c\left[1 + T\right] \|u\|_{C^2([0,T];W)}.$$

Proof. (1) Set $\eta_h(t) = u(t) - I_H u(t)$ and $e_h(t) = I_H u(t) - u_h(t)$. Note that $u - u_h = \eta_h + e_h$. The definition of $\eta_h(t)$ implies

$$\forall j \in \{0, 1, 2\}, \quad \|\eta_h\|_{C^j([0,T];L)} + H\|\eta_h\|_{C^j([0,T];V)} \leq cH^{k+1}\|u\|_{C^2([0,T];W)}.$$

Subtracting (6.98) from (6.66) yields the consistency equation

$$\forall v_h \in X_h, \quad (d_t e_h, v_h)_L + a(e_h, v_h) - b_h(u_h^H, v_h^H) = -(d_t \eta_h, v_h)_L - a(\eta_h, v_h).$$

Since X_H is invariant under P_H and P_H is linear,

$$u_h^H = u_h - P_H u_h = u_h - I_H u - P_H(u_h - I_H u) = -e_h + P_H e_h = -e_h^H.$$

Therefore, the equation controlling e_h takes the form

$$\forall v_h \in X_h, \quad (d_t e_h, v_h)_L + a_h(e_h, v_h) = -(d_t \eta_h, v_h)_L - a(\eta_h, v_h). \quad (6.101)$$

Proceeding similarly for $d_t e_h$ yields

$$\forall v_h \in X_h, \quad (d_{tt} e_h, v_h)_L + a_h(d_t e_h, v_h) = -(d_{tt} \eta_h, v_h)_L - a(d_t \eta_h, v_h). \quad (6.102)$$

(2) Let us now estimate $e_h(0)$ and $d_t e_h(0)$. It is clear that $e_h(0) = 0$. Moreover, using (6.101) at $t = 0$ yields $\|d_t e_h(0)\|_L \leq \|d_t \eta_h(0)\|_L + \|\eta_h(0)\|_V$. That is to say, at $t = 0$, the following estimates hold:

$$\|e_h(0)\|_L = 0, \qquad \|d_t e_h(0)\|_L \leq cH^k\|u\|_{C^2([0,T];W)}.$$

(3) Let us now estimate $\|d_t e_h\|_{L^\infty(]0,T[;L)}$. Using $d_t e_h$ to test (6.102) yields

$$\tfrac{1}{2}\tfrac{d}{dt}(\|d_t e_h\|_L^2) + a_h(d_t e_h, d_t e_h) \leq (\|d_{tt} \eta_h\|_L + |d_t \eta_h|_V)\|d_t e_h\|_L.$$

Since the bilinear form a_h is positive,

$$\tfrac{d}{dt}(\|d_t e_h\|_L^2) \leq 2(\|d_{tt} \eta_h\|_L + |d_t \eta_h|_V)\|d_t e_h\|_L.$$

Lemma 6.54 implies

$$\|d_t e_h\|_{L^\infty(]0,T[;L)}^2 \leq c(\|d_t e_h(0)\|_L^2 + T^2\|\eta\|_{C^2([0,T];V)}^2).$$

Hence,

$$\|d_t e_h\|_{L^\infty([0,T];L)} \le cH^k(1+T)\|u\|_{C^2([0,T];W)}. \qquad (6.103)$$

(4) To obtain an estimate on e_h, test (6.101) with e_h and use (A.3), yielding

$$\frac{1}{2}\frac{d}{dt}(\|e_h\|_L^2) + a_h(e_h, e_h) \le \|d_t\eta_h\|_L\|e_h\|_L + a(e_h, \eta_h) - 2a_s(e_h, \eta_h)$$
$$\le \|d_t\eta_h\|_L\|e_h\|_L + |e_h|_V\|\eta_h\|_L + \tfrac{1}{2}a_s(e_h, e_h) + 2a_s(\eta_h, \eta_h).$$

Since $a_h(e_h, e_h) \ge a_s(e_h, e_h)$,

$$\frac{d}{dt}(\|e_h\|_L^2) + a_h(e_h, e_h) \le 2\|e_h\|_L\|d_t\eta_h\|_L + c(|e_h|_V + \|\eta_h\|_V)\|\eta_h\|_L. \quad (6.104)$$

Note that the term $|e_h|_V\|\eta_h\|_L$ is not yet controlled. It is precisely at this point that the inf-sup condition (6.97) plays its role:

$$c_a|e_H|_V \le \sup_{\phi_h \in X_h} \frac{a(e_H, \phi_h)}{\|\phi_h\|_L} + c_\delta\|e_h\|_L$$

$$\le \sup_{\phi_h \in X_h} \frac{-(d_t e_h, \phi_h)_L - a_h(e_h^H, \phi_h) - (d_t\eta_h, \phi_h)_L - a(\eta_h, \phi_h)}{\|\phi_h\|_L} + c_\delta\|e_h\|_L$$

$$\le \|d_t e_h\|_L + |e_h^H|_V + c_b|e_h^H|_b + \|d_t\eta_h\|_L + |\eta_h|_V + c_\delta\|e_h\|_L$$

$$\le c(\|d_t e_h\|_L + |e_h^H|_b + \|e_h\|_L + \|d_t\eta_h\|_L + |\eta_h|_V).$$

The bound (6.103) together with the triangle inequality implies

$$|e_h|_V \le |e_H|_V + |e_h^H|_V \le c(|e_h^H|_b + \|e_h\|_L + (1+T)H^k\|u\|_{C^2([0,T];W)}). \quad (6.105)$$

(5) Inserting (6.105) into (6.104) leads to

$$\frac{d}{dt}(\|e_h\|_L^2) + a_h(e_h, e_h) \le 2\|e_h\|_L\|d_t\eta_h\|_L + cH^{2k+1}\|u\|_{C^2([0,T];W)}^2$$
$$+ c'\left[|e_h^H|_b + \|e_h\|_L + (1+T)H^k\|u\|_{C^2([0,T];W)}\right]\|\eta_h\|_L$$
$$\le c\|e_h\|_L(\|\eta_h\|_L + \|d_t\eta_h\|_L) + c'(1+T)H^{2k+1}\|u\|_{C^2([0,T];W)}^2$$
$$+ \tfrac{1}{2}c_{b1}H|e_h^H|_b^2 + c''H^{-1}\|\eta_h\|_L^2.$$

Since $a_h(e_h, e_h) \ge c_{b1}H|e_h^H|_b^2$,

$$\frac{d}{dt}(\|e_h\|_L^2) + a_h(e_h, e_h) \le c\|e_h\|_L\|\eta_h\|_{C^1([0,T];L)} + c'(1+T)H^{2k+1}\|u\|_{C^2([0,T];W)}^2.$$

Lemma 6.54 and the fact that $e_h(0) = 0$ imply

$$\|e_h\|_{C^0([0,T];L)}^2 + \|e_h\|_{L^2(]0,T[;V)}^2 \le cT[1+T]H^{2k+1}\|u\|_{C^2([0,T];W)}^2.$$

This estimate yields

$$\|u - u_h\|_{C^0([0,T];L)} \le cH^{k+\frac{1}{2}}[H + T(1+T)]^{\frac{1}{2}}\|u\|_{C^2([0,T];W)}.$$

(6) To derive an error estimate in the graph norm, use (6.105) as follows:

$$\int_0^T |e_h|_V^2 \le c \int_0^T [|e_h^H|_b^2 + \|e_h\|_L^2 + (1+T)^2 H^{2k} \|u\|_{C^2([0,T];W)}^2]$$

$$\le cT(1+T)^2 H^{2k} \|u\|_{C^2([0,T];W)}^2.$$

The conclusion is now straightforward. $\square$

Remark 6.68.

(i) The norms used to measure the error are exactly those in (6.67)–(6.68). The estimate (6.100) is optimal. The estimate (6.99) is identical to those that can be obtained by the DG technique; see [JoN84].

(ii) Note that when T is large, $c_1 = \mathcal{O}(T)$ and $c_2 = \mathcal{O}(T)$; that is to say, in the most unfavorable case, the error grows linearly with T. $\square$

A singular perturbation result. The subgrid viscosity technique developed in §6.3.4 is tailored for first-order differential operators. In practice, we have often to deal with situations where $B = A + \epsilon D$, where A is a first-order differential operator, and D is a coercive second-order differential operator. From the mathematical point of view, the coercivity of D implies that the evolution equation is parabolic in the sense of Definition 6.5. If ϵ is $\mathcal{O}(1)$, the theory presented in §6.1 fully applies. If ϵ is small, the coercivity is not strong enough to guarantee that the Galerkin approximation is satisfactory, since, in first approximation, one has $B \approx A$. We show in this section that the subgrid viscosity technique can easily be extended to treat this situation.

Use the notation introduced above. In addition to the two Hilbert spaces already considered, L and $D(A) = V$, we introduce a new Hilbert space, say X, with dense and continuous embedding in V. We also introduce a bilinear form $d \in \mathcal{L}(X \times X; \mathbb{R})$, and we assume that there is a seminorm $|\cdot|_X$ in X so that $d(u,v) \le c_d |u|_X |v|_X$ for all u, v in X. In practice, D can be a degenerate elliptic operator. We assume that $a + d$ is coercive with respect to the seminorm $|\cdot|_X$, i.e.,

$$\forall v \in X, \qquad |v|_X^2 \le a(v,v) + d(v,v). \qquad (6.106)$$

Consider the following problem: For $f \in C^1([0,T]; L)$, $\epsilon \ge 0$, and $u_0 \in X$,

$$\begin{cases} \text{Seek } u \text{ in } \mathcal{W}(X, X') \text{ such that} \\ (d_t u, v)_L + a(u,v) + \epsilon d(u,v) = (f,v)_L, & \forall v \in X, \ \forall t \ge 0, \qquad (6.107) \\ (u(0), v) = (u_0, v), & \forall v \in L. \end{cases}$$

Assume that (6.107) is normalized so that $\epsilon \le 1$ and that there is $c > 0$ so that $\|v\|_X \le c(\|v\|_L + |v|_X)$. The consequence of this hypothesis is that problem (6.107) is parabolic and has a unique solution; see Theorem 6.6. In the framework of unsteady advection–diffusion equations, ϵ is the reciprocal of the Péclet number; see §5.4.4.

We now turn our attention to the approximation of problem (6.107). We use the discrete setting of §6.3.4 to construct an approximate solution to (6.107). Introduce a sequence of pairs of finite-dimensional spaces

$\{X_H, X_h\}_{\{H,h\}}$, both conformal in X, and satisfying hypotheses (i)–(vi) stated in §6.3.4. Furthermore, assume that the following inverse inequality holds:

$$|v_h|_X \le cH^{-1}\|v_h\|_L. \tag{6.108}$$

In practice, hypothesis (6.108) means that the norms of X and V involve derivatives of the same order.

Assume $u_0 \in W$ so that $I_H u_0$ is an optimal approximation to u_0. The discrete problem is:

$$\begin{cases} \text{Seek } u_h \text{ in } C^1([0,T]; X_h) \text{ such that, for all } t \ge 0 \text{ and } v_h \in X_h, \\ (d_t u_h, v_h)_L + a(u_h, v_h) + \epsilon d(u_h, v_h) + b_h(u_h^H, v_h^H) = (f, v_h)_L, \\ u_h(0) = I_H u_0. \end{cases} \tag{6.109}$$

Problem (6.109) is well-posed since it is a linear system of ODEs.

Theorem 6.69. *In the framework of the above assumptions, if the solution u to (6.107) is in $C^2([0,T]; W)$, then the solution u_h to (6.109) satisfies*

$$\|u - u_h\|_{C^0([0,T];L)} + \epsilon^{\frac{1}{2}}\|u - u_h\|_{L^2([0,T];X)} \le c_1(T, u)(H^{k+\frac{1}{2}} + \epsilon^{\frac{1}{2}}H^k), \tag{6.110}$$

$$\frac{1}{\sqrt{T}}\|u - u_h\|_{L^2([0,T];V)} \le c_2(T, u)H^k, \tag{6.111}$$

where the constants c_1 and c_2 are bounded from above as follows:

$$c_1 \le c[H + T(1+T)]^{\frac{1}{2}}\|u\|_{C^2([0,T];W)}, \qquad c_2 \le c[1+T]\|u\|_{C^2([0,T];W)}.$$

Proof. Similar to that of Theorem 6.67; see also [Gue01^a]. □

6.3.5 Numerical tests on the subgrid viscosity method

We conclude this section by showing various numerical examples. Since the characteristics-Galerkin and the DG methods are more documented in the literature than the subgrid viscosity technique presented in §6.3.4, we shall restrict ourselves to the latter. We evaluate the performance of the method by testing it on problems of increasing difficulty.

Example 6.70 (One-dimensional advection). Let us first perform convergence tests in space on the one-dimensional linear advection problem $\partial_t u + \partial_x u = 0$ in the domain $\Omega = \,]0, +1[$ with periodic boundary conditions and initial data $u(0, x) = \sin(2\pi x^\alpha)$. The exact solution is $u(x, t) = \sin(2\pi(x - t)^\alpha)$. The problem falls within the framework developed above with $A = \partial_x$, $L = L^2(\Omega)$, and $V = \{v \in L^2(\Omega); \partial_x v \in L^2(\Omega); v_{|x=0} = v_{|x=1}\}$.

We approximate the solution by means of the two-level $\mathbb{P}_1$ setting defined in §5.5.6. Let $\mathcal{T}_h$ be a mesh of Ω. Setting $\beta = 1$, Lemma 5.61 guarantees that the discrete inf-sup condition (6.97) is satisfied with $c_\delta = 0$. We define b_h as in (5.91), i.e.,

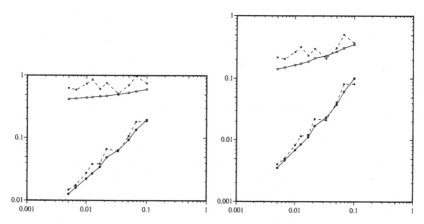

Fig. 6.2. L^2- and H^1-norms with respect to h: stabilized $\mathbb{P}_1$ solution (solid line); Galerkin $\mathbb{P}_1$ solution (discontinuous line); $u_0 = \sin(2\pi x^{0.6})$ (left); $u_0 = \sin(2\pi x^{0.8})$ (right).

$$b_h(u_h^H, v_h^H) = c_b \sum_{K \in T_h} \text{meas}(K)^{\frac{1}{d}} \int_K \nabla u_h^H \cdot \nabla v_h^H, \qquad (6.112)$$

with $d = 1$ for the present test case. Furthermore, we set $c_b = 0.1$. We consider shape-regular families of meshes, but to avoid super-convergence phenomena, each mesh is obtained by a random mapping of the uniform grid with the same number of nodes. To approximate the time-derivative, we use the BDF2 scheme. The time step Δt is chosen small enough to guarantee that the time error is much smaller than the space error, i.e., $\Delta t = 10^{-3}$. The total integration time is $T = 1$, i.e., the flow has crossed the domain once.

Convergence tests with $\alpha = 0.6$ and $\alpha = 0.8$ are reported in Figure 6.2. In both cases, the solution is in $C^0([0, +\infty[; H^1(\Omega))$. We present the L^2- and H^1-norms of the error as a function of h for the stabilized solution and the Galerkin solution. It is clear that the convergence properties of the stabilized solution are superior to that of the Galerkin solution.

To illustrate the convergence difficulties of the Galerkin approximation, we present in Figure 6.3 the stabilized solution and the Galerkin solution on three different meshes ($h = \frac{1}{60}$, $h = \frac{1}{100}$, and $h = \frac{1}{200}$). Note that for the three meshes considered, the Galerkin solution is polluted by spurious oscillations spreading all over the domain, whereas the stabilized solution exhibits some very localized oscillations only in the vicinity of the point where the first-order derivative is singular. $\qquad\qquad\Box$

Example 6.71 (Advection with non-smooth data). We consider the one-dimensional advection equation $\partial_t u + \partial_x u = 0$ in the domain $\Omega =]-1, +1[$ with periodic boundary conditions and initial data (see [ShO89]):

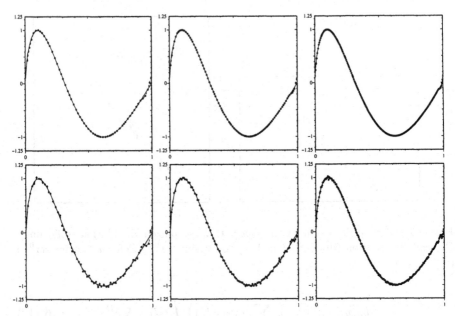

Fig. 6.3. Convergence tests with $u_0 = \sin(2\pi x^{0.6})$: stabilized $\mathbb{P}_1$ solution (top); Galerkin $\mathbb{P}_1$ solution (bottom). From left to right: $h = \frac{1}{60}$; $h = \frac{1}{100}$; and $h = \frac{1}{200}$.

$$
u_0(x) = \begin{cases}
e^{-300(x+0.7)^2} & \text{if } |x + 0.7| \le 0.25, \\
1 & \text{if } |x + 0.1| \le 0.2, \\
\left(1 - \left(\frac{x-0.6}{0.2}\right)^2\right)^{\frac{1}{2}} & \text{if } |x - 0.6| \le 0.2, \\
0 & \text{otherwise.}
\end{cases}
\tag{6.113}
$$

Similarly to time-independent first-order PDEs, the fact that the data u_0 is not in $D(A)$ triggers the *Gibbs phenomenon*; see Examples 5.65 and 5.66 in §5.5.8 for illustrations in a stationary setting. To limit the oscillations, we introduce a *shock-capturing* term based on the nonlinear form $c_h(u_h; \cdot, \cdot)$ defined in (5.93). This term is $\mathcal{O}(h^{k+1})$ when the solution is smooth, i.e., it does not modify the convergence order of the method. The approximate problem is modified accordingly:

$$
\begin{cases}
\text{Seek } u_h \text{ in } C^1([0, T]; X_h) \text{ such that, for all } t \ge 0 \text{ and } v_h \in X_h, \\
(d_t u_h, v_h)_L + a_h(u_h, v_h) + b_h(u_h^H, v_h^H) + c_h(u_h; u_h, v_h) = (f, v_h)_L, \\
u_h(0) = I_H u_0.
\end{cases}
$$

We perform numerical experiments with the two-level $\mathbb{P}_1$ and $\mathbb{P}_2$ settings on three different grids composed of 50, 100, and 200 nodes, respectively. We set $c_b = 0.05$ and $c_{sc} = 0.05$. We use BDF2 with $\Delta t = 10^{-3}$ to march in time so that the time error is negligible with respect to the space error (recall that we want to quantify the space error only). The standard $\mathbb{P}_1$ Galerkin

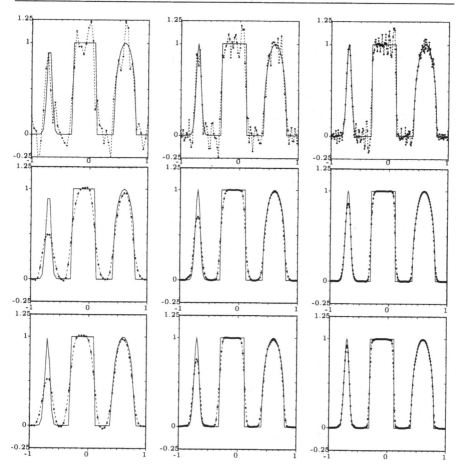

Fig. 6.4. One-dimensional advection problem with rough initial data: standard $\mathbb{P}_1$ Galerkin solution (top); $\mathbb{P}_1$ stabilized solution (center); and $\mathbb{P}_2$ stabilized solution (bottom). From left to right: 50 nodes; 100 nodes; and 200 nodes.

solution computed on the three meshes at $T = 4$ is shown in the top panels of Figure 6.4. It is clear that this type of approximation is useless for engineering purposes. The $\mathbb{P}_1$ and $\mathbb{P}_2$ subgrid viscosity solutions are reported in Figure 6.4. Both $\mathbb{P}_1$ and $\mathbb{P}_2$ approximations exhibit satisfactory convergence behavior when the mesh is refined. □

Example 6.72 (Shock tube problem). We now treat a one-dimensional shock tube problem. Consider an infinitely long tube containing a perfect inviscid gas in two different thermodynamic states on each side of a membrane at $x = 0$. The velocity, the pressure, the density, and the total energy of the gas are denoted by u, p, ρ, and e, respectively. Setting $\phi = (\rho, \rho u, e)$, perfect inviscid fluids are described by the Euler equations:

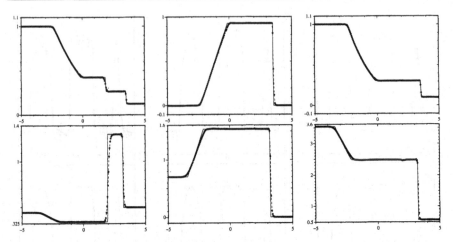

Fig. 6.5. $\mathbb{P}_1$ stabilized solution for the Sod tube at $T=2$ (top) and for the Lax tube at $T=1.3$ (bottom); 200 nodes. From left to right: density; velocity; and pressure.

$$\begin{cases} \partial_t \phi + \partial_x f(\phi) = 0, & \text{for } -\infty < x < +\infty \text{ and } t > 0, \\ \phi(0) = \phi_0, & \text{for } -\infty < x < +\infty, \end{cases}$$

where $f(\phi) = (\rho u, \rho u^2 + p, ue + pu)$. The conservation equations are completed by the state equation $p = (\gamma - 1)(e - \frac{1}{2}\rho u^2)$ with $\gamma = 1.4$.

We consider two test cases corresponding to the initial states:

$$\text{Sod Tube} \begin{cases} u_l = 0, & p_l = 1, & \rho_l = 1, \\ u_r = 0, & p_r = 0.1, & \rho_r = 0.125, \end{cases}$$

$$\text{Lax Tube} \begin{cases} u_l = 0.698, & p_l = 3.528, & \rho_l = 0.445, \\ u_r = 0, & p_r = 0.5, & \rho_r = 0.571, \end{cases}$$

with indices l and r referring to the initial left and right state, respectively. In the literature, the solution described by the first set of initial data is referred to as the Sod solution, and the second is called the Lax solution.

We consider the truncated domain $\Omega =]-5, +5[$. The march in time is performed using the BDF2 scheme with $\Delta t = 5 \times 10^{-3}$. The three evolution equations are solved independently after linearization and second-order extrapolation. More precisely, we use the mass conservation equation to update the density, the momentum conservation equation to update the velocity, and the energy conservation equation to update the total energy. Without subgrid viscosity, the standard Galerkin solution is polluted by strong oscillations. To implement the subgrid viscosity technique, we use the two-level $\mathbb{P}_1$ setting, and stabilize each equation by adding the forms b_h and c_h defined in (6.112) and (5.93), respectively.

Numerical experiments are performed with the two-level $\mathbb{P}_1$ setting on a mesh with 201 nodes. The results of the simulation at $T = 2$ for the Sod

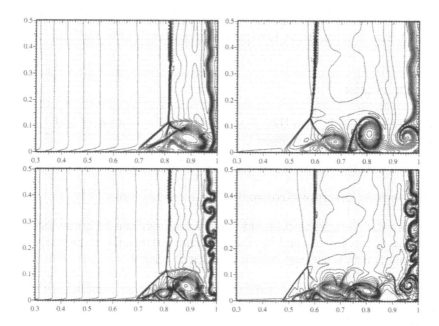

Fig. 6.6. Density isocontours for the shock box problem with a stabilized Galerkin approximation using a two-level $\mathbb{P}_1$ setting: Reynolds 200 (top); Reynolds 1000 (bottom); $t = 0.6$ (left); and $t = 1$ (right).

solution and at $T = 1.3$ for the Lax solution are reported in Figure 6.5. On the two considered solutions, we observe a rarefaction wave on the density distribution, a contact discontinuity, and a shock.   □

Example 6.73 (Compressible Navier–Stokes equations). To further illustrate the capability of the method, we solve a two-dimensional compressible Navier–Stokes problem proposed by Daru and Tenaud [DaT01]. We consider a box $\Omega = \,]0,1[^2$ filled with a viscous ideal gas. A diaphragm situated at $x - \frac{1}{2}$ separates the box into two parts. The fluid is initially at rest and in two different thermodynamic states on each side of the diaphragm. In the left part, we set $\rho_l = 120$ and $p_l = \frac{\rho_l}{\gamma}$, whereas in the right part, we set $\rho_r = 1.2$ and $p_r = \frac{\rho_r}{\gamma}$. The constant γ is set to 1.4. At $t = 0$ the diaphragm is broken. The shock moves to the right of the box, then reflects on the right side. When propagating back to the left, the shock strongly interacts with the boundary layer that it created at the bottom of the box. The interaction produces a λ-shock and a massive separation of the boundary layer.

The solution is assumed to be symmetric with respect to the axis $y = \frac{1}{2}$; as a result, the computational domain is restricted to $\Omega = \,]0,1[\times \,]0,\frac{1}{2}[$. We use a two-level $\mathbb{P}_1$ setting. Two Reynolds numbers are considered: $Re = 200$ and $Re = 1000$. The Prandtl number is 0.73. In Figure 6.6 we show density contours for these two Reynolds numbers at times $T = 0.6$ and $T = 1$. The contour step is $\Delta\rho = 5$, and the contour lines are shown from $\rho = 10$ to $\rho = 120$. The solution shown here compares quite well with [DaT01].   □

6.4 Exercises

Exercise 6.1. Let V be a Hilbert space. Using the notation of §6.1.1, show that $\mathcal{W}(V, V')$ is a Hilbert space.

Exercise 6.2. Prove that the solution sets of (6.2) and (6.4) are identical.

Exercise 6.3 (Discrete Gronwall). Prove Lemma 6.26.

Exercise 6.4. In Lemma 6.54, the main hypothesis can be recast into the form $d_t\phi \le \phi + \frac{1}{4}a^2 + b$ and Gronwall's Lemma 6.9 could be used. Explain why Lemma 6.54 is sharper, especially when T is large.

Exercise 6.5. Consider the sequences of non-negative numbers $\{a^n\}_{n\ge 0}$, $\{f^n\}_{n\ge 1}$, $\{g^n\}_{n\ge 1}$, and $\{h^n\}_{n\ge 1}$. Assume there is $\gamma \ge 0$ such that, for all $n \ge 0$,

$$a^{n+1} \le a^n + \gamma f^{n+1}\sqrt{a^n} + \gamma g^{n+1}\sqrt{a^{n+1}} + \gamma h^{n+1}.$$

Prove that there is c, independent of γ, such that, for all $N \ge 0$ and $1 \le p \le +\infty$,

$$a^N \le c\left(a^0 + (N\gamma)^2 \frac{1}{N^{\frac{2}{p}}}(\|f\|^2_{\ell^p[1,N]} + \|g\|^2_{\ell^p[1,N]}) + (N\gamma)\frac{1}{N^{\frac{1}{p}}}\|h\|_{\ell^p[1,N]}\right).$$

(*Hint*: Fix N, then use $f^{n+1}\sqrt{a^n} \le \frac{1}{N\gamma}a^n + \frac{N\gamma}{4}(f^{n+1})^2$, $\forall n \le N - 1$.)

Exercise 6.6 (Leap-frog). Let $A \in \mathcal{L}(V; L)$ be a maximal monotone operator, $D(A) = V \subset L$ where L is a Hilbert space; see §6.3.1 for notation. Let $u_0 \in V$, let $f \in L^2(]0, T[; L)$, and set $f^n = \frac{1}{\Delta t}\int_{t^{n-1}}^{t^{n+1}} f(t)\,dt$ for $n \ge 1$. Approximate the problem $d_t u + Au = f$, $u(0) = u_0$, by using the so-called *leap-frog scheme*:

$$\begin{cases} u^0 = u_0, \quad u^1 = u_0 - \Delta t(Au_0 - \frac{1}{\Delta t}\int_0^{\Delta t} f(t)\,dt), \\ \frac{1}{2\Delta t}(u^{n+1} - u^{n-1}) + Au^n = f^{n+1}. \end{cases}$$

(i) Prove $\|u^{n+1}\|_L^2 + \|u^n\|_L^2 \le \|u^n\|_L^2 + \|u^{n-1}\|_L^2 + 4\Delta t f^n(\|u^n\|_L^2 + \|u^{n-1}\|_L^2)^{\frac{1}{2}}$.
(ii) Why is this scheme appealing?
(iii) Prove $\|u_{\Delta t}\|_{\ell^\infty([0,T];L)} \le c(\|u_0\|_L + \Delta t\|Au_0\|_L + T\|f\|_{L^2(]0,T[;L)})$.
(iv) Complete the error analysis.

Exercise 6.7 (Explicit Euler). Use the notation of Exercise 6.6. Let $V_h \subset V$ be a finite-dimensional subspace of V. Let $c_i(h) = \max_{v_h \in V_h} \|Av_h\|_L / \|v_h\|_L$. Let $N \geq 1$, set $\Delta t = \frac{T}{N}$ and $t^n = n\Delta t$ for $0 \leq n \leq N$. Let $u_0 \in V$, $f \in L^2(]0, T[; L)$, and set $f^n = \frac{1}{\Delta t} \int_{t^{n-1}}^{t^n} f(t)\, dt$. Consider the *explicit Euler* scheme:

$$\begin{cases} (u_h^0, v_h)_L = (u_0, v_h)_L, & \forall v_h \in V_h, \\ \frac{1}{\Delta t}(u_h^{n+1} - u_h^n, v_h)_L + (Au_h^n, v_h)_L = (f^{n+1}, v_h)_L, & \forall v_h \in V_h. \end{cases}$$

(i) Give a bound on $c_i(h)$ when A is a first-order differential operator and V_h is a finite element space based on a quasi-uniform mesh.

(ii) Prove that if $\Delta t\, c_i(h)^2 \leq 1$, then

$$\|u_h^{n+1}\|_L^2 \leq \|u_h^n\|_L^2 + 2\Delta t\|u_h^{n+1}\|_L^2 + \Delta t\|f^{n+1}\|_L^2.$$

(*Hint*: Use $2p(p - p) = p^2 + (p - q)^2 - q^2$ and $2pq \leq \gamma p^2 + \frac{1}{\gamma}q^2$ for all $\gamma > 0$.)

(iii) Assuming $\Delta t\, c_i(h)^2 \leq 1$, prove the estimate

$$\|u_{h\Delta t}\|_{\ell^\infty([0,T];L)} \leq c(T)(\|f\|_{L^2(]0,T[;L)} + \|u_0\|_L).$$

Exercise 6.8 (Semi-explicit scheme). Let L, V, and X be three Hilbert spaces such that $X \subset V \subset L \equiv L' \subset V' \subset X'$. Let $A \in \mathcal{L}(V; L)$ and $D \in \mathcal{L}(X; X')$ be two operators as described in §6.3.1 and in the last part of §6.3.4. Assume that D is coercive with respect to the X-norm with coercivity constant equal to 1. Let $\mathfrak{c} = \max_{u \in V} \|Au\|_L / \|u\|_X$. Let $u_0 \in X$ and $f \in L^2(]0, T[; X')$. Let $N \geq 1$, set $\Delta t = \frac{T}{N}$, $t^n = n\Delta t$ for $0 \leq n \leq N$, and $f^n = \int_{t^{n-1}}^{t^n} f(t)\, dt$. Let $\alpha > 0$. Approximate the problem $d_t u + \alpha D u + A u = f$, $u(0) = u_0$, by using the scheme

$$\begin{cases} u_{|t=0} = u_0, \\ \frac{1}{\Delta t}(u^{n+1} - u^n) + \alpha D u^{n+1} + A u^n = f^{n+1}. \end{cases}$$

(i) Prove that if $\Delta t \leq \frac{\alpha}{2\mathfrak{c}^2}$, then

$$\|u^{n+1}\|_L^2 + \alpha \Delta t\|u^{n+1}\|_X^2 \leq \|u^n\|_L^2 + \alpha \frac{\Delta t}{2}\|u^n\|_X^2 + \frac{\Delta t}{\alpha}\|f^{n+1}\|_{X'}^2.$$

(ii) Derive the corresponding error estimates.

(iii) Redo questions (i) and (ii) using BDF2. (*Hint*: Consider

$$\begin{cases} u(0) = u_0, \\ \frac{1}{\Delta t}(\frac{3}{2}u^{n+1} - u^n + \frac{1}{2}u^{n-1}) + \alpha D u^{n+1} + A(2u^n - u^{n-1}) = f^{n+1}, \end{cases}$$

and use (6.33).)

Exercise 6.9 (CFL number). Use the notation of Exercise 6.8 and let d be the bilinear form associated with the coercive operator D. Let $X_h \subset X$

be a finite-dimensional subset of X. Let $c_i(h) = \max_{v_h \in X_h} \|v_h\|_X / \|v_h\|_L$. Approximate the problem $d_t u + Au = f$, $u(0) = u_0$ by using the scheme, $\forall v \in X_h$,

$$
\begin{cases}
d(u_h^0, v_h) = d(u_0, v_h), \\
\frac{1}{\Delta t}(u_h^{n+1} - u_h^n, v_h)_L + \alpha(h, \Delta t) d(u_h^n, v_h) + a(u^n, v_h) = \langle f^{n+1}, v_h \rangle_{X', X},
\end{cases}
$$

where $\alpha(h, \Delta t)$ is a so-called artificial viscosity.

   (i) Explain why the above scheme can be more attractive than the implicit Euler scheme (obtained with $\alpha = 0$)?
  (ii) Prove that if $2\frac{\Delta t}{\alpha}(\alpha^2 c_i(h)^2 + c^2) \le 1$, then

$$
\|u^{n+1}\|_L^2 + \alpha \Delta t \|u^{n+1}\|_X^2 \le \|u^n\|_L^2 + \alpha \frac{\Delta t}{2} \|u^n\|_X^2 + 2\frac{\Delta t}{\alpha} \|f^{n+1}\|_{X'}^2.
$$

 (iii) Prove that the above stability condition implies $4\Delta t c\, c_i(h) \le 1$. The constant $\Delta t c\, c_i(h)$ is called the Courant–Friedrichs–Levy (CFL) number. Determine the admissible range for $\alpha(h, \Delta t)$. Show that $4\Delta t c^2$ is an admissible value for α.
  (iv) Derive the corresponding error estimates.

Exercise 6.10. Prove Proposition 6.59. (*Hint*: Write the Taylor expansion of $u(X(x, t^{n+1}; t), t)$ at t^{n+1} with a second-order Lagrange remainder; then use Lemma 6.58(i).)

Exercise 6.11. The goal is to prove the stability inequality in Lemma 6.56.

   (i) Prove

$$
\|U_-^n\|_L^2 + \delta(h)\|d_t U + AU\|_{L^2(]0,t^n[;L)}^2
$$
$$
\le \|u_0\|_L^2 + \delta(h)\|f\|_{L^2(]0,t^n[;L)}^2 + 2\|f\|_{L^2(]0,t^n[;L)}\|U\|_{L^2(]0,t^n[;L)}.
$$

  (ii) Prove, for all $t \in \,]t^{n-1}, t^n[$,

$$
\|U(t)\|_L^2 \le \|U_-^n\|_L^2 + \frac{c}{\delta(h)} \int_t^{t^n} \|U\|_L^2 + \delta(h) \int_t^{t^n} \|d_t U + AU\|_L^2.
$$

      (*Hint*: $\|U\|_L^2 = \|U_-^n\|_L^2 - 2\int_t^{t^n}[U(d_t U + AU) - (U, AU)_L]$ and use an inverse inequality.)
 (iii) Deduce $\|U(t)\|_L^2 \le c(\|U_-^n\|_L^2 + \delta(h) \int_t^{t^n} \|d_t U + AU\|_L^2)$ for $t \in \,]t^{n-1}, t^n[$.
  (iv) Prove, for $t \in \,]0, T[$,

$$
\|U(t)\|_L^2 \le c(\|u_0\|_L^2 + \delta(h)\|f\|_{L^2(]0,T[;L)}^2 + \|f\|_{L^2(]0,T[;L)}\|U\|_{L^2(]0,t[;L)}).
$$

   (v) Conclude.

Part III

Implementation

Data Structuring and Mesh Generation

Since the notion of finite elements is often associated with that of unstructured meshes, ignoring how the latter are constructed and organized sometimes leads to the belief that the finite element method is an approximation technique far more complicated than finite differences. The first objective of this chapter is to demystify mesh generation techniques. The second is to introduce the basic notions addressed in the specialized literature, e.g., [Car97, FrG00, GeB98, KnS93, ThS98].

This chapter is organized into three sections. The first section presents enumeration principles and data structures to represent meshes. The second section describes a general algorithm on which most mesh generators are based. A list of mesh generators available on the World Wide Web is also given. The third section is more specific: it reviews Delaunay triangulations and describes some techniques that can be used to construct such triangulations.

7.1 Data Structuring

7.1.1 The cloud of points

A mesh is simply a cloud of points that are numbered and connected. These points are generated using a geometric reference finite element (see §1.3.2) and are called the *geometric nodes*.

Let N_{geo} be the total number of geometric nodes in the mesh ($N_{\text{geo}} = N_{\text{v}}$ if the geometric finite element is of degree 1). The geometric nodes are numbered from 1 to N_{geo}. This enumeration is said to be *global*. The geometric nodes are defined by their coordinates. These quantities are stored in a double-entry array of size $d \times N_{\text{geo}}$, where d is the space dimension. Henceforth, we denote this array by

$$\text{coord}(1{:}d, 1{:}N_{\text{geo}}),$$

and we say that coord is the *coordinate array* of the mesh. For $1 \leq k \leq d$ and $1 \leq n \leq N_{\text{geo}}$, coord$(k, n)$ is the k-th coordinate of the geometric node whose global index is n.

7.1.2 Connectivity

The geometric nodes of the mesh are organized into elements by means of a *connectivity array*. Let N_{el} be the number of elements in the mesh. We number the elements from 1 to N_{el}. Let $(\widehat{K}, \widehat{P}_{\text{geo}}, \widehat{\Sigma}_{\text{geo}})$ be the geometric reference finite element. Let n_{geo} be the number of Lagrange nodes in this finite element, and let us number these nodes from 1 to n_{geo}. Let K be an arbitrary element in the mesh and let $T_K : \widehat{K} \to K$ be the transformation that maps $\widehat{K}$ to K. We number the Lagrange nodes of K from 1 to n_{geo} so that the transformation $T_K : \widehat{K} \to K$ induced by this numbering (i.e., the transformation for which the numbering in K is the image under T_K of the numbering in $\widehat{K}$) is a $\mathcal{C}^1$-diffeomorphism; see §1.3.2 and Figure 1.12. The enumeration in K is said to be *local*. To materialize the existence of the geometric elements into a data structure, we define a connectivity array, say geo_cnty. We associate with each local node in an element its global index by means of a double-entry array of size $n_{\text{geo}} \times N_{\text{el}}$,

$$\texttt{geo\_cnty}(1 {:} n_{\text{geo}}, 1 {:} N_{\text{el}}).$$

For $1 \leq n \leq n_{\text{geo}}$ and $1 \leq m \leq N_{\text{el}}$, the integer geo_cnty$(n, m)$ is the global index of the n-th node in the m-th element.

To account for boundary conditions, it is necessary to define a connectivity array for the boundary nodes. If the intersection of an element with the domain boundary is a $(d-1)$-manifold, we refer to this intersection as a boundary face. We also call faces the sides of the reference element $\widehat{K}$ and, to simplify, we assume that all the faces of $\widehat{K}$ have the same number of geometric nodes, say $n_{\text{geo}}^{\partial}$. We number all the boundary faces from 1 to N_{f}^{∂}, and we locally number the geometric nodes of each face from 1 to $n_{\text{geo}}^{\partial}$. We associate with each local node in a face its global index by means of a double-entry array of size $n_{\text{geo}}^{\partial} \times N_{\text{f}}^{\partial}$, say

$$\texttt{geo\_cnty\_bnd}(1 {:} n_{\text{geo}}^{\partial}, 1 {:} N_{\text{f}}^{\partial}).$$

For $1 \leq n \leq n_{\text{geo}}^{\partial}$ and $1 \leq m \leq N_{\text{f}}^{\partial}$, the integer geo_cnty_bnd(n, m) is the global index of the n-th geometric node in the m-th face. The array geo_cnty_bnd is called the *boundary connectivity array*.

Example 7.1. For an affine mesh of simplices, $n_{\text{geo}} = d+1$ and $n_{\text{geo}}^{\partial} = d$. The local enumeration in a simplex is illustrated in Figure 7.1. Figure 7.2 presents an example of local and global enumerations. In this example, there are three triangles (i.e., elements) with global indices 56, 213, and 315, and the connectivity array takes the values geo_cnty$(1, 315) = 250$, geo_cnty$(2, 315) =$

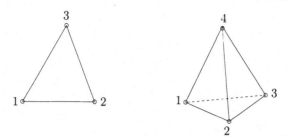

Fig. 7.1. Local enumeration of vertices in a simplex in two and three dimensions.

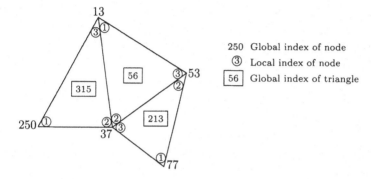

Fig. 7.2. Example of local and global enumerations for two-dimensional simplices.

371, geo_cnty(3, 315) = 13, geo_cnty(1, 56) = 13, geo_cnty(2, 56) = 371,
geo_cnty(3, 56) = 53, etc.   □

7.1.3 Other pointers

The total quantity of information required to fully describe a mesh depends on the application. To illustrate this point, let us go through three examples.

Partitioning of the boundary. Some applications involve a partition of the boundary on which various types of boundary conditions are enforced, e.g., essential or natural. Let I_b be the number of components of the boundary $\partial\Omega$. We number the boundary components from 1 to I_b, i.e., $\partial\Omega = \bigcup_{i=1}^{I_b} \partial\Omega_i$. To know on which component of $\partial\Omega$ a face is located, we define an array of dimension N_f^∂, say

$$\texttt{i\_bc}(1{:}N_f^\partial).$$

For $1 \leq m \leq N_f^\partial$, we assign to $\texttt{i\_bc}(m)$ the index of the boundary component on which the m-th boundary face is located.

Partitioning of the domain. In some applications, the computational domain is partitioned into I_d subdomains, say $\Omega = \bigcup_{i=1}^{I_d} \Omega_i$, and physical parameters are assigned different values, or different equations are solved depending

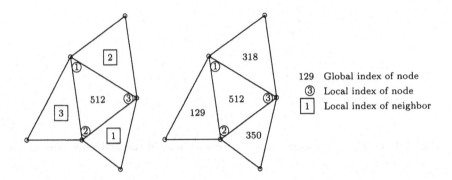

Fig. 7.3. Enumeration of the neighborhood of a two-dimensional simplex: local enumeration (left); global enumeration (right).

on the subdomain considered. To store this information, we define an array of size N_{el}, say

$$\texttt{i\_dom}(1\text{:}N_{\text{el}}).$$

For $1 \leq m \leq N_{\text{el}}$, we assign to $\texttt{i\_dom}(m)$ the index of the subdomain to which the m-th element belongs.

Neighborhood array. It is sometimes useful to identify the elements that are in the close neighborhood of a given element. Let us illustrate this idea with simplices. We call a neighbor of a simplex any of the $(d+1)$ simplices (at most) that share a face (edge in two dimensions) with the simplex in question. To distinguish the various neighbors, we must number them locally. One simple definition consists of assigning to each neighbor of the m-th simplex, say K_m with $1 \leq m \leq N_{\text{el}}$, the local index of the vertex of K_m that does not belong to the common interface. This definition is illustrated in Figure 7.3. To deduce from the local index of a neighbor its global index, we define a double-entry array of size $(d+1) \times N_{\text{el}}$, say

$$\texttt{neigh}(1\text{:}d+1, 1\text{:}N_{\text{el}}).$$

For $1 \leq n \leq d+1$ and $1 \leq m \leq N_{\text{el}}$, $\texttt{neigh}(n, m)$ stores the global index of the n-th neighbor of the m-th simplex. If the face opposite of the n-th node is on the domain boundary, we set $\texttt{neigh}(n, m) = 0$.

Example 7.2. Figure 7.3 illustrates the concept of the neighborhood array. In this example,

$$\texttt{neigh}(1, 512) = 330, \quad \texttt{neigh}(2, 512) = 318, \quad \texttt{neigh}(3, 512) = 129. \qquad \square$$

7.2 Mesh Generators

Mesh generation is an important aspect of finite element simulations since the quality of the results depends on mesh quality. Mesh generation is very often

time-consuming, especially for complex, three-dimensional, industrial configurations. Mesh generators involve two types of tasks. The first task is to represent geometrically the boundary of the domain using mapping techniques,
i.e., parametrization of paths or surfaces. This task is in direct connection
with Computer Assisted Design. The second task consists of meshing lines,
surfaces, and volumes. This section briefly describes the general organization
of mesh generators and gives a list of mesh generators available on the World
Wide Web.

7.2.1 An algorithm for two-dimensional mesh generators

Let us start from the following observations:

1. A two-dimensional domain is entirely defined by its one-dimensional
   boundary.
2. The one-dimensional boundary can be decomposed into its connected components.
3. Each connected component can be partitioned into a union of elementary
   paths.
4. Each elementary path can be assigned two extremities (possibly by cutting
   the path if it is closed). These points are referred to as the vertices of $\partial\Omega$.
5. Each path can be mapped to the interval $[0, 1]$.

By reading in reverse order the above list, a general algorithm for two-
dimensional mesh generators is obtained:

1. Locate the vertices of $\partial\Omega$ and partition $\partial\Omega = \bigcup_{e=1}^{E} \partial\Omega_e$ so that each
   elementary path $\partial\Omega_e$ is limited by two vertices (possibly identical). Here,
   E denotes the total number of elementary paths.
2. Connect the vertices by parameterized paths $\gamma_e : [0, 1] \to \partial\Omega_e, 1 \leq e \leq E$.
3. Divide the paths into small segments as follows: For $1 \leq e \leq E$, let
   $\bigcup_{i=0}^{I-1} [x_{e,i}, x_{e,i+1}]$ be a partition of $[0, 1]$; then, $\bigcup_{i=0}^{I-1} \gamma_e([x_{e,i}, x_{e,i+1}])$ is a
   partition of $\partial\Omega_e$. The partition $\bigcup_{e=1}^{E} \bigcup_{i=0}^{I-1} \gamma_e([x_{e,i}, x_{e,i+1}])$ is meant to be
   the trace of the mesh yet to be generated; it is called the *boundary mesh*.
4. Finally, mesh the interior of the domain by extending the boundary mesh
   to the interior of Ω. A practical example based on an advancing front
   method is presented in §7.3.

Example 7.3. Figure 7.4 illustrates the above algorithm. The domain to be
meshed has a boundary with two connected components. The external component is the union of the three paths AB, BC, and CA. The closed path
composing the internal boundary is transformed into a path that is homeomorphic to a segment by cutting it at D. In conclusion, the boundary of Ω is
decomposed into the union of four paths: $\partial\Omega_1 = AB, \partial\Omega_2 = BC, \partial\Omega_3 = CA$,
and $\partial\Omega_4 = DD$. To mesh Ω, one can proceed as follows:

1. Identify the four boundary vertices A, B, C, and D.

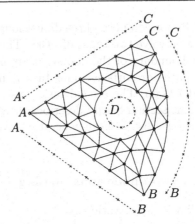

Fig. 7.4. Decomposition of the main steps to mesh a two-dimensional domain with triangles: (i) identification of the boundary paths; (ii) segmentation of the boundary paths; and (iii) triangulation of the interior of Ω by extending the boundary mesh.

2. Define parametrizations for the four paths AB, BC, CA, and DD.
3. Divide the paths into small segments of prescribed size; this yields the boundary mesh.
4. Construct the interior mesh by extending the boundary mesh.   □

7.2.2 An algorithm for cylindrical mesh generators

Numerous industrial applications use either cylinders or domains that are homeomorphic to cylinders. To mesh a cylinder-like domain, a possible strategy consists of meshing first the right section of the cylinder. Then, depending on the elements chosen to mesh the right section, the interior of the domain can be meshed with prisms of triangular or quadrangular base. The algorithm can be decomposed as follows:

1. Define the transformation that maps Ω to a reference cylinder.
2. Mesh the right section of the reference cylinder using the algorithm described in §7.2.1.
3. Translate the mesh of the right section of the reference cylinder along the cylinder generatrix using a prescribed space stepping. Repeat the operation until the reference cylinder is entirely meshed. This yields the reference mesh.
4. Define the mesh of Ω to be the image of the reference mesh by the inverse of the transformation that maps Ω to the reference cylinder.

Example 7.4. Figure 7.5 shows the mesh of a tank obtained with the above algorithm. Note that some cells have been removed from the cylinder to account for pipes that penetrate partially into the tank.   □

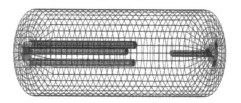

Fig. 7.5. Mesh of a tank. Courtesy of S. Chapuliot and J.-P. Magnaud (CEA).

7.2.3 An algorithm for three-dimensional mesh generators

The algorithm presented in §7.2.1 extends to three dimensions. As in two dimensions, the algorithm is deduced from the geometric description of three-dimensional domains:

1. A three-dimensional volume is entirely defined by its two-dimensional boundary.
2. The two-dimensional boundary can be decomposed into its connected components.
3. Each connected component can be decomposed into a union of elementary surfaces with edges, say $\partial\Omega = \bigcup_{e=1}^{E} \partial\Omega_e$. Here, E denotes the total number of such surfaces.
4. For each elementary surface with edges, there is a mapping $\gamma_e : \partial\Omega_e^{2D} \subset \mathbb{R}^2 \to \partial\Omega_e$, where $\partial\Omega_e^{2D}$ is a domain in $\mathbb{R}^2$.
5. Apply the algorithm described in §7.2.1 to generate a geometric description of each two-dimensional domain $\partial\Omega_e^{2D}$, $1 \le e \le E$.

An algorithm to mesh a three-dimensional domain is obtained by reading the above list from bottom to top:

1. For each two-dimensional domain $\partial\Omega_e^{2D}$, $1 \le e \le E$, construct a mesh by applying the algorithm presented in §7.2.1. Let $\mathcal{T}_{h,e}^{2D}$ be the mesh in question.
2. A mesh for $\partial\Omega_e$, $1 \le e \le E$, is defined to be $\mathcal{T}_{h,e}^{\partial} = \gamma_e(\mathcal{T}_{h,e}^{2D})$.
3. The union of the meshes of the elementary surfaces, $\bigcup_{e=1}^{E} \mathcal{T}_{h,e}^{\partial}$, yields the boundary mesh.
4. The interior of Ω is meshed by extending the boundary mesh.

Example 7.5. The above algorithm is illustrated in Figure 7.6. The domain to be meshed is a cone; its boundary is connected but has no edges. It can be decomposed into two simpler surfaces: the base and the lateral surface. The base is homeomorphic to a disk of boundary PP. By cutting the lateral surface along the segment PQ, the surface thus created is homeomorphic to a triangle. A boundary mesh is obtained by meshing the disk enclosed within the path $\partial\Omega_1^{2D} = PP$ and the triangle $\partial\Omega_2^{2D} = PQP$. Finally, the interior of the cone is meshed by extending the boundary mesh. $\square$

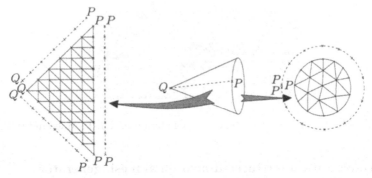

Fig. 7.6. Decomposition of the steps to mesh a three-dimensional domain: (i) identify the boundary vertices; (ii) identify the boundary paths; (iii) partition the boundary paths; (iv) triangulate the two-dimensional elementary surfaces by extending the partitioning of the edges; (v) compute the images of the two-dimensional triangulations; and (vi) mesh the interior of Ω by extending the boundary mesh.

7.2.4 Dividing cuboids and prisms into tetrahedra

When meshing a three-dimensional domain with tetrahedra, it may sometimes be easier to mesh it first with cuboids or with triangular prisms, and then to

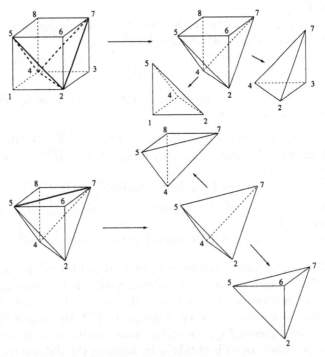

Fig. 7.7. Dividing a cuboid into five tetrahedra. First subdivision (top): two tetrahedra and a heptahedron. Second subdivision (bottom): the heptahedron is divided into three tetrahedra. The cutting lines are materialized by thick lines.

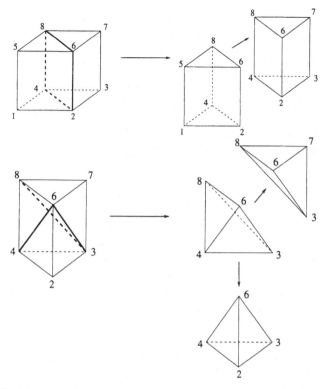

Fig. 7.8. Dividing a cuboid into six tetrahedra. The first subdivision yields two prisms (top). Second subdivision (bottom): divide each prism into three tetrahedra. The cutting lines are materialized by thick lines.

divide the cuboids or prisms into tetrahedra. In this section, we present two simple strategies to divide cuboids and prisms into tetrahedra.

Division into five tetrahedra. A cuboid can be divided into five tetrahedra as shown in Figure 7.7. The vertices of the cuboid are numbered from 1 to 8. A first subdivision yields two tetrahedra with vertices (1245) and (2347) plus a heptahedron. A second subdivision of the heptahedron yields three tetrahedra whose vertices are (2756), (4578), and (2457).

Division into six tetrahedra. A cuboid can also be divided into six tetrahedra as shown in Figure 7.8. A first subdivision yields two triangular prisms. Then, each triangular prism is divided into three tetrahedra. For instance, Figure 7.8 shows how to divide the prism (234678) into the tetrahedra (2346), (3867), and (3846).

7.2.5 Where to find a mesh generator?

Generating a mesh by hand each time one needs to perform a simulation in a given domain is out of the question. It is also unreasonable to try to program a mesh generator from scratch, unless one is already a skillful craftsman in this field. Hence, before performing any computation, the first task of an engineer is either to acquire a mesh generator or to subcontract the generation of a mesh of the domain for which a simulation is required.

A large number of robust two-dimensional mesh generators are freely accessible on the World Wide Web. Some efficient three-dimensional mesh generators are also freely accessible. Many are commercially distributed, but some can be freely acquired for academic purposes by special agreement with the developers. Mesh generators accessible on the World Wide Web include the following:

(i) **Mefisto**: This software has been developed by researchers of the Laboratoire d'Analyse Numérique of Université Paris VI (A. Perronnet, P. Joly, C. Doursat et al.). It is free and contains downloadable executable modules. The software consists of a two-dimensional and a three-dimensional mesh generator plus a set of finite element solvers for thermal and elasticity problems. It is accessible at the following address:

   http://www.ann.jussieu.fr/~perronne/mefisto.gene.html

(ii) **EMC2**: This acronym stands for *Editeur de Maillage et de Contours en deux dimensions* [Two-dimensional Mesh and Contour Editor]; **Bamg**: two-dimensional anisotropic mesh generator, acronym for Bidimensional Anisotropic Mesh Generator; **GHS3D**: powerful mesh generator for tetrahedral meshes. It is commercially distributed under the name **TetMesh**; **BL2D**: plane isotropic or anisotropic mesh generator. All these mesh generators have been developed at INRIA (P.L. George, F. Hecht, H. Borouchaki et al.) and can be found at the following address:

http://www-rocq1.inria.fr/gamma/cdrom/projs/gamma/sofware-eng.htm

(iii) Finally, an almost exhaustive list of either free or commercial mesh generators can be found at the following addresses:

http://www-users.informatik.rwth-aachen.de/~roberts/software.html

   or

             http://www.andrew.cmu.edu/user/sowen/mesh.html

where an updated list of literature on the subject can also be found.

7.3 Example: Delaunay Triangulations

Delaunay triangulations are endowed with interesting regularity properties; see, e.g., [Raj94, GeB98]. (Recall that the word triangulation refers to affine simplicial meshes in any space dimension; see Definition 1.53.) If the aspect ratio of an element is defined to be the ratio of the radius of the circumscribed sphere to that of the inscribed sphere, then, Delaunay triangulations are guaranteed to contain tetrahedra (triangles in two dimensions) of moderate aspect ratio. This fact is important since the interpolation properties of finite elements may not be optimal if the aspect ratio of the elements is too large; see

§1.5 and Exercise 7.12. Note, however, that for problems in which the Hessian of the exact solution is ill-conditioned (e.g., in the presence of sharp fronts), anisotropic meshes can be more computationally effective. For an introduction to the subject, see, e.g., to Habashi et al. [HaD00, AiB02, DoV02].

7.3.1 Definitions and elementary properties

Definition 7.6 (Delaunay triangulation). *A triangulation T is said to be Delaunay if, for every element K in T, the interior of the circumscribed sphere does not contain any vertex of the triangulation.*

In the specialized literature, this property is often referred to as the "empty circumcircle property." A sufficient condition for a triangulation to be Delaunay is stated in the following:

Proposition 7.7. *If all the simplices of a triangulation T contain the center of the circumscribed sphere, this triangulation is Delaunay.*

Proof. Left as an exercise. □

The following lemma due to Delaunay proves very useful in applications:

Lemma 7.8 (Delaunay). *If the empty circumcircle property is satisfied for every pair of simplices having a face in common, then T is a Delaunay triangulation.*

Proof. See [Del34] or [GeB98]. □

Many other properties hold in arbitrary dimensions, but to remain within the scope of this book, we only quote, without proof, two interesting properties of two-dimensional Delaunay triangulations.

Proposition 7.9. *A two-dimensional triangulation in which all the angles are acute is Delaunay.*

Proof. Left as an exercise. □

To state the second property, we introduce an order relation on triangulations. Let $\mathcal{P}$ be a set of points (vertices) in $\mathbb{R}^2$. We call a triangulation of $\mathcal{P}$ any set T of d-simplices (for example, tetrahedra in three dimensions) such that (see Definition 1.53):

(i) The set of vertices of T is $\mathcal{P}$.
(ii) The interiors of the simplices do not intersect each other.
(iii) The union of the simplices is the convex hull of $\mathcal{P}$.
(iv) Every d-simplex intersects $\mathcal{P}$ only at its vertices.

The set of all such triangulations is denoted by $\mathbb{T}_{\mathcal{P}}$. For $\mathcal{T} \in \mathbb{T}_{\mathcal{P}}$, denote by N_{el}, N_v, N_{ed}, and N_{ed}^{∂} the number of elements, the number of vertices, the number of edges, and the number of boundary edges, respectively. The integer N_v (i.e., the cardinal number of $\mathcal{P}$) does not depend on $\mathcal{T} \in \mathbb{T}_{\mathcal{P}}$. This holds also for N_{ed}^{∂}. The Euler relations (see Lemma 1.57), $N_{el} - N_{ed} + N_v = 1$ and $2N_{ed} - N_{ed}^{\partial} = 3N_{el}$, imply that N_{ed} and N_{el} do not depend on $\mathcal{T} \in \mathbb{T}_{\mathcal{P}}$. In conclusion, the number of elements, N_{el}, and the number of angles in the triangulation, $3N_{el}$, are independent of $\mathcal{T} \in \mathbb{T}_{\mathcal{P}}$.

For $\mathcal{T} \in \mathbb{T}_{\mathcal{P}}$, denote by $\alpha(\mathcal{T}) = (\alpha_1(\mathcal{T}), \dots, \alpha_{3N_{el}}(\mathcal{T}))$ the list of all the angles of the triangulation arranged in non-decreasing order. Similarly, denote by $r(\mathcal{T}) = (r_1(\mathcal{T}), \dots, r_{N_{el}}(\mathcal{T}))$ the list of the radii of the circumcircles of the triangulation, also arranged in non-decreasing order. Since N_{el} is independent of $\mathcal{T}$, the following total order relations in $\mathbb{T}_{\mathcal{P}}$ can be defined:

Definition 7.10. *Let $\mathcal{T}_1, \mathcal{T}_2 \in \mathbb{T}_{\mathcal{P}}$. Then, $\alpha(\mathcal{T}_1) < \alpha(\mathcal{T}_2)$ if and only if either there is an index j, $1 < j \leq 3N_{el}$, such that $\alpha_i(\mathcal{T}_1) = \alpha_i(\mathcal{T}_2)$ for $i \leq j - 1$ and $\alpha_j(\mathcal{T}_1) < \alpha_j(\mathcal{T}_2)$, or $\alpha_1(\mathcal{T}_1) < \alpha_1(\mathcal{T}_2)$. The total order relation $r(\mathcal{T}_1) < r(\mathcal{T}_2)$ in $\mathbb{T}_{\mathcal{P}}$ is defined similarly.*

The following remarkable property holds:

Proposition 7.11. *Among all the possible triangulations in $\mathbb{T}_{\mathcal{P}}$, a Delaunay triangulation maximizes $\alpha(\mathcal{T})$ and minimizes $r(\mathcal{T})$.*

Proof. See Exercise 7.10. □

In particular, we see that the first angle in $\alpha(\mathcal{T})$ (i.e., the smallest angle) is maximized. Unfortunately, this max-min angle property does not hold in three dimensions.

7.3.2 The Bowyer–Watson algorithm

There are many algorithms to construct Delaunay triangulations. The most used practical algorithm, applicable in any space dimension, is the Bowyer–Watson algorithm; see [Bow81, Wat81]. This algorithm constructs Delaunay triangulations by induction. It adds points sequentially into an existing Delaunay triangulation, starting usually from a very simple triangulation (e.g., one large triangle or rectangle) enclosing all the points to be triangulated.

Let $\mathcal{P}_i$ be a set of points in $\mathbb{R}^d$ and let $\mathcal{T}_i$ be a Delaunay triangulation of $\mathcal{P}_i$. Let P_{i+1} be a point in the convex hull $\text{conv}(\mathcal{P}_i)$ not belonging to the set $\mathcal{P}_i$. To insert the new point P_{i+1} into $\mathcal{T}_i$, the algorithm proceeds as follows:

1. Find all the existing simplices whose circumcircle contains the new point. Denote by $\mathcal{C}_{P_{i+1}}$ the union of the simplices in question. It is possible to show that $\mathcal{C}_{P_{i+1}}$ is star-shaped with respect to P_{i+1}.
2. Delete from $\mathcal{T}_i$ the simplices of $\mathcal{C}_{P_{i+1}}$.

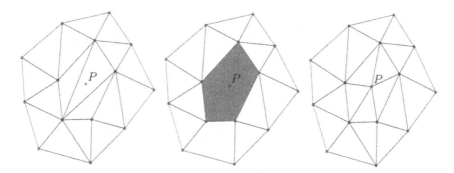

Fig. 7.9. Bowyer–Watson algorithm: insertion of point P (left); construction of the cavity $\mathcal{C}_P$ (shaded area, center); reconnection to form $\mathcal{B}_P$ (right).

3. Define $\mathcal{B}_{P_{i+1}}$ to be the set of the simplices constructed by joining the new point to all the vertices on the boundary of $\mathcal{C}_{P_{i+1}}$. This is always possible since $\mathcal{C}_{P_{i+1}}$ is star-shaped; see Figure 7.9.
4. Add $\mathcal{B}_{P_{i+1}}$ to $\mathcal{T}_i$.

This construction is based on the following:

Theorem 7.12. *The triangulation* $\mathcal{T}_{i+1} = (\mathcal{T}_i \backslash \mathcal{C}_{P_{i+1}}) \cup \mathcal{B}_{P_{i+1}}$ *is Delaunay.*

Proof. See [GeB98]; the proof is based on Delaunay's Lemma.   □

The Bowyer–Watson algorithm is the workhorse of most Delaunay mesh generators. Its complexity, $\mathcal{O}(N_v \log N_v)$, is quasi-optimal since the construction of the cavity $\mathcal{C}_{P_{i+1}}$ and the reconnections to form $\mathcal{B}_{P_{i+1}}$ are local operations.

7.3.3 Triangulating a two-dimensional domain

Let Ω be a polygonal domain in $\mathbb{R}^2$ (Ω is not assumed to be convex). The objective of this section is to show how to construct a Delaunay triangulation of Ω. The technique proceeds in three steps:

1. Form an initial triangulation of a box containing the domain and deduce from it a boundary mesh.
2. Construct a background mesh on which a user-defined function controlling locally the size of the triangles is specified.
3. Sequentially insert points into the domain using the Bowyer–Watson algorithm.

Step 1 (constructing the boundary mesh). Let $\mathcal{P}_{I_0} = \{P_i; 1 \leq i \leq I_0\}$ be a set of points belonging to the boundary $\partial\Omega$ of Ω. These points are created by the user when discretizing the set of elementary paths composing $\partial\Omega$ so that $\partial\Omega$ is the union of the segments P_iP_{i+1}; see Figure 7.10.

To apply the Bowyer–Watson algorithm, it is necessary to work in a convex domain. To this end, a simple convex polygon containing all the points P_i,

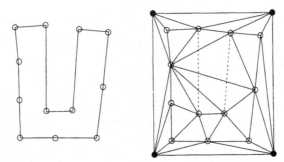

Fig. 7.10. Boundary triangulation of a non-convex domain: boundary and boundary points (left); boundary triangulation (right). Note that two boundary edges are missing. The four supplementary vertices used to embed Ω into a rectangle are shown in black.

$1 \le i \le I_0$, is introduced. Such a polygon can be, for instance, a rectangle large enough to contain the set of points $\mathcal{P}_{I_0}$. A first Delaunay triangulation, $\mathcal{T}_0$, is obtained by dividing the rectangle along one of its diagonals. Then, all the boundary points are inserted into this triangulation using the Bowyer–Watson algorithm.

The Bowyer–Watson algorithm does not guarantee that the boundary edges are preserved once all the points in $\mathcal{P}_{I_0}$ are inserted; see Figure 7.10. To ensure that $\partial\Omega$ is a union of edges of triangles belonging to $\mathcal{T}_{I_0}$, the following steps must be followed:

1. Find the missing edges $P_i P_{i+1}$.
2. Find the polygonal cavity $\mathcal{T}_{i,i+1}$ composed of all the triangles which intersect the open segment $P_i P_{i+1}$, i.e., excluding P_i and P_{i+1}.
3. Retriangulate the cavity $\mathcal{T}_{i,i+1}$ so that $P_i P_{i+1}$ is an edge of the new triangulation. For this purpose, one of the simplest techniques is to flip edges as shown in Figure 7.11.

After the boundary is reconstructed, there is no guarantee that the new triangulation is still Delaunay. However, if the segments composing the boundary $\partial\Omega$ are sufficiently small, the new triangulation is not far from being Delaunay. The terminology *quasi-Delaunay triangulation* is often employed in these circumstances. Another possibility of reconstructing the boundary is to insert new boundary points on the missing edges. It can be proved that, by repeated insertion of new points at the middle of the missing edges, a Delaunay triangulation preserving boundary edges is eventually constructed [Reb93].

Step 2 (defining a control function). The next step consists of adding internal points to the triangulation according to quality criteria set by the user. One way of measuring the quality of the mesh is to compare the diameter of the triangles to a user-defined control function $H(X)$ specifying the desired size of the triangles as a function of the position X in the domain. Since the boundary triangulation $\mathcal{T}_{I_0}$ is available, a continuous $\mathbb{P}_1$ approximation of the

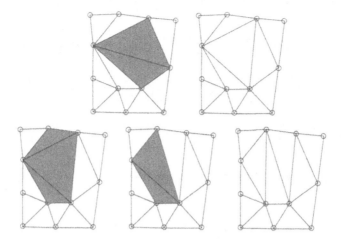

Fig. 7.11. Reconstruction of the boundary of the domain shown in Figure 7.10. The first missing edge is localized and the triangles which intersect this edge are flagged (shaded zone in top left and bottom left panels). The first missing edge is recovered after flipping one edge only (top row), whereas flipping two edges is required to recover the second missing edge (bottom row).

control function $H(X)$ is well-defined once the user specifies its value at the points $\{P_i; 1 \le i \le I_0\}$.

It happens quite often that in the neighborhood of particular points in the domain the user wants to specify the size of the triangles (i.e., to refine or unrefine locally). This is accomplished by specifying an additional set of points $\{P_i; I_0+1 \le i \le I_1\}$ together with the values of the control function at these points $\{H(P_i); I_0+1 \le i \le I_1\}$. The new points are inserted by means of the Bowyer–Watson algorithm and, after possibly flipping some edges, a new Delaunay triangulation T_{I_1} is constructed on which a new $\mathbb{P}_1$ approximation of $H(X)$ is defined. The T_{I_1} triangulation is called the *background mesh*. Its only purpose is to define a $\mathbb{P}_1$ approximation of the *control function $H(X)$*.

Step 3 (inserting points). Starting from T_{I_1}, there are many techniques to construct a Delaunay triangulation with triangles of appropriate size. We henceforth restrict ourselves to the so-called *advancing front method*. The material presented in this section is adapted from Rebay [Reb93].

First, the triangles of T_{I_1} are labeled to be internal or external depending on whether they are inside Ω or outside. Recall that the presence of external triangles results from the extension of Ω into a rectangular box so that the domain to be triangulated is convex. Then, internal triangles are further labeled to be either accepted or refused, depending on whether the radius of their circumcircle is smaller than $H(C_K)$, where C_K is the center of the circumcircle of the triangle K. The class of refused triangles is again divided into those that are active and those that are not. An active triangle is a refused triangle

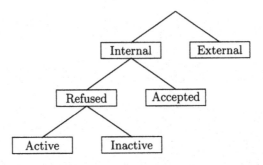

Fig. 7.12. Triangle hierarchy.

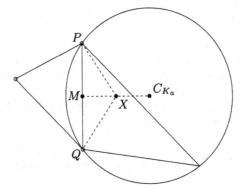

Fig. 7.13. Insertion technique: the accepted (or external) triangle is on the left; the active triangle K_a is on the right; X is the new point to be inserted.

that shares an edge with an accepted triangle or with an external triangle. An inactive triangle is a refused triangle that is not active; see Figure 7.12.

Let us now describe how internal points are inserted. Let K_a be an active triangle. Among the accepted neighbors of K_a, let us choose the one whose edge shared with K_a is the smallest; see Figure 7.13. Let PQ be the edge in question, let M be its middle, and let C_{K_a} be the center of the circumcircle of K_a. Set $p = \|PM\|$ and $q = \|C_{K_a}M\|$. The objective is to insert a new point X on the segment MC_{K_a}. Since X is inside the circumcircle of K_a, the triangle K_a will be deleted by the Bowyer–Watson algorithm. As a result, it is expected that a refused triangle will be replaced by a triangle of better aspect ratio. To set the position of the point X, it is sufficient to specify the radius of the circumcircle of the triangle PQX, say ρ_X, since the center of the circumcircle in question is necessarily on the line MX. A suitable value for ρ_X could be $\rho_M = H(M)$, but since any circle passing through P and Q necessarily has a radius larger than p, any acceptable radius must be larger than $\max(\rho_M, p)$. Moreover, to force X to be on the segment MC_{K_a}, it is necessary to take $\|MX\| \leq q$; that is, $\rho_X \leq \frac{p^2+q^2}{2q}$. In practice, it is advisable to choose

$$\rho_X = \min \left[\max(\rho_M, p), \frac{p^2 + q^2}{2q} \right]. \tag{7.1}$$

Denoting by n the unit vector $\frac{MC_{K_a}}{\|MC_{K_a}\|}$, the new point is defined to be

$$X = M + \left[\rho_X + \sqrt{\rho_X^2 - p^2} \right] n. \tag{7.2}$$

This point is inserted into the triangulation by means of the Bowyer–Watson algorithm and by testing only the internal triangles. This is to ensure that the newly inserted points are always inside Ω and to avoid losing boundary edges. Note that nothing guarantees that the triangle PQX will be part of the new triangulation; however, it is guaranteed that the new triangulation will contain two edges PX and QX having a length consistent with the control function. Likewise, the accepted triangle sharing the edge PQ with K_a (the left triangle in Figure 7.13) may not be part of the new triangulation since X may belong to the interior of its circumcircle.

Once X has been inserted and the triangulation regenerated, the status of the new triangles and their edge neighbors is reexamined. The new triangles are accepted or refused according to the following criterion: Let K_i be a new triangle and let M_i be the middle of the edge opposite to X; K_i is accepted if

$$\rho_{K_i} \le \tfrac{3}{2} H(M_i),$$

where the empirical factor $\frac{3}{2}$ is user-dependent. Then, the refused triangles, either newly created or old, are given an active or inactive status.

In summary, the advancing front method proceeds as follows:

1. Divide all the triangles into internal and external. Divide the internal triangles into accepted and refused on the basis of their circumcircle radius. Divide the refused triangles into active and inactive.
2. Order the active triangles with respect to their circumcircle radius.
3. Pick the active triangle with the largest circumcircle radius. Insert a point as described above.
4. Regenerate the triangulation by means of the Bowyer–Watson algorithm.
5. Divide the new triangles into accepted and refused. Divide the new refused triangles into active and inactive.
6. Divide the refused triangles that share an edge with the newly created triangles into active and inactive.
7. If the list of active triangles is empty stop, otherwise go to step 3.

The main advantage of the advancing front method is to create at each new insertion one or several suitable triangles. Furthermore, an interesting characteristic of this method is that it progressively replaces refused triangles by accepted ones. The process starts from the boundary and moves progressively into the interior of the domain in a way quite similar to that of a propagating front.

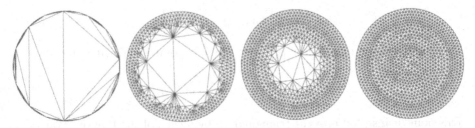

Fig. 7.14. Triangulation of a circle by the advancing front method. Note the propagation of the front separating the accepted and refused triangles. Boundary triangulation (left panel), and final mesh (right panel).

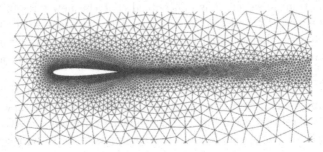

Fig. 7.15. Triangulation of the NACA0012 profile. Note the refinement near the airfoil and in the wake enforced by means of a control function.

Example 7.13. The propagating front phenomenon is illustrated in Figure 7.14. The domain is a circle. The figure shows various stages of the triangulation process. The left panel shows the boundary triangulation. The final triangulation is shown in the right panel. □

Example 7.14. An example of a mesh produced with a non-uniform control function is presented in Figure 7.15. The domain to be meshed is the outside of the NACA0012 profile. The refinement near the airfoil and in the wake is enforced by a control function and the use of a background mesh. □

For complements on other techniques and on the three-dimensional generalization of the advancing front method, the reader is referred to [Car97, GeB98, FrG00].

7.4 Exercises

Exercise 7.1. Consider a two-dimensional triangulation of a domain Ω. Given a point $m \in \Omega$ with coordinates (x, y), write an algorithm using the arrays coord and geo_cnty to find the triangle to which the point m belongs. Describe a modification of the algorithm that also uses the array neigh. Which algorithm do you expect to be more computationally effective?

Exercise 7.2. Let Ω be a two-dimensional polygon and let $\mathcal{T}_h$ be a triangulation Ω. Let $P^1_{c,h}$ and $P^2_{c,h}$ be the spaces spanned by piecewise linears and quadratics on this mesh, respectively; see (1.76).

(i) Express $\dim(P^1_{c,h} \cap H^1_0(\Omega))$ and $\dim(P^2_{c,h} \cap H^1_0(\Omega))$ in terms of N_v and N^i_{ed}.

(ii) Assume that the ratio N^∂_{ed}/N_v goes to zero when N_v becomes large. How does the ratio $\dim(P^2_{c,h} \cap H^1_0(\Omega))/\dim(P^1_{c,h} \cap H^1_0(\Omega))$ behave when N_v is large?

Exercise 7.3. Let $\Omega = \,]0,1[^2$, let $N \geq 2$ be an integer, and set $h = \frac{1}{N-1}$. Consider the quadrangular, uniform mesh of Ω with vertices a_{ij} having coordinates $((i-1)h, (j-1)h)$ for $1 \leq i,j \leq N$. Divide each quadrangular cell into two triangles by cutting it along its main diagonal; see left panel of Figure 7.16 for an example with $N = 4$. Write an algorithm to generate the arrays coord, geo_cnty, geo_cnty_bnd, and neigh. The vertices are numbered so that a_{ij} is assigned to the global number $(j-1)N + i$ for $1 \leq i,j \leq N$.

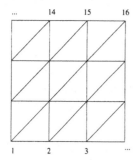 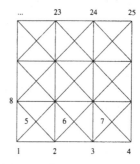

Fig. 7.16. Examples of quadrangular, uniform meshes.

Exercise 7.4. Redo Exercise 7.3 if the quadrangular cells are cut into four triangles along both diagonals; see right panel of Figure 7.16 for an example with $N = 4$.

Exercise 7.5. Let pqr be a triangle. Denote by h_p the distance from p to the line passing through q and r (and introduce a similar notation for h_q and h_r). Prove that the radius of the circumcircle R satisfies

$$R = \frac{\|p-q\|\,\|p-r\|}{2h_p} = \frac{\|q-p\|\,\|q-r\|}{2h_q} = \frac{\|r-p\|\,\|r-q\|}{2h_r}.$$

Exercise 7.6. Prove Propositions 7.7 and 7.9.

Exercise 7.7. Let $\mathcal{P}$ be a set of points in $\mathbb{R}^2$ containing at least three points and let $\mathcal{T}$ be a Delaunay triangulation of the convex hull of $\mathcal{P}$. Let p be any point in $\mathcal{P}$. Assume that there is a unique point q in $\mathcal{P}$ such that $\|p - q\|$ minimizes the Euclidean distance from p to $\mathcal{P}\backslash\{p\}$. Show that pq must be an edge of the triangulation $\mathcal{T}$. What happens if q is not unique?

Exercise 7.8. Adopt the notation of §7.3.2. Let $C_{P_{i+1}}$ be the cavity created by the Bowyer–Watson algorithm to insert P_{i+1}. Show that $C_{P_{i+1}}$ is star-shaped with respect to P_{i+1}.

Exercise 7.9. Let $\mathcal{P}$ be a set of points in $\mathbb{R}^2$ containing at least three points. Assume that there are no three collinear points and no four cocircular points. Two points p and q are said to form a Delaunay segment if there is a disk D having p and q on its boundary and such that $\mathcal{P} \cap D = \{p, q\}$. Let $\mathcal{T}$ be a Delaunay triangulation of $\mathcal{P}$.

   (i) Show that every edge of $\mathcal{T}$ is a Delaunay segment.
  (ii) Show that every Delaunay segment is an edge of $\mathcal{T}$.
 (iii) Prove that $\mathcal{T}$ is unique.
 (iv) What happens if $\mathcal{P}$ contains cocircular points?

Exercise 7.10. Let $\mathcal{P}$ be a set of points in $\mathbb{R}^2$ containing at least four points.

   (i) Consider four points in convex position (this means that for any triangle constructed using three of these points, the fourth point lies outside the triangle) and the two possible triangulations of these points. Using Definition 7.10 to order the angles of the triangulation, show that a Delaunay triangulation of these four points has the largest angle sequence.
  (ii) Let $\mathcal{T}$ be a Delaunay triangulation of $\mathcal{P}$. Prove that $\mathcal{T}$ maximizes the angle sequence.

Exercise 7.11. Let $\mathcal{P}$ be a set of points in $\mathbb{R}^2$ and let $\mathcal{T}$ be a Delaunay triangulation of $\mathcal{P}$. The *Gabriel Graph* is defined to be the set of edges pq, $p, q \in \mathcal{P}$, such that

$$\forall x \in \mathcal{P}\backslash\{p, q\}, \qquad \|p - q\|^2 < \|p - x\|^2 + \|q - x\|^2.$$

   (i) Prove that the Gabriel graph belongs to $\mathcal{T}$.
  (ii) Show that the Gabriel graph is connected.

Exercise 7.12. Show that the stiffness matrix of the Laplacian on a two-dimensional Delaunay mesh with $\mathbb{P}_1$ Lagrange finite elements is an *M-matrix*, i.e., all its off-diagonal entries are non-positive and its row-wise sums are non-negative. (*Hint*: Express the off-diagonal entries $\mathcal{A}_{ij}$ of the stiffness matrix in terms of the cotangent of the two angles facing the edge connecting the vertices S_i and S_j.)

8

Quadratures, Assembling, and Storage

This chapter deals with various aspects related to the implementation of the finite element method. It is organized into four sections. The first section is concerned with *quadratures*, i.e., methods to evaluate integrals approximately. These are almost unavoidable in applications since only a few academic problems involve integrals which can be evaluated analytically. We present various quadratures and evaluate the impact of this type of approximation on the accuracy of the finite element method. In the first section, we also list important arrays (Jacobian, shape functions, derivatives, etc.) which are required in a finite element code. The second section deals with *assembling* techniques for matrices and right-hand sides. The next section reviews some basic *storage* techniques for sparse matrices. In the last section, we briefly discuss how to deal with essential boundary conditions, whether homogeneous or not.

8.1 Quadratures: Theory and Implementation

Consider the discrete problem:

$$
\begin{cases}
\text{Seek } u_h \in W_h \text{ such that} \\
a_h(u_h, v_h) = f_h(v_h), \quad \forall v_h \in V_h,
\end{cases}
\tag{8.1}
$$

where W_h and V_h are finite-dimensional (finite element) spaces based on a mesh $\mathcal{T}_h$ of a domain Ω in $\mathbb{R}^d$, a_h is a bilinear form defined on $W_h \times V_h$, and f_h is a linear form defined on V_h. In general, a_h and f_h take the form

$$
a_h(u_h, v_h) = \int_\Omega A_h(x, u_h, v_h) \, \mathrm{d}x, \quad f_h(v_h) = \int_\Omega F_h(x, v_h) \, \mathrm{d}x + \int_{\partial\Omega_N} G_h(x, v_h) \, \mathrm{d}x,
$$

where A_h, F_h, and G_h are operators and $\partial\Omega_N$ is a part of the boundary (possibly of zero measure) where a non-homogeneous natural boundary condition is enforced. These formulas show that before solving the linear system associated with (8.1), it is necessary to evaluate integrals over Ω and $\partial\Omega_N$. The purpose of this section is to present some quadratures serving this purpose.

8.1.1 Principle of quadratures

Definition 8.1. *Let K be a non-empty, Lipschitz, compact, connected subset of $\mathbb{R}^d$. Let $l_q \geq 1$ be an integer. A quadrature on K with l_q points consists of:*

(i) *A set of l_q real numbers $\{\omega_1, \ldots, \omega_{l_q}\}$ called* quadrature weights.

(ii) *A set of l_q points $\{\xi_1, \ldots, \xi_{l_q}\}$ in K called* Gauß points *or* quadrature nodes.

The largest integer k such that

$$\forall p \in \mathbb{P}_k, \quad \int_K p(x)\,dx = \sum_{l=1}^{l_q} \omega_l\, p(\xi_l), \tag{8.2}$$

is called the quadrature order *and is denoted by k_q.*

Quadratures are used to approximate integrals over K. Indeed, setting $h_K = \operatorname{diam}(K)$ and using Taylor expansions, one easily verifies that

$$\forall \phi \in C^{k_q+1}(K), \quad \tfrac{1}{\operatorname{meas}(K)}\left|\int_K \phi(x)\,dx - \sum_{l=1}^{l_q} \omega_l\, \phi(\xi_l)\right| \leq c\, h_K^{k_q+1} \sup_{\substack{x \in K \\ |\alpha| = k_q+1}} |\partial^\alpha \phi(x)|.$$

Numerous quadrature formulas are available, and a large amount of literature is devoted to the subject. Many formulas frequently used in finite element codes can be found in [AbS72, HaS56, Str69, Str71]. We now give some examples in dimension 1, 2, and 3.

Examples in one dimension. The most frequently used one-dimensional quadratures are based on the Legendre polynomials introduced in §1.1.4; see Definition 1.14. We first consider quadratures on the reference interval $\widehat{K} = [0, 1]$.

Proposition 8.2. *Let $l_q \geq 1$, denote by $\widehat{\xi}_1, \ldots, \widehat{\xi}_{l_q}$ the l_q roots of the Legendre polynomial $\widehat{\mathcal{E}}_{l_q}(x)$, and set*

$$\widehat{\omega}_l = \int_0^1 \prod_{\substack{j=1 \\ j \neq l}}^{l_q} \frac{t - \widehat{\xi}_j}{\widehat{\xi}_l - \widehat{\xi}_j}\,dt. \tag{8.3}$$

Then, $\{\widehat{\xi}_1, \ldots, \widehat{\xi}_{l_q}; \widehat{\omega}_1, \ldots, \widehat{\omega}_{l_q}\}$ is a quadrature of order $k_q = 2l_q - 1$ on $[0, 1]$.

Proof. Let $\{\widehat{\mathcal{L}}_1^{l_q}, \ldots, \widehat{\mathcal{L}}_{l_q}^{l_q}\}$ be the set of Lagrange polynomials associated with the Gauß points $\{\widehat{\xi}_1, \ldots, \widehat{\xi}_{l_q}\}$. Then, $\widehat{\omega}_l = \int_0^1 \widehat{\mathcal{L}}_l^{l_q}(t)\,dt$ for $1 \leq l \leq l_q$. Let p be a polynomial of degree less than l_q. Owing to the identity

$$p(t) = \sum_{l=1}^{l_q} p(\widehat{\xi}_l)\widehat{\mathcal{L}}_l^{l_q}(t), \quad \forall t \in [0, 1],$$

k_q l_q	Nodes ξ_l	Weights ω_l
1 1	m	δ
3 2	$m \pm \frac{\delta}{2}\frac{\sqrt{3}}{3}$	$\frac{1}{2}\delta$
5 3	$m \pm \frac{\delta}{2}\sqrt{\frac{3}{5}}$	$\frac{5}{18}\delta$
	m	$\frac{8}{18}\delta$
7 4	$m \pm \frac{\delta}{2}\sqrt{\frac{1}{35}(15 + 2\sqrt{30})}$	$(\frac{1}{4} - \frac{1}{12}\sqrt{\frac{5}{6}})\delta$
	$m \pm \frac{\delta}{2}\sqrt{\frac{1}{35}(15 - 2\sqrt{30})}$	$(\frac{1}{4} + \frac{1}{12}\sqrt{\frac{5}{6}})\delta$

Table 8.1. Nodes and weights for quadratures on the interval $[a, b]$; $m = \frac{1}{2}(a + b)$ and $\delta = b - a$.

and integrating both sides of this equation, we deduce that the quadrature (8.2) is exact for p. Consider now a polynomial p of degree less than $2l_q$; write p in the form $p = q\widehat{\mathcal{E}}_{l_q} + r$ where both q and r are polynomials of degree less than l_q. Owing to the orthogonality property (1.25) of Legendre polynomials, we infer

$$\int_0^1 p(t)\, dt = \int_0^1 r(t)\, dt = \sum_{l=1}^{l_q} \widehat{\omega}_l r(\widehat{\xi}_l) = \sum_{l=1}^{l_q} \widehat{\omega}_l p(\widehat{\xi}_l),$$

since the points $\widehat{\xi}_l$ are the roots of $\widehat{\mathcal{E}}_{l_q}$. This completes the proof. $\qquad\square$

By a simple change of variables, the quadrature of Proposition 8.2 can be used to approximate integrals over the interval $[a, b]$. Table 8.1 gives some values for the weights ω_l and the points ξ_l.

Remark 8.3.

(i) The weights defined in (8.3) have a closed form in terms of the roots of the Legendre polynomials. However, these roots can be determined analytically only for the first few Legendre polynomials. When considering high-order quadratures, it is therefore necessary to evaluate the Gauß points using an algorithm that approximates the roots of the suitable Legendre polynomial. An example using the Newton–Raphson iteration and recursion relations for the polynomials is presented in [KaS99^b, p. 357].

(ii) Gauß points are often termed *Gauß–Legendre points* in one dimension since they are the roots of the Legendre polynomials. It is also possible to consider the roots of other polynomials to define quadratures; see, e.g., Exercise 8.1 for the Gauß–Lobatto quadrature in which the two endpoints of $[0, 1]$ are included among the Gauß points. $\qquad\square$

k_q l_q	Barycentric coord.	Multiplicity	Weights ω_l
1 1	$\left(\frac{1}{3},\frac{1}{3},\frac{1}{3}\right)$	1	S
2 3	$\left(\frac{1}{6},\frac{1}{6},\frac{2}{3}\right)$	3	$\frac{1}{3}S$
2 3	$\left(\frac{1}{2},\frac{1}{2},0\right)$	3	$\frac{1}{3}S$
3 4	$\left(\frac{1}{3},\frac{1}{3},\frac{1}{3}\right)$	1	$-\frac{9}{16}S$
	$\left(\frac{1}{5},\frac{1}{5},\frac{3}{5}\right)$	3	$\frac{25}{48}S$
3 7	$\left(\frac{1}{3},\frac{1}{3},\frac{1}{3}\right)$	1	$\frac{9}{20}S$
	$\left(\frac{1}{2},\frac{1}{2},0\right)$	3	$\frac{2}{15}S$
	$(1,0,0)$	3	$\frac{1}{20}S$
4 6	$(a_i,a_i,1-2a_i)$ for $i=1,2$	3	ω_i for $i=1,2$
	$a_1 = 0.445948490915965$		$\omega_1 = S \times 0.223381589678010$
	$a_2 = 0.091576213509771$		$\omega_2 = S \times 0.109951743655322$
5 7	$\left(\frac{1}{3},\frac{1}{3},\frac{1}{3}\right)$	1	$\frac{9}{40}S$
	$(a_i,a_i,1-2a_i)$ for $i=1,2$	3	
	$a_1 = \frac{6-\sqrt{15}}{21}$		$\frac{155-\sqrt{15}}{1200}S$
	$a_2 = \frac{6+\sqrt{15}}{21}$		$\frac{155+\sqrt{15}}{1200}S$
6 12	$(a_i,a_i,1-2a_i)$ for $i=1,2$	3	
	$a_1 = 0.063089014491502$		$S \times 0.050844906370206$
	$a_2 = 0.249286745170910$		$S \times 0.116786275726378$
	$(a,b,1-a-b)$	6	
	$a = 0.310352451033785$		$S \times 0.082851075618374$
	$b = 0.053145049844816$		

Table 8.2. Nodes and weights for quadratures on a triangle of area S.

Examples in two dimensions. Table 8.2 lists some quadratures on triangles in two dimensions. In this table, we call multiplicity the number of permutations that must be performed on the barycentric coordinates to obtain the list of all the Gauß points of the quadrature. For instance, the fourth-order formula has six Gauß points; the corresponding barycentric coordinates and weights are

$$\{a_1,a_1,1-2a_1;\omega_1\}, \ \{1-2a_1,a_1,a_1;\omega_1\}, \ \{a_1,1-2a_1,a_1;\omega_1\},$$
$$\{a_2,a_2,1-2a_2;\omega_2\}, \ \{1-2a_2,a_2,a_2;\omega_2\}, \ \{a_2,1-2a_2,a_2;\omega_2\},$$

where the weights of the first and second family of points are $\omega_1 \simeq S \times 0.223381589678010$ and $\omega_2 \simeq S \times 0.109951743655322$, respectively. Here, S denotes the surface of the triangle.

k_q l_q	Barycentric coord.	Multiplicity	Weights ω_l
1 1	$\left(\frac{1}{4}, \frac{1}{4}, \frac{1}{4}, \frac{1}{4}\right)$	1	V
2 4	$(a, a, a, 1 - 3a)$ $a = \frac{5 - \sqrt{5}}{20}$	4	$\frac{1}{4}V$
2 10	$\left(\frac{1}{2}, \frac{1}{2}, 0, 0\right)$ $(1, 0, 0, 0)$	6 4	$\frac{1}{5}V$ $-\frac{1}{20}V$
3 5	$\left(\frac{1}{4}, \frac{1}{4}, \frac{1}{4}, \frac{1}{4}\right)$ $\left(\frac{1}{6}, \frac{1}{6}, \frac{1}{6}, \frac{1}{2}\right)$	1 4	$-\frac{4}{5}V$ $\frac{9}{20}V$
5 15	$\left(\frac{1}{4}, \frac{1}{4}, \frac{1}{4}, \frac{1}{4}\right)$ $(a_i, a_i, a_i, 1 - 2a_i)$ for $i = 1, 2$ $a_1 = \frac{7 - \sqrt{15}}{34}$ $a_2 = \frac{7 + \sqrt{15}}{34}$ $(a, a, \frac{1}{2} - a, \frac{1}{2} - a)$ $a = \frac{10 - 2\sqrt{5}}{40}$	1 4 6	$\frac{16}{135}V$ $\frac{2665 + 14\sqrt{15}}{37800}V$ $\frac{2665 - 14\sqrt{15}}{37800}V$ $\frac{10}{189}V$

Table 8.3. Nodes and weights for quadratures on a tetrahedron of volume V.

Quadratures on rectangles are conveniently deduced from one-dimensional quadratures by taking the Gauß points in tensor product form.

Examples in three dimensions. Table 8.3 gives some quadratures on a tetrahedron in three dimensions. As in two dimensions, the multiplicity is the number of permutations to perform on the barycentric coordinates to obtain all the Gauß points of the quadrature. For instance, the third-order formula has five Gauß points which are the point $\left(\frac{1}{4}, \frac{1}{4}, \frac{1}{4}, \frac{1}{4}\right)$ with the weight $-\frac{4}{5}V$ and the four points $\left(\frac{1}{6}, \frac{1}{6}, \frac{1}{6}, \frac{1}{2}\right)$, $\left(\frac{1}{6}, \frac{1}{6}, \frac{1}{2}, \frac{1}{6}\right)$, $\left(\frac{1}{6}, \frac{1}{2}, \frac{1}{6}, \frac{1}{6}\right)$, $\left(\frac{1}{2}, \frac{1}{6}, \frac{1}{6}, \frac{1}{6}\right)$ with the weight $\frac{9}{20}V$.

Quadratures on cuboids are conveniently deduced from one-dimensional quadratures by taking the Gauß points in tensor product form.

8.1.2 Evaluation of an integral

Consider the integral $\int_\Omega \phi(x)\,dx$ where ϕ is a smooth function. Using the decomposition

$$\int_\Omega \phi(x)\,dx = \sum_{K \in \mathcal{T}_h} \int_K \phi(x)\,dx,$$

the computation of $\int_\Omega \phi(x)\,dx$ reduces to evaluating integrals over each element in the mesh. Let K be an arbitrary element in $\mathcal{T}_h$. Since the transformation T_K, mapping $\widehat{K}$ to K, is a $\mathcal{C}^1$-diffeomorphism, the change of variables $x = T_K(\widehat{x})$ yields

$$\int_K \phi(x)\, dx = \int_{\widehat{K}} \phi(T_K(\widehat{x}))\det(J_K(\widehat{x}))\, d\widehat{x},$$

where $J_K(\widehat{x}) = \frac{\partial T_K(\widehat{x})}{\partial \widehat{x}}$ is the Jacobian matrix of T_K at $\widehat{x}$. Hence, the whole problem reduces to evaluating integrals over the reference element $\widehat{K}$. Using a quadrature with l_q Gauß points on $\widehat{K}$, say $\{\widehat{\xi}_1, \ldots, \widehat{\xi}_{l_q}; \widehat{\omega}_1, \ldots, \widehat{\omega}_{l_q}\}$, yields

$$\int_K \phi(x)\, dx \approx \sum_{l=1}^{l_q} \widehat{\omega}_l \det(J_K(\widehat{\xi}_l))\, \phi(T_K(\widehat{\xi}_l)).$$

Let k_q be the order of the quadrature. Setting $\omega_{lK} = \widehat{\omega}_l \det(J_K(\widehat{\xi}_l))$ and $\xi_{lK} = T_K(\widehat{\xi}_l)$ leads to

$$\int_\Omega \phi(x)\, dx \approx \sum_{K \in \mathcal{T}_h} \sum_{l=1}^{l_q} \omega_{lK}\, \phi(\xi_{lK}),$$

i.e., $\{\xi_{1K}, \ldots, \xi_{l_q K}; \omega_{1K}, \ldots, \omega_{l_q K}\}$ is a quadrature on K. The purpose of the following lemma is to assess the accuracy of this quadrature:

Lemma 8.4. *Assume that $\{\mathcal{T}_h\}_{h>0}$ is a shape-regular family of affine meshes. Let $1 \le p \le +\infty$, let p' be such that $\frac{1}{p} + \frac{1}{p'} = 1$, assume that $k_q + 1 > \frac{d}{p}$, and let $s \in \mathbb{N}$ satisfy $\frac{d}{p} < s \le k_q + 1$. Then, there is $c(\widehat{K})$, independent of h, such that, for all $K \in \mathcal{T}_h$ and $\phi \in W^{s,p}(K)$,*

$$\left| \int_K \phi(x)\, dx - \sum_{l=1}^{l_q} \omega_{lK}\, \phi(\xi_{lK}) \right| \le c(\widehat{K})\, h_K^s \mathrm{meas}(K)^{\frac{1}{p'}} |\phi|_{s,p,K}. \tag{8.4}$$

Proof. Let $K \in \mathcal{T}_h$ and set $E(\phi) = \int_K \phi(x)\, dx - \sum_{l=1}^{l_q} \omega_{lK}\, \phi(\xi_{lK})$. Since the mesh is affine, $E(\phi) = |\det(J_K)|\, \widehat{E}(\widehat{\phi})$ with obvious notation. It is clear that $\widehat{E} : C^0(\widehat{K}) \to \mathbb{R}$ is linear and continuous. Moreover, since $sp > d$, the embedding $W^{s,p}(\widehat{K}) \subset C^0(\widehat{K})$ is continuous; see Corollary B.43 or Theorem B.46. As a result, $|\widehat{E}(\widehat{\phi})| \le c_1(\widehat{K}) \|\widehat{\phi}\|_{C^0(\widehat{K})} \le c_2(\widehat{K}) \|\widehat{\phi}\|_{W^{s,p}(\widehat{K})}$, i.e., the operator $\widehat{E}$ is bounded on $W^{s,p}(\widehat{K})$. Since $\widehat{E}(\widehat{p}) = 0$ for all $\widehat{p}$ in $\mathbb{P}_{s-1} \subset \mathbb{P}_{k_q}$, we deduce from the Bramble–Hilbert Lemma that

$$|\widehat{E}(\widehat{\phi})| \le c(\widehat{K}) |\widehat{\phi}|_{W^{s,p}(\widehat{K})}.$$

Since the mesh is affine,

$$|\widehat{\phi}|_{W^{s,p}(\widehat{K})} \le c(\widehat{K}) \|J_K\|_d^s |\det(J_K)|^{-\frac{1}{p}} |\phi|_{W^{s,p}(K)}.$$

Owing to the shape-regularity of $\{\mathcal{T}_h\}_{h>0}$ together with (1.94), we infer

$$|E(\phi)| \le c_3(\widehat{K}) h_K^s \mathrm{meas}(K)^{1-\frac{1}{p}} |\phi|_{W^{s,p}(K)},$$

whence (8.4) readily follows.   □

Theorem 8.5. *Under the hypotheses of Lemma 8.4, there is c, uniform in h and p, such that, for all $\phi \in W^{s,p}(\Omega)$,*

$$\left| \int_\Omega \phi(x)\,dx - \sum_{K \in \mathcal{T}_h} \sum_{l=1}^{l_q} \omega_{lK}\, \phi(\xi_{lK}) \right| \le c\, h^s |\phi|_{s,p,\Omega}. \tag{8.5}$$

Proof. Let R be the left-hand side in (8.5). Using (8.4), we infer

$$R \le \sum_{K \in \mathcal{T}_h} \left| \int_K \phi(x)\,dx - \sum_{l=1}^{l_q} \omega_{lK}\, \phi(\xi_{lK}) \right|$$

$$\le c(\widehat{K}) \sum_{K \in \mathcal{T}_h} h_K^s\, \operatorname{meas}(K)^{\frac{1}{p'}} |\phi|_{W^{s,p}(K)}$$

$$\le c(\widehat{K})\, h^s \left(\sum_{K \in \mathcal{T}_h} \operatorname{meas}(K) \right)^{\frac{1}{p'}} \left(\sum_{K \in \mathcal{T}_h} |\phi|_{W^{s,p}(K)}^p \right)^{\frac{1}{p}},$$

where the Hölder inequality in $\mathbb{R}^{N_{el}}$ is used to derive the last bound. Since $\sum_{K \in \mathcal{T}_h} \operatorname{meas}(K) = \operatorname{meas}(\Omega)$, (8.5) readily follows. $\square$

8.1.3 The discrete problem with quadratures

Let us now investigate the impact of quadratures on the stability and convergence properties of the finite element method. Using the above notation, the bilinear form a_h is approximated by

$$a_h(u_h, v_h) \approx a_{hQ}(u_h, v_h) = \sum_{K \in \mathcal{T}_h} \sum_{l=1}^{l_q} \omega_{lK}\, A_h(\xi_{lK}, u_h(\xi_{lK}), v_h(\xi_{lK})). \tag{8.6}$$

Likewise, the volume integral in the right-hand side is approximated by

$$\int_\Omega F_h(x, v_h)\,dx \approx \sum_{K \in \mathcal{T}_h} \sum_{l=1}^{l_q} \omega_{lK}\, F_h(\xi_{lK}, v_h(\xi_{lK})).$$

The surface integral in the right-hand side is approximated similarly. To simplify, assume that the mesh is such that $\partial\Omega_N$ is the union of faces of mesh elements. Denote by $\mathcal{N}_h$ the collection of those faces. Assume also that all the faces of $\widehat{K}$ are equivalent, up to invertible affine transformations, to a geometric reference face $\widehat{F} \subset \mathbb{R}^{d-1}$. For instance, if $\widehat{K}$ is the reference triangle in $\mathbb{R}^2$, we choose $\widehat{F}$ to be the unit segment $[0, 1]$. As a result, for all faces F in $\mathcal{N}_h$, there is a C^1-diffeomorphism T_F mapping $\widehat{F}$ to F. Let $g_F(\widehat{x}) = (J_F(\widehat{x}))^T J_F(\widehat{x})$ be the metric tensor where $J_F(\widehat{x}) = \frac{\partial T_F(\widehat{x})}{\partial \widehat{x}} \in \mathbb{R}^{d,d-1}$ is the Jacobian matrix of the mapping T_F at $\widehat{x}$. Consider a quadrature on $\widehat{F}$ defined by l_q^∂ Gauß points $\{\widehat{\xi}_1^\partial, \ldots, \widehat{\xi}_{l_q^\partial}^\partial\}$ and l_q^∂ weights $\{\widehat{\omega}_1^\partial, \ldots, \widehat{\omega}_{l_q^\partial}^\partial\}$. Setting

$\omega_{lF} = \widehat{\omega}_l^\partial \sqrt{\det(g_F(\widehat{x}))}$ and $\xi_{lF} = T_F(\widehat{\xi}_l^\partial)$, we can approximate the surface integral as follows:

$$\int_{\partial\Omega_N} g(x)\,v_h(x)\,\mathrm{d}x \approx \sum_{F\in\mathcal{N}_h} \sum_{l=1}^{l_q^\partial} \omega_{lF} G_h(\xi_{lF}, v_h(\xi_{lF})).$$

Finally, we introduce the approximate right-hand side

$$f_{hQ}(v_h) = \sum_{K\in\mathcal{T}_h} \sum_{l=1}^{l_q} \omega_{lK} F_h(\xi_{lK}, v_h(\xi_{lK})) + \sum_{F\in\mathcal{N}_h} \sum_{l=1}^{l_q^\partial} \omega_{lF} G_h(\xi_{lF}, v_h(\xi_{lF})). \quad (8.7)$$

Hence, the problem which is actually implemented in a computer code is:

$$\begin{cases} \text{Seek } u_h \in W_h \text{ such that} \\ a_{hQ}(u_h, v_h) = f_{hQ}(v_h), \quad \forall v_h \in V_h, \end{cases} \quad (8.8)$$

where a_{hQ} and f_{hQ} are defined in (8.6) and (8.7), respectively.

A natural question arising at this point is to determine whether the quadratures do not perturb the problem "too much." In particular, is the linear system (8.8) still invertible? How to choose the order of the quadratures so that the convergence rate of the approximation remains optimal? To give simple answers to these questions, we restrict ourselves to a scalar, second-order, elliptic problem. Set

$$\begin{cases} A_h(x, u_h, v_h) = \nabla v_h\cdot\sigma(x)\cdot\nabla u_h + v_h\left(\beta(x)\cdot\nabla u_h\right) + \mu(x)u_h v_h, \\ F_h(x, v_h) = f(x)v_h, \\ G_h(x, v_h) = g(x)v_h, \end{cases} \quad (8.9)$$

where σ, β, μ, f, and g satisfy suitable hypotheses so that the discrete problem (8.1) and its continuous version are well-posed; see §3.1.

Theorem 8.6. *Let $\{\mathcal{T}_h\}_{h>0}$ be a shape-regular family of affine meshes and let $V_h \subset V$ be a finite-dimensional space based on $\mathcal{T}_h$ and the reference finite element $\{\widehat{K}, \widehat{P}, \widehat{\Sigma}\}$. Assume that $\mathbb{P}_1 \subset \widehat{P}$ and that there exists $l \geq 1$ such that:*

(i) $\widehat{P} \subset \mathbb{P}_l$.

(ii) *The quadrature on $\widehat{K}$ is of order $2l - 2$.*

Then, problem (8.8) is well-posed if h is small enough.

Proof. See Exercise 8.6; see also, e.g., [StF73] and [Cia78, Theorem 4.1.2]. $\square$

Theorem 8.7. *Along with the hypotheses of Theorem 8.6, assume that there exists $k \geq 1$ such that:*

(i) $\mathbb{P}_k \subset \widehat{P}$.

(ii) *The surface quadrature is of order $k + l - 1$.*

Moreover, assume that $f \in W^{k,q}(\Omega)$ for $q \geq 2$ and $qk > d$, $g \in W^{k+1,s}(\partial\Omega_N)$ for $s \geq 2$, and that u, the solution to the continuous problem, is in $H^{k+1}(\Omega)$. Let u_h solve (8.8). Then, there is $h_0 > 0$ and c such that, for all $h \leq h_0$,

$$\|u - u_h\|_{1,\Omega} \leq ch^k(\|u\|_{k+1,\Omega} + \|f\|_{W^{k,q}(\Omega)} + h^{\frac{1}{2}}\|g\|_{W^{k+1,s}(\partial\Omega_N)}). \qquad (8.10)$$

Proof. See Exercise 8.8; see also, e.g., [CiR72^a, Cia78], [DaL90, Chap. 12], or [RaT83, p. 123]. □

Example 8.8. Consider an approximation with $\mathbb{P}_1$ Lagrange finite elements. To guarantee that the error in the H^1-norm is still $\mathcal{O}(h)$ despite quadratures, the volume quadrature must be of order 0 and the surface quadrature of order 1. The rule of thumb for the volume quadrature is that the highest order terms must be evaluated exactly if the coefficients of the PDE are constant. □

Remark 8.9. Theorems 8.6 and 8.7 can easily be extended to account for vector-valued PDEs. □

8.1.4 The quadrature module

In this section, we address practical aspects associated with the implementation of quadrature formulas. In particular, we enumerate a non-exhaustive list of arrays needed to compute the matrix and the right-hand side associated with (8.8). Throughout this section, we assume $W_h = V_h$ for the sake of simplicity.

The geometric transformation T_K. Let $\{\widehat{K}, \widehat{P}_{\text{geo}}, \widehat{\Sigma}_{\text{geo}}\}$ be the geometric reference finite element on which $\mathcal{T}_h$ is based; see Definition 1.50 in §1.3.2. Let n_{geo} be the number of degrees of freedom in $\widehat{\Sigma}_{\text{geo}}$ and let $\{\widehat{g}_1, \ldots, \widehat{g}_{n_{\text{geo}}}\}$ be the geometric nodes of $\widehat{K}$. Denote by $\{\widehat{\psi}_1, \ldots, \widehat{\psi}_{n_{\text{geo}}}\}$ the geometric reference shape functions on $\widehat{K}$.

Given a Gauß point $\widehat{\xi}_l$ in $\widehat{K}$ and an element K in $\mathcal{T}_h$, we want to evaluate the coordinates of $\xi_{lK} = T_K(\widehat{\xi}_l)$. Recall that coord denotes the coordinate array of the geometric nodes of the mesh and that geo_cnty denotes the connectivity array of the mesh; see §7.1. These arrays are usually provided by the mesh generator that created the mesh $\mathcal{T}_h$. Let N_{el} be the number of elements in the mesh. Then, according to (1.51), the k-th Cartesian component of $T_{K^m}(\widehat{\xi}_l)$, $1 \leq m \leq N_{\text{el}}$, $1 \leq k \leq d$, is given by

$$\left(T_{K^m}(\widehat{\xi}_l)\right)_k = \sum_{n=1}^{n_{\text{geo}}} \text{coord}(k, \text{geo\_cnty}(n, m))\, \text{psi}(n, l),$$

where we have introduced the double-entry array $\text{psi}(1{:}n_{\text{geo}}, 1{:}l_{\text{q}})$ such that

$$\text{psi}(n, l) = \widehat{\psi}_n(\widehat{\xi}_l), \quad 1 \leq n \leq n_{\text{geo}}, 1 \leq l \leq l_{\text{q}}.$$

Jacobians and weights. Let $J_{K^m}(\widehat{x}) \in \mathbb{R}^{d,d}$ be the Jacobian matrix of the transformation T_{K^m} at point $\widehat{x}$ in $\widehat{K}$. Using the above notation, the (k_1, k_2)-th entry of $J_K(\widehat{x})$ at $\widehat{\xi}_l$, $1 \leq k_1, k_2 \leq d$, is given by

$$\left(J_{K^m}(\widehat{\xi}_l)\right)_{k_1, k_2} = \sum_{n=1}^{n_{\text{geo}}} \text{coord}(k_1, \text{geo\_cnty}(n,m)) \, \text{dpsi\_dhatK}(k_2, n, l),$$

where we have introduced the triple-entry array $\text{dpsi\_dhatK}(1{:}d, 1{:}n_{\text{geo}}, 1{:}l_q)$ such that

$$\text{dpsi\_dhatK}(k, n, l) = \frac{\partial \widehat{\psi}_n}{\partial \widehat{x}_k}(\widehat{\xi}_l), \quad 1 \leq k \leq d, \ 1 \leq n \leq n_{\text{geo}}, \ 1 \leq l \leq l_q.$$

Since the determinant of $J_{K^m}(\widehat{\xi}_l)$ is always multiplied by the weight $\widehat{\omega}_l$ in the quadratures, it is advisable to store this product once and for all in a double-entry array of weights:

$$\text{weight\_K}(l, m) = \widehat{\omega}_l \det\left(J_{K^m}(\widehat{\xi}_l)\right), \quad 1 \leq l \leq l_q, \ 1 \leq m \leq N_{\text{el}}.$$

Remark 8.10. In the case of affine meshes, the partial derivatives of $\widehat{\psi}_n$ are constant on each simplex; as a consequence, the size of the array dpsi_dhatK can be reduced to $d \times n_{\text{geo}}$. □

Local shape functions. Let $\{\widehat{K}, \widehat{P}, \widehat{\Sigma}\}$ be the reference finite element. Let n_{sh} be the number of degrees of freedom of $\widehat{\Sigma}$ and let $\{\widehat{\theta}_1, \ldots, \widehat{\theta}_{n_{\text{sh}}}\}$ be the local shape functions on $\widehat{K}$. Since the value of the local shape functions at the Gauß points is required many times in the quadrature formulas, it is advisable to compute all those values once and for all and store them in the array

$$\text{theta}(n, l) = \widehat{\theta}_n(\widehat{\xi}_l), \quad 1 \leq n \leq n_{\text{sh}}, \ 1 \leq l \leq l_q.$$

Likewise, it is generally worthwhile to store the values of the first-order derivatives of the local shape functions in the array

$$\text{dtheta\_dhatK}(k, n, l) = \frac{\partial \widehat{\theta}_n}{\partial \widehat{x}_k}(\widehat{\xi}_l), \quad 1 \leq k \leq d, \ 1 \leq n \leq n_{\text{sh}}, \ 1 \leq l \leq l_q.$$

Global shape functions: Enumeration and connectivity. To simplify, assume that the linear bijective mapping ψ_K used to generate the approximation space V_h is such that $\psi_K(v) = v \circ T_K$. This assumption holds, for instance, if the reference finite element is a Lagrange finite element; see §1.4.

Let $N = \dim(V_h)$. The global shape functions are numbered from 1 to N, say $\{\varphi_1, \ldots, \varphi_N\}$. To materialize the relation between local and global shape functions, we define a double-entry connectivity array of size $n_{\text{sh}} \times N_{\text{el}}$,

$$\text{pde\_cnty}(1{:}n_{\text{sh}}, 1{:}N_{\text{el}}).$$

This connectivity array is such that, for $1 \leq n \leq n_{\mathrm{sh}}$ and $1 \leq m \leq N_{\mathrm{el}}$,

$$\forall x \in K_m, \quad \varphi_{\texttt{pde\_cnty}(n,m)}(x) = \widehat{\theta}_n \circ T_{K_m}^{-1}(x). \tag{8.11}$$

That is, $\theta_n = \widehat{\theta}_n \circ T_{K_m}^{-1}$ is the restriction to K_m of the $\texttt{pde\_cnty}(n,m)$-th global shape function.

Derivatives of global shape functions. Let us now evaluate the first-order derivatives of the global shape functions at the Gauß points. Let K_m be an element in the mesh and let φ_i be a global shape function whose support has a non-empty intersection with K_m. Let $1 \leq n \leq n_{\mathrm{sh}}$ be such that $i = \texttt{pde\_cnty}(n,m)$, i.e., $\varphi_{i|K_m} = \theta_n = \widehat{\theta}_n \circ T_{K_m}^{-1}$. Using (8.11) and the chain rule yields

$$\frac{\partial \theta_n}{\partial x_{k_1}}(\xi_{lK_m}) = \frac{\partial \widehat{\theta}_n \circ T_{K_m}^{-1}}{\partial x_{k_1}}(\xi_{lK_m}) = \sum_{k_2=1}^{d} \frac{\partial \widehat{\theta}_n}{\partial \widehat{x}_{k_2}}(\widehat{\xi}_l) \left(\left[J_{K_m}(\widehat{\xi}_l) \right]^{-1} \right)_{k_2,k_1}.$$

Let us distinguish two cases:

(i) If the transformation T_{K_m} is affine (this is often the case for subparametric transformations), the Jacobian matrix J_{K_m} and its inverse are constants on the element, i.e., they do not depend on the Gauß points. In this case, one may store the inverse of the Jacobian matrix in an array

$$\texttt{inv\_jac\_K}(k_1, k_2, m) = \left(\left[J_{K_m} \right]^{-1} \right)_{k_1,k_2},$$

and each time the quantity $\frac{\partial \theta_n}{\partial x_{k_1}}(\xi_{lK_m})$ is needed, the following operations must be performed:

$$\frac{\partial \theta_n}{\partial x_{k_1}}(\xi_{lK_m}) = \sum_{k_2=1}^{d} \texttt{dtheta\_dhatK}(k_2, n, l)\,\texttt{inv\_jac\_K}(k_2, k_1, m).$$

The memory space required to store $\texttt{inv\_jac\_K}$ is proportional to $d \times d \times N_{\mathrm{el}}$. If $d \ll n_{\mathrm{sh}} \times l_{\mathrm{q}}$, this storage space is less than $d \times n_{\mathrm{sh}} \times l_{\mathrm{q}} \times N_{\mathrm{el}}$ which is that required to store all the derivatives of the shape functions. This technique proves useful if the available memory space is limited. It trades substantial memory savings against moderate additional computations. Note that memory allocation may be a severe limiting factor for large-scale simulations in three dimensions.

(ii) If the mesh is not affine, it is advisable to evaluate the quantities $\frac{\partial \theta_n}{\partial x_k}(\xi_{lK_m})$ once and for all and to store them in an array

$$\texttt{dtheta\_dK}(k, n, l, m) = \frac{\partial \theta_n}{\partial x_k}(\xi_{lK_m}), \quad \begin{cases} 1 \leq k \leq d,\ 1 \leq n \leq n_{\mathrm{sh}}, \\ 1 \leq l \leq l_{\mathrm{q}},\ 1 \leq m \leq N_{\mathrm{el}}. \end{cases}$$

As mentioned above, the size of this array, $d \times n_{\mathrm{sh}} \times l_{\mathrm{q}} \times N_{\mathrm{el}}$, can be extremely large. If memory space is at a premium, strategy (i) is preferable.

Boundary integrals. Surface quadratures can be evaluated in a fashion very similar to that of volume quadratures. The details are left to the reader.

8.2 Assembling

Assembling refers to the phase in a finite element program where the entries of the stiffness matrix $\mathcal{A}$ and those of the right-hand side vector F are computed. To give an idea on how assembling can be implemented, we consider the approximate problem (8.8) in which the bilinear form a_{hQ} and the linear form f_{hQ} are defined in (8.6) and (8.7), respectively, with A_h, F_h, and G_h defined in (8.9). For the sake of simplicity, assume that $W_h = V_h$ in (8.8) so that the same array pde_cnty can be used to span the trial space and the test space.

8.2.1 Assembling of the stiffness matrix

The assembling of the stiffness matrix $\mathcal{A}$ can be implemented as shown in Algorithm 8.1. Note the use of the connectivity array pde_cnty.

Algorithm 8.1. Assembling of $\mathcal{A}$.

$\mathcal{A} = 0$
for $1 \leq m \leq N_{\mathrm{el}}$ do
  for $1 \leq l \leq l_{\mathrm{q}}$ do
    for $1 \leq ni \leq n_{\mathrm{sh}}$ do; $i = $ pde_cnty(ni, m)
      for $1 \leq nj \leq n_{\mathrm{sh}}$ do; $j = $ pde_cnty(nj, m)
        $\mathcal{A}_{ij} = \mathcal{A}_{ij} + $ weight_K$(l, m) * A_h(\xi_{lK_m}, \varphi_j(\xi_{lK_m}), \varphi_i(\xi_{lK_m}))$
      end for
    end for
  end for
end for

Let us now review some implementation details. To evaluate the quantity $A_h(\xi_{lK_m}, \varphi_j(\xi_{lK_m}), \varphi_i(\xi_{lK_m}))$, we distinguish two cases depending on whether the functions σ, β, and μ are known analytically or are already discretized.

Case 1: Functions that are known analytically. We assume here that the functions σ, β, and μ are known analytically. Since we need the values of σ, β, and μ at the Gauß point $\xi_{lK_m} = T_{K_m}(\widehat{\xi_l})$, it is necessary to evaluate ξ_{lK_m} in the l-loop. The details are shown in Algorithm 8.2. Note the use of both the geometric connectivity array geo_cnty and the connectivity array for the global shape functions pde_cnty.

Algorithm 8.2. Assembling of $\mathcal{A}$ when σ, β, and μ are known analytically.

$\mathcal{A} = 0$

for $1 \leq m \leq N_{\mathrm{el}}$ do

  for $1 \leq l \leq l_{\mathrm{q}}$ do

    ======Evaluate the Cartesian coordinates of Gauß point ξ_{lK_m}

    for $1 \leq k \leq d$ do

$$\texttt{xi\_l}(k) = \sum_{n=1}^{n_{\mathrm{geo}}} \texttt{coord}(k, \texttt{geo\_cnty}(n, m)) * \texttt{psi}(n, l)$$

    end for

    for $1 \leq ni \leq n_{\mathrm{sh}}$ do; $i = \texttt{pde\_cnty}(ni, m)$

      for $1 \leq nj \leq n_{\mathrm{sh}}$ do; $j = \texttt{pde\_cnty}(nj, m)$

      ======Evaluate $\nabla\varphi_j \cdot \sigma \cdot \nabla\varphi_i$

$$x_1 = \sum_{k_1, k_2=1}^{d} \texttt{dtheta\_dK}(k_1, ni, l, m) * \sigma_{k_1, k_2}(\texttt{xi\_l}) * \texttt{dtheta\_dK}(k_2, nj, l, m)$$

      ======Evaluate $\varphi_i\,(\beta \cdot \nabla\varphi_j)$

$$x_2 = \texttt{theta}(ni, l) \sum_{k_1=1}^{d} \beta_{k_1}(\texttt{xi\_l}) * \texttt{dtheta\_dK}(k_1, nj, l, m)$$

      ======Evaluate $\varphi_i\,\mu\,\varphi_j$

$$x_3 = \texttt{theta}(ni, l) * \mu(\texttt{xi\_l}) * \texttt{theta}(nj, l)$$

      ======Accumulate

$$\mathcal{A}(i, j) = \mathcal{A}(i, j) + [x_1 + x_2 + x_3] * \texttt{weight\_K}(l, m)$$

      end for

    end for

  end for

end for

Case 2: Discrete functions. We now assume that the functions σ, β, and μ are in V_h or in a Cartesian product of V_h. Let $\texttt{sigma}$, $\texttt{beta}$, and $\texttt{mu}$ be the corresponding coordinate vectors. To implement quadratures involving these functions, we must compute their values at the Gauß points on every element K_m. For instance, for the scalar-valued function μ,

$$\mu(\xi_{lK_m}) = \sum_{n=1}^{n_{\mathrm{sh}}} \texttt{mu}(\texttt{pde\_cnty}(n, m)) * \texttt{theta}(n, l).$$

To obtain the value of σ and β at ξ_{lK_m}, we apply the above formula to their Cartesian components. This situation is described in Algorithm 8.3.

Remark 8.11.

(i) Of course, it is possible that among the three fields σ, β, and μ, some are discrete and some continuous. Then, one or the other of the two assembling techniques presented above must be used depending on the circumstances.

(ii) Note that the loop on m in Algorithms 8.2 and 8.3 is easily parallelizable. However, the inner loops are generally too short to be worth vectorizing.

(iii) To optimize memory access time, it is sometimes worthwhile to modify

Algorithm 8.3. Assembling of $\mathcal{A}$ when σ, β, and μ are discrete.

$\mathcal{A} = 0$
for $1 \leq m \leq N_{\mathrm{el}}$ **do**
  for $1 \leq l \leq l_{\mathrm{q}}$ **do**
    ======Evaluate σ, β and μ at Gauß point ξ_{lK_m}
    for $1 \leq k_1 \leq d$ **do**
      for $1 \leq k_2 \leq d$ **do**

$$\texttt{sigma\_l}(k_1, k_2) = \sum_{n=1}^{n_{\mathrm{sh}}} \texttt{sigma}(k_1, k_2, \texttt{pde\_cnty}(n, m)) * \texttt{theta}(n, l)$$

      end for

$$\texttt{beta\_l}(k_1) = \sum_{n=1}^{n_{\mathrm{sh}}} \texttt{beta}(k_1, \texttt{pde\_cnty}(n, m)) * \texttt{theta}(n, l)$$

    end for

$$\texttt{mu\_l} = \sum_{n=1}^{n_{\mathrm{sh}}} \texttt{mu}(\texttt{pde\_cnty}(n, m)) * \texttt{theta}(n, l)$$

    for $1 \leq ni \leq n_{\mathrm{sh}}$ **do**; $\ i = \texttt{pde\_cnty}(ni, m)$
      for $1 \leq nj \leq n_{\mathrm{sh}}$ **do**; $\ j = \texttt{pde\_cnty}(nj, m)$
      ======Evaluate $\nabla\varphi_j \cdot \sigma \cdot \nabla\varphi_i$

$$x_1 = \sum_{k_1, k_2 = 1}^{d} \texttt{dtheta\_dK}(k_1, ni, l, m) * \texttt{sigma\_l}(k_1, k_2) * \texttt{dtheta\_dK}(k_2, nj, l, m)$$

      ======Evaluate $\varphi_i \left(\beta \cdot \nabla\varphi_j \right)$

$$x_2 = \texttt{theta}(ni, l) * \sum_{k_1 = 1}^{d} \texttt{beta\_l}(k_1) * \texttt{dtheta\_dK}(k_1, nj, l, m)$$

      ======Evaluate $\varphi_i \, \mu \, \varphi_j$
$$x_3 = \texttt{mu\_l} * \texttt{theta}(ni, l) * \texttt{theta}(nj, l)$$
      ======Accumulation
$$\mathcal{A}_{ij} = \mathcal{A}_{ij} + [x_1 + x_2 + x_3] * \texttt{weight\_K}(l, m)$$
      end for
    end for
  end for
end for

the order of the loops and to introduce temporary arrays. For instance, instead of accumulating the quantities $[x_1 + x_2 + x_3] * \texttt{weight\_K}(l, m)$ into $\mathcal{A}_{ij}$ in the l-loop, it may be preferable to introduce a temporary array $\texttt{temp}(1:n_{\mathrm{sh}}, 1:n_{\mathrm{sh}})$ and perform the accumulation into $\texttt{temp}(ni, nj)$. Then, when the l-loop is finished, the quantities stored in $\texttt{temp}(ni, nj)$ are accumulated into $\mathcal{A}_{ij}$. This type of optimization avoids unnecessary memory access. Its efficiency is, of course, hardware-dependent. $\square$

8.2.2 Assembling of the right-hand side

Assuming for the sake of simplicity that the boundary integral in (8.7) is zero, the assembling of the right-hand side vector can be implemented as in Algorithm 8.4.

Algorithm 8.4. Assembling of F.

$b = 0$
for $1 \leq m \leq N_{\mathrm{el}}$ do
  for $1 \leq l \leq l_{\mathrm{q}}$ do
    for $1 \leq ni \leq n_{\mathrm{sh}}$ do; $i = \mathtt{pde\_cnty}(ni, m)$
      $b(i) = b(i) + \mathtt{weight\_K}(l, m) * F_h(\xi_{lK_m}, \varphi_i(\xi_{lK_m}))$
    end for
  end for
end for

Algorithm 8.5 illustrates the situation where $F_h(x, v) = \beta(x) \cdot \nabla v + f(x) v$ when β and f are analytically known functions.

Algorithm 8.5. Assembling of F when $F_h(x, v) = \beta(x) \cdot \nabla v + f(x) v$.

$b = 0$
for $1 \leq m \leq N_{\mathrm{el}}$ do
  for $1 \leq l \leq l_{\mathrm{q}}$ do
    ======Evaluate the Cartesian coordinates of Gauß point ξ_{lK_m}
    for $1 \leq k_1 \leq d$ do
$$\mathtt{xi\_1}(k_1) = \sum_{n=1}^{n_{\mathrm{geo}}} \mathtt{coord}(k_1, \mathtt{geo\_cnty}(n, m)) \, \mathtt{psi}(n, l)$$
    end for
    for $1 \leq ni \leq n_{\mathrm{sh}}$ do; $i = \mathtt{pde\_cnty}(ni, m)$
      ======Evaluate $f \varphi_i$
      $x_1 = f(\mathtt{xi\_1}) * \mathtt{theta}(ni, l)$
      ======Evaluate $\beta \cdot \nabla \varphi_i$
$$x_2 = \sum_{k_1=1}^{d} \beta_{k_1}(\mathtt{xi\_1}) * \mathtt{dtheta\_dK}(k_1, ni, l, m)$$
      ======Accumulate
      $b(i) = b(i) + [x_1 + x_2] * \mathtt{weight\_K}(l, m)$
    end for
  end for
end for

Remark 8.12. Variations on this theme are numerous. For instance, if f or β are discrete fields, one must compute their value at ξ_{lK_m} as in the preceding section. One could also assume that $F_h(x, v_h)$ is of the form $\nabla w_h \cdot \sigma(x) \cdot \nabla v_h + v_h (\beta(x) \cdot \nabla w_h)$, where w_h is a discrete field. This case arises when one uses an explicit or semi-implicit march in time to approximate a time-dependent problem; see Chapter 6. □

8.3 Storage of Sparse Matrices

Definition 8.13 (Sparse matrix). *Let $A \in \mathbb{R}^{N,N'}$. Denote by nnz the number of non-zero entries in A. The matrix A is said to be* sparse *if*

$$nnz \ll N \times N'.$$

In general, matrices associated with finite element discretizations of PDEs are sparse because the global shape functions have small supports. Let $\{\mathcal{T}_h\}_{h>0}$ be a shape-regular family of meshes of Ω, let V_h be an approximation space of dimension N based on $\mathcal{T}_h$, and let $\{\varphi_1, \ldots, \varphi_N\}$ be the global shape functions. Let a_h be a bilinear form defined on $V_h \times V_h$ and let $A = (A_{ij})_{1 \leq i,j \leq N}$ be the stiffness matrix with entries $A_{ij} = a_h(\varphi_j, \varphi_i)$. (Again, for the sake of simplicity, a standard Galerkin setting is considered.) Denote by N_{row} the maximum number of non-zero entries per row in the matrix A and let us estimate an upper bound on N_{row}. Since the family $\{\mathcal{T}_h\}_{h>0}$ is shape-regular, the upper bound on the number of elements that are common to one degree of freedom is independent of h; let nc be this number. Let n_{sh} be the number of local shape functions per element. Given a shape function, say φ_i, there are at most $nc \times n_{\text{sh}}$ other shape functions whose support has an intersection of non-zero measure with the support of φ_i. Hence, the estimate $N_{\text{row}} \leq nc \times n_{\text{sh}}$ holds. This upper bound is overestimated, since some shape functions may have a support included in one or two elements only.

Example 8.14.
    (i) In two dimensions, for $\mathbb{P}_1$ finite elements ($n_{\text{sh}} = 3$), each shape function interacts with at most $nc+1$ other shape functions (i.e., $N_{\text{row}} \leq nc+1$); for $\mathbb{P}_2$ finite elements ($n_{\text{sh}} = 6$), each shape function interacts with at most $3nc + 1$ other shape functions (i.e., $N_{\text{row}} \leq 3nc + 1$); see also Exercise 8.10 for further bounds.
    (ii) A matrix with a sparse profile is shown in the left panel of Figure 8.1. Black squares indicate non-zero entries. The matrix is obtained from the discretization of a homogeneous Dirichlet problem in two dimensions with $\mathbb{P}_1$ Lagrange finite elements and the mesh shown in the right panel of Figure 8.1. The boundary nodes have been eliminated in this example; see §8.4.2. The matrix is 16×16 and contains at most seven non-zero entries per row ($nc = 6$, $N_{\text{row}} = 7$).   □

8.3.1 CSR and CSC formats

One of the most frequently used storage techniques for sparse matrices is probably the so-called *Compressed Sparse Rows* (resp., *Columns*) format, usually referred to as the CSR (resp., CSC) format. Both formats are very similar, the role played by rows and columns being simply interchanged. For the sake of brevity, we only present the CSR format.

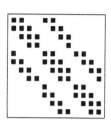 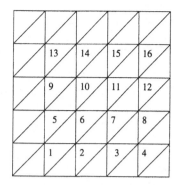

Fig. 8.1. Example of sparse matrix: sparsity pattern (left) and underlying mesh (right).

Let $\mathcal{A}(1{:}N, 1{:}N')$ be a sparse matrix not necessarily square and containing nnz non-zero entries. To store this matrix in the CSR format, define three arrays denoted by $\mathtt{ia}(1{:}N{+}1)$, $\mathtt{ja}(1{:}nnz)$, and $\mathtt{aa}(1{:}nnz)$.

1. The integer array $\mathtt{ia}$ stores the number of non-zero entries in each row. More precisely, set $\mathtt{ia}(1) = 1$ and choose the value of $\mathtt{ia}(i+1)$ such that $\mathtt{ia}(i+1) - \mathtt{ia}(i)$ is equal to the number of non-zero entries in the i-th row of the matrix $\mathcal{A}$, $1 \leq i \leq N$. Obviously, the following identity holds: $nnz = \mathtt{ia}(N+1) - \mathtt{ia}(1)$.
2. The integer array $\mathtt{ja}$ gives the column indices of the non-zero entries. More precisely, for $1 \leq i \leq N$, the list $(\mathtt{ja}(p))_{\mathtt{ia}(i) \leq p \leq \mathtt{ia}(i+1)-1}$ contains all the column indices of the non-zero entries in row i. Conventionally, the array $\mathtt{ja}$ is ordered such that, for a given row, the column indices are in increasing order.
3. The real array $\mathtt{aa}$ contains the non-zero entries of the matrix. For a given row i, the list $(\mathtt{aa}(p))_{\mathtt{ia}(i) \leq p \leq \mathtt{ia}(i+1)-1}$ contains all the non-zero entries of the row in question. The same ordering is used for $\mathtt{ja}$ and $\mathtt{aa}$, so that $\mathtt{aa}(p) = \mathcal{A}_{i,\mathtt{ja}(p)}$.

Example 8.15. For the following 5×5 matrix:

$$
\mathcal{A} = \begin{bmatrix}
1. & 0. & 0. & 2. & 0. \\
3. & 4. & 0. & 5. & 0. \\
6. & 0. & 7. & 8. & 9. \\
0. & 0. & 10. & 11. & 0. \\
0. & 0. & 0. & 0. & 12.
\end{bmatrix},
\tag{8.12}
$$

its storage in the CSR format yields the arrays

$\mathtt{aa}$ =	1.	2.	3.	4.	5.	6.	7.	8.	9.	10.	11.	12.

$\mathtt{ja}$ =	1	4	1	2	4	1	3	4	5	3	4	5

$\mathtt{ia}$ =	1	3	6	10	12	13

Note that $nnz = \text{ia}(6) - \text{ia}(1) = 12$.   □

Let us now see how to use the CSR format in practical situations.

Retrieving a non-zero entry in $\mathcal{A}$. Assuming that for a given pair of indices (i, j), the entry $\mathcal{A}_{ij}$ is non-zero, a practical problem is to retrieve the value of $\mathcal{A}_{ij}$ from the array **aa**. Using the CSR storage technique, this operation is described in Algorithm 8.6.

Algorithm 8.6. Retrieving $\mathcal{A}_{ij} \neq 0$ in CSR format.

---

for $\text{ia}(i) \leq p \leq \text{ia}(i+1) - 1$ **do**
  if $\text{ja}(p) = j$ **then**
    value $= \text{aa}(p)$; Exit loop on p
  end if
end for

Assembling. Using the notation of §8.2, the assembling of a matrix $\mathcal{A}$ stored in the CSR format is described in Algorithm 8.7. Note the use of the temporary array d which avoids unnecessary memory access.

Algorithm 8.7. Matrix assembling in CSR format.

$\text{aa} = 0$
for $1 \leq m \leq N_{\text{el}}$ **do**
  for $1 \leq l \leq l_{\text{q}}$ **do**
    $d = 0$
    for $1 \leq ni \leq n_{\text{sh}}$ **do**; $i = \text{pde\_cnty}(ni, m)$
      for $1 \leq nj \leq n_{\text{sh}}$ **do**; $j = \text{pde\_cnty}(nj, m)$
        $d(ni, nj) = d(ni, nj) + \text{weight\_K}(l, m) * A_h(\xi_{lK_m}, \varphi_j(\xi_{lK_m}), \varphi_i(\xi_{lK_m}))$
      end for
    end for
    end for
    for $1 \leq ni \leq n_{\text{sh}}$ **do**; $i = \text{pde\_cnty}(ni, m)$
      for $1 \leq nj \leq n_{\text{sh}}$ **do**; $j = \text{pde\_cnty}(nj, m)$
        for $\text{ia}(i) \leq p \leq \text{ia}(i+1) - 1$ **do**
          if $\text{ja}(p) = j$ **then**
            $\text{aa}(p) = \text{aa}(p) + d(ni, nj)$; Exit loop on p
          end if
        end for
      end for
    end for
  end for
end for

Matrix–vector multiplication. For iterative solution methods (see §9.3), matrix–vector multiplication is an operation which is repeated a large number of times. Algorithm 8.8 shows how to perform the matrix–vector multiplication when the matrix is stored in CSR format. The technique is optimal in the sense that it involves only the number of operations that are strictly necessary.

Algorithm 8.8. Matrix–vector multiplication in CSR format.

---

for $1 \leq i \leq N$ do
  $yi = 0$
  for $\mathtt{ia}(i) \leq p \leq \mathtt{ia}(i+1) - 1$ do
    $yi = yi + \mathtt{aa}(p) * x(\mathtt{ja}(p))$
  end for
  $y(i) = yi$
end for

8.3.2 Ellpack–Itpack format

The CSR format has some drawbacks, in particular when used on parallel computers. Indeed, since the rows of the compressed matrix are not of constant length, it may be difficult to distribute the entries of the matrix to the different processors of the computer. Similarly, the fact that the index of the first non-zero entry has to be computed for each row may hamper vectorization.

The Ellpack–Itpack format partially solves these difficulties. This format is based on the hypothesis that each row of the matrix contains almost the same number of non-zero entries. Let $n_{\text{row}}(i)$ be the number of non-zero entries in row i, $1 \leq i \leq N$, of the matrix $\mathcal{A} \in \mathbb{R}^{N \times N'}$. Let $N_{\text{row}} = \max_{1 \leq i \leq N} n_{\text{row}}(i)$ be the maximum number of non-zero entries per row in $\mathcal{A}$. Define two arrays $\mathtt{aa}(1{:}N, 1{:}N_{\text{row}})$ and $\mathtt{ja}(1{:}N, 1{:}N_{\text{row}})$ as follows:

1. The real array $\mathtt{aa}$ contains the non-zero entries of $\mathcal{A}$. For a given row i, $\mathtt{aa}(i, 1{:}n_{\text{row}}(i))$ contains all the non-zero entries in row i, and if $n_{\text{row}}(i) < N_{\text{row}}$, the entries $\mathtt{aa}(i, (n_{\text{row}}(i) + 1){:}N_{\text{row}})$ are set to zero.
2. The integer array $\mathtt{ja}$ contains the column indices of the non-zero entries in the matrix $\mathcal{A}$. For a given row i, $\mathtt{ja}(i, 1{:}n_{\text{row}}(i))$ contains all the column indices of the non-zero entries in row i. The ordering of the indices in $\mathtt{ja}$ is the same as that in $\mathtt{aa}$. The simplest convention consists of ordering the column indices in increasing order. If $n_{\text{row}}(i) < N_{\text{row}}$, the entries $\mathtt{ja}(i, (n_{\text{row}}(i) + 1){:}N_{\text{row}})$ are given an arbitrary value, say $\mathtt{ja}(i, n_{\text{row}}(i))$.

The Ellpack–Itpack format is well-adapted to meshes that are almost structured. For instance, for continuous $\mathbb{Q}_1$ finite elements, $N_{\text{row}} = 9$ in two dimensions and $N_{\text{row}} = 27$ in three dimensions.

Example 8.16. For the 5×5 matrix shown in (8.12), $N_{\text{row}} = 4$ and

$$
\mathbf{aa} = \begin{bmatrix} 1. & 2. & 0. & 0. \\ 3. & 4. & 5. & 0. \\ 6. & 7. & 8. & 9. \\ 10. & 11. & 0. & 0. \\ 12. & 0. & 0. & 0. \end{bmatrix}, \quad
\mathbf{ja} = \begin{bmatrix} 1 & 4 & 4 & 4 \\ 1 & 2 & 4 & 4 \\ 1 & 3 & 4 & 5 \\ 3 & 4 & 4 & 4 \\ 5 & 5 & 5 & 5 \end{bmatrix}.
\qquad \square
$$

Matrix–vector multiplication. To evaluate the matrix–vector multiplication $y = \mathcal{A}x$ in Ellpack–Itpack format, one proceeds as in Algorithm 8.9.

Algorithm 8.9. Matrix–vector multiplication in Ellpack–Itpack format.

for $1 \le i \le N$ **do**
  $yi = 0$
  for $1 \le p \le N_{\text{row}}$ **do**
    $yi = yi + \mathbf{aa}(i,p) * x(\mathbf{ja}(i,p))$
  end for
  $y(i) = yi$
end for

8.4 Non-Homogeneous Dirichlet Boundary Conditions

The goal of this section is to discuss from an algorithmic viewpoint various options to deal with essential (Dirichlet) boundary conditions when assembling the stiffness matrix and the right-hand side vector. For the sake of simplicity, we restrict the discussion to a scalar, second-order, elliptic PDE. All the results are easily extendable to other settings, e.g., the Stokes equations or first-order PDEs. Consider the problem:

$$
\begin{cases}
\text{Seek } u \in H^1(\Omega) \text{ such that} \\
a(u,v) = \int_\Omega fv, & \forall v \in H_0^1(\Omega), \\
\gamma_0(u) = g, & \text{in } H^{\frac{1}{2}}(\partial\Omega),
\end{cases}
\tag{8.13}
$$

where γ_0 is the trace operator defined in §B.3.5. We assume that $f \in L^2(\Omega)$, $g \in \mathcal{C}^{0,1}(\partial\Omega)$ (recall that $\mathcal{C}^{0,1}(\partial\Omega) \subset H^{\frac{1}{2}}(\partial\Omega)$ with continuous embedding; see Example B.32(ii)) and that the bilinear form a satisfies the assumptions of the BNB Theorem on $H_0^1(\Omega) \times H_0^1(\Omega)$. As a result, problem (8.13) is well-posed; see §2.1.4.

We seek an approximate solution to (8.13) in a H^1-conformal approximation space V_h of dimension N using Lagrange finite elements. Denote by

$\{\varphi_1, \ldots, \varphi_N\}$ the nodal basis of V_h and by $\{a_1, \ldots, a_N\}$ the associated nodes. Up to renumbering, we can assume without loss of generality that the nodes located at the boundary are $\{a_1, \ldots, a_{N_D}\}$; hence,

$$V_{h0} = \operatorname{span}\{\varphi_{N_D+1}, \ldots, \varphi_N\} = \{v_h \in V_h; \gamma_0(v_h) = 0\}.$$

Since the essential data g is continuous, it can be approximated by its Lagrange interpolant, i.e.,

$$g_h = \sum_{i=1}^{N_D} g_h(a_i)\, \gamma_0(\varphi_i).$$

The discrete problem we consider is the following:

$$\begin{cases} \text{Seek } u_h \in V_h \text{ such that} \\ a_h(u_h, v_h) = f_h(v_h), \quad \forall v_h \in V_{h0}, \\ \gamma_0(u_h) = g_h. \end{cases} \qquad (8.14)$$

See §3.2.2 for the analysis of (8.14) together with error estimates. The bilinear form a_h and the linear form f_h are approximations of a and f, respectively, e.g., resulting from the use of quadratures.

8.4.1 The linear system

Setting $u_h = \sum_{i=1}^{N} U_i\varphi_i$, the equation $\gamma_0(u_h) = g_h$ is equivalent to $U_i = g_h(a_i)$ for $1 \le i \le N_D$. As a result, $U_i = g_h(a_i)$, $1 \le i \le N_D$, and $\sum_{j=1}^{N} a_h(\varphi_j, \varphi_i)U_j = f_h(\varphi_i)$, $N_D + 1 \le i \le N$. To distinguish the interior degrees of freedom from those associated with the Dirichlet data, we introduce the following notation: For all $X \in \mathbb{R}^N$, set $X = (\widehat{X}, \widetilde{X})$ with $\widehat{X} = (X_1, \ldots, X_{N_D})$ and $\widetilde{X} = (X_{N_D+1}, \ldots, X_N)$. On the same token, write

$$\widehat{F} = (g_h(a_1), \ldots, g_h(a_{N_D})) \quad \text{and} \quad \widetilde{F} = (f_h(\varphi_{N_D+1}), \ldots, f_h(\varphi_N)).$$

Furthermore, define

$$B_{ij} = a_h(\varphi_j, \varphi_i) \quad \text{for } N_D + 1 \le i \le N, \quad 1 \le j \le N_D,$$
$$\widetilde{A}_{ij} = a_h(\varphi_j, \varphi_i) \quad \text{for } N_D + 1 \le i, j \le N.$$

Note that the matrix $\widetilde{A}$ involves only the global shape functions associated with the nodes located inside Ω. Using the above notation, the linear system to be solved takes the partitioned form

$$\begin{bmatrix} \mathcal{I} & 0 \\ \hline B & \widetilde{A} \end{bmatrix} \begin{bmatrix} \widehat{U} \\ \widetilde{U} \end{bmatrix} = \begin{bmatrix} \widehat{F} \\ \widetilde{F} \end{bmatrix}. \qquad (8.15)$$

8.4.2 Eliminating the essential nodes

The subsystem associated with the Dirichlet unknowns is trivial: $\widehat{U} = \widehat{F}$.
Hence, it may seem natural to eliminate the unknown $\widehat{U}$, yielding

$$\widetilde{\mathcal{A}}\widetilde{U} = \widetilde{F} - \mathcal{B}\widehat{F}.$$

This technique has two advantages: First, the possible symmetry properties of
the bilinear form a are inherited by the matrix $\widetilde{\mathcal{A}}$, whereas the system (8.15)
is not symmetric. Second, the final size of the linear system is optimal since
only the internal degrees of freedom are unknown. However, this technique
requires assembling two matrices, $\widetilde{\mathcal{A}}$ and $\mathcal{B}$, instead of one. The matrix $\widetilde{\mathcal{A}}$ is
square and of order $N - N_{\mathrm{D}}$, whereas $\mathcal{B} \in \mathbb{R}^{N-N_{\mathrm{D}}, N_{\mathrm{D}}}$ is rectangular. Note
also that the sparsity profiles of the two matrices are different.

Remark 8.17. Consider the lifting $u_{hg} = \sum_{i=1}^{N_{\mathrm{D}}} g(a_i)\varphi_i$ of the approximate
essential boundary condition. Setting $u_h = \phi_h + u_{hg}$, the new unknown $\phi_h \in$
V_{h0} clearly solves the problem $a_h(\phi_h, v_h) = f_h(v_h) - a_h(u_{hg}, v_h)$ for all $v_h \in$
V_{h0}. Introducing the vector $\Phi \in \mathbb{R}^{N-N_{\mathrm{D}}}$ such that $\phi_h = \sum_{i=N_{\mathrm{D}}+1}^{N} \Phi_{i-N_{\mathrm{D}}}\varphi_i$,
it is readily deduced that $\widetilde{\mathcal{A}}\Phi = \widetilde{F} - \mathcal{B}\widehat{F}$, i.e., $\Phi = \widetilde{U}$.   □

8.4.3 Keeping the essential nodes

To avoid assembling two matrices and manipulating two different sparsity
formats, the trivial equations associated with the Dirichlet nodes are retained.
Writing (8.15) in the condensed form $\mathcal{A}U = F$, a very simple technique to
assemble the matrix $\mathcal{A}$ without having to assemble $\mathcal{B}$ and $\widetilde{\mathcal{A}}$ is the following:

(i) Assemble the matrix without making any distinction between the nodes.
   This yields the matrix associated with the Neumann problem. Of course,
   the rows corresponding to essential nodes are not correctly assembled.

(ii) To correct the rows $1 \leq i \leq N_{\mathrm{D}}$, it is sufficient to set all the matrix
   entries to 0, except those on the diagonal which are set to 1.

To assemble the right-hand side, one can proceed similarly: First, assemble
the vector without taking care of the Dirichlet data; then, for $1 \leq i \leq N_{\mathrm{D}}$,
replace the entry of index i by the essential data $g_h(a_i)$.

   The above technique is simple. It slightly increases the number of unknowns by keeping the essential degrees of freedom. If the bilinear form a is
symmetric, this technique has the apparent drawback of breaking the symmetry of the stiffness matrix. Actually, symmetry is not lost if an iterative method
is used to solve the linear system. To illustrate this point in the framework of
the iterative solution methods presented in §9.3, we introduce the notion of
Krylov space. Let $r \in \mathbb{R}^N$ and set

$$\mathcal{K}(\mathcal{A}, r, k) = \mathrm{span}\{r, \mathcal{A}r, \ldots, \mathcal{A}^{k-1}r\}.$$

The vector space $\mathcal{K}(\mathcal{A}, r, k)$ is the so-called Krylov space of order k generated by r and associated with the matrix $\mathcal{A}$. The Krylov space technique is at the origin of numerous iterative solution methods.

Proposition 8.18. *Let $\mathcal{A} \in \mathbb{R}^{N \times N}$ be a partitioned matrix with the following block structure:*

$$\mathcal{A} = \left[\begin{array}{c|c} \mathcal{I} & 0 \\ \hline \mathcal{B} & \tilde{\mathcal{A}} \end{array} \right].$$

Let $r_0 \in \mathbb{R}^N$ be such that $\hat{r}_0 = 0$. Then, for all $y \in \mathcal{K}(\mathcal{A}, r_0, k)$, $\hat{y} = 0$. Moreover, if $\tilde{\mathcal{A}}$ is symmetric, the restriction of $\mathcal{A}$ to $\mathcal{K}(\mathcal{A}, r_0, k)$ is symmetric.

Proof. A simple computation yields $\mathcal{A}y = (\hat{y}, \mathcal{B}\hat{y} + \tilde{\mathcal{A}}\tilde{y})$, whence we deduce $\widehat{\mathcal{A}y} = 0$ if $\hat{y} = 0$. This property extends by induction on k to $\mathcal{K}(\mathcal{A}, r_0, k)$ if $\hat{r}_0 = 0$. Moreover, let x and y be two vectors in $\mathcal{K}(\mathcal{A}, r_0, k)$. From what precedes we infer $\hat{y} = 0$ and $\hat{x} = 0$. As a result, $(y, \mathcal{A}x)_N = (\tilde{y}, \tilde{\mathcal{A}}\tilde{x})_{N-N_{\mathrm{D}}}$. This implies that the restriction of $\mathcal{A}$ to the space $\mathcal{K}(\mathcal{A}, r_0, k)$ has the same symmetry properties as $\tilde{\mathcal{A}}$. $\qquad \square$

An important consequence of Proposition 8.18 is that, if a Krylov-based iterative solution method is used to solve (8.15) and if the initial guess satisfies the Dirichlet data, the following properties hold:

(i) The residual is always zero at the essential degrees of freedom during the iterations.
(ii) The iterative algorithm behaves exactly as if the Dirichlet unknowns had been eliminated.

To conclude, retaining the Dirichlet unknowns is slightly more expensive than eliminating them, but the matrix is simpler to assemble and iterative solution algorithms behave as if these unknowns had been eliminated.

8.4.4 Penalty method

Another technique to enforce Dirichlet boundary values without eliminating the associated degrees of freedom consists of using a penalty method: First, one assembles the matrix and the right-hand side as above, i.e., without taking into account the essential boundary conditions. One obtains the matrix and the right-hand side associated with the homogeneous Neumann problem. Then, one adds the equation

$$\epsilon^{-1} U_i = \epsilon^{-1} g_{hi},$$

to each row i corresponding to a Dirichlet node. More precisely, for each row i such that $1 \le i \le N_{\mathrm{D}}$, one adds ϵ^{-1} to the diagonal entry of the matrix and $\epsilon^{-1} g_{hi}$ to the right-hand side. When the matrix of the Neumann problem is symmetric, symmetry is preserved by the penalty method. This technique, which is very simple to implement, is useful when one wishes to utilize a direct solution method based on the symmetry of the matrix such as the Choleski or the LDL^T factorization; see §9.2.1.

The mathematical analysis of the above penalty method can be found in [Lio68, Aub70, Bab73^b]. If ϵ^{-1} is not large enough, the method suffers from a lack of consistency. This problem can be avoided by adding extra boundary terms ensuring consistency as shown by Nitsche [Nit71]. For the Laplace operator, the suitable bilinear form is $\int_\Omega \nabla u \cdot \nabla v - \int_{\partial\Omega}(v\partial_n u + u\partial_n v) + \epsilon^{-1}\int_{\partial\Omega}(u-g)$ and $\epsilon \sim h^{-1}$.

8.5 Exercises

Exercise 8.1 (Gauß–Lobatto quadrature). Let $k \geq 2$. The $(k+1)$ Gauß–Lobatto points are the two endpoints of $\widehat{K} = [0,1]$, i.e., $\widehat{\xi}_0 = 0$ and $\widehat{\xi}_k = 1$, and the $(k-1)$ roots of $\widehat{\mathcal{E}}'_k$, $\{\widehat{\xi}_1, \ldots, \widehat{\xi}_{k-1}\}$, where $\widehat{\mathcal{E}}_k$ is the Legendre polynomial of degree k. Show that if the weights are evaluated from $\widehat{\omega}_l = \int_0^1 \widehat{\theta}_l(t)\,dt$, where $\{\widehat{\theta}_0, \ldots, \widehat{\theta}_k\}$ are the nodal ($\mathcal{C}^0$-continuous) basis functions defined in (1.33), the resulting quadrature is exact for polynomials up to degree $2k-1$.

Exercise 8.2. Let $k \geq 1$ and let $\{\widehat{\theta}_0, \ldots, \widehat{\theta}_k\}$ be the polynomials defined in (1.31). Consider the matrices $\mathcal{M}$ and $\mathcal{A}$ of order $k+1$ with entries

$$\mathcal{M}_{mn} = \int_0^1 \widehat{\theta}_m(t)\widehat{\theta}_n(t)\,dt, \quad \mathcal{A}_{mn} = \int_0^1 \tfrac{d}{dt}\widehat{\theta}_m(t)\tfrac{d}{dt}\widehat{\theta}_n(t)\,dt,$$

for $0 \leq m, n \leq k$. Using the conventional representation of Figure 8.1, draw the sparsity pattern of the matrices $\mathcal{M}$ and $\mathcal{A}$ for $k=9$.

Exercise 8.3. Let K be a simplex in two dimensions. Let a_0 be the barycenter of K, let a_1, a_2, and a_3 be the vertices of K, and let a_4, a_5, and a_6 be the midpoints of the edges of K. Prove that the following quadratures:

$$\{a_0; \text{meas}(K)\},$$
$$\{a_1, a_2, a_3; \tfrac{1}{3}\text{meas}(K), \tfrac{1}{3}\text{meas}(K), \tfrac{1}{3}\text{meas}(K)\},$$
$$\{a_4, a_5, a_6; \tfrac{1}{3}\text{meas}(K), \tfrac{1}{3}\text{meas}(K), \tfrac{1}{3}\text{meas}(K)\},$$

are of order zero, one, and two, respectively.

Exercise 8.4. Let K_1 and K_2 be two triangles in $\mathbb{R}^2$ with one face in common, say F. Let $\{x_1, \ldots, x_k; \omega_1, \ldots, \omega_k\}$ be a quadrature of degree $2k-1$ on F; see Proposition 8.2. Let $v_h : K_1 \cup K_2 \to \mathbb{R}$ be a function such that $v_{h|K_1} \in \mathbb{P}_k(K_1)$, $v_{h|K_2} \in \mathbb{P}_k(K_2)$, and $\int_F (v_{h|K_1} - v_{h|K_2})\psi = 0$ for all ψ in $\mathbb{P}_{k-1}(F)$. Prove that v_h is continuous at the Gauß points.

Exercise 8.5. Let $\{\widehat{K}, \widehat{P}, \widehat{\Sigma}\}$ be a reference finite element and let $\{K, P, \Sigma\}$ be a finite element such that the mapping $T_K : \widehat{K} \to K$ is affine. Assume $\mathbb{P}_k \subset \widehat{P} \subset \mathbb{P}_l$, $l \geq k \geq 1$. Let $\{\widehat{\xi}_1, \ldots, \widehat{\xi}_{l_q}; \widehat{\omega}_1, \ldots, \widehat{\omega}_{l_q}\}$ be a quadrature on $\widehat{K}$. Let $\widehat{E}$ be the linear form such that $\widehat{E}(\widehat{\phi}) = \int_{\widehat{K}} \widehat{\phi}(\widehat{x}) \, d\widehat{x} - \sum_{l=1}^{l_q} \widehat{\omega}_l \widehat{\phi}(\widehat{\xi}_l)$ for $\widehat{\phi} \in \mathcal{C}^0(\widehat{K})$. Let $a \in W^{k,\infty}(K)$.

(i) Using norm equivalences in $\mathbb{P}_{l-1}$, prove that, for all $\widehat{w} \in \mathbb{P}_{l-1}$ and $\widehat{\phi} \in W^{k,\infty}(\widehat{K})$, $|\widehat{E}(\widehat{\phi}\widehat{w})| \leq \widehat{c}\|\widehat{\phi}\|_{k,\infty,\widehat{K}}\|\widehat{w}\|_{0,\widehat{K}}$.

(ii) Prove that if the quadrature is of order $k+l-2$, then, for all $\widehat{w} \in \mathbb{P}_{l-1}$ and $\widehat{\phi} \in W^{k,\infty}(\widehat{K})$, $|\widehat{E}(\widehat{\phi}\widehat{w})| \leq \widehat{c}|\widehat{\phi}|_{k,\infty,\widehat{K}}\|\widehat{w}\|_{0,\widehat{K}}$. (*Hint*: Use Lemma B.68.)

(iii) For $\phi \in \mathcal{C}^0(K)$, set $E(\phi) = \widehat{E}(\phi \circ T_K)$. Deduce that if the quadrature is of order $k+l-2$,

$$\forall p, q \in P, \quad |E(a\nabla p \cdot \nabla q)| \leq c\, h_K^k \|a\|_{k,\infty,K} \|p\|_{k,K} |q|_{1,K}. \tag{8.16}$$

(iv) Likewise, prove that if the quadrature is of order $2l-2$, then

$$\forall p, q \in P, \quad |E(a\nabla p \cdot \nabla q)| \leq c\, h_K \|a\|_{1,\infty,K} |p|_{1,K} |q|_{1,K}. \tag{8.17}$$

Exercise 8.6. The goal of this exercise is to prove Theorem 8.6.

(i) Using (8.17), prove that $|a(u_h, v_h) - a_{hQ}(u_h, v_h)| \leq c\, h \|u_h\|_{1,\Omega} \|v_h\|_{1,\Omega}$.

(ii) Assuming that a satisfies a uniform inf-sup condition on $V_h \times V_h$, $V_h \subset H^1(\Omega)$, prove that a_{hQ} also satisfies a uniform inf-sup condition.

(iii) Explain why the above argument can be extended to the Crouzeix–Raviart non-conforming setting considered in §3.2.3.

Exercise 8.7. Use the same notation and hypotheses as in Exercise 8.5. Assume that the quadrature is of order $k+l-2$.

(i) Let $q \geq \max(1, \frac{d}{k})$ where d is the space dimension. Let π be the L^2-projection from $L^2(\widehat{K})$ to $\mathbb{P}_1(\widehat{K})$. Show that

$$\forall \widehat{\phi} \in W^{k,q}(\widehat{K}), \; \forall \widehat{p} \in \widehat{P}, \quad |\widehat{E}(\widehat{\phi}\pi\widehat{p})| \leq \widehat{c} \sum_{j=0}^1 |\widehat{\phi}|_{k-j,q,\widehat{K}} |\widehat{p}|_{j,\widehat{K}}.$$

(ii) Assume $k \geq 2$ and let r be a number such that the embeddings $W^{k,q}(\Omega) \subset W^{k-1,r}(\Omega) \subset \mathcal{C}^0(K)$ are continuous, i.e., $\frac{1}{r} = \frac{1}{q} - \frac{1}{d}$ and $(k-1)r > d$; see Corollary B.43. Show that for all $\widehat{\phi} \in W^{k,q}(\widehat{K})$ and $\widehat{p} \in \widehat{P}$,

$$|\widehat{E}(\widehat{\phi}(\widehat{p} - \pi\widehat{p}))| \leq \widehat{c}|\widehat{\phi}|_{k-1,r,\widehat{K}}\|\widehat{p} - \pi\widehat{p}\|_{0,\widehat{K}} \leq \widehat{c}' \sum_{j=0}^1 |\widehat{\phi}|_{k-j,q,\widehat{K}} |\widehat{p}|_{j,\widehat{K}}.$$

(iii) Assume $k = 1$. Show that $|\widehat{E}(\widehat{\phi}(\widehat{p} - \pi\widehat{p}))| \leq \widehat{c} \sum_{j=0}^1 |\widehat{\phi}|_{k-j,q,\widehat{K}} |\widehat{p}|_{j,\widehat{K}}$ for all $\widehat{\phi} \in W^{k,q}(\widehat{K})$ and $\widehat{p} \in \widehat{P}$.

(iv) Using the above results, prove that for all $\phi \in W^{k,q}(K)$ and $p \in P$,

$$|E(\phi p)| \leq c\, h_K^k \operatorname{meas}(K)^{\frac{1}{2} - \frac{1}{q}} \|\phi\|_{k,q,K} \|p\|_{1,K}. \tag{8.18}$$

Exercise 8.8. Using the hypotheses of Theorem 8.7, the notation of Exercises 8.5 and 8.7, and assuming that there is no boundary integral in (8.8), i.e., $g = 0$ in (8.9), prove the error estimate (8.10). (*Hint*: Use the First Strang Lemma 2.27 together with (8.16) and (8.18).)

Exercise 8.9. Consider the mesh in the right panel of Figure 8.1 and let V_h^1 be the H^1-conformal finite element space based on this mesh. Let the domain be $]0, 1[\times]0, 1[$.

(i) Calculate the entries of the stiffness matrix, i.e., $\mathcal{A}_{ij} = \int_\Omega \nabla \varphi_j \cdot \nabla \varphi_i$.
(ii) Calculate the entries of the mass matrix, i.e., $\mathcal{M}_{ij} = \int_\Omega \varphi_j \varphi_i$.
(iii) Write the arrays needed to store $\mathcal{A}$ and $\mathcal{M}$ in the CSR format.
(iv) Write the arrays needed to store $\mathcal{A}$ and $\mathcal{M}$ in the Ellpack–Itpack format.

Exercise 8.10. Let $\mathcal{A}$ be the stiffness matrix for the Laplacian based on the two-dimensional Lagrange finite element $\mathbb{P}_k$. Denote by n_v, n_e, and n_t the number of local degrees of freedom located at a vertex, an edge, and inside the reference triangle, respectively. Let c_v be the maximum number of triangles sharing a given vertex in the mesh.

(i) Derive an upper bound for N_{row}, the maximum number of non-zero entries per row in $\mathcal{A}$, in terms of n_v, n_e, n_t, and c_v. Compute this upper bound for $k = 1, 2$, and 3.
(ii) Derive an upper bound for N_{row} if the reference finite element is the Crouzeix–Raviart finite element. What is the advantage of this finite element in terms of the structure of $\mathcal{A}$?

Exercise 8.11 (Coordinate format). Consider the following storage format: Let $\mathcal{A}$ be a $N \times N$ sparse matrix. Store the non-zero entries $\mathcal{A}_{ij}$ in the real array $\mathbf{aa}(1{:}nnz)$ and store in the same order the row and columns indices i and j in the integer arrays $\mathbf{ia}(1{:}nnz)$ and $\mathbf{ja}(1{:}nnz)$, respectively. Use this format to store the matrix (8.12). Write an algorithm to perform a matrix–vector product in this format. Write an algorithm to extract the entry $\mathcal{A}_{ij}$ from the array $\mathbf{aa}$. Compare this format with the CSR format.

Exercise 8.12. Consider the following storage format for sparse $N \times N$ matrices: Store the non-zero entries $\mathcal{A}_{ij}$ in the real array $\mathbf{aa}(1{:}nnz)$ and store in the same order the integer $(i-1)N + j$ in the integer array $\mathbf{ja}(1{:}nnz)$. Use this format to store the matrix (8.12). What are the advantages and disadvantages of this data structure? Write an algorithm to perform matrix–vector products in this format.

Linear Algebra

The goal of this chapter is to investigate efficient methods to solve linear systems of the form $\mathcal{A}U = F$ where the matrix $\mathcal{A}$ and the right-hand side F result from a finite element approximation to a linear model problem satisfying the well-posedness conditions of the BNB Theorem. The first section is concerned with the concept of *matrix conditioning*. The idea is to evaluate a real number that quantifies the stability of the linear system with respect to perturbations. In particular, we estimate the condition number of the mass matrix and that of the stiffness matrix. The second section deals with *reordering techniques* for sparse matrices. These techniques are particularly useful when solving sparse linear systems using direct methods. The third section reviews elementary properties of some widely used *iterative solution methods*: the Conjugate Gradient algorithm for symmetric positive definite systems and, more generally, projection based Krylov-type methods. We investigate the convergence rate of these methods and show how this rate can be improved using *preconditioning techniques*. The last section presents a brief introduction to the *parallel implementation* of iterative solution methods. For the sake of brevity, relaxation methods and multigrid methods are not discussed herein. The reader is referred, e.g., to [BrS94, GoV89, LaT93, Ort87, QuV97, Saa96] for further insight.

9.1 Conditioning

This section investigates the concept of matrix conditioning. The main result is that, under reasonable assumptions, the condition number of the mass matrix is controlled independently of the mesh size h, while the condition number of the stiffness matrix associated with a second-order differential operator explodes as $\frac{1}{h^2}$. However, in the framework of the BNB Theorem, stability is recovered in spite of the ill-conditioning of the stiffness matrix.

9.1.1 Preliminary definitions and results

Let $\| \cdot \|_N$ be the Euclidean norm in $\mathbb{R}^N$. The associated matrix norm, also denoted by $\| \cdot \|_N$, is defined to be

$$\forall \mathcal{Z} \in \mathbb{R}^{N,N}, \quad \|\mathcal{Z}\|_N = \max_{V \in \mathbb{R}^N} \frac{\|\mathcal{Z}V\|_N}{\|V\|_N}.$$

A classical result [GoV89] is that

$$\|\mathcal{Z}\|_N = \rho(\mathcal{Z}^T \mathcal{Z})^{\frac{1}{2}} = \rho(\mathcal{Z}\mathcal{Z}^T)^{\frac{1}{2}},$$

where $\rho(\mathcal{Z})$ denotes the spectral radius of $\mathcal{Z}$ and $\mathcal{Z}^T$ denotes the transpose of $\mathcal{Z}$. If the matrix $\mathcal{Z}$ is *symmetric*, $\|\mathcal{Z}\|_N = \lambda_{\max}(\mathcal{Z})$ where $\lambda_{\max}(\mathcal{Z})$ is the largest eigenvalue of $\mathcal{Z}$ in absolute value.

Definition 9.1 (Condition number). *Let $\mathcal{Z}$ be an invertible matrix of order N. Its* condition number *is defined as*

$$\kappa(\mathcal{Z}) = \|\mathcal{Z}\|_N \|\mathcal{Z}^{-1}\|_N.$$

Proposition 9.2. *Let $\mathcal{Z}$ be an invertible matrix of order N. Then, the following properties hold:*

(i) $\kappa(\mathcal{Z}) \geq 1$.

(ii) *If $\mathcal{Z}$ is symmetric,*

$$\kappa(\mathcal{Z}) = \frac{\lambda_{\max}(\mathcal{Z})}{\lambda_{\min}(\mathcal{Z})},$$

where $\lambda_{\max}(\mathcal{Z})$ and $\lambda_{\min}(\mathcal{Z})$ denote the largest and smallest eigenvalue of $\mathcal{Z}$ in absolute value, respectively.

(iii) *Denoting by $\mathfrak{I}(\mathcal{Z})$ the set of matrices $\mathcal{E} \in \mathbb{R}^{N,N}$ such that $\mathcal{Z} + \mathcal{E}$ is singular,*

$$\frac{1}{\kappa(\mathcal{Z})} = \min_{\mathcal{E} \in \mathfrak{I}(\mathcal{Z})} \frac{\|\mathcal{E}\|_N}{\|\mathcal{Z}\|_N}. \tag{9.1}$$

Proof. The proof of (i) and (ii) is straightforward. For the proof of (iii), see [Kah66]. $\qquad\square$

Definition 9.3. *An invertible matrix $\mathcal{Z} \in \mathbb{R}^{N,N}$ is said to be* ill-conditioned *if*

$$\kappa(\mathcal{Z}) \gg 1.$$

Remark 9.4.

(i) The characterization (9.1) shows that $\kappa(\mathcal{Z})$ can be interpreted as the reciprocal of the relative distance of $\mathcal{Z}$ to the set of singular matrices. Thus, when $\kappa(\mathcal{Z})$ is large, the matrix $\mathcal{Z}$ is "almost singular."

(ii) More generally, if $\| \cdot \|$ denotes a matrix norm associated with some vector norm, items (i) and (iii) in Proposition 9.2 remain valid for the condition number defined as $\kappa(\mathcal{Z}) = \|\mathcal{Z}\| \|\mathcal{Z}^{-1}\|$; see Exercise 9.1 for an example. $\qquad\square$

9.1.2 Ill-conditioning and linear system stability

Let $\mathcal{A}$ be an invertible matrix of order N and consider the linear system

$$\mathcal{A}U = F. \tag{9.2}$$

In this section, we are interested in assessing the impact of perturbations in the system matrix $\mathcal{A}$ and in the right-hand side F on the solution U.

Proposition 9.5. *Let U be the solution to (9.2) and let $U+\delta U$ be the solution to the perturbed system $\mathcal{A}(U + \delta U) = F + \delta F$. Assume $F \neq 0$. Then,*

$$\frac{\|\delta U\|_N}{\|U\|_N} \leq \kappa(\mathcal{A})\frac{\|\delta F\|_N}{\|F\|_N}, \tag{9.3}$$

and this inequality is sharp.

Proof. The estimate results from

$$\|\delta U\|_N \leq \|\mathcal{A}^{-1}\|_N \|\delta F\|_N \qquad \text{and} \qquad \|F\|_N \leq \|\mathcal{A}\|_N \|U\|_N.$$

Furthermore, compacity of the unit ball in finite dimension implies that there exist U_0 and F_0 such that

$$\|\mathcal{A}U_0\|_N = \|\mathcal{A}\|_N \|U_0\|_N \qquad \text{and} \qquad \|\mathcal{A}^{-1}F_0\|_N = \|\mathcal{A}^{-1}\|_N \|F_0\|_N,$$

and this implies that estimate (9.3) is sharp. □

Proposition 9.6. *Let U be the solution to (9.2) and let $U+\delta U$ be the solution to the perturbed system $(\mathcal{A} + \delta\mathcal{A})(U + \delta U) = F$. Assume $F \neq 0$. Then,*

$$\frac{\|\delta U\|_N}{\|U + \delta U\|_N} \leq \kappa(\mathcal{A})\frac{\|\delta\mathcal{A}\|_N}{\|\mathcal{A}\|_N}, \tag{9.4}$$

and this inequality is sharp.

Proof. Similar to the proof of Proposition 9.5. ⊔

Estimates (9.3) and (9.4) show that if $\mathcal{A}$ is ill-conditioned, small perturbations (due, for instance, to roundoff errors) in $\mathcal{A}$ or in F can induce significant variations in the solution.

9.1.3 Conditioning of the mass matrix

Let Ω be a polyhedron in $\mathbb{R}^d$. Let $\{\mathcal{T}_h\}_{h>0}$ be a family of affine meshes of Ω, let $\{\widehat{K}, \widehat{P}, \widehat{\Sigma}\}$ be the reference finite element, and let $\psi_K : V(K) \to V(\widehat{K})$ be the linear bijective mapping introduced in §1.4 to construct the approximation space, say V_h. For the sake of simplicity, we assume that $\psi_K(v) = v \circ T_K$ where $T_K : \widehat{K} \to K$ is the geometric transformation mapping $\widehat{K}$ to K. This assumption holds, for instance, if the reference finite element is a Lagrange finite element.

Denote by $\{\varphi_1, \ldots, \varphi_N\}$ the global shape functions in V_h with $N = \dim V_h$. For a function $v_h \in V_h$, denote by $V \in \mathbb{R}^N$ the coordinate vector of v_h relative to the basis $\{\varphi_1, \ldots, \varphi_N\}$, i.e., $v_h = \sum_{i=1}^{N} V_i \varphi_i \in V_h$. The goal of this section is to estimate the condition number of the mass matrix $\mathcal{M} = (\int_\Omega \varphi_i \varphi_j)_{1 \leq i,j \leq N}$. Observe that $\mathcal{M}$ is symmetric positive definite and denote by $\mu_{\min}$ and $\mu_{\max}$ its smallest and largest eigenvalue, respectively. Then, it is clear that

$$\forall v_h \in V_h, \quad \mu_{\min}^{\frac{1}{2}} \|V\|_N \leq \|v_h\|_{0,\Omega} \leq \mu_{\max}^{\frac{1}{2}} \|V\|_N. \tag{9.5}$$

This shows that the spectrum of $\mathcal{M}$ is closely related to the equivalence of the norms $\|V\|_N$ and $\|v_h\|_{0,\Omega}$ in V_h. The main result is that if the family $\{\mathcal{T}_h\}_{h>0}$ is *quasi-uniform* (see Definition 1.140), the condition number of $\mathcal{M}$ is controlled *uniformly in h*.

Lemma 9.7. *If $\{\mathcal{T}_h\}_{h>0}$ is* quasi-uniform, *there exist $c_1 > 0$, $c_2 > 0$ such that*

$$\forall h, \ \forall v_h \in V_h, \quad c_1 h^d \|V\|_N^2 \leq \|v_h\|_{0,\Omega}^2 \leq c_2 h^d \|V\|_N^2. \tag{9.6}$$

Proof. (1) Let $\{\widehat{\theta}_1, \ldots, \widehat{\theta}_{n_{\mathrm{sh}}}\}$ be the local shape functions for the reference finite element. Denote by $\mathcal{S}^{n_{\mathrm{sh}}}$ the unit sphere in $\mathbb{R}^{n_{\mathrm{sh}}}$ and define the operator

$$\psi : \mathcal{S}^{n_{\mathrm{sh}}} \ni \eta \longmapsto \left\| \sum_{k=1}^{n_{\mathrm{sh}}} \eta_k \widehat{\theta}_k \right\|_{0,\widehat{K}}^2 \in \mathbb{R}.$$

The operator ψ is clearly continuous. Moreover, since $\mathcal{S}^{n_{\mathrm{sh}}}$ is compact, ψ reaches its minimum and its maximum, say $\widehat{c}_1$ and $\widehat{c}_2$, respectively. Assume that $\widehat{c}_1 = 0$. Then, there exists $\eta \in \mathcal{S}^{n_{\mathrm{sh}}}$ such that $\psi(\eta) = 0$, yielding $\sum_{k=1}^{n_{\mathrm{sh}}} \eta_k \widehat{\theta}_k = 0$. Since $\{\widehat{\theta}_1, \ldots, \widehat{\theta}_{n_{\mathrm{sh}}}\}$ is a basis, this implies $\eta_1 = \ldots = \eta_{n_{\mathrm{sh}}} = 0$, contradicting the fact that $\eta \in \mathcal{S}^{n_{\mathrm{sh}}}$. Therefore, $0 < \widehat{c}_1 \leq \widehat{c}_2$. Consider now $\widehat{V} \in \mathbb{R}^{n_{\mathrm{sh}}}$ with $\widehat{V} \neq 0$. Let $\widehat{v} = \sum_{i=1}^{n_{\mathrm{sh}}} \widehat{V}_i \widehat{\theta}_i$ and $\eta_i(\widehat{v}) = \frac{\widehat{V}_i}{\|\widehat{V}\|_{n_{\mathrm{sh}}}}$ for $1 \leq i \leq n_{\mathrm{sh}}$. Clearly, $\eta(\widehat{v}) = (\eta_i(\widehat{v}))_{1 \leq i \leq n_{\mathrm{sh}}}$ is in $\mathcal{S}^{n_{\mathrm{sh}}}$. Since

$$\psi(\eta(\widehat{v})) = \frac{\|\widehat{v}\|_{0,\widehat{K}}^2}{\|\widehat{V}\|_{n_{\mathrm{sh}}}^2},$$

the following inequalities hold:

$$\forall \widehat{V} \in \mathbb{R}^{n_{\mathrm{sh}}}, \quad \widehat{c}_1 \|\widehat{V}\|_{n_{\mathrm{sh}}}^2 \leq \|\widehat{v}\|_{0,\widehat{K}}^2 \leq \widehat{c}_2 \|\widehat{V}\|_{n_{\mathrm{sh}}}^2. \tag{9.7}$$

(2) Consider now an arbitrary element K in the mesh. Denote by $T_K : \widehat{K} \to K$ the corresponding transformation and by $\{\theta_1, \ldots, \theta_{n_{\mathrm{sh}}}\}$ the local shape functions. For $V \in \mathbb{R}^{n_{\mathrm{sh}}}$, set $v = \sum_{i=1}^{n_{\mathrm{sh}}} V_i \theta_i$ and $\widehat{v} = v \circ T_K$. Changing variables in the integral in (9.7) yields

$$\forall V \in \mathbb{R}^{n_{\text{sh}}}, \quad \frac{\text{meas}(K)}{\text{meas}(\widehat{K})} \widehat{c}_1 \|V\|_{n_{\text{sh}}}^2 \leq \|v\|_{0,K}^2 \leq \frac{\text{meas}(K)}{\text{meas}(\widehat{K})} \widehat{c}_2 \|V\|_{n_{\text{sh}}}^2.$$

Clearly, $\frac{\text{meas}(K)}{\text{meas}(\widehat{K})} \leq ch_K^d \leq ch^d$. Furthermore, the quasi-uniformity of the mesh implies $c'h^d \leq \frac{\text{meas}(K)}{\text{meas}(\widehat{K})}$. As a result, there are $c_1 > 0$, $c_2 > 0$ such that

$$\forall h, \ \forall K \in \mathcal{T}_h, \ \forall V \in \mathbb{R}^{n_{\text{sh}}}, \quad c_1 h^d \|V\|_{n_{\text{sh}}}^2 \leq \|v\|_{0,K}^2 \leq c_2 h^d \|V\|_{n_{\text{sh}}}^2.$$

(3) For $1 \leq i \leq N$, denote by ζ_i the number of elements $K' \in \mathcal{T}_h$ whose intersection with the support of φ_i is of non-zero measure. Since $\{\mathcal{T}_h\}_{h>0}$ is quasi-uniform, $\max_{1 \leq i \leq N} \zeta_i$ is controlled uniformly with respect to h.

(4) To conclude the proof, consider an arbitrary function $v_h = \sum_{i=1}^{N} V_i \varphi_i$ in V_h. Step 2 shows that

$$\forall h, \ \forall K \in \mathcal{T}_h, \quad c_1 h^d \sum_{i \in \Upsilon_K} V_i^2 \leq \|v_h\|_{0,K}^2 \leq c_2 h^d \sum_{i \in \Upsilon_K} V_i^2,$$

where Υ_K is the set of indices i such that the intersection of K with the support of φ_i has non-zero measure. Summing over the elements yields

$$c_1 h^d \sum_{K \in \mathcal{T}_h} \sum_{i \in \Upsilon_K} V_i^2 \leq \|v_h\|_{0,\Omega}^2 \leq c_2 h^d \sum_{K \in \mathcal{T}_h} \sum_{i \in \Upsilon_K} V_i^2.$$

Since $\sum_{K \in \mathcal{T}_h} \sum_{i \in \Upsilon_K} V_i^2 = \sum_{i=1}^{N} \zeta_i V_i^2$, (9.6) readily results from step 3. $\qquad \square$

Theorem 9.8 (Mass matrix condition number). *If the affine family $\{\mathcal{T}_h\}_{h>0}$ is quasi-uniform, there exist $c_1 > 0$ and $c_2 > 0$ such that every eigenvalue μ of the mass matrix verifies*

$$\forall h, \quad c_1 h^d \leq \mu \leq c_2 h^d.$$

As a result, there exists c, independent of h, such that

$$\kappa(\mathcal{M}) \leq c.$$

Proof. Since the mass matrix is symmetric, its eigenvalues are real. Let μ be one of them and let $V \in \mathbb{R}^N$ be an associated eigenvector. Setting $v_h = \sum_{i=1}^{N} V_i \varphi_i$ yields $\|v_h\|_{0,\Omega}^2 = (V, \mathcal{M}V)_N = \mu(V, V)_N = \mu\|V\|_N^2$. Conclude using Lemma 9.7. $\qquad \square$

Example 9.9. In one dimension, the mass matrix obtained with the $\mathbb{P}_1$ Lagrange finite element on a uniform mesh with step size $h = \frac{1}{N+1}$ is

$$\mathcal{M} = \tfrac{h}{6} \text{tridiag}(1, 4, 1),$$

and its eigenvalues are $\{\frac{h}{3}(2 + \cos(ih\pi))\}_{1 \leq i \leq N}$. A straightforward calculation shows that the ratio of the largest to the smallest eigenvalue is controlled independently of h. $\qquad \square$

Remark 9.10.

(i) A practical consequence of Theorem 9.8 is that when $\{\mathcal{T}_h\}_{h>0}$ is quasi-uniform, the linear system $\mathcal{M}U = G$ is relatively inexpensive to solve approximately using the Conjugate Gradient algorithm; see §9.3.2.

(ii) If $\{\mathcal{T}_h\}_{h>0}$ is not quasi-uniform, the condition number of the mass matrix can be estimated by $h_{\max}^d h_{\min}^{-d}$, where $h_{\max}$ and $h_{\min}$ are the largest and smallest element diameters in the mesh, respectively.   □

9.1.4 Conditioning of the stiffness matrix

The goal of this section is to estimate the condition number of the stiffness matrix associated with a finite element approximation to a PDE-based problem. Use the notation introduced in the previous section and let the assumptions stated therein hold. Consider the discrete problem (2.11) and use the notation of §2.2. Assume $\dim(W_h) = \dim(V_h)$ and that the discrete inf-sup condition (BNB1$_h$) holds, i.e.,

$$\exists \alpha_h > 0, \quad \inf_{u_h \in W_h} \sup_{v_h \in V_h} \frac{a_h(u_h, v_h)}{\|u_h\|_{W_h} \|v_h\|_{V_h}} \geq \alpha_h. \tag{9.8}$$

Moreover, we make the following technical assumptions:

$$\exists c_{sP}, \quad \forall w_h \in W_h, \quad c_{sP}\|w_h\|_{0,\Omega} \leq \|w_h\|_{W_h}, \tag{9.9}$$

$$\exists c_{tP}, \quad \forall v_h \in V_h, \quad c_{tP}\|v_h\|_{0,\Omega} \leq \|v_h\|_{V_h}, \tag{9.10}$$

$$\exists s > 0, \exists c_{sI}, \quad \forall w_h \in W_h, \quad \|w_h\|_{W_h} \leq c_{sI}h^{-s}\|w_h\|_{0,\Omega}, \tag{9.11}$$

$$\exists t > 0, \exists c_{tI}, \quad \forall v_h \in V_h, \quad \|v_h\|_{V_h} \leq c_{tI}h^{-t}\|v_h\|_{0,\Omega}. \tag{9.12}$$

Estimates (9.9) and (9.10) are Poincaré-like inequalities expressing the fact that the norms equipping W_h and V_h control the L^2-norm. Furthermore, (9.11) and (9.12) are inverse inequalities. When the mesh family $\{\mathcal{T}_h\}_{h>0}$ is quasi-uniform, the constants s and t can be interpreted as the order of the differential operator used to defined the norms in W_h and V_h, respectively.

Let $\mathcal{M}_s$ and $\mathcal{M}_t$ be the mass matrices associated with the discrete spaces W_h and V_h, respectively. Denote by $\mu_{t,\min}$ and $\mu_{s,\min}$ (resp., $\mu_{t,\max}$ and $\mu_{s,\max}$) the smallest (resp., largest) eigenvalue of $\mathcal{M}_s$ and $\mathcal{M}_t$, respectively. Let $\{\psi_1, \dots, \psi_N\}$ and $\{\varphi_1, \dots, \varphi_N\}$ be the global shape functions in W_h and V_h, respectively. For $u_h \in W_h$ and $v_h \in V_h$, denote by $U = (U_i)_{1 \leq i \leq N}$ and $V = (V_i)_{1 \leq i \leq N}$ the coordinate vectors relative to these bases, i.e., $u_h = \sum_{i=1}^N U_i\psi_i$ and $v_h = \sum_{i=1}^N V_i\varphi_i$. Recalling that the stiffness matrix $\mathcal{A}$ has entries $\left(a_h(\psi_j, \varphi_i)\right)_{1 \leq i,j \leq N}$, we infer $(V, \mathcal{A}U)_N = a_h(u_h, v_h)$.

Theorem 9.11 (Stiffness matrix condition number). *Under the above assumptions,*

$$\forall h, \quad \kappa(\mathcal{A}) \leq \kappa(\mathcal{M}_s)^{\frac{1}{2}}\kappa(\mathcal{M}_t)^{\frac{1}{2}} \frac{c_{sI}c_{tI}}{c_{sP}c_{tP}} \frac{\|a_h\|}{\alpha_h} h^{-s-t}. \tag{9.13}$$

Moreover, if $\{\mathcal{T}_h\}_{h>0}$ is quasi-uniform and a_h is uniformly bounded and stable with respect to h,

$$\kappa(\mathcal{A}) \le ch^{-s-t}. \tag{9.14}$$

Proof. (1) Let us bound $\|\mathcal{A}\|_N$. Consider $U \in \mathbb{R}^N$; then,

$$\|\mathcal{A}U\|_N = \sup_{V \in \mathbb{R}^N} \frac{(V, \mathcal{A}U)_N}{\|V\|_N} = \sup_{V \in \mathbb{R}^N} \frac{a_h(u_h, v_h)}{\|V\|_N}$$

$$\le \|a_h\| \, \|u_h\|_{W_h} \left(\sup_{V \in \mathbb{R}^N} \frac{\|v_h\|_{V_h}}{\|V\|_N} \right).$$

Using the inverse inequalities (9.11)–(9.12) yields

$$\|\mathcal{A}U\|_N \le c_{sI} c_{tI} h^{-s-t} \|a_h\| \, \|u_h\|_{0,\Omega} \left(\sup_{V \in \mathbb{R}^N} \frac{\|v_h\|_{0,\Omega}}{\|V\|_N} \right)$$

$$\le c_{sI} c_{tI} h^{-s-t} \|a_h\| \, \|u_h\|_{0,\Omega} \, \mu_{t,\max}^{\frac{1}{2}}$$

$$\le c_{sI} c_{tI} \mu_{s,\max}^{\frac{1}{2}} \mu_{t,\max}^{\frac{1}{2}} h^{-s-t} \|a_h\| \, \|U\|_N.$$

That is to say,

$$\|\mathcal{A}\|_N \le \|a_h\| \, c_{sI} c_{tI} \mu_{s,\max}^{\frac{1}{2}} \mu_{t,\max}^{\frac{1}{2}} h^{-s-t}.$$

(2) Let us bound $\|\mathcal{A}^{-1}\|_N$. A straightforward calculation yields

$$\|U\|_N \le \mu_{s,\min}^{-\frac{1}{2}} \|u_h\|_{0,\Omega} \quad \text{and} \quad \|V\|_N \le \mu_{t,\min}^{-\frac{1}{2}} \|v_h\|_{0,\Omega}.$$

The discrete stability condition (9.8) together with the Poincaré-like inequalities (9.9)–(9.10) implies

$$c_{sP} \mu_{s,\min}^{\frac{1}{2}} \alpha_h \|U\|_N \le c_{sP} \alpha_h \|u_h\|_{0,\Omega} \le \alpha_h \|u_h\|_{W_h}$$

$$\le \sup_{v_h \in V_h} \frac{a_h(u_h, v_h)}{\|v_h\|_{V_h}} \le \frac{1}{c_{tP}} \sup_{V \in \mathbb{R}^N} \frac{(V, \mathcal{A}U)_N}{\|v_h\|_{0,\Omega}}$$

$$\le \frac{1}{c_{tP}} \|\mathcal{A}U\|_N \sup_{V \in \mathbb{R}^N} \frac{\|V\|_N}{\|v_h\|_{0,\Omega}} \le \frac{1}{c_{tP}} \mu_{t,\min}^{-\frac{1}{2}} \|\mathcal{A}U\|_N.$$

Hence,

$$\|\mathcal{A}^{-1}\|_N \le \frac{1}{\alpha_h} \frac{1}{c_{sP} c_{tP}} \mu_{s,\min}^{-\frac{1}{2}} \mu_{t,\min}^{-\frac{1}{2}}.$$

The upper bound (9.13) is a direct consequence of the above estimates.
(3) If the family $\{\mathcal{T}_h\}_{h>0}$ is quasi-uniform, $\kappa(\mathcal{M}_s)$ and $\kappa(\mathcal{M}_s)$ are bounded uniformly with respect to h, yielding (9.14). $\square$

Remark 9.12. If $\{\mathcal{T}_h\}_{h>0}$ is not quasi-uniform, the estimate on the condition number of the stiffness matrix takes the form $\kappa(\mathcal{A}) \le c_2 h_{\max}^d h_{\min}^{-d-s-t}$ provided a_h is uniformly bounded and stable with respect to h. $\square$

Example 9.13. For first-order PDEs, the exponents s and t in (9.11)–(9.12) are $s = 1$ and $t = 0$; hence, under the assumptions for (9.14), $\kappa(\mathcal{A}) \leq ch^{-1}$. For second-order PDEs, $s = 1$ and $t = 1$; hence, the upper bound is $\kappa(\mathcal{A}) \leq ch^{-2}$. Finally, for fourth-order PDEs, $s = 2$ and $t = 2$ yielding $\kappa(\mathcal{A}) \leq ch^{-4}$.   □

It is possible to sharpen (9.13) when a_h is symmetric and coercive. Assume that $W = V$ and $W_h = V_h$ and denote by α_h the coercivity constant of a_h in V_h. Assume that there is $c_{\text{loc}} > 0$ such that, for all h, there is a non-zero function $u_h \in V_h$ satisfying

$$c_{\text{loc}} \|u_h\|_{0,\Omega} \leq h^s \|u_h\|_{V_h}. \tag{9.15}$$

This hypothesis is a localized Poincaré inequality. For instance, if $\| \cdot \|_{V_h}$ is a (possibly broken) H^1-norm, (9.15) holds with $s = 1$ for every function whose support has a diameter of order h; see Remark B.62(iii). In particular, (9.15) holds for global shape functions in finite element spaces.

We also assume that there is c_P, independent of h, such that

$$\forall z \in V, \quad c_P \|z\|_{0,\Omega} \leq \|z\|_{V(h)}. \tag{9.16}$$

Recall that the (possibly broken) norm $\| \cdot \|_{V(h)}$ equipping $V(h) := V + V_h$ is such that $\| \cdot \|_{V(h)} = \| \cdot \|_{V_h}$ on V_h, and V is uniformly continuously embedded in $V(h)$. Denote by c_{inj} the embedding constant, i.e., $\|z\|_{V(h)} \leq c_{\text{inj}}\|z\|_V$ for all $z \in V$. Define

$$c_{\min} = \inf_{h>0} \inf_{z \in V} \frac{\|z\|_{V(h)}}{\|z\|_{0,\Omega}}. \tag{9.17}$$

Clearly, $c_{\min} \leq c_{\text{inj}} \inf_{z \in V} \frac{\|z\|_V}{\|z\|_{0,\Omega}} < \infty$, i.e., $c_{\min}$ is finite. Moreover, owing to (9.16), we infer $c_{\min} \geq c_P$, i.e., $c_{\min}$ is positive.

Theorem 9.14. *Assume (9.15), (9.16), (9.17), and that V is uniformly continuously embedded in $V(h)$. Assume that the discrete setting has the approximability property and that a_h is symmetric and coercive. Then,*

$$\exists h_0, \ \forall h \leq h_0, \quad \kappa(\mathcal{A}) \geq \frac{1}{4} \frac{c_{\text{loc}}^2}{c_{\min}^2} \frac{\alpha_h}{\|a_h\|} \kappa(\mathcal{M})^{-1} h^{-2s}. \tag{9.18}$$

Proof. We derive lower bounds for $\|\mathcal{A}\|_N$ and $\|\mathcal{A}^{-1}\|_N$.
(1) Let u_h be one of the functions satisfying (9.15). Then,

$$\|\mathcal{A}U\|_N \|U\|_N \geq a_h(u_h, u_h) \geq \alpha_h \|u_h\|_{V_h}^2 \geq \alpha_h c_{\text{loc}}^2 h^{-2s} \|u_h\|_{0,\Omega}^2.$$

This yields the estimate $\|\mathcal{A}U\|_N \geq \alpha_h c_{\text{loc}}^2 h^{-2s} \mu_{\min} \|U\|_N$, i.e.,

$$\|\mathcal{A}\|_N \geq \alpha_h c_{\text{loc}}^2 h^{-2s} \mu_{\min}.$$

(2) Owing to the approximability property, it is possible to find $z_h \in V_h$, $z_h \neq 0$, such that

$$\|z_h\|_{V_h} = \|z_h\|_{V(h)} \leq 2c_{\min}\|z_h\|_{0,\Omega},$$

if h is small enough. Denoting by $Z \in \mathbb{R}^N$ the coordinate vector of z_h relative to the basis $\{\varphi_1, \ldots, \varphi_N\}$ yields

$$\lambda_{\min}\|Z\|_N^2 \leq a_h(z_h, z_h) \leq 4c_{\min}^2\|a_h\|\|z_h\|_{0,\Omega}^2 \leq 4c_{\min}^2\|a_h\|\mu_{\max}\|Z\|_N^2,$$

where $\lambda_{\min}$ is the smallest eigenvalue of $\mathcal{A}$. As a result,

$$\lambda_{\min} \leq 4c_{\min}^2\|a_h\|\mu_{\max}.$$

Since $\mathcal{A}$ is symmetric positive definite, $\lambda_{\min}^{-1} = \|\mathcal{A}^{-1}\|_N$, and the lower bound for $\kappa(\mathcal{A})$ follows readily. $\qquad\square$

Example 9.15. Set $\Omega = \,]0,1[$ and consider the Laplace equation in Ω supplemented with homogeneous Dirichlet boundary conditions. The stiffness matrix obtained with continuous $\mathbb{P}_1$ Lagrange finite elements on a uniform mesh with step size $h = \frac{1}{N+1}$ is

$$\mathcal{A} = \tfrac{1}{h}\mathrm{tridiag}(-1, 2, -1),$$

and the eigenvalues are $\{\frac{2}{h}(1-\cos(ih\pi))\}_{1 \leq i \leq N}$. One readily verifies that $\kappa(\mathcal{A})$ is asymptotically of order h^{-2} as $h \to 0$. $\qquad\square$

Remark 9.16. The reader must bear in mind that the condition number of the stiffness matrix and that of the mass matrix depend on the choice made for the global shape functions. In particular, the fact that the condition number of the stiffness matrix explodes when $h \to 0$ results from the global shape functions having a localized support. An alternative possibility is to choose hierarchical bases for V_h and W_h. If those hierarchical bases are well-chosen, it is possible to bound the condition number of the stiffness matrix uniformly with respect to h; see, e.g., [Hac85, BrP00]. Note, however, that (9.18) implies that if $\kappa(\mathcal{A})$ is bounded, then $\kappa(\mathcal{M})$ must explode, i.e., it is not possible to find bases for which both $\kappa(\mathcal{A})$ and $\kappa(\mathcal{M})$ are bounded uniformly. $\qquad\square$

9.1.5 Ill-conditioning and discrete PDE stability

Use the notation of the previous section and let $u_h \in W_h$ solve

$$a_h(u_h, v_h) = f_h(v_h), \quad \forall v_h \in V_h. \tag{9.19}$$

Let U^m be an approximate solution to the linear system

$$\mathcal{A}U = F, \tag{9.20}$$

resulting, for instance, from the use of an iterative method. Let $E^m = U - U^m$ be the error and let $R^m = F - \mathcal{A}U^m$ be the residual. The quality of the approximation provided by U^m is assessed by the error E^m. However, the only quantity accessible to actual computation is the residual R^m. It is therefore important to determine whether the error is controlled by the residual.

A naive analysis yields rather discouraging results. Indeed, since $\mathcal{A}E^m = R^m$, Proposition 9.5 implies an estimate of the form

$$\frac{\|E^m\|_N}{\|U\|_N} \le \kappa(\mathcal{A}) \frac{\|R^m\|_N}{\|F\|_N},$$

whence we deduce using Theorem 9.11 that

$$\frac{\|E^m\|_N}{\|U\|_N} \le c\,h^{-s-t} \frac{\|R^m\|_N}{\|F\|_N},$$

i.e., the error in the Euclidean norm cannot be controlled by the residual on fine meshes.

However, a sharper estimate can be derived if the stability properties of the underlying PDE are taken into account. We first point out that we are not interested in the coordinate vector $U^m \in \mathbb{R}^N$ but in the reconstructed function instead, say

$$u_h^m = \sum_{i=1}^N U_i^m \psi_i \in W_h.$$

This means that our goal is to control $\|u_h - u_h^m\|_{W_h}$ instead of $\|E^m\|_N$. To this end, we use the matrix interpretation of the discrete stability inequality (9.8). Equipping the vector space $\mathbb{R}^N$ with the norm

$$\forall Z \in \mathbb{R}^N, \qquad \|Z\|_\star = \sup_{Y \in \mathbb{R}^N} \frac{(Y, Z)_N}{\|y_h\|_{V_h}},$$

the following result holds:

Lemma 9.17. *Assume* (9.8). *Then, for all* $Z = (Z_1, \dots, Z_N) \in \mathbb{R}^N$, *setting* $z_h = \sum_{i=1}^N Z_i \psi_i$,

$$\alpha_h \|z_h\|_{W_h} \le \|\mathcal{A}Z\|_\star.$$

Proof. Since $z_h \in W_h$, (9.8) yields

$$\alpha_h \|z_h\|_{W_h} \le \sup_{y_h \in V_h} \frac{a_h(z_h, y_h)}{\|y_h\|_{V_h}} = \sup_{Y \in \mathbb{R}^N} \frac{(Y, \mathcal{A}Z)_N}{\|y_h\|_{V_h}} = \|\mathcal{A}Z\|_\star. \qquad \square$$

Proposition 9.18. *Assume* (9.8). *Then,*

$$\frac{\|u_h - u_h^m\|_{W_h}}{\|u_h\|_{W_h}} \le \frac{\|a_h\|}{\alpha_h} \frac{\|R^m\|_\star}{\|F\|_\star}. \tag{9.21}$$

Proof. Since $F = (f_h(\varphi_1), \dots, f_h(\varphi_N))$,

$$\|F\|_\star = \sup_{Y \in \mathbb{R}^N} \frac{(Y, F)_N}{\|y_h\|_{V_h}} = \sup_{y_h \in V_h} \frac{f_h(y_h)}{\|y_h\|_{V_h}} = \sup_{y_h \in V_h} \frac{a_h(u_h, y_h)}{\|y_h\|_{V_h}} \le \|a_h\| \|u_h\|_{W_h}.$$

Furthermore, owing to Lemma 9.17, $\alpha_h \|u_h - u_h^m\|_{W_h} \le \|\mathcal{A}E^m\|_\star = \|R^m\|_\star$, completing the proof. $\qquad \square$

A practical consequence of (9.21) is that an appropriate convergence criterion for an iterative method appears to be the relative residual in the $\|\cdot\|_*$-norm, i.e., the ratio $\frac{\|R^m\|_*}{\|F\|_*}$. However, the $\|\cdot\|_*$-norm is rather difficult to evaluate in practice. It is instead preferable to evaluate quantities involving Euclidean and L^2-norms. This motivates the following:

Proposition 9.19. *Assume* (9.8) *and that there is* $f \in L^2(\Omega)$ *such that* $|f_h(v_h)| \leq \|f\|_{0,\Omega}\|v_h\|_{0,\Omega}$. *Then,*

$$\frac{\|u_h - u_h^m\|_{W_h}}{\|f\|_{0,\Omega}} \leq \frac{\kappa(\mathcal{M}_t)^{\frac{1}{2}}}{c_{tP}\alpha_h}\frac{\|R^m\|_N}{\|F\|_N}. \tag{9.22}$$

Proof. Denote by $\mu_{t,\max}$ and $\mu_{t,\min}$ the largest and the smallest eigenvalue of the mass matrix $\mathcal{M}_t$, respectively. Then,

$$\|F\|_N = \sup_{Y \in \mathbb{R}^N} \frac{(Y, F)_N}{\|Y\|_N} = \sup_{Y \in \mathbb{R}^N} \frac{f_h(y_h)}{\|Y\|_N}$$

$$\leq \mu_{t,\max}^{\frac{1}{2}} \sup_{y_h \in V_h} \frac{f(y_h)}{\|y_h\|_{0,\Omega}} \leq \mu_{t,\max}^{\frac{1}{2}}\|f\|_{0,\Omega}.$$

Moreover, since $\mu_{t,\min}^{\frac{1}{2}}\|Y\|_N \leq \|y_h\|_{0,\Omega} \leq \frac{1}{c_{tP}}\|y_h\|_{V_h}$, inequality (9.8) implies

$$\|R^m\|_N = \|AE^m\|_N = \sup_{Y \in \mathbb{R}^N} \frac{(Y, AE^m)_N}{\|Y\|_N} \geq \mu_{t,\min}^{\frac{1}{2}}c_{tP} \sup_{Y \in \mathbb{R}^N} \frac{a_h(e_h^m, y_h)}{\|y_h\|_{V_h}}$$

$$\geq c_{tP}\,\mu_{t,\min}^{\frac{1}{2}}\alpha_h\|e_h^m\|_{W_h},$$

with $e_h^m = u_h - u_h^m$. The conclusion follows readily. $\qquad\square$

Estimate (9.22) has important practical consequences. It means that if we control the relative residual in the Euclidean norm, then we control the relative error measured in the W_h-norm. Furthermore, the control is uniform with respect to h if the family $\{\mathcal{T}_h\}_{h>0}$ is quasi-uniform and if the bilinear form a_h is uniformly bounded and stable with respect to h. In conclusion, *if the approximate problem is well-posed, the linear system is stable in spite of the possible ill-conditioning of the system matrix*, i.e., perturbations (measured in the natural stability norm) of the right-hand side induce perturbations on the solution that are of the same order of magnitude.

9.2 Reordering

This section deals with reordering techniques for sparse matrices. We begin with a brief discussion on direct solution methods for linear systems to illustrate the fact that for sparse systems, reordering techniques can improve significantly the computational efficiency of direct solution methods. We then present various reordering techniques based on the concept of adjacency graphs: level-set orderings; independent-set orderings; and multicolor orderings.

9.2.1 Direct solution methods

A first approach that can be considered to solve a linear system is to employ a *direct solution method*. The best-known example consists of constructing the *LU factorization* with complete pivoting of the matrix $\mathcal{A}$ in the form

$$\mathcal{P}\mathcal{A}\mathcal{Q} = \mathcal{T}_{\text{lo}}\mathcal{T}_{\text{up}}, \qquad\qquad (9.23)$$

where $\mathcal{P}$ and $\mathcal{Q}$ are permutation matrices, $\mathcal{T}_{\text{lo}}$ is lower triangular, and $\mathcal{T}_{\text{up}}$ is upper triangular. These matrices are constructed using *Gaussian elimination*; see, e.g., [GoV89, pp. 96–119] or [BuF93, p. 323] for a detailed presentation. The permutation matrices $\mathcal{P}$ and $\mathcal{Q}$ are needed to avoid divisions by zero and, more generally, by very small numbers compared to the matrix entries. In some cases, partial pivoting is considered instead of complete pivoting, e.g., only row permutations are performed and $\mathcal{Q}$ is the identity matrix.

Once the LU factorization of $\mathcal{A}$ has been constructed, the linear system $\mathcal{A}U = F$ is readily solved in three steps:

  (i) Solve the lower triangular system $\mathcal{T}_{\text{lo}}U' = \mathcal{P}F$.
 (ii) Solve the upper triangular system $\mathcal{T}_{\text{up}}U'' = U'$.
(iii) Set $U = \mathcal{Q}U''$.

In computer implementations, an important criterion when selecting a solution algorithm is its asymptotic cost.

Definition 9.20 (Asymptotic cost). *Consider an algorithm to solve a linear system of order N. Let $\omega(N)$ be the number of* floating point operations (flops) *performed, one flop being typically an addition or a multiplication. The algorithm is said to have* asymptotic cost $\chi(N)$ *if*

$$\frac{\omega(N)}{\chi(N)} \to 1 \qquad \text{as} \quad N \to \infty.$$

It is useful to bear in mind that the asymptotic cost of a direct solution based on LU factorization without pivoting is $\frac{2N^3}{3}$ and that this cost is dominated by that incurred when forming the LU factorization. The asymptotic cost of the lower and upper triangular solves scales as N^2 only.

When the system matrix is *symmetric*, algorithms exploiting symmetry are usually employed. For instance, a symmetric matrix can be factored into the form $\mathcal{A} = \mathcal{T}_{\text{lo}}\mathcal{D}\mathcal{T}_{\text{lo}}^T$, where $\mathcal{T}_{\text{lo}}$ is lower triangular with diagonal entries equal to one and $\mathcal{D}$ is diagonal. This form is the so-called LDL^T *factorization*; see, e.g., [GoV89, p. 137] or [BuF93, p. 376] for a detailed presentation. Because it exploits the symmetry of $\mathcal{A}$, the LDL^T factorization has an asymptotic cost scaling as $\frac{N^3}{3}$, i.e., half the cost of the LU factorization.

Remark 9.21. If the matrix $\mathcal{A}$ is *symmetric positive definite*, the Choleski algorithm can be employed; see, e.g., [GoV89, p. 141] or [BuF93, p. 376]. The

Choleski factorization of a symmetric positive definite matrix takes the form $\mathcal{A} = \mathcal{T}_{\text{lo}} \mathcal{T}_{\text{lo}}^T$, where $\mathcal{T}_{\text{lo}}$ is lower triangular. Owing to the absence of the diagonal matrix $\mathcal{D}$ considered in the LDL^T factorization, the Choleski algorithm requires the evaluation of N square roots, which can be computationally expensive. $\qquad\qquad\square$

Linear systems resulting from finite element discretizations of PDEs generally involve sparse matrices; see Definition 8.13. In practical simulations, the size of matrices and therefore the level of accuracy that can be achieved (resulting from grid refinement or high-order polynomial interpolation) is very often limited by computer memory. In this context, the major disadvantage of direct solution methods is that they *fill the sparse structure of the system matrix*. To see this, consider a sparse matrix $\mathcal{A} \in \mathbb{R}^{N,N}$. For $1 \le i \le N$, set

$$\begin{aligned}
l_i &= \min\{j \le i; \mathcal{A}_{ij} \ne 0\}, \\
u_i &= \max\{j \ge i; \mathcal{A}_{ij} \ne 0\}.
\end{aligned} \tag{9.24}$$

The number $\max_{1 \le i \le N}(u_i - l_i)$ is called the *bandwidth* of $\mathcal{A}$. Denote by $n_{\text{row}}(i)$ the number of non-zero entries in the i-th row of $\mathcal{A}$. Note that the total number of non-zero entries in $\mathcal{A}$ is $nnz_{\mathcal{A}} = \sum_{i=1}^{N} n_{\text{row}}(i)$. An important observation is that matrices resulting from finite element discretizations are such that $n_{\text{row}}(i) \ll u_i - l_i$ for large N even if the enumeration of the degrees of freedom is carefully done. Consider now the LU factorization of $\mathcal{A}$. One can show that in the absence of pivoting, the lower triangular matrix $\mathcal{T}_{\text{lo}}$ is such that $\mathcal{T}_{\text{lo},ij} = 0$ for $j < l_i$, but $\mathcal{T}_{\text{lo},ij} \ne 0$ for most $j \in \{l_i, \ldots, i\}$. Similarly, the upper triangular matrix $\mathcal{T}_{\text{up}}$ is such that $\mathcal{T}_{\text{up},ij} = 0$ for $j > u_i$, but $\mathcal{T}_{\text{up},ij} \ne 0$ for most $j \in \{i, \ldots, u_i\}$. As a result, the total number of non-zero entries generated by the LU factorization is of the order of $\sum_{i=1}^{N}(u_i - l_i) \gg \sum_{i=1}^{N} n_{\text{row}}(i) = nnz_{\mathcal{A}}$ and is therefore much larger than $nnz_{\mathcal{A}}$. Exercise 9.3 deals with an example of storage overhead incurred by the LU factorization.

An important technique to *reduce the level of fill-in* in an LU factorization is to reorder the entries of the matrix. Consider, for instance, the square matrix $\mathcal{A} \in \mathbb{R}^{8,8}$ whose sparsity pattern is shown in the left panel of Figure 9.1. It is clear that the LU factorization of $\mathcal{A}$ results in disastrous fill-ins, i.e., $\mathcal{T}_{\text{lo}}$ and $\mathcal{T}_{\text{up}}$ are dense matrices. Actually, all the entries are filled after the first step of the Gaussian elimination. Now, let us consider the permutation $\sigma : \{1, 2, 3, 4, 5, 6, 7, 8\} \mapsto \{8, 7, 6, 5, 4, 3, 2, 1\}$. Let $\mathcal{B}$ be the matrix obtained by permuting the rows and the columns of $\mathcal{A}$ using σ, i.e., $\mathcal{B}_{ij} = \mathcal{A}_{\sigma(i)\sigma(j)}$. The sparsity pattern of $\mathcal{B}$ is shown in the right panel of Figure 9.1. It is straightforward to check that no fill-ins occur when forming the LU factorization of $\mathcal{B}$. This example illustrates the fact that significant savings in memory and computational time can be achieved by enumerating properly the degrees of freedom in a finite element code.

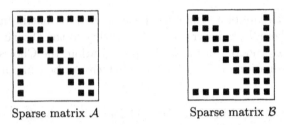

Sparse matrix $\mathcal{A}$ Sparse matrix $\mathcal{B}$

Fig. 9.1. Two different orderings for a sparse matrix.

9.2.2 Adjacency graphs

As demonstrated by the above example, it is often important to reorder the unknowns and the equations before solving a linear system. Of course, the reordering technique to be used depends on the strategy chosen to solve the linear system (direct, iterative, parallel, etc.). Choosing optimal reordering strategies is a difficult branch of graph theory. See [Saa96] for a review of reordering techniques in the context of iterative solution methods and to [GeL81, GeG93] for a review in the context of direct solution methods.

To better understand the enumeration issue, it is convenient to introduce the notion of adjacency graph. Let V be a set, let $\Re$ be a binary relation on V, and denote $E = \{(x, y) \in V \times V; x\Re y\}$. The pair $(V, E) = G$ is called a *graph*. The elements of V are the *graph vertices* or nodes and those of E are the *graph edges*. The graph G is said to be an *undirected graph* if $\Re$ is symmetric. A vertex y is said to be adjacent to x if $(x, y) \in E$. For a subset $X \subset V$, the *adjacent set* of X is defined as $\text{Adj}(X) = \{y \in V \backslash X; \exists x \in X, (x, y) \in E\}$. The set $\text{Adj}(x)$, defined as $\text{Adj}(\{x\})$, is called the *neighborhood* of x. The cardinal number of $\text{Adj}(x)$ is called the *degree* of x. A common way of representing graphs is to associate with each vertex in V a point in the plane and to draw a directed line between two points (possibly identical) whenever their associated vertices are in E.

Let $\mathcal{A} \in \mathbb{R}^{N,N}$ be a square matrix. The *adjacency graph* of $\mathcal{A}$ is the pair (V, E) where $V = \{1, 2, \dots, N\}$ and $(i, j) \in E$ iff $\mathcal{A}_{ij} \neq 0$, $i, j \in V$. Clearly, the graph is undirected iff ($\mathcal{A}_{ij} \neq 0$ iff $\mathcal{A}_{ji} \neq 0$, $i, j \in V$). Figure 9.2 displays the adjacency graph of an 8×8 sparse matrix. A circle around a number means that the corresponding diagonal entry in the matrix is not zero.

9.2.3 Level-set orderings

Assume that V is finite, let $G = (V, E)$ be a graph, and let $x \in V$ be a vertex. The elements of an indexed collection of disjoint subsets of V, say $L_1, L_2, L_3, \dots$, are said to be *level-sets* associated with x if these subsets satisfy the following recursive property: $L_1 = \{x\}$ and $L_{k+1} = \text{Adj}(L_k) \backslash (\bigcup_{l=1}^{k} L_l)$. L_k is said to be the k-th level-set. Since V is finite, the list $L_1, L_2, L_3, \dots$ is finite. Moreover, if G is a *strongly connected graph*, i.e., if there exists a path

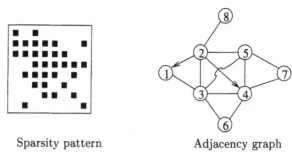

Sparsity pattern   Adjacency graph

Fig. 9.2. Adjacency graph of an 8×8 sparse matrix.

from each vertex to every other vertex, then $L_1, L_2, L_3, \ldots$ forms a partition of V.

For every vertex y in V, define the distance from x to y, $d_x(y)$, as follows: If there is k such that $y \in L_k$, then set $d_x(y) = k - 1$, otherwise set $d_x(y) = +\infty$ (note that $d_x(y) < +\infty$ if the graph is strongly connected). In general $d_x(y) \neq d_y(x)$ unless the graph is undirected; think of $V = \{x, y\}$ and $E = \{(x, y)\}$ where $d_x(y) = 1$ and $d_y(x) = +\infty$.

Algorithm 9.1. Evaluate the level-sets of i_1.

Input: i_1
Output: perm, max_levelset, stride
$k = 2$; count $= 1$; virgin$(1 : N) =$.true.
perm$(1) = i_1$; stride$(1) = 1$; stride$(2) = 2$
loop
  nb_vert_in_Lk $= 0$
  for stride$(k-1) \leq l \leq$ stride$(k) - 1$ **do**
    for all $j \in$ Adj(perm(l)) **do**
      if (virgin(j)) **then**
        virgin$(j) = $.false.; nb_vert_in_Lk $=$ nb_vert_in_Lk $+ 1$
        count $=$ count $+ 1$; perm(count) $= j$
      end if
    end for
  end for
  if (nb_vert_in_Lk $= 0$) **then**
    max_levelset $= k - 1$; exit loop
  end if
  stride$(k+1) =$ stride$(k) +$ nb_vert_in_Lk; $k = k + 1$
end loop
if count $\neq N$ **then** G is not strongly connected

Algorithm 9.1 shows a possible way to evaluate the level-sets associated with a vertex i_1 in the adjacency graph of a sparse matrix $A \in \mathbb{R}^{N,N}$. The integer max_levelset is the number of level-sets associated with the vertex

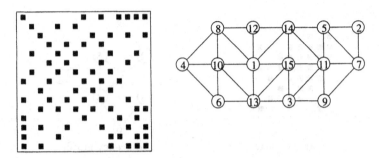

Fig. 9.3. Sparsity pattern and adjacency graph.

i_1. The level-sets are deduced from the arrays `perm` and `stride` as follows:

$$\underbrace{\{\texttt{perm}(1)\}}_{=L_1}, \ldots, \underbrace{\{\texttt{perm}(\texttt{stride}(k)), \ldots, \texttt{perm}(\texttt{stride}(k+1)-1)\}}_{=L_k}, \ldots,$$

If there is a vertex i_2 which is not in any of the level-sets associated with i_1, i.e., i_1 is not connected to i_2, then the level-sets associated with i_2 are constructed. The process is repeated until the union of all the level-sets forms a partition of V. At the end, all the permutation arrays are collected in a single array still denoted by `perm`.

Example 9.22. To illustrate the level-set concept, let us consider the undirected adjacency graph shown in Figure 9.3. The level-sets associated with vertex 2 are

$$L_1 = \{2\}, \quad L_2 = \{5, 7\}, \quad L_3 = \{9, 11, 14\}, \quad L_4 = \{1, 3, 12, 15\},$$
$$L_5 = \{8, 10, 13\}, \quad L_6 = \{4, 6\}.$$

Clearly, `max_levelset` $= 6$, `stride` $= (1, 2, 4, 7, 11, 14, 16)$, and a possible choice for `perm` is `perm` $= (2, 5, 7, 9, 11, 14, 1, 3, 12, 15, 8, 10, 13, 4, 6)$. □

The simplest reordering strategy for $\mathcal{A}$ consists of setting

$$\mathcal{B}_{ij} = \mathcal{A}_{\texttt{perm}(i)\texttt{perm}(j)}.$$

This technique is known as the *Breadth-First-Search* (BFS) reordering. One interest of this reordering is the following result:

Proposition 9.23. *Assume G is undirected and* `max_levelset` ≥ 3. *The array* `stride` *defines a tridiagonal block structure of $\mathcal{B}$, i.e.,*

$$\mathcal{B}_{ij} = 0 \quad \text{if} \quad \begin{cases} \texttt{stride}(k) \leq i \leq \texttt{stride}(k+1)-1, \text{ i.e., } \texttt{perm}(i) \in L_k, \\ \texttt{stride}(k') \leq j \leq \texttt{stride}(k'+1)-1, \text{ i.e., } \texttt{perm}(j) \in L_{k'}, \\ |k-k'| \geq 2. \end{cases}$$

Proof. By contradiction. Assume $\mathcal{B}_{ij} \neq 0$ and $|k - k'| \geq 2$ with $\texttt{perm}(i) \in L_k$ and $\texttt{perm}(j) \in L_{k'}$; then, $\mathcal{A}_{\texttt{perm}(i)\texttt{perm}(j)} \neq 0$. This means $\texttt{perm}(i) \in \text{Adj}(\texttt{perm}(j))$ and $\texttt{perm}(j) \in \text{Adj}(\texttt{perm}(i))$, since G is undirected and $i \neq j$. Assume further that $k' \geq k$. Then, $\texttt{perm}(j) \notin \bigcup_{l=1}^{k} L_l$, since the level-sets are disjoint. Moreover, $\texttt{perm}(j) \in \text{Adj}(\texttt{perm}(i))$ and $\texttt{perm}(i) \in L_k$ means $\texttt{perm}(j) \in \text{Adj}(L_k)$. Combining the above two statements yields $\texttt{perm}(j) \in L_{k+1} = \text{Adj}(L_k) \backslash (\bigcup_{l=1}^{k} L_l)$. This means $k' = k + 1$, which contradicts $|k - k'| \geq 2$. The argument applies also if $k' \leq k$ since the graph is undirected. $\qquad\square$

Proposition 9.23 shows that choosing level-sets with $\texttt{max\_levelset}$ as large as possible minimizes the bandwidth of $\mathcal{B}$. This can be achieved by picking the initial vertex i_1 such that $\max_{y \in V} d_{i_1}(y)$ is maximal.

The ordering depends on the way the vertices are traversed in each level-set. In the BFS reordering, the vertices are traversed in the natural order. Another strategy consists of ordering the vertices in each level-set by increasing degree. This ordering technique is known as the *Cuthill–McKee* (CMK) ordering. Another popular strategy consists of reversing the CMK ordering. It has been observed that the reversing strategy yields a better scheme for sparse Gaussian elimination; see George et al. [GeL81, GeG93] for further insight into these techniques and their many generalizations.

Example 9.24. Figure 9.4 displays the adjacency graph and the sparsity pattern of the CMK-reordered matrix corresponding to the matrix shown in Figure 9.3. The level-sets associated with vertex 2 have been used, and in each level-set, the nodes are ordered by increasing degree. The resulting permutation array is $\texttt{perm} = (2, 5, 7, 9, 14, 11, 12, 3, 15, 1, 8, 10, 13, 4, 6)$. The reordered matrix has a tridiagonal block structure, and the size of the k-th block is $\texttt{stride}(k + 1) - \texttt{stride}(k)$ where the array $\texttt{stride}$ is evaluated in Example 9.22. $\qquad\square$

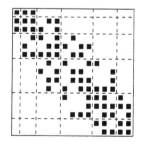 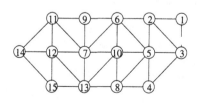

Fig. 9.4. Sparsity pattern and adjacency graph after reordering the matrix shown in Figure 9.3.

9.2.4 Independent Set Orderings (ISO)

The aim of ISO techniques is to find a permutation of the vertices such that the reordered matrix has the following 2×2 block structure:

$$\mathcal{B} = \begin{bmatrix} \mathcal{D} & \mathcal{E} \\ \hline \mathcal{F} & \mathcal{H} \end{bmatrix},$$

where $\mathcal{D}$ is diagonal and as large as possible. For this purpose, we introduce the notion of *independent set*. Let $G = (V, E)$ be a graph. $S \subset V$ is said to be an independent set if for all $x, y \in V$, the edge (x, y) is not in S whenever $x \neq y$. An independent set is said to be *maximal* if S is maximal with respect to the inclusion order.

Assume that G is the adjacency graph of a square matrix $\mathcal{A}$. Let S be an independent set. Let **perm** be any permutation array of $\{1, \ldots, N\}$ such that $S = \{\text{perm}(1), \ldots, \text{perm}(\text{card}(S))\}$. Define the reordered matrix $\mathcal{B}$ such that $\mathcal{B}_{ij} = \mathcal{A}_{\text{perm}(i)\text{perm}(j)}$ for all $1 \leq i, j \leq N$. We readily infer the following:

Proposition 9.25. *The triplet* $\{1, \text{card}(S), N\}$ *defines a* 2×2 *block structure of* $\mathcal{B}$ *where the top left block is diagonal.*

Let **traverse** be a permutation array of $\{1, \ldots, N\}$. Algorithm 9.2 presents a simple strategy to construct an independent set. A possible choice for **traverse** consists of setting $\text{traverse}(i) = i$ but, in general, **traverse** is set to maximize the cardinal number of S. Since the cardinal number of S is equal to the number of times the statement $(\text{virgin}(j))$ is true in Algorithm 9.2, a possible technique to maximize this number is to choose j so that $\text{card}(\text{Adj}(j))$ is small, i.e., among all the nodes left, j must be one of the lowest degree. Then, a crude strategy consists of using **traverse** to sort the nodes in increasing degree.

Algorithm 9.2. Independent set ordering.

```
S = ∅; virgin =.true.
for i = 1, ..., N do
    j = traverse(i)
    if (virgin(j)) then
        S = S ∪ {j}; virgin(j) =.false.
        for all k ∈ Adj(j) do
            virgin(k) =.false.
        end for
    end if
end for
```

9.2.5 Multicolor orderings

A third standard reordering method is based on so-called *graph coloring* techniques. A coloring process is a transformation $C : V \rightarrow \mathbb{N}$ such that, if

$(x, y) \in E$, $C(x) \neq C(y)$ whenever $x \neq y$. For $x \in V$, $C(x)$ is referred to as the color of x. The goal of graph coloring is to find a coloring process C such that the cardinal number of the range of C is as small as possible, i.e., the number of colors to color V is as small as possible.

In the context of linear algebra, optimality is not a major issue and one is usually satisfied by using simple heuristics. For instance, given a permutation array $\texttt{traverse}$ of $\{1, \ldots, N\}$, Algorithm 9.3 describes a basic coloring strategy. The simplest choice for the array $\texttt{traverse}$ consists of setting $\texttt{traverse}(i) = i$, but more sophisticated choices are possible. For instance, it can be shown that if the graph can be colored with two colors only and if BFS is used to initialize $\texttt{traverse}$, then Algorithm 9.3 finds a two-color partitioning; see Exercise 9.6. Finally, independently of $\texttt{traverse}$, the number of colors found by Algorithm 9.3 is at most equal to 1 plus the largest degree in the graph; see Exercise 9.7.

Algorithm 9.3. Multicoloring algorithm.

```
color = 0
for i = 1, ..., N do
  j = traverse(i)
  color(j) = min{k > 0; k ∉ color(Adj(j))}
end for
```

Assume that we have colored the adjacency graph of a matrix $\mathcal{A}$. Denote by $\texttt{k\_max}$ the number of colors used, and let $C : V \to \{1, 2, \ldots, \texttt{k\_max}\}$ be the corresponding color mapping. Let $\texttt{col\_part}(1:\texttt{k\_max})$ be the array such that $\texttt{col\_part}(1) = 1$ and $\texttt{col\_part}(k + 1) = \texttt{col\_part}(k) + \text{card}(C^{-1}(k))$ for $1 \leq k \leq \texttt{k\_max}$. Let $\texttt{perm}$ be any permutation array so that the color of the vertices in the set $\{\texttt{perm}(\texttt{col\_part}(k)), \ldots, \texttt{perm}(\texttt{col\_part}(k + 1) - 1)\}$ is k. Define the reordered matrix $\mathcal{B}$ such that $\mathcal{B}_{ij} = \mathcal{A}_{\texttt{perm}(i)\texttt{perm}(j)}$. Then, multicoloring ordering partially finds its justification in the following:

Proposition 9.26. *The array* $\texttt{col\_part}$ *defines a* $\texttt{k\_max} \times \texttt{k\_max}$ *block structure of* $\mathcal{B}$ *where the diagonal blocks are diagonal.*

Proof. Left as an exercise. □

9.3 Iterative Solution Methods

The main motivation for using an iterative method to solve (approximately) a linear system is that it takes advantage of the *sparsity* of the system matrix. In this section, the emphasis is set on one important class of iterative methods, the so-called *projection-based iterative methods*. This class contains several examples of widely used methods, such as the conjugate gradient algorithm for symmetric positive definite systems and its generalizations to non-symmetric systems. When dealing with ill-conditioned matrices, preconditioning is critical to enhance the convergence rate of iterative methods; this section contains a brief introduction to these techniques.

9.3.1 Projection-based iterative methods

Consider the linear system $\mathcal{A}U = F$ and assume that $\mathcal{A}$ is non-singular. Let $V \in \mathbb{R}^N$ be an approximation to its solution U. Denote by $E(V) = U - V$ the error and by $R(V) = F - \mathcal{A}V$ the residual. Given two subspaces $\mathcal{K}$ and $\mathcal{L}$ of $\mathbb{R}^N$ having the same dimension, a projection method attempts to improve the approximate solution V by a vector W that solves the problem:

$$
\begin{cases}
\text{Seek } W \in V + \mathcal{K} \text{ such that} \\
R(W) \perp \mathcal{L}.
\end{cases}
\tag{9.25}
$$

Since $\mathcal{K}$ and $\mathcal{L}$ have the same dimension and $\mathcal{A}$ is invertible, (9.25) is well-posed.

Two particular choices for the spaces $\mathcal{K}$ and $\mathcal{L}$ lead to remarkable optimality properties satisfied by the solution W.

Residual minimization in Euclidean norm. Set $\mathcal{L} = \mathcal{A}\mathcal{K}$. Then, one readily verifies that the solution W to (9.25) minimizes the Euclidean norm of the residual over the affine space $V + \mathcal{K}$ and that

$$
R(W) = (\mathcal{I} - P_{\mathcal{A}\mathcal{K},(\mathcal{A}\mathcal{K})^\perp})R(V),
\tag{9.26}
$$

where $\mathcal{I}$ is the identity matrix and $P_{\mathcal{A}\mathcal{K},(\mathcal{A}\mathcal{K})^\perp}$ is the Euclidean orthogonal projector onto $\mathcal{A}\mathcal{K}$.

Error minimization in energy norm. Assume that the matrix $\mathcal{A}$ is symmetric positive definite so that it induces the scalar product

$$
(V, W)_{\mathcal{A}} = (\mathcal{A}V, W)_N,
\tag{9.27}
$$

and the so-called *energy norm* $\|V\|_{\mathcal{A}} = (V, V)_{\mathcal{A}}^{\frac{1}{2}}$. Set $\mathcal{L} = \mathcal{K}$. Then, one readily verifies that the solution W to (9.25) minimizes the energy norm of the error over the affine space $V + \mathcal{K}$ and that

$$
E(W) = (I - P_{\mathcal{K},\mathcal{K}^{\perp_{\mathcal{A}}}})E(V),
\tag{9.28}
$$

where $P_{\mathcal{K},\mathcal{K}^{\perp_{\mathcal{A}}}}$ denotes the orthogonal projector onto $\mathcal{K}$ with respect to the scalar product (9.27).

Iterative methods based on projection methods. A wide range of iterative methods can be designed on the basis of the projection step (9.25). For instance, consider an initial iterate $U^0 \in \mathbb{R}^N$ and two sequences of spaces $\{\mathcal{K}_m\}_{m\geq 1}$ and $\{\mathcal{L}_m\}_{m\geq 1}$. Then, the mth step of the iterative process is as follows:

$$\begin{cases} \text{Seek } U^m \in U^0 + \mathcal{K}_m \text{ such that} \\ R(U^m) \perp \mathcal{L}_m. \end{cases} \qquad (9.29)$$

Remark 9.27. It is also possible to design iterative methods based on the projection (9.25) in which the dimension of spaces $\mathcal{K}$ and $\mathcal{L}$ is kept constant at each iteration, but where the vector V is repeatedly modified. An example is proposed in Exercise 9.11. □

9.3.2 The Conjugate Gradient (CG) algorithm

The CG algorithm is a well-established method for solving iteratively linear systems involving a *symmetric positive definite matrix*. Designed by Hestenes and Stiefel in 1952 [HeS52], it can be formulated in the framework of iterative projection methods [Saa96]. Given $U^0 \in \mathbb{R}^N$, let $R^0 = F - \mathcal{A}U^0$ be the initial residual. Note that $R^0 \neq 0$ unless U^0 is the solution. For an integer $m \geq 1$, consider the *Krylov space*

$$\mathcal{K}_m = \text{span}\{R^0, \mathcal{A}R^0, \dots, \mathcal{A}^{m-1}R^0\}. \qquad (9.30)$$

At step m, CG performs the projection (9.25) with $V = U^0$ and $\mathcal{L} = \mathcal{K} = \mathcal{K}_m$. We infer from §9.3.1 that the iterate U^m minimizes the energy norm of the error over the affine space $U^0 + \mathcal{K}_m$.

Owing to the symmetry of $\mathcal{A}$, several simplifications occur in the CG algorithm. The most remarkable is that although U^m satisfies an optimality property over the whole affine space $U^0 + \mathcal{K}_m$, an entire basis of $\mathcal{K}_m$ need not be stored. The implementation CG is shown in Algorithm 9.4. Observe that *only three vectors* are needed to evaluate U^m, namely U^{m-1}, R^{m-1}, and P^{m-1}.

Remark 9.28. In practice, one often chooses a relative convergence criterion in Algorithm 9.4, i.e., one sets $\texttt{tol} = \texttt{tol}_r \|R^0\|_N$ where $\texttt{tol}_r$ is a user-defined relative tolerance. This remark also holds for the other algorithms presented in the rest of this section. □

Proposition 9.29. *The following properties hold:*

(i) $\mathcal{K}_m = \text{span}\{R^0, \dots, R^{m-1}\} = \text{span}\{P^0, \dots, P^{m-1}\}$.
(ii) *The set* $\{P^0, \dots, P^{m-1}\}$ *is an $\mathcal{A}$-orthogonal basis of $\mathcal{K}_m$.*
(iii) *The set* $\{R^0, \dots, R^{m-1}\}$ *is an orthogonal basis of $\mathcal{K}_m$.*
(iv) $\forall m, n \in \mathbb{N}$, $(R^m, P^n)_N = 0$ *if* $m > n$, *and* $(R^m, \mathcal{A}P^n)_N = 0$ *if* $m > n+1$.
(v) *The CG algorithm converges in at most N iterations.*

Proof. Left as an exercise. □

Algorithm 9.4. CG.

---

choose U^0, set $R^0 = F - \mathcal{A}U^0$ and $P^0 = R^0$
choose a tolerance `tol`
set $m = 0$
while $\|R^m\|_N > $ `tol` **do**
$\quad \alpha^m = (R^m, R^m)_N / (\mathcal{A}P^m, P^m)_N$
$\quad U^{m+1} = U^m + \alpha^m P^m$
$\quad R^{m+1} = R^m - \alpha^m \mathcal{A}P^m$
$\quad \beta^m = (R^{m+1}, R^{m+1})_N / (R^m, R^m)_N$
$\quad P^{m+1} = R^{m+1} + \beta^m P^m$
$\quad m \leftarrow m + 1$
end while

Asymptotic cost. The asymptotic cost of one CG iteration is dominated by the cost of computing the matrix–vector product $\mathcal{A}P^m$. If the matrix $\mathcal{A}$ is dense, this cost scales as $2N^2$. It is, however, much lower for a sparse matrix. Assume that the number of non-zero entries per row in $\mathcal{A}$ is bounded by $N_{\text{row}} \ll N$. The asymptotic cost of a matrix–vector product then scales as $2N_{\text{row}}N$. In addition, the asymptotic cost of a scalar product scales as $2N$ and so does the asymptotic cost of the "vector update" $x \leftarrow x + \alpha y$, where x and y are vectors and α is a scalar. Since one CG iteration includes two scalar products and three vector updates, the total asymptotic cost per iteration scales as $2(N_{\text{row}} + 5)N$. Let M be the number of CG iterations performed to achieve convergence. The total asymptotic cost is then $2(N_{\text{row}} + 5)NM$. As a result, the CG algorithm is computationally competitive with respect to a direct solution method if M is not too large and, ideally, if M is significantly lower than N. Clearly, M depends on the convergence rate of the algorithm, i.e., the rate at which the error is reduced from one iteration to the next. An estimate of the convergence rate of CG is given by the following:

Proposition 9.30 (Convergence rate). *Let $\mathcal{A} \in \mathbb{R}^{N,N}$ be a symmetric positive definite matrix and let $\kappa(\mathcal{A})$ be its condition number. Let $\{U^m\}_{m\geq 0}$ be the sequence of iterates generated by the CG algorithm. Then,*

$$\|U^m - U\|_{\mathcal{A}} \leq 2 \left(\frac{\kappa(\mathcal{A})^{\frac{1}{2}} - 1}{\kappa(\mathcal{A})^{\frac{1}{2}} + 1} \right)^m \|U^0 - U\|_{\mathcal{A}}.$$

Proof. See [Saa96, p. 193] and [GoV89, p. 525]. $\qquad\qquad\qquad\qquad\qquad \Box$

Remark 9.31. If $\mathcal{A}$ is the stiffness matrix associated with a finite element approximation of a second-order PDE, say the Laplace equation, (9.13) implies that the number of CG iterations required to reduce the initial error by a factor h^k scales as $k\,h^{-1}|\log h|$. $\qquad\qquad\qquad\qquad\qquad\qquad\qquad\qquad \Box$

9.3.3 The non-symmetric case (1): GMRes

This section is concerned with one possible generalization of the CG algorithm to solve iteratively *non-symmetric* linear systems, namely the "Generalized Minimal Residual" (GMRes) algorithm designed by Saad and Schultz in 1986 [SaS86]. Given an initial iterate U^0 with corresponding residual R^0, GMRes performs the projection (9.25) with $V = U^0$, $\mathcal{K} = \mathcal{K}_m$ defined in (9.30), and $\mathcal{L} = \mathcal{A}\mathcal{K}$. We infer from §9.3.1 that the iterate U^m minimizes the Euclidean norm of the residual over the affine space $U^0 + \mathcal{K}_m$.

Because of the lack of symmetry of $\mathcal{A}$, a basis of the Krylov space $\mathcal{K}_m$ must be constructed and stored. One way of constructing an orthogonal basis for $\mathcal{K}_m$ is to use a clever form of the Gram–Schmidt process proposed by Arnoldi [Arn51]. This technique is described in Algorithm 9.5. Note that if the algorithm breaks down for some j, i.e., $h_{j+1,j} = 0$, then $\mathcal{K}_j$ is maximal, i.e., $\mathcal{K}_j$ is invariant under $\mathcal{A}$. In this event, the projection algorithm gives the exact solution. This unlikely occurrence is termed a lucky breakdown.

Algorithm 9.5. Arnoldi's algorithm.

set $\beta = \|R^0\|_N$ and $V^1 = R^0/\beta$
set $j = 1$
while $j < m$ **do**
  $h_{i,j} = (\mathcal{A}V_j, V_i)_N$ for $i = 1, \ldots, j$
  $\widehat{V}_{j+1} = \mathcal{A}V_j - \sum_{i=1}^{j} h_{i,j} V_i$
  $h_{j+1,j} = \|\widehat{V}_{j+1}\|_N$; **if** $h_{j+1,j} = 0$ **stop**
  $V_{j+1} = \widehat{V}_{j+1}/h_{j+1,j}$
  $j \leftarrow j + 1$
end while

For an integer j, denote by $\mathcal{V}_j \in \mathbb{R}^{N,j}$ the matrix whose columns are the first j orthogonal vectors $\{V_1, \ldots, V_j\}$ generated by Algorithm 9.5 and denote by $\mathcal{H}_j \in \mathbb{R}^{j+1,j}$ the matrix whose non-zero entries are the corresponding quantities $h_{i,j}$. Note that $\mathcal{H}_j$ is an upper *Hessenberg matrix*, i.e., $\mathcal{H}_{j,kl} = 0$ if $k > l + 1$, for $1 \leq k \leq j + 1$ and $1 \leq l \leq j$. Then, one readily checks that

$$\mathcal{A}\mathcal{V}_j = \mathcal{V}_{j+1}\mathcal{H}_j. \tag{9.31}$$

Consider now the m-th iterate of the GMRes algorithm. We want to solve the minimization problem

$$\min_{Z \in \mathcal{K}_m} \|F - \mathcal{A}(U^0 + Z)\|_N = \min_{Z \in \mathcal{K}_m} \|R^0 - \mathcal{A}Z\|_N.$$

We expand Z with respect to the Arnoldi basis $\{V_1, \ldots, V_m\}$, i.e., we write $Z = \mathcal{V}_m Y$ for some $Y \in \mathbb{R}^m$. Since $R^0 = \beta V_1$, we infer

$$\|R^0 - \mathcal{A}Z\|_N = \|\beta V_1 - \mathcal{A}\mathcal{V}_m Y\|_N = \|\mathcal{V}_{m+1}(\beta e_1 - \mathcal{H}_m Y)\|_N,$$

where e_1 is the first vector of the canonical basis of $\mathbb{R}^{m+1}$. Since the basis $\{V_1, \ldots, V_{m+1}\}$ is orthonormal, we obtain the minimization problem

$$\min_{Y \in \mathbb{R}^m} \|\beta e_1 - \mathcal{H}_m Y\|_{m+1}. \tag{9.32}$$

This is a standard Least-Squares problem. Since the rank of $\mathcal{H}_m$ is maximal, the Least-Squares solution is unique. To compute the solution, one performs a QR factorization of the matrix $\mathcal{H}_m$. Using the fact that $\mathcal{H}_m$ is a Hessenberg matrix, it is possible to find (as explained below) a unitary matrix $\mathcal{Q}_m \in \mathbb{R}^{m+1, m+1}$ such that the matrix $\mathcal{R}_m = \mathcal{Q}_m \mathcal{H}_m \in \mathbb{R}^{m+1, m}$ has the following 2×1 block structure:

$$\mathcal{R}_m = \begin{pmatrix} \mathcal{U}_m \\ \hline 0 \end{pmatrix},$$

where $\mathcal{U}_m \in \mathbb{R}^{m, m}$ is upper triangular and invertible. As a result,

$$\|\beta e_1 - \mathcal{H}_m Y\|_{m+1} = \|\mathcal{Q}_m(\beta e_1 - \mathcal{H}_m Y)\|_{m+1} = \|G_m - \mathcal{R}_m Y\|_{m+1},$$

where $G_m = \beta \mathcal{Q}_m e_1$. Owing to the particular structure of the matrix $\mathcal{R}_m$, it is now straightforward to solve (9.32). Indeed, first invert the square upper triangular system

$$\mathcal{U}_m \begin{pmatrix} Y_1 \\ \vdots \\ Y_m \end{pmatrix} = \begin{pmatrix} G_{m,1} \\ \vdots \\ G_{m,m} \end{pmatrix}. \tag{9.33}$$

Then, the m-th iterate in the GMRes algorithm is simply

$$U^m = U^0 + \sum_{i=1}^m Y_m V_m,$$

and the residual is $|G_{m,m+1}| = \|F - \mathcal{A}U^m\|_N$. A very handy feature of the algorithm is that U^m need not be reconstructed to evaluate the residual. This leads to Algorithm 9.6.

We now turn our attention to the computation of the *QR factorization* of $\mathcal{H}_m$. The unitary matrix $\mathcal{Q}_m$ is decomposed into a product of elementary rotation matrices, $\mathcal{Q}_m = \prod_{j=1}^m \mathcal{F}_j$ such that

$$\mathcal{F}_j = \begin{pmatrix} \mathcal{I}_{j-1} & 0 & 0 \\ \hline 0 & \begin{matrix} c_j & s_j \\ -s_j & c_j \end{matrix} & 0 \\ \hline 0 & 0 & \mathcal{I}_{m-j} \end{pmatrix},$$

where $\mathcal{I}_l$ is the identity matrix in $\mathbb{R}^{l,l}$ and $c_j^2 + s_j^2 = 1$. Set $\mathcal{H}_m^{(0)} = \mathcal{H}_m$ and $\mathcal{H}_m^{(j)} = \mathcal{F}_j \ldots \mathcal{F}_1 \mathcal{H}_m$. It is clear that, whatever c_j and s_j, $\mathcal{H}_m^{(j)}$ remains a Hessenberg matrix. In particular, setting

---

Algorithm 9.6. GMRes.

---

choose U^0 and a tolerance `tol`
set $m = 0$, $R^0 = F - \mathcal{A}U^0$, $\beta = \|R^0\|_N$, and $V^1 = R^0/\beta$
while $\|R^m\|_N >$ **tol do**
$\quad m \leftarrow m + 1$
$\quad h_{i,m} = (\mathcal{A}V_m, V_i)_N$ for $i = 1, \ldots, m$
$\quad \widehat{V}_{m+1} = \mathcal{A}V_m - \sum_{i=1}^{m} h_{i,m} V_i$
$\quad h_{m+1,m} = \|\widehat{V}_{m+1}\|_N$
$\quad V_{m+1} = \widehat{V}_{m+1}/h_{m+1,m}$
$\quad$ compute the matrices $\mathcal{H}_m$, $\mathcal{Q}_m$, and $\mathcal{R}_m$
$\quad$ evaluate the vector $G_m = \mathcal{Q}_m \beta e_1$
$\quad$ set $\|R^m\|_N = |G_{m,m+1}|$
end while
solve upper triangular system (9.33)
set $U^m = U^0 + \sum_{k=1}^{m+1} Y_k V_k$

$$c_j = \frac{h_{j,j}^{(j-1)}}{\left(\left(h_{j,j}^{(j-1)}\right)^2 + \left(h_{j+1,j}^{(j-1)}\right)^2\right)^{\frac{1}{2}}} \quad \text{and} \quad s_j = \frac{h_{j+1,j}^{(j-1)}}{\left(\left(h_{j,j}^{(j-1)}\right)^2 + \left(h_{j+1,j}^{(j-1)}\right)^2\right)^{\frac{1}{2}}},$$

and left-multiplying $\mathcal{H}_m^{(j-1)}$ by $\mathcal{F}_j$, the entry $h_{j+1,j}^{(j-1)}$ is replaced by 0. More generally, $\mathcal{H}_{m,kl}^{(j)} = 0$, for $1 \le l \le j$ and $k > l$. As a result, setting $\mathcal{R}_m = \mathcal{H}_m^{(m)}$ yields $\mathcal{Q}_m \mathcal{H}_m = \mathcal{R}_m$.

In practice, the algorithm is implemented as follows: At iteration j, the sequences c_j and s_j are stored, and the upper triangular matrix $\mathcal{U}_m$ is stored in place of $\mathcal{H}_m$. At the next iteration, the last column of $\mathcal{H}_{m+1}$ is evaluated. Denote by $C_{m+1} = (h_{1,m+1}, \ldots, h_{m+2,m+1})^T$ this column. Then, compute $R_{m+1} = \mathcal{F}_{m+1} \ldots \mathcal{F}_1 C_{m+1}$. Note that the last entry of R_{m+1} is zero. Finally, set $\widehat{R}_m = (R_{m+1,1}, \ldots, R_{m+1,m})^T$ where $R_{m+1,l}$, $1 \le l \le m+2$, are the entries of R_{m+1}, and

$$\mathcal{U}_{m+1} = \left(\begin{array}{c|c} \mathcal{U}_m & \widehat{R}_m \\ \hline 0 & R_{m+1,m+1} \end{array}\right).$$

Since the entire Arnoldi basis is needed to compute U^m, computational costs and storage requirements in the GMRes algorithm grow linearly with the iteration number. In practice, it is often preferable to impose a maximal dimension n for the Krylov space, with n ranging typically between 10 and 50. If convergence is not achieved once the maximum dimension has been reached, the algorithm is re-initialized by setting $U^0 = U^n$ and a new Krylov space is generated. The resulting algorithm is called GMRes(n).

Proposition 9.32 (Convergence). *Let $\mathcal{A} \in \mathbb{R}^{N,N}$ be a nonsingular matrix.*

(i) *If not restarted, Algorithm 9.6 converges in at most N iterations.*
(ii) *If A is positive definite, GMRes(n) converges for all $n \ge 1$.*

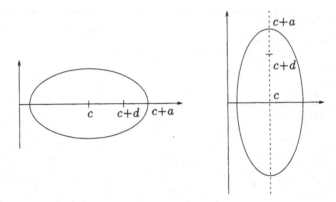

Fig. 9.5. Ellipse $E(c, d, a)$ containing the spectrum of the matrix $\mathcal{A}$; d is either real (left) or imaginary (right).

Proof. Property (i) is a consequence of the fact that GMRes minimizes the Euclidean norm of the residual over the affine space $U_0 + \mathcal{K}_m$. Therefore, in the absence of restarts, the algorithm must converge in at most N iterations. Property (ii) is proved in [Saa96, p. 195]. $\qquad\qquad\qquad\square$

The convergence rate of Algorithm 9.6 depends on the spectrum of the matrix $\mathcal{A}$. Let $E(c, d, a)$ be the ellipse in the complex plane with center $c \in \mathbb{R}$, focal distance d, and large half-axis a; see Figure 9.5.

Proposition 9.33 (Convergence rate). *Let $\mathcal{A} \in \mathbb{R}^{N,N}$. Assume that $\mathcal{A}$ can be diagonalized in the form $\mathcal{A} = T \Lambda T^{-1}$ where $\Lambda = \mathrm{diag}(\lambda_1, \ldots, \lambda_N)$ with eigenvalues λ_i arranged in increasing order. Assume that all the eigenvalues are contained in the ellipse $E(c, d, a)$ and that $0 \notin E(c, d, a)$. Then, the residual R^m at the m-th step in the GMRes algorithm satisfies*

$$\|R^m\|_N \leq \kappa(T) \frac{C_m(\frac{a}{d})}{C_m(\frac{c}{d})} \|R^0\|_N, \qquad (9.34)$$

where $C_m(z) = (z + \sqrt{z^2 - 1})^m + (z + \sqrt{z^2 - 1})^{-m}$.

Proof. See [Saa96, p. 196]. $\qquad\qquad\qquad\qquad\qquad\qquad\qquad\qquad\square$

Estimate (9.34) can be used to assess the convergence rate of the GMRes algorithm provided the matrix T is reasonably well-conditioned. In practice, one can use the approximate expression

$$\frac{C_m(\frac{a}{d})}{C_m(\frac{c}{d})} \simeq \left(\frac{a + \sqrt{a^2 - d^2}}{c + \sqrt{c^2 - d^2}} \right)^m.$$

When the matrix $\mathcal{A}$ is ill-conditioned and has real spectrum, $a \simeq d$, $c+a = \lambda_N$, $c - a = \lambda_1$, yielding

$$\frac{C_m(\frac{a}{d})}{C_m(\frac{c}{d})} \simeq \left(1 - \frac{2}{\kappa(\mathcal{A})^{\frac{1}{2}}}\right)^m.$$

i.e., the convergence rates of GMRes and CG are asymptotically the same.

9.3.4 The non-symmetric case (2): Bi-CGStab

We have seen in §9.3.2 that the CG algorithm is endowed with the remarkable feature that the m-th iterate U^m satisfies an optimality property over the m-dimensional affine space $U^0 + \mathcal{K}_m$ and, at the same time, U^m can be evaluated using three-term recursions without making any explicit reference to a complete basis of the Krylov space $\mathcal{K}_m$. Unfortunately, this feature no longer holds for non-symmetric systems, as shown by Faber and Manteuffel; see [Saa96, p. 186]. Iterative methods such as the GMRes algorithm guarantee an optimality property (minimization of the Euclidean norm of the residual), but need to store a basis for $\mathcal{K}_m$. In this section, we briefly describe some generalizations of the CG algorithm to non-symmetric linear systems that give up the optimality property, but employ simple recursions to compute the iterates.

A first generalization, the Bi-Conjugate Gradient (Bi-CG) method, is (9.25) with the Krylov spaces $\mathcal{K}_m = \text{span}\{R^0, \mathcal{A}R^0, \ldots, \mathcal{A}^{m-1}R^0\}$ and $\mathcal{L}_m = \text{span}\{\overline{R}^0, \mathcal{A}^T \overline{R}^0, \ldots, (\mathcal{A}^T)^{m-1}\overline{R}^0\}$, where R^0 and $\overline{R}^0$ are the residuals associated with two given vectors. Unfortunately, neither the residual at the m-th iteration nor the error satisfy the optimality properties stated in §9.3.1. A further modification of the algorithm has been introduced in the Conjugate Gradient Squared (CGS) method proposed by Sonneveld in 1989 [Son89]. If the m-th residual in the Bi-CG algorithm is given in the form $R^m = p_m(\mathcal{A})R^0$ where $p_m(\mathcal{A})$ is a polynomial of degree m in $\mathcal{A}$, the m-th residual of the CGS algorithm is such that $\widehat{R}^m = p_m^2(\mathcal{A})R^0$. Because of the lack of an optimality property, the Euclidean norm of the residual is non-monotonic, i.e., erratic convergence behavior where the residual makes large jumps in the "wrong" direction is sometimes observed. As a remedy, van der Vorst proposed in 1992 a stabilized version of the method, called the Bi-Conjugate Gradient Stabilized (Bi-CGStab) algorithm [vaV92], an implementation of which is shown in Algorithm 9.7.

Algorithm 9.7 involves only two matrix–vector products per iteration so that the asymptotic cost per iteration remains moderate when compared to that of GMRes. Moreover, only six vectors need to be stored, an attractive feature if memory is at premium. The drawback is that neither the residual nor the error satisfy an optimality property so that there is no theoretical guarantee that the algorithm does not breakdown before reaching the solution. Furthermore, even if no breakdown occurs, divisions by very small numbers may take place, hampering the stability of the algorithm. Nevertheless, the Bi-CGStab algorithm usually works fairly well on large linear systems arising in engineering applications. As for the previous algorithms, the *convergence rate* of Bi-CGStab depends on the condition number of the matrix $\mathcal{A}$.

Algorithm 9.7. Bi-CGStab.

---

choose U^0, set $R^0 = F - \mathcal{A}U^0$ and $P^0 = R^0$

choose $\overline{R}^0 \in \mathbb{R}^N$ (e.g., $\overline{R}^0 = R^0$)

choose a tolerance `tol`

set $m = 0$

while $\|R^m\|_N > $ `tol` **do**

$\quad \alpha^m = (R^m, \overline{R}^0)_N / (\mathcal{A}P^m, \overline{R}^0)_N$

$\quad S^m = R^m - \alpha^m \mathcal{A}P^m$

$\quad \omega^m = (\mathcal{A}S^m, S^m)_N / (\mathcal{A}S^m, \mathcal{A}S^m)_N$

$\quad U^{m+1} = U^m + \alpha^m P^m + \omega^m S^m$

$\quad R^{m+1} = S^m - \omega^m \mathcal{A}S^m$

$\quad \beta^m = (\alpha^m / \omega^m) \times (R^{m+1}, \overline{R}^0)_N / (R^m, \overline{R}^0)_N$

$\quad P^{m+1} = R^{m+1} + \beta^m (P^m - \omega^m \mathcal{A}P^m)$

$\quad m \leftarrow m + 1$

end while

9.3.5 Preconditioning

When the matrix $\mathcal{A}$ is ill-conditioned, Proposition 9.30 shows that the CG algorithm exhibits an "agonizingly slow" convergence rate. The same conclusion holds for the GMRes algorithm and for the Bi-CGStab algorithm. A remedy to this problem consists of using a *preconditioner*. The idea is to replace the linear system $\mathcal{A}U = F$ by a new linear system $\widetilde{\mathcal{A}}\widetilde{U} = \widetilde{F}$ for which the matrix $\widetilde{\mathcal{A}}$ is better conditioned. Let $\mathcal{P}$ be a non-singular matrix in the form $\mathcal{P} = \mathcal{P}_L \mathcal{P}_R$ and assume that linear systems of the form $\mathcal{P}_L X = Y$ and $\mathcal{P}_R X' = Y'$ are relatively inexpensive to solve. Using the matrix $\mathcal{P}$ as a split-preconditioner yields the linear system

$$\underbrace{(\mathcal{P}_L^{-1} \mathcal{A} \mathcal{P}_R^{-1})}_{\widetilde{\mathcal{A}}} \underbrace{(\mathcal{P}_R U)}_{\widetilde{U}} = \underbrace{\mathcal{P}_L^{-1} F}_{\widetilde{F}}. \tag{9.35}$$

As particular cases, we can consider a *left-preconditioner*, for which $\mathcal{P} = \mathcal{P}_L$ and $\mathcal{P}_R = I$, yielding

$$\underbrace{(\mathcal{P}_L^{-1} \mathcal{A})}_{\widetilde{\mathcal{A}}} U = \underbrace{\mathcal{P}_L^{-1} F}_{\widetilde{F}},$$

as well as a *right-preconditioner*, for which $\mathcal{P} = \mathcal{P}_R$ and $\mathcal{P}_L = I$, yielding

$$\underbrace{(\mathcal{A} \mathcal{P}_R^{-1})}_{\widetilde{\mathcal{A}}} \underbrace{(\mathcal{P}_R U)}_{\widetilde{U}} = F.$$

Preconditioned CG. To derive a preconditioned version of CG, assume that $\mathcal{P}$ is *symmetric positive definite* and consider a symmetric split-preconditioner in the form $\mathcal{P} = \mathcal{P}_L \mathcal{P}_L^T$. This yields Algorithm 9.8. Note that the algorithm is implemented without any explicit reference to the matrix $\mathcal{P}_L$. Setting $\widetilde{U}^m =$

$\mathcal{P}_L U^m$, $\widetilde{P}^m = \mathcal{P}_L P^m$, and $\widetilde{R}^m = \mathcal{P}_L R^m$, it is readily verified that $\widetilde{U}^m$, $\widetilde{P}^m$, and $\widetilde{R}^m$ are the iterates generated by the CG algorithm when applied to (9.35). As a result, the convergence rate of Algorithm 9.8 depends on the condition number of the preconditioned matrix $\widetilde{\mathcal{A}}$.

Algorithm 9.8. Preconditioned CG.

---

choose U^0, set $R^0 = F - \mathcal{A}U^0$ and $P^0 = \mathcal{P}^{-1} R^0$
choose a tolerance `tol`
set $m = 0$
while $\|R^m\|_N > $ `tol` **do**
$\quad \alpha^m = (\mathcal{P}^{-1} R^m, R^m)_N / (\mathcal{A}P^m, P^m)_N$
$\quad U^{m+1} = U^m + \alpha^m P^m$
$\quad R^{m+1} = R^m - \alpha^m \mathcal{A}P^m$
$\quad \beta^m = (\mathcal{P}^{-1} R^{m+1}, R^{m+1})_N / (\mathcal{P}^{-1} R^m, R^m)_N$
$\quad P^{m+1} = \mathcal{P}^{-1} R^{m+1} + \beta^m P^m$
$\quad m \leftarrow m + 1$
end while

Choosing a preconditioner is a *compromise* between the computational cost incurred per iteration and the convergence rate that can be achieved. The former depends on the cost of inverting $\mathcal{P}$ while the latter depends on the condition number of $\widetilde{\mathcal{A}}$. Taking a very simple preconditioner such as $\mathcal{P} = \text{diag}(\mathcal{A})$ does not modify the asymptotic cost incurred per iteration. However, in most cases, the convergence rate is not substantially improved. Conversely, taking a preconditioner $\mathcal{P}$ very close to $\mathcal{A}$ yields relatively high convergence rates, but each iteration is computationally expensive. Two preconditioners often considered for the CG algorithm are:

(i) **Symmetric Gauß–Seidel (SGS).** Split the matrix $\mathcal{A}$ into

$$\mathcal{A} = \mathcal{D} - \mathcal{L} - \mathcal{L}^T,$$

where $\mathcal{D}$ is diagonal and $\mathcal{L}$ is strictly lower triangular. The SGS preconditioner is defined by

$$\mathcal{P} = (\mathcal{D} - \mathcal{L})\mathcal{D}^{-1}(\mathcal{D} - \mathcal{L})^T. \tag{9.36}$$

Note that linear systems of the form $\mathcal{P}X = Y$ are relatively easy to solve since they only involve forward and backward substitutions.

(ii) **Incomplete LDL^T factorization.** The idea is to form the LDL^T factorization of $\mathcal{A}$ and to discard any entry in the lower and upper triangular matrices that would not match the sparse structure of $\mathcal{A}$. This yields a preconditioner of the form $\mathcal{P} = \mathcal{L}\mathcal{D}\mathcal{L}^T$, where $\mathcal{L}$ is a lower triangular matrix and $\mathcal{D}$ is diagonal. Again, linear systems of the form $\mathcal{P}X = Y$ are relatively easy to solve.

Preconditioned GMRes and Bi-CGStab. Because there is no symmetry property to preserve, it is possible to implement a left- or a right-preconditioned version of GMRes and Bi-CGStab. *Left-preconditioned GMRes* generates the Krylov space

$$\mathcal{K}_m^{\mathcal{P}} = \text{span}\{R^0, (\mathcal{P}^{-1}\mathcal{A})R^0, \ldots, (\mathcal{P}^{-1}\mathcal{A})^{m-1}R^0\},$$

yielding Algorithm 9.9. Likewise, *left-preconditioned Bi-CGStab* is obtained by replacing $\mathcal{A}$ by $\mathcal{P}^{-1}\mathcal{A}$ and F by $\mathcal{P}^{-1}F$ in Algorithm 9.7.

Algorithm 9.9. Left-preconditioned GMRes.

---

choose U^0 and a tolerance tol
set $m = 0$, $R^0 = \mathcal{P}^{-1}(F - \mathcal{A}U^0)$, $\beta = \|R^0\|_N$, and $V^1 = R^0/\beta$
while $\|R^m\|_N > $ tol **do**
 $m \leftarrow m + 1$
 $h_{i,m} = (\mathcal{P}^{-1}\mathcal{A}V_m, V_i)_N$ for $i = 1, \ldots, m$
 $\widehat{V}_{m+1} = \mathcal{P}^{-1}\mathcal{A}V_m - \sum_{i=1}^m h_{i,m}V_i$
 $h_{m+1,m} = \|\widehat{V}_{m+1}\|_N$
 $V_{m+1} = \widehat{V}_{m+1}/h_{m+1,m}$
 compute the matrices $\mathcal{H}_m$, $\mathcal{Q}_m$, and $\mathcal{R}_m$
 evaluate the vector $G_m = \mathcal{Q}_m\beta e_1$
 set $\|R^m\|_N = |G_{m,m+1}|$
end while
solve upper triangular system (9.33)
set $U^m = U^0 + \sum_{k=1}^{m+1} Y_k V_k$

Two preconditioners often considered for GMRes and Bi-CGStab are:

(i) **Symmetric Successive Over-Relaxation (SSOR).** Split the matrix $\mathcal{A}$ into

$$\mathcal{A} = \mathcal{D} - \mathcal{L} - \mathcal{U},$$

where $\mathcal{D}$ is diagonal, $\mathcal{L}$ is strictly lower triangular, and $\mathcal{U}$ is strictly upper triangular. The SSOR preconditioner is defined by

$$\mathcal{P} = (\mathcal{D} - \omega\mathcal{L})\mathcal{D}^{-1}(\mathcal{D} - \omega\mathcal{U}), \tag{9.37}$$

where ω is a real parameter usually taking values between 0 and 2. For $\omega = 1$ and a symmetric matrix, the SGS preconditioner is recovered. Linear systems of the form $\mathcal{P}X = Y$ are relatively easy to solve since they involve only lower and upper triangular solves.

(ii) **Incomplete LU (ILU) factorization.** The idea is to form the LU factorization of $\mathcal{A}$ and to discard any entry in the lower and upper triangular matrices that does not match the sparsity pattern of $\mathcal{A}$. This yields a preconditioner of the form $\mathcal{P} = \mathcal{LU}$, where $\mathcal{L}$ is lower triangular and $\mathcal{U}$ is upper triangular. Again, linear systems of the form $\mathcal{P}X = Y$

are relatively easy to solve. Algorithm 9.10 produces the incomplete LU factorization of a matrix $\mathcal{A}$ of order N. Note that the loops are ordered in such a way that rows can be computed one at a time. This is particularly convenient when the matrix $\mathcal{A}$ is stored in CSR format; see §8.3.1. Several variants of Algorithm 9.10 can be considered. For instance, some level of fill-in can be allowed. One can also attempt to compensate for discarded entries in the ILU factorization by subtracting them from the diagonal entry, leading to the Modified ILU (MILU) factorization; see [Saa96, p. 286].

Algorithm 9.10. ILU factorization.

> **for** $i = 2 \ldots N$ **do**
>> **for** $k = 1 \ldots i - 1$ **do**
>>> **if** $\mathcal{A}_{ik} \neq 0$ **then**
>>>> $\mathcal{A}_{ik} \leftarrow \mathcal{A}_{ik}/\mathcal{A}_{kk}$
>>>> **for** $j = k + 1 \ldots N$ **do**
>>>>> **if** $\mathcal{A}_{ij} \neq 0$ **then**
>>>>>> $\mathcal{A}_{ij} \leftarrow \mathcal{A}_{ij} - \mathcal{A}_{ik}\mathcal{A}_{kj}$
>>>>> **end if**
>>>> **end for**
>>> **end if**
>> **end for**
> **end for**

Remark 9.34. Many other preconditioning techniques can be considered. A particularly important class is that of multilevel preconditioners. The main idea behind these techniques is to make a change of basis so that the solution is expanded with respect to a more or less hierarchical basis. See [Hac85, BrP90, BrP00] for an introduction to this very important topic.   □

9.3.6 Where to find a linear solver?

Over the last decades, a wide range of computationally efficient iterative methods and preconditioners have been designed for solving large, sparse linear systems such as those arising in finite element discretizations of PDEs. An extensive list of linear solvers freely available on the World Wide Web can be found at the address:

http://www.netlib.org/utk/people/JackDongarra/la-sw.html

9.4 Introduction to Parallel Implementation

Progress in high performance scientific computing is sustained in part by the steady improvements achieved in computer technology, among which parallelism is one of the most important. The purpose of this section is to discuss

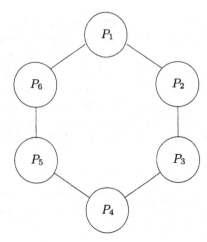

Fig. 9.6. A simplified example of distributed-memory parallel architecture.

briefly some issues related to the parallel implementation of a linear solver. The section is staged at an introductory level and the reader is referred, e.g., to [Saa96, pp. 324–380] for further insight.

9.4.1 Parallel architectures

Several types of parallel architectures are available. Broadly speaking, one can distinguish between *shared-memory* and *distributed-memory* architectures. The former usually consists of a single supercomputer that contains several processors sharing a common memory space. The latter consists of a network of processors, each one having its own memory. Distributed-memory architectures are relatively simple to set up from a cluster of workstations. The number of processors typically ranges between ten and a few hundreds. Processors communicate by passing messages to other processors. A simplified example of a distributed-memory architecture is illustrated in Figure 9.6. In this example, processors only communicate with their two neighbors. In general, communications among all the processors are possible. Henceforth, we restrict the discussion to distributed-memory architectures.

In distributed-memory architectures, there is no global synchronization of the tasks performed by the various processors in parallel. Each processor performs its own tasks as soon as the necessary data are made available. Global task management can be assigned to one particular processor (the "master" processor) that passes messages to the other processors (the "slave" processors) and controls task synchronization whenever necessary.

9.4.2 Parallel efficiency

Let T be the runtime (or wall-clock time) of the code when executed on a single processor. Let $\tau(p)$ be the runtime of the code when executed on a distributed-memory architecture with p processors. Parallel efficiency is assessed by the ratio

$$S = \frac{p\tau(p)}{T}.$$

In principle, this ratio should always be lower than one. Values of the order of 0.9 usually indicate a relatively high level of parallelization. Low levels of parallelization are observed for tasks that require a fairly important amount of communication among processors or whenever the workload is unevenly distributed among processors.

9.4.3 Domain decomposition

Domain decomposition is a relatively straightforward paradigm to parallelize codes solving PDEs by the finite element method. The idea is to take advantage of the fact that in a finite element code, most operations are localized in space. For instance, evaluating a finite element residual involves computing integrals that can be split among the mesh cells and then reassembled.

A simple and convenient way to exploit this idea is to split the computational domain Ω into as many subdomains Ω_i as processors available. Each subdomain is assigned to a processor P_i. Then, the global degrees of freedom, whose support is entirely in Ω_i are assigned to Ω_i. Those degrees of freedom whose support is cut by the interface between two subdomains, say Ω_i and Ω_j, are assigned to either Ω_i or Ω_j. This strategy induces a partitioning of the undirected adjacency graph of the stiffness matrix $\mathcal{A}$. Denote by Ξ_i the list of degrees of freedom assigned to Ω_i. A degree of freedom assigned to Ω_i, say l, is said to be internal if $\mathrm{Adj}(l) \subset \Xi_i$. Otherwise, there is at least one subdomain Ω_j such that $\mathrm{Adj}(l) \cap \Xi_j \neq \emptyset$, and l is said to be at the interface between Ω_i and Ω_j; see Figure 9.7. The list of the degrees of freedom in Ω_i and at the interface between Ω_i and Ω_j is denoted by Ξ_{ij}. Note that if l is internal to Ω_i and m is internal to Ω_j, $i \neq j$, then $\mathrm{Adj}(l) \cap \mathrm{Adj}(m) = \emptyset$, i.e., $\mathcal{A}_{lm} = 0 = \mathcal{A}_{ml}$.

9.4.4 Parallelizing a linear solver

We briefly discuss the parallel implementation of a linear solver based on preconditioned iterative methods such as those discussed in §9.3. The main tasks to be parallelized are scalar products, matrix–vector products, and the preconditioning.

Scalar products. Consider two vectors U and $V \in \mathbb{R}^N$. Assume that the scalar product $(U,V)_N$ must be made available to all the processors. Since

$$(U,V)_N = \sum_{l=1}^{N} U_l V_l = \sum_{i=1}^{p} \sum_{l \in \Xi_i} U_l V_l,$$

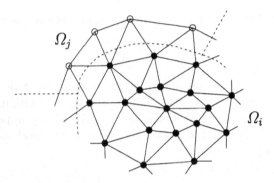

Fig. 9.7. Partitioning of the nodes in the adjacency graph; black and grey nodes are in Ξ_i, gray nodes are in Ξ_{ij} , and whites nodes are in Ξ_{ji}.

where p is the number of processors, the sums $\sum_{l \in \Xi_i} U_l V_l$ can be performed by the processors in parallel. The results must then be assembled. This requires communication and synchronization among processors.

Matrix–vector products. Consider a matrix–vector product $\mathcal{A}U$. Each processor knows only a part of the matrix $\mathcal{A}$ and of the vector U. For instance, assume that the memory of processor P_i stores the components $(U_l)_{l \in \Xi_i}$ of U and the rows $(\mathcal{A}_{lk})_{k \in \Xi_i \cup_{s \in \Lambda_i} \Xi_{si}}$, $l \in \Xi_i$, where Λ_i is the list of the indices of the processors that share an interface with P_i. The components $(V_l)_{l \in \Xi_i}$ of $V = \mathcal{A}U$ are computed in two steps. First, all the processors exchange the appropriate components U_l of U where l is an interface degree of freedom; see Figure 9.7. That is, P_i sends the list $(U_l)_{l \in \Xi_{ij}}$ to P_j and P_j sends the list $(U_m)_{m \in \Xi_{ji}}$ to P_i. Then, each processor computes independently the sums

$$V_l = \sum_{k \in \Xi_i \cup_{s \in \Lambda_i} \Xi_{si}} \mathcal{A}_{lk} U_k \qquad \text{for} \quad l \in \Xi_i. \tag{9.38}$$

Thus, the processors communicate only if they share interface degrees of freedom and, in this case, they only exchange information relevant to those degrees of freedom. A global synchronization is needed at the end of the communication step before the sums (9.38) are assembled locally on each processor.

Preconditioning. The parallelization of a preconditioner is not a simple problem. One possible strategy is to employ preconditioners locally defined on each subdomain Ω_j and to include some level of communication among processors before or after the local preconditioners are applied. Numerous examples of preconditioners implemented on parallel architectures are presented in [Saa96, p. 353]; see also [BrP90, ErD95].

9.4.5 Where to find a parallel iterative library?

Over the last decade, several state-of-the-art iterative methods that perform well on parallel computers have been developed. Since parallelizing a finite element code from scratch requires a substantial amount of work, it may be advisable to use libraries developed by specialists. Among the various parallelization tools available freely on the World Wide Web, we give two examples. The Aztec iterative solver package can be found at the address:

http://www.cs.sandia.gov/CRF/aztec1.html

It is a library of iterative linear system solvers using distributed memory techniques. It is intended as a software tool for those who want to avoid parallel programming details. The second software is the Portable, Extensible Toolkit for Scientific Computation (PETSc) which can be found at the address:

http://www-unix.mcs.anl.gov/petsc/petsc-2/

It includes a long list of parallel linear and nonlinear equation solvers and is intended to be easy to use for beginners.

9.5 Exercises

Exercise 9.1. Consider the norm $\|X\|_\infty = \max_{1\le i\le N} |X_i|$ for $X \in \mathbb{R}^N$.

(i) Verify that the associated matrix norm is, for a matrix $\mathcal{Z} \in \mathbb{R}^{N,N}$, $\|\mathcal{Z}\|_\infty = \max_{1\le i\le N} \sum_{1\le j\le N} |\mathcal{Z}_{ij}|$.

(ii) Let $\{e_1,\dots,e_N\}$ be the canonical basis of $\mathbb{R}^N$. Let $\alpha \in \mathbb{R}$ and consider the matrix $\mathcal{Z} = \mathcal{I} + \alpha e_1 \otimes e_N$ with entries $\mathcal{Z}_{ij} = \delta_{ij} + \alpha \delta_{i1}\delta_{jN}$. Verify that $\mathcal{Z}^{-1} = \mathcal{I} - \alpha e_1 \otimes e_N$.

(iii) Evaluate the condition number of the matrix $\mathcal{Z}$ in the $\|\cdot\|_\infty$-norm. What happens if α is large?

Exercise 9.2. Let $\mathcal{Z} \in \mathbb{R}^{N,N}$ be the upper triangular matrix such that $\mathcal{Z}_{ii} = 1$ for $1 \le i \le N$, and $\mathcal{Z}_{ij} = -1$ for $1 \le i < j \le N$. Let $X \in \mathbb{R}^N$ have coordinates $X_i = 2^{1-i}$ for $1 \le i \le N$.

(i) Compute $\mathcal{Z}X$, $\|\mathcal{Z}X\|_N$, and $\|X\|_N$.
(ii) Show that $\|\mathcal{Z}\|_N \ge 1$.
(iii) Derive a lower bound for $\kappa(\mathcal{Z})$ in the Euclidean norm.

Exercise 9.3. Let M be a positive integer and set $h = \frac{1}{M+1}$. Consider a triangulation of $\Omega = \,]0,1[\,\times\,]0,1[$ with vertices $\{(kh, lh); 0 \le k, l \le M+1\}$ and constructed by cutting each square cell along one of its diagonals. The degrees of freedom are numbered from bottom to top and from left to right. Let $\mathcal{A}$ be the stiffness matrix resulting from the $\mathbb{P}_1$ Lagrange discretization of the Laplacian with homogeneous boundary conditions.

(i) For $1 \le i \le M$, compute the quantities n_i, l_i, and u_i defined in (9.24).
(ii) Evaluate $nnz_\mathcal{A}$, the total number of non-zero entries in $\mathcal{A}$.

(iii) Verify that $nnz_A \ll \sum_{i=1}^{M^2}(u_i - l_i)$ for large M.

Exercise 9.4 (Crout's factorization). Consider a tridiagonal matrix $A \in \mathbb{R}^{N,N}$ and set $c_i = A_{i-1,i}$ for $2 \le i \le N$, $a_i = A_{i,i}$ for $1 \le i \le N$, and $b_i = A_{i,i+1}$ for $1 \le i \le N - 1$. Verify that Algorithm 9.11 produces the LU factorization of A where the two bidiagonal matrices $\mathcal{L}$ and $\mathcal{U}$ are such that

$$\mathcal{L}_{i-1,i} = m_i, \ 2 \le i \le N, \qquad \mathcal{L}_{i,i} = l_i, \ 1 \le i \le N,$$
$$\mathcal{U}_{i,i+1} = u_i, \ 1 \le i \le N - 1, \qquad \mathcal{U}_{i,i} = 1, \ 1 \le i \le N.$$

Algorithm 9.11. Crout's factorization of a tridiagonal matrix.

$l_1 = a_1$ and $u_1 = \frac{b_1}{l_1}$
for $i = 2 \ldots N - 1$ **do**
    set $m_i = c_i$, $l_i = a_i - m_i u_{i-1}$, and $u_i = \frac{b_i}{l_i}$
end for
set $m_N = c_N$ and $l_N = a_N - m_N u_{N-1}$

Exercise 9.5 (Permutation matrix). Let σ be a permutation of $\{1, \ldots, N\}$. Let $\mathcal{P} \in \mathbb{R}^{N,N}$ be such that $\mathcal{P}_{ij} = \mathcal{I}_{\sigma(i)j}$, where $\mathcal{I}$ is the identity matrix.

(i) Prove that $\mathcal{P}^T = \mathcal{P}^{-1}$.
(ii) Let $A \in \mathbb{R}^{N,N}$, and set $\mathcal{B}_{ij} = \mathcal{B}_{\sigma(i)\sigma(j)}$. Prove that $\mathcal{B} = \mathcal{P}A\mathcal{P}^T$.
(iii) Show that the condition number of A is reordering-invariant.
(iv) Show that the spectrum of A is reordering-invariant.

Exercise 9.6. Prove that the total number of colors found by Algorithm 9.3 is at most equal to 1 plus the largest degree in the graph. (*Hint*: By induction.)

Exercise 9.7. Assume that a graph G can be colored with two colors only. Prove that if the BFS reordering is used to initialize **perm**, Algorithm 9.3 finds a two-color partitioning.

Exercise 9.8. Prove Proposition 9.26.

Exercise 9.9. Let W be the solution of (9.25) with $\mathcal{L} = A\mathcal{K}$.

(i) Show that $R(W)$ minimizes the Euclidean norm of the residual over the affine space $V + \mathcal{K}$.
(ii) Conversely, show that if $R(W)$ minimizes the Euclidean norm of the residual over the affine space $V + \mathcal{K}$, then W solves (9.25).
(iii) Prove (9.26).

Exercise 9.10. Let W be the solution of (9.25) with $\mathcal{L} = \mathcal{K}$ and assume that A is symmetric positive definite.

(i) Show that $R(W)$ minimizes the energy norm of the error over the affine space $V + \mathcal{K}$.

(ii) Conversely, show that if $R(W)$ minimizes the energy norm of the error over the affine space $V + \mathcal{K}$, then W solves (9.25).

(iii) Prove (9.28).

Exercise 9.11 (Gauß–Seidel). Consider the following iterative method: Given $U^m \in \mathbb{R}^N$, the next iterate U^{m+1} is obtained by performing N projection steps (9.25) in the following way: First, set $U^{m+1,0} = U^m$; then, for $1 \leq i \leq N$, $U^{m+1,i}$ is obtained from (9.25) by taking $V = U^{m+1,i-1}$ and $\mathcal{K} = \mathcal{L} = \text{span}\{e_i\}$ where e_i is the i-th vector of the canonical basis of $\mathbb{R}^N$; finally, set $U^{m+1} = U^{m+1,N}$. Show that

$$U_i^{m+1} = \frac{1}{\mathcal{A}_{ii}} \left(F_i - \sum_{j=1}^{i-1} \mathcal{A}_{ij} U_j^{m+1} - \sum_{j=i+1}^{N} \mathcal{A}_{ij} U_j^m \right), \quad 1 \leq i \leq N.$$

This algorithm is known as the *Gauß-Seidel method*.

Exercise 9.12. Prove Proposition 9.29.

Exercise 9.13. Let $\mathcal{A}$ be a symmetric positive definite matrix of order N.

(i) Let $\{Z^1, \ldots, Z^N\}$ be an $\mathcal{A}$-orthogonal basis of $\mathbb{R}^N$ and let $U \in \mathbb{R}^N$ be such that $\mathcal{A}U = F$. Show that $U = \sum_{n=1}^{N} \frac{(F, Z^n)}{(Z^n, \mathcal{A}Z^n)} Z^n$.

(ii) Show that for the CG algorithm, $U^m = U^0 + \sum_{n=0}^{m-1} \frac{(R^0, P^n)_N}{(P^n, \mathcal{A}P^n)_N} P^n$. (*Hint*: Prove first by induction on n that $(R^n, R^n)_N = (R^0, P^n)_N$.) Comment on this expression for $m = N$.

Exercise 9.14. Let $\mathcal{A} \in \mathbb{R}^{N,N}$ be symmetric positive definite matrix and let

$$J : \mathbb{R}^N \ni V \longmapsto J(V) = \tfrac{1}{2}(\mathcal{A}V, V)_N - (F, V)_N \in \mathbb{R}.$$

Let U^m be the m-th iterate of the CG algorithm.

(i) Show that U^m minimizes J over the affine space $U^0 + \mathcal{K}_m$.

(ii) Show that

$$J(U^m) = \min_{W \in U^0 + \mathcal{K}_{m-1}} J(W) + \min_{\alpha \in \mathbb{R}} \left(-\alpha (R^{m-1}, R^{m-1})_N + \tfrac{\alpha^2}{2}(P^m, \mathcal{A}P^m)_N \right).$$

(iii) Show that the above minimization problems have respective solutions given by $W = U^{m-1}$ and $\alpha = \frac{(R^{m-1}, R^{m-1})_N}{(\mathcal{A}P^m, P^m)_N}$. Comment on this result in view of Algorithm 9.4.

Exercise 9.15. Prove (9.31).

Exercise 9.16. Write the right-preconditioned version of GMRes and Bi-CGStab.

Exercise 9.17. Write Algorithm 9.10 in terms of the arrays ia, ja, and aa used to store the stiffness matrix in the CSR format; see §8.3.1.

A Posteriori Error Estimates and Adaptive Meshes

The goal of *a posteriori error estimates* is to assess the error between the exact solution u and its finite element approximation u_h in terms of known quantities only, i.e., the size of the mesh cells, the problem data, and the approximate solution. Such estimates differ from *a priori error estimates* of the form $\|u - u_h\|_{1,\Omega} \le ch^k \|u\|_{k+1,\Omega}$ in that the upper bound does not depend on the unknown quantity $\|u\|_{k+1,\Omega}$.

Since the early work of Babuška and Rheinbolt [BaR78^a, BaR78^b], the practical importance of a posteriori error estimates has been widely recognized. In addition to assessing the quality of the approximate solution, these estimates provide valuable information to construct a new mesh on which the approximate solution will (hopefully) be more accurate. This process can be repeated several times, thereby generating a sequence of so-called *adaptive meshes*.

This chapter is organized into four sections. The first section introduces the setting for a posteriori error estimates and presents a first type of error estimates, the so-called *residual-based error estimates*. The next section studies a posteriori error estimates based on the use of *hierarchical bases*. Both sections deal with estimates that control the error in the natural stability norm, i.e., the norm for which the conditions of the BNB Theorem hold. The third section investigates error estimates based on *duality techniques* designed to control arbitrary output functionals of the error. The last section addresses some implementation aspects of *adaptive mesh generation*. This chapter is set at an introductory level; see [AiO00, BeR01, JoR95, JoS95, OdD91, Ver96] for further aspects of the material covered herein.

10.1 Residual-Based Error Estimates

In this section, we derive residual-based error estimates and investigate their optimality properties. For the sake of brevity, we restrict the presentation to the Laplacian and to the Stokes problem both supplemented with homogeneous Dirichlet conditions. The techniques presented hereafter can be extended to more general PDEs (such as convection–diffusion–reaction equations) and to more general boundary conditions. An example is proposed in Exercise 10.1.

10.1.1 The setting

Consider the following problem:

$$
\begin{cases}
\text{Seek } u \in W \text{ such that} \\
a(u, v) = f(v), \quad \forall v \in V,
\end{cases}
\tag{10.1}
$$

where W and V are Hilbert spaces of functions defined over a domain Ω, $f \in V'$, and $a \in \mathcal{L}(W \times V; \mathbb{R})$. We assume that the bilinear form a satisfies the conditions of the BNB Theorem. Problem (10.1) is therefore well-posed.

Let W_h and V_h be two approximation spaces constructed from a family $\{\mathcal{T}_h\}_{h>0}$ of meshes of Ω and a reference finite element $\{\widehat{K}, \widehat{P}, \widehat{\Sigma}\}$. Set $W(h) = W + W_h$ and equip this space with an extended norm denoted by $\| \cdot \|_{W(h)}$. Consider the approximate problem:

$$
\begin{cases}
\text{Seek } u_h \in W_h \text{ such that} \\
a_h(u_h, v_h) = f_h(v_h), \quad \forall v_h \in V_h,
\end{cases}
\tag{10.2}
$$

where a_h and f_h are approximations to the bilinear form a and the linear form f, respectively. We assume that (10.2) is well-posed (see Theorem 2.22) and that suitable consistency and approximability properties hold to ensure that u_h converges to the exact solution u as $h \to 0$; see §2.3.1.

Definition 10.1. *A function $e(h, u_h, f)$ is said to be an* a posteriori *error estimate if*

$$
\|u - u_h\|_{W(h)} \leq e(h, u_h, f).
\tag{10.3}
$$

Furthermore, if $e(h, u_h, f)$ can be localized in the form

$$
e(h, u_h, f) = \left(\sum_{K \in \mathcal{T}_h} e_K(u_h, f)^2 \right)^{\frac{1}{2}},
\tag{10.4}
$$

the quantities $e_K(u_h, f)$ are called local error indicators.

Remark 10.2. The estimate (10.3) is sometimes called a *reliability property* since it shows that $e(h, u_h, f)$ controls the error $u - u_h$ in the natural stability norm. □

Local error indicators provide a basis for adaptive mesh refinement. Heuristically, if $e_K(u_h, f)$ is large, the mesh must be refined locally, whereas if $e_K(u_h, f)$ is very small, the mesh cell K can be recombined with some of its neighbors to form a larger mesh cell. This procedure can be repeated several times, leading (hopefully) to a mesh for which:

(i) The total error estimate $e(h, u_h, f)$ is less than a prescribed tolerance.
(ii) The local error indicators $e_K(u_h, f)$ is balanced over the mesh cells.

Clearly, computational efficiency requires that the local error indicators also satisfy some *optimality property* guaranteeing that the upper bound (10.4) is sharp enough. To this end, assume that the extended norm $\|\cdot\|_{W(h)}$ can be localized in the form $\|\cdot\|_{W(h)}^2 = \sum_{K \in \mathcal{T}_h} \|\cdot\|_{W(h),K}^2$. Ideally, one would like to derive local error indicators such that

$$\forall h, \ \forall K \in \mathcal{T}_h, \quad c_1 e_K(u_h, f) \le \|u - u_h\|_{W(h),K} \le c_2 e_K(u_h, f), \qquad (10.5)$$

where c_1 and c_2 are mesh-independent constants. Inequalities (10.5) imply that $e_K(u_h, f)$ is *equivalent* to the local error $\|u - u_h\|_{W(h),K}$. Unfortunately, it is not possible in general to derive (10.5). Instead, one often proves a global upper bound $\sum_{K \in \mathcal{T}_h} \|u - u_h\|_{W(h),K}^2 \le c_2 \sum_{K \in \mathcal{T}_h} e_K(u_h, f)^2$, while the lower bound takes the form

$$c_1 e_K(u_h, f) \le \|u - u_h\|_{W(h),\Delta_K} + \Pi(h_K, \Delta_K, f), \qquad (10.6)$$

where Δ_K is a patch of elements around K and $\Pi(h_K, \Delta_K, f)$ is a perturbation that is either negligible or is asymptotically of the same order as the error $\|u - u_h\|_{W(h),\Delta_K}$.

10.1.2 The Laplacian

Let Ω be a polyhedron in $\mathbb{R}^d$, $f \in L^2(\Omega)$, and consider the problem:

$$\begin{cases} \text{Seek } u \in H_0^1(\Omega) \text{ such that} \\ a(u, v) = \int_\Omega fv, \quad \forall v \in H_0^1(\Omega), \end{cases} \qquad (10.7)$$

where $a(u, v) = \int_\Omega \nabla u \cdot \nabla v$. Leaving room for generalizations, we shall not use the coercivity of the bilinear form a but only assume that a satisfies the conditions of BNB Theorem. In particular, we assume that there exists $\alpha > 0$ such that

$$\inf_{u \in H_0^1(\Omega)} \sup_{v \in H_0^1(\Omega)} \frac{a(u, v)}{\|u\|_{1,\Omega} \|v\|_{1,\Omega}} \ge \alpha. \qquad (10.8)$$

For the sake of simplicity, we restrict the presentation to simplicial, affine mesh families, say $\{\mathcal{T}_h\}_{h>0}$. Let V_h be a $H_0^1(\Omega)$-conformal approximation space based on $\mathcal{T}_h$ and a Lagrange finite element of degree k. This yields the approximate problem:

$$\begin{cases} \text{Seek } u_h \in V_h \text{ such that} \\ a(u_h, v_h) = \int_\Omega f v_h, \quad \forall v_h \in V_h. \end{cases} \tag{10.9}$$

Assuming that the exact solution is smooth enough, the following *a priori error estimate* holds (see Chapter 3):

$$\|u - u_h\|_{1,\Omega} \le c \inf_{v_h \in V_h} \|u - v_h\|_{1,\Omega} \le c' \left(\sum_{K \in T_h} h_K^{2k} \|u\|_{k+1,K}^2 \right)^{\frac{1}{2}}.$$

To derive an *a posteriori error estimate*, we use the stability property (10.8) to obtain

$$\alpha \|u - u_h\|_{1,\Omega} \le \sup_{v \in H_0^1(\Omega)} \frac{a(u - u_h, v)}{\|v\|_{1,\Omega}} \le \sup_{v \in H_0^1(\Omega)} \frac{\langle \Delta(u - u_h), v \rangle_{H^{-1}, H_0^1}}{\|v\|_{1,\Omega}}$$

$$\le \|f + \Delta u_h\|_{-1,\Omega}.$$

This yields our first a posteriori error estimate.

Proposition 10.3. *Let u solve (10.7) and u_h solve (10.9). Then,*

$$\|u - u_h\|_{1,\Omega} \le \tfrac{1}{\alpha} \|f + \Delta u_h\|_{-1,\Omega}. \tag{10.10}$$

The main difficulty with the a posteriori estimate (10.10) is that the norm $\|\cdot\|_{-1,\Omega}$ cannot be localized. To derive a local error indicator, we still use the idea of integration by parts to eliminate the exact solution, but we perform it elementwise. Let $\mathcal{F}_h^i$ be the set of interior faces. For $F \in \mathcal{F}_h^i$ with $F = K_1 \cap K_2$, denote by n_1 and n_2 the outward normal to K_1 and K_2, respectively. Let $[\![\partial_n u_h]\!]$ be the jump of the normal derivative of u_h across F, i.e., $[\![\partial_n u_h]\!] = \nabla u_{h|K_1} \cdot n_1 + \nabla u_{h|K_2} \cdot n_2$. The main result is stated in the following:

Theorem 10.4. *Let u solve (10.7) and u_h solve (10.9). Assume that the family $\{T_h\}_{h>0}$ is shape-regular. Then, there is c such that*

$$\forall h, \quad \|u - u_h\|_{1,\Omega} \le c \left(\sum_{K \in T_h} e_K(u_h, f)^2 \right)^{\frac{1}{2}}, \tag{10.11}$$

with local error indicators

$$e_K(u_h, f) = h_K \|f + \Delta u_h\|_{0,K} + \tfrac{1}{2} \sum_{F \in \mathcal{F}_K} h_F^{\frac{1}{2}} \|[\![\partial_n u_h]\!]\|_{0,F}, \tag{10.12}$$

where $\mathcal{F}_K$ is the set of faces of K that are not on $\partial\Omega$ and $h_F = \mathrm{diam}(F)$.

Proof. Since $a(u - u_h, v_h) = 0$ for all $v_h \in V_h$, (10.8) yields

$$\forall v_h \in V_h, \quad \|u - u_h\|_{1,\Omega} \le \frac{1}{\alpha} \sup_{v \in H_0^1(\Omega)} \frac{a(u - u_h, v - v_h)}{\|v\|_{1,\Omega}}. \tag{10.13}$$

The numerator in the right-hand side is expanded as follows:

$$a(u - u_h, v - v_h) = \int_{\Omega} -\Delta u(v - v_h) - \nabla u_h \cdot \nabla(v - v_h)$$

$$= \sum_{K \in \mathcal{T}_h} \left(\int_K (f + \Delta u_h)(v - v_h) - \sum_{F \in \partial K} \int_F \partial_n u_h(v - v_h) \right),$$

where F denotes a face of the element K. Since $v - v_h$ vanishes on $\partial \Omega$, the summation over F involves only interior faces. Using the continuity of $v - v_h$ across F yields

$$a(u - u_h, v - v_h) \leq \sum_{K \in \mathcal{T}_h} \left(\|f + \Delta u_h\|_{0,K} \|v - v_h\|_{0,K} \right.$$

$$\left. + \sum_{F \in \mathcal{F}_K} \tfrac{1}{2} \|[\partial_n u_h]\|_{0,F} \|v - v_h\|_{0,F} \right)$$

$$\leq \sum_{K \in \mathcal{T}_h} e_K(u_h, f) \eta_K(v - v_h),$$

where $\eta_K(v) = \max(h_K^{-1} \|v\|_{0,K}, h_F^{-\frac{1}{2}} \max_{F \in \mathcal{F}_K} \|v\|_{0,F})$ for $v \in H_0^1(\Omega)$. Taking $v_h = \mathcal{SZ}_h v$ in the above equation, where $\mathcal{SZ}_h$ is the Scott–Zhang interpolation operator introduced in Lemma 1.130, yields

$$a(u - u_h, v - \mathcal{SZ}_h v) \leq \sum_{K \in \mathcal{T}_h} e_K(u_h, f) \eta_K(v - \mathcal{SZ}_h v).$$

For an element K and a face F, let Δ_K be the set of simplices that share at least one point with K, and let Δ_F be the set of simplices that share at least one point with F; see Figure 1.25. Define the integers

$$M = \max_{K \in \mathcal{T}_h} \operatorname{card}\{K' \in \mathcal{T}_h;\ K \in \Delta_{K'}\},$$

$$N = \max_{F \in \mathcal{F}_h} \operatorname{card}\{F \in \mathcal{F}_h;\ K \in \Delta_F\},$$

where $\mathcal{F}_h$ denotes the set of mesh faces. Clearly, the integers M and N only depend on the shape-regularity of the mesh, and are bounded by a constant independent of h. Owing to the interpolation properties of $\mathcal{SZ}_h$,

$$\sum_{K \in \mathcal{T}_h} \eta_K(v - \mathcal{SZ}_h v)^2 \leq \max(M, N) \|v\|_{1,\Omega}^2,$$

yielding

$$\frac{a(u - u_h, v - \mathcal{SZ}_h v)}{\|v\|_{1,\Omega}} \leq c \left(\sum_{K \in \mathcal{T}_h} e_K(u_h, f)^2 \right)^{\frac{1}{2}}.$$

Conclude using (10.13).   □

Remark 10.5.

(i) In the literature, $e_K(u_h, f)$ is called a *residual-based error indicator* because the quantity $f + \Delta u_h$ is the residual of the original PDE, $f = -\Delta u$. This type of error indicator has been introduced and analyzed in one dimension in [BaR78^a]. The proof presented above is adapted from [Ver94].

(ii) If V_h is constructed using $\mathbb{P}_1$ Lagrange finite elements, $\Delta u_{h|K} = 0$ and $[\![\partial_n u_h]\!]_{|F}$ is constant. This yields the simplified expression

$$e_K(u_h, f) = h_K \|f\|_{0,K} + \frac{1}{2} \sum_{F \in \mathcal{F}_K} h_F^{\frac{1}{2}} \operatorname{meas}(F)^{\frac{1}{2}} [\![\partial_n u_h]\!]_{|F}. \tag{10.14}$$

Note that in two dimensions, $h_F = \operatorname{meas}(F)$.

(iii) The coercivity of the bilinear form a has not been used. Instead, the stability inequality (10.8) has played a central role. Therefore, the above analysis can be generalized to problems that enter the framework of the BNB Theorem without being necessarily endowed with a coercivity property.

(iv) It is remarkable that in the a posteriori analysis, it is the stability of the continuous problem that comes into play, i.e., equation (10.8) instead of the inf-sup condition associated with the discrete problem. The latter is invoked only to guarantee that the discrete problem is well-posed. □

Optimality of the error indicator. We now want to determine whether the residual-based error indicator defined in (10.12) satisfies some optimality property, i.e., whether the lower bound in (10.5) also holds. Before stating the main result, we introduce some notation and present two technical lemmas.

Lemma 10.6 (Verfürth 1). *Let $b_K \in H_0^1(K)$ be a function such that:*

(i) $0 \le b_K \le 1$.
(ii) $\exists D \subset K$ with $\operatorname{meas}(D) > 0$ and $b_{K|D} \ge \frac{1}{2}$.

Let $m \in \mathbb{N}$. Then, there exist c_1 and c_2 such that,

$$\forall K \in \mathcal{T}_h, \ \forall \phi \in \mathbb{P}_m(K), \quad \begin{cases} \|b_K \phi\|_{0,K} \le \|\phi\|_{0,K} \le c_1 \|b_K^{\frac{1}{2}} \phi\|_{0,K}, \\ |b_K \phi|_{1,K} \le c_2 h_K^{-1} \|\phi\|_{0,K}. \end{cases}$$

Proof. The above inequalities are first established on the reference simplex using the equivalence of norms in finite-dimensional vector spaces and then transferred back to the simplex K by a change of variables. The complete proof is presented in [Ver94] and [Ver96, p. 10]. □

Example 10.7. An example of function b_K satisfying the assumptions of Lemma 10.6 is the H^1-conformal *cell bubble function*

$$b_K = (d+1)^{d+1} \lambda_0 \dots \lambda_d,$$

where $\{\lambda_0, \dots, \lambda_d\}$ are the barycentric coordinates of K. □

Fig. 10.1. Left: The shaded zone represents the set D_K defined as the union of the simplices sharing a face with the simplex K. Center: The shaded zone represents the set D_F defined as the union of the two simplices sharing the interface F. Right: Level-sets of the lifting $P_F(\phi)$ in D_F.

We now introduce a lifting operator for functions defined on an interface. Consider first the reference simplex $\widehat{K}$, e.g., the unit simplex of $\mathbb{R}^d$. Denote by $\widehat{F}_1$ the face of $\widehat{K}$ corresponding to the hyperplane $\{\widehat{x}_1 = 0\}$. Let $\widehat{\Pi}_1$ be the orthogonal projection from $\mathbb{R}^d$ to this hyperplane, and introduce the lifting operator $\widehat{P}_{\widehat{F}_1} : \mathbb{P}_k(\widehat{F}_1) \rightarrow \mathbb{P}_k(\widehat{K})$ defined as $\widehat{P}_{\widehat{F}_1}(\widehat{\phi}) = \widehat{\phi} \circ \widehat{\Pi}_1$ for all $\widehat{\phi} \in \mathbb{P}_k(\widehat{F}_1)$. Let $F \in \mathcal{F}_h$ be an interior mesh face. Let K be one of the two simplices of which F is a face and let $T_K : \widehat{K} \rightarrow K$ be the corresponding affine transformation. Without loss of generality, we assume that $F = T_K(\widehat{F}_1)$. The lifting operator $P_{F,K} : \mathbb{P}_k(F) \rightarrow \mathbb{P}_k(K)$ is defined as

$$\forall \phi \in \mathbb{P}_k(F), \quad P_{F,K}(\phi) = \widehat{P}_{\widehat{F}_1}(\phi \circ T_K) \circ T_K^{-1}.$$

Let D_F be the union of the two simplices K and K' sharing the interface F; see the central panel in Figure 10.1. The lifting operator P_F is defined as

$$\forall \phi \in \mathbb{P}_k(F), \quad P_F(\phi) = \begin{cases} P_{F,K}(\phi) & \text{on } K, \\ P_{F,K'}(\phi) & \text{on } K'. \end{cases} \tag{10.15}$$

$P_F(\phi)$ is supported on D_F; level-sets are shown in the right panel of Figure 10.1.

Lemma 10.8 (Verfürth 2). *Let b_F be a function in $H_0^1(D_F)$ such that:*

(i) $0 \leq b_F \leq 1$.
(ii) $\exists D \subset D_F$ with $\text{meas}(D) > 0$ such that $b_{F|D} \geq \frac{1}{2}$.
(iii) $b_{F|F} \in H_0^1(F)$.
(iv) $\exists D' \subset F$ with $\text{meas}(D') > 0$ (surface measure) such that $b_{F|D'} \geq \frac{1}{2}$.

Let $k \in \mathbb{N}$. Then, there exist c_1, c_2, c_3, and c_4 such that

$$\forall F \in \mathcal{F}_h, \ \forall \phi \in \mathbb{P}_k(F), \quad \begin{cases} \|b_F\phi\|_{0,F} \leq \|\phi\|_{0,F} \leq c_1 \|b_F^{\frac{1}{2}}\phi\|_{0,F}, \\ c_2 h_F^{\frac{1}{2}} \|\phi\|_{0,F} \leq \|b_F P_F(\phi)\|_{0,D_F} \leq c_3 h_F^{\frac{1}{2}} \|\phi\|_{0,F}, \\ |b_F P_F(\phi)|_{1,D_F} \leq c_4 h_F^{-\frac{1}{2}} \|\phi\|_{0,F}. \end{cases}$$

Proof. See [Ver94] and [Ver96, p. 10]. $\square$

Example 10.9. Let us give an example of a *bubble function* b_F satisfying the assumptions of Lemma 10.8 in dimension 2. Without loss of generality, assume that the vertices of the segment $F = K \cap K'$ are the vertices 1 and 2 in triangles K and K'. Denote by $\lambda_{i,K}$ the barycentric coordinate associated with the i-th vertex of K. Define the function b_F as

$$b_F = \begin{cases} 4\lambda_{1,K}\lambda_{2,K} & \text{on } K, \\ 4\lambda_{1,K'}\lambda_{2,K'} & \text{on } K'. \end{cases}$$

Clearly, $b_F \in H_0^1(D_F)$ and the restriction of b_F to F is in $H_0^1(F)$. The other assumptions of Lemma 10.8 are readily checked. $\square$

Theorem 10.10 (Optimality). *Let $l \in \mathbb{N}$ and set $Z_{lh} = \{v_h \in L^2(\Omega); \forall K \in \mathcal{T}_h, v_{h|K} \in \mathbb{P}_l(K)\}$. Then, there exists c, depending only on the shape-regularity of the mesh, the reference finite element, and l, such that*

$$e_K(u_h, f) \le c \left(|u - u_h|_{1,D_K} + h_K \inf_{v_h \in Z_{lh}} \|f - v_h\|_{0,D_K} \right). \tag{10.16}$$

Proof. (1) Let us first bound $\|f + \Delta u_h\|_{0,K}$. Clearly, for all $v_h \in Z_{lh}$,

$$\|f + \Delta u_h\|_{0,K} \le \|f - v_h\|_{0,K} + \|v_h + \Delta u_h\|_{0,K},$$

showing that the quantity $\|v_h + \Delta u_h\|_{0,K}$ must be controlled. Since $v_h + \Delta u_h$ is in $\mathbb{P}_{\max(k-2,l)}(K)$, Lemma 10.6 yields

$$c\|v_h + \Delta u_h\|_{0,K}^2 \le \|b_K^{\frac{1}{2}}(v_h + \Delta u_h)\|_{0,K}^2 = \int_K (v_h + \Delta u_h)b_K(v_h + \Delta u_h)$$

$$= \int_K (f + \Delta u_h)b_K(v_h + \Delta u_h) + \int_K (v_h - f)b_K(v_h + \Delta u_h)$$

$$\le \int_K \nabla(u - u_h)\cdot\nabla(b_K(v_h + \Delta u_h)) + \|v_h - f\|_{0,K}\|v_h + \Delta u_h\|_{0,K}.$$

Note that the integration by parts in the last inequality uses the fact that b_K vanishes on the boundary of K. The inverse inequality established in Lemma 10.6 implies

$$c\|v_h + \Delta u_h\|_{0,K}^2 \le |u - u_h|_{1,K}|b_K(v_h + \Delta u_h)|_{1,K} + \|v_h - f\|_{0,K}\|v_h + \Delta u_h\|_{0,K}$$

$$\le (ch_K^{-1}|u - u_h|_{1,K} + \|v_h - f\|_{0,K})\|v_h + \Delta u_h\|_{0,K}.$$

As a result,

$$\|f + \Delta u_h\|_{0,K} \le c\,(h_K^{-1}|u - u_h|_{1,K} + \|v_h - f\|_{0,K}).$$

(2) Let us now bound $\|[\![\partial_n u_h]\!]\|_{0,F}$. Since $[\![\partial_n u_h]\!]_{|F}$ is in $\mathbb{P}_{k-1}(F)$, Lemma 10.8 yields

$$c\,\|[\![\partial_n u_h]\!]\|_{0,F}^2 \leq \int_F [\![\partial_n u_h]\!]b_F[\![\partial_n u_h]\!] \leq \int_F [\![\partial_n(u_h - u)]\!]b_F P_F([\![\partial_n u_h]\!]).$$

In the last inequality, the fact that $[\![\partial_n u]\!]$ vanishes on F has been used, and the lifting operator defined in (10.15) has been introduced. Since the function $b_F P_F([\![\partial_n u_h]\!])$ vanishes on the boundary of D_F and is continuous across F, an integration by parts yields

$$c\,\|[\![\partial_n u_h]\!]\|_{0,F}^2 \leq \int_{D_F} \nabla(u - u_h)\cdot\nabla(b_F P_F([\![\partial_n u_h]\!]))$$

$$+ \sum_{K\in D_F} \int_K b_F P_F([\![\partial_n u_h]\!])\Delta(u - u_h)$$

$$\leq |u - u_h|_{1,D_F} |b_F P_F([\![\partial_n u_h]\!])|_{1,D_F}$$

$$+ \sum_{K\in D_F} \|b_F P_F([\![\partial_n u_h]\!])\|_{0,K}\|f + \Delta u_h\|_{0,K}.$$

The second estimate and the inverse inequality of Lemma 10.8 imply

$$\|[\![\partial_n u_h]\!]\|_{0,F}^2 \leq c_1 h_F^{-\frac{1}{2}}|u - u_h|_{1,D_F}\|[\![\partial_n u_h]\!]\|_{0,F}$$

$$+ c_2 \sum_{K\in D_F} h_F^{\frac{1}{2}}\|[\![\partial_n u_h]\!]\|_{0,F}\|f + \Delta u_h\|_{0,K}$$

$$\leq c\left(h_F^{-\frac{1}{2}}|u - u_h|_{1,D_F} + h_F^{\frac{1}{2}}\sum_{K\in D_F}\|f + \Delta u_h\|_{0,K}\right)\|[\![\partial_n u_h]\!]\|_{0,F}.$$

Use the estimate of $\|f + \Delta u_h\|_{0,K}$ derived in the first part of the proof and the shape-regularity assumption implying that $h_F \leq h_K \leq ch_F$ to obtain

$$\|[\![\partial_n u_h]\!]\|_{0,F} \leq c\left(h_F^{-\frac{1}{2}}|u - u_h|_{1,D_F}\right.$$

$$\left. + h_F^{\frac{1}{2}}\sum_{K\in D_F}(h_K^{-1}|u - u_h|_{1,K} + \|v_h - f\|_{0,K})\right)$$

$$\leq c(h_F^{-\frac{1}{2}}|u - u_h|_{1,D_F} + h_F^{\frac{1}{2}}\|v_h - f\|_{0,D_F}).$$

Finally, combine the two above estimates to infer

$$e_K(u_h, f) \leq c_1(|u - u_h|_{1,K} + h_K\|v_h - f\|_{0,K})$$

$$+ c_2 \sum_{F\in\mathcal{F}_K}(|u - u_h|_{1,D_F} + h_F\|v_h - f\|_{0,D_F})$$

$$\leq c(|u - u_h|_{1,D_K} + h_K\|v_h - f\|_{0,D_K}).$$

Conclude by taking the infimum on v_h. $\qquad\qquad\qquad\qquad\qquad\square$

Remark 10.11.

(i) The proof of Theorem 10.10 is adapted from [Ver96, pp. 10–18] for $l = 0$. The advantage of considering an arbitrary l will appear more clearly at the next item, where the impact of quadratures is discussed. Note that the constant c in (10.16) explodes with l.

(ii) In many applications, the quantity $\|f + \Delta u_h\|_{0,K}$ cannot be computed exactly, i.e., approximate integration using quadratures must be performed; see §8.1. Let k be the degree of the Lagrange finite element on which the approximation is based. Let k' be the largest polynomial degree for which the quadrature is exact. Let f_{lh} be a polynomial in $\mathbb{P}_l(K)$ approximating f over K. The quantity $\|f_{lh} + \Delta u_h\|_{0,K}$ can be computed exactly if $k' \geq \max(2l, 2(k-2))$. As a result, the following approximate error indicators can be computed:

$$e_{lK}(u_h, f) = h_K \|f_{lh} + \Delta u_h\|_{0,K} + \tfrac{1}{2} \sum_{F \in \mathcal{F}_K} h_F^{\frac{1}{2}} \|[\![\partial_n u_h]\!]\|_{0,F}.$$

Assuming $f \in W^{l+1,\infty}(K)$, we infer

$$|e_K(u_h, f) - e_{lK}(u_h, f)| = h_K \|f - f_{lh}\|_{0,K} \leq c\, h_K^{l+2} \|f\|_{W^{l+1,\infty}(K)}.$$

Therefore, the error associated with quadratures does not spoil the error indicator provided $h_K^{l+2} \ll h_K^{k+1}$, i.e., $l \geq k$. This condition means that error indicators should be evaluated using a quadrature that is exact for polynomials of degree $2k$.

(iii) It is possible to derive other types of local error indicators for the Poisson problem; see, e.g., [BaW85] for error indicators based on a local Neumann problem and [BaR78^b] for those based on a local Dirichlet problem. The residual-based error indicator (10.12) is generally cheaper to evaluate. Furthermore, one can show that all these error indicators are more or less equivalent; see [Ver96, pp. 19–28] for a thorough discussion.   □

10.1.3 The Stokes problem

In this section, we generalize to the Stokes problem the residual-based error indicator derived in the previous section for the Laplacian. Let Ω be a polyhedron in $\mathbb{R}^d$, and let $f \in [L^2(\Omega)]^d$. For the sake of simplicity, we consider the Stokes problem with homogeneous Dirichlet conditions:

$$\begin{cases} \text{Seek } (u, p) \in [H_0^1(\Omega)]^d \times L_{j=0}^2(\Omega) \text{ such that} \\ \int_\Omega \nabla u : \nabla v - \int_\Omega p \nabla \cdot v = \int_\Omega f \cdot v, & \forall v \in [H_0^1(\Omega)]^d, \\ \int_\Omega q \nabla \cdot u = 0, & \forall q \in L_{j=0}^2(\Omega). \end{cases}$$

See Chapter 4 for the mathematical analysis and the finite element approximation of this problem.

Let $\{\mathcal{T}_h\}_{h>0}$ be a shape-regular family of simplicial meshes of Ω. Consider mixed finite element approximation spaces $X_h \subset H_0^1(\Omega)$ and $M_h \subset L_{j=0}^2(\Omega)$

such that the compatibility condition derived in §4.2 holds. The approximate Stokes problem is then well-posed, and we denote by $(u_h, p_h) \in X_h \times M_h$ its unique solution. For an interior face $F \in \mathcal{F}_h^i$ with $F = K_1 \cap K_2$, set $[\![\partial_n u_h - p_h n]\!] = (\nabla u_{h|K_1} \cdot n_1 - p_{h|K_1} n_1) + (\nabla u_{h|K_2} \cdot n_2 - p_{h|K_2} n_2)$ with obvious notation.

Theorem 10.12. *Under the above assumptions, the quantity*

$$e_K(u_h, p_h, f) = h_K \| f + \Delta u_h - \nabla p_h \|_{0,K} + \| \nabla \cdot u_h \|_{0,K}$$
$$+ \tfrac{1}{2} \sum_{F \in \mathcal{F}_K} h_F^{\frac{1}{2}} \| [\![\partial_n u_h - p_h n]\!] \|_{0,F},$$

is a residual-based error indicator *for the Stokes problem.*

Proof. The Stokes problem is rewritten in the compact form:

$$\begin{cases} \text{Seek } u \in V \text{ such that} \\ a(u, v) = \int_\Omega f \cdot v, \quad \forall v \in V, \end{cases}$$

with $u = (u, p)$, $V = [H_0^1(\Omega)]^d \times L_{j=0}^2(\Omega)$, $v = (v, q)$, $f = (f, 0)$, and

$$a(u, v) = \int_\Omega \nabla u : \nabla v - \int_\Omega p \nabla \cdot v + \int_\Omega q \nabla \cdot u.$$

Equipped with the norm $\| u \|_V = \| u \|_{1,\Omega} + \| p \|_{0,\Omega}$, V is a Hilbert space, $a \in \mathcal{L}(V \times V; \mathbb{R})$, and a satisfies on $V \times V$ the stability inequality

$$\exists \alpha > 0, \ \forall u \in V, \quad \alpha \| u \|_V \leq \sup_{v \in V} \frac{a(u, v)}{\| v \|_V}.$$

Set $V_h = X_h \times M_h$ and let $u_h = (u_h, p_h)$ be the solution to the approximate Stokes problem in V_h. The orthogonality relation $a(u - u_h, v_h) = 0$ and the above stability inequality imply

$$\forall v_h \in V_h, \quad \alpha \| u - u_h \|_V \leq \sup_{v \in V} \frac{a(u - u_h, v - v_h)}{\| v \|_V}.$$

Choose $v_h = (v_h, 0)$ and expand the numerator in the right-hand side using $(v - v_h)_{|\partial \Omega} = 0$ and $\nabla \cdot u = 0$ to obtain

$$a(u - u_h, v - v_h) = \int_\Omega (-\Delta u + \nabla p) \cdot (v - v_h) - \sum_{K \in \mathcal{T}_h} \int_K \nabla u_h : \nabla (v - v_h)$$

$$+ \sum_{K \in \mathcal{T}_h} \int_K p_h \nabla \cdot (v - v_h) + \sum_{K \in \mathcal{T}_h} \int_K q \nabla \cdot (u - u_h)$$

$$= \sum_{K \in \mathcal{T}_h} \int_K (f + \Delta u_h - \nabla p_h) \cdot (v - v_h)$$

$$- \sum_{F \in \mathcal{F}_h} \int_F [\![\partial_n u_h - p_h n]\!] \cdot (v - v_h) - \sum_{K \in \mathcal{T}_h} \int_K q \nabla \cdot u_h.$$

This yields

$$a(u - u_h, v - v_h) \leq \sum_{K \in \mathcal{T}_h} \|f + \Delta u_h - \nabla p_h\|_{0,K} \|v - v_h\|_{0,K}$$

$$+ \sum_{F \in \mathcal{F}_h} \|[\![\partial_n u_h - p_h n]\!]\|_{0,F} \|v - v_h\|_{0,F} + \sum_{K \in \mathcal{T}_h} \|q\|_{0,K} \|\nabla \cdot u_h\|_{0,K}.$$

Set $v_h = (\mathcal{SZ}_h v_1, \ldots, \mathcal{SZ}_h v_d)$, where $\mathcal{SZ}_h$ is the Scott–Zhang interpolation operator, and conclude as in the proof of Theorem 10.4. $\qquad \square$

Optimality of the error indicator. The error indicator derived in Theorem 10.12 is quasi-optimal, as shown in the following:

Proposition 10.13. *For $l \in \mathbb{N}$, set $Z_{lh} = \{v_h \in [L^2(\Omega)]^d; \forall K \in \mathcal{T}_h, v_{h|K} \in [\mathbb{P}_l(K)]^d\}$. Then, there exists c, depending only on the shape-regularity of the mesh, the reference elements for velocity and pressure, and l, such that*

$$e_K(u_h, p_h, f) \leq c \left(|u - u_h|_{1, D_K} + \|p - p_h\|_{0, D_K} + h_K \inf_{v_h \in Z_{lh}} \|f - v_h\|_{0, D_K} \right).$$

Proof. The proof is only sketched since it is similar to that of Theorem 10.10. An upper bound for $\|f + \Delta u_h - \nabla p_h\|_{0,K}$ is derived as in the first step of the proof of Theorem 10.10: introduce $v_h \in [\mathbb{P}_l(K)]^d$ and $b_K(v_h + \Delta u_h - \nabla p_h)$; then, use the vector version of Lemma 10.6. This yields the estimate

$$\|f + \Delta u_h - \nabla p_h\|_{0,K} \leq c \left(h_K^{-1}(|u - u_h|_{1,K} + \|p - p_h\|_{0,K}) \right.$$

$$\left. + \inf_{v_h \in [\mathbb{P}_l(K)]^d} \|f - v_h\|_{0,K} \right).$$

To control $\|\nabla \cdot u_h\|_{0,K}$, simply note that $\|\nabla \cdot u_h\|_{0,K} = \|\nabla \cdot (u - u_h)\|_{0,K} \leq c |u - u_h|_{1,K}$. Finally, to control the interface terms, proceed as in the second step of the proof of Theorem 10.10. For an interior face $F \in \mathcal{F}_h^i$, introduce $b_F P_F([\![\partial_n u_h - p_h n]\!])$ where P_F is the vector version of the lifting operator defined in Lemma 10.8. Observe that $\partial_n u - pn$ is continuous across F, and use the vector version of Lemma 10.8. This yields the estimate

$$\|[\![\partial_n u_h - p_h n]\!]\|_{0,F} \leq c \left(h_K^{-1}(|u - u_h|_{1,D_F} + \|p - p_h\|_{0,D_F}) \right.$$

$$\left. + \inf_{v_h \in Z_{lh}} \|f - v_h\|_{0,D_F} \right),$$

and the conclusion readily follows. $\qquad \square$

Remark 10.14.

(i) The above results can be extended to more general versions of the Stokes problem; see Exercise 10.2 for an example.

(ii) In the same spirit as Remark 10.5(iv), we point out that no discrete inf-sup condition is invoked in the analysis. □

10.1.4 Non-conformal Crouzeix–Raviart approximation

Residual-based a posteriori error estimates in a non-conformal setting have been investigated in [AcB01, Ang02, Sch02, ElE03]. For the sake of brevity, we discuss only the non-conformal approximation of the Laplacian with homogeneous Dirichlet conditions. Let Ω be a polyhedron in $\mathbb{R}^d$, $f \in L^2(\Omega)$, and let $u \in H_0^1(\Omega)$ be the solution to (10.7). Let $\{\mathcal{T}_h\}_{h>0}$ be a family of meshes of Ω and denote by $P_{c,h,0}^1$ the H_0^1-conformal approximation space based on the $\mathbb{P}_1$ Lagrange finite element (recall that $P_{c,h,0}^1 = P_{c,h}^1 \cap H_0^1(\Omega)$ with $P_{c,h}^1$ defined in (1.76)). Let $P_{pt,h,0}^1$ be the non-conformal approximation space based on the Crouzeix–Raviart finite element

$$P_{pt,h,0}^1 = \{v_h \in L^1(\Omega_h); \forall K \in \mathcal{T}_h, v_{h|K} \in \mathbb{P}_1; \forall F \in \mathcal{F}_h, \int_F [\![v_h]\!] = 0\}, \quad (10.17)$$

with the convention that a zero outer value is taken for $F \in \mathcal{F}_h^\partial$. Functions in $P_{pt,h,0}^1$ are piecewise linear, their mean value is continuous across interior faces (edges in two dimensions), and their mean value vanishes at boundary faces. Consider the approximate problem:

$$\begin{cases} \text{Seek } u_h \in P_{pt,h,0}^1 \text{ such that} \\ \sum_{K \in \mathcal{T}_h} \int_K \nabla u_h \cdot \nabla v_h = \int_\Omega f v_h, \quad \forall v_h \in P_{pt,h,0}^1, \end{cases} \quad (10.18)$$

and define $a_h(u_h, v_h) = \sum_{K \in \mathcal{T}_h} \int_K \nabla u_h \cdot \nabla v_h$. Our goal is to derive an a posteriori error estimate for the error $u - u_h$ evaluated in the broken energy norm $|v|_{h,1,\Omega} = a_h(v,v)^{\frac{1}{2}}$. Recall that this norm controls uniformly the L^2-norm on $H_0^1(\Omega) + P_{pt,h,0}^1$; see Lemma 3.35. For a region that is a collection of elements, say R, define $|v|_{h,1,R} = (\sum_{K \in R} \|\nabla v\|_{0,K}^2)^{\frac{1}{2}}$.

Proposition 10.15. *Assume that the family $\{\mathcal{T}_h\}_{h>0}$ is shape-regular. Then, there is c, independent of h, such that*

$$|u - u_h|_{h,1,\Omega} \le c \left(\sum_{K \in \mathcal{T}_h} e_K(u_h, f)^2 \right)^{\frac{1}{2}} + \inf_{v_h \in P_{c,h,0}^1} |u_h - v_h|_{h,1,\Omega}, \quad (10.19)$$

where the residual-based error indicators $e_K(u_h, f)$ are defined in (10.14).

Proof. Since $P_{c,h,0}^1 \subset P_{pt,h,0}^1$, Galerkin orthogonality implies

$$\forall w_h \in P_{c,h,0}^1, \quad a_h(u - u_h, w_h) = 0.$$

Consider an arbitrary $w \in H_0^1(\Omega)$. Take $w_h = S\mathcal{Z}_h w$ in the above equation, where $S\mathcal{Z}_h$ is the Scott–Zhang interpolation operator introduced in Lemma 1.130, to obtain

$$a_h(u - u_h, w) = a_h(u - u_h, w - S\mathcal{Z}_h w).$$

Using the same techniques as in the proof of Theorem 10.4 yields

$$a_h(u - u_h, w) \leq c \left(\sum_{K \in \mathcal{T}_h} e_K(u_h, f)^2 \right)^{\frac{1}{2}} |w|_{h,1,\Omega}.$$

Now let v_h be an arbitrary function in $P_{c,h,0}^1$. Setting $w = u - v_h$ leads to

$$|u - v_h|_{h,1,\Omega}^2 = a_h(u - u_h, w) + a_h(u_h - v_h, w).$$

Hence,

$$|u - v_h|_{h,1,\Omega} \leq c \left(\sum_{K \in \mathcal{T}_h} e_K(u_h, f)^2 \right)^{\frac{1}{2}} + |u_h - v_h|_{h,1,\Omega}.$$

Conclude using the triangle inequality and taking the infimum over v_h. $\qquad\square$

Estimate (10.19) consists of the local error indicators associated with the conformal approximation plus an additional term resulting from nonconformity. This term can be controlled by taking some particular $v_h \in P_{c,h,0}^1$. Consider, for instance, the *Oswald interpolation operator* [Osw93] $\mathcal{O}_h : P_{pt,h,0}^1 \to P_{c,h,0}^1$ defined for $w_h \in P_{pt,h,0}^1$ as follows:

$$\forall a \in \mathcal{V}_h, \quad \mathcal{O}_h w_h(a) = \frac{1}{n_a} \sum_{K \in \Omega_a} w_{h|K}(a) \qquad \text{if} \quad a \notin \partial\Omega, \tag{10.20}$$

and $\mathcal{O}_h w_h(a) = 0$ if $a \in \partial\Omega$. Here, $\mathcal{V}_h$ denotes the set of vertices in the mesh, Ω_a the set of elements containing the vertex a, and n_a the cardinal number of this set. For $F \in \mathcal{F}_h$ and $w_h \in P_{pt,h,0}^1$, denote by $[\![w_h]\!]_F$ the jump of w_h across F (the subscript is dropped when there is no ambiguity), it being understood that a zero outer value is taken for $F \subset \partial\Omega$. The following interpolation result holds provided the family $\{\mathcal{T}_h\}_{h>0}$ is shape-regular (see Exercise 10.3 for a proof): There exists c such that, for all $w_h \in P_{pt,h,0}^1$ and $K \in \mathcal{T}_h$,

$$|w_h - \mathcal{O}_h w_h|_{1,K} \leq c \sum_{\substack{F \in \mathcal{F}_h \\ F \cap K \neq \emptyset}} h_F^{-\frac{1}{2}} \|[\![w_h]\!]\|_{0,F}, \tag{10.21}$$

where h_F is the diameter of F. This yields the following:

Corollary 10.16. *Assume that the family* $\{\mathcal{T}_h\}_{h>0}$ *is shape-regular. Then, there exists* c *such that, for all* h,

$$|u - u_h|_{h,1,\Omega} \le c \left(\sum_{K \in \mathcal{T}_h} e_K(u_h, f)^2 + \sum_{\substack{F \in \mathcal{F}_h \\ F \cap K \neq \emptyset}} h_F^{-1} \|[\![u_h]\!]\|_{0,F}^2 \right)^{\frac{1}{2}}. \quad (10.22)$$

Remark 10.17. The quantity $e_K(u_h, f) + \sum_{F \in \partial K} h_F^{-\frac{1}{2}} \|[\![u_h]\!]\|_{0,F}$ is a *residual-based error indicator* for non-conformal settings. Other a posteriori estimators for non-conformal finite element methods are presented in [HoW96]. □

The non-conformal residual-based a posteriori error estimator defined in (10.22) is endowed with an optimality property, as stated in the following:

Proposition 10.18. *Let* $l \in \mathbb{N}$ *and set* $Z_{lh} = \{v_h \in L^2(\Omega); \forall K \in \mathcal{T}_h, v_{h|K} \in \mathbb{P}_l(K)\}$. *Then, there exists* c, *depending only on the shape-regularity of the mesh, the reference finite element, and* l, *such that, for all* $K \in \mathcal{T}_h$,

$$e_K(u_h, f) + \sum_{F \in \partial K} h_F^{-\frac{1}{2}} \|[\![u_h]\!]\|_{0,F} \le c \Big(|u - u_h|_{h,1,D_K} \tag{10.23}$$
$$+ h_K \inf_{v_h \in Z_{lh}} \|f - v_h\|_{0,D_K} \Big).$$

Proof. (1) To bound $e_K(u_h, f)$, apply the same proof as that of Theorem 10.10 with the modification that integrals are split elementwise.
(2) To bound the second term in the left-hand side of (10.23), consider $K_1 \in \mathcal{T}_h$ and $F \in \partial K_1$. When F is an interior face, there is $K_2 \in \mathcal{T}_h$ such that $F = K_1 \cap K_2$. For $i = 1, 2$, let n_i be the normal to F pointing outward of K_i. Recall that $[\![u_h]\!]_F = u_{h|K_1} n_1 + u_{h|K_2} n_2$. Consider the Neumann problems:

$$\begin{cases} \text{Seek } \psi_i \in H^1_{\int = 0}(K_i) \text{ such that} \\ \int_K \nabla \psi_i \cdot \nabla \varphi = - \int_F [\![u_h]\!]_F \cdot n_i \, \varphi, \quad \forall \varphi \in H^1_{\int = 0}(K_i), \end{cases}$$

where $H^1_{\int = 0}(K_i) = \{\varphi \in H^1(K_i); \int_{K_i} \varphi = 0\}$. A scaling argument shows that $|\psi_i|_{1,K_i} \le c h_F^{\frac{1}{2}} \|[\![u_h]\!]\|_{0,F}$. Moreover, the identity

$$\|[\![u_h]\!]\|_{0,F}^2 = \sum_{i=1,2} \int_{K_i} \nabla \psi_i \cdot \nabla (u - u_h),$$

implies $h_F^{-\frac{1}{2}} \|[\![u_h]\!]\|_{0,F} \le c \sum_{i=1,2} |u - u_h|_{1,K_i}$, completing the proof. □

10.2 Hierarchical Error Estimates

A posteriori error estimates of hierarchical type have been introduced by Bank and Weiser [BaW85]. The analysis was first carried out for standard

Galerkin methods in a consistent and conformal setting [BaW85, BaS93] and then extended to non-standard Galerkin methods in a non-consistent and non-conformal setting [AcA98]. An efficient tool for deriving hierarchical error estimators is the concept of hierarchical basis functions (see §1.1.5), an idea originally brought forward in [BaS93]. The basic concept underlying hierarchical a posteriori error estimates is to enrich the discrete problem by introducing an additional set of fluctuating basis functions. The estimate is then obtained by solving a new discrete problem where the data is the residual of the approximate solution tested against the fluctuating basis functions. Hierarchical error estimates can be extended to multilevel settings provided sequences of enriched spaces are properly introduced. This technique is not covered herein; see, e.g., [AiO00, p. 100].

10.2.1 The setting

Consider the model problem (10.1) with Hilbert spaces W and V, a bilinear form $a \in \mathcal{L}(W \times V; \mathbb{R})$, and a linear form $f \in \mathcal{L}(V; \mathbb{R})$. Denote by $\|a\|$ the continuity constant of a in $\mathcal{L}(W \times V; \mathbb{R})$. As in the previous section, we assume that a satisfies the conditions of the BNB Theorem. This implies the existence of $\alpha > 0$ such that

$$\inf_{w \in W} \sup_{v \in V} \frac{a(w, v)}{\|w\|_W \|v\|_V} \geq \alpha. \tag{10.24}$$

Let $\{\mathcal{T}_h\}_{h>0}$ be a family of meshes of Ω and let W_h and V_h be approximation spaces constructed on $\mathcal{T}_h$. Consider the discrete problem (10.2). As in the previous section, we assume that (10.2) is well-posed, implying that the bilinear form a_h satisfies a discrete inf-sup condition on $W_h \times V_h$ with associated constant α_h.

Now, assume that the discrete spaces W_h and V_h can be enriched as follows:

$$W_h^e = W_h \oplus W_h^f, \qquad V_h^e = V_h \oplus V_h^f,$$

where W_h^e and V_h^e are the *enriched spaces* and W_h^f and V_h^f are the *fluctuation spaces*. The enriched spaces are used to define an enriched discrete problem:

$$\begin{cases} \text{Seek } u_h^e \in W_h^e \text{ such that} \\ a_h^e(u_h^e, v_h^e) = f_h^e(v_h^e), \quad \forall v_h^e \in V_h^e, \end{cases} \tag{10.25}$$

while the fluctuation spaces are used to define a fluctuation discrete problem:

$$\begin{cases} \text{Seek } e_h^f \in W_h^f \text{ such that} \\ a_h^f(e_h^f, v_h^f) = f_h^f(v_h^f) - a_h^e(u_h, v_h^f), \quad \forall v_h^f \in V_h^f. \end{cases} \tag{10.26}$$

Note that the right-hand side in (10.26) is the residual of the approximate solution u_h tested against the fluctuation space. The solution e_h^f to problem (10.26) serves to construct the hierarchical a posteriori error estimate;

see Theorem 10.19 below. In actual computations, the enriched solution u_h^e is not evaluated; it is used only in the theoretical analysis. We make the following assumptions:

(a₁) $W(h) = W + W_h^e$ is a Hilbert space when equipped with an inner product, say $(\cdot,\cdot)_{W(h)}$. Denote by $\|\cdot\|_{W(h)}$ the corresponding norm and assume that W_h^e is continuously embedded into $W(h)$ for this norm. Equip the spaces V_h^e, V_h^f, and V_h with the same inner product, say $(\cdot,\cdot)_{V_h}$, and corresponding norm, say $\|\cdot\|_{V_h}$.

(a₂) The bilinear forms a_h^f and a_h^e are uniformly bounded on $W(h) \times V_h^e$ (the uniform boundedness of a_h^f can be weakened to $(W + W_h^f) \times V_h^f$), and the linear forms f_h^f and f_h^e are continuous on V_h^f and V_h^e, respectively.

(a₃) The following inf-sup conditions are satisfied:

$$
\begin{cases}
\displaystyle \inf_{w_h^e \in W_h^e} \sup_{v_h^e \in V_h^e} \frac{a_h^e(w_h^e, v_h^e)}{\|w_h^e\|_{W(h)} \|v_h^e\|_{V_h}} \geq \alpha_h^e > 0, \\[3ex]
\displaystyle \inf_{w_h^f \in W_h^f} \sup_{v_h^f \in V_h^f} \frac{a_h^f(w_h^f, v_h^f)}{\|w_h^f\|_{W(h)} \|v_h^f\|_{V_h}} \geq \alpha_h^f > 0.
\end{cases}
\tag{10.27}
$$

(a₄) $\dim(W_h^e) = \dim(V_h^e)$ and $\dim(W_h^f) = \dim(V_h^f)$.

The well-posedness of problems (10.25) and (10.26) then results from Theorem 2.22.

To compare $\|e_h^f\|_{W(h)}$ with the norm of the approximation error $u - u_h$, we require the hierarchical setting to satisfy two assumptions:

(sat) *Saturation property*: there exists $\beta < 1$ such that

$$
\|u - u_h^e\|_{W(h)} \leq \beta \|u - u_h\|_{W(h)}.
\tag{10.28}
$$

(scs) *Strong Cauchy–Schwarz inequality*: there exists $\gamma < 1$ such that

$$
\forall v_h \in V_h, \ \forall v_h^f \in V_h^f, \quad (v_h, v_h^f)_{V_h} \leq \gamma \|v_h\|_{V_h} \|v_h^f\|_{V_h}.
\tag{10.29}
$$

In the literature, this type of inequality is often called the *Cauchy–Buniakowski–Schwarz (CBS) inequality*.

Assumption (sat) simply means that the enriched discrete solution u_h^e converges more rapidly to u than u_h. Assumption (scs) can be interpreted as a lower bound, uniform in h, of the angle between the test spaces V_h and V_h^f, thus avoiding that these spaces become asymptotically collinear. Assumptions (sat) and (scs) have been introduced in [BaW85, BaS93] in a simplified setting; see also [AiM94] for the use of the strong Cauchy–Schwarz inequality in the framework of nonlinear Galerkin methods.

Theorem 10.19. *Assume* (a₁)–(a₄). *Let u solve* (10.1), *let u_h solve* (10.2), *and let e_h^f solve* (10.26).

(i) *Set* $c_1 = \frac{\|a_h^e\|(1+\beta)}{\alpha_h^f}$ *and* $c_2 = \frac{1}{\alpha_h^f}$. *Under the saturation assumption* (sat),

$$\|e_h^f\|_{W(h)} \leq c_1 \|u - u_h\|_{W(h)} + c_2 \sup_{v_h^f \in V_h^f} \frac{f_h^f(v_h^f) - f_h^e(v_h^f)}{\|v_h^f\|_{V_h}}. \qquad (10.30)$$

(ii) *Set* $c_3 = \frac{1}{\alpha_h^e(1-\beta)(1-\gamma^2)^{\frac{1}{2}}}$ *and* $c_4 = \|a_h^f\|$. *Under the saturation assumption* (sat) *and the strong Cauchy–Schwarz inequality* (scs),

$$\|u - u_h\|_{W(h)} \leq c_3 \Bigg(c_4 \|e_h^f\|_{W(h)} + \sup_{v_h^f \in V_h^f} \frac{f_h^f(v_h^f) - f_h^e(v_h^f)}{\|v_h^f\|_{V_h}}$$
$$+ \sup_{v_h \in V_h} \frac{f_h^e(v_h) - a_h^e(u_h, v_h)}{\|v_h\|_{V_h}} \Bigg). \qquad (10.31)$$

Proof. The proof is adapted from [AcA98].
(1) The triangle inequality $\|u - u_h\|_{W(h)} \leq \|u - u_h^e\|_{W(h)} + \|u_h - u_h^e\|_{W(h)}$ and the saturation property (sat) imply

$$(1 - \beta)\|u - u_h\|_{W(h)} \leq \|u_h - u_h^e\|_{W(h)} \leq (1 + \beta)\|u - u_h\|_{W(h)}. \qquad (10.32)$$

(2) Let us prove estimate (10.30). Use assumptions (a_2) and (a_3) to obtain

$$\alpha_h^f \|e_h^f\|_{W(h)} \leq \sup_{v_h^f \in V_h^f} \frac{a_h^f(e_h^f, v_h^f)}{\|v_h^f\|_{V_h}} = \sup_{v_h^f \in V_h^f} \frac{f_h^f(v_h^f) - a_h^e(u_h, v_h^f)}{\|v_h^f\|_{V_h}}$$

$$= \sup_{v_h^f \in V_h^f} \frac{a_h^e(u_h^e, v_h^f) - a_h^e(u_h^e, v_h^f) + f_h^f(v_h^f) - a_h^e(u_h, v_h^f)}{\|v_h^f\|_{V_h}}$$

$$= \sup_{v_h^f \in V_h^f} \frac{f_h^f(v_h^f) - a_h^e(u_h^e, v_h^f) + a_h^e(u_h^e - u_h, v_h^f)}{\|v_h^f\|_{V_h}}$$

$$\leq \|a_h^e\| \, \|u_h^e - u_h\|_{W(h)} + \sup_{v_h^f \in V_h^f} \frac{f_h^f(v_h^f) - f_h^e(v_h^f)}{\|v_h^f\|_{V_h}}.$$

Estimate (10.30) then results from (10.32).
(3) Let us prove estimate (10.31). Let $v_h \in V_h$ and $v_h^f \in V_h^f$ with $v_h + v_h^f \neq 0$. Assumption (scs) yields

$$\|v_h + v_h^f\|_{V_h}^2 = \|v_h\|_{V_h}^2 + \|v_h^f\|_{V_h}^2 + 2(v_h, v_h^f)_{V_h}$$
$$\geq \|v_h\|_{V_h}^2 + \|v_h^f\|_{V_h}^2 - 2\gamma\|v_h\|_{V_h}\|v_h^f\|_{V_h}$$
$$= (\|v_h\|_{V_h} - \gamma\|v_h^f\|_{V_h})^2 + (1 - \gamma^2)\|v_h^f\|_{V_h}^2$$
$$= (\|v_h^f\|_{V_h} - \gamma\|v_h\|_{V_h})^2 + (1 - \gamma^2)\|v_h\|_{V_h}^2.$$

Hence,

$$\max\{\|v_h\|_{V_h}, \|v_h^f\|_{V_h}\} \leq \frac{1}{(1-\gamma^2)^{\frac{1}{2}}}\|v_h + v_h^f\|_{V_h}. \qquad (10.33)$$

As a result,

$$
\begin{aligned}
\alpha_h^e \|u_h^e - u_h\|_{W(h)} &\leq \sup_{v_h + v_h^f \in V_h^e} \frac{a_h^e(u_h^e - u_h, v_h + v_h^f)}{\|v_h + v_h^f\|_{V_h}} \\
&= \sup_{v_h + v_h^f \in V_h^e} \frac{a_h^e(u_h^e, v_h + v_h^f) - a_h^e(u_h, v_h + v_h^f)}{\|v_h + v_h^f\|_{V_h}} \\
&= \sup_{v_h + v_h^f \in V_h^e} \frac{f_h^e(v_h + v_h^f) - a_h^e(u_h, v_h) - f_h^f(v_h^f) + f_h^f(v_h^f) - a_h^e(u_h, v_h^f)}{\|v_h + v_h^f\|_{V_h}} \\
&= \sup_{v_h + v_h^f \in V_h^e} \frac{a_h^f(e_h^f, v_h^f) + f_h^e(v_h) - a_h^e(u_h, v_h) + f_h^e(v_h^f) - f_h^f(v_h^f)}{\|v_h + v_h^f\|_{V_h}} \\
&\leq \frac{1}{(1-\gamma^2)^{\frac{1}{2}}} \Bigg(\|a_h^f\| \, \|e_h^f\|_{W(h)} + \sup_{v_h \in V_h} \frac{f_h^e(v_h) - a_h^e(u_h, v_h)}{\|v_h\|_{V_h}} \\
&\quad + \sup_{v_h^f \in V_h^f} \frac{f_h^e(v_h^f) - f_h^f(v_h^f)}{\|v_h^f\|_{V_h}} \Bigg),
\end{aligned}
$$

the last inequality resulting from (10.33). Finally, estimate (10.31) results from (10.32). $\qquad\square$

Remark 10.20. If a localization property $\|\cdot\|_{W(h)}^2 = \sum_{K \in T_h} \|\cdot\|_{W(h),K}^2$ holds, the quantities $\|e_h^f\|_{W(h),K}$ are called *hierarchical error indicators*. $\qquad\square$

In practical simulations, it is often useful to approximate the bilinear form a_h^f arising in the left-hand side of (10.26) by a bilinear form, say $\tilde{a}_h^f$, yielding linear systems that are easier to solve. To evaluate the corresponding error estimate, the fluctuation problem we consider is thus:

$$
\begin{cases}
\text{Seek } \tilde{e}_h^f \in W_h^f \text{ such that} \\
\tilde{a}_h^f(\tilde{e}_h^f, v_h^f) = f_h^f(v_h^f) - a_h^e(u_h, v_h^f), \quad \forall v_h^f \in V_h^f.
\end{cases}
\tag{10.34}
$$

Corollary 10.21. *Along with the hypotheses of Theorem 10.19, assume that the bilinear form $\tilde{a}_h^f$ is continuous on $(W + W_h^f) \times V_h^f$ and satisfies*

$$
\inf_{w_h^f \in W_h^f} \sup_{v_h^f \in V_h^f} \frac{\tilde{a}_h^f(w_h^f, v_h^f)}{\|w_h^f\|_{W(h)} \|v_h^f\|_{V_h}} \geq \tilde{\alpha}_h^f > 0.
$$

Let $\tilde{e}_h^f$ solve (10.34) and let e_h^f solve (10.26). Then,

$$
c_5 \|e_h^f\|_{W(h)} \leq \|\tilde{e}_h^f\|_{W(h)} \leq c_6 \|e_h^f\|_{W(h)},
\tag{10.35}
$$

with constants $c_5 = \frac{\alpha_h^f}{\|\tilde{a}_h^f\|}$ and $c_6 = \frac{\|a_h^f\|}{\tilde{\alpha}_h^f}$.

Proof. We deduce from (10.34) that

$$\tilde{a}_h^{\mathrm{f}}(\tilde{e}_h^{\mathrm{f}}, v_h^{\mathrm{f}}) = a_h^{\mathrm{f}}(e_h^{\mathrm{f}}, v_h^{\mathrm{f}}), \quad \forall v_h^{\mathrm{f}} \in V_h^{\mathrm{f}}.$$

Therefore,

$$\tilde{\alpha}_h^{\mathrm{f}} \|\tilde{e}_h^{\mathrm{f}}\|_{W(h)} \le \sup_{v_h^{\mathrm{f}} \in V_h^{\mathrm{f}}} \frac{\tilde{a}_h^{\mathrm{f}}(\tilde{e}_h^{\mathrm{f}}, v_h^{\mathrm{f}})}{\|v_h^{\mathrm{f}}\|_{V_h}} = \sup_{v_h^{\mathrm{f}} \in V_h^{\mathrm{f}}} \frac{a_h^{\mathrm{f}}(e_h^{\mathrm{f}}, v_h^{\mathrm{f}})}{\|v_h^{\mathrm{f}}\|_{V_h}} \le \|a_h^{\mathrm{f}}\| \, \|e_h^{\mathrm{f}}\|_{W(h)},$$

yielding the upper bound in (10.35). The lower bound is proved similarly. $\square$

The reliability and the optimality of the hierarchical error estimate depends on the value of the parameters $c_1, \dots, c_4$ in (10.30) and (10.31). These parameters depend in turn on the value of the saturation constant β and the strong Cauchy–Schwarz constant γ. The constant β depends on the exact solution u so that one has little a priori control over it in general. One favorable situation is when it can be proved that the enriched solution u_h^{e} converges to the exact solution u at a higher order rate than the approximate solution u_h. In these circumstances, it is reasonable to expect that (10.28) holds for sufficiently small h. At variance with the saturation constant, the strong Cauchy–Schwarz constant γ only depends on the discrete setting and can be computed explicitly in many situations. To this end, the following lemmas are useful:

Lemma 10.22. *Let $\Pi_h \in \mathcal{L}(V_h^{\mathrm{e}}; V_h)$ be the projector based on the decomposition $V_h^{\mathrm{e}} = V_h^{\mathrm{f}} \oplus V_h$. Then, $\|\Pi_h\|_{\mathcal{L}(V_h^{\mathrm{e}}; V_h)}$ is uniformly bounded with respect to h if and only if the strong Cauchy–Schwarz inequality (10.29) holds with*

$$\gamma = \left(1 - \frac{1}{\|\Pi_h\|_{\mathcal{L}(V_h^{\mathrm{e}}; V_h)}^2} \right)^{\frac{1}{2}}. \tag{10.36}$$

Proof. (1) Assume that $\|\Pi_h\|_{\mathcal{L}(V_h^{\mathrm{e}}; V_h)}$ is uniformly bounded with respect to h. Let $v_h \in V_h$, $v_h^{\mathrm{f}} \in V_h^{\mathrm{f}}$, and set $v_h^{\mathrm{e}} = v_h - \frac{(v_h, v_h^{\mathrm{f}})_{V_h}}{\|v_h^{\mathrm{f}}\|_{V_h}^2} v_h^{\mathrm{f}}$. Clearly, $v_h = \Pi_h v_h^{\mathrm{e}}$ implying that $\|v_h\|_{V_h} \le \|\Pi_h\|_{\mathcal{L}(V_h^{\mathrm{e}}; V_h)} \|v_h^{\mathrm{e}}\|_{V_h}$. Dividing by $\|\Pi_h\|_{\mathcal{L}(V_h^{\mathrm{e}}; V_h)}$, taking the square of the resulting inequality, and developing $\|v_h^{\mathrm{e}}\|_{V_h}^2$ yields

$$\frac{(v_h, v_h^{\mathrm{f}})_{V_h}^2}{\|v_h^{\mathrm{f}}\|_{V_h}^2} - 2\frac{(v_h, v_h^{\mathrm{f}})_{V_h}^2}{\|v_h^{\mathrm{f}}\|_{V_h}^2} + \left(1 - \frac{1}{\|\Pi_h\|_{\mathcal{L}(V_h^{\mathrm{e}}; V_h)}^2} \right) \|v_h\|_{V_h}^2 \ge 0,$$

leading to inequality (10.29) with γ given by (10.36).
(2) Conversely, assume that inequality (10.29) holds for some $\gamma < 1$. Let $v_h^{\mathrm{e}} \in V_h^{\mathrm{e}}$. Set $v_h = \Pi_h v_h^{\mathrm{e}}$ and $v_h^{\mathrm{f}} = v_h^{\mathrm{e}} - v_h$. Use to the identity

$$\|v_h^{\mathrm{e}}\|_{V_h}^2 - (1 - \gamma^2) \|v_h\|_{V_h}^2 = \gamma^2 \|v_h\|_{V_h}^2 - \frac{(v_h, v_h^{\mathrm{f}})_{V_h}^2}{\|v_h^{\mathrm{f}}\|_{V_h}^2}$$

$$+ \left(\|v_h^{\mathrm{f}}\|_{V_h} + \frac{(v_h, v_h^{\mathrm{f}})_{V_h}}{\|v_h^{\mathrm{f}}\|_{V_h}} \right)^2 \ge 0,$$

to infer $\|\Pi_h\|_{\mathcal{L}(V_h^e;V_h)} \leq (1 - \gamma^2)^{-\frac{1}{2}}$. $\square$

To evaluate the constant γ, we introduce bases for the discrete spaces V_h and V_h^f, say $\{\varphi_1, \dots, \varphi_N\}$ and $\{\varphi_1^f, \dots, \varphi_{N^f}^f\}$, and the Gram matrices $G_{11} \in \mathbb{R}^{N,N}$, $G_{21}^T = G_{12} \in \mathbb{R}^{N,N^f}$, and $G_{22} \in \mathbb{R}^{N^f,N^f}$ with entries

$$G_{11,ij} = (\varphi_i, \varphi_j)_{V_h}, \quad 1 \leq i,j \leq N,$$

$$G_{12,ij} = (\varphi_i, \varphi_j^f)_{V_h}, \quad 1 \leq i \leq N, \; 1 \leq j \leq N^f,$$

$$G_{22,ij} = (\varphi_i^f, \varphi_j^f)_{V_h}, \quad 1 \leq i,j \leq N^f.$$

Lemma 10.23. *The constant γ arising in the strong Cauchy–Schwarz inequality is given by*

$$\gamma^2 = \max_{X_1 \in \mathbb{R}^N} \frac{X_1^T G_{12} G_{22}^{-1} G_{21} X_1}{X_1^T G_{11} X_1}. \tag{10.37}$$

Proof. Consider the projector introduced in Lemma 10.22. Clearly,

$$\|\Pi_h\|_{\mathcal{L}(V_h^e;V_h)}^2 = \max_{X_1 \in \mathbb{R}^N} \max_{X_2 \in \mathbb{R}^{N^f}} \frac{X_1^T G_{11} X_1}{(X_1, X_2)^T G(X_1, X_2)},$$

with the global Gram matrix

$$G = \left[\begin{array}{c|c} G_{11} & G_{12} \\ \hline G_{21} & G_{22} \end{array} \right].$$

A straightforward calculation shows that the supremum over X_2 is attained for $X_2 = -G_{22}^{-1} G_{21} X_1$. Therefore, introducing the Schur complement matrix $S_{11} = G_{11} - G_{12} G_{22}^{-1} G_{21}$ readily yields

$$\|\Pi_h\|_{\mathcal{L}(V_h^e;V_h)}^2 = \max_{X_1 \in \mathbb{R}^N} \frac{X_1^T G_{11} X_1}{X_1^T S_{11} X_1}.$$

Conclude using (10.36). $\square$

Lemma 10.23 is generally used as follows: Assume that the inner product $(\cdot,\cdot)_{V_h}$ can be localized in the form $\sum_{K \in \mathcal{T}_h} (\cdot,\cdot)_{V_h,K}$ and consider the associated localized norm $\|\cdot\|_{V_h,K}$. For $K \in \mathcal{T}_h$, define γ_K by

$$\gamma_K = \max_{v_h \in V_h} \max_{v_h^f \in V_h^f} \frac{(v_h, v_h^f)_{V_h,K}}{\|v_h\|_{V_h,K} \|v_h^f\|_{V_h,K}}.$$

Then, letting N_{el} be the number of elements in $\mathcal{T}_h$, the Cauchy–Schwarz inequality in $\mathbb{R}^{N_{el}}$ yields

$$(v_h, v_h^f)_{V_h} \leq \left(\max_{K \in \mathcal{T}_h} \gamma_K \right) \|v_h\|_{V_h} \|v_h^f\|_{V_h}.$$

A simple way to proceed is thus to establish the strong Cauchy–Schwarz inequality on the reference element $\widehat{K}$ and then to extend it to an arbitrary element K. The derivation of an upper bound on γ_K uniform in h usually requires that the mesh family $\{\mathcal{T}_h\}_{h>0}$ be shape-regular; see the examples below.

10.2.2 Example 1 (Consistent and conformal approximation)

In this section, we simplify the framework of Theorem 10.19 by assuming that the setting is consistent and conformal. In particular, we assume that the spaces W_h, W_h^f, and W_h^e are subsets of W and that the spaces V_h, V_h^f, and V_h^e are subsets of V. All the discrete spaces are equipped with the corresponding inner products of W and V and associated norms $\|\cdot\|_W$ and $\|\cdot\|_V$. Furthermore, because the setting is consistent, all the bilinear forms coincide, i.e., $a_h = a_h^f = a_h^e = a$, and the same conclusion holds for the linear forms $f_h = f_h^f = f_h^e = f$. To further simplify the setting, all the inf-sup constants are regrouped into a single constant denoted by α.

With the above simplifications, the enriched problem becomes:

$$\begin{cases} \text{Seek } u_h^e \in W_h^e \text{ such that} \\ a(u_h^e, v_h^e) = f(v_h^e), \quad \forall v_h^e \in V_h^e, \end{cases} \tag{10.38}$$

while the fluctuation problem becomes:

$$\begin{cases} \text{Seek } e_h^f \in W_h^f \text{ such that} \\ a(e_h^f, v_h^f) = f(v_h^f) - a(u_h, v_h^f), \quad \forall v_h^f \in V_h^f. \end{cases} \tag{10.39}$$

The conclusion of Theorem 10.19 simplifies into

$$\tfrac{\alpha}{\|a\|}(1 - \beta)(1 - \gamma^2)^{\frac{1}{2}} \|u - u_h\|_W \le \|e_h^f\|_W \le \tfrac{\|a\|}{\alpha}(1 + \beta)\|u - u_h\|_W. \tag{10.40}$$

The factor $(1 + \beta)$ can be eliminated from the upper bound in (10.40) since

$$\alpha\|e_h^f\|_W \le \sup_{v_h^f \in V_h^f} \frac{a(e_h^f, v_h^f)}{\|v_h^f\|_{V_h}} = \sup_{v_h^f \in V_h^f} \frac{a(u - u_h, v_h^f)}{\|v_h^f\|_{V_h}} \le \|a\| \, \|u - u_h\|_W. \tag{10.41}$$

The lower bound in (10.40) and the upper bound (10.41) have been derived in [BaS93] in a standard setting, i.e., with the further assumption that $W = V$ and $W_h = V_h$.

As a simple application, consider an elliptic PDE supplemented with homogeneous Dirichlet conditions and a H^1-conformal approximation space based on the $\mathbb{P}_1$ Lagrange finite element and a mesh family $\{\mathcal{T}_h\}_{h>0}$ of a polyhedron Ω in $\mathbb{R}^d$. For $h > 0$, denote by $V_h = P_{c,h,0}^1$ the corresponding approximation space. To enrich $P_{c,h,0}^1$, consider the vector space $V_h^f = B_E$ spanned by H^1-conformal edge *bubble functions*. Let e be an edge in the mesh and let Δ_e be the set of the mesh cells to which e belongs. For $K \in \Delta_e$, denote by $\lambda_{1,e,K}$ and $\lambda_{2,e,K}$ the barycentric coordinates in K associated with the two vertices of K that lie in e. Then, define the edge bubble function

$$b_{e|K} = 4\lambda_{1,e,K}\lambda_{2,e,K}. \tag{10.42}$$

Note that b_e has support in Δ_e. Set $B_E = \text{span}\{b_e\}_{e \in \mathcal{E}_h}$. Clearly, B_E is also the span of continuous, piecewise quadratic polynomials that are zero at the vertices of $\mathcal{T}_h$. The enriched space $V_h^e = P_{c,h,0}^1 \oplus B_E$ is equal to $P_{c,h,0}^2$, the H^1-conformal approximation space based on the $\mathbb{P}_2$ Lagrange finite element. Provided the exact solution is smooth enough, the approximate solution in $P_{c,h,0}^2$ converges to second-order to u, whereas the approximate solution in $P_{c,h,0}^1$ converges to first-order. It is therefore reasonable to expect that the saturation property is satisfied provided h is small enough. Finally, assuming that the family $\{\mathcal{T}_h\}_{h>0}$ is shape-regular, it can be proven that the pair of spaces $\{P_{c,h,0}^1, B_E\}$ satisfies the strong Cauchy–Schwarz inequality; see [Bra81, MaM82] and Exercises 5.14 and 10.5.

10.2.3 Example 2 (Non-conformal approximation)

Consider the PDE $-\Delta u = f$ supplemented with Dirichlet conditions and use the Crouzeix–Raviart finite element space $W_h = P_{pt,h,0}^1$ defined in (10.17) to approximate this problem. To enrich $P_{pt,h,0}^1$, two approaches can be considered based on either H^1-conformal face bubbles or non-conformal cell bubbles; see [ElE03]. In both cases, the space $W(h) = H_0^1(\Omega) + W_h^e$ is equipped with the scalar product $(u,v)_{h,1,\Omega} = \sum_{K \in \mathcal{T}_h} \int_K \nabla u \cdot \nabla v$. Recall that the associated norm $| \cdot |_{h,1,\Omega}$ controls uniformly the L^2-norm on $H_0^1(\Omega) + P_{pt,h,0}^1$; see Lemma 3.35.

Enrichment by H^1-conformal face bubbles. Let $F \in \mathcal{F}_h^i$ be an interior face in the mesh. The corresponding face bubble, say b_F, has support in D_F which is defined to be the set of the mesh cells containing F. In three dimensions, for $K \in D_F$, denote by $\lambda_{1,F,K}$, $\lambda_{2,F,K}$, and $\lambda_{3,F,K}$ the barycentric coordinates in K associated with the three vertices of K that lie in F. Then,

$$b_{F|K} = 27\lambda_{1,F,K}\lambda_{2,F,K}\lambda_{3,F,K}. \tag{10.43}$$

In two dimensions, the face bubble function coincides with the edge bubble function defined in (10.42).

Set $B_F = \mathrm{span}\{b_F\}_{F \in \mathcal{F}_h^i}$. The enriched space $V_h^e = P_{pt,h,0}^1 \oplus B_F$ has dimension $2N_f^i$, where N_f^i is the number of interior faces in the mesh. In two dimensions, the pair of spaces $\{P_{pt,h,0}^1, B_F\}$ can be shown to satisfy the strong Cauchy–Schwarz inequality provided the mesh family $\{\mathcal{T}_h\}_{h>0}$ is shape-regular; see [ElE03]. Furthermore, it is reasonable to expect that the enriched space V_h^e has better interpolation properties than $P_{pt,h,0}^1$, but since the patch-test (see §1.4.3) is not satisfied for the first-order moments, the enriched solution u_h^e may converge to u to first-order only. Therefore, it is not possible to infer a priori the validity of the saturation assumption.

Enrichment by non-conformal cell bubbles. Another possibility to enrich the Crouzeix–Raviart finite element space is to consider the space B_C spanned by the *Fortin–Soulié bubbles* defined in (1.72). For a cell $K \in \mathcal{T}_h$, the Fortin–Soulié bubble b_K has support in K. Therefore, B_C has dimension N_{el}, the number of elements in the mesh, and the enriched space $V_h^e = P_{pt,h,0}^1 \oplus B_C$ has dimension $N_f^i + N_{el}$. Because V_h^e has no better interpolation properties than $P_{pt,h,0}^1$, it is not possible to infer a priori the validity of the saturation assumption. Numerical experiments indicate that the saturation assumption is satisfied on model elliptic problems [ElE03].

One original feature of non-conformal cell bubbles is that $P_{pt,h,0}^1 \perp B_C$ for the inner product $(\cdot, \cdot)_{h,1,\Omega}$. As a result, the pair of spaces $\{P_{pt,h,0}^1, B_C\}$ satisfies the strong Cauchy–Schwarz inequality with constant $\gamma = 0$. In these circumstances, the hierarchical error estimate takes the simple form

$$\|e_h^f\|_{h,1,\Omega} = \left(\sum_{K \in \mathcal{T}_h} \frac{(f, b_K)_{0,K}^2}{\|\nabla b_K\|_{0,K}^2} \right)^{\frac{1}{2}}, \tag{10.44}$$

where $f \in L^2(\Omega)$ is the data. Therefore, $\|e_h^f\|_{h,1,\Omega}$ does not depend on the approximate solution u_h; see Exercise 10.6.

Circumventing the saturation property. Because the saturation property is often difficult to prove, it is interesting to derive an error estimate that does not rely on this assumption.

Proposition 10.24. *Let u be the exact solution to the homogeneous Dirichlet problem with data $f \in L^2(\Omega)$. Let $P_{c,h,0}^1$ be the H_0^1-conformal approximation space based on the $\mathbb{P}_1$ Lagrange finite element. Let u_h be the discrete solution obtained using an approximation space V_h. Assume that $P_{c,h,0}^1 \subset V_h$ and that the functions in V_h are piecewise linear. Define e_h^f to be the solution of the fluctuation problem:*

$$\begin{cases} \text{Seek } e_h^f \in B_F \text{ such that} \\ (e_h^f, v_h^f)_{h,1,\Omega} = (f, v_h^f)_{0,\Omega} - (u_h, v_h^f)_{h,1,\Omega}, \quad \forall v_h^f \in B_F, \end{cases} \tag{10.45}$$

where B_F is the H^1-conformal space spanned by face bubbles. Then, if the family $\{\mathcal{T}_h\}_{h>0}$ is shape-regular, there is c such that, for all h,

$$|u - u_h|_{1,\Omega} \leq c \left(|e_h^f|_{1,\Omega} + h\|f\|_{0,\Omega} + \inf_{v_h \in P_{c,h,0}^1} |u_h - v_h|_{1,\Omega} \right). \tag{10.46}$$

Proof. (1) Let $\mathcal{F}_h$ be the set of faces in $\mathcal{T}_h$ and for $F \in \mathcal{F}_h$, let b_F be the associated face bubble. Recall that $B_F = \text{span}\{b_F\}_{F \in \mathcal{F}_h}$. Define the interpolation operator

$$\Pi_B : H^1(\Omega) \ni v \longmapsto \sum_{F \in \mathcal{F}_h} \left(\frac{\int_F v}{\int_F b_F} \right) b_F \in B_F.$$

Clearly, Π_B is uniformly stable in $L^2(\Omega)$, and using elementary properties of face bubble functions, it is straightforward to verify the stability property

$$\forall v \in H_0^1(\Omega), \ \forall K \in \mathcal{T}_h, \quad \|\nabla \Pi_B v\|_{0,K} \leq c\, h_K^{-\frac{1}{2}} \sum_{F \in \partial K} \|v\|_{0,F}.$$

Furthermore, for all $v_h \in V_h$ and $v \in H_0^1(\Omega)$, an integration by parts yields $(v_h, v)_{h,1,\Omega} = (v_h, \Pi_B v)_{h,1,\Omega}$ since v_h is piecewise linear.

(2) Let $w \in H_0^1(\Omega)$ and set $z = w - \mathcal{SZ}_h w$, where $\mathcal{SZ}_h$ is the Scott–Zhang interpolation operator. Use the fact that $\mathcal{SZ}_h w \in P_{c,h,0}^1$ and $P_{c,h,0}^1 \subset V_h$ to obtain

$$(u - u_h, w)_{h,1,\Omega} = (u - u_h, z)_{h,1,\Omega} = (f, z)_{0,\Omega} - (u_h, \Pi_B z)_{h,1,\Omega}$$
$$= (f, z - \Pi_B z)_{0,\Omega} + (e_h^{\mathrm{f}}, \Pi_B z)_{h,1,\Omega}.$$

The stability properties of Π_B derived in step 1 and the interpolation properties of $\mathcal{SZ}_h$ imply

$$\|z - \Pi_B z\|_{0,\Omega} \leq c \|z\|_{0,\Omega} = c \|w - \mathcal{SZ}_h w\|_{0,\Omega} \leq c\, h |w|_{1,\Omega},$$

$$|\Pi_B z|_{h,1,\Omega} \leq c \sum_{K \in \mathcal{T}_h} \left(\sum_{F \in \partial K} h_K^{-1} \|w - \mathcal{SZ}_h w\|_{0,F}^2 \right)^{\frac{1}{2}} \leq c |w|_{1,\Omega}.$$

As a result,

$$(u - u_h, w)_{h,1,\Omega} \leq c \left(|e_h^{\mathrm{f}}|_{h,1,\Omega} + h \|f\|_{0,\Omega} \right) |w|_{1,\Omega}.$$

(3) Let $v_h \in P_{c,h,0}^1$ and take $w = u - v_h$ in step 2. This yields

$$|w|_{1,\Omega}^2 = (u_h - v_h, w)_{h,1,\Omega} + (u - u_h, w)_{h,1,\Omega}$$
$$\leq c \left(|e_h^{\mathrm{f}}|_{h,1,\Omega} + h \|f\|_{0,\Omega} + |u_h - v_h|_{h,1,\Omega} \right) |w|_{1,\Omega}.$$

and, hence, an estimate on $|u - v_h|_{h,1,\Omega}$. Conclude using the triangle inequality and taking the infimum over v_h. $\qquad\square$

Remark 10.25.

(i) The proof above is adapted from [AcA03].

(ii) Neither the saturation assumption nor the strong Cauchy–Schwarz inequality are needed in Proposition 10.24.

(iii) When the approximation space V_h is based on the Crouzeix–Raviart finite element, one can use the Oswald interpolation operator introduced in §10.1.4 to control $\inf_{v_h \in P_{c,h,0}^1} |u_h - v_h|_{h,1,\Omega}$ by the jumps of u_h over the faces in the mesh; see [ElE03]. $\qquad\square$

10.3 Duality-Based Error Estimates

This section discusses a posteriori error estimates by duality techniques. It is set at an introductory level; see [BeR96, BeR01] for further theoretical insight and numerous examples.

10.3.1 The main ideas

To introduce the main ideas, consider a simple problem. Given a matrix $\mathcal{A} \in \mathbb{R}^{N,N}$ and a vector $F \in \mathbb{R}^N$, assume that we have computed an approximate solution to the linear system $\mathcal{A}U = F$, say U_*. Let $E = U - U_*$ be the error and let $R = F - \mathcal{A}U_*$ be the residual. Since the residual is obtained by inserting the approximate solution into the original problem $\mathcal{A}U = F$, it is often termed the *truncation error* or the *consistency error*. Note that the error and the residual are related as follows:

$$\mathcal{A}E = R. \tag{10.47}$$

This relation provides the basis for most a posteriori error estimates. For instance, an error estimate in the Euclidean norm readily results from (10.47) in the form $\|E\|_N \leq \|\mathcal{A}\|_N^{-1}\|R\|_N$.

In applications, it often happens that the user wishes to control the error in a different way. Suppose, for the sake of illustration, that one wishes to control the i-th component of the error ($1 \leq i \leq N$). Then, one expects that E_i results from two contributions, namely the local truncation error (which can be assessed by the residual component R_i) and a transmitted error produced by the coupling with the other components, i.e.,

$$E_i = \sum_{j=1}^{N} R_j Z_j,$$

where the quantities Z_j, $1 \leq j \leq N$, account for error propagation. To compute these quantities, one introduces the *dual problem*

$$\mathcal{A}^T Z = \Phi_i,$$

where $\{\Phi_1, \ldots, \Phi_N\}$ is the canonical basis of $\mathbb{R}^N$. Indeed, one readily verifies that $E_i = (E, \Phi_i)_N = (E, \mathcal{A}^T Z)_N = (R, Z)_N$ owing to (10.47).

To generalize the above approach, consider a nonlinear operator $\Psi : \mathbb{R}^N \to \mathbb{R}$. In the context of error estimates, the operator Ψ is often termed the *(error) output functional*. To control the error $\Psi(U) - \Psi(U_*)$, assume that Ψ is smooth enough to define its gradient $\Psi'(x) \in \mathbb{R}^N$ for all $x \in \mathbb{R}^N$, and introduce the *dual problem*

$$\mathcal{A}^T Z = \int_0^1 \Psi'(U_* + sE)\,\mathrm{d}s. \tag{10.48}$$

Then, a straightforward calculation yields

$$\Psi(U) - \Psi(U_*) = \int_0^1 (\Psi'(U_* + sE), E)_N\,\mathrm{d}s = \left(\int_0^1 \Psi'(U_* + sE)\,\mathrm{d}s, E\right)_N$$

$$= (\mathcal{A}^T Z, E)_N = (R, Z)_N,$$

owing to (10.47).

Instead of controlling the error in the output functional, one can also control the output functional of the error. Obviously, these two estimates coincide if Ψ is linear. To control $\Psi(U - U_*)$, assume that Ψ admits a density function $\psi : \mathbb{R}^N \to \mathbb{R}^N$ such that $\Psi(x) = (\psi(x), x)_N$ for all $x \in \mathbb{R}^N$, and introduce the *dual problem*

$$\mathcal{A}^T Z = \psi(E). \tag{10.49}$$

Then, one readily verifies that

$$\Psi(U - U_*) = (\psi(E), E)_N = (\mathcal{A}^T Z, E)_N = (R, Z)_N.$$

In summary, let $\mathcal{E}$ be a user-specified error measure, i.e., either $\Psi(U) - \Psi(U_*)$ or $\Psi(U - U_*)$. The first important idea in duality-based error estimates is to introduce an appropriate dual problem, either (10.48) or (10.49), and to derive the *error representation* formula

$$\mathcal{E} = (R, Z)_N. \tag{10.50}$$

Example 10.26. Consider the output functional $\Psi(x) = (x, \mathcal{P}x)_N$, where $\mathcal{P}$ is a given symmetric matrix. Note that $\Psi'(x) = 2\mathcal{P}x$ and that Ψ admits the density functional $\psi(x) = \mathcal{P}x$. Then, one readily verifies that to control $\Psi(U) - \Psi(U_*)$, the corresponding dual problem is $\mathcal{A}^T Z = \mathcal{P}U + \mathcal{P}U_*$, whereas to control $\Psi(U - U_*)$, the dual problem is $\mathcal{A}^T Z = \mathcal{P}U - \mathcal{P}U_*$. □

The second important idea underlying duality-based error estimates is to exploit the optimality properties satisfied by the approximate solution U_*. For instance, assume that U_* has been obtained from a projection method in which the residual is made orthogonal to a given subspace $\mathcal{K}$ of $\mathbb{R}^N$; see §9.3.1. Let $\Pi_{\mathcal{K}} Z$ be the orthogonal projection of the dual solution Z onto $\mathcal{K}$. Owing to (10.50), we readily infer

$$\mathcal{E} = (R, Z)_N = (R, Z - \Pi_{\mathcal{K}} Z)_N \leq \|R\|_N \inf_{x \in \mathcal{K}} \|Z - x\|_N.$$

The ideas presented in this section are now extended to finite element approximations of PDEs.

10.3.2 The finite element setting

Let W and V be two Hilbert spaces, $f \in V'$, $a \in \mathcal{L}(W \times V; \mathbb{R})$, and assume that problem (10.1) is well-posed. Consider a consistent and conformal approximation to (10.1): Given two subspaces $W_h \subset W$ and $V_h \subset V$,

$$\begin{cases} \text{Seek } u_h \in W_h \text{ such that} \\ a(u_h, v_h) = f(v_h), \quad \forall v_h \in V_h. \end{cases} \tag{10.51}$$

Assume that (10.51) is well-posed. Let $e = u - u_h$ be the error and define the residual $\rho(u_h; \cdot) \in \mathcal{L}(V; \mathbb{R})$ to be $\rho(u_h; v) = f(v) - a(u_h, v), \forall v \in V$.

Let $\Psi : W \rightarrow \mathbb{R}$ be a nonlinear operator which shall play the role of the output functional. Define the error measure $\mathcal{E}$ as follows:

$$\mathcal{E} = \begin{cases} \Psi(u) - \Psi(u_h), & \text{or} \\ \Psi(u - u_h). \end{cases} \tag{10.52}$$

In the first case, we assume that Ψ is smooth enough so that, for all $x \in W$, the first-order derivative $\Psi'(x; \cdot) \in \mathcal{L}(W; \mathbb{R})$ is well-defined, and we introduce the following dual problem:

$$\begin{cases} \text{Seek } z \in V \text{ such that} \\ a(w, z) = \int_0^1 \Psi'(u_h + se; w) \, ds, \quad \forall w \in W. \end{cases} \tag{10.53}$$

In the second case, we assume that Ψ admits a density functional $\psi : W \rightarrow W$ such that $\Psi(x) = (\psi(x), x)_W$ for all $x \in W$, and introduce the following dual problem:

$$\begin{cases} \text{Seek } z \in V \text{ such that} \\ a(w, z) = (\psi(e), w)_W, \quad \forall w \in W. \end{cases} \tag{10.54}$$

In both cases, we readily deduce from (10.53) and (10.54) the following *error representation* formula:

$$\mathcal{E} = a(e, z) = \rho(u_h; z),$$

and using Galerkin orthogonality yields the following:

Proposition 10.27 (Error representation).

$$\forall z_h \in V_h, \quad \mathcal{E} = \rho(u_h; z - z_h). \tag{10.55}$$

Equation (10.55) shows that in duality-based error estimates, the interpolation properties of the test space, rather than those of the solution space, play a role. We also point out that as for residual-based error estimates, the discrete inf-sup condition is not directly used in the analysis.

Remark 10.28.

(i) It is evident that (10.55) yields $|\mathcal{E}| = |\rho(u_h; z - z_h)|$, but in general the absolute value is not necessary unless the right-hand side is bounded further by applying the Cauchy–Schwarz inequality.

(ii) Assume that $W \subset L$ with continuous embedding where L is a Hilbert space with scalar product $(\cdot, \cdot)_L$. It is then possible to consider an error output functional Ψ which admits a density functional $\psi : L \rightarrow L$ such that $\Psi(x) = (\psi(x), x)_L$ for all $x \in W$. The corresponding dual problem is (10.54) in which the right-hand side is replaced by $(\psi(e), w)_L$, and this leads again to the error representation (10.55). This formalism is, for instance, useful to represent the error in the $\| \cdot \|_L$ norm. Indeed, taking the density functional to be $\psi(x) = x$ and solving the corresponding dual problem with solution z yields $\|e\|_L^2 = a(e, z - z_h), \forall z_h \in V_h$. □

Localization. In the context of finite element approximations to PDEs, the discrete spaces W_h and V_h are constructed using a mesh $\mathcal{T}_h$ of the domain Ω on which the PDE is posed. Let $m \geq 1$, $L = [L^2(\Omega)]^m$, $f \in L$, and consider the PDE $\mathcal{L}u = f$ in Ω, where $\mathcal{L}$ is an $\mathbb{R}^m$-valued differential operator. We make the following localization hypothesis: for all $v \in V$,

$$\rho(u_h; v) = \sum_{K \in \mathcal{T}_h} \rho_K(u_h; v), \tag{10.56}$$

$$\rho_K(u_h; v) = (f - \mathcal{L}u_h, v)_{[L^2(K)]^m} + ((\Lambda u_h) \cdot n_K, v)_{[L^2(\partial K)]^m}, \tag{10.57}$$

where $\rho_K(u_h; \cdot)$ represents the local residual, n_K is the outward normal of the mesh element K, and Λ is an $\mathbb{R}^{m,d}$-valued trace operator; see the two examples below. Let $F \in \mathcal{F}_h^i$ be an interior face in the mesh such that $F = K_1 \cap K_2$. Denote by n_1 and n_2 the outward normal to K_1 and K_2 on F, respectively. Use subscripts 1 and 2 to denote restrictions to K_1 and K_2, respectively. For $u_h \in W_h$ and $v \in V$, define the following jump and average operators on F:

$$[\![\Lambda u_h \cdot n]\!] = (\Lambda u_h)_1 \cdot n_1 + (\Lambda u_h)_2 \cdot n_2 \in \mathbb{R}^m,$$
$$\{\Lambda u_h\} = \tfrac{1}{2}((\Lambda u_h)_1 + (\Lambda u_h)_2) \in \mathbb{R}^{m,d},$$
$$[\![v]\!] = v_1 \otimes n_1 + v_2 \otimes n_2 \in \mathbb{R}^{m,d},$$
$$\{v\} = \tfrac{1}{2}(v_1 + v_2) \in \mathbb{R}^m.$$

Then, using (10.55), (10.56), and (10.57), we infer the following:

Proposition 10.29 (Localized error representation). *In the above framework, for all* $z_h \in V_h$,

$$\mathcal{E} = \sum_{K \in \mathcal{T}_h} (f - \mathcal{L}u_h, z - z_h)_{[L^2(K)]^m} + \sum_{F \in \mathcal{F}_h^\partial} ((\Lambda u_h) \cdot n, z - z_h)_{[L^2(F)]^m}$$
$$+ \sum_{F \in \mathcal{F}_h^i} ([\![\Lambda u_h \cdot n]\!], \{z - z_h\})_{[L^2(F)]^m} + (\{\Lambda u_h\}, [\![z - z_h]\!])_{[L^2(F)]^{m,d}}, \tag{10.58}$$

where n *is the outward normal to* Ω.

We now review two examples: second-order elliptic PDEs and the Stokes problem (both with homogeneous Dirichlet conditions). Duality techniques can be generalized to other boundary conditions; see, e.g., Exercise 10.7. They can also be generalized to nonlinear problems; see [BeR01] and Exercise 10.11. See also [KaS99^a] for duality-based error estimates in a non-conformal setting.

Second-order elliptic PDEs. Consider the PDE $\mathcal{L}u = f \in L^2(\Omega)$ with

$$\mathcal{L}u = -\nabla \cdot (\sigma \cdot \nabla u) + \beta \cdot \nabla u + \mu u, \tag{10.59}$$

supplemented with homogeneous Dirichlet conditions. Recall that the weak formulation of this problem is as follows:

$$\begin{cases} \text{Seek } u \in H_0^1(\Omega) \text{ such that} \\ a(u,v) = \int_\Omega fv, \quad \forall v \in H_0^1(\Omega), \end{cases} \tag{10.60}$$

with the bilinear form $a(u,v) = \int_\Omega \nabla v \cdot \sigma \cdot \nabla u + v(\beta \cdot \nabla u) + \mu uv$. We assume that problem (10.60) is well-posed; see Theorem 3.8 for sufficient conditions. Let V_h be a H^1-conformal approximation space constructed from an affine mesh $\mathcal{T}_h$ of Ω. Denote by u_h the corresponding approximate solution. We assume that V_h is endowed with the following optimal interpolation property: There exists an integer $k \geq 1$ and an interpolation constant c_i, independent of h, such that, for all $v \in H^{k+1}(\Omega)$ and $K \in \mathcal{T}_h$,

$$\inf_{v_h \in V_h} \max(\|v - v_h\|_{0,K}, h_K^{\frac{1}{2}}\|v - v_h\|_{0,\partial K}) \leq c_i \, h_K^{k+1}|v|_{k+1,K}. \tag{10.61}$$

This property holds, for instance, if Lagrange finite elements of degree $k \geq 1$ and a shape-regular family of meshes are considered.

Proposition 10.30. *In the above framework, assume that the dual solution z of either (10.53) or (10.54) is in $H^{k+1}(\Omega)$. For $K \in \mathcal{T}_h$, define the residuals ς_K and the weights η_K to be*

$$\varsigma_K = h_K\|f - \mathcal{L}u_h\|_{0,K} + \tfrac{1}{2}h_K^{\frac{1}{2}}\|[\![n \cdot \sigma \cdot \nabla u_h]\!]\|_{0,\partial K},$$
$$\eta_K = c_i \, h_K^k|z|_{k+1,K}.$$

Then,

$$|\mathcal{E}| \leq \sum_{K \in \mathcal{T}_h} \varsigma_K \eta_K. \tag{10.62}$$

The quantities $\varsigma_K \eta_K$ are called dual-weighted error indicators.

Proof. Direct consequence of Proposition 10.29. Take $m = 1$ and the boundary operator $\Lambda u = \sigma \cdot \nabla u$. The contribution of boundary faces vanishes owing to the homogeneous Dirichlet condition. Moreover, because of H^1-conformity, $\{z - z_h\} = z - z_h$ and $[\![z - z_h]\!] = 0$. To conclude, use (10.61). □

Remark 10.31. Estimate (10.62) is not an posteriori estimate according to Definition 10.1 since the dual solution z depends on the exact solution. Practical means of approximating the dual solution are briefly discussed in §10.4.1; see [BeR01] for further insight. □

The Stokes problem. For the sake of simplicity, we consider the Stokes problem with homogeneous Dirichlet conditions. Let Ω be a domain in $\mathbb{R}^d$. Consider a pair of approximation spaces $\{X_h, M_h\}$ such that $X_h \subset H_0^1(\Omega)$, $M_h \subset L_{\int=0}^2(\Omega)$, and such that the compatibility condition derived in §4.2 is satisfied. Assume that the spaces X_h and M_h are endowed with the following optimal interpolation property: There exists an integer $k \geq 1$ and an interpolation constant c_i, independent of h, such that, for all $K \in \mathcal{T}_h$, $v \in [H^{k+1}(K)]^d$, and $q \in H^k(K)$,

$$\begin{cases} \inf_{v_h \in V_h} (\|v - v_h\|_{0,K} + h_K^{\frac{1}{2}} \|v - v_h\|_{0,\partial K}) \le c_i \, h_K^{k+1} |v|_{k+1,K}, \\ \\ \inf_{q_h \in M_h} \|q - q_h\|_{0,K} \le c_i \, h_K^k |q|_{k,K}. \end{cases}$$

Proposition 10.32. *In the above framework, assume that the dual solution* $z = (v, q)$ *of either (10.53) or (10.54) is in* $[H^{k+1}(\Omega)]^d \times H^k(\Omega)$. *Define the residuals* ς_K *and the weights* η_K *to be to be*

$$\varsigma_K = h_K(\|f + \Delta u_h - \nabla p_h\|_{0,K} + \|\nabla \cdot u_h\|_{0,K}) + \tfrac{1}{2} h_K^{\frac{1}{2}} \|[\![\partial_n u_h - p_h n]\!]\|_{0,\partial K},$$
$$\eta_K = c_i \, h_K \max(|v|_{2,K}, |q|_{1,K}).$$

Then,

$$|\mathcal{E}| \le \sum_{K \in \mathcal{T}_h} \varsigma_K \eta_K.$$

Proof. Direct consequence of Proposition 10.29. Take $m = d + 1$ and define the $\mathbb{R}^{d+1,d}$-valued boundary operator to be

$$\Lambda(u_h, p_h) = \begin{bmatrix} \partial_1 u_{h,1} & \cdots & \partial_d u_{h,1} \\ \vdots & \ddots & \vdots \\ \partial_1 u_{h,d} & \cdots & \partial_d u_{h,d} \\ \hline 0 & \cdots & 0 \end{bmatrix} - p_h \begin{bmatrix} 1 & \cdots & 0 \\ \vdots & \ddots & \vdots \\ 0 & \cdots & 1 \\ 0 & \cdots & 0 \end{bmatrix}.$$

The contribution of boundary faces vanishes owing to the homogeneous Dirichlet condition. Moreover, because of H^1-conformity of the velocity space, $\{v - v_h\} = v - v_h$ and $[\![v - v_h]\!] = 0$. The quantities $\{q - q_h\}$ and $[\![q - q_h]\!]$ do not yield any contribution owing to the fact that the last line of $\Lambda(u_h, p_h)$ is identically zero. To conclude, use (10.61). $\qquad\square$

10.3.3 A posteriori estimates of modeling and discretization errors

In many engineering applications, it often happens that the most accurate and validated model cannot be used in numerical simulations since it requires prohibitive computational costs. Simpler and less accurate models are then preferred, at least in some parts of the computational domain. However, it is not known a priori where the accurate model is needed. It is also desirable to avoid a situation where the accuracy of the solution is affected by a poor model although a very fine mesh is used. In other words, one wishes to design a strategy to balance adaptively modeling and discretization errors. The material presented henceforth is adapted from [BrE03].

Let W and V be two Hilbert spaces, $f \in V'$, $a, \delta \in \mathcal{L}(W \times V; \mathbb{R})$, and consider the following problem:

$$\begin{cases} \text{Seek } u \in W \text{ such that} \\ a(u, v) + \delta(u, v) = f(v), \quad \forall v \in V. \end{cases} \tag{10.63}$$

Here, the bilinear form δ stands for the part of the model one wishes to neglect, and a represents the coarse model. Considering solely the coarse model yields the problem:

$$\begin{cases} \text{Seek } u_m \in W \text{ such that} \\ a(u_m, v) = f(v), \quad \forall v \in V. \end{cases} \qquad (10.64)$$

We assume that problems (10.63) and (10.64) are well-posed. Note that for the modeling error $u - u_m$, the following perturbed Galerkin orthogonality holds

$$\forall v \in V, \quad a(u - u_m, v) = -\delta(u, v).$$

However, the problem that is actually solved is not (10.64) but its finite element approximation:

$$\begin{cases} \text{Seek } u_{hm} \in W_h \text{ such that} \\ a(u_{hm}, v_h) = f(v_h), \quad \forall v_h \in V_h. \end{cases} \qquad (10.65)$$

For the sake of simplicity, we consider a conformal setting in which $W_h \subset W$ and $V_h \subset V$. We also assume that (10.65) is well-posed.

Let $\Psi \in \mathcal{L}(W; \mathbb{R})$ and let $\mathcal{E} = \Psi(u) - \Psi(u_{hm}) = \Psi(u - u_{hm})$ be the error measure. Note that, for the sake of simplicity, we consider a linear operator Ψ; see [BrE03] for the nonlinear case. Consider the dual problem:

$$\begin{cases} \text{Seek } z \in V \text{ such that} \\ a(w, z) + \delta(w, z) = \Psi(w), \quad \forall w \in W, \end{cases} \qquad (10.66)$$

as well as its discrete counterpart with the coarse model only:

$$\begin{cases} \text{Seek } z_{hm} \in V_h \text{ such that} \\ a(w_h, z_{hm}) = \Psi(w_h), \quad \forall w_h \in W_h. \end{cases} \qquad (10.67)$$

Proposition 10.33. *Using the above notation, the following identity holds: For all $(u_h, z_h) \in W_h \times V_h$,*

$$\begin{aligned} \Psi(u) - \Psi(u_{hm}) = \ & -\delta(u_{hm}, z_{hm}) \\ & + \tfrac{1}{2}\big(\rho(u_{hm}; z - z_h) + \rho^*(z_{hm}; u - u_h)\big) \\ & - \tfrac{1}{2}\big(\delta(u_{hm}, z - z_{hm}) + \delta(u - u_{hm}, z_{hm})\big) \end{aligned} \qquad (10.68)$$

with residuals $\rho(u_{hm}; v) = f(v) - a(u_{hm}, v)$ for all $v \in V$, and $\rho^*(z_{hm}, w) = \Psi(w) - a(w, z_{hm})$ for all $w \in W$.

Proof. Left as an exercise.   □

In practice, the error representation (10.68) can be used as follows: The first term in the right-hand side, $-\delta(u_{hm}, z_{hm})$, measures the modeling error. The terms in the second line can be used to estimate the discretization error by localizing the interpolation errors $u - u_h$ and $z - z_h$ and using patchwise reconstructions. Finally, the terms in the third line are one order smaller than $-\delta(u_{hm}, z_{hm})$ in the modeling error. These terms can be neglected in a first approximation; see [BrE03] for more details on adaptive algorithms balancing modeling and discretization errors.

10.4 Adaptive Mesh Generation

This section briefly discusses how the local error indicators derived in the previous sections can be used to design an algorithm for *adaptive mesh generation*; see, e.g., [BeR01] for further insight into practical aspects of adaptive mesh generation.

10.4.1 Strategies for mesh adaption

A general flowchart for adaptive mesh generation is shown in Algorithm 10.1. In this section, we are mainly concerned with the implementation of step 5.

Algorithm 10.1. Adaptive mesh generation.

1: Construct a relatively coarse mesh $\mathcal{T}_{(0)}$. Set $i = 0$.
2: Solve the approximate problem on $\mathcal{T}_{(i)}$. Denote by $u_{(i)}$ the solution.
3: For each element $K_{(i)}$ in $\mathcal{T}_{(i)}$, compute a local error indicator $e_{K_{(i)}}(u_{(i)}, f)$.
4: Compute an a posteriori estimate for the global error. If the approximate solutions meets the prescribed accuracy level, stop.
5: Otherwise deduce from the local error indicators $e_{K_{(i)}}(u_{(i)}, f)$ a new mesh. Set $i \leftarrow i + 1$ and return to step 2.

Evaluation of the error indicators. The a posteriori error estimates derived in §10.1, §10.2, and §10.3 can be recast in the form

$$|\mathcal{E}| \leq \left(\sum_{K \in \mathcal{T}_h} \eta_K(u_h, f)^2 \right)^{\frac{1}{2}}.$$

The quantities $\eta_K(u_h, f)$ are the error indicators. When residual-based or hierarchical error estimates are considered, the quantities $\eta_K(u_h, f)$ can be readily evaluated.

When using duality techniques, the computation of $\eta_K(u_h, f)$ requires solving a dual problem posed on an infinite-dimensional space. The simplest approach is to evaluate an approximate dual solution $z_{(i)}$ on the computational mesh $\mathcal{T}_{(i)}$ and then to estimate second derivatives of z on an element $K_{(i)}$ using the values of $z_{(i)}$ on a macroelement centered at $K_{(i)}$. For instance, one can consider the patch of mesh cells sharing at least one point with $K_{(i)}$; see [BeR01] for the case where triangulations based on hierarchical quadrangular meshes are used, and [BuE04] for an application to flame simulation on fully unstructured triangulations. Furthermore, since the data ψ for the dual problem usually depends on the exact solution u, it must be approximated somehow, e.g., using the approximate solutions obtained on previous meshes.

Constructing a new mesh. Once the error indicators have been evaluated, the next question to address is how to use this information to construct a new mesh. Suppose that a tolerance `tol` for the error is prescribed, i.e., the goal is to construct a mesh $\mathcal{T}_{(i)}$ such that $\mathcal{E} \leq$ `tol`. Clearly, a sufficient condition for this is that

$$\left(\sum_{K_{(i)} \in \mathcal{T}_{(i)}} \eta_{K_{(i)}}(u_{(i)}, f)^2 \right)^{\frac{1}{2}} \leq \texttt{tol}.$$

If the above inequality is not satisfied, a new mesh $\mathcal{T}_{(i+1)}$ must be constructed. To this end, one of the simplest and most popular strategies is to equidistribute the cell contributions, yielding the so-called *error-balancing strategy*; see, e.g., [BaR78^b]. More specifically, we wish to construct a new mesh $\mathcal{T}_{(i+1)}$ and a new approximate solution $u_{(i+1)}$ such that

$$\forall K_{(i+1)} \in \mathcal{T}_{(i+1)}, \quad \eta_{K_{(i+1)}}(u_{(i+1)}, f)^2 \simeq \frac{\texttt{tol}^2}{N_{\mathrm{el}(i+1)}},$$

where $N_{\mathrm{el}(i+1)}$ is the total number of cells in $\mathcal{T}_{(i+1)}$. In practice, these equations are used as a guideline to generate the new mesh.

A relatively straightforward way to proceed is as follows: First, define the function $\varphi : \mathbb{R}_+ \times \mathbb{R}_+ \to \mathbb{R}_+$ such that

$$\varphi(x, \delta) = \begin{cases} \beta & \text{if } x < \gamma\delta, \\ 1 & \text{if } \gamma\delta \leq x \leq \delta, \\ \alpha & \text{if } \delta < x, \end{cases}$$

with parameters $\alpha < 1 < \beta$ and $\gamma < 1$. The parameter α controls the ratio of mesh refinement, β that of mesh derefinement, and γ the threshold for derefinement. Then, introduce the $N_{\mathrm{el}(i)}$ auxiliary quantities

$$\forall K_{(i)} \in \mathcal{T}_{(i)}, \quad h'_{K_{(i)}} = \varphi\left(\eta_{K_{(i)}}(u_{(i)}, f), \frac{\texttt{tol}}{N_{\mathrm{el}(i)}^{\frac{1}{2}}} \right) h_{K_{(i)}}. \tag{10.69}$$

Depending on the mesh structure, several algorithms can be considered to construct $\mathcal{T}_{(i+1)}$ from the auxiliary quantities defined in (10.69).

(i) If the meshes are endowed with some hierarchical structure allowing for local cell refinement and cluster cell derefinement, one can, for instance, decide to refine the cells for which $h_{K_{(i)}} > h'_{K_{(i)}}$, and to unrefine cell patches, say $\mathcal{P}$, for which $\max_{K_{(i)} \in \mathcal{P}} h_{K_{(i)}} < \min_{K_{(i)} \in \mathcal{P}} h'_{K_{(i)}}$ holds; see [BeR01] for numerous examples where the underlying structure is that of quadrangular or hexahedral meshes.

(ii) If a Delaunay mesh generator is used (see §7.3.3), the quantities $h'_{K_{(i)}}$ can be used to define the control function, and it is also possible to refine cell patches locally using the Bowyer–Watson algorithm described in §7.3.2; see [BuE04] for an application to reactive flow simulation on unstructured triangulations.

An alternative strategy to error balancing is to order the mesh cells according to the size of the error indicator η_K and to refine the cells that make up a certain percentage of the total error estimate. One can also choose to refine a given percentage of the cells whose error indicator is the largest. The main interest of this type of strategies is that they yield a relatively moderate increase of the total number of cells from one mesh to the next finer one.

Remark 10.34.

(i) One can consider weighting coefficients to enhance mesh refinement in some specific regions of the computational domain Ω. Let $g : \Omega \to \mathbb{R}_+$ be a user-designed function such that $\int_\Omega g = 1$. On the mesh $\mathcal{T}_{(i)}$, evaluate the $N_{\mathrm{el}(i)}$ weighting coefficients $g_{K_{(i)}} = \int_{K_{(i)}} g$. Note that these coefficients are positive and sum up to unity. The auxiliary quantities $h'_{K_{(i)}}$ are then evaluated from (10.69) with the right-hand side multiplied by $g_{K_{(i)}}$.

(ii) It is also possible to consider optimization strategies to decide which cells should be refined; see [BeR01] and Exercise 10.8. □

10.4.2 Numerical illustration

This section briefly presents an example of adaptive mesh generation for the simulation of a reactive flow problem; see [Bra98, BuE04] for other combustion problems.

Consider a stationary, axisymmetric, methane/air Bunsen flame. Fuel and oxidizer are premixed in stoichiometric proportions and flown into ambient air through a cylindrical tube. The tube radius is 4 mm and its width is 0.5 mm. The injection velocity profile is parabolic with an average velocity of 50 cm/s. Under such experimental conditions, it is possible to stabilize a steady-state flame of conical shape above the burner. The distance between the flame cone vertex and the injection plane is defined as the flame length. In the present setup, the flame length is approximately 8 mm.

The equations governing the reactive flow express the conservation of species mass, momentum, and energy in the form

$$\nabla \cdot (\rho_i v) + \nabla \cdot F_i = S_i,$$
$$\nabla \cdot (\rho_i v \otimes v) + \nabla p + \nabla \cdot P = \rho g,$$
$$\nabla \cdot ((\rho e + p)v) + \nabla \cdot (Q + P \cdot v) = \rho v \cdot g.$$

In the above equations, the fundamental unknowns (the dependent variables) are the species densities ρ_i, $1 \leq i \leq n_S$, where n_S is the total number of species, the hydrodynamic velocity v, and the total specific energy e. The diffusion fluxes F_i, P, and Q are the species mass diffusion fluxes, the viscous stress tensor, and the heat flux vector, respectively. These fluxes are expressed in terms of the gradients of the dependent variables and multicomponent transport coefficients which also depend on the dependent variables. Moreover, S_i is the source term accounting for chemical reactions, $\rho = \sum_{i=1}^{n_S} \rho_i$ is the mixture density, p is the pressure given by the ideal gas law, and g is the gravitational constant. See, e.g., [ErG94, Gio99] for an introduction to multicomponent flow modeling and [ErG98] for Bunsen flame modeling.

An approximate solution to the flame equations is obtained using $\mathbb{P}_1$ Lagrange finite elements for each dependent unknown. A GaLS formulation is implemented; see §4.3 and §5.4, respectively. The computational domain is $\Omega = [0, 5] \times [0, 20]$ in cm. Given a triangulation $\mathcal{T}_{(i)}$ of Ω, the discretized equations form a system of nonlinear equations which is solved approximately using Newton's method. Linear systems are solved iteratively using a preconditioned GMRes or Bi-CGStab algorithm; see §9.3. Once an approximate solution is obtained on the triangulation $\mathcal{T}_{(i)}$, the L^1-norm of the error on the methane density over the control domain $\omega = [0, 1] \times [0, 1]$ (in cm) is evaluated using the duality techniques described in §10.3. Isolines for the methane concentration are depicted in Figure 10.2 on two adaptively refined meshes. More details on the solution algorithms can be found in [BuE03].

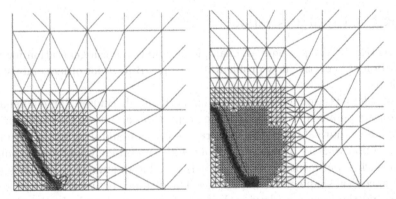

Fig. 10.2. Isolines for the methane concentration on two adaptively refined meshes.

10.5 Exercises

Exercise 10.1. The goal of this exercise is to derive a residual-based error indicator for a convection–diffusion–reaction problem with mixed Dirichlet–Neumann conditions. Consider a partition $\partial\Omega = \partial\Omega_D \cup \partial\Omega_N$ and assume $\text{meas}(\partial\Omega_D) > 0$. Let $f \in L^2(\Omega)$, $g \in L^2(\partial\Omega_N)$, and consider the problem:

$$\begin{cases} \text{Seek } u \in V \text{ such that} \\ a(u,v) = \int_\Omega fv + \int_{\partial\Omega_N} gv, \quad \forall v \in V, \end{cases}$$

with $V = \{v \in H^1(\Omega); v_{|\partial\Omega_D} = 0\}$ and $a(u,v) = \int_\Omega \nabla v \cdot \sigma \cdot \nabla u + v(\beta \cdot \nabla u) + \mu uv$. Consider a V-conformal approximation space based on a shape-regular family of meshes $\{\mathcal{T}_h\}_{h>0}$ and a Lagrange finite element of degree k. Let $\mathcal{N}_K$ denote the set (possibly empty) of faces of K located on $\partial\Omega_N$. Prove that the quantity

$$e_K(u_h, f, g) = h_K \|f - \mathcal{L}u_h\|_{0,K} + \tfrac{1}{2} \sum_{F \in \mathcal{F}_K} h_F^{\frac{1}{2}} \|[\![n \cdot \sigma \cdot \nabla u_h]\!]\|_{0,F}$$

$$+ \sum_{F \in \mathcal{N}_K} h_F^{\frac{1}{2}} \|g - n \cdot \sigma \cdot \nabla u_h\|_{0,F},$$

with $\mathcal{L}$ defined in (10.59) is an error indicator.

Exercise 10.2. Set $r(u) = \beta \cdot \nabla u + \mu u$ and consider the following PDE system with boundary conditions:

$$\begin{cases} -\nabla \cdot (\sigma \cdot \nabla u) + r(u) + \nabla p = f, & \text{a.e. in } \Omega, \\ \nabla \cdot u = s, & \text{a.e. in } \Omega, \\ u = 0, & \text{a.e. on } \partial\Omega_D, \\ n \cdot (\sigma \cdot \nabla u - pI) = g, & \text{a.e. on } \partial\Omega_N. \end{cases}$$

(i) Write a weak formulation for this problem and specify reasonable assumptions on the data σ, β, μ, f, s, and g for well-posedness.

(ii) Let $\mathcal{N}_K$ denote the set (possibly empty) of faces of K located on $\partial\Omega_N$. Prove that the following quantity is an error indicator:

$$e_K(u_h, p_h, f, g, s) = h_K \|f + \nabla \cdot (\sigma \cdot \nabla u_h) - r(u_h) - \nabla p_h\|_{0,K} + \|s - \nabla \cdot u_h\|_{0,K}$$

$$+ \tfrac{1}{2} \sum_{F \in \mathcal{F}_K} h_F^{\frac{1}{2}} \|[\![n \cdot \sigma \cdot \nabla u_h - p_h n]\!]\|_{0,F} + \tfrac{1}{2} \sum_{F \in \mathcal{N}_K} h_F^{\frac{1}{2}} \|g - n \cdot \sigma \cdot \nabla u_h + p_h n\|_{0,F}.$$

Exercise 10.3. Let $\{\mathcal{T}_h\}_{h>0}$ be a shape-regular family of meshes of the polyhedron $\Omega \subset \mathbb{R}^d$ and consider the Oswald interpolation operator defined in (10.20). The goal of this exercise is to prove the estimate (10.21).

(i) Let $K \in \mathcal{T}_h$. Denote $h_K = \text{diam}(K)$ and let $\mathcal{V}_K$ be the set of the $(d+1)$ vertices of K. Show that there exists c such that, for all $w_h \in P^1_{\text{pt},h,0}$ and $K \in \mathcal{T}_h$,

$$|w_h - \mathcal{O}_h w_h|_{1,K} \leq c h_K^{\frac{d}{2}-1} \sum_{a \in \mathcal{V}_K} |(w_{h|K} - \mathcal{O}_h w_h)(a)|.$$

(*Hint*: Use the barycentric coordinates of K.)

(ii) For a vertex a in the mesh, let n_a be the number of mesh cells to which a belongs. Using the shape-regularity of $\{\mathcal{T}_h\}_{h>0}$, prove that $\forall K \in \mathcal{T}_h$ and $\forall a \in \mathcal{V}_K$, there exist constants c_F such that

$$\forall w_h \in P^1_{\text{pt},h,0}, \quad (w_{h|K} - \mathcal{O}_h w_h)(a) = \sum_{\substack{F \in \mathcal{F}_h \\ F \cap K \neq \emptyset}} c_F [\![w_h]\!]_F(a).$$

(iii) Using an inverse inequality, show that for all faces $F \in \mathcal{F}_h$ with diameter h_F, the following estimate holds:

$$\forall w_h \in P^1_{\text{pt},h,0}, \quad \|[\![w_h]\!]\|_{L^\infty(F)} \leq c h_F^{-\frac{d-1}{2}} \|[\![w_h]\!]\|_{L^2(F)},$$

for a constant independent of F. Conclude.

Exercise 10.4. Consider the general setting for hierarchical error estimates introduced in §10.2.1. Assume that the constant γ in the strong Cauchy–Schwarz inequality is zero. Prove that the enriched solution u_h^e of problem (10.25) admits the decomposition $u_h^e = u_h + u_h^f$ with $u_h^f \in W_h^f$ and u_h is the discrete solution to (10.2).

Exercise 10.5. Let $\widehat{K}$ be the unit simplex in $\mathbb{R}^2$. Let $B_{\widehat{K}}$ be the space spanned by the three nodal $\mathbb{P}_2$ shape functions associated with the midpoints of the three edges of K. Consider the scalar product $(\widehat{u}, \widehat{v})_{\widehat{K}} = \int_{\widehat{K}} \nabla \widehat{u} \cdot \nabla \widehat{v}$. Show that the pair of spaces $\{\mathbb{P}_1, B_{\widehat{K}}\}$ satisfies the strong Cauchy–Schwarz inequality with constant $\gamma_{\widehat{K}} = (\frac{2}{3})^{\frac{1}{2}}$. (*Hint*: Select barycentric coordinates and edge bubble functions for the basis functions and use Lemma 10.23.)

Exercise 10.6. Consider the Crouzeix–Raviart finite element space $P^1_{\text{pt},h,0}$ defined in (10.17) and the Fortin–Soulié bubble space B_C introduced in §10.2.3.

(i) Prove $P^1_{\text{pt},h,0} \perp B_C$ for the inner product $(u, v)_{h,1,\Omega} = \sum_{K \in \mathcal{T}_h} \int_K \nabla u \cdot \nabla v$.

(ii) Prove (10.44).

Exercise 10.7. Consider the model problem of Exercise 10.1. Given two functions $\psi_1 \in L^2(\Omega)$ and $\psi_2 \in L^2(\partial\Omega_N)$, define the functional

$$\Psi(u - u_h) = \int_\Omega \psi_1(u - u_h) + \int_{\partial\Omega_N} \psi_2(u - u_h).$$

(i) Introduce an appropriate dual problem to represent the error. Comment on the direction of the advection velocity and on the natural boundary conditions considered in the dual problem.

(ii) Prove the a posteriori error estimate (10.62) with the residuals

$$\rho_K = h_K \|f - \mathcal{L}u_h\|_{0,K} + \tfrac{1}{2}h_K^{\frac{1}{2}}\|[\![n \cdot \sigma \cdot \nabla u_h]\!]\|_{0,\partial K}$$
$$+ h_K^{\frac{1}{2}}\|g - n \cdot \sigma \cdot u_h\|_{0,\partial K \cap \partial \Omega_N}.$$

Exercise 10.8. Given two functions $h : \Omega \to \mathbb{R}_+$ and $w : \Omega \to \mathbb{R}_+$, define

$$N(h) = \int_\Omega h^{-d} \qquad \text{and} \qquad \eta(h) = \int_\Omega w h^\alpha.$$

The function h represents the local size of mesh cells, the parameter α the order of the error indicators, the quantity $N(h)$ the total number of mesh cells, and $\eta(h)$ the global error estimate. Consider the following constrained optimization problems:

(i) Find h that minimizes $N(h)$ under the constraint $\eta(h) \le \texttt{tol}$.
(ii) Find h that minimizes $\eta(h)$ under the constraint $N(h) \le N_{\max}$, where the parameter $N_{\max}$ results, for instance, from computer memory limitations.

Assume $w^{\frac{d}{\alpha+d}} \in L^1(\Omega)$ and set $W = \|w^{\frac{d}{\alpha+d}}\|_{L^1(\Omega)}$. Prove that both optimization problems have a solution given by

$$h = h_0 w^{-\frac{1}{\alpha+d}},$$

with $h_0 = (\frac{\texttt{tol}}{W})^{\frac{1}{\alpha}}$ for problem (i) and $h_0 = (\frac{W}{N_{\max}})^{\frac{1}{d}}$ for problem (ii).

Exercise 10.9. Prove Proposition 10.33.

Exercise 10.10. Let X be a Hilbert space and $A : X \to \mathbb{R}$ be a *nonlinear* operator from X to $\mathbb{R}$. Assume that A is smooth enough to define, for all $x \in X$, its first-order derivative $A'(x; \cdot) \in X'$ and its second-order derivative $A''(x; \cdot, \cdot) \in \mathcal{L}(X^2; \mathbb{R})$. Consider the problem of finding the stationary points of A, i.e., the points $x \in X$ such that $A'(x; \varphi) = 0$, $\forall \varphi \in X$. Consider a standard, conformal Galerkin method to approximate x: given a subspace $X_h \subset X$, seek $x_h \in X_h$ such that $A'(x_h; \varphi_h) = 0$, $\forall \varphi_h \in X_h$.

(i) Assume that A is convex on X, i.e., for all $(y, t) \in X^2$, $A''(y; t, t) \ge 0$. Prove that

$$0 \le Ax_h - Ax \le -\tfrac{1}{2}\min_{\varphi_h \in X_h} A'(x_h; x - \varphi_h).$$

(*Hint*: Set $e = x - x_h$ and use the identity $Ax - Ax_h = \tfrac{1}{2}A'(x_h; e) + \tfrac{1}{2}A'(x; e) + \tfrac{1}{2}\int_0^1 A''(x_h + se; e, e)\,ds$.)
(ii) Consider the general case when the nonlinear operator A is not necessarily convex. Assume that A is smooth enough to define for all $x \in X$ the third-order derivative $A'''(x; \cdot, \cdot, \cdot) \in \mathcal{L}(X^3; \mathbb{R})$. Prove the *error representation* formula

$$Ax - Ax_h = \frac{1}{2} \min_{\varphi_h \in X_h} A'(x_h; x - \varphi_h) + \rho,$$

with remainder term given by $\rho = \frac{1}{2} \int_0^1 A'''(x_h + se; e, e, e) s(s-1) \, ds$.

Exercise 10.11. Let W and V be two Hilbert spaces, $f \in V'$, $a \in \mathcal{L}(W \times V; \mathbb{R})$, and assume that problem (10.1) is well-posed. Let $\Psi : W \to \mathbb{R}$ be a nonlinear operator. Given two subspaces $W_h \subset W$ and $V_h \subset V$, consider the approximate problem:

$$\begin{cases} \text{Seek } u_h \in W_h \text{ such that} \\ a(u_h, v_h) = f(v_h), \quad \forall v_h \in V_h. \end{cases} \quad (10.70)$$

Set $X = W \times V$ and introduce the *Lagrangian functional* on X defined by

$$\mathcal{L}(x) = \Psi(u) + f(z) - a(u, z), \quad x = (u, z).$$

Assume that $\mathcal{L}$ admits a unique stationary point $x = (u, z) \in X$.

(i) Set $X_h = W_h \times V_h$. Show that $x_h = (u_h, z_h)$ is an approximate stationary point of $\mathcal{L}$, i.e., $\mathcal{L}'(x_h; \varphi_h) = 0 \; \forall \varphi_h \in X_h$, if and only if u_h solves (10.70) and z_h solves the dual problem $a(w_h, z_h) = \Psi'(u_h; w_h)$, $\forall w_h \in W_h$.

(ii) Prove the error representation formula

$$\Psi(u) - \Psi(u_h) = \frac{1}{2} \min_{v_h \in V_h} r(u_h, z - v_h) + \frac{1}{2} \min_{w_h \in W_h} r^*(u_h; z_h, u - w_h) + \rho,$$

with residuals $r(u_h, v) = f(v) - a(u_h, v)$, $r^*(u_h; z_h, w) = \Psi'(u_h; w) - a(w, z_h)$, and remainder term $\rho = \frac{1}{2} \int_0^1 \Psi'''(u_h + se_u; e_u, e_u, e_u) s(s-1) \, ds$ where $e_u = u - u_h$. (*Hint*: Use Exercise 10.10.)

(iii) Extend the above result to the case where the form a is nonlinear in its first argument.

Part IV

Appendices

A

Banach and Hilbert Spaces

The goal of this appendix is to restate some fundamental results on Banach and Hilbert spaces. The emphasis is set on the characterization of bijective Banach operators. The results collected herein provide a theoretical framework for the mathematical analysis of the finite element method. Most of classical results are stated without proof; see [Aub00, Bre91, OdD96, Rud66, Sho96, Yos80, Zei95] for further insight.

A.1 Basic Definitions and Results

For the sake of simplicity, we consider real vector spaces. All the material presented herein can be easily extended to complex vector spaces.

A.1.1 Norm and scalar product

Let V be a real vector space.

Definition A.1 (Norm). *A norm on V is a mapping*

$$\| \cdot \|_V : V \ni v \longmapsto \|v\|_V \in \mathbb{R}_+,$$

satisfying the three following properties:

(i) $\|v\|_V = 0 \iff v = 0$.
(ii) $\forall c \in \mathbb{R}, \ \forall v \in V, \ \|cv\|_V = |c| \, \|v\|_V$.
(iii) *Triangle inequality:* $\forall v, w \in V, \ \|v + w\|_V \leq \|v\|_V + \|w\|_V$.

Moreover, a seminorm on V is a mapping from V to $\mathbb{R}_+$ which satisfies only properties (ii) and (iii).

Definition A.2 (Equivalent norms). *Two norms $\| \cdot \|_{V,1}$ and $\| \cdot \|_{V,2}$ are said to be equivalent if there exist two positive constants c_1 and c_2 such that*

$$\forall v \in V, \qquad c_1 \|v\|_{V,2} \leq \|v\|_{V,1} \leq c_2 \|v\|_{V,2}.$$

Remark A.3. If the normed space V is finite-dimensional, all the norms in V are equivalent. This result is false in infinite-dimensional normed spaces. $\square$

Definition A.4 (Inner product). *An* inner product *or* scalar product *on V is a bilinear mapping*

$$(\cdot,\cdot)_V : V \times V \ni (v,w) \longmapsto (v,w)_V \in \mathbb{R},$$

satisfying the three following properties:

(i) *Symmetry:* $\forall v, w \in V$, $(v,w)_V = (w,v)_V$.
(ii) *Positivity:* $\forall v \in V$, $(v,v)_V \geq 0$.
(iii) $(v,v)_V = 0 \iff v = 0$.

Proposition A.5. *Let $(\cdot,\cdot)_V$ be a scalar product on V. By setting*

$$\forall v \in V, \qquad \|v\|_V = (v,v)_V^{\frac{1}{2}},$$

one defines a norm on V. Moreover, the Cauchy–Schwarz inequality *holds:*

$$\forall v, w \in V, \qquad (v,w)_V \leq \|v\|_V \|w\|_V. \tag{A.1}$$

Remark A.6. The Cauchy–Schwarz[1] inequality can be seen as a consequence of the identity

$$\|v\|_V \|w\|_V - (v,w)_V = \frac{\|v\|_V \|w\|_V}{2} \left\| \frac{v}{\|v\|_V} - \frac{w}{\|w\|_V} \right\|_V^2,$$

valid for all non-zero v and w in V. This identity clearly shows that equality holds in (A.1) if and only if v and w are collinear. $\square$

Proposition A.7 (Arithmetic–geometric inequality). *Let $x_1, \ldots, x_n$ be non-negative numbers. Then,*

$$(x_1 x_2 \ldots x_n)^{\frac{1}{n}} \leq \frac{1}{n}(x_1 + \ldots + x_n). \tag{A.2}$$

This inequality is frequently used in conjunction with the Cauchy–Schwarz inequality. In particular, it implies

$$\forall \gamma > 0, \ \forall v, w \in V, \quad (v,w)_V \leq \frac{\gamma}{2}\|v\|_V^2 + \frac{1}{2\gamma}\|w\|_V^2. \tag{A.3}$$

A.1.2 Operators in Banach spaces

Definition A.8 (Banach spaces). Banach spaces *are complete normed vector spaces. This means that a Banach space is a vector space V equipped with a norm $\|\cdot\|_V$ such that every Cauchy sequence (with respect to the metric $d(x,y) = \|x-y\|_V$) in V has a limit in V.*

---

[1] Augustin-Louis Cauchy (1789–1857) and Herman Schwarz (1843–1921)

Definition A.9. *Let V and W be two normed vector spaces. $\mathcal{L}(V;W)$ is the vector space of continuous linear mappings from V to W. The mapping A is also called an* operator.

Proposition A.10. *Let V be a normed vector space and let W be a Banach space. Equip $\mathcal{L}(V;W)$ with the norm*

$$\forall A \in \mathcal{L}(V;W), \quad \|A\|_{\mathcal{L}(V;W)} = \sup_{v \in V} \frac{\|Av\|_W}{\|v\|_V}.$$

Then, $\mathcal{L}(V;W)$ is a Banach space.

Proof. See [Rud66, p. 87] or [Yos80, p. 111]. □

Remark A.11. In this book, we shall systematically abuse the notation by writing $\sup_{v \in V} \frac{\|Av\|_W}{\|v\|_V}$ instead of $\sup_{\substack{v \in V \\ v \neq 0}} \frac{\|Av\|_W}{\|v\|_V}$. □

Definition A.12 (Compact operator). *Let V and W be two Banach spaces. $A \in \mathcal{L}(V;W)$ is called a* compact operator *if from every bounded sequence $\{v_n\}_{n \geq 0}$ in V, one can extract a subsequence $\{v_{n_k}\}$ such that the sequence $\{Av_{n_k}\}$ converges in W.*

Remark A.13. If $V \subset W$ and the injection of V into W is compact, then from every bounded sequence $\{v_n\}$ in V, one can extract a subsequence that converges in W. □

A.1.3 Duality

Definition A.14 (Dual space, Continuous linear forms). *Let V be a normed vector space. The* dual space *of V is defined to be $\mathcal{L}(V;\mathbb{R})$ and is denoted by V'. An element $A \in V'$ is called a* continuous linear form. *Its action on an element $v \in V$ is often denoted by means of duality brackets $\langle \cdot, \cdot \rangle_{V',V}$ so that $\langle A, v \rangle_{V',V} = Av$.*

Remark A.15. Owing to Proposition A.10, V' is a Banach space when equipped with the norm

$$\forall A \in V', \quad \|A\|_{V'} = \sup_{v \in V} \frac{\langle A, v \rangle_{V',V}}{\|v\|_V}.$$ □

Theorem A.16 (Hahn–Banach). *Let V be a normed vector space and let G be a subspace of V equipped with the same norm. Let $B \in G' = \mathcal{L}(G;\mathbb{R})$ be a linear continuous mapping with norm*

$$\|B\|_{G'} = \sup_{g \in G} \frac{\langle B, g \rangle_{G',G}}{\|g\|_V}.$$

Then, there exists $A \in V'$ with the following properties:

(i) A is an extension of B, i.e., $Ag = Bg$ for all $g \in G$.

(ii) $\|A\|_{V'} = \|B\|_{G'}$.

Proof. See [Rud66, p. 56], [Yos80, p. 102], or [Bre91, p. 4]. The statement is actually a simplified version of the Hahn–Banach Theorem.   □

Corollary A.17. *Let V be a normed vector space. For all $v \in V$,*

$$\|v\|_V = \sup_{A \in V', \|A\|_{V'}=1} |\langle A, v \rangle_{V',V}| = \max_{A \in V', \|A\|_{V'}=1} |\langle A, v \rangle_{V',V}|.$$

Proof. Assume $v \neq 0$. Clearly, $\sup_{A \in V', \|A\|_{V'}=1} |\langle A, v \rangle_{V',V}| \leq \|v\|_V$. Let $G = \operatorname{span}(v)$ and $B \in G'$ defined as $\langle B, tv \rangle_{G',G} = t\|v\|_V^2$ for $t \in \mathbb{R}$. Owing to the Hahn–Banach Theorem, there exists $Z \in V'$ such that $\|Z\|_{V'} = \|B\|_{G'} = \|v\|_V$ and $\langle Z, v \rangle_{V',V} = \|v\|_V^2$. Conclude by taking $A = \|v\|_V^{-1} Z$.   □

Corollary A.18. *Let V be a normed space and let $F \subset V$ be subspace. Assume $(\forall f \in V', \ f(F) = 0) \Rightarrow (f = 0)$. Then, $\overline{F} = V$.*

Proof. See [Rud66, Theorem 5.19] or [Bre91, p. 7].   □

Definition A.19 (Dual operator). *Let V and W be two normed vector spaces and let $A \in \mathcal{L}(V; W)$. The dual operator $A^T : W' \to V'$ is defined by*

$$\forall v \in V, \ \forall w' \in W', \quad \langle A^T w', v \rangle_{V',V} = \langle w', Av \rangle_{W',W}.$$

Definition A.20 (Continuous bilinear forms). *Let Z_1 and Z_2 be two normed vector spaces. $\mathcal{L}(Z_1 \times Z_2; \mathbb{R})$ denotes the vector space of continuous bilinear forms on $Z_1 \times Z_2$. It is a Banach space when equipped with the norm*

$$\|a\|_{Z_1,Z_2} = \sup_{z_1 \in Z_1, z_2 \in Z_2} \frac{a(z_1, z_2)}{\|z_1\|_{Z_1} \|z_2\|_{Z_2}}.$$

Proposition A.21. *Let Z_1 and Z_2 be two Banach spaces and let $a \in \mathcal{L}(Z_1 \times Z_2; \mathbb{R})$. Then, the mapping $A : Z_1 \to Z_2'$ defined by*

$$\forall z_1 \in Z_1, \ \forall z_2 \in Z_2, \quad \langle A z_1, z_2 \rangle_{Z_2',Z_2} = a(z_1, z_2),$$

is in $\mathcal{L}(Z_1; Z_2')$ and $\|A\|_{\mathcal{L}(Z_1;Z_2')} = \|a\|_{Z_1,Z_2}$.

Definition A.22 (Double dual). *The double dual of a Banach space V is the dual of V' and is denoted by V''.*

Remark A.23. Owing to Proposition A.10, V'' is a Banach space.   □

Proposition A.24. *Let V be a Banach space and let $J_V : V \to V''$ be the linear mapping defined by*

$$\forall u \in V, \ \forall v' \in V', \quad \langle J_V u, v' \rangle_{V'',V'} = \langle v', u \rangle_{V',V}.$$

Then, J_V is an isometry.

Proof. J_V is an isometry since

$$\|J_V u\|_{V''} = \sup_{\substack{v' \in V' \\ \|v'\|_{V'}=1}} \langle J_V u, v' \rangle_{V'',V'} = \sup_{\substack{v' \in V' \\ \|v'\|_{V'}=1}} \langle v', u \rangle_{V',V} = \|u\|_V,$$

where the last equality results from Corollary A.17.   □

Remark A.25.

(i) Since the mapping J_V is an isometry, it is injective. As a result, V can be identified with the subspace $J_V(V) \subset V''$.

(ii) It may happen that the mapping J_V is not surjective. In this case, the space V is strictly included in V''. For instance, $L^\infty(\Omega) = L^1(\Omega)'$ but $L^1(\Omega) \subsetneq L^\infty(\Omega)'$ with strict inclusion; see §B.1.2 or [Bre91, pp. 63–66] for the definition of these spaces.   □

Definition A.26 (Reflexive Banach spaces). *Let V be a Banach space. V is said to be* reflexive *if J_V is an isomorphism.*

A.1.4 Hilbert spaces

Definition A.27 (Hilbert spaces). *A Hilbert space is an inner product space that is complete with respect to the norm defined by the inner product (and is hence a Banach space). A Hilbert space is said to be* separable *if it admits a countable and dense subset.*

Theorem A.28 (Riesz–Fréchet). *Let V be a Hilbert space. For each $v' \in V'$, there exists a unique $u \in V$ such that*

$$\forall w \in V, \quad \langle v', w \rangle_{V',V} = (u, w)_V.$$

Moreover, the mapping $v' \in V' \mapsto u \in V$ is an isometric isomorphism.

Proof. See [Yos80, p. 90] or [Bre91, p. 81].   □

An important consequence of the Riesz–Fréchet Theorem is the following:

Proposition A.29. *Hilbert spaces are reflexive.*

Proof. Let V be a Hilbert space. The Riesz–Fréchet Theorem implies that V can be identified with V'; similarly, V' can be identified with V''.   □

Definition A.30 (Orthogonal projection). *Let H be a Hilbert space with scalar product $(\cdot, \cdot)_H$ and associated norm $\|\cdot\|_H$. Let S be a closed subspace of H. The* orthogonal projection *from H to S is defined to be the operator $P_{H,S} \in \mathcal{L}(H; S)$ such that*

$$\forall w \in S, \quad (P_{H,S}(v), w)_H = (v, w)_H. \tag{A.4}$$

Proposition A.31. *$P_{H,S}$ is characterized by the property*

$$\forall v \in H, \quad \|v - P_{H,S}(v)\|_H = \min_{w \in S} \|v - w\|_H. \tag{A.5}$$

Proof. See Exercise 2.2.   □

A.2 Bijective Banach Operators

This section presents classical results to characterize bijective linear Banach operators [Aub00, Bre91, Yos80]. Some of the material presented herein is adapted from [Aze95, GuQ97]. Henceforth, V and W are real Banach spaces, and operators in $\mathcal{L}(V; W)$ are called *Banach operators*.

A.2.1 Fundamental results

For $A \in \mathcal{L}(V; W)$, we denote by $\text{Ker}(A)$ its kernel and by $\text{Im}(A)$ its range. The operator A being continuous, $\text{Ker}(A)$ is closed in V. Hence, the quotient of V by $\text{Ker}(A)$, $V/\text{Ker}(A)$, can be defined. This space is composed of equivalence classes $\bar{v}$ such that v and w are in the same class $\bar{v}$ if and only if $v - w \in \text{Ker}(A)$.

Theorem A.32. *Equipped with the norm $\|\bar{v}\| = \inf_{v \in \bar{v}} \|v\|_V$, $V/\text{Ker}(A)$ is a Banach space. Moreover, defining $\overline{A} : V/\text{Ker}(A) \to \text{Im}(A)$ by $\overline{A}\bar{v} = Av$ for all v in $\bar{v}$, $\overline{A}$ is an isomorphism.*

Proof. See [Yos80, p. 60]. $\qquad\qquad\qquad\qquad\qquad\qquad\qquad\qquad\qquad$ □

For $M \subset V$, $N \subset V'$, we introduce the so-called *annihilator* of M and N,

$$M^{\perp} = \{v' \in V'; \forall m \in M, \langle v', m \rangle_{V',V} = 0\},$$
$$N^{\perp} = \{v \in V; \forall n' \in N, \langle n', v \rangle_{V',V} = 0\}.$$

A characterization of $\text{Ker}(A)$ and $\text{Im}(A)$ is given by the following:

Lemma A.33. *For A in $\mathcal{L}(V; W)$, the following properties hold:*

  (i) $\text{Ker}(A) = (\text{Im}(A^T))^{\perp}$.
  (ii) $\text{Ker}(A^T) = (\text{Im}(A))^{\perp}$.
  (iii) $\overline{\text{Im}(A)} = (\text{Ker}(A^T))^{\perp}$.
  (iv) $\overline{\text{Im}(A^T)} \subset (\text{Ker}(A))^{\perp}$.

Proof. See [Yos80, pp. 202–209] or [Bre91, p. 28]. $\qquad\qquad\qquad\qquad\qquad$ □

Lemma A.33(i) shows that the characterization of operators with closed range is important to characterize surjective operators. This is the purpose of the following fundamental theorem:

Theorem A.34 (Banach or Closed Range). *Let $A \in \mathcal{L}(V; W)$. The following statements are equivalent:*

  (i) $\text{Im}(A)$ *is closed.*
  (ii) $\text{Im}(A^T)$ *is closed.*
  (iii) $\text{Im}(A) = (\text{Ker}(A^T))^{\perp}$.
  (iv) $\text{Im}(A^T) = (\text{Ker}(A))^{\perp}$.

Proof. See [Yos80, p. 205] or [Bre91, p. 29]. □

We now put in place the second keystone of the edifice:

Theorem A.35 (Open Mapping). *If $A \in \mathcal{L}(V; W)$ is surjective and U is an open set in V, then $A(U)$ is open in W.*

Proof. See [Rud66, pp. 47–48], [Yos80, p. 75], or [Bre91, p. 18]. □

Theorem A.35, also due to Banach, has far-reaching consequences. In particular, we deduce the following:

Lemma A.36. *Let $A \in \mathcal{L}(V; W)$. The following statements are equivalent:*

(i) $\mathrm{Im}(A)$ *is closed.*
(ii) *There exists $\alpha > 0$ such that*

$$\forall w \in \mathrm{Im}(A), \ \exists v_w \in V, \quad Av_w = w \quad \text{and} \quad \alpha \|v_w\|_V \le \|w\|_W. \quad \text{(A.6)}$$

Proof. The implication (i) $\Rightarrow$ (ii). Since $\mathrm{Im}(A)$ is closed in W, $\mathrm{Im}(A)$ is a Banach space. Applying the Open Mapping Theorem to $A : V \to \mathrm{Im}(A)$ and $U = B_V(0, 1)$ (the unit ball in V) yields that $A(B_V(0, 1))$ is open in $\mathrm{Im}(A)$. Since $0 \in A(B_V(0, 1))$, there is $\gamma > 0$ such that $B_W(0, \gamma) \subset A(B_V(0, 1))$. Let $w \in \mathrm{Im}(A)$. Since $\frac{\gamma}{2} \frac{w}{\|w\|_W} \in B_W(0, \alpha)$, there is $z \in B_V(0, 1)$ such that $Az = \frac{\gamma}{2} \frac{w}{\|w\|_W}$. In other words, setting $v = \frac{2\|w\|_W}{\gamma} z$, $Av = w$ and $\frac{\gamma}{2} \|v\|_V \le \|w\|_W$. The implication (ii) $\Rightarrow$ (i). Let $\{w_n\}$ be a sequence in $\mathrm{Im}(A)$ that converges to some $w \in W$. Using (A.6), we infer that there exists a sequence $\{v_n\}$ in V such that $Av_n = w_n$ and $\alpha \|v_n\|_V \le \|w_n\|_W$. Since $\{v_n\}$ is a Cauchy sequence in V and V is a Banach space, v_n converges to a certain $v \in V$. Owing to the continuity of A, Av_n converges to Av. Hence, $w = Av \in \mathrm{Im}(A)$, proving statement (i). □

Remark A.37. A first consequence of Lemma A.36 is that if $A \in \mathcal{L}(V; W)$ is bijective, then its inverse is necessarily continuous. Indeed, the fact that A is bijective implies that A is injective and $\mathrm{Im}(A)$ is closed. Lemma A.36 implies that there is $\alpha > 0$ such that $\|A^{-1} w\|_V \le \frac{1}{\alpha} \|w\|_W$, i.e., A^{-1} is continuous. □

Let us finally give a sufficient condition for the image of an injective operator to be closed.

Lemma A.38 (Petree–Tartar). *Let X, Y, Z be three Banach spaces. Let $A \in \mathcal{L}(X; Y)$ be an injective operator and let $T \in \mathcal{L}(X; Z)$ be a compact operator. If there is $c > 0$ such that $c\|x\|_X \le \|Ax\|_Y + \|Tx\|_Z$, then $\mathrm{Im}(A)$ is closed; equivalently, there is $\alpha > 0$ such that*

$$\forall x \in X, \quad \alpha \|x\|_X \le \|Ax\|_Y.$$

Proof. By contradiction. Assume that there is a sequence $\{x_n\}$ of X such that $\|x_n\|_X = 1$ and $\|Ax_n\|_Y$ converges to zero when n goes to infinity. Since T is compact and the sequence $\{x_n\}$ is bounded, there is a subsequence $\{x_{n_k}\}$ such that $\{Tx_{n_k}\}$ is a Cauchy sequence in Z. Owing to the inequality

$$\alpha\|x_{n_k} - x_{m_k}\|_X \leq \|Ax_{n_k} - Ax_{m_k}\|_Y + \|Tx_{n_k} - Tx_{m_k}\|_Z,$$

$\{x_{n_k}\}$ is a Cauchy sequence in X. Let x be the limit of the subsequence $\{x_{n_k}\}$ in X. The continuity of A implies $Ax_{n_k} \to Ax$ and $Ax = 0$ since $Ax_{n_k} \to 0$. Since A is injective $x = 0$, which contradicts the hypothesis $\|x_{n_k}\|_X = 1$. □

A.2.2 Characterization of surjective operators

As a consequence of the Closed Range Theorem together with Lemma A.36, which is a rephrasing of the Open Mapping Theorem, we deduce two lemmas characterizing surjective operators. The proofs are left as an exercise.

Lemma A.39. *Let $A \in \mathcal{L}(V; W)$. The following statements are equivalent:*

(i) $A^T : W' \to V'$ *is surjective.*
(ii) $A : V \to W$ *is injective and* $\mathrm{Im}(A)$ *is closed in* W.
(iii) *There exists $\alpha > 0$ such that*

$$\forall v \in V, \qquad \|Av\|_W \geq \alpha\|v\|_V. \tag{A.7}$$

(iv) *There exists $\alpha > 0$ such that*

$$\inf_{v \in V} \sup_{w' \in W'} \frac{\langle w', Av \rangle_{W',W}}{\|w'\|_{W'}\|v\|_V} \geq \alpha. \tag{A.8}$$

Lemma A.40. *Let $A \in \mathcal{L}(V; W)$. The following statements are equivalent:*

(i) $A : V \to W$ *is surjective.*
(ii) $A^T : W' \to V'$ *is injective et* $\mathrm{Im}(A^T)$ *is closed in* V'.
(iii) *There exists $\alpha > 0$ such that*

$$\forall w' \in W', \qquad \|A^T w'\|_{V'} \geq \alpha\|w'\|_{W'}. \tag{A.9}$$

(iv) *There exists $\alpha > 0$ such that*

$$\inf_{w' \in W'} \sup_{v \in V} \frac{\langle A^T w', v \rangle_{V',V}}{\|w'\|_{W'}\|v\|_V} \geq \alpha. \tag{A.10}$$

Remark A.41. The statement (i) $\Leftrightarrow$ (iv) in Lemma A.40 is sometimes referred to as Lions' Theorem; see, e.g., [Sho96, LiM68]. Establishing the a priori estimate (A.10) is a necessary and sufficient condition to prove that the problem $Au = f$ has at least one solution u in V for all f in W. □

One easily verifies (see Lemma A.42) that (A.6) implies the *inf-sup condition* (A.10). In practice, however, it is often easier to check condition (A.10) than to prove that for all $w \in \text{Im}(A)$, there exists an inverse image v_w satisfying (A.6). At this point, the natural question that arises is to determine whether the constant α in (A.10) is the *same* as that in (A.6). The answer to this question is the purpose of the next lemma which is due to Azerad [Aze95, Aze99]. This lemma will be used in the study of saddle-point problems; see Theorem 2.34.

Lemma A.42. *Let V and W be two Banach spaces and let $A \in \mathcal{L}(V; W)$ be a surjective operator. Let $\alpha > 0$. The property*

$$\forall w \in \text{Im}(A), \ \exists v_w \in V, \quad Av_w = w \quad and \quad \alpha\|v_w\|_V \leq \|w\|_W,$$

implies

$$\inf_{w' \in W'} \sup_{v \in V} \frac{\langle A^T w', v \rangle_{V',V}}{\|w'\|_{W'}\|v\|_V} \geq \alpha.$$

The converse is true if V is reflexive.

Proof. (1) The implication. By definition of the norm in W',

$$\forall w' \in W', \qquad \|w'\|_{W'} = \sup_{\substack{w \in W \\ \|w\|_W \leq 1}} \langle w', w \rangle_{W',W}.$$

For all w in W, there is $v_w \in V$ such that $Av_w = w$ and $\alpha\|v_w\|_V \leq \|w\|_W$. Let w' in W'. Therefore,

$$\langle w', w \rangle_{W',W} = \langle w', Av_w \rangle_{W',W} = \langle A^T w', v_w \rangle_{V',V} \leq \tfrac{1}{\alpha}\|A^T w'\|_{V'}\|w\|_W.$$

Hence,

$$\|w'\|_{W'} = \sup_{\substack{w \in W \\ \|w\|_W \leq 1}} \langle w', w \rangle_{W',W} \leq \tfrac{1}{\alpha}\|A^T w'\|_{V'}.$$

The desired inequality follows from the definition of the norm in V'.

(2) Let us prove the converse statement by assuming that V is reflexive. The inf-sup inequality being equivalent to $\|A^T w'\|_{V'} \geq \alpha\|w'\|_{W'}$ for all $w' \in W'$, A^T is injective. Let $v' \in \text{Im}(A^T)$ and define $z'(v') \in W'$ such that $A^T(z'(v')) = v'$. Note that $z'(v')$ is unique since A^T is injective. Hence, $z'(\cdot) : \text{Im}(A^T) \subset V' \to W'$ is a mapping. Let $w \in W$ and let us construct an inverse image for w, say v_w, satisfying (A.6). We first define the linear form $\tilde{w} : \text{Im}(A^T) \to \mathbb{R}$ by

$$\forall v' \in \text{Im}(A^T), \quad \tilde{w}(v') = \langle z'(v'), w \rangle_{W',W},$$

that is, $\tilde{w}(A^T z') = \langle z', w \rangle_{W',W}$ for all $z' \in W'$. Hence,

$$|\tilde{w}(v')| \leq \|z'(v')\|_{W'}\|w\|_W \leq \tfrac{1}{\alpha}\|A^T z'(v')\|_{V'}\|w\|_W$$
$$\leq \tfrac{1}{\alpha}\|v'\|_{V'}\|w\|_W.$$

This means that $\tilde{w}$ is continuous on $\mathrm{Im}(A^T)$ equipped with the norm of V'. Owing to the Hahn–Banach Theorem, $\tilde{w}$ can be extended to V' with the same norm. Let $\tilde{\tilde{w}} \in V''$ be the extension in question with $\|\tilde{\tilde{w}}\|_{V''} \leq \frac{1}{\alpha}\|w\|_W$. Since V is assumed to be reflexive, there is $v_w \in V$ such that $J_V(v_w) = \tilde{\tilde{w}}$. As a result,

$$\forall z' \in W', \qquad \langle z', Av_w \rangle_{W',W} = \langle A^T z', v_w \rangle_{V',V} = \langle J_V(v_w), A^T z' \rangle_{V'',V'}$$
$$= \langle \tilde{\tilde{w}}, A^T z' \rangle_{V'',V'} = \langle z', w \rangle_{W',W},$$

showing that $Av_w = w$. Hence, v_w is an inverse image of w and

$$\|v_w\|_V = \|J_V(v_w)\|_{V''} = \|\tilde{\tilde{w}}\|_{V''} \leq \frac{1}{\alpha}\|w\|_W. \qquad \square$$

A.2.3 Characterization of bijective Banach operators

Theorem A.43. *Let $A \in \mathcal{L}(V; W)$. A is bijective if and only if $A^T : W' \to V'$ is injective and there exists $\alpha > 0$ such that*

$$\forall v \in V, \qquad \|Av\|_W \geq \alpha\|v\|_V. \tag{A.11}$$

Proof. (1) The implication. Since A is surjective, $\mathrm{Ker}(A^T) = \mathrm{Im}(A)^{\perp} = \{0\}$, i.e., A^T is injective. Since $\mathrm{Im}(A) = W$ is closed and A is injective, we deduce from Lemma A.39 that there exists $\alpha > 0$ such that $\underline{\|Av\|_W \geq \alpha\|v\|_V}$.
(2) The converse. The injectivity of A^T implies $\overline{\mathrm{Im}(A)} = (\mathrm{Ker}(A^T))^{\perp} = W$, i.e., $\mathrm{Im}(A)$ is dense in W. Let us prove that $\mathrm{Im}(A)$ is closed. Let $\{v_n\}$ be a sequence in V such that $\{Av_n\}$ is a Cauchy sequence in W. The inequality $\|Av_n\|_W \geq \alpha\|v_n\|_V$ implies that $\{v_n\}$ is a Cauchy sequence in V. Let v be its limit. The continuity of A implies $Av_n \to Av$; hence, $\mathrm{Im}(A)$ is closed. Therefore, $\mathrm{Im}(A) = W$, i.e., A is surjective. Finally, the injectivity of A is a direct consequence of inequality (A.11). $\square$

Remark A.44. The interpretation of Theorem A.43 is that a Banach operator is bijective if and only if it is injective, its range is closed, and its dual operator is injective. $\square$

Corollary A.45. *Let $A \in \mathcal{L}(V; W)$. The following statements are equivalent:*

(i) *A is bijective.*
(ii) *There exists a constant $\alpha > 0$ such that*

$$\forall v \in V, \quad \|Av\|_W \geq \alpha\|v\|_V, \tag{A.12}$$
$$\forall w' \in W', \quad (A^T w' = 0) \implies (w' = 0). \tag{A.13}$$

(iii) *There exists a constant $\alpha > 0$ such that*

$$\inf_{v \in V} \sup_{w' \in W'} \frac{\langle w', Av \rangle_{W',W}}{\|w'\|_{W'}\|v\|_V} \geq \alpha, \tag{A.14}$$
$$\forall w' \in W', \quad (\langle w', Av \rangle_{W',W} = 0, \ \forall v \in V) \implies (w' = 0). \tag{A.15}$$

Proof. Condition (A.13) is equivalent to stating that A^T is injective. Therefore, Theorem A.43 shows that the bijectivity of A is equivalent to the conditions of statement (ii). Furthermore, statements (ii) and (iii) are clearly equivalent since the inf-sup condition (A.14) is a simple reformulation of (A.12) and since (A.13) and (A.15) are clearly equivalent. □

Now, let us assume that $A \in \mathcal{L}(V; W)$ is associated with a bilinear form $a \in \mathcal{L}(Z_1 \times Z_2; \mathbb{R})$ such that $\langle Az_1, z_2 \rangle_{Z_2', Z_2} = a(z_1, z_2)$, i.e., $V = Z_1$ and $W = Z_2'$.

Corollary A.46. *If Z_2 is reflexive, the following statements are equivalent:*

(i) *For all $f \in Z_2'$, there is a unique $u \in Z_1$ such that $a(u, z_2) = \langle f, z_2 \rangle_{Z_2', Z_2}$ for all $z_2 \in Z_2$.*
(ii) *There is $\alpha > 0$ such that*

$$\inf_{z_1 \in Z_1} \sup_{z_2 \in Z_2} \frac{a(z_1, z_2)}{\|z_1\|_{Z_1} \|z_2\|_{Z_2}} \geq \alpha, \tag{A.16}$$

$$\forall z_2 \in Z_2, \quad (\forall z_1 \in Z_1, a(z_1, z_2) = 0) \implies (z_2 = 0). \tag{A.17}$$

Proof. Item (i) amounts to stating that A is bijective. Owing to Corollary A.45, the bijectivity of A is equivalent to conditions (A.12) and (A.13). Clearly, (A.12) is equivalent to $\sup_{z_2 \in Z_2} \frac{a(z_1, z_2)}{\|z_2\|_{Z_2}} \geq \alpha \|z_1\|_{Z_1}$, which is (A.16). Furthermore, (A.15) is clearly equivalent to (A.17) since Z_2 is reflexive. □

Remark A.47. Corollary A.46 is the BNB Theorem; see §2.1.3. □

A.2.4 Coercive operators

In this section, we focus on a smaller class of operators, that of coercive operators.

Definition A.48 (Coercive operator). $A \in \mathcal{L}(V; V')$ *is said to be a coercive operator if there is a constant $\alpha > 0$ such that*

$$\forall v \in V, \quad \langle Av, v \rangle_{V', V} \geq \alpha \|v\|_V^2.$$

The following proposition shows that the notion of coercivity is relevant only in *Hilbert spaces*:

Proposition A.49. *Let V be a Banach space. V can be equipped with a Hilbert structure with the same topology if and only if there is a coercive operator in $\mathcal{L}(V; V')$.*

Proof. See Exercise 2.8. □

Corollary A.50. *Coercivity is a sufficient condition for an operator $A \in \mathcal{L}(V; V')$ to be bijective.*

Proof. See Lemma 2.8.   □

Remark A.51. Corollary A.50 is the *Lax–Milgram Lemma*; see §2.1.2.   □

We now introduce the class of self-adjoint operators.

Definition A.52 (Self-adjoint operator). *Let V be a reflexive Banach space, so that V and V'' are identified. $A \in \mathcal{L}(V; V')$ is said to be a self-adjoint operator if $A^T = A$.*

Self-adjoint bijective operators are characterized as follows:

Corollary A.53. *Let V be a reflexive Banach space and let $A \in \mathcal{L}(V; V')$ be a self-adjoint operator. Then, A is bijective if and only if there is $\alpha > 0$ such that*

$$\forall v \in V, \qquad \|Av\|_{V'} \geq \alpha \|v\|_V. \tag{A.18}$$

Proof. Owing to Theorem A.43, the bijectivity of A implies that A satisfies inequality (A.18). Conversely, inequality (A.18) means that A is injective. It follows that A^T is injective since $A^T = A$ by hypothesis. The conclusion is then a consequence of Theorem A.43.   □

We finally introduce the concept of monotonicity.

Definition A.54 (Monotone operator). *$A \in \mathcal{L}(V; V')$ is said to be monotone operator if*

$$\forall v \in V, \qquad \langle Av, v \rangle_{V',V} \geq 0.$$

Corollary A.55. *Let V be a reflexive Banach space and let $A \in \mathcal{L}(V; V')$ be a monotone self-adjoint operator. Then, A is bijective if and only if A is coercive.*

Proof. See Exercise 2.9.   □

A.2.5 Application to saddle-point problems

Let X and M be two real Banach spaces. Consider two continuous linear operators $A : X \to X'$ and $B : X \to M$, and let $B^T : M' \to X'$. The goal of this section is to study the following problem: Given $f \in X'$ and $g \in M$,

$$\begin{cases} \text{Seek } u \in X \text{ and } p \in M' \text{ such that} \\ Au + B^T p = f, \\ Bu = g. \end{cases} \tag{A.19}$$

Denote by $\mathrm{Ker}(B)$ the kernel of the operator B and define the operator $\pi A : \mathrm{Ker}(B) \to \mathrm{Ker}(B)'$ such that $\langle \pi A u, v \rangle = \langle Au, v \rangle$ for all $u, v \in \mathrm{Ker}(B)$.

Theorem A.56. *Problem (A.19) is well-posed if and only if:*

(i) $\pi A : \mathrm{Ker}(B) \to \mathrm{Ker}(B)'$ *is an isomorphism.*

(ii) $B : X \to M$ *is surjective.*

Proof. (1) Let us first show that if problem (A.19) is well-posed, then statements (i) and (ii) are satisfied.

1.a Let h be in M, denote by (u,p) the solution to problem (A.19) with $f = 0$ and $g = h$. It is clear that u satisfies $Bu = h$. Hence, B is surjective.

1.b Let us show that πA is surjective. Let $h \in \mathrm{Ker}(B)'$. Owing to the Hahn–Banach Theorem, there is an extension $\tilde h \in X'$ such that $\langle \tilde h, v \rangle = \langle h, v \rangle$ for all v in $\mathrm{Ker}(B)$ and $\|\tilde h\|_{X'} = \|h\|_{\mathrm{Ker}(B)'}$. Let (u,p) be the solution to (A.19) with $f = \tilde h$ and $g = 0$. It is clear that u is in $\mathrm{Ker}(B)$. Since $\langle B^T p, v \rangle = \langle p, Bv \rangle = 0$ for all v in $\mathrm{Ker}(B)$, we infer $\langle \pi Au, v \rangle = \langle Au, v \rangle = \langle \tilde h, v \rangle = \langle h, v \rangle$ for all v in $\mathrm{Ker}(B)$. As a result, $\pi Au = h$.

1.c Let us show that πA is injective. Let u in $\mathrm{Ker}(B)$ be such that $\pi Au = 0$. Then, $\langle Au, v \rangle = 0$ for all v in $\mathrm{Ker}(B)$; as a result, Au is in $\mathrm{Ker}(B)^\perp$. B being surjective, $\mathrm{Im}(B)$ is closed and owing to Banach's Theorem, $\mathrm{Im}(B^T) = \mathrm{Ker}(B)^\perp$. As a result, $Au \in \mathrm{Im}(B^T)$, i.e., there is $p \in M'$ such that $Au = -B^T p$. Hence, $Au + B^T p = 0$ and $Bu = 0$, which shows that (u,p) is the solution to problem (A.19) with $f = 0$ and $g = 0$. The solution to (A.19) being unique, we infer $u = 0$.

(2) Conversely, assume that statements (i) and (ii) are satisfied.

2.a For $f \in X'$ and $g \in M$, let us show that there is at least one solution to problem (A.19). B being surjective, there is u_g in X such that $Bu_g = g$. The linear form $f - Au_g$ is continuous on X and $\mathrm{Ker}(B)$. Denote by $h_{f,g}$ the linear form on $\mathrm{Ker}(B)$ such that $\langle h_{f,g}, v \rangle = \langle f, v \rangle - \langle Au_g, v \rangle$ for all v in $\mathrm{Ker}(B)$. Let $\phi \in \mathrm{Ker}(B)$ be the solution to the problem $\pi A\phi = h_{f,g}$ and set $u = \phi + u_g$. The linear form $f - Au$ is clearly in $\mathrm{Ker}(B)^\perp$. Since B is surjective, $\mathrm{Ker}(B)^\perp = \mathrm{Im}(B^T)$; that is, there is $p \in M'$ such that $B^T p = f - Au$. Moreover, $Bu = B(\phi + u_g) = Bu_g = g$. Hence, we have constructed a solution to problem (A.19).

2.b Let us show that the solution is unique. Let (u,p) be such that $Bu = 0$ and $Au + B^T p = 0$. Clearly, $u \in \mathrm{Ker}(B)$ and $\pi Au = 0$. Since πA is injective, $u = 0$. As a result $B^T p = 0$. Since B is surjective, B^T is necessarily injective, which implies $p = 0$.

$\square$

Remark A.57. Problem (A.19) is studied in detail in §2.4.1 and §4.1. This problem can be interpreted as a saddle-point problem when the operator A is self-adjoint and monotone; see §2.4.1. $\square$

B

Functional Analysis

This appendix collects the functional analysis concepts used in this book. We first restate fundamental results on Lebesgue integration together with the main properties of the $L^p(\Omega)$ spaces. Then, we introduce the concept of distributions and distributional derivatives. We review the main properties of the Sobolev spaces $W^{s,p}(\Omega)$. Most of the results are stated without proof, the reader being referred to specialized textbooks for complements [Ada75, Bar01, Bre91, MaZ97, Rud87, Sob63, Yos80].

Henceforth, Ω is a (measurable) open set of $\mathbb{R}^d$ with boundary $\partial\Omega$. Whenever it is well-defined, its outward normal is denoted by n.

B.1 Lebesgue and Lipschitz Spaces

B.1.1 Measurable functions and the Lebesgue integral

Let $\mathrm{M}(\Omega)$ be the space of scalar-valued functions on Ω that are *Lebesgue-measurable*. In particular, $\mathrm{M}(\Omega)$ contains functions that are piecewise continuous and, more generally, all the functions that are integrable in the Riemann sense. All the functions used in this book are measurable.

Measurable functions are defined up to sets of zero measure. In other words, $\mathrm{M}(\Omega)$ is a space of equivalence classes of functions, that is, two functions belong to the same equivalence class if they coincide *almost everywhere* (henceforth, a.e.), i.e., everywhere but on a set of zero Lebesgue measure.

Example B.1. The function $\phi :]0,1[\rightarrow \{0,1\}$ which takes the value 1 on rational numbers and which is zero otherwise belongs to the same equivalence class as the zero function. Hence, $\phi = 0$ a.e. on $]0,1[$. □

B.1.2 Lebesgue spaces

Definition B.2. $L^1(\Omega)$ *is the space of the scalar-valued functions that are Lebesgue-integrable. The space of* locally *integrable functions is denoted by*

$L^1_{\mathrm{loc}}(\Omega)$ and is defined as

$$L^1_{\mathrm{loc}}(\Omega) = \{f \in \mathbb{M}(\Omega); \forall \mathrm{compact}\ K \subset \Omega,\ f \in L^1(K)\}. \qquad (\mathrm{B.1})$$

Throughout this book, integrals are always understood in the Lebesgue sense. Whenever no confusion may arise and to alleviate the notation, we omit the Lebesgue measure under the integral sign; hence, we write $\int_\Omega f$ instead of $\int_\Omega f(x)\,\mathrm{d}x$.

Theorem B.3 (Lebesgue's Dominated Convergence). *Let* $\{f_n\}_{n\geq 0}$ *be a sequence of functions in* $L^1(\Omega)$ *such that:*

(i) $f_n(x) \to f(x)$ *a.e. in* Ω.
(ii) *There is* $g \in L^1(\Omega)$ *such that, for all* n, $|f_n(x)| \leq g(x)$ *a.e. in* Ω.

Then, $f \in L^1(\Omega)$ *and* $f_n \to f$ *in* $L^1(\Omega)$.

Proof. See [Bar01, p. 123], [Bre91, p. 54], or [Rud66]. □

Definition B.4. *For* $1 \leq p \leq +\infty$, *let*

$$L^p(\Omega) = \{f \in \mathbb{M}(\Omega);\ \|f\|_{0,p,\Omega} < +\infty\}, \qquad (\mathrm{B.2})$$

where

$$\|f\|_{0,p,\Omega} = \left(\int_\Omega |f|^p\right)^{\frac{1}{p}}, \quad \text{for } 1 \leq p < +\infty, \qquad (\mathrm{B.3})$$

$$\|f\|_{0,\infty,\Omega} = \underset{x \in \Omega}{\mathrm{ess\ sup}}\,|f(x)| = \inf\{M \geq 0;\ |f(x)| \leq M \text{ a.e. on } \Omega\}. \qquad (\mathrm{B.4})$$

Theorem B.5 (Fischer–Riesz). *Let* $1 \leq p \leq +\infty$. *Equipped with the* $\|\cdot\|_{0,p,\Omega}$ *norm, the vector space* $L^p(\Omega)$ *is a Banach space.*

Proof. See [Bar01, p. 142], [Sob63, §I.2.2], [Bre91, p. 57], or [Rud66]. □

In this book, we also employ the notation $\|f\|_{L^p(\Omega)} = \|f\|_{0,p,\Omega}$. For $1 \leq p \leq +\infty$, we denote by p' its *conjugate*, i.e., $\frac{1}{p} + \frac{1}{p'} = 1$ with the convention that $p' = 1$ if $p = +\infty$ and $p' = +\infty$ if $p = 1$.

Theorem B.6 (Hölder). *Let* $f \in L^p(\Omega)$ *and* $g \in L^{p'}(\Omega)$ *with* $1 \leq p \leq +\infty$ *and* $\frac{1}{p} + \frac{1}{p'} = 1$. *Then,* $fg \in L^1(\Omega)$ *and*

$$\int_\Omega |fg| \leq \|f\|_{L^p(\Omega)} \|g\|_{L^{p'}(\Omega)}. \qquad (\mathrm{B.5})$$

Proof. See [Bar01, p. 404], [Bre91, p. 56], or [Rud66]. □

The following corollary is an easy consequence of Hölder's inequality:

Corollary B.7 (Interpolation inequality). *Let* $1 \leq p \leq q \leq +\infty$ *and* $0 \leq \alpha \leq 1$. *Let* r *be such that* $\frac{1}{r} = \frac{\alpha}{p} + \frac{1-\alpha}{q}$. *Then,*

$$\forall f \in L^p(\Omega) \cap L^q(\Omega), \quad \|f\|_{L^r(\Omega)} \leq \|f\|_{L^p(\Omega)}^\alpha \|f\|_{L^q(\Omega)}^{1-\alpha}. \qquad (\mathrm{B.6})$$

Theorem B.8 (Riesz' Representation Theorem). *Let $1 \leq p < +\infty$. The dual space of $L^p(\Omega)$ can be identified with $L^{p'}(\Omega)$.*

Proof. See [Sob63, §I.3.3], [Bre91, p. 61], or [Rud66]. □

One important consequence of Theorem B.8 is that $L^p(\Omega)$ is reflexive if $1 < p < +\infty$. However, $L^1(\Omega)$ and $L^\infty(\Omega)$ are not reflexive. The dual of $L^1(\Omega)$ is $L^\infty(\Omega)$, but the dual of $L^\infty(\Omega)$ strictly contains $L^1(\Omega)$; see [Rud66] or [Bre91, p. 65]. Among all the Lebesgue spaces, $L^2(\Omega)$ plays a particular role owing to the following important consequence of the Fischer–Riesz Theorem:

Theorem B.9. $L^2(\Omega)$ *is a Hilbert space when equipped with the scalar product*

$$(f,g)_{L^2(\Omega)} = \int_\Omega fg.$$

Henceforth, we denote by $(\cdot\,,\cdot)_{0,\Omega}$ the scalar product in $L^2(\Omega)$, and when no confusion is possible, we denote the corresponding norm by

$$\|f\|_{0,\Omega} = \left(\int_\Omega f^2 \right)^{\frac{1}{2}}.$$

In $L^2(\Omega)$, the Hölder inequality becomes the *Cauchy–Schwarz inequality:*

$$\forall f, g \in L^2(\Omega), \quad (f,g)_{0,\Omega} \leq \|f\|_{0,\Omega} \|g\|_{0,\Omega}.$$

B.1.3 Spaces of continuous functions

Definition B.10 (Hölder spaces).

(i) *Let E be a subset of $\overline{\Omega}$. $C^0(E)$ denotes the space of functions that are continuous on E, and for every integer $k \geq 1$, $C^k(E)$ is the space of functions that are k-times continuously Fréchet-differentiable on E.*

(ii) *For $0 < \alpha \leq 1$, $C^{0,\alpha}(E)$ is the space of functions that are Hölder of exponent α on E, i.e., $f \in C^{0,\alpha}(E)$ if f is continuous on E and*

$$\sup_{x,y \in E} \frac{|f(x) - f(y)|}{|x - y|^\alpha} < +\infty.$$

For every integer $k \geq 1$, $C^{k,\alpha}(E)$ is the space of the functions f in $C^k(E)$ such that $\partial^\gamma f \in C^{0,\alpha}(E)$ for all multi-index γ of length $|\gamma| = k$.

(iii) *For $E = \overline{\Omega}$ (resp., $E = \Omega$), $C^{0,1}(E)$ is called the space of globally (resp., locally) Lipschitz functions.*

Proposition B.11. *Let $k \geq 0$ be an integer and $0 < \alpha \leq 1$. If Ω is bounded, $C^{k,\alpha}(\overline{\Omega})$ is a Banach space when equipped with the norm*

$$\|f\|_{C^{k,\alpha}(\overline{\Omega})} = \max_{|\gamma| \leq k} \sup_{x \in \overline{\Omega}} |\partial^\gamma f(x)| + \max_{|\gamma| = k} \sup_{x,y \in \overline{\Omega}} \frac{|\partial^\gamma f(x) - \partial^\gamma f(y)|}{|x - y|^\alpha}.$$

Remark B.12. The above definitions extend to functions that are defined only at the boundary $\partial\Omega$. In particular, we denote by $C^{0,1}(\partial\Omega)$ the space of Lipschitzian functions on $\partial\Omega$. □

B.2 Distributions

B.2.1 Preliminary definitions

Definition B.13. $\mathcal{D}(\Omega)$ *is the vector space of C^∞ functions whose support in Ω is compact.*

Theorem B.14. *Let $1 \le p < +\infty$. Then, $\mathcal{D}(\Omega)$ is dense in $L^p(\Omega)$.*

Proof. See [MaZ97, p. 5], [Bre91, p. 61], or [Rud66]. □

Lemma B.15. *Let $f \in L^1_{\mathrm{loc}}(\Omega)$ be such that $\int_\Omega f\varphi = 0$, $\forall \varphi \in \mathcal{D}(\Omega)$. Then, $f = 0$ a.e. in Ω.*

Proof. See [MaZ97, p. 6], [Bre91, p. 61], or [Rud66]. □

Definition B.16 (Distributions). *A linear mapping*

$$u : \mathcal{D}(\Omega) \ni \varphi \longmapsto \langle u, \varphi \rangle_{\mathcal{D}',\mathcal{D}} \in \mathbb{R} \ or \ \mathbb{C},$$

is said to be a distribution *on Ω if and only if the following property holds: For all compact K in Ω, there is an integer p and a constant c such that*

$$\forall \varphi \in \mathcal{D}(\Omega),\ \mathrm{supp}(\varphi) \subset K, \quad |\langle u, \varphi \rangle_{\mathcal{D}',\mathcal{D}}| \le c \sup_{x \in K, |\alpha| \le p} |\partial^\alpha \varphi(x)|.$$

Remark B.17. $\mathcal{D}(\Omega)$ is not a normed vector space. As a result, $\mathcal{D}'(\Omega)$ is not the dual of $\mathcal{D}(\Omega)$ in the sense of Definition A.14. We shall nevertheless use the notation $\mathcal{D}'(\Omega)$. □

Example B.18.
  (i) Every function f in $L^1_{\mathrm{loc}}(\Omega)$ can be identified with the distribution

$$\tilde{f} : \mathcal{D}(\Omega) \ni \varphi \longmapsto \langle \tilde{f}, \varphi \rangle_{\mathcal{D}',\mathcal{D}} = \int_\Omega f\varphi.$$

This identification is possible owing to Lemma B.15 since, for $f, g \in L^1_{\mathrm{loc}}(\Omega)$,

$$(f = g \ \text{a.e. in } \Omega) \iff \left(\forall \varphi \in \mathcal{D}(\Omega), \int_\Omega f\varphi = \int_\Omega g\varphi \right).$$

We will constantly abuse the notation by identifying $f \in L^1_{\mathrm{loc}}(\Omega)$ with the associated distribution $\tilde{f} \in \mathcal{D}'(\Omega)$.
  (ii) Every function of $C^{k,\alpha}(\Omega)$, $k \ge 0$, can be identified with a distribution.
  (iii) Let a be a point in Ω. The *Dirac measure* at a is the distribution

$$\delta_{x=a} : \mathcal{D}(\Omega) \ni \varphi \longmapsto \langle \delta_{x=a}, \varphi \rangle_{\mathcal{D}',\mathcal{D}} = \varphi(a).$$

Note that $\delta_{x=a} \notin L^1(\Omega)$, i.e., there is no function $f \in L^1(\Omega)$ such that $\varphi(a) = \int_\Omega f\varphi$, $\forall \varphi \in \mathcal{D}(\Omega)$. Otherwise, one would have $0 = \int_\Omega f\varphi$ for all functions $\varphi \in \mathcal{D}(\Omega \backslash \{a\})$, which, owing to Lemma B.15, would imply $f = 0$ a.e. in $\Omega \backslash \{a\}$, i.e., $f = 0$ a.e. in Ω. □

B.2.2 The distributional derivative

The key to the distribution theory is that every distribution is differentiable in the following sense:

Definition B.19. *Let $u \in \mathcal{D}'(\Omega)$ be a distribution and let $1 \leq i \leq d$. The distributional derivative $\partial_i u \in \mathcal{D}'(\Omega)$ is defined as follows:*

$$\partial_i u : \mathcal{D}(\Omega) \ni \varphi \longmapsto \langle \partial_i u, \varphi \rangle_{\mathcal{D}',\mathcal{D}} = -\langle u, \partial_i \varphi \rangle_{\mathcal{D}',\mathcal{D}}.$$

More generally, for a multi-index α, the distribution $\partial^\alpha u$ is defined such that

$$\partial^\alpha u : \mathcal{D}(\Omega) \ni \varphi \longmapsto \langle \partial^\alpha u, \varphi \rangle_{\mathcal{D}',\mathcal{D}} = (-1)^{|\alpha|} \langle u, \partial^\alpha \varphi \rangle_{\mathcal{D}',\mathcal{D}}.$$

Hereafter, we set conventionally $\partial^0 u = u$ and $\nabla u = (\partial_1 u, \ldots, \partial_d u)$.

Remark B.20. The notion of distributional derivative is the extension of that of pointwise (or classical) derivative. This notion allows for the differentiation of functions that are not derivable in the classical sense. When $u \in L^1_{loc}(\Omega)$, distributional derivatives of u are sometimes called *weak derivatives*. □

Proposition B.21. *Let α be a multi-index and let $u \in C^{|\alpha|}(\Omega)$. Then, up to the order $|\alpha|$, the weak derivatives and the classical derivatives of u coincide.*

Proof. Let $(\partial^\alpha u)_{cl}$ denote the pointwise derivative; then,

$$\forall \varphi \in \mathcal{D}(\Omega), \quad \langle (\partial^\alpha u)_{cl}, \varphi \rangle_{\mathcal{D}',\mathcal{D}} = \int_\Omega (\partial^\alpha u)_{cl} \, \varphi \quad (\text{since } (\partial^\alpha u)_{cl} \in L^1_{loc}(\Omega))$$

$$= (-1)^{|\alpha|} \int_\Omega u \, \partial^\alpha \varphi \quad (\text{integration by parts})$$

$$= (-1)^{|\alpha|} \langle u, \partial^\alpha \varphi \rangle_{\mathcal{D}',\mathcal{D}} \quad (\text{since } u \in L^1_{loc}(\Omega)).$$

Note that in the integration by parts there are no boundary terms since $\partial^\beta \varphi_{|\partial\Omega} = 0$ for all multi-index β. □

Theorem B.22 (Rademacher). *Let $u \in C^{0,1}(\Omega)$ be a locally Lipschitz function on Ω. Then, u is differentiable a.e. in Ω and the pointwise derivative coincides a.e. with the weak derivative.*

Proof. See [MaZ97, p. 44]. □

Example B.23.
  (i) Take $\Omega = \,]-1, 1[$ and $u = 1 - |x|$. The weak derivative of u is

$$\partial_x u = \begin{cases} 1 & \text{if } x < 0, \\ -1 & \text{if } x > 0. \end{cases}$$

This example is fundamental. Its generalization to higher space dimension is used repeatedly in the book.

(ii) The *Heavyside function* (or step function) on $\Omega = \,]-1, 1[$ is defined by

$$H(x) = \begin{cases} 0 & \text{if } x < 0, \\ 1 & \text{if } x \geq 0. \end{cases}$$

It is clear that H is not derivable in the classical sense. However, a simple computation shows that $\partial_x H = \delta_{x=0}$. $\square$

Lemma B.24. *Let $\Omega = \,]0, 1[$. For all f in $L^1(\Omega)$,*

$$\partial_x \left(\int_0^x f \right) = f \qquad \text{in } \mathcal{D}'(\Omega). \tag{B.7}$$

Proof. We use a density argument. Since $\mathcal{D}(\Omega)$ is dense in $L^1(\Omega)$, there is a sequence $\{f_n\}_{n \geq 0}$ in $\mathcal{D}(\Omega)$ that converges to f in $L^1(\Omega)$ and such that $\|f_n\|_{1,\Omega} \leq 2\|f\|_{1,\Omega}$. For all ϕ in $\mathcal{D}(\Omega)$, it is clear that $|\int_0^1 f_n \phi - \int_0^1 f\phi| \leq (\sup_{x \in \Omega} |\phi(x)|) \int_0^1 |f_n - f| \to 0$. Likewise, $(\int_0^x f_n)\phi'(x) \to (\int_0^x f)\phi'(x)$ a.e. in Ω, and $|(\int_0^x f_n)\phi'(x)| \leq 2|\phi'(x)| \int_0^1 |f|$. Lebesgue's Dominated Convergence Theorem implies that $\int_0^1 (\int_0^x f_n)\phi'(x)\,\mathrm{d}x \to \int_0^1 (\int_0^x f)\phi'(x)\,\mathrm{d}x$. Passing to the limit in the relation

$$\int_0^1 \left(\int_0^x f_n \right) \phi'(x)\,\mathrm{d}x = -\int_0^1 f_n(x)\phi(x)\,\mathrm{d}x,$$

yields

$$\forall \phi \in \mathcal{D}(\Omega), \qquad \int_0^1 \left(\int_0^x f \right) \phi'(x)\,\mathrm{d}x = -\int_0^1 f(x)\phi(x)\,\mathrm{d}x.$$

This shows that (B.7) holds in the distribution sense. $\square$

B.3 Sobolev Spaces

B.3.1 The $W^{s,p}(\Omega)$ spaces

Definition B.25 (Sobolev spaces). *Let s and p be two integers with $s \geq 0$ and $1 \leq p \leq +\infty$. The so-called Sobolev space $W^{s,p}(\Omega)$ is defined as*

$$W^{s,p}(\Omega) = \{u \in \mathcal{D}'(\Omega); \partial^\alpha u \in L^p(\Omega), |\alpha| \leq s\}, \tag{B.8}$$

where the derivatives are understood in the distribution sense.

Proposition B.26. $W^{s,p}(\Omega)$ *is a Banach space when equipped with the norm*

$$\|u\|_{W^{s,p}(\Omega)} = \sum_{|\alpha| \le s} \|\partial^\alpha u\|_{L^p(\Omega)}. \tag{B.9}$$

The case $p = 2$ is particularly interesting since the spaces $W^{s,2}(\Omega)$ have a Hilbert structure, and we henceforth denote them by $H^s(\Omega)$.

Theorem B.27 (Hilbert Sobolev spaces). *Let* $s \ge 0$. *The space* $H^s(\Omega) = W^{s,2}(\Omega)$ *is a Hilbert space when equipped with the scalar product*

$$(u, v)_{s,\Omega} = \sum_{|\alpha| \le s} \int_\Omega \partial^\alpha u \, \partial^\alpha v.$$

The associated norm is denoted by $\| \cdot \|_{s,\Omega}$.

Sometimes we shall also make use of the following notation:

$$\|v\|_{s,p,\Omega} = \|v\|_{W^{s,p}(\Omega)}, \qquad |v|^2_{s,p,\Omega} = \sum_{|\alpha|=s} \|D^\alpha v\|^2_{L^p(\Omega)}, \tag{B.10}$$

$$\|v\|_{s,\Omega} = \|v\|_{H^s(\Omega)}, \qquad |v|^2_{s,\Omega} = \sum_{|\alpha|=s} \|D^\alpha v\|^2_{L^2(\Omega)}. \tag{B.11}$$

Example B.28.

(i) $H^1(\Omega) = \{u \in L^2(\Omega); \partial_i u \in L^2(\Omega), 1 \le i \le d\}$ is a Hilbert space for the scalar product

$$(u, v)_{1,\Omega} = \int_\Omega uv + \int_\Omega \nabla u \cdot \nabla v = \int_\Omega uv + \sum_{i=1}^d \int_\Omega \partial_i u \, \partial_i v.$$

The associated norm is

$$\|u\|^2_{1,\Omega} = \int_\Omega u^2 + \int_\Omega (\nabla u)^2 = \|u\|^2_{0,\Omega} + |u|^2_{1,\Omega}.$$

(ii) Let $\Omega = \,]0, 1[$ and consider the function $u(x) = x^\alpha$ where $\alpha \in \mathbb{R}$. One easily verifies that $u \in L^2(\Omega)$ if $\alpha > -\frac{1}{2}$, $u \in H^1(\Omega)$ if $\alpha > \frac{1}{2}$, and, more generally $u \in H^s(\Omega)$ if $\alpha > s - \frac{1}{2}$.

(iii) The function shown in the left panel of Figure B.1 belongs to $L^2(\Omega)$ but not to $\mathcal{C}^0(\overline{\Omega})$. This function does not belong to $H^1(\Omega)$ either, since its distributional first derivative is the sum of two Dirac measures supported at the two discontinuity points. The function shown in the right panel of Figure B.1 belongs to $H^1(\Omega)$. However, it is neither in $\mathcal{C}^1(\overline{\Omega})$ nor in $H^2(\Omega)$ since its first derivative is discontinuous.

(iv) Let $\Omega = B(0, \frac{1}{2}) \subset \mathbb{R}^2$ be the ball centered at 0 and of radius $\frac{1}{2}$. A simple computation shows that the function

$$u(x_1, x_2) = \log\big(-\log(x_1^2 + x_2^2)\big)$$

is in $H^1(\Omega)$. This example shows that in two dimensions, functions in $H^1(\Omega)$ are neither necessarily continuous nor bounded. □

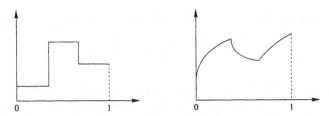

Fig. B.1. The function on the left is not $H^1(]0, 1[)$, that on the right is.

The following lemma characterizes the nullspace of ∇; see [MaZ97, p. 24].

Lemma B.29. *Assume that Ω is an open connected set. Let $1 \leq p \leq +\infty$. Let u in $W^{1,p}(\Omega)$ such that $\nabla u = 0$ a.e. on Ω; then, u is constant.*

Definition B.30 (Fractional Sobolev spaces). *For $0 < s < 1$ and $1 \leq p < +\infty$, the so-called Sobolev space with fractional exponent is defined as*

$$W^{s,p}(\Omega) = \left\{ u \in L^p(\Omega); \frac{u(x) - u(y)}{\|x - y\|^{s + \frac{d}{p}}} \in L^p(\Omega \times \Omega) \right\}. \qquad (B.12)$$

Furthermore, when $s > 1$ is not integer, letting $\sigma = s - [s]$ where $[s]$ is the integer part of s, $W^{s,p}(\Omega)$ is defined as

$$W^{s,p}(\Omega) = \{ u \in W^{[s],p}(\Omega); \partial^\alpha u \in W^{\sigma,p}(\Omega), \forall \alpha, |\alpha| = [s] \}. \qquad (B.13)$$

When $p = 2$, we denote $H^s(\Omega) = W^{s,2}(\Omega)$.

Remark B.31.

(i) The Sobolev spaces with fractional exponent can also be introduced by interpolation between $L^p(\Omega)$ and $W^{1,p}(\Omega)$ or by Fourier transform if $p = 2$ and Ω is either $\mathbb{R}^d$ or the d-torus; see [Ada75, LiM68].

(ii) Using mappings, it is possible to define $W^{s,p}(\partial\Omega)$ whenever $\partial\Omega$ is a smooth manifold.   □

Example B.32.

(i) It can be easily verified that the function shown in the left panel of Figure B.1 is in $W^{s,p}(]0, 1[)$ for all $0 < s$ and $1 \leq p < +\infty$ such that $sp < 1$.

(ii) One can also verify that if Ω is bounded and $1 \leq p < +\infty$, $C^{0,\alpha}(\overline{\Omega}) \subset W^{s,p}(\Omega)$ provided $0 \leq s < \alpha \leq 1$.   □

B.3.2 Density

As an alternative to Definition B.25, one can also define Sobolev spaces as follows: Let $X^{s,p}(\Omega)$ be the closure of $\mathcal{C}^s(\Omega) \cap W^{s,p}(\Omega)$ with respect to the Sobolev norm $\| \cdot \|_{W^{s,p}(\Omega)}$. Observe that $X^{s,\infty}(\Omega) \neq W^{s,\infty}(\Omega)$ since $X^{s,\infty}(\Omega) = \mathcal{C}_b^s(\Omega)$, where $\mathcal{C}_b^s(\Omega)$ is the Banach space composed of the functions whose derivatives up to order s are continuous and bounded on Ω. However, for $1 \leq p < +\infty$, $X^{s,p}(\Omega) = W^{s,p}(\Omega)$. More precisely, the following result is proved in [MeS64]:

Theorem B.33 (Meyers–Serrin). *Let Ω be any open set. Then $C^\infty(\Omega) \cap W^{s,p}(\Omega)$ is dense in $W^{s,p}(\Omega)$ for $p < +\infty$.*

Theorem B.34 (Friedrichs). *Let Ω be any open set and let $u \in W^{1,p}(\Omega)$ with $1 \le p < +\infty$. Then, there exists a sequence $\{u_n\}_{n\ge0}$ in $\mathcal{D}(\mathbb{R}^d)$ such that*

(i) $u_n \to u$ in $L^p(\Omega)$.
(ii) $\nabla u_{n|\omega} \to \nabla u_{|\omega}$ in $[L^p(\omega)]^d$ for all ω such that $\bar\omega \subset \Omega$ and $\bar\omega$ compact.

Proof. See [Bre91, p. 151]. □

Remark B.35. In general, $C^\infty(\overline\Omega)$ is not dense in $W^{s,p}(\Omega)$. For this stronger density result to hold, some additional regularity hypotheses on Ω must be assumed. In particular, this result cannot hold whenever Ω lies on both sides of parts of its boundary. For instance, think of the function $u(x_1, x_2) = r\cos(\frac{1}{2}\theta)$ in $\Omega = \{0 \le r < 1,\ 0 < \theta < 2\pi\}$, where (r, θ) are the cylindrical coordinates. Let B be the closed ball of radius 1 centered at 0. Note that $B = \overline\Omega$. It is clear that u is in $W^{1,p}(\Omega)$, but u is not in $W^{1,p}(B)$. Indeed, $\partial_\theta u = -\frac{1}{2}r\sin(\frac{1}{2}\theta) + 2r\delta_{\theta=0}$ where $\delta_{\theta=0}$ is the Dirac measure supported by the segment $\{0 < x_1 < 1,\ x_2 = 0\}$, and $\delta_{\theta=0}$ cannot be identified with any function in $L^p(\Omega)$; see Example B.18(iii). Since $C^1(\overline\Omega) \subset W^{1,p}(B)$ and $W^{1,p}(B)$ is complete, the closure of $C^1(\overline\Omega)$ in $W^{1,p}(B)$ is a subspace of $W^{1,p}(B)$. Since $u \notin W^{1,p}(B)$, u cannot be a limit of functions in $C^1(\overline\Omega)$. □

To extend the previous density results, we introduce the following:

Definition B.36 ((s,p)-extension property). *Let $1 \le p \le +\infty$ and let $s \ge 0$. Ω is said to have the (s,p)-extension property if there is a bounded linear operator $L : W^{s,p}(\Omega) \to W^{s,p}(\mathbb{R}^d)$ such that $Lu_{|\Omega} = u$ for all $u \in W^{s,p}(\Omega)$. For $u \in W^{s,p}(\Omega)$, Lu is called (s,p)-extension of u.*

A fundamental result of Calderón–Stein [Cal61, Ste70] is the following:

Theorem B.37. *Every open set with a Lipschitz boundary has the $(1,p)$-extension property.*

Corollary B.38. *Let Ω be a bounded open set having the $(1,p)$-extension property. The restriction to Ω of functions in $\mathcal{D}(\mathbb{R}^d)$ span a dense subspace of $W^{1,p}(\Omega)$.*

Proof. Apply Theorem B.34 to a $(1,p)$-extension of $u \in W^{1,p}(\Omega)$. □

B.3.3 Embedding and compacity

Proposition B.39. *Let Ω be an open bounded set. Then, for $1 \le p < q \le +\infty$, the embedding $L^q(\Omega) \subset L^p(\Omega)$ is continuous.*

Proof. Easy consequence of Hölder's inequality. □

One of the key arguments in the embedding theory is the following:

Theorem B.40 (Sobolev). *Let $1 \leq p < d$ and denote by p^* the number such that $\frac{1}{p^*} = \frac{1}{p} - \frac{1}{d}$. Then,*

$$\exists c = \frac{p^*}{1^*}, \ \forall u \in W^{1,p}(\mathbb{R}^d), \quad \|u\|_{L^{p^*}(\mathbb{R}^d)} \leq c \|\nabla u\|_{L^p(\mathbb{R}^d)}. \qquad (B.14)$$

Proof. See [MaZ97, p. 32], [Sob63, §I.7.4], or [Bre91, p. 162]. $\qquad\square$

Corollary B.41. *Let $1 \leq p, q \leq +\infty$. The following embeddings are continuous:*

$$W^{1,p}(\mathbb{R}^d) \subset L^q(\mathbb{R}^d) \ \text{if} \ \begin{cases} \text{either } 1 \leq p < d \text{ and } p \leq q \leq p^*, \\ \text{or } p = d \text{ and } p \leq q < +\infty. \end{cases} \qquad (B.15)$$

Proof. See [MaZ97, p. 34], [Sob63, §I.8.2], or [Bre91, p. 165]. $\qquad\square$

Theorem B.42 (Morrey). *Let $d < p \leq +\infty$ and $\alpha = 1 - \frac{d}{p}$. The following embedding is continuous:*

$$W^{1,p}(\mathbb{R}^d) \subset L^\infty(\mathbb{R}^d) \cap C^{0,\alpha}(\mathbb{R}^d). \qquad (B.16)$$

Proof. See [MaZ97, p. 37] or [Bre91, p. 166]. $\qquad\square$

Corollary B.43. *Let $1 \leq p, q \leq +\infty$. Let $s \geq 1$ be an integer. Let Ω be a bounded open set having the $(1, p)$-extension property. The following embeddings are continuous:*

$$W^{s,p}(\Omega) \subset \begin{cases} L^q(\Omega) & \text{if } 1 \leq p < \frac{d}{s} \text{ and } p \leq q \leq p^*, \\ L^q(\Omega) & \text{if } p = \frac{d}{s} \text{ and } p \leq q < +\infty, \\ L^\infty(\Omega) \cap C^{0,\alpha}(\overline{\Omega}) & \text{if } p > \frac{d}{s} \text{ and } \alpha = 1 - \frac{d}{sp}. \end{cases} \qquad (B.17)$$

Remark B.44. This theorem implies that in one dimension, the functions of $H^1(\Omega)$ are continuous, whereas in dimension 2 or 3, this may not be the case; see Example B.28(iv). If $d = 2$ or 3, functions in $H^2(\Omega)$ are continuous. $\qquad\square$

Example B.45. Let $1 < \alpha$, $1 \leq p < 2$, and $\Omega = \{(x_1, x_2) \in \mathbb{R}^2; 0 < x_1 < 1, 0 < x_2 < x_1^\alpha\}$. Let $u(x_1, x_2) = x_1^\beta$ with $1 - \frac{1+\alpha}{p} < \beta < 0$. Then, $u \in W^{1,p}(\Omega)$ and $u \in L^q(\Omega)$ for all q such that $1 \leq q < p_\alpha$ where $\frac{1}{p_\alpha} = \frac{1}{p} - \frac{1}{1+\alpha}$. Set $\frac{1}{p^*} = \frac{1}{p} - \frac{1}{2}$ and $\epsilon = \frac{\beta - 1}{1 + \alpha} + \frac{1}{p} > 0$; one can choose β so that ϵ is arbitrarily small. Set $\frac{1}{p_\beta} = \frac{1}{p_\alpha} - \epsilon$. If ϵ is small enough, $p_\alpha < p_\beta < p^*$. Then $u \notin L^q(\Omega)$ for $p_\beta \leq q \leq p^*$, which would contradict Corollary B.43 if the $(1, p)$-extension hypothesis had been omitted. The hypothesis $1 < \alpha$ means that Ω has a cusp at the origin; hence, Ω is not Lipschitz. This counterexample shows that some regularity on Ω is needed for Corollary B.43 to hold. $\qquad\square$

We conclude this section by stating a very useful compacity result.

Theorem B.46 (Rellich–Kondrachov). *Let $1 \le p \le +\infty$ and let $s \ge 0$. Let Ω be a bounded open set having the (s,p)-extension property. The following injections are compact:*

(i) *If $sp \le d$, $W^{s,p}(\Omega) \subset L^q(\Omega)$ for all $1 \le q < p^*$ where $\frac{1}{p^*} = \frac{1}{p} - \frac{s}{d}$.*

(ii) *If $sp > d$, $W^{s,p}(\Omega) \subset C^0(\overline{\Omega})$.*

Proof. See [MaZ97], [BrS94, Chap. 1], or [Bre91, Chap. 8]. $\qquad\square$

B.3.4 $W_0^{s,p}(\Omega)$ and its dual

Definition B.47. *For $1 \le p < +\infty$ and $s \ge 0$, set*

$$W_0^{s,p}(\Omega) = \overline{\mathcal{D}(\Omega)}^{W^{s,p}(\Omega)}, \qquad (B.18)$$

and let $W^{-s,p'}(\Omega) = \left(W_0^{s,p}(\Omega)\right)'$ be the dual of $W_0^{s,p}(\Omega)$ with the norm

$$\forall f \in W^{-s,p'}(\Omega), \qquad \|f\|_{W^{-s,p'}(\Omega)} = \sup_{u \in W_0^{s,p}(\Omega)} \frac{\langle f, u \rangle_{W^{-s,p'},W_0^{s,p}}}{\|u\|_{W^{s,p}(\Omega)}}.$$

For $p = 2$, set $H_0^s(\Omega) = W_0^{s,2}(\Omega)$ and $H^{-s}(\Omega) = (W_0^{s,2}(\Omega))'$.

Proposition B.48. *Let $1 \le p < +\infty$ and let Ω be an open set. If u is in $W_0^{1,p}(\Omega)$, the zero-extension of u is in $W^{1,p}(\mathbb{R}^d)$.*

Corollary B.49. *The conclusions of Corollary B.43 and Theorem B.46 hold in $W_0^{1,p}(\Omega)$ without the $(1,p)$-extension hypothesis on Ω.*

The distributions in $W^{-1,p'}(\Omega)$ are characterized by the following:

Proposition B.50. *Let $1 \le p < +\infty$, let Ω be an open set, and let $F \in W^{-1,p'}(\Omega)$. Then, there are $f_0, f_1, \ldots, f_d \in L^{p'}(\Omega)$ such that*

$$\forall v \in W_0^{1,p}(\Omega), \quad \langle F, v \rangle_{W^{-1,p'},W^{1,p}} = \int_\Omega f_0 v + \sum_{k=1}^d \int_\Omega f_k \partial_{x_k} v,$$

and $\|F\|_{W^{-1,p'}(\Omega)} = \max_{0 \le k \le d} \|f_k\|_{L^{p'}(\Omega)}$. One can set $f_0 = 0$ if Ω is bounded.

Proof. See [Rud66] or [Bre91, p. 175]. $\qquad\square$

Remark B.51.

(i) The notation for the dual of $W^{s,p}(\Omega)$ has been chosen so that when $s = 0$, it is coherent with the Riesz Theorem.

(ii) It is remarkable that some of the objects contained in $W^{-s,p'}(\Omega)$ are not functions but distributions. For instance, if $sp > d$, the Dirac measure at $x \in \Omega$ is in $W^{-s,p'}(\Omega)$. For a function u in $L^p(\Omega)$, all the first-order distributional derivatives of u are in $W^{-1,p'}(\Omega)$. $\qquad\square$

B.3.5 The trace theory

In general, it is meaningless to speak of the value at $\partial\Omega$ of a function in $L^p(\Omega)$. For instance, take $\Omega =]0,1[^2$ and $u(x_1, x_2) = x_1^{-\frac{\alpha}{p}}$ with $0 < \alpha < 1$. It is clear that $u \in L^p(\Omega)$, but $u_{|x_1=0} = +\infty$. Let $\gamma_0 : \mathcal{C}^0(\overline{\Omega}) \to \mathcal{C}^0(\partial\Omega)$ map functions in $\mathcal{C}^0(\overline{\Omega})$ to their trace on $\partial\Omega$. If Ω is a Lipschitz bounded open set, γ_0 can be continuously extended to $W^{1,p}(\Omega)$; see, e.g., [Ada75]. Let us abuse the notation by still denoting the extension in question by γ_0.

Theorem B.52 (Trace Theorem 1). *Let $1 \le p < +\infty$ and Ω be a Lipschitz bounded open set. Then,*

(i) $\gamma_0 : W^{1,p}(\Omega) \to W^{\frac{1}{p'},p}(\partial\Omega)$ *is surjective.*
(ii) *The kernel of γ_0 is $W_0^{1,p}(\Omega)$.*

Statement (ii) in this theorem means that the functions in $W_0^{1,p}(\Omega)$ are those in $W^{1,p}(\Omega)$ that are zero at $\partial\Omega$.

When $p = 2$, we set $H^{\frac{1}{2}}(\partial\Omega) = W^{\frac{1}{2},2}(\partial\Omega)$. Statement (i) in Theorem B.52 implies that every function in $H^{\frac{1}{2}}(\partial\Omega)$ is the trace of a function in $H^1(\Omega)$. More precisely, from the Open Mapping Theorem we deduce the following:

Corollary B.53. *Let $1 \le p < +\infty$ and Ω be a Lipschitz bounded open set. Then, there exists a constant c such that, $\forall g \in W^{\frac{1}{p'},p}(\partial\Omega)$, there exists $u_g \in W^{1,p}(\Omega)$ satisfying*

$$\gamma_0(u_g) = g \qquad and \qquad \|u\|_{W^{1,p}(\Omega)} \le c\,\|g\|_{W^{\frac{1}{p'},p}(\partial\Omega)}.$$

The function u_g is said to be a lifting of g in $W^{1,p}(\Omega)$.

Likewise $\partial_n u = n\cdot\nabla u$ is meaningful for functions in $W^{2,p}(\Omega)$ since, for such functions, $\nabla u \in W^{\frac{1}{p'},p}(\partial\Omega)$. Actually, provided Ω is smooth enough, the mapping $\gamma_1 : \mathcal{C}^1(\overline{\Omega}) \to \mathcal{C}^0(\partial\Omega)$ such that $\gamma_1(u) = \partial_n u$ can be continuously extended from $W^{2,p}(\Omega)$ to $W^{\frac{1}{p'},p}(\Omega)$.

Theorem B.54 (Trace Theorem 2). *Let $1 \le p < +\infty$ and Ω be a bounded open set of class $\mathcal{C}^2$. Then:*

(i) $(\gamma_0, \gamma_1) : W^{2,p}(\Omega) \times W^{2,p}(\Omega) \to W^{1+\frac{1}{p'},p}(\partial\Omega) \times W^{\frac{1}{p'},p}(\partial\Omega)$ *is surjective.*
(ii) *The kernel of (γ_0, γ_1) is $W_0^{2,p}(\Omega)$.*

Item (ii) in this theorem says that the functions in $W_0^{2,p}(\Omega)$ are those in $W^{2,p}(\Omega)$ that are zero at $\partial\Omega$ and whose normal derivative on $\partial\Omega$ is zero.

Remark B.55. If Ω is a polyhedron, the outward normal vector n is discontinuous at the edges. As a result, the normal derivative $\partial_n u$ cannot be smooth whatever the regularity of u. It is nevertheless possible to extend the results of Theorem B.54 to this situation [GiR86, p. 9]. For example, when Ω is a polygon in $\mathbb{R}^2$ with sides $\partial\Omega_j$ and normal vectors n_j, $1 \le j \le J$, the mapping $u \mapsto (\partial_{n_1}u, \ldots, \partial_{n_J}u)$ is continuous and surjective from $W^{2,p}(\Omega)$ to $\prod_{j=1}^J W^{\frac{1}{p'},p}(\partial\Omega_j)$. $\qquad\square$

B.3.6 The divergence formula and its consequences

For brevity, we henceforth omit the symbols γ_0 and γ_1 to denote the trace of a distribution or that of its normal derivative. In this section, p is a number in $[1, +\infty[$ and Ω denotes a Lipschitz bounded open set.

Lemma B.56 (Divergence formula). *Let u be a smooth vector field. Then,*

$$\int_\Omega \nabla \cdot u = \int_{\partial\Omega} u \cdot n. \tag{B.19}$$

Corollary B.57. *Let $u \in [L^p(\Omega)]^d$ such that $\nabla \cdot u \in L^p(\Omega)$. Then, $u \cdot n \in W^{-\frac{1}{p},p}(\partial\Omega)$ and*

$$\forall q \in W^{1,p'}(\Omega), \qquad \int_\Omega q\nabla \cdot u = -\int_\Omega u \cdot \nabla q + \int_{\partial\Omega} q\, u \cdot n. \tag{B.20}$$

Proof. Use (B.19) for smooth functions, together with $\nabla \cdot (qu) = q\nabla \cdot u + u \cdot \nabla q$. Conclude using a density argument. □

Corollary B.58. *Let $u \in [L^p(\Omega)]^3$ such that $\nabla \times u \in [L^p(\Omega)]^3$. Then, $u \times n \in [W^{-\frac{1}{p},p}(\partial\Omega)]^3$ and*

$$\forall v \in W^{1,p'}(\Omega), \qquad \int_\Omega (\nabla \times u) \cdot v = \int_\Omega u \cdot (\nabla \times v) - \int_{\partial\Omega} (u \times n) \cdot v. \tag{B.21}$$

Proof. Use $\nabla \cdot (u \times v) = (\nabla \times u) \cdot v - u \cdot (\nabla \times v)$ together with (B.19) and a density argument. □

Corollary B.59 (Green's formula). *Let $\sigma \in [L^\infty(\Omega)]^{d \times d}$ and let $u \in W^{1,p}(\Omega)$ be such that $\nabla \cdot (\sigma \cdot \nabla u) \in L^p(\Omega)$. Then, $n \cdot \sigma \cdot \nabla u \in W^{-\frac{1}{p},p}(\partial\Omega)$ and*

$$\forall v \in W^{1,p'}(\Omega), \qquad -\int_\Omega \nabla \cdot (\sigma \cdot \nabla u)\, v = \int_\Omega \nabla v \cdot \sigma \cdot \nabla u - \int_{\partial\Omega} (n \cdot \sigma \cdot \nabla u)v. \tag{B.22}$$

Proof. Use $\nabla \cdot ((\sigma \cdot \nabla u)\, v) = \nabla v \cdot \sigma \cdot \nabla u + \nabla \cdot (\sigma \cdot \nabla u)\, v$ together with (B.19) and a density argument. □

Remark B.60. To be rigorous, the boundary integrals in (B.20), (B.21), and (B.22) should be written in the form of duality products. For instance, in (B.20), we should write $\langle u \cdot n, q \rangle_{W^{-\frac{1}{p},p}(\partial\Omega), W^{\frac{1}{p},p'}(\partial\Omega)}$. □

B.3.7 Poincaré-like inequalities

Lemma B.61 (Poincaré). *Let $1 \le p < +\infty$ and let Ω be a bounded open set. Then, there exists $c_{p,\Omega} > 0$ such that*

$$\forall v \in W_0^{1,p}(\Omega), \qquad c_{p,\Omega}\|v\|_{L^p(\Omega)} \le \|\nabla v\|_{L^p(\Omega)}. \tag{B.23}$$

For $p = 2$, we denote $c_\Omega = c_{2,\Omega}$.

Proof. We only give the proof for $p < d$. Let $\tilde{v} \in W^{1,p}(\mathbb{R}^d)$ be the zero-extension of v; see Proposition B.48. Theorem B.40 implies $\|\tilde{v}\|_{L^{p^*}(\mathbb{R}^d)} \leq c\|\nabla\tilde{v}\|_{L^p(\mathbb{R}^d)}$. Since Ω is bounded and $p^* \geq p$, we infer $\|v\|_{L^p(\Omega)} = \|\tilde{v}\|_{L^p(\mathbb{R}^d)} \leq c\|\tilde{v}\|_{L^{p^*}(\mathbb{R}^d)}$, yielding (B.23). $\qquad\square$

Remark B.62.

(i) The Poincaré inequality shows that on $W_0^{1,p}(\Omega)$, the seminorm $v \mapsto |v|_{1,p,\Omega}$ is equivalent to the usual norm $v \mapsto \|v\|_{1,p,\Omega}$. For instance, in the Hilbertian case ($p = 2$),

$$\forall v \in H_0^1(\Omega), \quad \frac{c_\Omega}{(1 + c_\Omega^2)^{\frac{1}{2}}}\|v\|_{1,\Omega} \leq |v|_{1,\Omega} \leq \|v\|_{1,\Omega}.$$

(ii) To give some further insight into (B.23), we give a proof in one dimension using $\Omega =]0,1[$. For $\varphi \in \mathcal{D}(\Omega)$, use $\varphi(0) = 0$ to write $\varphi(x) = \int_0^x \varphi'(y)\,dy$ for all $x \in \Omega$. Hence,

$$
\begin{aligned}
\|\varphi\|_{0,p,\Omega}^p = \int_0^1 |\varphi(x)|^p\,dx &= \int_0^1 \left|\int_0^x \varphi'(y)\,dy\right|^p\,dx \\
&\leq \int_0^1 \left(\int_0^x dy\right)^{\frac{p}{p'}}\left(\int_0^x |\varphi'(y)|^p\,dy\right)dx \\
&\leq \int_0^1 |\varphi'(y)|^p\,dy = |\varphi|_{1,p,\Omega}^p.
\end{aligned}
$$

Conclude using the density of $\mathcal{D}(\Omega)$ in $W_0^{1,p}(\Omega)$.

(iii) Let B_R be the ball with radius R. Using a scaling argument, one readily infers that $Rc_{p,B_R} = c_{p,B_1}$. $\qquad\square$

Lemma B.63. *Let $1 \leq p < +\infty$ and Ω be a bounded connected open set having the $(1,p)$-extension property. Let f be a linear form on $W^{1,p}(\Omega)$ whose restriction on constant functions is not zero. Then, there is $c_{p,\Omega} > 0$ such that*

$$\forall v \in W^{1,p}(\Omega), \quad c_{p,\Omega}\|v\|_{W^{1,p}(\Omega)} \leq \|\nabla v\|_{L^p(\Omega)} + |f(v)|. \tag{B.24}$$

Proof. Use the Petree–Tartar Lemma. To this end, set $X = W^{1,p}(\Omega)$, $Y = [L^p(\Omega)]^d \times \mathbb{R}$, $Z = L^p(\Omega)$, and $A : X \ni v \mapsto (\nabla v, f(v)) \in Y$. Owing to Lemma B.29 and the hypotheses on f, A is continuous and injective. Moreover, the injection $X \subset Z$ is compact owing to Theorem B.46. $\qquad\square$

Example B.64. Let E be a subset of Ω of non-zero measure and set $f(v) = \text{meas}(E)^{-1}\int_E v$. It is clear that f is continuous, and if c is a constant function, $f(c)$ is zero if and only if c is zero. A second possibility consists of setting $f(v) = \text{meas}(\partial\Omega_1)^{-1}\int_{\partial\Omega_1} v$ where $\partial\Omega_1$ is a subset of $\partial\Omega$ of non-zero $(d-1)$-measure. The continuity of f is a consequence of the Trace Theorem B.52. $\qquad\square$

Corollary B.65. *Under the hypotheses of Lemma B.63, assume $f(1_\Omega) = 1$, where 1_Ω is the constant function equal to 1 on Ω. Then,*

$$\forall v \in W^{1,p}(\Omega), \quad c\|v - f(v)\|_{W^{1,p}(\Omega)} \leq \|\nabla v\|_{L^p(\Omega)}. \tag{B.25}$$

As an easy consequence of Lemma B.63, we infer a useful generalization of the Poincaré inequality.

Lemma B.66 (Poincaré–Friedrichs). *Under the above hypotheses, let $W = \{v \in W^{1,p}(\Omega); \ f(v) = 0\}$. Then, W is a closed subspace of $W^{1,p}(\Omega)$ and*

$$\forall v \in W, \quad c\|v\|_{W^{1,p}(\Omega)} \leq \|\nabla v\|_{L^p(\Omega)}. \tag{B.26}$$

We conclude this section with two lemmas that play an important role in numerical analysis. In particular, they simplify considerably the error analysis of the finite element method. Recall that $\mathbb{P}_l$ is the vector space of real-valued polynomials in the variables $x_1, \ldots, x_d$ and of global degree at most l.

Lemma B.67 (Deny–Lions). *Let $1 \leq p \leq +\infty$ and let Ω be a connected bounded open set having the $(1,p)$-extension property. Let $l \geq 0$. There exists $c > 0$ such that*

$$\forall v \in W^{l+1,p}(\Omega), \qquad \inf_{\pi \in \mathbb{P}_l} \|v + \pi\|_{l+1,p,\Omega} \leq c\,|v|_{l+1,p,\Omega}. \tag{B.27}$$

Proof. Let $N_l = \dim \mathbb{P}_l$, and introduce the N_l continuous linear forms

$$f_\alpha : W^{l+1,p}(\Omega) \ni v \longmapsto f_\alpha(v) = \int_\Omega \partial^\alpha v \in \mathbb{R}, \qquad |\alpha| \leq l,$$

where $\alpha = (\alpha_1, \ldots, \alpha_d)$ is a multi-index. Define $X = W^{l+1,p}(\Omega)$, $Y = [L^p(\Omega)]^{N_{l+1} - N_l} \times \mathbb{R}^{N_l}$, and $Z = W^{l,p}(\Omega)$. Define the operator

$$A : X \ni v \longmapsto ((D^\alpha v)_{|\alpha|=l+1}, (f_\alpha(v))_{|\alpha| \leq l}) \in Y.$$

A is clearly continuous. Let $v \in W^{l+1,p}(\Omega)$ and assume that $Av = 0$. Repeated application of Lemma B.29 yields $v \in \mathbb{P}_l$ and $f_\alpha(v) = 0$ for all $|\alpha| \leq l$. Owing to the definition of the linear forms f_α, $v = 0$. That is to say, A is injective. Owing to the Rellich–Kondrachov Theorem, the injection $X = W^{l+1,p}(\Omega) \subset W^{l,p}(\Omega) = Z$ is compact. Then, the Petree–Tartar Lemma implies that there is $c > 0$ such that

$$\forall v \in W^{l+1,p}(\Omega), \qquad c\|v\|_{l+1,p,\Omega} \leq |v|_{l+1,p,\Omega} + \sum_{|\alpha| \leq l} |f_\alpha(v)|. \tag{B.28}$$

Let $\pi(v) \in \mathbb{P}_l$ be such that $f_\alpha(v \mid \pi(v)) = 0$ for $|\alpha| \leq l$. Then,

$$\inf_{\pi \in \mathbb{P}_l} \|v + \pi\|_{l+1,p,\Omega} \leq \|v + \pi(v)\|_{l+1,p,\Omega} \leq c\,|v + \pi(v)|_{l+1,p,\Omega} = c\,|v|_{l+1,p,\Omega},$$

and this completes the proof; see [DeL55]. $\qquad\square$

Lemma B.68 (Bramble–Hilbert). *Assume the hypotheses of the Deny–Lions Lemma hold. Then, there is $c > 0$ such that, for all $f \in (W^{k+1,p}(\Omega))'$ vanishing on $\mathbb{P}_k$,*

$$\forall v \in W^{k+1,p}(\Omega), \quad |f(v)| \leq c\|f\|_{(W^{k+1,p}(\Omega))'}|v|_{k+1,p,\Omega}. \tag{B.29}$$

Proof. For all $\pi \in \mathbb{P}_k$, $|f(v)| = |f(v + \pi)| \leq \|f\|_{(W^{k+1,p}(\Omega))'}\|v + \pi\|_{k+1,p,\Omega}$, that is, $|f(v)| \leq \|f\|_{(W^{k+1,p}(\Omega))'} \inf_{\pi \in \mathbb{P}_k} \|v + \pi\|_{k+1,p,\Omega}$. Then, conclude using the Deny–Lions Lemma; see [BrH70]. $\qquad\square$

B.3.8 The right inverse of the divergence operator

This section investigates the surjectivity of the divergence operator. An open bounded set Ω is said to be star-shaped with respect to a ball B if for any $x \in \Omega$ and $z \in B \subset \Omega$, the segment joining x and z is contained in Ω.

Lemma B.69 (Bogovskiĭ). *Assume $d \geq 2$ and let Ω be a bounded open set in $\mathbb{R}^d$ star-shaped with respect to a ball B. Let $1 < p < +\infty$ and set*

$$L^p_{\int=0}(\Omega) = \left\{ v \in L^p(\Omega);\ \int_\Omega v = 0 \right\}.$$

Then, the operator $\nabla \cdot : [W^{1,p}_0(\Omega)]^d \to L^p_{\int=0}(\Omega)$ is surjective.

Proof. See [Bog80], [Gal94, Lemma 3.1, Chap. III], or [DuM01]. ☐

Remark B.70. The hypothesis "Ω is star-shaped" can be replaced by "the boundary of Ω is Lipschitz" when $p = 2$; see [GiR86, pp. 18–26]. ☐

Owing to Lemma A.42 and Theorem B.8, Lemma B.69 implies:

Corollary B.71. *Under the hypotheses of Lemma B.69, there is $\beta > 0$ such that*

$$\inf_{q \in L^{p'}_0(\Omega)}\ \sup_{v \in [W^{1,p}_0(\Omega)]^d}\ \frac{\int_\Omega q\nabla \cdot v}{\|v\|_{[W^{1,p}(\Omega)]^d}\|q\|_{L^{p'}(\Omega)}} \geq \beta. \tag{B.30}$$

Remark B.72. Lemma B.69 has been extended to fractional Sobolev spaces $[W^{s,p}(\Omega)]^d$ by Solonnikov [Sol01, Prop. 2.1]. ☐

As a consequence of Lemma B.69, we deduce a "coarse" version of de Rham's Theorem in $L^p(\Omega)$.

Theorem B.73 (de Rham). *Under the hypotheses of Lemma B.69, the continuous linear forms on $[W^{1,p}_0(\Omega)]^d$ that are zero on $\mathrm{Ker}(\nabla \cdot)$ are gradients of functions in $L^{p'}_{\int=0}(\Omega)$.*

Proof. This a simple consequence of the Closed Range Theorem. Indeed, consider the weak gradient operator $\nabla : L^{p'}_{\int=0}(\Omega) \to [W^{-1,p'}(\Omega)]^d$. It is clear that $-\nabla = (\nabla \cdot)^T$. Since $\nabla \cdot$ is surjective by Lemma B.69, the Closed Range Theorem implies $[\mathrm{Ker}(\nabla \cdot)]^\perp = \mathrm{Im}(\nabla)$. ☐

Nomenclature

General Convention

Throughout the book, c denotes a constant that does not depend on h, unless specified otherwise, and whose value may change at each occurrence.

Basic Notation

$\text{card}(E)$	Cardinal number of the set E
$\text{meas}(E)$	Lebesgue measure of $E \subset \mathbb{R}^d$
$u_{|E}$	Restriction of the function u to the set E
$\text{span}\{v_1, \ldots, v_n\}$	Vector space spanned by the vectors $v_1, \ldots, v_n$
$\dim(V)$	Dimension of the vector space V
δ_{ij}	Kronecker symbol: $\delta_{ij} = 1$ if $i = j$ and 0 otherwise

Vectors and Matrices

$(u_1, \ldots, u_n)$	Cartesian components of the vector u in $\mathbb{R}^n$
$(u, v)_n$ or $u \cdot v$	Euclidean scalar product in $\mathbb{R}^n$: $u \cdot v = \sum_{i=1}^n u_i v_i$
$\|u\|_n$	Euclidean norm in $\mathbb{R}^n$: $\|u\|_n = (u \cdot u)^{\frac{1}{2}}$
$u \times v$	Cross-product in $\mathbb{R}^3$:
	$\quad u \times v = (u_2 v_3 - u_3 v_2, u_3 v_1 - u_1 v_3, u_1 v_2 - u_2 v_1)^T.$
$\mathbb{R}^{m,n}$	Vector space $m \times n$ matrices with real-valued entries
$\mathcal{A}, \mathcal{M}$	Matrices
$\mathcal{I}$	Identity matrix
$\mathcal{A}_{ij}$	Entry of $\mathcal{A}$ in the ith row and the jth column
$\mathcal{A}^T$	Transpose of the matrix $\mathcal{A}$
$\text{tr}(\mathcal{A})$	Trace of $\mathcal{A}$: For $\mathcal{A} \in \mathbb{R}^{n,n}$, $\text{tr}(\mathcal{A}) = \sum_{i=1}^n \mathcal{A}_{ii}$
$\det(\mathcal{A})$	Determinant of $\mathcal{A}$
$\text{rank}(\mathcal{A})$	Rank of $\mathcal{A}$

$\mathrm{diag}(\mathcal{A})$	Diagonal of $\mathcal{A}$: For $\mathcal{A} \in \mathbb{R}^{n,n}$, $\mathrm{diag}(\mathcal{A})_{ij} = \delta_{ij}\mathcal{A}_{ij}, 1 \le i,j \le n$
$\mathcal{A}{\cdot}u$	Matrix–vector product: For $\mathcal{A} \in \mathbb{R}^{m,n}$ and for $1 \le i \le m$, $(\mathcal{A}{\cdot}u)_i = \sum_{j=1}^{n} \mathcal{A}_{ij}\,u_j$
$v{\cdot}\mathcal{A}{\cdot}u$	For $\mathcal{A} \in \mathbb{R}^{m,n}$, $v{\cdot}\mathcal{A}{\cdot}u = \sum_{i=1}^{m}\sum_{j=1}^{n} v_i \mathcal{A}_{ij} u_j$
$\mathcal{A}{:}\mathcal{M}$	Double contraction: For $\mathcal{A} \in \mathbb{R}^{m,n}$ and $\mathcal{M} \in \mathbb{R}^{m,n}$, $\mathcal{A}{:}\mathcal{M} = \sum_{i=1}^{m}\sum_{j=1}^{n} \mathcal{A}_{ij}\mathcal{M}_{ij}$
$u{\otimes}v$	Tensor product: For $u \in \mathbb{R}^m$ and $v \in \mathbb{R}^n$, $u{\otimes}v \in \mathbb{R}^{m,n}$ and $(u{\otimes}v)_{ij} = u_i\,v_j,\ 1 \le i \le m,\ 1 \le j \le n$

Differential Operators

$(x_1,\dots,x_d)$	Cartesian coordinates in $\mathbb{R}^d$
$d_t u$	Time derivative of u
$\partial_i u$	Distributional derivative of u with respect to x_i
$\partial_\xi u$	Distributional derivative of u with respect to ξ
$\partial_{ij} u$	Second-order derivative of u with respect to x_i and x_j
$\partial^\alpha u$	$\partial_{x_1}^{\alpha_1} \dots \partial_{x_d}^{\alpha_d} u$ where $\alpha = (\alpha_1,\dots,\alpha_d) \in \mathbb{N}^d$ is a *multi-index*
$|\alpha|$	Length of $\alpha = (\alpha_1,\dots,\alpha_d) \in \mathbb{N}^d$: $|\alpha| = \alpha_1 + \dots + \alpha_d$
∇u	Gradient: $\nabla u = (\partial_1 u,\dots,\partial_d u)^T \in \mathbb{R}^d$ if u is $\mathbb{R}$-valued or $\nabla u = (\partial_j u_i)_{1 \le i \le m,\, 1 \le j \le d} \in \mathbb{R}^{m,d}$ if u is $\mathbb{R}^m$-valued
$\nabla{\cdot}u$	Divergence: $\nabla{\cdot}u = \sum_{i=1}^{d} \partial_i u_i$ if u is $\mathbb{R}^d$-valued or $\nabla{\cdot}u = (\sum_{j=1}^{d} \partial_j u_{ij})_{1 \le i \le m}^T \in \mathbb{R}^m$ if u is $\mathbb{R}^{m,d}$-valued
$\nabla{\times}u$	Curl: $\nabla{\times}u = (\partial_2 u_3 - \partial_3 u_2, \partial_3 u_1 - \partial_1 u_3, \partial_1 u_2 - \partial_2 u_1)^T$ if u is $\mathbb{R}^3$-valued or $\nabla{\times}u = \partial_1 u_2 - \partial_2 u_1$ if u is $\mathbb{R}^2$-valued
$\beta{\cdot}\nabla u$	Advection operator: β is $\mathbb{R}^d$-valued and $\beta{\cdot}\nabla u = \sum_{i=1}^{d} \beta_i \partial_i u$
Δu	Laplace operator: $\Delta u = \sum_{i=1}^{d} \partial_{ii} u$ if u is $\mathbb{R}$-valued or $\Delta u = (\sum_{j=1}^{d} \partial_{jj} u_i)_{1 \le i \le m}^T \in \mathbb{R}^m$ if u is $\mathbb{R}^m$-valued
$D^2 u$	Hessian operator: $D^2 u = (\partial_{ij} u)_{1 \le i,j \le d}$, if u is $\mathbb{R}$-valued. $(D^2 u)^T = D^2 u$ and $\mathrm{tr}(D^2 u) = \Delta u$

Function Spaces

$\mathcal{L}(E;F)$	Vector space of the bounded linear operators from E to F
$\mathrm{Ker}\,A$	Kernel (i.e., nullspace) of the linear operator A
$\mathrm{Im}\,A$	Range of the linear operator A
X'	Topological dual of the topological space X
A^T	Dual operator of A: if $A \in \mathcal{L}(E;F)$, $A^T \in \mathcal{L}(F',E')$
$\|u\|_X$	Norm of u in the normed space X
$\mathbb{P}_k$	Vector space of polynomials in the variables $x_1,\dots,x_d$ of global degree at most k
$\mathbb{Q}_k$	Vector space of polynomials in the variables $x_1,\dots,x_d$ of partial degree at most k
$\mathcal{D}(\Omega)$	Infinitely differentiable functions compactly supported in Ω

$\mathcal{C}^0(E; F)$	Space of continuous functions from the topological space E to the topological space F
$\mathcal{C}^0(\Omega)$, $\mathcal{C}^k(\Omega)$	Space of continuous functions on $\Omega \subset \mathbb{R}^d$, and space of k times continuously differentiable functions on Ω
$\mathcal{C}^{k,\alpha}(\Omega)$ (resp., $\mathcal{C}^{k,\alpha}(\overline{\Omega})$)	Space of functions whose derivatives up to order k are locally (resp., globally) α-Hölder continuous
$\delta_{x=a}$	Dirac measure supported at a
$L^p(\Omega)$	Functions whose p-th power is Lebesgue integrable on Ω
$L^p_{\int=0}(\Omega)$	$\{v \in L^p(\Omega); \int_\Omega v = 0\}$
p'	Conjugate of p, $\frac{1}{p} + \frac{1}{p'} = 1$
p^*	$\frac{1}{p^*} = \frac{1}{p} - \frac{1}{d}$, and $p^* = +\infty$ if $p = d$ (d is the space dimension)
$W^{s,p}(\Omega)$	Functions whose derivatives up to order s are in $L^p(\Omega)$
$W^{s,p}_0(\Omega)$	Closure of $\mathcal{D}(\Omega)$ in $W^{s,p}(\Omega)$
$W^{-s,p'}(\Omega)$	Dual of $W^{s,p}_0(\Omega)$
$[W^{s,p}(\Omega)]^m$	$\mathbb{R}^m$-valued functions whose components are in $W^{s,p}(\Omega)$
$\|u\|_{0,p,\Omega}$	Norm in $L^p(\Omega)$: $\|u\|_{0,p,\Omega} = (\int_\Omega |u|^p)^{\frac{1}{p}}$.
$|u|_{s,p,\Omega}$	Seminorm in $W^{s,p}(\Omega)$: $|u|_{s,p,\Omega} = \sum_{|\alpha|=s} \|\partial^\alpha u\|_{0,p,\Omega}$
$\|u\|_{s,p,\Omega}$	Norm in $W^{s,p}(\Omega)$: $\|u\|_{s,p,\Omega} = \sum_{l \le s} |u|_{l,p,\Omega}$
$H^s(\Omega)$, $H^s_0(\Omega)$	$W^{s,2}(\Omega)$, $W^{s,2}_0(\Omega)$
$H^s_{\int=0}(\Omega)$	$\{v \in H^s(\Omega); \int_\Omega v = 0\}$
$\|u\|_{s,\Omega}$, $|u|_{s,\Omega}$	$\|u\|_{s,2,\Omega}$, $|u|_{s,2,\Omega}$
$(u, v)_{0,E}$	Scalar product on $L^2(E)$: $\int_E uv$
$H(\mathrm{div}; \Omega)$	$\{v \in [L^2(\Omega)]^d; \nabla\cdot v \in L^2(\Omega)\}$
$H_0(\mathrm{div}; \Omega)$	$\{v \in [L^2(\Omega)]^d; \nabla\cdot v \in L^2(\Omega); v\cdot n_{|\partial\Omega} = 0\}$
$H(\mathrm{curl}; \Omega)$	$\{v \in [L^2(\Omega)]^3; \nabla\times v \in [L^2(\Omega)]^3\}$
$H_0(\mathrm{curl}; \Omega)$	$\{v \in [L^2(\Omega)]^3; \nabla\times v \in [L^2(\Omega)]^3; v\times n_{|\partial\Omega} = 0\}$
$\mathcal{C}^j([0, T]; V)$	V-valued functions that are of class $\mathcal{C}^j$ with respect to t
$L^p(]0, T[; V)$	V-valued functions whose norm in V is in $L^p(]0, T[)$
$\mathcal{W}(B_0, B_1)$	$\{v :]0, T[\to B_0; v \in L^{p_1}(]0, T[; B_0); d_t v \in L^{p_2}(]0, T[; B_1)\}$

Mesh-Related Symbols

$h_K = \mathrm{diam}(K)$	Diameter of $K \subset \mathbb{R}^d$
N_{geo}	Number of geometrical nodes
N_{el}	Number of cells (or elements) in the mesh
N_{ed}, $N^{\mathrm{i}}_{\mathrm{ed}}$, $N^{\partial}_{\mathrm{ed}}$	Number of edges, internal edges, and edges at the boundary
N_{f}, $N^{\mathrm{i}}_{\mathrm{f}}$, $N^{\partial}_{\mathrm{f}}$	Number of faces, internal faces, and faces at the boundary
N_{v}, $N^{\mathrm{i}}_{\mathrm{v}}$, $N^{\partial}_{\mathrm{v}}$	Number of vertices, internal vertices, and vertices at the boundary
$\{\mathcal{T}_h\}_{h>0}$	Family of meshes
$\mathcal{E}_h$, $\mathcal{E}^{\mathrm{i}}_h$, $\mathcal{E}^{\partial}_h$	Set of edges, internal edges, and edges at the boundary
$\mathcal{F}_h$, $\mathcal{F}^{\mathrm{i}}_h$, $\mathcal{F}^{\partial}_h$	Set of faces ($(d-1)$-manifolds), internal faces, and faces at the boundary

k_q Quadrature order

l_q, l_q^∂ Number of Gauß points and Gauß points at the boundary

Reference Finite Elements

$\{\widehat{K}, \widehat{P}_\mathrm{geo}, \widehat{\Sigma}_\mathrm{geo}\}$ Geometric reference finite element

$\{\widehat{K}, \widehat{P}, \widehat{\Sigma}\}$ Reference finite element

n_geo Number of shape functions for the geometric reference FE

n_sh Number of shape functions for the reference FE

$V(\widehat{K})$ The linear forms in $\widehat{\Sigma}$ extend continuously to $V(\widehat{K})$.

T_K Transformation that maps the reference element $\widehat{K}$ to K

ψ_K Mapping from $V(K)$ to $V(\widehat{K})$

Finite Element Spaces

$P_{\mathrm{c},h}^k$, $Q_{\mathrm{c},h}^k$ Vector space of functions that are piecewise in $\mathbb{P}^k$ or in $\mathbb{Q}^k$ and are continuous

$P_{\mathrm{pt},h}^k$ Vector space of functions that are piecewise in $\mathbb{P}^k$ and satisfy the patch-test condition

$P_{\mathrm{td},h}^k$ Vector space of (totally discontinuous) functions that are piecewise in $\mathbb{P}^k$

$P_{\mathrm{pt},h}^1$ Crouzeix–Raviart approximation space

D_h Raviart–Thomas approximation space

R_h Nédélec approximation space

Algorithmic Symbols

`coord`   Coordinate array of geometric nodes

`geo_cnty`   Connectivity array of geometric nodes

`geo_cnty_bnd`   Connectivity array of geometric nodes at the boundary

`neigh`   Neighborhood connectivity array of geometric elements

`i_bc`   Boundary index array

`i_dom`   Domain index array

`pde_cnty`   Connectivity array of global d.o.f.

`pde_cnty_bnd`   Connectivity array of global d.o.f. at the boundary

`psi`   Array of local geometric shape functions

`dpsi_dhatK`   Array of gradients of local geometric shape functions in $\widehat{K}$

`theta`   Array of local shape functions

`dtheta_dhatK`   Array of gradients of local shape functions in $\widehat{K}$

`dtheta_dK`   Array of gradients of local shape functions in K

`weight_K`   Array of quadrature weights

`xi_l`   Array of Gauß points in K

References

[AbS72] ABRAMOWITZ, M., AND STEGUN, I. *Handbook of Mathematical Functions*, 9th ed. Dover, New York, 1972.

[AcA03] ACHCHAB, B., ACHCHAB, S., AND AGOUZAL, A. Hierarchical robust a posteriori error estimator for a singularly perturbed problem. *C. R. Acad. Sci. Paris, Sér. I* **336**:1 (2003) 95–100.

[AcA98] ACHCHAB, B., AGOUZAL, A., BARANGER, J., AND MAITRE, J. Estimateur d'erreur a posteriori hiérarchique. Application aux éléments finis mixtes. *Numer. Math.* **80** (1998) 159–179.

[AcB01] ACHDOU, Y., AND BERNARDI, C. Adaptive finite volume or finite element discretization of Darcy's equations with variable permeability. *C. R. Acad. Sci. Paris, Sér. I* **333**:7 (2001) 693–698.

[Ada75] ADAMS, R. *Sobolev Spaces*, Vol. 65, Pure and Applied Mathematics, Academic Press, New York, 1975.

[AiO00] AINSWORTH, M., AND ODEN, J. *A Posteriori Error Estimation in Finite Element Analysis*. Wiley, New York, 2000.

[AiB02] AIT-ALI-YAHIA, D., BARUZZI, G., HABASHI, W., FORTIN, M., DOMPIERRE, J., AND VALLET, M.-G. Anisotropic mesh adaptation: Towards user-independent, mesh-independent and solver-independent CFD. II: Structured grids. *Int. J. Numer. Methods Fluids* **39**:8 (2002) 657–673.

[AiM94] AIT OU AMMI, A., AND MARION, M. Nonlinear Galerkin methods and mixed finite elements: Two-grid algorithms for the Navier Stokes equations. *Numer. Math.* **68**:2 (1994) 189–213.

[Ama00] AMANN, H. Compact embeddings of vector-valued Sobolev and Besov spaces. *Glas. Mat. Ser. III* **35 (55)**:1 (2000) 161–177.

[AmB98] AMROUCHE, C., BERNARDI, C., DAUGE, M., AND GIRAULT, V. Vector potentials in three-dimensional non-smooth domains. *Math. Methods Appl. Sci.* **21**:9 (1998) 823–864.

[AmG91] AMROUCHE, C., AND GIRAULT, V. On the existence and regularity of the solution of the Stokes problem in arbitrary dimension. *Proc. Japan Acad.* **67** (1991) 171–175.

[Ang02] ANGERMANN, L. A posteriori error estimates for FEM with violated Galerkin orthogonality. *Numer. Methods Partial Differential Equations* **18**:2 (2002) 241–259.

[ArK67]   ARGYRIS, J., AND KELSEY, S. *Energy Theorems and Structural Analysis*,
          3rd ed. Buttelworths, London, 1967. Initially published in Aircraft Engrg.,
          26, pp. 347–356, 383–387, 394, 410–422, (1954) and **27**, pp. 42–58, 80–94,
          125–134, 145–158 (1955).

[ArB01]   ARNOLD, D., BREZZI, F., COCKBURN, B., AND MARINI, L. Unified anal-
          ysis of discontinuous Galerkin methods for elliptic problems. *SIAM J.
          Numer. Anal.* **39**:5 (2001/02) 1749–1779.

[ArB84]   ARNOLD, D., BREZZI, F., AND FORTIN, M. A stable finite element for the
          Stokes equations. *Calcolo* **21** (1984) 337–344.

[Arn51]   ARNOLDI, W. The principle of minimized iteration in the solution of the
          matrix eigenvalue problem. *Quart. Appl. Math.* **9** (1951) 17–29.

[Aub63]   AUBIN, J.-P. Un théorème de compacité. *C. R. Acad. Sci. Paris, Sér. I*
          256 (1963) 5042–5044.

[Aub70]   AUBIN, J.-P. Approximation des problèmes aux limites non homogènes
          pour des opérateurs non linéaires. *J. Math. Anal. Appl.* **30** (1970) 510–
          521.

[Aub87]   AUBIN, J.-P. *Analyse fonctionnelle appliquée*, Vols. 1 and 2. Presses Uni-
          versitaires de France, Paris, 1987.

[Aub00]   AUBIN, J.-P. *Applied Functional Analysis*, 2nd ed. Pure and Applied
          Mathematics. Wiley-Interscience, New York, 2000.

[Aze95]   AZERAD, P. *Analyse des équations de Navier–Stokes en bassin peu profond
          et de l'équation de transport*. Thèse de l'Univeristé de Neuchâtel, 1995.

[Aze99]   AZERAD, P. 1999. Personnal communication.

[AzK85]   AZIZ, A., KELLOGG, R., AND STEPHENS, A. Least-Squares methods for
          elliptic systems. *Math. Comp.* **44**:169 (1985) 53–70.

[Bab73^a] BABUŠKA, I. The finite element method with Lagrangian multipliers.
          Numer. Math. **20** (1973) 179–192.

[Bab73^b] BABUŠKA, I. The finite element method with penalty. *Math. Comp.* **27**
          (1973) 221–228.

[BaA72]   BABUŠKA, I., AND AZIZ, A. Survey lectures on the mathematical founda-
          tion of the finite element method. In *The Mathematical Foundations of the
          Finite Element Method with Applications to Partial Differential Equations*,
          pp. 1–359. Academic Press, New York, 1972.

[BaD81]   BABUŠKA, I., AND DORR, M.R. Error estimates for the combined h and p
          versions of the finite element method. *Numer. Math.* **37**:2 (1981) 257–277.

[BaR78^a] BABUŠKA, I., AND RHEINBOLT, W. Error estimates for adaptive finite
          element method computations. *SIAM J. Numer. Anal.* **15** (1978) 736–754.

[BaR78^b] BABUŠKA, I., AND RHEINBOLT, W. A posteriori error estimates for the
          finite element method. *Int. J. Numer. Methods Engrg.* **12** (1978) 1597–
          1615.

[BaS81]   BABUŠKA, I., SZABO, B.A., AND KATZ, I.N. The p-version of the finite
          element method. *SIAM J. Numer. Anal.* **18**:3 (1981) 515–545.

[BaB93]   BAIOCCHI, C., BREZZI, F., AND FRANCA, L. Virtual bubbles and Galerkin-
          Least-Squares type methods (GaLS). *Comput. Methods Appl. Mech. Engrg.*
          105 (1993) 125–141.

[BaS93]   BANK, R., AND SMITH, R. A posteriori error estimates based on hierar-
          chical bases. *SIAM J. Numer. Anal.* **30**:4 (1993) 921–935.

[BaW85]   BANK, R., AND WEISER, A. Some a posteriori error estimators for elliptic
          partial differential equations. *Math. Comp.* **44** (1985) 283–301.

[Bar01] BARTLE, R. *A Modern Theory of Integration*. Graduate Studies in Mathematics, Vol. 32. American Mathematical Society, Providence, RI, 2001.

[Bat96] BATHE, K. *Finite Element Procedures*. Prentice-Hall, Englewood Cliffs, NJ, 1996.

[BaO99] BAUMANN, C., AND ODEN, J. A discontinuous hp finite element method for the Euler and Navier–Stokes equations. *Int. J. Numer. Methods Engrg.* **31** (1999) 79–95.

[BeB01] BECKER, R., AND BRAACK, M. A finite element pressure gradient stabilization for the Stokes equations based on local projections. *Calcolo* **38**:4 (2001) 173–199.

[BeR96] BECKER, R., AND RANNACHER, R. A feed-back approach to error control in finite element methods: Basic analysis and examples. *East-West J. Numer. Math.* **4**:4 (1996) 237–264.

[BeR01] BECKER, R., AND RANNACHER, R. An optimal control approach to a posteriori error estimation in finite element methods. *Acta Numerica* **10** (2001) 1–102.

[BeP79] BERCOVIER, M., AND PIRONNEAU, O. Error estimates for finite element solution of the Stokes problem in the primitive variables. *Numer. Math.* **33** (1979) 211–224.

[Ber89] BERNARDI, C. Optimal finite element interpolation on curved domains. *SIAM J. Numer. Anal.* **26** (1989) 1212–1240.

[BeG98] BERNARDI, C., AND GIRAULT, V. A local regularization operator for triangular and quadrilateral finite elements. *SIAM J. Numer. Anal.* **35**:5 (1998) 1893–1916.

[Ber99] BERTOLUZZA, S. The discrete commutator property of approximation spaces. *C. R. Acad. Sci. Paris, Sér. I* **329**:12 (1999) 1097–1102.

[BeO96] BEY, K.S., AND ODEN, J.T. hp-version discontinuous Galerkin methods for hyperbolic conservation laws. *Comput. Methods Appl. Mech. Engrg.* **133**:3-4 (1996) 259–286.

[Boc97] BOCHEV, P. Experiences with negative norm least-squares methods for the Navier–Stokes equations. *Electron. Trans. Numer. Anal.* **6** (1997) 44–62.

[Boc99] BOCHEV, P. Negative norm least-squares methods for the velocity-vorticity-pressure Navier–Stokes equations. *Numer. Methods Partial Differential Equations* **15**:2 (1999) 237–256.

[Bof01] BOFFI, D. A note on the de Rahm complex and a discrete compactness property. *Appl. Math. Lett.* **14** (2001) 33–38.

[Bog80] BOGOVSKIĬ, M. Solutions of some problems of vector analysis, associated with the operators div and grad. In *Theory of Cubature Formulas and the Application of Functional Analysis to Problems of Mathematical Physics*, Trudy Sem. S. L. Soboleva, No. 1, Vol. 149, pp. 5–40. Akad. Nauk SSSR Sibirsk. Otdel. Inst. Mat., Novosibirsk, Russia, 1980.

[BoN85] BOLAND, J., AND NICOLAIDES, R. Stability and semistable low-order finite elements for viscous flows. *SIAM J. Numer. Anal.* **22** (1985) 474–492.

[BoT01] BOS, L., TAYLOR, M.A., AND WINGATE, B.A. Tensor product Gauss-Lobatto points are Fekete points for the cube. *Math. Comp.* **70**:236 (2001) 1543–1547.

[Bos93] BOSSAVIT, A. *Electromagnétisme en vue de la modélisation*, SMAI Series on Mathematics and Applications, Vol. 14. Springer-Verlag, Paris, 1993. See also: *Computational Electromagnetism, Variational Formulations, Complementary, Edge Elements*, Academic Press, New York, 1998.

[BoM97] BOUKIR, K., MADAY, Y., MÉTIVET, B., AND RAZAFINDRAKOTO, E. A
         high-order characteristics/finite element method for the incompressible
         Navier–Stokes equations. *Int. J. Numer. Methods Fluids* **25**:12 (1997)
         1421–1454.

[Bow81] BOWYER, A. Computing Dirichlet tesselations. *Comput. J.* **24** (1981)
         162–167.

[Bra98] BRAACK, M. *An adaptive finite element method for reactive flow problems.*
         PhD thesis, University of Heidelberg, 1998.

[BrE03] BRAACK, M., AND ERN, A. A posteriori control of modeling errors and
         discretization errors. *Multiscale Model. Simul.* **1** (2003) 221–238.

[Bra81] BRAESS, D. The contraction number of a multigrid method for solving the
         Poisson equation. *Numer. Math.* **37** (1981) 387–404.

[Bra97] BRAESS, D. *Finite Elements. Theory, Fast Solvers, and Applications in
         Solid Mechanics,* 2nd ed. Cambridge University Press, Cambridge, 1997.

[BrH70] BRAMBLE, J.H. AND HILBERT, S.R. Estimation of linear functionals on
         Sobolev spaces with application to Fourier transforms and spline interpo-
         lation. *SIAM J. Numer. Anal.* **7** (1970) 112–124.

[BrL97] BRAMBLE, J.H., LAZAROV, R., AND PASCIAK, J. A least-squares approach
         based on a discrete minus one inner product for first-order systems. *Math.
         Comp.* **66**:219 (1997) 935–955.

[BrL01] BRAMBLE, J.H., LAZAROV, R., AND PASCIAK, J. Least-squares methods
         for linear elasticity based on a discrete minus one inner product. *Comput.
         Methods Appl. Mech. Engrg.* **191**:8–10 (2001) 727–744.

[BrP96] BRAMBLE, J.H., AND PASCIAK, J. Least-squares methods for Stokes equa-
         tions based on a discrete minus one inner product. *J. Comput. Appl. Math.*
         74:1–2 (1996) 155–173.

[BrP00] BRAMBLE, J.H., PASCIAK, J.E., AND VASSILEVSKI, P.S. Computational
         scales of Sobolev norms with application to preconditioning. *Math. Comp.*
         69:230 (2000) 463–480.

[BrP90] BRAMBLE, J.H., PASCIAK, J.E., AND XU, J. Parallel multilevel precondi-
         tioners. *Math. Comp.* **55**:191 (1990) 1–22.

[BrS70] BRAMBLE, J.H., AND SCHATZ, A. Rayleigh–Ritz–Galerkin-methods for
         Dirichlet's problem using subspaces without boundary conditions. *Comm.
         Pure Appl. Math.* **23** (1970) 653–675.

[BrS71] BRAMBLE, J.H., AND SCHATZ, A. Least squares for 2mth-order elliptic
         boundary-value problems. *Math. Comp.* **25** (1971) 1–32.

[BrS98] BRAMBLE, J.H., AND SUN, T. A negative-norm least squares method for
         Reissner–Mindlin plates. *Math. Comp.* **67**:223 (1998) 901–916.

[BrS94] BRENNER, S., AND SCOTT, R. *The Mathematical Theory of Finite Element
         Methods.* Texts in Applied Mathematics, Vol. 15. Springer, New York, 1994.

[Bre91] BREZIS, H. *Analyse fonctionnelle. Théorie et applications. [Functional
         Analysis. Theory and applications.]* Applied Mathematics Series for the
         Master's Degree. Masson, Paris, 1983.

[Bre74] BREZZI, F. On the existence, uniqueness and approximation of saddle-
         point problems arising from Lagrange multipliers. *RAIRO Anal. Numér.*
         (1974) 129–151.

[BrB92] BREZZI, F., BRISTEAU, M., FRANCA, L., MALLET, M., AND ROGÉ, G.
         A relationship between stabilized finite element methods and the Galerkin
         method with bubble functions. *Comput. Methods Appl. Mech. Engrg.* **96**
         (1992) 117–129.

[BrD87] BREZZI, F., DOUGLAS JR., J., DURÁN, R.G., AND FORTIN, M. Mixed finite elements for second order elliptic problems in three variables. *Numer. Math.* **51**:2 (1987) 237–250.

[BrD86] BREZZI, F., DOUGLAS JR., J., AND MARINI, L.D. Recent results on mixed finite element methods for second order elliptic problems In *Vistas in applied mathematics* (Transl. Ser. Math. Engrg.) pp. 25–43. Optimization Software, New York, 1986.

[BrF91^a] BREZZI, F., AND FALK, R. Stability of a higher-order Hood–Taylor method. *SIAM J. Numer. Anal.* **28** (1991) 581–590.

[BrF91^b] BREZZI, F., AND FORTIN, M. *Mixed and Hybrid Finite Element Methods.* Springer-Verlag, New York, 1991.

[BrF97] BREZZI, F., FRANCA, L., HUGHES, T.J., AND RUSSO, A. $b = \int g$. *Comput. Methods Appl. Mech. Engrg.* **145** (1997) 329–364.

[BrH82] BROOKS, A., AND HUGHES, T.J. Streamline Upwind/Petrov–Galerkin formulations for convective dominated flows with particular emphasis on the incompressible Navier–Stokes equations. *Comput. Methods Appl. Mech. Engrg.* **32** (1982) 199–259.

[BrC01] BROWN, D., CORTEZ, R., AND MINION, M. Accurate projection methods for the incompressible Navier–Stokes equations. *J. Comput. Phys.* **168**:2 (2001) 464–499.

[BuF93] BURDEN, R., AND FAIRES, J. *Numerical Analysis*, 5th ed. PWS Publishing, Boston, MA, 1993.

[BuE02] BURMAN, E., AND ERN, A. Nonlinear diffusion and discrete maximum principle for stabilized Galerkin approximations of the convection-diffusion-reaction equation. *Comput. Methods Appl. Mech. Engrg.* **191** (2002) 3833–3855.

[BuE03] BURMAN, E., ERN, A., AND GIOVANGIGLI, V. Adaptive finite element methods for low Mach, steady, laminar combustion. *J. Comput. Phys.* **188**:2 (2003) 472–492.

[BuE04] BURMAN, E., ERN, A., AND GIOVANGIGLI, V. Bunsen flame simulation by finite elements on adaptively refined, unstructured triangulations. *Combust. Theory Modelling* (2004).

[Cal61] CALDERÓN, A. Lebesgue spaces of differentiable functions and distributions. *Proc. Sympos. Pure Math.* **4** (1961) 33–49.

[Car97] CAREY, G. *Computational Grids. Generation, Adaptation, and Solution Strategies.* Series in Computational and Physical Processes in Mechanics and Thermal Sciences. Taylor & Francis, Washington, DC, 1997.

[CaP91] CARRIER, G., AND PEARSON, C. *Ordinary Differential Equations.* SIAM, Philadelphia, PA, 1991.

[Cat61] CATTABRIGA, L. Su un problema al contorno relativo al sistema di equazioni di Stokes. *Rend. Sem. Mat. Univ. Padova* **31** (1961) 308–340.

[Céa64] CÉA, J. Approximation variationnelle des problèmes aux limites. *Ann. Inst. Fourier, Grenoble* **14** (1964) 345–444.

[ChB95] CHEN, Q., AND BABUŠKA, I. Approximate optimal points for polynomial interpolation of real functions in an interval and in a triangle. *Comput. Methods Appl. Mech. Engrg.* **128**:3-4 (1995) 405–417.

[Cho68] CHORIN, A.J. Numerical solution of the Navier–Stokes equations. *Math. Comp.* **22** (1968) 745–762.

[Cho69] CHORIN, A.J. On the convergence of discrete approximations to the Navier-Stokes equations. *Math. Comp.* **23** (1969) 341–353.

[Cia78]   CIARLET, P. *The Finite Element Method for Elliptic Problems*. North-Holland, Amsterdam, 1978.

[Cia91]   CIARLET, P. *Basic Error Estimates for Elliptic Problems*, Vol. II : Finite Element Methods, ch. 2. Handbook of Numerical Analysis, P.G. Ciarlet and J.-L. Lions, editors. North-Holland, Amsterdam, 1991.

[Cia97]   CIARLET, P. *Mathematical Elasticity II: Theory of Plates*. Studies in Mathematics and its Applications, Vol. 27. Elsevier, Amsterdam, 1997.

[CiR72^a]   CIARLET, P., AND RAVIART, P.-A. The combined effect of curved boundaries and numerical integration in isoparametric finite element methods. In *The mathematical foundations of the finite element method with applications to partial differential equations* (Proc. Sympos., Univ. Maryland, Baltimore, MD, 1972) pp. 409–474. Academic Press, New York, 1972.

[CiR72^b]   CIARLET, P., AND RAVIART, P.-A. Interpolation theory over curved elements, with applications to finite element methods. *Comput. Methods Appl. Mech. Engrg.* **1** (1972) 217–249.

[Clé75]   CLÉMENT, P. Approximation by finite element functions using local regularization. *RAIRO Anal. Numér.* **9** (1975) 77–84.

[Clo60]   CLOUGH, R. The finite element method in plane stress analysis. In *Proc. 2nd ASCE Conference on Electronic Computation* Pittsburgh, PA, 1960.

[CoJ98]   COCKBURN, B., JOHNSON, C., SHU, C., AND TADMOR, E. *Advanced Numerical Approximation of Nonlinear Hyperbolic Equations*. Lecture Notes in Mathematics, Vol. 1697. Springer-Verlag, New York, 1998.

[CoK00]   COCKBURN, B., KARNIADAKIS, G., AND SHU, C. *Discontinuous Galerkin Methods–Theory, Computation and Applications*. Lecture Notes in Computer Science and Engineering, Vol. 11. Springer, New York, 2000.

[CoS98]   COCKBURN, B., AND SHU, C. The local discontinuous Galerkin method for time-dependent convection–diffusion systems. *SIAM J. Numer. Anal.* **35** (1998) 2440–2463.

[Cod98]   CODINA, R. Comparison of some finite element methods for solving the diffusion–convection–reaction equations. *Comput. Methods Appl. Mech. Engrg.* **156** (1998) 185–210.

[Cos91]   COSTABEL, M. A coercive bilinear form for Maxwell's equations. *J. Math. Anal. Appl.* **157**:2 (1991) 527–541.

[CoD02]   COSTABEL, M., AND DAUGE, M. Crack singularities for general elliptic systems. *Math. Nachr.* **235** (2002) 29–49.

[Cro00]   CROISILLE, J.-P. Finite volume box schemes and mixed methods. *Math. Model. Numer. Anal. (M2AN)* **34**:2 (2000) 1087–1106.

[CrG02]   CROISILLE, J.-P., AND GREFF, I. Some nonconforming mixed box schemes for elliptic problems. *Numer. Methods Partial Differential Equations* **18**:3 (2002) 355–373.

[CrM84]   CROUZEIX, M., AND MIGNOT, A. *Analyse numérique des équations différentielles*. Masson, Paris, 1984.

[CrR73]   CROUZEIX, M., AND RAVIART, P.-A. Conforming and nonconforming finite element methods for solving the stationnary Stokes equations. *RAIRO Anal. Numér.* **3** (1973) 33–75.

[DaT01]   DARU, V., AND TENAUD, C. Evaluation of TVD high-resolution schemes for unsteady viscous shocked flows. *Comput. Fluids* **30**:1 (2001) 89–113.

[DaL88]   DAUTRAY, R., AND LIONS, J.-L. *Functional and Variational Methods*. Vol. 2. Mathematical Analysis and Numerical Methods for Science and Technology. Springer-Verlag, Berlin, 1988.

[DaL90] DAUTRAY, R., AND LIONS, J.-L. *Integral Equations and Numerical Methods*. Vol. 4. Mathematical Analysis and Numerical Methods for Science and Technology. Springer-Verlag, Berlin, 1990.

[DaL93] DAUTRAY, R., AND LIONS, J.-L. *Evolution Problems, II.* Vol. 6. Mathematical Analysis and Numerical Methods for Science and Technology. Springer-Verlag, Berlin, 1993.

[Del34] DELAUNAY, B. Sur la sphère vide. *Bul. Acad. Sci. URSS, Class. Sci. Nat.* (1934) 793–800.

[DeM00] DEMKOWICZ, L., MONK, P., VARDAPETYAN, L., AND RACHOWICZ, W. de Rahm diagram for hp finite element spaces. *Comput. Math. Appl.* **39** (2000) 29–38.

[DeL55] DENY, J. AND LIONS, J.-L. Les espaces du type de Beppo Levi. *Ann. Inst. Fourier, Grenoble* **5** (1955) 305–370.

[Des86] DESTUYNDER, P. *Une théorie asymptotique des plaques minces en élasticité linéaire.* Masson, Paris, 1986.

[DoV02] DOMPIERRE, J., VALLET, M.-G., BOURGAULT, Y., FORTIN, M., AND HABASHI, W. Anisotropic mesh adaptation: Towards user-independent, mesh-independent and solver-independent CFD. III: Unstructured meshes. *Int. J. Numer. Methods Fluids* **39**:8 (2002) 675–702.

[DoR82] DOUGLAS JR., J., AND RUSSELL, T. Numerical methods for convection-dominated diffusion problems based on combining the method of characteristics with finite element or finite difference procedures. *SIAM J. Numer. Anal.* **19** (1982) 871–885.

[DuM01] DURÁN, R.G., AND MUSCHIETTI, M.A. An explicit right inverse of the divergence operator which is continuous in weighted norms. *Studia Math.* **148**:3 (2001) 207–219.

[DuL72] DUVAUT, G., AND LIONS, J.-L. *Les inéquations en mécanique et en physique.* Dunod, Paris, 1972.

[Dži68] DŽIŠKARIANI, A.V. The least squares and Bubnov–Galerkin methods. *Ž. Vyčisl. Mat. i Mat. Fiz.* **8** (1968) 1110–1116.

[EwL95] E, W., AND LIU, J. Projection method I: Convergence and numerical boundary layers. *SIAM J. Numer. Anal.* **32** (1995) 1017–1057.

[ElE03] EL ALAOUI, L., AND ERN, A. Residual based and hierarchical a posteriori estimates for nonconforming mixed finite element methods. *Math. Model. Numer. Anal. (M2AN)* (2003). Submitted.

[ErE96] ERIKSSON, K., ESTEP, D., HANSBO, P., AND JOHNSON, C. *Computational Differential Equations.* University Press, Cambridge, 1996.

[ErD95] ERN, A., DOUGLAS, C.C., AND SMOOKE, M.D. Detailed chemistry modeling of laminar diffusion flames on parallel computers. *Int. J. Supercomput. Appl.* **9** (1995) 167–186.

[ErG94] ERN, A., AND GIOVANGIGLI, V. *Multicomponent Transport Algorithms.* Lecture Notes in Physics, Vol. 24. Springer-Verlag, Heidelberg, 1994.

[ErG98] ERN, A., AND GIOVANGIGLI, V. Thermal diffusion effects in hydrogen/air and methane/air flames. *Combust. Theory Modelling* **2** (1998) 349–372.

[ErG02] ERN, A., AND GUERMOND, J.-L. *Elements finis: théorie, applications, mise en œuvre.* SMAI Series on Mathematics and Applications, Vol. 36. Springer-Verlag, Paris, 2002.

[Fal91] FALK, R. Nonconforming finite element methods for the equations of linear elasticity. *Math. Comp.* **57** (1991) 529–550.

[For77]   FORTIN, M. An analysis of the convergence of mixed finite element methods. *RAIRO Anal. Numér.* **11** (1977) 341–354.

[FoS83]   FORTIN, M., AND SOULIÉ, M. A non-conforming piecewise quadratic finite element on triangles. *Int. J. Numer. Methods Engrg.* **19** (1983) 505–520.

[FrF92]   FRANCA, L., AND FREY, S. Stabilized finite element methods: II. The incompressible Navier–Stokes equations. *Comput. Methods Appl. Mech. Engrg.* **99** (1992) 209–233.

[FrG00]   FREY, P., AND GEORGE, P.-L. *Mesh Generation. Application to Finite Elements.* Hermes Science, Oxford, 2000.

[Fre58]   FRIEDRICHS, K. Symmetric positive linear differential equations. *Comm. Pure Appl. Math.* **11** (1958) 333–418.

[Gal94]   GALDI, G. *An Introduction to the Mathematical Theory of the Navier–Stokes Equations.* Vol. I. Springer Tracts in Natural Philosophy, Vol. 38. Springer-Verlag, New York, 1994.

[Gea71]   GEAR, C. *Numerical Initial Value Problems in Ordinary Differential Equations.* Prentice-Hall, Englewoods Cliffs, NJ, 1971.

[GeB98]   GEORGE, P.-L., AND BOROUCHAKI, H. *Delaunay Triangulation and Meshing. Application to Finite Elements.* Editions Hermès, Paris, 1998.

[GeG93]   GEORGE, A., GILBERT, J., AND LIU, J., editors. *Graph Theory and Sparse Matrix Computation.* The IMA Volumes in Mathematics and its Applications, Vol. 56. Springer-Verlag, New York, 1993.

[GeL81]   GEORGE, A., AND LIU, J. *Computer solution of large sparse positive definite systems.* Prentice-Hall Series in Computational Mathematics. Prentice-Hall, Englewood Cliffs, NJ, 1981.

[Gio99]   GIOVANGIGLI, V. *Multicomponent Flow Modeling.* Modeling and Simulation in Science, Engineering and Technology. Birkhäuser, Boston, MA, 1999.

[GiR86]   GIRAULT, V., AND RAVIART, P.-A. *Finite Element Methods for Navier–Stokes Equations. Theory and Algorithms.* Springer Series in Computational Mathematics, Vol. 5. Springer-Verlag, Berlin, 1986.

[God79]   GODA, K. A multistep technique with implicit difference schemes for calculating two- or three-dimensional cavity flows. *J. Comput. Phys.* **30** (1979) 76–95.

[God71]   GODBILLION, C. *Eléments de topologie algébrique.* Hermann, Paris, 1971.

[GoV89]   GOLUB, G., AND VAN LOAN, C. *Matrix Computations,* 2nd ed. The John Hopkins University Press, Baltimore, MD, 1989.

[GoW96]   GORDON, C., AND WEBB, D. You can't hear the shape of a drum. *American Scientist* (1996).

[Gri85]   GRISVARD, P. *Boundary Value Problems in Non-Smooth Domains.* Pitman, London, 1985.

[Gri92]   GRISVARD, P. *Singularities in Boundary Value Problems.* Masson, Paris, 1992.

[Gue96]   GUERMOND, J.-L. Some practical implementations of projection methods for Navier–Stokes equations. *Math. Model. Numer. Anal. (M2AN)* **30** (1996) 637–667. Also in *C. R. Acad. Sci. Paris, Sér. I* **319** (1994) 887–892.

[Gue99^a]   GUERMOND, J.-L. Stabilization of Galerkin approximations of transport equations by subgrid modeling. *Math. Model. Numer. Anal. (M2AN)* **33**:6 (1999) 1293–1316. Also in *C. R. Acad. Sci. Paris, Sér. I* **328** (1999) 617–622.

[Gue99^b] GUERMOND, J.-L. Un résultat de convergence d'ordre deux en temps pour l'approximation des équations de Navier–Stokes par une technique de projection incrémentale. *Math. Model. Numer. Anal. (M2AN)* **33**:1 (1999) 169–189. Also in *C. R. Acad. Sci. Paris, Sér. I* **325** (1997) 1329–1332.

[Gue01^a] GUERMOND, J.-L. Subgrid stabilization of Galerkin approximations of linear contraction semi-groups of class C^0 in Hilbert spaces. *Numer. Methods Partial Differential Equations* **17** (2001) 1–25.

[Gue01^b] GUERMOND, J.-L. Subgrid stabilization of Galerkin approximations of linear monotone operators. *IMA J. Numer. Anal.* **21** (2001) 165–197.

[GuQ97] GUERMOND, J.-L., AND QUARTAPELLE, L. On sensitive vector Poisson and Stokes problems. *Math. Model. Methods Appl. Sci. (M3AS)* **7**:5 (1997) 681–698.

[GuQ98] GUERMOND, J.-L., AND QUARTAPELLE, L. On the approximation of the unsteady Navier–Stokes equations by finite element projection methods. *Numer. Math.* **80**:5 (1998) 207–238.

[GuS01] GUERMOND, J.-L., AND SHEN, J. Quelques résultats nouveaux sur les méthodes de projection. *C. R. Acad. Sci. Paris, Sér. I* **333** (2001) 1111–1116.

[GuS04] GUERMOND, J.-L., AND SHEN, J. On the error estimates for the rotational pressure–correction projection methods. *Math. Comp.* (2004). To appear.

[Gun89] GUNZBURGER, M. *Finite Element Methods for Viscous Incompressible Flows. A Guide to Theory, Practice, and Algorithms.* Computer Science and Scientific Computing. Academic Press, Boston, MA, 1989.

[HaD00] HABASHI, W., DOMPIERRE, J., BOURGAULT, Y., AIT-ALI-YAHIA, D., FORTIN, M., AND VALLET, M.-G. Anisotropic mesh adaption: Towards user-independent, mesh-independent and solver-independent CFD. I: General principles. *Int. J. Numer. Methods Fluids* **32**:6 (2000) 725–744.

[Hac85] HACKBUSCH, W. *Multigrid Methods and Applications*, Springer Series in Computational Mathematics, Vol. 4. Springer-Verlag, Berlin, 1985.

[Had32] HADAMARD, J. *Le problème de Cauchy et les équations aux dérivées partielles linéaires hyperboliques.* Hermann, Paris, 1932.

[HaW91] HAIRER, E., AND WANNER, G. *Solving Ordinary Differential Equations II: Stiff and Differential-Algebraic Problems.* Springer-Verlag, New York, 1991.

[HaS56] HAMMER, P., AND STROUD, A. Numerical integration over simplexes. *Math. Tables Aids Comput.* **10** (1956) 137–139.

[Hec81] HECHT, F. Construction d'une base de fonctions $\mathbb{P}_1$ non conforme à divergence nulle dans $\mathbb{R}^3$. *RAIRO Anal. Numér.* **15**:2 (1981) 119–150.

[Hec84] HECHT, F. Construction of a basis at free divergence in finite element and application to the Navier–Stokes equations. In *Numerical Solutions of Nonlinear Problems* (Rocquencourt, 1983) pp. 284–297. INRIA, Rocquencourt, France, 1984.

[HeS52] HESTENES, M., AND STIEFEL, E. Method of conjugate gradients for solving linear systems. *J. Res. Nat. Bur. Standards* **49** (1952) 409–436.

[HoW96] HOPPE, R., AND WOHLMUTH, B. Element-oriented and edge-oriented local error estimators for nonconforming finite element methods. *Math. Model. Numer. Anal. (M2AN)* **39** (1996) 237–263.

[HoS00] HOUSTON, P., SCHWAB, C., AND SÜLI, E. Stabilized hp-finite element methods for first-order hyperbolic problems. *SIAM J. Numer. Anal.* **37**:5 (2000) 1618–1643.

[HuG98] HU, N., GUO, X., AND KATZ, I. Bounds for eigenvalues and condition numbers in the p-version of the finite element method. *Math. Comp.* **67**:224 (1998) 1423–1450.

[HuF89] HUGHES, T.J., FRANCA, L., AND HULBERT, G. A new finite element formulation for computational fluid dynamics: VIII. The Galerkin/Least-Squares method for advection–diffusive equations. *Comput. Methods Appl. Mech. Engrg.* **73** (1989) 173–189.

[HuM86] HUGHES, T.J., AND MALLET, M. A new finite element formulation for computational fluid dynamics. IV: A discontinuity-capturing operator for multidimensional advective-diffusive systems. *Comput. Methods Appl. Mech. Engrg.* **58** (1986) 329–336.

[IrR72] IRONS, B.M., AND RAZZAQUE, A. Experience with the patch test for convergence of finite elements. In *The Mathematical Foundations of the Finite Element Method with Applications to Partial Differential Equations*, pp. 557–587. Academic Press, New York, 1972.

[Jam78] JAMET, P. Galerkin-type approximations which are discontinuous in time for parabolic equations in a variable domain. *SIAM J. Numer. Anal.* **15** (1978) 912–928.

[Jia89] JIANG, B. *The Least-Squares Finite Element Method.* Scientific Computation. Springer-Verlag, New York, 1998.

[Joh87] JOHNSON, C. *Numerical Solution of Partial Differential Equations by the Finite Element Method.* Cambridge University Press, Cambridge, 1987.

[JoN84] JOHNSON, C., NÄVERT, U., AND PITKÄRANTA, J. Finite element methods for linear hyperbolic equations. *Comput. Methods Appl. Mech. Engrg.* **45** (1984) 285–312.

[JoP86] JOHNSON, C., AND PITKÄRANTA, J. An analysis of the discontinuous Galerkin method for a scalar hyperbolic equation. *Math. Comp.* **46**:173 (1986) 1–26.

[JoR95] JOHNSON, C., RANNACHER, R., AND BOMAN, M. Numerics and hydrodynamic stability: Toward error control in computational fluid dynamics. *SIAM J. Numer. Anal.* **32**:4 (1995) 1058–1079.

[JoS87] JOHNSON, C., AND SZEPESSY, A. On the convergence of a finite element method for a nonlinear hyperbolic conservation law. *Math. Comp.* **49**:180 (1987) 427–444.

[JoS95] JOHNSON, C., AND SZEPESSY, A. Adaptive finite element methods for conservation laws based on a posteriori error estimates. *Comm. Pure Appl. Math.* **48** (1995) 199–234.

[JoS90] JOHNSON, C., SZEPESSY, A., AND HANSBO, P. On the convergence of shock-capturing streamline diffusion finite element methods for hyperbolic conservation laws. *Math. Comp.* **54**:189 (1990) 107–129.

[Kah66] KAHAN, W. Numerical linear algebra. *Canad. Math. Bull.* **9** (1966) 757–801.

[KaS99^a] KANSCHAT, G., AND SUTTMEIER, F. A posteriori error estimates for nonconforming finite element schemes. *Calcolo* **36** (1999) 129–141.

[KaS99^b] KARNIADAKIS, G., AND SPENCER, J. *Spectral/hp Element Methods for CFD.* Numerical Mathematics and Scientific Computation. Oxford University Press, New York, 1999.

[KnL02] KNOPP, T., LUBE, G., AND RAPIN, G. Stabilized finite element methods with shock capturing for advection–diffusion problems. *Comput. Methods Appl. Mech. Engrg.* **191**:27-28 (2002) 2997–3013.

[KnS93] KNUPP, P., AND STEINBERG, S. *Fundamentals of Grid Generation*. CRC Press, Boca Raton, FL, 1993.

[Lam91] LAMBERT, J. *Numerical Methods for Ordinary Differential Systems*. Wiley, New York, 1991.

[LaT93] LASCAUX, P., AND THEODOR, R. *Analyse numérique matricielle appliquée à l'art de l'ingénieur*, 2nd ed. Vol. I and II. Masson, Paris, 1993.

[LaM54] LAX, P.D., AND MILGRAM, A.N. Parabolic equations. In *Contributions to the theory of partial differential equations*, Annals of Mathematics Studies, no. 33, pp. 167–190. Princeton University Press, Princeton, NJ, 1954.

[Len86] LENOIR, M. Optimal isoparametric finite elements and error estimates for domains involving curved boundaries. *SIAM J. Numer. Anal.* **23**:3 (1986) 562–580.

[Les75] LESAINT, P. *Sur la résolution des sytèmes hyperboliques du premier ordre par des méthodes d'éléments finis*. PhD thesis, University of Paris VI, 1975.

[LeR74] LESAINT, P., AND RAVIART, P.-A. On a finite element method for solving the neutron transport equation. In *Mathematical aspects of Finite Elements in Partial Differential Equations*, pp. 89–123. C. de Boors, editor. Academic Press, New York, 1974.

[Lev53] LEVY, S. Structural analysis and influence coefficients for delta wings. *J. Aeronaut. Sci.* **20** (1953).

[Lio59] LIONS, J.-L. Quelques résultats d'existence dans des équations aux dérivées partielles non linéaires. *Bull. Soc. Math. France* **87** (1959) 245–273.

[Lio68] LIONS, J.-L. Problèmes aux limites non homogènes à données irrégulières: Une méthode d'approximation. In *Numerical Analysis of Partial Differential Equations* (C.I.M.E. 2 Ciclo, Ispra, 1967) pp. 283–292. Edizione Cremonese, Rome, 1968.

[Lio69] LIONS, J.-L. *Quelques méthodes de résolution des problèmes aux limites non linéaires*, Vol. 1. Dunod, Paris, 1969.

[LiM68] LIONS, J.-L., AND MAGENES, E. *Problèmes aux limites non homogènes et applications*, Vol. 1. Dunod, Paris, 1968.

[Luč69] LUČKA, A. The rate of convergence to zero of the residual and the error for the Bubnov–Galerkin method and the method of least squares. In *Proc. Sem. Differential and Integral Equations*, No. I, pp. 113–122 (Russian). Akad. Nauk Ukrain. SSR Inst. Mat., Kiev, Ukraine, 1969.

[LuP98] LUCQUIN, B., AND PIRONNEAU, O. *Introduction to Scientific Computing*. Wiley, New York, 1998.

[MaM82] MAITRE, J., AND MUSY, F. The contraction number of a class of two level methods; an exact evaluation for some finite element subspaces and model problems. In *Multigrid Methods: Proceedings, Cologne 1981* No. 960, pp. 535–544. Lecture Notes in Mathematics, Springer-Verlag, Heidelberg, 1982.

[MaZ97] MALÝ, J., AND ZIEMER, W. *Fine Regularity of Solutions of Elliptic Partial Differential Equations*. Mathematical Surveys and Monograph, Vol. 51. American Mathematical Society, Providence, RI, 1997.

[MeS64] MEYERS, N.G., AND SERRIN, J. $H = W$. *Proc. Nat. Acad. Sci. USA* **51** (1964) 1055–1056.

[Mon92] MONK, P. Analysis of a finite element method for Maxwell's equations. *SIAM J. Numer. Anal.* **29**:3 (1992) 714–729.

[MoP88] MORTON, K., PRIESTLEY, A., AND SÜLI, E. Stability of the Lagrange-Galerkin method with non-exact integration. *RAIRO Anal. Numér.* **22**:4 (1988) 625–653.

[Neč62] NEČAS, J. Sur une méthode pour résoudre les équations aux dérivées partielles de type elliptique, voisine de la variationnelle. *Ann. Scuola Norm. Sup. Pisa* **16** (1962) 305–326.

[Néd80] NÉDÉLEC, J.-C. Mixed finite elements in $\mathbf{R}^3$. *Numer. Math.* **35**:3 (1980) 315–341.

[Néd86] NÉDÉLEC, J.-C. A new family of mixed finite elements in $\mathbb{R}^3$. *Numer. Math.* **50** (1986) 57–81.

[Néd91] NÉDÉLEC, J.-C. *Notions sur les techniques d'éléments finis.* SMAI Series on Mathematics and Applications, Vol. 7. Ellipses, Paris, 1991.

[Nit71] NITSCHE, J. Über ein Variationsprinzip zur Lösung von Dirichlet-Problemen bei Verwendung von Teilräumen, die keinen Randbedingungen unterworfen sind. *Abh. Math. Sem. Univ. Hamburg* **36** (1971) 9–15.

[Nit76] NITSCHE, J. Über L_∞-Abschätzungen von Projektionen auf finite Elemente. In *Finite Elemente* (Tagung, Inst. Angew. Math., Univ. Bonn, Bonn, 1975). *Bonn. Math. Schrift.* **89** (1976) 13–30.

[Ode91] ODEN, J. *Finite Elements: An Introduction,* Vol. II : Finite Element Methods, ch. 1. Handbook of Numerical Analysis. P.G. Ciarlet and J.-L. Lions, editors. North-Holland, Amsterdam, 1991.

[OdB98] ODEN, J., BABUŠKA, I., AND BAUMANN, C. A discontinuous hp finite element method for diffusion problems. *J. Comput. Phys.* **146**:2 (1998) 491–519.

[OdD91] ODEN, J., AND DEMKOWICZ, L. h-p adaptive finite element methods in computational fluid dynamics. *Comput. Methods Appl. Mech. Engrg.* **89** (1991) 11–40.

[OdD96] ODEN, J., AND DEMKOWICZ, L. *Applied Functional Analysis.* CRC Series in Computational Mechanics and Applied Analysis. CRC Press, Boca Raton, FL, 1996.

[OlD95] OLSEN, E., AND DOUGLAS JR., J. Bounds on spectral condition numbers of matrices arising in the p-version of the finite element method. *Numer. Math.* **69**:3 (1995) 333–352.

[Ort87] ORTEGA, J. *Matrix Theory, a Second Course.* Plenum, New York, 1987.

[Osw93] OSWALD, P. On a BPX-preconditioner for $\mathbb{P}_1$ elements. *Computing* **51** (1993) 125–133.

[Pat84] PATERA, A. A spectral method for fluid dynamics: Laminar flow in a channel expansion. *J. Comput. Phys.* **54** (1984) 468–488.

[PeC94] PEHLIVANOV, A., CAREY, G., AND LAZAROV, R. Least-Squares mixed finite elements for second-order elliptic problems. *SIAM J. Numer. Anal.* **31**:5 (1994) 1368–1377.

[Pir82] PIRONNEAU, O. On the transport-diffusion algorithm and its applications to the Navier–Stokes equations. *Numer. Math.* **38** (1982) 309–332.

[Pir83] PIRONNEAU, O. *Méthodes des éléments finis pour les fluides.* Masson, Paris, 1983.

[QuV97] QUARTERONI, A., AND VALLI, A. *Numerical Approximation of Partial Differential Equations,* 2nd ed. Springer Series in Computational Mathematics, Vol. 23. Springer-Verlag, New York, 1997.

[Raj94] RAJAN, V. Optimality of the Delaunay triangulation in R^d. *Discrete Comput. Geom.* **12** (1994) 189–202.

[Ran92] RANNACHER, R. On Chorin's projection method for the incompressible Navier–Stokes equations. In *The Navier–Stokes Equations II—Theory and Numerical Methods* (Oberwolfach, 1991). Lecture Notes in Mathematics, Vol. 1530, pp. 167–183. Springer-Verlag, Berlin, 1992.

[RaS82] RANNACHER, R., AND SCOTT, R. Some optimal error estimates for piecewise linear finite element approximations. *Math. Comp.* **38**:158 (1982) 437–445.

[RaT92] RANNACHER, R., AND TUREK, S. Simple nonconforming quadrilateral Stokes element. *Numer. Methods Partial Differential Equations* **8**:2 (1992) 97–111.

[RaT77] RAVIART, P.-A., AND THOMAS, J.-M. A mixed finite element method for second-order elliptic problems. In *Mathematical Aspects of the Finite Element Method*. E. Magenes, I. Galligani, editors. Lecture Notes in Mathematics, Vol. 606. Springer–Verlag, New York, 1977.

[RaT83] RAVIART, P.-A., AND THOMAS, J.-M. *Introduction à l'analyse numérique des équations aux dérivées partielles*. Masson, Paris, 1983.

[Reb93] REBAY, S. Efficient unstructured mesh generation by means of Delaunay triangulation and the Bowyer-Watson algorithm. *J. Comput. Phys.* **106** (1993) 125–138.

[RiW99] RIVIÈRE, B., WHEELER, M., AND GIRAULT, V. Improved energy estimates for interior penalty, constrained and discontinuous Galerkin methods for elliptic problems, I. *Comput. Geosci.* **8** (1999) 337–360.

[RoT91] ROBERTS, J., AND THOMAS, J.-M. *Mixed and Hybrid Methods*. Vol. II: Finite Element Methods, chap. 4. Handbook of Numerical Analysis. P.G. Ciarlet and J.-L. Lions, editors. North-Holland, Amsterdam, 1991.

[Rua96] RUAS, V. Circumventing discrete Korn's inequalities in convergence analyses of nonconforming finite element approximations of vector fields. *Z. Angew. Math. Mech.* **76**:8 (1996) 483–484.

[Rud66] RUDIN, W. *Real and Complex Analysis*. McGraw-Hill, New York, 1966.

[Rud87] RUDIN, W. *Analyse réelle et complexe*, 4th ed. Masson, Paris, 1987.

[Rus85] RUSSELL, T. Time stepping along characteristics with incomplete iteration for a Galerkin approximation of miscible displacement in porous media. *SIAM J. Numer. Anal.* **22** (1985) 970–1013.

[Saa96] SAAD, Y. *Iterative Methods for Sparse Linear Systems*. PWS Publishing, Boston, MA, 1996.

[SaS86] SAAD, Y., AND SCHULTZ, M. GMRES: A generalized minimal residual algorithm for solving nonsymmetric linear systems. *SIAM J. Sci. Statist. Comput.* **7** (1986) 856–869.

[Sch02] SCHIEWECK, F. A posteriori error estimates with post-processing for nonconforming finite elements. *Math. Model. Numer. Anal. (M2AN)* **36**:3 (2002) 489–503.

[Sco76] SCOTT, R. Optimal L^∞ estimates for the finite element method on irregular meshes. *Math. Comp.* **30**:136 (1976) 681–697.

[ScZ90] SCOTT, R., AND ZHANG, S. Finite element interpolation of nonsmooth functions satisfying boundary conditions. *Math. Comp.* **54**:190 (1990) 483–493.

[She92^a] SHEN, J. On error estimates of projection methods for the Navier–Stokes equations: First-order schemes. *SIAM J. Numer. Anal.* **29** (1992) 57–77.

[She92^b] SHEN, J. On pressure stabilization method and projection method for unsteady Navier–Stokes equations. In *Advances in Computer Methods for Partial Differential Equations*, R. Vichnevetsky, D. Knight, G. Richter, editors, pp. 658–662. IMACS, New Brunswick, NJ, 1992.

[She93] SHEN, J. A remark on the projection-3 method. *Int. J. Numer. Methods Fluids* **16**:3 (1993) 249–253.

[She96] SHEN, J. On error estimates of projection methods for the Navier–Stokes equations: Second-order schemes. *Math. Comp.* **65**:215 (1996) 1039–1065.

[Sho96] SHOWALTER, R. *Monotone Operators in Banach Spaces and Nonlinear Partial Differential Equations*. Mathematical Surveys and Monographs, Vol. 49. American Mathematical Society, Providence, RI, 1996.

[ShO89] SHU, C., AND OSHER, S. Efficient implementation of essentially non-oscillatory shock-capturing schemes, ii. *J. Comput. Phys.* **83** (1989) 32–78.

[Sim87] SIMON, J. Compact sets in the space $L^p(0, T; B)$. *Ann. Mat. Pura Appl.* **146** (1987) 65–96.

[Sob63] SOBOLEV, S. *Applications of Functional Analysis in Mathematical Physics*, 2nd ed. Translations of Mathematical Monographs, Vol. VII. American Mathematical Society, Providence, RI, 1963.

[Sol01] SOLONNIKOV, V.A. L^p-estimates for solutions of the heat equation in a dihedral angle. *Rend. Mat. Appl.* **21** (2001) 1–15.

[Son89] SONNEVELD, P. CGS, a fast Lanczos-type solver for nonsymmetric linear systems. *SIAM J. Sci. Statist. Comput.* **10**:1 (1989) 36–52.

[Ste70] STEIN, E. *Singular Integrals and Differentiability Properties of Functions*. Princeton University Press, Princeton, NJ, 1970.

[StB80] STOER, J., AND BULIRSCH, R. *Introduction to Numerical Analysis*. Springer-Verlag, New York, 1980.

[Str72] STRANG, G. Variational crimes in the finite element method. In *The Mathematical Foundations of the Finite Element Method with Applications to Partial Differential Equations*, A. Aziz, editor. Academic Press, New York, 1972.

[StF73] STRANG, G., AND FIX, G. *An Analysis of the Finite Element Method*. Prentice-Hall Series in Automatic Computation. Prentice-Hall, Englewood Cliffs, NJ, 1973.

[StL99] STRIKWERDA, J., AND LEE, Y. The accuracy of the fractional step method. *SIAM J. Numer. Anal.* **37**:1 (1999) 37–47.

[Str69] STROUD, A. A fifth degree integration formula for the n-simplex. *SIAM J. Numer. Anal.* **6** (1969) 90–98.

[Str71] STROUD, A. *Approximate Calculation of Multiple Integrals*. Prentice-Hall, Englewood Cliffs, NJ, 1971.

[Sül88] SÜLI, E. Convergence and nonlinear stability of the Lagrange–Galerkin method for the Navier–Stokes equations. *Numer. Math.* **53**:4 (1988) 459–483.

[Tad89] TADMOR, E. Convergence of spectral methods for nonlinear conservation laws. *SIAM J. Numer. Anal.* **26**:1 (1989) 30–44.

[TaW00] TAYLOR, M.A., WINGATE, B.A., AND VINCENT, R.E. An algorithm for computing Fekete points in the triangle. *SIAM J. Numer. Anal.* **38**:5 (2000) 1707–1720.

[Tem69] TEMAM, R. Sur l'approximation de la solution des équations de Navier–Stokes par la méthode des pas fractionnaires ii. *Arch. Rational Mech. Anal.* **33** (1969) 377–385.

[Tem77] TEMAM, R. *Navier–Stokes Equations*. Studies in Mathematics and its Applications, Vol. 2. North-Holland, Amsterdam, 1977.

[Tho97] THOMÉE, V. *Galerkin Finite Element Methods for Parabolic Problems*. Springer Series in Computational Mathematics, Vol. 25. Springer-Verlag, New York, 1997.

[ThS98] THOMPSON, J., SONI, B., AND WEATHERILL, N. *Handbook of Grid Generation*. CRC Press, Boca Raton, FL, 1998.

[TiM96] TIMMERMANS, L., MINEV, P., AND VAN DE VOSSE, F. An approximate projection scheme for incompressible flow using spectral elements. *Int. J. Numer. Methods Fluids* **22** (1996) 673–688.

[ToV96] TOBISKA, L., AND VERFÜRTH, R. Analysis of a streamline diffusion finite element method for the Stokes and Navier–Stokes equations. *SIAM J. Numer. Anal.* **33**:1 (1996) 107–127.

[Tur99] TUREK, S. *Efficient Solvers for Incompressible Flow Problems. An Algorithmic and Computational Approach*. Lecture Notes in Computational Science and Engineering, Vol. 6. Springer-Verlag, Berlin, 1999.

[TuC56] TURNER, M., CLOUGH, R., MARTIN, H., AND TOPP, L. Stiffness and deflection analysis of complex structures. *J. Aero. Sci.* **23** (1956) 805–823.

[vaV92] VAN DER VORST, H. Bi-CGStab: A more stably converging variant of CG-S for the solution of nonsymmetric linear systems. *SIAM J. Sci. Statist. Comput.* **13** (1992) 631–644.

[vaK86] VAN KAN, J. A second-order accurate pressure–correction scheme for viscous incompressible flow. *SIAM J. Sci. Statist. Comput.* **7**:3 (1986) 870–891.

[Ver84] VERFÜRTH, R. Error estimates for a mixed finite element approximation of the Stokes equation. *RAIRO Anal. Numér.* **18** (1984) 175–182.

[Ver94] VERFÜRTH, R. A posteriori error estimations and adaptive mesh-refinement techniques. *J. Comput. Appl. Math.* **50** (1994) 67–83.

[Ver96] VERFÜRTH, R. *A Review of a Posteriori Error Estimation and Adaptive Mesh-Refinement Techniques*. Wiley, Chichester, UK, 1996.

[Wat81] WATSON, D. Computing the n-dimensional Delaunay tesselation with applications to Voronoï polytopes. *Comput. J.* **24** (1981) 167–172.

[Whe73] WHEELER, M. A priori L_2 error estimates for Galerkin approximations to parabolic partial differential equations. *SIAM J. Numer. Anal.* **10** (1973) 723–759.

[Whi57] WHITNEY, H. *Geometric Integration Theory*. Princeton University Press, Princeton, NJ, 1957.

[Yos80] YOSIDA, K. *Functional Analysis*. Classics in Mathematics. Springer-Verlag, Berlin, 1995. Reprint of the sixth 1980 edition.

[Zei95] ZEIDLER, E. *Applied Functional Analysis*. Applied Mathematical Sciences, Vol. 108. Springer-Verlag, New York, 1995.

[Zha95] ZHANG, S. Successive subdivisions of tetrahedra and multigrid methods on tetrahedral meshes. *Houston J. Math.* **21** (1995) 541–555.

[Zho97] ZHOU, G. How accurate is the streamline diffusion finite element method? *Math. Comp.* **66** (1997) 31–44.

[Zlá73] ZLÁMAL, M. Curved elements in the finite element method. I. *SIAM J. Numer. Anal.* **10** (1973) 229–240.

[Zlá74] ZLÁMAL, M. Curved elements in the finite element method. II. *SIAM J. Numer. Anal.* **11** (1974) 347–362.

Author Index

Subject Index

Applied Mathematical Sciences

(continued from page ii)

(continued on next page)

Applied Mathematical Sciences

(continued from previous page)